SYSTEMS ANALYSIS AND DESIGN METHODS

FIFTH EDITION

SYSTEMS ANALYSIS AND DESIGN METHODS

JEFFREY L. WHITTEN
Professor

LONNIE D. BENTLEY
Professor

KEVIN C. DITTMAN
Assistant Professor

All at Purdue University
West Lafayette, IN

Boston   Burr Ridge, IL   Dubuque, IA   Madison, WI   New York   San Francisco   St. Louis
Bangkok   Bogotá   Caracas   Lisbon   London   Madrid
Mexico City   Milan   New Delhi   Seoul   Singapore   Sydney   Taipei   Toronto

McGraw-Hill Higher Education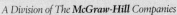

A Division of The **McGraw-Hill** *Companies*

SYSTEMS ANALYSIS AND DESIGN METHODS

Published by Irwin/McGraw-Hill, an imprint of The McGraw-Hill Companies, Inc. 1221 Avenue of the Americas, New York, NY, 10020. Copyright © 2000, 1998, 1994, 1989, 1986, by The McGraw-Hill Companies, Inc. All rights reserved. No part of this publication may be reproduced or distributed in any form or by any means, or stored in a database or retrieval system, without the prior written consent of The McGraw-Hill Companies, Inc., including, but not limited to, in any network or other electronic storage or transmission, or broadcast for distance learning.

Some ancillaries, including electronic and print components, may not be available to customers outside the United States.

This book is printed on acid-free paper.

domestic   1 2 3 4 5 6 7 8 9 0 WCK/WCK 0 9 8 7 6 5 4 3 2 1 0
international   1 2 3 4 5 6 7 8 9 0 WCK/WCK 0 9 8 7 6 5 4 3 2 1 0

ISBN   0-07-231539-3

Vice President/Editor-in-chief: *Michael W. Junior*
Publisher: *David Kendric Brake*
Senior sponsoring editor: *Rick Williamson*
Developmental editor: *Christine Wright*
Senior marketing manager: *Jeff Parr*
Senior project manager: *Susan Trentacosti*
Manager, new book production: *Melonie Salvati*
Senior designer: *Kiera Cunningham*
Senior supplement coordinator: *Marc Mattson*
Cover image: *©Linda Bleck/SIS*
Compositor: *GTS Graphics, Inc.*
Typeface: *10/12 Garamond Light*
Printer: *Quebecor World/Versailles*

Library of Congress Cataloging-in-Publication Data

Whitten, Jeffrey L.
    Systems analysis and design methods / Jeffrey L. Whitten, Lonnie D. Bentley, Kevin C. Dittman. —5th ed.
      p. cm.
      ISBN 0-07-231539-3 (alk. paper)
      1. System design. 2. System analysis. I. Bentley, Lonnie D. II. Dittman, Kevin C. III. Title.

QA76.9.S88 W48 2001
004.2′1—dc21   00-040694

INTERNATIONAL EDITION ISBN 0-07-118070-2
Copyright © 2001. Exclusive rights by The McGraw-Hill Companies, Inc. for manufacture and export. This book cannot be re-exported from the country to which it is sold by McGraw-Hill. The International Edition is not available in North America.

www.mhhe.com

WHY WE WROTE THIS BOOK

More than ever, today's students are "consumer-oriented," due in part to the changing world economy that promotes quality, competition, and professional currency. They expect to walk away from a course with more than a grade and a promise that they'll someday appreciate what they've learned. They want to "practice" the application of concepts, not just study applications of concepts. As with the previous editions of this book, we wrote it to: (1) balance the coverage of concepts, tools, techniques, and their application, (2) provide the most examples of system analysis and design deliverables available in any book, and (3) balance the coverage of classical methods (such as *structured analysis* and *information engineering*) and emerging methods (e.g., *object-oriented analysis* and *rapid application development*). Additionally, we wrote the textbook to serve the reader as a post-course, professional reference for best current practices.

Consistent with the first four editions, we have written the book using a lively, conversational tone. Our experience suggests that the more traditional, academic tone detracts from student interest. The "talk with you—not at you" style seems to work well with a wider variety of students. We hope that our style does not offend or patronize any specific audience. We apologize if it does.

INTENDED AUDIENCE

Systems Analysis and Design Methods, fifth edition, is intended to support one or more practical courses in information systems development. These courses are normally taught at the sophomore, junior, senior, or graduate level. They are taught in vocational trade schools, junior colleges, colleges, and universities. The courses are taught to both information systems and business majors.

We recommend that students should have taken a computer- and information systems-literacy course. While **not** required or assumed, a programming course can significantly enhance the learning experience provided by this textbook.

ORGANIZATION

Systems Analysis and Design Methods, fifth edition, is divided into five parts. Past experience indicates that instructors can omit and resequence chapters as they feel is important to their audience. Every effort has been made to decouple chapters from one another as much as possible to assist in resequencing the material—even to the extent of reintroducing selected concepts and terminology.

Part One, The Context of Systems Analysis and Design, presents the information systems development scenario and process. The chapters introduce the student to systems analysts, other project team members (including users and management), information systems building blocks (based on the Zachman framework), a contemporary systems development life cycle, and project management. Part One can be covered relatively quickly. Some adopters may prefer to omit project management, or delay it until the end of the book.

Part Two, Systems Analysis Methods, covers the front-end life cycle activities, tools, and techniques for analyzing business problems, specifying business requirements for an information system, and proposing a business and system solution. Coverage includes requirements gathering, data modeling with entity-relationship diagrams, process modeling with data flow diagrams, requirements specification in a data dictionary, and solution identification and the system proposal.

Part Three, Systems Design Methods, covers the middle life cycle activities, tools, and techniques. It includes coverage of both general and detailed design with a particular emphasis on application architecture, rapid development and prototyping, external design (outputs, inputs, and interfaces), and internal design.

Part Four, Beyond Systems Analysis and Design, is a capstone unit that places systems analysis and design into perspective by surveying the back-end life cycle activities. Specifically, chapters examine system implementation, support, maintenance, and reengineering.

Part Five, Advanced Analysis and Design Methods, teaches object-oriented analysis and design methods using the Unified Modeling Language. The two modules could be integrated into the analysis and design units respectively, omitted altogether, or introduced at the end of the course (or at the beginning of an advanced course).

CHANGES FOR THE FIFTH EDITION

In this edition, we continue to react to changes and expected changes in the information technology domain. Our industry faced many exciting problems including Year 2000 (Y2K) compatibility and the single European currency (called the *euro*). And there are even more opportunities as client/server computing meets the Internet, intranets, and extranets for electronic business and commerce applications. Finally, we see exciting systems analysis and design challenges with *Enterprise Resource Planning* (ERP) applications (such as *SAP*), systems integration, and business process redesign (BPR).

We believe that we have preserved the features adopters liked in the previous editions. And in the spirit of continuous improvement we have made the following changes:

- The information system development, systems analysis, systems design, and systems implementation chapters have been structurally simplified.
- At the request of adopters, the cross-life cycle modules (e.g., project management, interpersonal skills, fact finding and JAD, and feasibility analysis) have been updated and integrated into the mainstream chapters of the book.

- The use of automated tools (such as CASE and RAD) for systems analysis, design, and construction is once again reinforced throughout the book. Some of the tools demonstrated in the fifth edition include *Visio Professional, System Architect 2001, Project 2000,* and *Visual Basic.*
- The fifth edition continues the pedagogical use of full-color applied to an adaptation of Zachman's *Framework for Information Systems Architecture.* The Information Systems Architecture matrix uses these colors to introduce recurring concepts. System models then reinforce those concepts with a consistent use of the same colors.
- The *matrix* framework based on Zachman's *Framework for Information Systems Architecture* continues to organize the subject's conceptual foundations. The fifth edition framework has been updated (and simplified!) to reflect contemporary technologies and methods. The framework has been visually integrated into both the textbook's system development methodology and the beginning of every chapter as a chapter map, showing which aspects of the framework are relevant to that chapter.
- The SoundStage Entertainment Club chapter-opening case study has been enhanced and updated to reflect the advent of Web-centric applications of the Internet, intranets, and extranets.

Specific chapter and module enhancements include:

- Chapter 1, the *modern systems analyst,* has been renamed to **players in the systems game** to reflect a new emphasis on systems analysis and design as a "team sport." Consistent with the textbook's title and subject, the

systems analyst is still emphasized; however, the framework is introduced to help students better appreciate the roles of the management, user, and technical communities.

- The revamped matrix framework that will be used to organize the rest of the text is introduced in Chapter 2.
- The impact of contemporary techniques (such as **model-driven development, rapid application development,** and **commercial off-the-shelf software integration**) and automated tools (such as **CASE** and **ADEs**) is introduced in the **information systems development** chapter (Chapter 3).
- Immediately after the information systems development chapter, **project management** is introduced in Chapter 4. This chapter has been significantly updated to focus on the activities of project management while retaining (and improving) the demonstration of Microsoft *Project.* The Capability Maturity Model (CMM) drives our coverage of project management.
- The **systems analysis** chapter (Chapter 5) includes new material on the subject of **business process analysis and redesign.** All information systems must be integrated into the business processes of an organization. This is especially true when software applications are procured instead of being built in-house.
- The former fact-finding techniques and joint application development modules have been merged into a single **requirements discovery** chapter (Chapter 6) that is now part of the systems analysis unit.
- Based on encouragement from several adopters, we returned

normalization to the **data modeling** chapter (Chapter 7).

— By popular demand, we provide a complete set of leveled data flow diagrams in the **process modeling** chapter (Chapter 8) (perceived as a strength in the first three editions). Coverage of the bottom-up (Yourdon *modern structured analysis*) approach is clarified from the fourth edition.

— The *network modeling* chapter was deleted since its modeling paradigm has not come into mainstream practice; however, **distribution analysis** coverage has been fully integrated into the data and process modeling material, Chapters 7 and 8.

— The analysis-to-design transition coverage is improved by combining **feasibility analysis** (formerly a module) with coverage of preparing a physical/technical **system proposal** in Chapter 9.

— The **systems design** overview chapter (Chapter 10) offers improved coverage of **commercial off-the-shelf software (COTS)** as an alternative to designing and developing an in-house solution. This "route" introduces issues of both procurement and system integration. This changes the rules of engagement for system design.

— The **application architecture** chapter (Chapter 11) has been updated to reflect the latest in client/server, Web, and other information technologies applicable to information systems. Physical data flow diagrams are used throughout the chapter to demonstrate modern architectures.

— The database design chapter (Chapter 12) has been simplified and updated to include coverage of data distribution analysis.

— The **output, input,** and **graphical user interface design**

chapters (Chapters 13, 14, and 15) have been further updated to reflect design considerations for both client/server ("fat client") and Web-based ("thin client") applications.

— The **system construction and implementation** chapter (Chapter 16) provides improved emphasis on system testing, conversion, and user training for distributed information systems.

— The **systems operation and support** chapter (Chapter 17) has been updated to reflect contemporary maintenance and reengineering issues.

— Also at the request of adopters and reviewers, the **object-oriented analysis and design** chapters have been relocated to the end of the book as Modules A and B. Many adopters told us that they omit this advanced material or cover it at the end of the course for transition to an advanced course. The modules have been significantly updated to reflect the official emergence of the *Unified Modeling Language* (UML) which evolved from the collaboration of three OOA experts: Grady Booch, Ivar Jacobson ("use case"), and James Rumbaugh ("object modeling technique" or OMT).

INSTRUCTIONAL RESOURCES AND SUPPLEMENTS

It has always been our intent to provide our adopters with a complete course, not just a textbook. We are especially excited about this edition's comprehensive support package. It includes Web-hosted support, software bundles, and other resources for both the student and instructor. Most have been developed in parallel with this edition. The supplements for the fifth edition of *Systems Analysis and Design Methods* include the following components.

For the Instructor
Instructor's Resource CD-ROM.
A presentation manager shell allows you to organize and customize the following components to the needs of your course:

— **Instructor's Guide with Electronic Transparencies and PowerPoint.** This instructor's guide includes course planning materials, teaching guidelines and transparencies, templates, and answers to end-of-chapter problems, exercises, and minicases.

The transparency repository on the CD-ROM includes many more slides than could be offered in a traditional printed book. All slides are in Microsoft's *PowerPoint* format (complete with instructor notes that provide teaching guidelines and tips). Instructors can (1) pick-and-choose those slides they wish to use, (2) customize slides to their own preferences, and (3) add new slides. Slides can (a) be organized into electronic presentations, or (b) printed as transparencies or transparency masters.

The slides are also provided in *Adobe Acrobat* format for non-PowerPoint users.

— **Test Bank/Computerized Test Bank.** A test bank and Brownstone Diploma test generation software for online or traditional testing contain questions in the following formats: true/false, multiple choice, sentence completion, and matching. The answers are cross-referenced to the page numbers in the text.

— **Projects and Cases Solutions.** Suggested solutions and supporting material for the optional projects and cases are provided.

Website and Online Learning Center.
With the previous edition, we provided the first comprehensive website for adopters of a systems analysis and design textbook. The new website at www.mhhe.com/whitten (URL is also

on front cover of the text) provides a password-protected instructor section for downloading the latest supplemental resources and updates, as well as an Online Learning Center with additional lecture notes, material on chapters not found in the text, and solutions to extra projects and cases. Instructors can also contact the authors from this location.

For the Student

Website and Online Learning Center.
The student side of the website contains downloadable templates in *Visio, System Architect,* and Microsoft *Word, Excel,* and *Project,* as well as links to interesting and relevant information regarding systems work. The student Online Learning Center includes additional material such as supplemental

chapters and modules, projects and cases, and self-assessment quizzes for each chapter.

Projects and Cases to Accompany Systems Analysis and Design Methods.
This casebook, available with this edition as an optional CD-ROM or on the student Online Learning Center, has been updated to include new semester project case studies that can be completed in conjunction with the textbook. A *build your own project* model is retained for those instructors and students who want to maximize value by leveraging students' past and current work experience, or for use with a live-client project.

System Architect 2001 Student Edition.
In an exclusive agreement between Irwin/McGraw-Hill and Popkin Soft-

ware & Systems, a student edition of *System Architect 2001* is also available on CD-ROM as a packaging option with the text. *SA 2001* supports all of the diagrams covered in the textbook and includes instructions and tutorials on the CD.

Visible Analyst Workbench Student Edition.
In cooperation with Visible Systems, Irwin/McGraw-Hill offers Visible Analyst Workbench as a second CASE tool packaging option with the text. Visible Analyst has recently been updated to support 32-bit technology and incorporates the UML methodology.

Jeffrey L. Whitten
Lonnie D. Bentley
Kevin C. Dittman

ACKNOWLEDGMENTS

We are indebted to many individuals who contributed to the development of five editions of this textbook.

We wish to thank the reviewers and critics of this and prior editions:

Jeanne M. Alm, *Moorhead State University*

Charles P. Bilbrey, *James Madison University*

Ned Chapin, *California State University–Hayward*

Carol Clark, *Middle Tennessee State University*

Gail Corbitt, *California State University–Chico*

Larry W. Cornwell, *Bradley University*

Barbara B. Denison, *Wright State University*

Linda Duxbury, *Carleton University*

Dana Edberg, *University of Nevada–Reno*

Craig W. Fisher, *Marist College*

Raoul J. Freeman, *California State University–Dominguez Hills*

Dennis D. Gagnon, *Santa Barbara City College*

Abhijit Gopal, *University of Calgary*

Patricia J. Guinan, *Boston University*

Bill C. Hardgrave, *University of Arkansas–Fayetteville*

Alexander Hars, *University of Southern California*

Richard C. Housley, *Golden Gate University*

Constance Knapp, *Pace University*

Riki S. Kuchek, *Orange Coast College*

Thom Luce, *Ohio University*

Charles M. Lutz, *Utah State University*

Ross Malaga, *University of Maryland–Baltimore County*

Chip McGinnis, *Park College*

William H. Moates, *Indiana State University*

Ronald J. Norman, *San Diego State University*

Charles E. Paddock, *University of Nevada–Las Vegas*

June A. Parsons, *Northern Michigan University*

Harry Reif, *James Madison University*

Gail L. Rein, *SUNY–Buffalo*

Rebecca H. Rutherfoord, *Southern College of Technology*

Jerry Sitek, *Southern Illinois University–Edwardsville*

Craig W. Slinkman, *University of Texas–Arlington*

John Smiley, *Holy Family College*

Mary Thurber, *Northern Alberta Institute of Technology*

Jerry Tillman, *Appalachian State University*

Jonathan Trower, *Baylor University*

Margaret S. Wu, *University of Iowa*

Jacqueline E. Wyatt, *Middle Tennessee State University*

Vincent C. Yen, *Wright State University*

Ahmed S. Zaki, *College of William and Mary*

Your patience and constructive criticism were essential to shaping this edition and appreciated.

Special thanks are offered to Dorothy Jane Miller who provided special assistance throughout the project. We also thank our students and alumni. You make teaching worthwhile.

Finally, we acknowledge the contributions, encouragement, and patience of the staff at McGraw-Hill/Irwin. For this edition, special thanks to Rick Williamson for never losing his cool in the wake of deadlines missed; to Christine Wright, developmental editor; to Susan Trentacosti, who as project manager for three editions has established the benchmark of excellence in that role. We also thank Keri Johnson, Kiera Cunningham, Melonie Salvati, Marc Mattson, and Merrily Mazza. We hope we haven't forgotten anyone.

To those who used our previous four editions, thank you for your continued support. And for new adopters, we hope you'll see a difference in this text. We eagerly await your reactions, comments, and suggestions.

Jeffrey L. Whitten
Lonnie D. Bentley
Kevin C. Dittman

BRIEF CONTENTS

CONTENTS

PART TWO　SYSTEMS ANALYSIS METHODS　　　161

PART THREE   SYSTEMS DESIGN METHODS   391

SYSTEMS ANALYSIS AND DESIGN METHODS

THE CONTEXT OF SYSTEMS ANALYSIS AND DESIGN

This is a practical book about information systems development methods. All businesses and organizations develop information systems. You can be assured that you will play some role in the systems analysis and design for those systems—either as a customer or user of those systems or as a developer of those systems. Systems analysis and design is about business problem solving and computer applications. The methods you will learn in this book can be applied to a wide variety of problem domains, not just those involving the computer.

Before we begin, we assume you've completed an introductory course in computer-based information systems. Many of you have also completed one or more programming courses (using technologies such as *Access, COBOL,* C/C++, or *Visual Basic*). That will prove helpful, since systems analysis and design precede and/or integrate with those activities. But don't worry—we'll review all the necessary principles on which systems analysis and design is based.

Part One focuses on the big picture. Before you learn about specific activities, tools, techniques, methods, and technology, you need to understand this big picture. As you explore the context of systems analysis and design, we will introduce many ideas,

tools, and techniques that are not explored in great detail until later in the book. Try to keep that in mind as you explore the big picture.

Systems development isn't magic. There are no secrets for success, no perfect tools, techniques, or methods. To be sure, there are skills that can be mastered. But the complete and consistent application of those skills is still an art.

We start in Part One with fundamental concepts, philosophies, and trends that provide the context of systems analysis and design methods—in other words, the basics! If you understand these basics, you will be better able to apply, with confidence, the practical tools and techniques you will learn in Parts Two through Five. You will also be able to adapt to new situations and methods.

Four chapters make up this part. Chapter 1, Players in the Systems Game, introduces you to the PARTICIPANTS in systems analysis and design with special emphasis on the modern systems analyst as the facilitator of systems work. You'll also learn about the relationships among systems analysts, end-users, managers, and other information systems professionals. Finally, you'll learn to prepare yourself for a career as an analyst (if that is your goal). Regardless, you will understand how you will interact with this important professional.

Chapter 2, Information System Building Blocks, introduces the PRODUCT we will teach you how to build—*information systems.* Specifically, you will learn to examine information systems in terms of common building blocks: DATA, PROCESSES, and INTERFACES— each from the perspective of different participants or stakeholders. A visual matrix framework will help you organize these building blocks so that you can see them applied in the subsequent chapters.

Chapter 3, Information Systems Development, introduces a high-level (meaning general) process for information systems development. This is called a *systems development life cycle.* We will present the life cycle in a form in which most of you will experience it—a *systems development methodology.* This methodology will be the context in which you will learn to use and apply the systems analysis and design methods taught in the remainder of the book.

Chapter 4, Project Management, introduces project management techniques. All systems projects are dependent on the principles that are surveyed. This chapter introduces two modeling techniques for project management: *Gantt* and *PERT.* These tools help you schedule activities, evaluate progress, and adjust schedules.

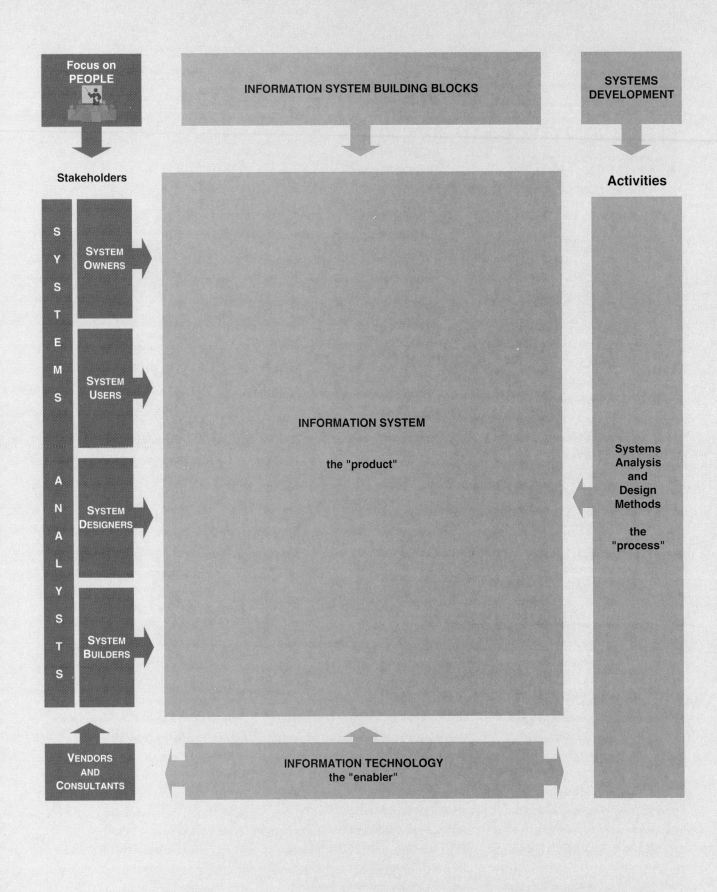

1

PLAYERS IN THE SYSTEMS GAME

CHAPTER PREVIEW AND OBJECTIVES

This is a book about information systems analysis and design. No matter what your occupation or position in any business, you will likely participate in systems analysis and design. Some of you will become *systems analysts,* key players in systems analysis and design. The rest of you will work *with* systems analysts as projects come and go in your organizations. This chapter introduces you to all the players in the information systems game, but it emphasizes systems analysts, the people who usually facilitate systems development through the methods described in this book. The chapter will address the following questions:

— What are information systems, and who are the stakeholders in the information systems game?

—What role will you play in the development and use of information systems?

— Who are information systems users? How is the definition changing in a remote computing and Internet-centric world?

— What is a systems analyst, and why are these individuals the key players in the development and implementation of information systems? Where do systems analysts work?

— What are systems analysis and design?

— What modern business and technology trends are affecting information systems development?

— What are the career prospects for systems analysts?

— If you want to pursue a career as a systems analyst, what knowledge and skills do you need? (The complete set of knowledge and skills is beyond the scope of any single book.)

SOUNDSTAGE

SOUNDSTAGE ENTERTAINMENT CLUB

CASE STUDY

This is the continuing story of Sandra Shepherd and Bob Martinez, systems analysts for SoundStage Entertainment Club. SoundStage Entertainment Club is one of the fastest growing music and video clubs in America. The company headquarters, central region warehouse, and sales office are located in Indianapolis, Indiana. Other sales offices and warehouses are located in Baltimore and Seattle.

Systems analysis and design is more than concepts, tools, techniques, and methods. It is about people working with people. Although experience is the best teacher, you can learn a great deal by observing other systems analysts in action. Nancy Picard, chief information officer (CIO) and vice president of Information Services (IS) for SoundStage Entertainment Club, consented to let you watch two of her analysts on a typical project.

Sandra Shepherd, a senior systems analyst and project manager, volunteered for this demonstration. She has successfully implemented several information systems for SoundStage and should be able to provide you with a valuable learning experience. Bob Martinez, Sandra's partner, is a new analyst/ programmer at SoundStage. In fact, today is his first day!

Bob has to go through orientation today, and Nancy has invited you to observe the orientation. It'll be a good way for you and Bob to get acquainted with SoundStage.

SCENE

In Nancy's office where Bob has been directed after completing his employment paperwork.

NANCY

Hi, Bob! It's great to have you aboard. My name's Nancy Picard, and I'm the CIO and vice president of Information Services. I didn't get a chance to meet you when you interviewed last month. Why don't you tell me a little about yourself?

BOB

Well, I just received my bachelor's degree in computer information systems and technology from Purdue University. I was president of my local student chapter of the Association of Information Technology Professionals and hope to get involved in the Indianapolis professional chapter as well. My career goals are oriented toward applying the systems analysis and design skills I learned in college. That's the main reason I accepted this job. It looked like I'd get a chance to do some analysis and design here—not just programming.

NANCY

That's why we hired you, Bob. You may be interested to know that we were especially impressed by your classroom experience with computer tools for systems analysis and design. We just bought a package called *System Architect 2001* from Popkin Software & Systems. We want you to learn and use that package on your first project and report your experiences back to the management staff. You'll learn more about this technology from your partner, Sandra Shepherd, a senior systems analyst who is very familiar with the tools and techniques you learned in college.

BOB

I'm familiar with *System Architect*. We used it in my systems analysis courses at Purdue.

NANCY

Great! Sandra will show you the ropes and help you learn about SoundStage and our way of doing things. You'll meet with Sandra soon. OK, let's take a tour of the building.

[Walking up to the second floor]

NANCY

Did you hear the big news? *[pause]* We just acquired controlling interest in Private Screenings Video Club. Along with our recent acquisition of GameScreen, the electronic games club, we can now offer comprehensive entertainment merchandise to our members. Of course,

consolidating our operations and information services will be a real challenge. Your first assignment will directly address that challenge.

Here we are. The second floor houses several departments including Personnel, Building Services, Accounting, Marketing, and Finance. These outside wall offices belong to executive managers, staff, and assistants.

I don't know how much you know or remember about SoundStage, so I'll give you a quick overview. We used to be called SoundStage Record and Tape Club—an LP record and cassette tape subscription service. As you know, LP records are virtually obsolete, and audio entertainment alternatives have been expanded to include compact discs and minidisks. The GameScreen merger will add video games to our inventory. And the Private Screenings merger will add videotapes and discs to the product mix. Our management believes the new DVD videodiscs will become a big market. This new product mix is why we changed our name to SoundStage *Entertainment* Club.

Customers join the club through advertisements and member referrals. We want to start looking at electronic commerce options on the Internet. Our advertisements typically dangle a carrot such as, "Choose any 10 CDs for a penny and agree to buy 5 more within two years at regular club prices." I'm sure you've seen such offers.

BOB

Yes, I've been a SoundStage member for three years.

NANCY

Club members receive monthly promotions and catalogs that offer a selection of the month. Currently, they must respond to the offer within a few weeks or that selection will automatically be shipped and billed to their account. We have decided to change that business model because customers hate those return cards. Customers can also order alternative selections and

SOUNDSTAGE

S O U N D S T A G E   E N T E R T A I N M E N T   C L U B

special merchandise from the catalogs. After members fulfill their original subscription agreement, they are eligible for bonus coupons that may be redeemed for free merchandise from our catalogs.

BOB

How many members are there?

NANCY

The last number I heard was 340,000. Of that total, about 180,000 accounts are active, having purchased merchandise in the last 12 months. But that doesn't include the GameScreen and Private Screenings members that we will soon inherit.

[Walking along the hallway of executive offices.]

NANCY

The office on your left belongs to our president and chief executive officer, Steven Short. This is our organization chart, Bob (see Figure A). As you can see, SoundStage is divided into three main divisions besides the Information Services division. The offices you walked through on your way to this office belong to the Business Services division.

On the first floor we have the Operations division, which handles day-to-day operations including purchasing, inventory control, warehousing, and shipping and receiving. I will soon take you through those facilities.

In the basement, you'll find the Information Services division, where you'll have your office. Our computer operations are located down there, along with my staff, which totals about 35 people.

FIGURE A   *SoundStage Entertainment Club Organization Chart*

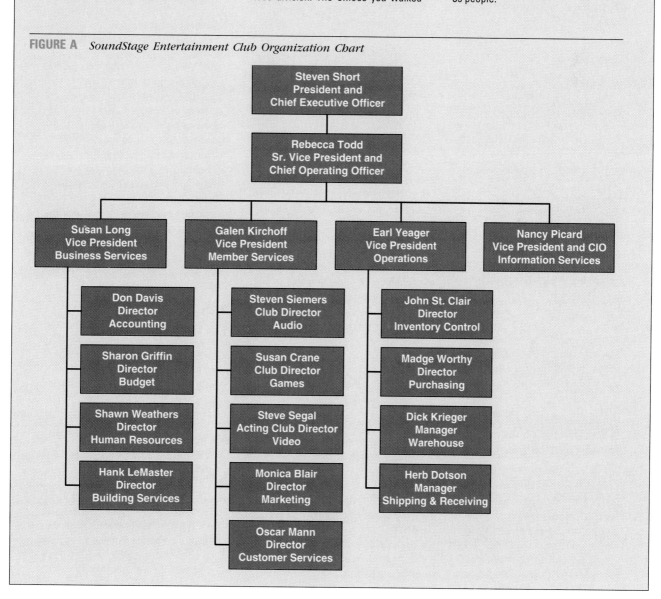

[Down to the first floor; another maze of offices.]

NANCY

These are the offices for the Operations division, including Purchasing and Inventory Control. They buy the merchandise we resell to the customers.

And this office area we are approaching is the Member Services division. Those clerks are processing orders, back orders, follow-ups, and other customer transactions. That's where your first project is going to be. Sandra will be taking you there to meet your customers.

BOB

Customers?

NANCY

We refer to the users of our services as our customers. The customer focus is a cornerstone of our total quality culture (TQC). You'll be indoctrinated to the TQC in your corporate orientation next week. It is taken very seriously here. Promotions and pay increases are driven entirely by your contribution to the company's bottom line, which is a direct function of our overall quality.

BOB

Doesn't bother me a bit. Where to next?

[Through a set of double doors into a huge warehouse.]

NANCY

This is the warehouse. This is where the action is! Over there are the shipping and receiving offices and docks.

BOB

I saw this before. I'm still taken aback by the size and activity. This has to be a very large operation to coordinate.

NANCY

I wouldn't want to do it. And it will get larger with the mergers. There is already talk of expanding the facility by 50 percent. We haven't done much in the way of information services for the warehouse. We do support Purchasing and Inventory Control, but that is as close as we come to supporting the warehouse. Our information strategy plan, however, has placed great importance on information services for optimizing this operation. The first project, based on bar-coding of inventory, is slated to begin next quarter.

Well, let's go downstairs and tour our own Information Services division, and then I'll take you to Sandra's office.

[Downstairs, in the basement facilities, beginning with the Computer Operations Center.]

NANCY

This is an organization chart for Information Services, Bob (see Figure B). You might want to refer to it as we complete this part of the tour. I now report directly to Rebecca Todd. That reflects the recent reorganization. I used to report to the vice president of Business Services, but it caused some problems with prioritizing requests from the other divisions.

This is the Computer Operations Center. You are looking at our IBM AS/400 computer. Historically, this midrange computer has supported most of our computer-processing needs. But our strategic information technology architecture will migrate us from the centralized AS/400 solution to a distributed computing solution based on servers, workstations, and the Internet.

BOB

Is that one of them? *[pointing to a small rack unit]*

NANCY

Yes, that is our new Compaq communications server. It runs Windows NT Server and Microsoft *Exchange Server,* our electronic mail, and fax system. Our distributed systems will use a combination of Windows NT and UNIX servers, plus Windows NT client workstations. That is the information technology architecture to be used for your first project.

BOB

This is pretty impressive stuff. You've definitely piqued my interest.

[Through the double doors (using Nancy's security badge) and into an adjacent open office area.]

NANCY

And this is where you'll be working. We call it the Development Center. All of our systems analysts, contract programmers, and technical specialists work here.

BOB

Contract programmers?

NANCY

Yes. We've moved away from *COBOL* and RPG programming to *Visual Basic* and *C++* programming, and we are looking seriously at *Java.* We've refocused our existing programmers to become either analyst programmers or other technical specialists. To meet most of our programming needs, we contract for programmers on a project-by-project basis. Of course, we keep a few good programmers around for management, technical support, and maintenance.

BOB

Interesting approach! Does that mean I won't program?

NANCY

Actually, you will program in the rapid application development or prototyping sense of the word. But you will have contract programmers on your team to help you, and they will do the heavy-duty programming that transforms prototypes into working applications.

BOB

What's this area with the gigantic diagram on the wall?

NANCY

These offices are assigned to our data and database administrators. They manage our corporate data model and our database management systems. You'll work with the data administrator's staff

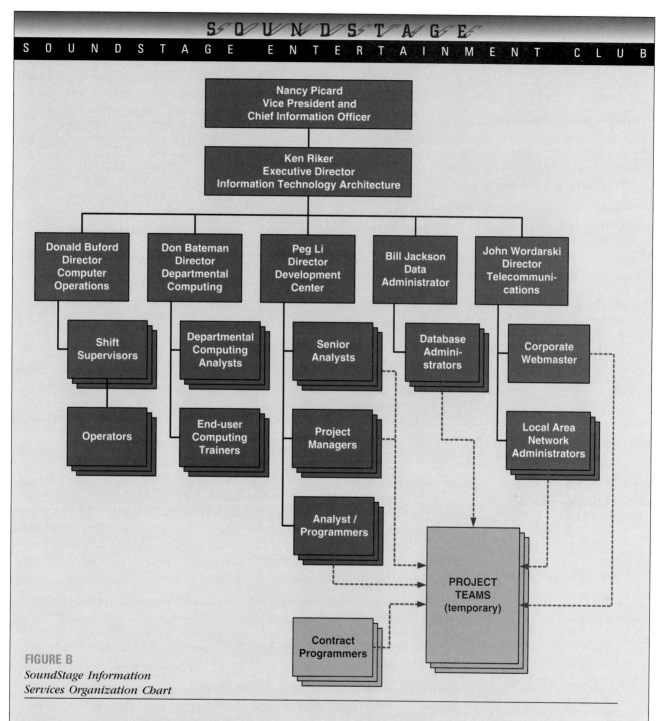

FIGURE B
*SoundStage Information
Services Organization Chart*

on any information systems projects that use the database. We are mandated by policy to implement all new systems using that database; therefore, I'm sure you'll soon be visiting his staff.

BOB

What methodology do you use to develop your databases and applications?

NANCY

We use a methodology called *FAST,* a *Framework for Application of Systems Techniques.* It is part methodology and part framework. For example, it provides frameworks for flavors of systems development such as "structured," "rapid development," "purchased applications," and "object-oriented."

Your partner, Sandra, is an experienced *FAST* developer.

Our methodology analyst, Susan Clark, coordinates *FAST* training and helps our analysts with their *FAST* questions. She has two tools analysts who help our analysts apply tools like *System Architect* and *Visual Basic* within the *FAST* methodology.

BOB

Obviously, I've got a lot to learn.

[Into an empty office.]

NANCY

And this is your office. I wonder where . . . here she is! Sandra, I want you to meet Bob Martinez, your new partner.

SANDRA

Bob and I met during his interview, Nancy. Welcome, Bob.

NANCY

Terrific! I didn't know you had already met. I don't know what you've told Bob about yourself, Sandra, but I'd like to do a little biographical sketch.

Sandra has been with us for seven years. She was recently promoted to senior systems analyst because she has proven herself to be one of our most competent, progressive, and personable analysts. Sandra has a

bachelor's degree in business. She had little formal computing education, but she has done well because she always seeks out opportunities to learn more through reading, seminars, and company training courses. Sandra's credentials also include recognition as a Certified Systems Professional or CSP.

Bob, you'll be learning from one of our best people! Once again, Bob, welcome! I'll leave you with Sandra now. She'll help you get organized and start teaching you all the things you'll need to start learning. Bye!

HOW TO USE THE DEMONSTRATION CASE

You've just been introduced to a case study that will be continued throughout this book. The continuing case shows you that tools and techniques alone do not make a systems analyst. Systems analysis and design

involve a commitment to work for and with a number of people.

When we started writing this book, we wanted to make sure the chapters would teach you the important concepts, tools, and techniques. But we were also afraid that you might begin to believe that, if you knew those tools and techniques, you'd have all the knowledge necessary to be a systems analyst.

Each chapter of this book begins with a SoundStage episode that introduces new ideas, tools, and/or techniques that will be examined in that chapter. In almost all cases, you will see Sandra and Bob working closely with their team, which includes management, system users, and technical specialists. This self-managed team will build a quality- and improvement-driven system solution for SoundStage, consistent with the corporate total quality culture initiative described in this episode.

WHY STUDY SYSTEMS ANALYSIS AND DESIGN METHODS?

Today, it is hard to imagine any industry or business that has not been affected by computer-based information systems and computer applications. Many businesses consider management of their information resource to be equal in importance to managing their other key resources: property, facilities, equipment, employees, and capital. This book is about "analyzing" business requirements for information systems and "designing" information systems that fulfill business requirements and use information technology.

> An **information system (IS)** is an arrangement of people, data, processes, information presentation, and information technology that interact to support and improve day-to-day operations in a business as well as support the problem-solving and decision-making needs of management and users.

> **Information technology (IT)** is a contemporary term that describes the combination of computer technology (hardware and software) with telecommunications technology (data, image, and voice networks).

Many organizations consider information systems and information technology to be essential to their ability to compete or gain a competitive advantage. Most businesses realize that all workers need to participate in the development of information systems—not just the information technology specialists.

Let's assume you want to build an information system. Who are the stakeholders in the system you want to build?

> A **stakeholder** is any person who has an interest in an existing or new information system. Stakeholders can be technical or nontechnical workers.

The stakeholders for information systems can be broadly classified into six groups:

— *System owners* pay for the system to be built and maintained. They own the system, set priorities for the system, and determine policies for its use. In some cases, system owners may also be system users.

— *System users* actually use the system to perform or support the work to be completed. System users define the business requirements and performance expectations for the system to be built.

— *System designers* design the system to meet the users' requirements. In many cases, these technical specialists may also be system builders.

— *System builders* construct, test, and deliver the system into operation.

— *Systems analysts* facilitate the development of information systems and computer applications by bridging the communications gap that exists between nontechnical system owners and users and technical system designers and builders.

— *IT vendors and consultants* sell hardware, software, and services to businesses for incorporation into their information systems.

Representatives of all six stakeholders usually staff today's information system development teams. They are the players in the systems game!

Regardless of your academic field of study or chosen career, you will almost certainly participate in one or more of the above roles during an information system development project. Consequently, systems analysis and design is not a subject just for computer professionals anymore! That is why you should become familiar with, or master, the methods taught in this book.

INFORMATION WORKERS

All the above stakeholders have one thing in common—they are what the U.S. Department of Labor calls information workers.

> The term **information worker** was coined to describe those people whose jobs involve the creation, collection, processing, distribution, and use of information.

The livelihoods of information workers depend on decisions made from information. Some information workers (such as systems analysts and programmers) create information systems that process and distribute information. Others (such as clerks, secretaries, and managers) primarily capture, distribute, and use data and information. Today, more than 60 percent of the U.S. labor force is involved in the production, distribution, and use of information. Not surprisingly, an information technology industry (which includes hardware, software, networks, and consulting) has developed to support the growing information needs of businesses.

Let's examine the six groups of information workers in greater detail.

System Owners

For any information system, large or small, there will be one or more system owners. System owners usually come from the ranks of management. For medium-to-large information systems, the system owners are usually middle or executive managers. For smaller systems, the system owners may be middle managers or supervisors.

> **System owners** are the information system's sponsors and chief advocates. They are usually responsible for funding the project to develop, operate, and maintain the information system.

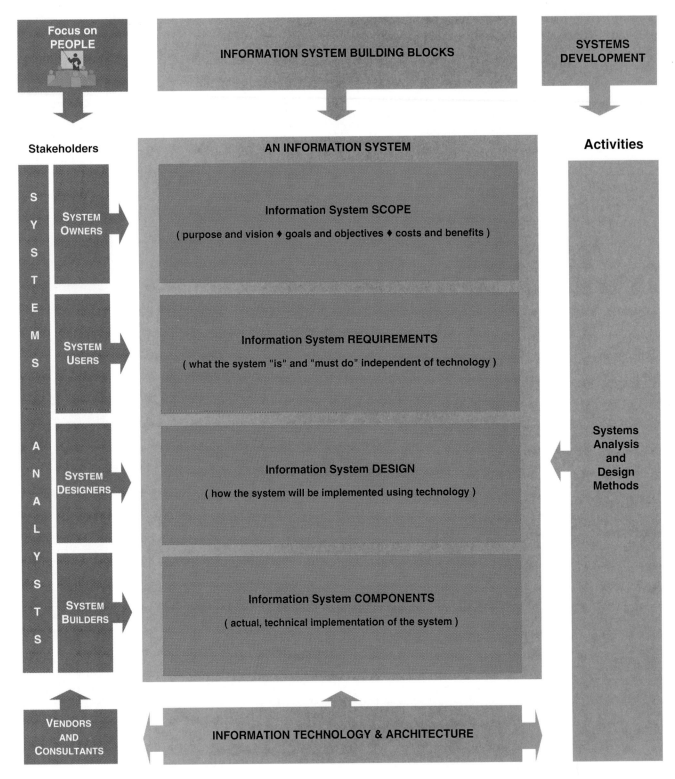

FIGURE 1.1   *People Perspectives for Systems Development*

As shown in Figure 1.1, system owners tend to be interested in the bottom line—how will the system return value or advantage to the business? Value is measured in different ways. What is the *purpose* of the system? What is the vision of the system—goals and objectives? How much will the system cost to build? How much will the system cost to operate? Will those costs be offset by measurable benefits? What about intangible benefits?

System users make up the vast majority of the information workers in any information system. **System Users**

> **System users** are the people who use or are affected by the information system on a regular basis—capturing, validating, entering, responding to, storing, and exchanging data and information. A common synonym is **client.**

Users don't always refer to themselves as users; some take offense at the term. According to Ed Yourdon, a noted systems author and consultant,

> The user is the "customer" in two important respects: (1) as in many other professions, the customer is always right, regardless of how demanding, unpleasant, or irrational he or she may seem; and (2) the customer is ultimately the person paying for the system and usually has the right or ability to refuse to pay if he or she is unhappy with the product received.[1]

Unlike system owners, system users tend to be less concerned with costs and benefits of the system. Instead, as illustrated in Figure 1.1, they are concerned with business requirements. Although users have become more technology-literate over the years, their primary concern is to get the job done. Consequently, discussions with most users need to be kept at the business level as opposed to the technical level. Much of this book is dedicated to teaching you how to effectively communicate business requirements for an information system.

There are many classes of system users. Each class should be directly involved in any information system development project that affects them. Let's briefly examine these classes.

Internal Users   Internal users are employees of the business for which an information system is built. Internal users make up the largest percentage of information system users in most businesses. Examples include clerical and service staff, technical and professional staff, supervisors, middle managers, and executive managers.

Clerical and service workers perform most of the day-to-day transaction processing in the average business. They process orders, invoices, payments, and the like. They type and file correspondence. They fill orders in the warehouse. And they manufacture goods on the shop floor. Most of the fundamental data in any business is captured or created by these workers, many of whom perform manual labors in addition to processing of data. The volume of data captured or created by these workers in the average organization is staggering. Information systems that target these workers tend to focus on transaction processing speed and accuracy.

Technical and professional staff consists largely of business and industrial specialists who perform highly skilled and specialized work. Examples include lawyers, accountants, engineers, scientists, market analysts, advertising designers, and statisticians. Their work is based on well-defined bodies of knowledge; hence, they are sometimes called knowledge workers.

> **Knowledge workers** are a subset of information workers whose responsibilities are based on a specialized body of knowledge.

Most knowledge workers are college educated. Their jobs depend not only on information, but also on their ability to properly use and react to that information. By way of their education and professional activities, technical and professional staff tend to be demanding users. Information systems that target these knowledge works tend to focus on data analysis as well as generating timely information for problem solving.

---

[1] Edward Yourdon, *Modern Structured Analysis* (Englewood Cliffs, NJ: Prentice Hall, Yourdon Press Computing Series, 1989), p. 41.

Supervisors, middle managers, and **executive managers** are all decision makers. Supervisors tend to focus on day-to-day management issues. Middle managers are more concerned with tactical or short-term management plans and problems. Executive managers are concerned with overall business performance, any strategic or long-term planning, and problem solving. Information systems for managers tend to focus on information access. Managers need the right information at the right time to solve problems and make good decisions.

Remote and Mobile Users A relatively new class of system users deserves special mention—remote and mobile users. Like internal users, they are employees of the business for which you are building information systems. Unlike internal users, they are geographically separated from the business.

The classic example is sales and service representatives. These **mobile users** live their professional lives on the road—customer to customer, client to client, buyer to buyer, and so forth. Historically, they have been second-class information system users, receiving little value from those systems. But with the advent of global telecommunications technology, and global competition, these so-called road warriors must be included in the new mix of information system users. Information systems that support mobile users are complex, but they also offer significant potential value to the modern business. Sales representatives who can directly tap databases and order-filling systems have a competitive advantage over those who cannot.

The new kid in the system user community is the **remote user.** Many businesses are looking to **telecommuting** to reduce costs and improve worker productivity. Telecommuting, stated simply, is working from home. Evidence suggests that many employees can be just as productive working at home if they can be connected to the company's information system through modern telecommunications technology.

External Users Modern information systems are now reaching beyond the boundaries of the traditional business to include customers and other businesses as system users. Businesses are redesigning their information systems to directly connect to and interoperate with other businesses, trading partners, suppliers, customers, and even consumers. For example, businesses increasingly connect their purchasing systems directly to the order processing systems of their suppliers to eliminate paperwork and more quickly replenish inventories of products, raw materials, and supplies. Accordingly, in business-to-business information systems, each business becomes an external user of the other business's information systems.

This concept is naturally extended to the consumer. When you directly purchase a product via the Internet, *you* become an external user of the retailer's order processing information system. You eliminate the necessity to phone or mail your order and the retailer's necessity to input your order into the information system.

System Designers

System designers are the technology specialists for information systems.

> **System designers** translate system users' business requirements and constraints into technical solutions. They design the computer files, databases, inputs, outputs, screens, networks, and programs that will meet the system users' requirements.

Again referring to Figure 1.1, system designers are interested in information technology choices and the design of systems within the constraints of the chosen technology. Today's system designers tend to focus on technical specialties such as databases, networks, user interfaces, or software. Some of you may be educating yourselves to specialize in one of these technical specialties.

System builders (see Figure 1.1) represent another category of information system development roles.

System Builders

> **System builders** construct the information system components based on the design specifications from the system designers. In many cases, the system designer and builder for a component are one and the same.

The applications programmer is the classic example of a system builder. However, other technical specialists may also be involved, such as systems programmers, database programmers, network administrators, and microcomputer software specialists. Although this book is not directly intended to educate or train the system builder, it is intended to teach system designers how to better communicate with system builders.

One knowledge worker plays a special role in information systems development, the systems analyst.

Systems Analysts

> A **systems analyst** *facilitates* the development of information systems and computer applications.

In this capacity, and as illustrated in Figure 1.1, the systems analyst must interact with each of the other stakeholders in the system. For the system owners and users, the analyst typically identifies and validates their business problems and needs. For the system designers and builders, the analyst ensures that the technical solution fulfills the business needs and then integrates the technical solution into the business. As part of the facilitation, the systems analyst performs *systems analysis* and *design*.

> **Systems analysis** is the study of a business problem domain to recommend improvements and specify the business requirements for the solution.

> **Systems design** is the specification or construction of a technical, computer-based solution for the business requirements identified in a systems analysis. (Note: Increasingly, the design takes the form of a working prototype.)

Some of you will routinely work with systems analysts. The rest of you will be *customers* of systems analysts who will try to help you solve your business and industrial problems by creating and improving your access to the data and information needed to do your job.

Most information systems are dependent on information technology that must be selected, installed and customized, integrated into the business, and technically supported. This technology is developed, sold, and supported by IT vendors. Increasingly, many businesses purchase software applications whenever possible, especially to support common business functions that don't create any competitive advantage (such as payroll and financial management). Information technology vendors have become more than just players in the information systems game; they have become partners of the businesses that purchase their products and services.

Information Technology Vendors and Consultants

Similarly, many businesses rely on external consultants to help them develop or acquire information systems and technology. The use of consultants may be driven by the need for specialized knowledge or skills or by an immediate need for additional analysts and programmers to complete a project. Regardless of the rationale, external consultants are a stakeholder in many IT projects.

The systems analyst facilitates most of the activities and methods taught in this book. In this section, we will examine the origins of the systems analyst. Then we will look at a more formal definition of the systems analyst and classify the roles and activities performed by the systems analyst. Finally, we will describe where systems analysts work.

THE MODERN SYSTEMS ANALYST

Why Do Businesses Need Systems Analysts?

The first systems analysts were born out of the industrial revolution. They were industrial engineers whose responsibilities centered on the design of efficient and effective manufacturing systems. Information systems analysts evolved from the need to improve the use of computer resources for the information processing needs of business applications.

Despite all its current and future technological capabilities, the computer still owes its power and usefulness to people. Businesspeople define the business problems to be solved by the computer. Computer programmers and technicians apply information technology to build information systems that solve those problems.

Information technology offers the opportunity to collect and store enormous volumes of data, process business transactions with great speed and accuracy, and provide timely and relevant information for management. Unfortunately, this potential has never been fully or even adequately realized in many businesses. Why? Business users may not fully understand the capabilities and limitations of modern information technology. Similarly, computer programmers and information technologists frequently do not understand the business applications they are trying to computerize or support. Worse still, some computer professionals become overly preoccupied with the technology.

A communications gap has always existed between those who need computer-based business solutions and those who understand information technology. The systems analyst bridges that gap. You can (and probably will) play a role as either a systems analyst or someone who works with systems analysts.

What Is a Systems Analyst?

In simple terms, systems analysts are people who understand both business and computing. They study business problems and opportunities and then transform business and information requirements into the computer-based information systems that are implemented by various technical specialists, including computer programmers. A more formal definition follows.

> A **systems analyst** studies the problems and needs of an organization to determine how people, data, processes, communications, and information technology can best accomplish improvements for the business.

When *information technology* is used, the analyst is responsible for the efficient capture of data from its business source, the flow of that data to the computer, the processing and storage of that data by the computer, and the flow of useful and timely information back to the business and its people.

Ultimately, a systems analyst is a business problem solver. Computers and information systems are of value to a business only if they help solve problems or affect improvements. Accordingly, a systems analyst helps the business by solving its problems through system concepts and information technology.

Systems analysts sell business management and computer users the services of information technology. More importantly, they sell *change*. Every new system changes the business. Increasingly, the very best systems analysts change their organizations—providing information that can be used for competitive advantage, finding new markets and services, and even dramatically changing and improving the way the organization does business.

In some organizations, system users are temporarily assigned to serve as business analysts.

> A **business analyst** is a systems analyst that specializes in business problem analysis and technology-independent requirements analysis.

Typically recruited from the user community, business analysts focus on business and nontechnical aspects of system problem solving. As experts in their business area, they help define system requirements for business problems and coordinate interactions between other business users and technical staff. Business analysts are usually appointed to a specific project or for a fixed duration.

There are also several legitimate, but often confusing, variations on the job title we are calling systems analyst. A *programmer/analyst* (or analyst/programmer) includes the responsibilities of both the computer programmer and the systems analyst. (Most systems analysts do some programming and most programmers do some systems analysis.) Other synonyms for systems analyst include *systems consultant, systems architect, systems engineer, information engineer, information analyst,* and *systems integrator.*

The systems analyst is basically a *problem solver.* Throughout this book, the term *problem* will be used to describe many situations including:

What Does a Systems Analyst Do?

- True *problem* situations, either real or anticipated, that require corrective action.
- *Opportunities* to improve a situation despite the absence of complaints.
- *Directives* to change a situation regardless of whether anyone has complained.

Most systems analysts use some variation of a problem-solving approach, which usually incorporates the following general steps:

1. Identify the problem.
2. Analyze and understand the problem.
3. Identify solution requirements or expectations.
4. Identify alternative solutions and decide a course of action.
5. Design and implement the "best" solution.
6. Evaluate the results. If the problem is not solved, return to Step 1 or 2 as appropriate.

The systems analyst's job presents a fascinating and exciting challenge. It offers high management visibility and opportunities for important decision making and creativity that may affect an entire organization. Furthermore, this job can offer these benefits relatively early in your career (compared to other jobs and careers at the same level).

Systems analysts can be found in most businesses. In this section, we'll examine where analysts work in both traditional and modern organizations. Then we'll examine two trends, *outsourcing* and *integration,* and discuss their implications for systems analysts. Finally, we'll examine alternatives to working in either a traditional or modern business information services organization.

Where Do Systems Analysts Work?

The Systems Analyst in the Traditional Business   Every business organizes itself uniquely. But certain patterns of organization seem to recur. In the traditional business, *information services* are centralized for the entire organization (or for a specific location). Figure 1.2 is an organization chart for a traditional information services unit.

In this organization, Information Services reports directly to the chief executive officer (CEO). The highest ranking information officer is sometimes called a **chief information officer (CIO).** This places Information Services at the same level of importance as the traditional business functions (e.g., Finance, Human Resources, Sales and Marketing, Production, and so on). The rest of Information Services is organized according to the following functions or centers:

- *Systems development.* Most systems analysts and programmers work here. The systems analysts and programmers are organized into permanent teams that support the information systems and applications for specific business functions. Each development team derives its budget and priorities from its corresponding business function. Notice that our systems development unit includes a **center for excellence.** The center for excellence is a group of

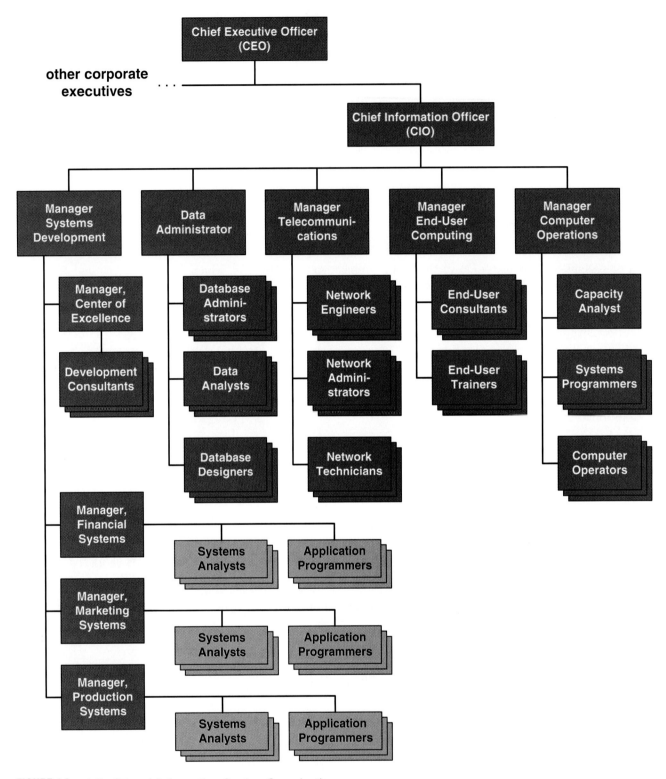

FIGURE 1.2   *A Traditional Information Services Organization*

experts who establish and enforce methods, tools, techniques, and quality for all system development projects. Experienced systems analysts, system designers, and system builders staff the center. They provide consulting and mentoring services for all projects.

- *Data administration.* This center manages the data and information resource in the organization. This includes the databases that are (will be) used by system developers to support applications. Data administration usually employs several systems analyst-like specialists called *data analysts* who analyze database requirements and design and construct the corresponding databases. To learn more about data administration and data analysts, refer to your local database management course or textbook.

- *Telecommunications.* This center designs, constructs, and manages the computer networks that have become essential to most businesses. The center does not employ traditional systems analysts; however, *network analysts* perform many of the same tasks as applied to designing local and wide area networks that will ultimately be used by systems and applications.

- *End-user computing.* This center supports the growing base of personal computers and local area networks in the end-user community. It provides installation services, training, and help-desk services (call-in help for various PC-related problems). In mature businesses, it also provides standards and consulting to end-users that develop their own systems with PC power tools such as spreadsheets and PC database management systems. In this latter role, the center employs analyst-like *end-user computing consultants.*

- *Computer operations.* This center runs all of the *shared* computers including mainframes, minicomputers, and nondepartmental servers. This unit rarely employs systems analysts.

The Systems Analyst in a Contemporary Information Services Organization   Many medium-to-large information services units for the modern business have reorganized to be somewhat *decentralized* with a focus on *empowerment* and *dynamic teams.* As shown in Figure 1.3, the resulting information services organization might best be described as a federation of competency centers that do not have any permanent assignment to any functional business unit. Each competency center might be thought of as a pool of developers who are assigned to projects based on the priorities of the business.

Accordingly, a systems analyst may be assigned to a financial systems project for the duration of that project. When the project is completed, the systems analyst returns to the pool and is subsequently reassigned to a different project (not necessarily in financial systems). During the project, the systems analyst and other team members are directly accountable to the business unit for which the system is being developed. Through this type of organization, Information Services tries to get closer to users and management to improve services and value.

Outsourcing   Another very significant trend in businesses that will affect many systems analysts is outsourcing.

> **Outsourcing** is the act of contracting with an outside vendor to assume responsibility for one or more IT functions or services. In many cases, ownership of IT assets (including technology and employees) is transferred to the outsourcer.

Representative IT outsourcers are listed in the margin.

As many as 70 percent of medium-to-large businesses have already outsourced some or all of their information services. The initial business driver was *cost reduction.* Outsourcers were able to sell and/or demonstrate that they could provide the information technology and services cheaper than the current in-house information services unit. Despite some well-publicized horror stories, the data suggest that most businesses are satisfied with their outsourcing deals. The Gartner Group predicts that 50 percent or more of businesses will be outsourcing information systems development by the year 2003. Also, future outsourcing will be based more on *value added*

——————— ✓ ———————

Major IT Outsourcers

Andersen Consulting
Cap/Gemini
Computer Sciences Corporation (CSC)
Electronic Data Systems (EDS)
IBM Corporation

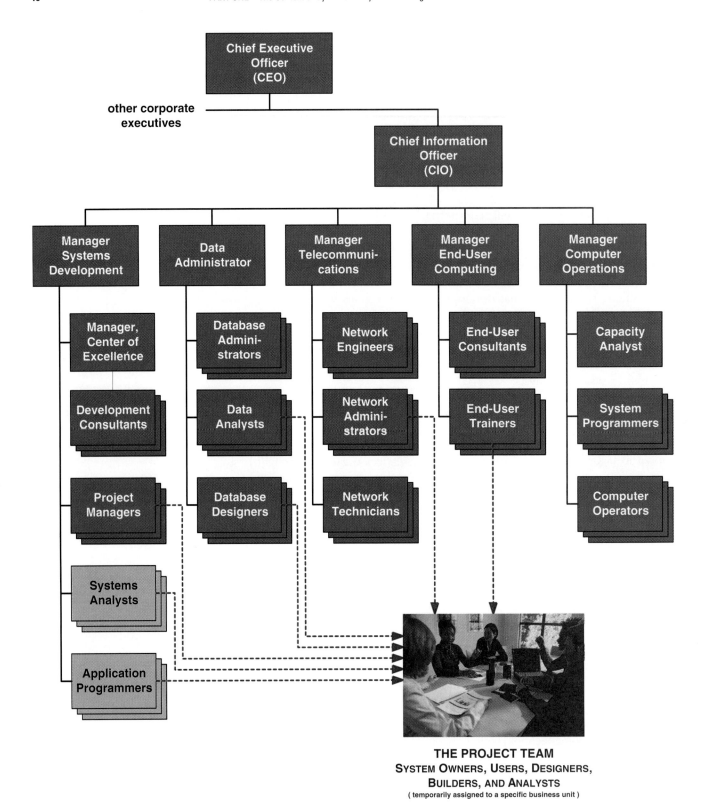

FIGURE 1.3 *A Contemporary Information Services Organization*

to the business than on cost reduction. In other words, the outsourcer will be contracted for technology or expertise that will return real value to the business.

How will outsourcing impact the systems analyst? First, many systems analysts (and other computer professionals) can expect to be absorbed by outsourcers that assume responsibility for an organization's information systems development. When an information services unit or function is outsourced, its employees typically become new employees of the outsourcing vendor. Over time, the outsourcer may retain the employee, terminate the employee, or reassign the employee to a different customer (meaning a different company). While you can understand the anxiety this may cause outsourced employees, Gartner Group data suggest the experience for most information technology professionals has been positive.

A second implication is the employment opportunity afforded by outsourcing vendors. These outsourcers must retain a high-quality, technically competent workforce of information technology managers and specialists, including systems analysts. They must also invest in that workforce to keep it current and maintain their business. Otherwise, their customers will not renew their outsourcing agreements. Thus, outsourcers offer systems analysts many of the advantages of *consulting* (see below), but with one major difference—less travel. Because outsourcing agreements tend to be long term, job assignments last longer than a single project.

Consulting   Another significant trend in information systems development is the use of consultants.

> **Consulting** is the act of contracting with an outside vendor to assume responsibility for or participate in one or more IT projects.

This differs from outsourcing in that the consulting engagement typically ends when the project is completed. It is a shorter-term obligation. Also, the consultants work directly with their client's IT staff on the project. The IT staff members continue to be employees of the client organization, unlike in outsourcing. Examples of well-known management consulting firms are listed in the margin. (Some of the outsourcers previously listed also provide consulting services.)

These management and systems consulting firms build information systems and applications for other organizations. Why wouldn't an organization build all systems through its own information systems unit? Perhaps the information systems unit is understaffed. Perhaps the unit's management is looking for technical expertise that its own staff doesn't (yet) possess. Perhaps management is looking for an unbiased opinion and fresh ideas. The list of reasons is endless.

Systems analysts employed by consulting firms are usually called **systems consultants.** They are lent (for a fee) to the *client* for *engagements* (a consulting term that means "project") that result in a new system for the client. Once the engagement is completed, they are reassigned to a new engagement, usually for a different organization. IT consulting firms represent an attractive employment option for aspiring systems analysts. The engagements tend to be very challenging and provide a wide variety of exposure and experiences. Also, consulting firms tend to keep their consultants on the cutting edge of technology and techniques to better compete for business. For college graduates who are particularly well schooled in the latest systems analysis and design methods, consulting firms represent an interesting and challenging employment alternative.

Independent Software Vendors   One systems analysis opportunity often overlooked by college graduates is the application software market. Independent software vendors (ISVs) build information systems and applications for resale to other businesses. Software vendors may specialize in specific business applications (called *horizontal applications,* such as accounting or human resources), applications for specific industries (called *vertical applications,* such as health care or higher education), or integrated enterprisewide applications (called *enterprise resource planning*).

Software vendors thrive because many businesses have a policy of not building any system they can purchase. Software packages are typically written to the most common business requirements of their intended market—that is, they are designed to meet general requirements and offer limited customizability.

———— ✓ ————

Major IT Consultants

American Management Systems
Andersen Consulting
Ernst & Young
IBM AD Consulting
Perot Systems
Price Waterhouse

The development of packaged software follows a problem-solving approach similar to that of developing custom information systems within an organization. In some ways, their development is more challenging because the software vendor wants its package to appeal to as large a market as possible. Software vendors usually hire two types of systems analysts. The first, a **software engineer,** is responsible for designing (and programming) the software package itself. The second, sometimes called a **sales engineer,** is responsible for helping customers that purchase the package integrate it into their business operations.

MODERN BUSINESS TRENDS AND IMPLICATIONS

Before we begin our study of information systems analysis and design, let's briefly examine the most important business and technical trends that will affect information systems development and stakeholders in the coming decade. Many trends quickly become fads, but here are some we believe will influence systems development in the coming years. Many of these trends are related and integrated such that they form a new business philosophy that will impact the way everyone works in the next century.

Total Quality Management

One major business trend of the 1990s was total quality management (TQM).

> **Total quality management (TQM)** is a comprehensive approach to facilitating quality improvements and management within a business and its products and services.

The key word is *comprehensive.* Businesses have learned that quality is critical for success. They have also learned that quality management does not begin and end with the products and services sold by the business. Instead, it begins with a culture that recognizes that everyone in the business is responsible for quality. TQM commitments require every business function, including information services, to identify quality indicators, measure quality, and make appropriate changes to improve quality.

Many businesses are attempting to formally certify their quality achievements through the *International Standards Organization's (ISO) 9000*-series standards. These standards provide quality requirements that must be met when designing, producing, and delivering products and services to customers. Information technology and services groups participate by becoming certified according to *ISO 9001, Quality systems—Model for quality assurance in design/development, production, installation, and servicing.* This model provides a collection of 20 standards for certifying software and information systems quality.

Another popular framework for software and systems quality certification is the Software Engineering Institute's (SEI) Capability Maturity Model (CMM). Many businesses are beginning to use the CMM within the context of ISO 9001 Certification.

> The **Capability Maturity Model (CMM)** is a framework to assess the maturity level of an organization's information systems development and management processes and products. It consists of five levels of maturity as measured by a set of guidelines called the key process areas.

Our discussions with college graduate recruiters suggest that an "obsessive" attitude toward quality management will become an essential characteristic of successful systems analysts (and all information technology professionals). Throughout this book, quality management, including CMM, will be a theme.

Business Process Redesign

Total quality management has forced many businesses to radically rethink and redesign their fundamental business processes.

> **Business process redesign (BPR)** is the study, analysis, and redesign of fundamental business processes to reduce costs and/or improve value added to the business.

Economic hardship and increased competition initiated the BPR phenomenon. Most businesses are learning that their fundamental business processes have not changed in decades, and that those business processes are grossly inefficient and/or costly. Many processes are overly bureaucratic and do not truly contribute value to the business. While computers have automated many of these business processes, in many cases the inherent inefficiencies in the original business system have not been addressed. The tired excuse for these problems was "we've always done it that way."

Enter business process redesign! A BPR project begins with identification of a *value chain,* a combination of processes that should result in some value added to the business. For example, a value chain might be all those processes that respond to a customer order and result in a satisfied customer. These business processes usually cut across departmental boundaries. Every facet of every process is analyzed for timeliness, bottlenecks, costs, and whether it truly adds value to the organization (or only adds bureaucracy). Business processes are subsequently redesigned for maximum efficiency. Finally, new business processes are analyzed for opportunities for further improvement through information technology.

Some studies suggest that a majority of new applications developments in this decade will be initiated by business process redesign. Systems owners, users, and analysts figure prominently in BPR. System users are the subject matter experts. Systems analysts provide process analysis and system problem-solving skills. This textbook will teach you some introductory business process analysis and design techniques.

Another TQM-related trend is continuous process improvement.

Continuous Process Improvement

> **Continuous process improvement (CPI)** is the continuous monitoring of business processes to effect small but measurable improvements in cost reduction and value added.

In a sense, CPI is the opposite of BPR. Whereas BPR is intended to implement dramatic change, CPI implements a continuous series of smaller changes. Continuous improvement contributes to cost reductions, improved efficiencies, and increased value and profit. Systems analysts may be called on to participate in CPI initiatives for any business process, including the design and implementation of improvements to information technology and applications that support the process.

Globalization of the Economy

The 1990s will be remembered as the era of economic globalization. Competition became global with emerging industrial nations offering lower-cost or higher-quality alternatives to many products. American businesses suddenly found themselves with new international competitors. On the other hand, many U.S. businesses also discovered new and expanded international markets for their own goods and services. Most businesses were forced to reorganize to operate in a global economy.

A related phenomenon has been the trend toward industrial consolidation. News of corporate mergers, acquisitions, and partnerships that cross national boundaries dominate business headlines. In many cases, these acquisitions were intended to stimulate internationalization of products and services.

Economic globalization affects the players in the systems game. Information systems and computer applications must be internationalized. They must support multiple languages, currency exchange rates, international trade regulations, accepted business practices (which differ in different countries), and so forth. Also, most information systems ultimately require information consolidation for performance analysis and decision making. Language barriers, currency exchange rates, transborder information regulations, and the like complicate such consolidation. Finally, there exists a demand for players who can communicate, orally and in writing, with management and users that speak different languages, dialects, and slang. Opportunities for international employment of systems analysts should continue to expand.

Information Technology Trends and Drivers

Advances in information technology can also be business drivers. In some cases, outdated technologies can become significant problems that drive business decisions. In other cases, newer technologies present business opportunities. Three examples of information technology driving business decisions are (1) the year 2000 and euro currency conversions, (2) enterprise resource planning, and (3) electronic commerce.

Year 2000 and the Euro Conversion   Over more than 30 years of developing information systems, most organizations have accumulated a large inventory of legacy systems.

> **Legacy systems** are older information system applications that are crucial to the day-to-day operation of a business, but which may use technologies considered old or outdated by current standards.

Legacy systems were typically developed using technologies that were popular at the time (e.g., *COBOL, CICS, IMS*). Many of the more critical and complex legacy systems have yet to be upgraded to newer technologies because of the cost, skills, and people required. Yet these systems are frequently modified to reflect new or changing business requirements. Occasionally, these changes occur on a large and complex scale. Two such examples include the year 2000 problem and the euro conversion requirement.

You are probably familiar with the widely publicized **year 2000** (Y2K) problem. Analysts believe that latent examples of Y2K problems may occur for several years after the turn of the century. Many computers and computer applications were designed to store dates with only two digits (e.g., *98* = 1998) to reduce storage space required. This presented problems when the millennium changed. For example, if you were born in 1981, and assuming the current year is 2000, a non-Y2K compliant computer application might calculate your age as follows:

$$\text{Age} = \text{Current year} - \text{Year of birth}$$
$$\text{thus}$$
$$\text{Your age} = 00 - 81 = -81 \text{ years old!}$$

How applications would react to such a result vary. Some applications simply crash. Others interpret that you are 0 years old or that you are 81 years old. Because legacy applications included many dates, calculations using dates, and logic based on dates, making those applications Y2K compliant consumed enormous resources during the late-1990s. In other words, a legacy technology problem initiated many information system projects.

Looking ahead, one significant international challenge is currently dominating legacy systems projects—the **European Economic and Monetary Union (EMU).** The EMU is a group of European nations that have agreed to replace their existing European currencies by June 2002 with a single currency called the euro. This could effectively lead to the member countries competing as a single economic power. In the nearer term, existing information systems of other nationalities must be redesigned to also support the euro, especially if they want to do business in the European markets. Many analysts consider this euro-conversion problem to be on the same scale as the aforementioned Y2K compatibility problem!

Enterprise Resource Planning   Historically, information systems in most businesses were developed in-house incrementally over many years. Each system had its own files and databases with loose and awkward integration. During the 1990s, businesses tried to integrate their legacy information systems, usually with poor results. Ideally, an organization would have preferred to redevelop its core business applications (e.g., finance, sales, human resources, and so on) from scratch as a single *integrated* information system. Unfortunately, few if any businesses had enough resources to try this. Recognizing that the basic applications needed by most businesses were similar, the software industry developed a solution to this problem—enterprise resource planning.

An **enterprise resource planning (ERP)** software product is a fully integrated information system that spans most basic business functions required by a major corporation.

An ERP product is built around a common database shared by the basic business functions. Examples of ERP software vendors are listed in the margin.

An ERP solution provides the core information system functions for the entire business. An organization must redesign its business processes to integrate with the ERP solution. Most organizations must still supplement the ERP solution to fulfill business requirements that are unique to the industry or business. An ERP implementation and integration usually represents the largest information system project ever undertaken by a company. It can cost tens of millions of dollars and require an army of managers, users, analysts, technical specialists, programmers, and consultants. ERP implementation and integration requires a new type of systems analyst called a **systems integrator.**

Electronic Commerce   Few technologies have enjoyed as rapid acceptance as the Internet. Internet technologies are changing the rules by which business is conducted. These technologies are also changing the fundamental technical architecture on which many organizations are building their internal information systems. Most organizations are using Internet technologies to build a private internal network called an intranet. And many organizations are using the same technologies to build secure, business-to-business networks called extranets. These technology shifts are driving a significant business trend, electronic commerce.

Electronic commerce (e-commerce or EC) involves conducting both internal and external business over the Internet, intranets, and extranets. Electronic commerce includes the buying and selling of goods and services, the transfer of funds, and the simplification of day-to-day business processes—all through digital communications.

There are three basic types of electronic-commerce-enabled applications.

— *Marketing* of corporate image, products, and services is the simplest form of electronic commerce application. Most businesses have achieved this level of electronic commerce in which the World Wide Web is used merely to "inform" customers about products, services, and policies.

— *Business-to-consumer (B2C)* electronic commerce attempts to offer new, Web-based channels of distribution for traditional products and services. The typical consumer can research, order, and pay for products directly via the Internet. Examples include Amazon.com (for books and music) and E-trade.com (for stocks and bonds). Both companies are examples of businesses created on the Web. Their competition, however, includes traditional businesses that have added Web-based electronic commerce front ends as an alternative consumer option (such as Barnes and Noble and Merrill Lynch).

— *Business-to-business (B2B)* electronic commerce is the real future! This is the most complex form of electronic commerce and could ultimately evolve into electronic business—the complete paperless and digital processing of virtually all business transactions that occur within and between businesses.

One example of B2B electronic commerce is electronic procurement. All businesses purchase raw materials, equipment, and supplies, frequently tens or hundreds of millions of dollars worth per year. EC-based procurement allows employees to browse electronic storefronts and catalogs, initiate purchase requisitions and work orders, route requisitions and work orders electronically for expenditure approvals, order the goods and services, and pay for the delivered goods and completed services—all without the traditional time-consuming and costly paperwork and bureaucracy.

✓

Enterprise Resource Planning Vendors

Baan
J. D. Edwards
Oracle
PeopleSoft
SAP AG

The transition to Web-based information systems and electronic commerce will take several years; however, no business, large or small, will be immune from the transition and an economy based on Internet technology.

PREPARING FOR A CAREER AS A SYSTEMS ANALYST

The principal role in systems projects is played by the systems analyst. What does it take to become a successful systems analyst? One writer suggests the following timeless answer:

> I submit that systems analysts are people who communicate with management and users at the management/user level; document their experience; understand problems before proposing solutions; think before they speak; facilitate systems development, not originate it; are supportive of the organization in question and understand its goals and objectives; use good tools and approaches to help solve systems problems; and enjoy working with people.[2]

This seems like a tall order. It is often difficult to pinpoint those skills and attributes necessary to succeed. However, the following subsections describe the skills most frequently cited by practicing systems analysts.

Working Knowledge of Information Technology

———— ✓ ————

Current Information Technologies

Automatic data capture
Client/server architectures
Component programming languages (e.g., *Visual Basic*)
Electronic commerce
Enterprise resource planning
Graphical user interfaces
Internet, intranet, and extranet
Object programming languages (e.g., *Java*)
Rapid application development
Relational database management systems
Sales force automation
Telecommunications and networking

The systems analyst is an agent of change. He or she is responsible for showing end-users and management how new technologies can benefit their business and its operations. To that end, the analyst must be aware of both existing and emerging information technologies and techniques. Such knowledge can be acquired in college courses, professional development seminars/courses, and in-house corporate training programs. Some technologies and topics that you should be studying today are listed in the margin.

One good way to keep up on what's happening is to regularly skim and read various trade periodicals about computers and information systems. Examples of helpful trade publications include *InfoWeek* and *ComputerWorld*. *InfoWeek* is an outstanding periodical that summarizes the week's news, includes numerous "trend" and "applications" articles, and surveys other periodicals and articles of interest to the busy IS professional. Several trade periodicals also focus specifically on information system development. Examples include *Application Development Trends, Intelligent Enterprise, Component Strategies,* and *Software Development.* Web-based journals such as *Database Programming and Design* are also becoming numerous.

Another way to keep current is through professional association. Consider joining a professional association such as the Association for Information Technology Professionals (AITP), the Association for Computing Machinery (ACM), the Association for Systems Management (ASM), and the Society for Information Management (SIM). Most of these organizations have Web sites.

Computer Programming Experience and Expertise

Whether or not systems analysts write programs, they must know how to program because they are the principal link between business users and computer programmers. Many, but not all, organizations consider experience in computer programming to be a prerequisite to systems analysis and design.

You should not, however, assume that a good programmer will become a good analyst or that a bad programmer could not become a good analyst. There is no such correlation. Unfortunately, many organizations insist on promoting good programmers who become poor or mediocre systems analysts. Worse still, mediocre programmers are often passed over in the belief that they cannot become good analysts.

It is difficult to imagine how systems analysts could adequately prepare business and technical specifications for a programmer if they didn't have *some* programming experience. This experience can be obtained at virtually any college or vocational school.

---

[2] Michael Wood, "Systems Analyst Title Most Abused in Industry: Redefinition Imperative," *ComputerWorld,* April 30, 1979, pp. 24, 26.

Most systems analysts need to be proficient in one or more high-level programming languages. The choice of language is subject to some debate. Most legacy systems were written in *COBOL;* however, few new systems are being designed for that language. Instead, many newer network-based applications are being written in graphical languages such as *Visual Basic.* More sophisticated applications are being built with object-oriented languages such as *C++* or *Java.* In particular, *Java* has become a popular language because of its portability to different computer platforms (e.g., *Macintosh, Windows,* and *UNIX*) as well as its applicability to the Internet, intranets, and extranets (in support of e-commerce).

Increasingly, systems analysts are expected to immerse themselves in the business. They are expected to be able to specify and defend technical solutions that address the bottom-line value returned to the business. This ability comes with experience, but aspiring analysts should take advantage of all opportunities to improve their business literacy (just as aspiring business and management students must improve their computer literacy).

Systems analysts should be able to communicate with business experts to gain knowledge of problems and needs. We strongly suggest you include courses in as many of the subjects listed in the margin as possible. Other specialized courses such as business process redesign and total quality management are highly valued by today's businesses.

By working with business experts, systems analysts gradually acquire business expertise. It is not uncommon for analysts to develop so much expertise over time they move out of information systems and into the user community.

The systems analyst must be able to take a large business problem, break down that problem into its component parts, analyze the various aspects of the problem, and then assemble an improved system to solve the problem. Engineers call this problem-solving process *analysis and synthesis.* The analyst must learn to analyze problems in terms of causes and effects rather than in terms of simple remedies. Being well organized is also part of developing good problem-solving skills.

Analysts must be able to creatively define alternative solutions to problems and needs. Creativity and insight are more likely to be gifts than skills, although they can be developed to some degree. Perhaps the best inspiration for students and young analysts comes from the late USN Rear Admiral Grace Hopper, mother of the *COBOL* language. She suggests, "The most damaging phrase in the English language is, 'We've always done it that way.'" Always be willing to look beyond your first idea for other solutions.

If you're looking for a general education or liberal arts elective, research the offerings of your philosophy department, which may include courses on general problem solving and logic. This information can be adapted to the systems problem solving addressed in this book.

Without exception, an analyst must be able to communicate effectively, both orally and in writing. The analyst should actively seek help or training in business writing, technical writing, interviewing, presentations, and listening. These skills can be learned, but most of us must force ourselves to seek help and work hard to improve them. College recruiters and business managers emphasize that communication skills are the single most important ingredient for success. Many college seniors lose job offers because of their inability to write and speak at the expected level, and others face slow career development because of inadequate communication skills. Almost without exception, your communication skills, not your technical skills, will be the single biggest factor in your career success or failure. Most schools offer several courses that can help you improve your interpersonal communication skills. Representative titles are listed in the margin.

General Business Knowledge

——————— ✓ ———————

Business Literacy Subjects
Accounting
Business law and ethics
Economics
Finance
Manufacturing
Marketing
Operations management
Organizational behavior

Problem-Solving Skills

Interpersonal Communication Skills

——————— ✓ ———————

Interpersonal Communication Subjects
Business speaking
Business writing
Interviewing
Listening
Persuasion
Technical discussion
Technical writing

Interpersonal Relations Skills

As illustrated in Figure 1.4, systems analysts are the facilitators of information systems development. The analyst may be the only individual who sees the system as a whole, the "big picture." As a facilitator, the systems analyst needs "to exercise the boldness of Lady Godiva, the introspection of Sherlock Holmes, the methodology of Andrew Carnegie, and the down-home common sense of Will Rogers."[3] Systems work is people-oriented and systems analysts must be extroverted. Interpersonal skills help us work effectively with people. Although these skills can be developed, some people simply do not possess the necessary outgoing personality. The interpersonal nature of systems work is demonstrated in the SoundStage case study that appears throughout this book.

Interpersonal skills are also important because of the political nature of the systems analyst's job. The analyst's first responsibility is to the business, its management, and its workers. Individuals in an organization frequently have conflicting goals and needs. They have personality clashes. They fight turf battles over who should be responsible for what and who should have decision authority over what. The analyst must mediate such problems and achieve benefits for the business as a whole.

Another aspect of interpersonal relations is the analyst's role as an agent of change. The systems analyst is frequently as welcome as an IRS auditor is! Many individuals feel comfortable with the status quo and resent the change the systems analyst brings. An analyst should study the theory and techniques of effecting change. Persuasion is an art that can be learned. Begin by studying sales techniques—after all, systems analysts sell change.

FIGURE 1.4
The Systems Analyst as a Facilitator

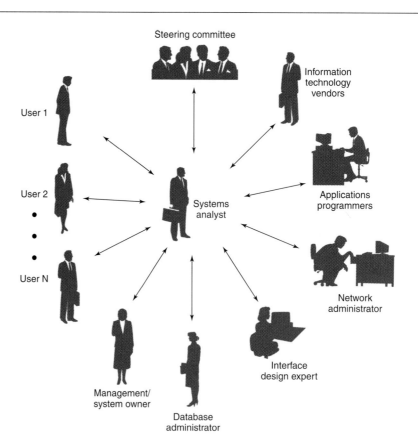

---

[3] Kenniston W. Lord, Jr., and James B. Steiner, *CDP Review Manual: A Data Processing Handbook,* 2nd ed. (New York: Van Nostrand Reinhold, 1978), p. 349.

Finally, systems analysts work in teams composed of IS professionals, end-users, and management. Being able to cooperate, to compromise, and to function as part of a team is critical for success in most projects. Because development teams include people with dramatically different levels of education and experience, group dynamics is an important skill to develop.

Many business or organizational leadership schools offer valuable courses on topics such as change management, team dynamics, leadership, and conflict resolution. If available, these courses can contribute significantly to the career growth of the systems analyst.

Flexibility and Adaptability

No two systems development projects encountered by a systems analyst are identical. Each project offers its own unique challenges. Thus, there is no single, magical approach or solution applicable to systems development. Successful systems analysts recognize this and learn to be flexible and adapt to special challenges or situations.

Many organizations have standards that dictate specific approaches, tools, and techniques that must be adhered to when developing a system. Although these standards should be followed as closely as possible, the systems analyst must be able to recognize when variations on (or exceptions to) those standards are necessary and beneficial to a particular project. At the same time, the analyst must be aware of the implications of not following the standards. It's a balancing act that usually improves with experience.

Character and Ethics

The nature of the systems analyst's job requires a strong character and sense of ethics.

> **Ethics** is a personal character trait in which an individual understands the difference between "right" and "wrong" and acts accordingly.

Systems analysts often encounter sensitive information when developing systems. It could be a file of a company's pricing structure for a bid or employee profile data that may include salaries, performance history, medical history, and career plans. Analysts must protect the security and confidentiality of any data they have been entrusted with.

Systems analysts also frequently uncover dissent in the ranks of employees and gain access to sensitive and private data and information (through sampling of memos, files, and forms) about customers, suppliers, employees, and the like. The analyst must be very careful not to share such feelings or information with the wrong people. Trust is sacred! Confidence is earned!

Systems analysts also design systems and write programs. But who owns such intellectual property? In most cases, the design and programs are the property of the organization because it paid for the services of the analyst and programmers. It would be unethical to take (or sell) such designs and programs to another company.

Most computer professional societies such as AITP have a code of conduct and code of ethics for their members to adhere to in conducting business. The following paragraphs are an excerpt from AITP's Code of Ethics relating to the protection of information:

> Further, I shall not use knowledge of a confidential nature to further my personal interest, nor shall I violate the privacy and confidentiality of information entrusted to me or to which I may gain access.
>
> That I have an obligation to my employer whose trust I hold, therefore, I shall endeavor to discharge this obligation to the best of my ability, to guard my employer's interest, and to advise him or her wisely and honestly.[4]

Many corporations require their employees to attend annual training seminars on company ethics. Code of conduct and ethics statements may be displayed in visible places in the workplace. Ethics manuals may be stored in convenient

---

[4] From the website of AITP: www.aitp.org/about/code_of_ethics.html

locations or on the company intranet. Violations of ethics statements can lead to disciplinary action or even termination.

The Computer Ethics Institute is a nonprofit research, education, and policy study organization based in Washington, D.C. It strives to make people more aware of computer ethics and to use computers more responsibly; a primary goal is to make computer ethics part of the standard school curriculum. To promote computer ethics awareness, the institute has published "The Ten Commandments of Computer Ethics" which appears in Figure 1.5.

Systems Analysis and Design Skills

All systems analysts need thorough and ongoing training in systems analysis and design. Systems analysis and design skills can be conveniently factored into three subsets—concepts and principles, tools, and techniques. This book begins that training.

When all else fails, the systems analyst that remembers the basic concepts and principles of systems work will still succeed. No tool, technique, process, or methodology is perfect in all situations! Concepts and principles will help you adapt to new and different situations and methods as they become available. We have purposefully emphasized applied concepts and principles in this book. This is not a mechanical, "monkey see, monkey do" book! We believe that if you carefully study the concepts presented in Part One, you will be better able to communicate with potential employers, business users, and computer programmers alike. Also note the references at the end of this chapter for books that emphasize problem solving.

Not too long ago, it was thought that the systems analyst's only tools were paper, pencil, and flowchart template. Over the years, several tools and techniques have been developed to help the analyst. Today, a new generation of computer-based tools is emerging. Tools and their associated techniques help the analyst build systems faster and with greater reliability. This book comprehensively covers the modern tools and techniques of the trade.

Techniques are specific approaches for applying tools in a disciplined manner to successfully develop systems, and we will present the most popular through-

---

FIGURE 1.5
Ethics for Systems Analysts

The Ten Commandments of Computer Ethics

1. Thou shalt not use a computer to harm other people.

2. Thou shalt not interfere with other people's computer work.

3. Thou shalt not snoop around in other people's computer files.

4. Thou shalt not use a computer to steal.

5. Thou shalt not use a computer to bear false witness.

6. Thou shalt not copy or use proprietary software for which you have not paid.

7. Thou shalt not use other people's computer resources without authorization or proper compensation.

8. Thou shalt not appropriate other people's intellectual output.

9. Thou shalt think about the social consequences of the program you are writing or the system you are designing.

10. Thou shalt always use a computer in ways that insure consideration and respect for your fellow humans.

Source: Computer Ethics Institute.

out this book. You will learn to use and integrate these techniques and avoid the pitfall of blind devotion to any one technique.

Systems analysis and design techniques are constantly evolving, so sensible systems analysts take advantage of any opportunity to improve their skills. Books provide the easiest source of self-improvement. Forward-thinking organizations are willing to invest in courses and seminars to keep their analysts current.

THE NEXT GENERATION

The life of a systems analyst is both challenging and rewarding. But what are the prospects for the future? Do organizations need systems analysts? Will they need them into the foreseeable future? Is the job changing for the future, and if so, how? These questions are addressed in this section.

Career Prospects

According to the U.S. Office of Employment Predictions, demand for information technology workers such as systems analysts, programmers, and other computer specialists is far exceeding the supply of workers available:

> Reports by groups such as the Information Technology Association and the U.S. Department of Commerce's Office of Technology identify what they consider strong evidence of the United States' inability to keep up with the high demand for information technology workers. As a result, highly qualified personnel enjoy rising starting salaries, multiple job offers, creative recruiting efforts, and a hiring climate that some have equated with a pro sports draft.[5]

By some counts, there exist more than 300,000 vacant IT jobs in the United States alone. This deficit is growing because the need is expanding faster than colleges are producing new qualified workers.

For systems analysts, the job outlook is especially bright. According to the Bureau of Labor Statistics, opportunities for systems analysts are expected to increase much faster than the average for all professions, significantly more than for programmers. Between 1996 and 2006, businesses will need as many as 520,000 new systems analysts, an increase of 103 percent since 1996.[6] During this same period, the overall job force is expected to increase only 8 to 22 percent. The demand is increasing because industry needs systems analysts to meet the seemingly endless need for more computer-based systems. Opportunities for success will be the greatest for the most qualified, skilled, and experienced analysts.

There are many career paths for the successful systems analyst. Some analysts leave the information systems field and join the user community. Their experience in developing business applications, combined with their total systems perspective, can make experienced analysts unique business specialists. Alternatively, analysts could become project managers, information systems managers, or technical specialists (for database, telecommunications, microcomputers, and so forth). Finally, skilled systems analysts are often recruited by the consulting and outsourcing industries. The career path opportunities are virtually limitless.

Predictions

As with any profession, systems analysts can expect change. We believe that organizations will become increasingly dependent on external sources, such as consultants and outsourcers, for their analysts. This will be driven by such factors as the complexity and rapid change of technology, the desire to accelerate systems development, and the continued difficulty in recruiting, retaining, and retraining skilled systems analysts (and other information technology professionals). In many cases, internally employed systems analysts will manage projects and consulting or outsourcing agreements.

---

[5] Carolyn Veneri, "Here Today, Jobs of Tomorrow: Opportunities in Information Technology," *Occupational Outlook Quarterly,* Fall 1998, p. 45.

[6] Ibid, pp. 47–53.

WHERE DO YOU GO FROM HERE?

Each chapter will provide guidance for self-paced instruction under the heading, "Where Do You Go from Here?" Recognizing that different students and readers have different backgrounds and interests, we will propose appropriate learning paths—most within this book, but some beyond the scope of this book.

Most readers should proceed directly to Chapter 2. The first four chapters provide much of the context for the remainder of the book. Several recurring themes, models, and terms are introduced in those chapters to allow you to define your own learning path from that point forward. This chapter focused on the *people* who develop information systems. Chapter 2 will take a closer look at the *product* itself—information systems—from an architectural perspective appropriate for system development. Chapter 3 examines the *process* of system development. Chapter 4 completes the foundation with an introduction to the *management* of system development.

We believe that an increasing percentage of tomorrow's systems analysts will not work in the information systems department. Instead, they will work directly for a business unit. This will enable them to better serve their users. It will also give users more power over what systems are built and supported.

Finally, we also believe that a greater percentage of systems analysts will come from noncomputing backgrounds. At one time most analysts were computer specialists. Today's computer graduates are becoming more business literate. Similarly, today's business and noncomputing graduates are becoming more computer literate. Their full-time help and insight will be needed to meet demand and to provide the business background necessary for tomorrow's more complex applications.

SUMMARY

1. An information system is an arrangement of people, data, processes, information presentation, and information technology that interact to support and improve the operations in a business and to support the problem-solving and decision-making needs of management and users.

2. Information technology is a contemporary term that describes the combination of computer technology (hardware and software) with telecommunications technology (data, image, and voice networks).

3. Information workers are the stakeholders in information systems. Information workers include those people whose jobs involve the creation, collection, processing, distribution, and use of information. They include:
   a. System owners, the sponsors and chief advocates of information systems.
   b. System users, the people who use or are affected by the information system on a regular basis. Geographically, system users may be internal, mobile, remote, or external.
   c. System designers, technology specialists who translate system users' business requirements and constraints into technical solutions.
   d. System builders, technology specialists who construct the information system components based on the design specifications.

   e. Systems analysts, the people who facilitate the development of information systems and computer applications. They coordinate the efforts of the owners, users, designers, and builders. Frequently, they may play one of those roles as well. Systems analysts perform systems analysis and design.
      i) Systems analysis is the study of a business problem domain to recommend improvements and specify the business requirements for the solution.
      ii) Systems design is the specification or construction of a technical, computer-based solution for the business requirements identified in a systems analysis.
   f. Information technology vendors and consultants, the people who bring external expertise, experience, and human resource capacity to many projects.

4. The systems analyst facilitates most of the activities to develop or acquire an information system. A systems analyst studies the problems and needs of an organization to determine how people, data, processes, communication, and information technology can best accomplish improvements for the business.

5. A business analyst is a systems analyst that specializes in business problem analysis and technology-independent requirements analysis.

6. Most systems analysts work for the systems development group in the information services group or department of a business. Alternatively systems analysts (under different titles) work for IT consulting firms, outsourcers, and application software vendors.

7. Players in the information systems game are being affected by a number of business and technology trends including:

   a. Total quality management, a comprehensive approach to facilitating quality improvements and management within a business.

   b. Business process redesign, the study, analysis, and redesign of fundamental business processes to reduce costs and/or improve value added to the business.

   c. Continuous process improvement, the continuous monitoring of business processes to effect small but measurable improvements to cost reduction and value added.

   d. The year 2000 compatibility problem and the euro currency conversion directive.

   e. Enterprise resource planning, the selection and implementation of a single vendor's fully integrated information system that spans most basic business functions required by a major corporation.

   f. Electronic commerce, a new way of doing business that involves conducting both internal and external business over the Internet, intranets, and extranets.

8. The systems analyst is a principal player in the information systems game. To prepare for a career as a systems analyst you will need to gain:

   a. A current working knowledge of information technology, which you must keep current.

   b. Some computer programming experience and expertise.

   c. General business knowledge and, if possible, business experience.

   d. Better than average problem-solving skills.

   e. Better than average interpersonal communication skills.

   f. Strong interpersonal relations and teamwork skills.

   g. Flexibility and adaptability to change as the problem and politics of business change.

   h. Good character and strong ethics, necessary because analysts gain access to sensitive and confidential data, facts, and opinions.

   i. Systems analysis and design skills—the subject of this book!

9. Prospects for a career as a systems analyst look bright through 2006 and beyond. A shortage of analysts will fuel the prospects for those experienced analysts who have remained current with the fast-changing world of information technology.

KEY TERMS

business analyst, p. 14
business process redesign (BPR), p. 20
Capability Maturity Model (CMM), p. 20
center for excellence, p. 15
chief information officer (CIO), p. 15
clerical and service worker, p. 11
client, p. 11
consulting, p. 19
continuous process improvement (CPI), p. 21
electronic commerce (e-commerce, EC), p. 23
enterprise resource planning (ERP), p. 23

ethics, p. 27
European Economic and Monetary Union (EMU), p. 22
executive manager, p. 12
information system (IS), p. 8
information technology (IT), p. 8
information worker, p. 9
knowledge worker, p. 11
legacy system, p. 22
middle manager, p. 12
mobile user, p. 12
outsourcing, p. 17
remote user, p. 12
sales engineer, p. 20
software engineer, p. 20

stakeholder, p. 8
supervisor, p. 12
system builder, p. 13
system designer, p. 12
system owner, p. 9
system user, p. 11
systems analysis, p. 13
systems analyst, pp. 13, 14
systems consultants, p. 19
systems design, p. 13
systems integrator, p. 23
technical and professional staff, p. 11
telecommuting, p. 12
total quality management (TQM), p. 20
year 2000 (Y2K), p. 22

REVIEW QUESTIONS

1. What is the difference between information systems and information technology?

2. Who are the six stakeholder groups in information systems?

3. Differentiate between information workers and knowledge workers.

4. What is a system owner's role in information systems development?

5. Differentiate between system owners and system users?

6. Differentiate between internal, remote and mobile, and external system users.

7. What are a system designer's and system builder's roles in information systems development?

8. What is the systems analyst role in systems development as it relates to the other stakeholders?

9. Differentiate between systems analysis and systems design.

10. What roles do information technology vendors and consultants play in the systems game?

11. What is a systems analyst, and what is his or her responsibility with respect to information technology?

12. Differentiate among a systems analyst, business analyst, and programmer/analyst.

13. Differentiate among three types of information systems problems encountered by analysts and other stakeholders.

14. List the six steps of the general problem-solving approach.

15. With respect to the systems analyst role, what is the difference between the traditional and modern information services organizations.

16. What is the purpose of a "systems development" center for excellence?

17. Differentiate between outsourcing and consulting with respect to systems analysts.

18. What is total quality management and how does it affect players in the systems game?

19. List two types of quality certification applicable to information systems.

20. What is business process redesign and how does it affect systems analysts?

21. Differentiate between business process redesign and continuous process improvement.

22. What are legacy systems? Name two types of problems encountered in legacy systems.

23. What is enterprise resource planning and what is its relationship to information systems and development?

24. What is electronic commerce? List types of electronic commerce information systems applications.

25. What is ethics and how does it affect systems analysts?

PROBLEMS AND EXERCISES

1. For each of the following classifications of system users, identify the individual as clerical staff, technical and professional staff, supervisory staff, middle management, or executive management. Defend your answer.
   a. Receptionist
   b. Shop floor foreman
   c. Financial manager
   d. Assistant store manager
   e. Chief operating officer
   f. Manufacturing control manager
   g. Terminal operator/data entry
   h. Applications programmer
   i. Programmer/analyst
   j. Warehouse clerk who fills orders
   k. Stockholder
   l. Product engineer
   m. Consultant
   n. Broker

2. Give three examples of system users from each of the following classifications: clerical workers, technical and professional staff, supervisory staff, middle management, and executive management. Explain the job responsibilities of each example you provided and state why your example represents that particular classification of system user. For each system user, describe a situation that would make that worker a client of a systems analyst.

3. Explain how the existence of remote users might complicate information system design. Describe a system in which *you* might be considered a remote user.

4. Consider an organization by which you were, or are, employed in any capacity. Identify the information workers at each level in the organization according to their classification of system user. (Alternative: Substitute your school for the organization. Students are part of the clerical staff classification. Do you see why?)

5. Consider your college course registration system. List as many stakeholders as possible in that system. Classify each as owners, users, designers, builders, and analysts.

6. Write a job description for a systems analyst.

7. Based on this chapter, write a 100-word advertisement for a systems analyst. Compare your advertisement with actual job ads. (Do not be concerned about significant differences, but do attempt to explain why those differences might occur.)

8. Using the general problem-solving steps described in this chapter, write a letter to your instructor that proposes development of an improved personal financial management system (similar to *Quicken* or *Money*). Tell your instructor what has to be done. Assume your instructor knows nothing about computers or systems development. In other words, be careful with your use of new terms.

9. Kathy Thomas has been asked to reclassify her systems analysts. Virtually all her analysts perform systems analysis, systems design, and application programming. However, depending on their experience, the percentage of time in these activities varies. Younger analysts do 80 percent programming, whereas the most experienced analysts do 80 percent systems analysis—largely because of their greater understanding of the business and its users and management. How should Kathy reclassify her personnel?

10. Construct a matrix that maps the roles of the systems analyst versus the programmers in terms of the general problem-solving methodology described in this chapter. Next, add a column for the roles of system owners and system users.

11. Diversified Plastics, Inc., has adopted an unusual information systems organization. The Management Systems group consists of systems analysts that perform only systems analysis and general systems design. The Technical Systems group consists of *application developers* that

perform detailed systems design, application programming and testing, and systems implementation and maintenance. History and unwritten policy dictates that all systems analysts must come from a business, engineering, or management education and backgrounds—with no computer experience required. Application developers must come from a computing education or background. Transfers between the groups are discouraged and in many cases not allowed. What are the advantages and disadvantages of such an organization and its policies?

12. Federated Mortgages' corporate information officer (CIO) is facing a budget dilemma. End-users have been buying personal computers (PCs) at an alarming rate. In a sense, the business users of these PCs, who have little or no formal information systems or technology education, are developing their own small-scale information systems. Because of this, the CIO has been asked to justify the continued growth of his own information systems and technology budget, especially the growth of his programming and systems analysis staff. Some believe that the number of programmers and analysts should be reduced. How can the CIO justify his staff? Will the roles of his programmers and systems analysts change? If so, how? Can the users completely replace the programmers and analysts?

13. Which of the following environments would you prefer to work in and why? An information services unit of a business? An outsourcer? A consulting firm? A software vendor?

14. You work for Total Recall, Inc. As part of an information systems project, you discover that offers to new employees require four signatures: the employee, the employee's manager, the manager's director, and the vice president of human resources. Having secured those four signatures, the employee's manager initiates a process that authorizes the employee to be paid via the payroll system. To complete this step, the employee must secure seven signatures, three of which were on the original contract. Comment on this process with respect to business process redesign.

15. Why would a minor in international studies (consisting of courses in a foreign language, cultural issues, international economics, and international business) add value to a degree in systems analysis and design?

16. How might a company's decision to use enterprise resource planning applications affect information systems development?

17. How will the Internet, intranets, and extranets change the rules of information systems development?

18. The students in an introductory programming course would like to know how systems analysis and systems design differ from computer programming. Specifically, they want to know how to choose between the two careers. Help them by explaining the differences between the two and pointing out factors that might influence their decisions.

19. Your company has given your team funds to attend a professional development course on some technology. One team member misses half the sessions in order spend time with his spouse on local sightseeing tours. Comment on the ethics of this situation.

20. You are assigned to the new payroll system project. You are aware of your best friend's dissatisfaction with his salary in the procurement department. Because you have read-access privileges to all human resource files, he asks you to provide salary (or average salary) information for all procurement staff at the same level and experience as his own. Comment on the ethics of this situation.

PROBLEMS AND RESEARCH

1. Make an appointment to visit with a systems analyst or programmer/analyst in a local business. Try to obtain a job description from the analyst. Compare that job description with the roles and responsibilities described in this chapter.

2. Visit a local information systems/services department in your business community. Compare its organization with the organizations described in this chapter. Are the five centers of activity present? What are they called? How is its organization different? How is it better? Do you see any disadvantages? Where do the systems analysts fit in?

3. Visit a local information systems/services department in your business community. How are its project teams formed? How are those teams organized? How does this compare with the organization described in this chapter?

4. Based on the systems analyst's job characteristics and requirements described in this chapter, evaluate your own skills and personality traits. In what areas would you need to improve?

5. Your library probably subscribes to at least one big-city newspaper. Additionally, your library, academic department, or instructor may subscribe to an information technology periodical such as *ComputerWorld*. Study the job advertisements for systems analysts and programmer/analysts. What skills are being sought? What experience is being required? How are those skills and experiences important to the role of the analyst as described in this chapter?

6. Prepare a curriculum plan of study for your education as a prospective systems analyst. If you are already working, prepare a statement that expresses your personal need for continuing education to become a systems analyst.

7. A systems analyst applies new technologies to business and industrial problems. As a prerequisite to this "technology transfer," the analyst must keep abreast of the latest trends and techniques. The best way to accomplish this is to develop a disciplined reading program. This extended project will help you develop this program.
   a. Visit your local school, community, or business library. Make a list of all computing, information technology, and information systems-oriented publications.
   b. Skim two or three issues of each publication to get a feel for their contents and orientation. Select the five

periodicals that you find most interesting and helpful. We recommend that you select five publications that address the areas of personal computing, enterprisewide computing, networking, software development, and data management.

c. Set up a browsing schedule. This should consist of one or two hours a week that you will spend browsing the journals. You should try to maintain this schedule for 10 to 15 weeks. If you miss a week, make it up within one week.

d. Set up a journal to track your progress. Record the date, the journals browsed, the title or subject of the cover story or headlines, and the title of one other article that caught your eye.

e. Learn to browse. You won't have time to read. If you try to read everything, you will get discouraged and

quit the program. Instead, study the table of contents. Read only the first paragraph or two of each article along with any highlighted text in the article. Read the conclusion or last paragraph of the article. Then move on to the next article no matter how interesting the present one. Note any article that you want to fully read after browsing your reading list.

f. After browsing each of your selected publications, select at least two articles to read thoroughly. The number you read is limited only by your interest and available time. Record these articles in your journal.

This project will show you how to keep up with a rapidly changing technological world without consuming excessive time and effort.

MINICASES

1. For whom should the systems analyst work? Rolland Industries is facing an information services reorganization dilemma. Non-IS management is pressing for a new structure whereby most systems analysts would directly report to their business user group (such as Accounting, Finance, Manufacturing, Personnel) as opposed to reporting to the Information Services group. Non-IS management believes that, in the existing structure, systems analysts are too influenced to "change everything" for the sake of computing because they report to Information Services. To ensure that systems meet IS standards, a small contingent of analysts would remain in IS as a quality assurance group that has final sign-off on all systems projects.

IS managers are resisting this change. They believe systems analysts will become technologically "out of tune" if removed from IS. They also feel that separating the systems analysts from one another will result in less sharing of ideas and, subsequently, reduce innovation. They also think data files and programs will be unnecessarily duplicated. Conflicts between analysts and programmers, who will remain in IS, will likely increase. IS also believes users will hire new systems analysts without regard to programming and technical experience or familiarity with IS's technical environment.

Systems analysts are split on the issue. They see the benefits of users being more directly in control of their own systems' destinies; however, they are concerned that users and user management will be less forgiving when faced with budget overruns and schedule delays that have historically plagued projects. Analysts are also concerned that they will become more prone to technological obsolescence if they are physically relocated outside of IS and its more technically oriented staff.

The decision will likely be made at a higher level than IS. What do you think should be done?

2. The following fable, which appears in a book by Gerald Weinberg called *Rethinking Systems Analysis and Design,*

is an entertaining yet effective minicase. Read and enjoy the fable, but pay attention to its moral. Many systems analysts have discovered the moral of this fable the hard way!

Three ostriches had a running argument over the best way for an ostrich to defend himself. Although they were brothers, their mother always said she couldn't understand how three eggs from the same nest could be so different. The youngest brother practiced biting and kicking incessantly and held the black belt. He asserted, "The best defense is a good offense." The middle brother, however, lived by the maxim that "he who fights and runs away, lives to fight another day." Through arduous practice, he had become the fastest ostrich in the desert, which you must admit is rather fast. The eldest brother, being wiser and more worldly, adopted the typical attitude of mature ostriches: "What you don't know can't hurt you." He was far and away the best head-burier that any ostrich could recall.

One day a feather hunter came to the desert and started robbing ostriches of their precious tail feathers. Now, an ostrich without his tail feather is an ostrich without pride, so most ostriches came to the three brothers for advice on how best to defend their family honor. "You three have practiced self-defense for years," said their spokesman. "You have the know-how to save us, if you will teach it to us." And so each of the three brothers took on a group of followers for instruction in the proper method of self-defense—according to each one's separate gospel.

Eventually, the feather hunter turned up outside the camp of the youngest brother, where he heard the grunts and snorts of all the disciples who were busily practicing kicking and biting. The hunter was on foot but armed with an enormous club, which he brandished menacingly as the youngest brother went out undaunted to engage him in combat. Yet fearless as he was, the ostrich was no match for the hunter, because the club was much longer than an ostrich's leg or neck. After taking many lumps and bumps, and not getting in a single kick or bite, the ostrich fell exhausted to the ground. The hunter casually plucked his precious tail feather, after which all his disciples gave up without a fight.

When the youngest ostrich told his brothers how his feather had been lost, they both scoffed at him. "Why didn't you run?"

demanded the middle one. "A man cannot catch an ostrich."

"If you had put your head in the sand and ruffled your feathers properly," chimed the eldest, "he would have thought you were a yucca and passed you by."

The next day the hunter left his club at home and went out hunting on a motorcycle. When he discovered the middle brother's training camp, all the ostriches began to run, the brother in the lead. But the motorcycle was much faster, and the hunter simply sped up alongside each ostrich and plucked his tail feather on the run.

That night the other two brothers had the last word. "Why didn't you turn on him and give him a good kick?" asked the youngest. "One solid kick and he would have fallen off that bike and broken his neck."

"No need to be so violent," added the eldest. "With your head buried and your body held low, he would have gone past you so fast he would have thought you were a sand dune."

A few days later, the hunter was out walking without his club when he came upon the eldest brother's camp. "Eyes under!" the leader ordered and was instantly obeyed. The hunter was unable to believe his luck, for all he had to do was walk slowly among the ostriches and pluck an enormous supply of tail feathers.

When the younger brothers heard this story, they felt impelled to remind their supposedly more mature sibling of their advice. "He was unarmed," said the youngest. "One good bite on the neck and you'd never have seen him again."

"And he didn't even have that infernal motorcycle," added the middle brother. "Why, you could have outdistanced him at half a trot."

But the brothers' arguments had no more effect on the eldest than his had on them, so they all kept practicing their own methods while they patiently grew new tail feathers.[7]

a. What is the moral of the fable?
b. This book will teach you a variety of tools, techniques, and methodologies for developing information systems and computer applications. How might the fable and its moral relate to your study of systems analysis and design?

SUGGESTED READINGS

Dejoie, Roy; George Fowler; and David Paradice. *Ethical Issues in Information Systems*. Boston: Boyd and Fraser Publishing Company, 1991. This is an excellent textbook for teaching ethics in an MIS curriculum. It is a collection of readings that should be must reading for any capstone MIS graduate.

Gartner Group IT Symposium and Expo (annual). Our institution's management information unit has long subscribed to the Gartner Group's service that reports on industry trends, the probabilities for success of trends and technologies, and suggested strategies for information technology transfer. Members of the author team have been fortunate to be able to attend the annual symposium. Gartner Group reports and symposiums have influenced each edition of the book.

Gause, Donald, and Gerald Weinberg. *Are Your Lights On? How to Figure Out What the Problem REALLY Is*. New York: Dorset House Publishing, 1990. Here's a short and easy-to-read book about general problem solving. You can probably read the entire book in one night, and it could profoundly improve your problem-solving potential as a systems analyst (or any other profession). We consider Dr. Weinberg one of the most influential systems analysis educators anywhere.

MacDonald, Robert. *Intuition to Implementation: Communicating about Systems Toward a Language of Structure in Data Processing System Development*. Englewood Cliffs, NJ: Prentice Hall, Yourdon Press Computing Series, 1987. This is a good conceptual systems problem-solving textbook. Although copyrighted in 1987, its timeless wisdom demonstrates that system concepts have withstood the test of time.

Weinberg, Gerald. *Rethinking Systems Analysis and Design*. New York: Dorset House Publishing, 1988. Don't let the date fool you. This is one of the best and most important books on this subject ever written. This book may not teach any of the popular systems analysis and design methods of our day, but it challenges the reader to leap beyond those methods to consider something far more important—the people side of systems work. Dr. Weinberg's theories and concepts are presented in the context of dozens of delightful fables and short experiential stories. We are grateful to him for our favorite systems analysis fable of all time, *The Three Ostriches*, as reprinted in the last minicase of this chapter.

Yourdon, Edward. *Modern Structured Analysis*. Englewood Cliffs, NJ: Prentice Hall, Yourdon Press Computing Series, 1989. This book was the initial source of our classification scheme for end-users. Yourdon's coverage is in Chapter 3, "Players in the Systems Game."

---

[7] Gerald M. Weinberg, *Rethinking Systems Analysis and Design,* pp. 23–24. Copyright © 1988, 1982 by Gerald M. Weinberg. Reprinted by permission of Dorset House Publishing, 353 W. 12th ST, New York, NY 10014 (212-620-4053/1-800-DH-BOOKS/www.dorsethouse.com).

Focus on PEOPLE

Focus on DATA

Focus on PROCESSES

Focus on INTERFACES

SYSTEMS DEVELOPMENT

Stakeholders

BUILDING BLOCKS OF AN INFORMATION SYSTEM

Activities

S Y S T E M S

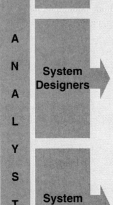

System Owners

List of business entities and rules ...

Business Knowledge

List of business events and responses ...

Business Functions

List of business locations and systems ...

Business Locations

System Users

Data Requirements

Process Requirements

Interface Requirements

A N A L Y S T S

System Designers

Database Schema

Application Schema & Specs

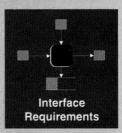

Interface Specifications

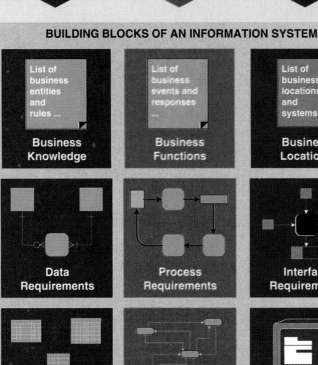

Systems Analysis and Design Methods

the "process"

System Builders

```
CREATE TABLE tblOrders
  colOrderNot CHAR(5) NOT
    NULL
  colOrderDate  DATE/TIME
NOT
    NULL ...
```

Database Programs

```
Sub Main ()
  frmLogin.Display
  If frmLogin.Cancel Then
    MsgBox "User clicked on
      exit"
  Else ...
```

Application Programs

```
<html>
<head>
<title> Order Entry Form </
title> ...
```

Interface Programs

Vendors and Consultants

INFORMATION TECHNOLOGY
the "enabler"

2

INFORMATION SYSTEM BUILDING BLOCKS

CHAPTER PREVIEW AND OBJECTIVES

Systems analysis and design methods are used to develop information systems for organizations. Before learning the *process* of building systems, you need a clear understanding of the *product* you are trying to build. This chapter takes an architectural look at information systems and applications. We will build a framework for information systems architecture that will subsequently be used to organize and relate all the chapters in this book. The chapter will address the following areas:

- Describe the difference between data and information.
- Define the product called an *information system,* and describe the role of information technology in information systems.
- Differentiate between *front-* and *back-office* information systems.
- Describe five classes of information system applications (transaction processing, management information, decision support, expert, and office automation systems) and how they interoperate.
- Describe the role of information systems architecture in systems development.
- Name six groups of stakeholders in information systems development.
- Name three focuses for information systems.
- Describe four perspectives of the DATA focus for an information system.
- Describe four perspectives of the PROCESS focus for an information system.
- Describe four perspectives of the INTERFACE focus for an information system.
- Describe the role of a computer network as it relates to DATA, PROCESSES, and INTERFACES.

SOUNDSTAGE

SOUNDSTAGE ENTERTAINMENT CLUB

SCENE

A conference room where Sandra Shepherd has called a project launch meeting. Sitting around the table are Bob Martinez (a teammate), Galen Kirchoff (vice president of Member Services), Steven Siemers (club director of the Audio Club), Susan Crane (club director of the Game Club), Steve Segal (interim club director of the Video Club), Monica Blair (director of Marketing), Oscar Mann (director of Customer Services), and Dick Krieger (director of Warehouse Operations).

SANDRA

Good morning! I see that you all found the coffee, juice, and rolls. Thank you for coming. As you know, we are launching a new systems project this morning. This is the launch meeting, and our goal is simply to orient everyone so that we can assemble the team and establish a vision.

First, I'd like to introduce you to two new faces. On my left is Bob Martinez. Bob has just joined Information Services as a systems analyst and will be assigned to the project. On my right is Steve Segal who has joined us from our new Silver Screenings Video Club acquisition. His title is acting club director of the Video Club. Welcome to both of you. I know everybody introduced one another before the meeting, so let's get started.

First, it is required in our methodology that every project has an executive sponsor. I'd like to turn the floor over to our executive sponsor, Mr. Galen Kirchoff.

GALEN

Good morning! I'll make my remarks brief. This morning, it is my privilege to empower this steering body to begin a long-anticipated project, the reengineering of our member services information system.

[Galen distributes copies of an administrative memorandum, Figure A.]

This is a project charter given to us by Rebecca Todd, our executive vice president and chief operating officer, in her capacity as chairperson of the Strategic Information Technology Planning group. That group hired a consulting firm to help us develop a strategic plan for business process redesign and information strategy planning. As part of that plan, they documented management's business plan and then developed an overall information systems architecture for our future databases, networks, and applications. They also developed a prioritized list of information systems development projects based on perceived value to the business plan. Well, the order fulfillment and member services system is first on the list. That's why you are all here.

I have personally committed this entire group at least one-quarter time to direct this project. This group will assemble a full-time project team to complete the project. Sandra and Bob have been committed full time.

From Information Services, Sandra will serve as project manager. Bob and various other IS staff will provide technical services. The three club directors have agreed to appoint one experienced assistant director full time to this project as a business analyst. That assistant director, Sarah Hartman, will serve a two-year appointment to Information Services as this business analyst. Additionally, I am asking that each of you designate one individual to work with the team one-quarter time as necessary to complete various aspects of the project.

Folks, this project is important. The business plan suggested a major expansion of marketing and member services. The member services system must be completely overhauled to achieve our strategic business vision. I know that I can count on each of you for your full cooperation. Thank you, Sandra.

SANDRA

Thank you, Galen. Everybody except Bob should be somewhat familiar with our FAST methodology for continuous systems improvement. Bob, FAST is an acronym for *Framework for the Application of Systems Techniques*. We want a well-controlled FAST project on this one—it'll be a great learning experience for Bob. The strategic planning library includes several high-level system models that should provide us with existing documentation.

GALEN

In a nutshell, we want to see what you can do to improve our order fulfillment and member systems. As you know, our product mix is rapidly changing, especially with the addition of games and videos. As part of the business process redesign, we want to disband the current club membership structure that ties members to a particular medium such as discs, cassettes, or videotapes. In its place, we want a flexible membership club that is not dependent on type of merchandise. Marketing has the details.

MONICA

Maybe we should skip the details until later. Suffice to say we have developed a business model for an integrated club that would allow a member to purchase from a wide variety of our consolidated product lines and count them all toward fulfillment of the member agreements. We've even discussed eliminating the passive order fulfillment.

BOB

What is passive order fulfillment?

MONICA

That's where a member's lack of response to a promotion triggers the automatic placement of an order... not one of our more popular business practices according to a recent survey of our members.

GALEN

The business plan suggested the development of cross-functional information systems. We're ignoring organizational boundaries and territories. The goal is to design systems across multiple organization boundaries according to common data needs and functional efficiency. Clearly, Marketing, each club's order entry services, and the Customer Services functions need to be integrated, regardless of where we place them in the organization chart. I would think we want Inventory Control, Warehousing, Purchasing, and Shipping & Receiving to be involved as well. Monica, why don't you describe the marketing dimensions?

SOUNDSTAGE
SOUNDSTAGE ENTERTAINMENT CLUB

Project Charter
Information Services

SoundStage Entertainment Club

Project Name:	2000-001	Member Services Information System
Version Control:	v1.02	Last revised January 3, 2000
File Location:	H:\information services\strategic plan\projects\2000-001MSIS Charter.doc	

Project Objectives

This project will develop new business processes and supporting information system processes and services to support the strategic vision for SoundStage products and member services. It is anticipated that the resulting system will provide for highly integrated processes and services that cross many internal business functions and reach out directly to customers. It is anticipated that this project will result in one of the following (listed in order of expected likelihood).

1. Develop an in-house information system that results in significant competitive advantage for SoundStage in a highly competitive market.

2. Establish a partnership with an IT vendor that involves purchasing a software solution, installing that solution, and customizing that solution to create competitive advantage. It is recognized that this solution may require business processes to be redesigned to fit the solution. This alternative supposes that modifications to the package will be done in cooperation with the vendor and in such a way as to permit the vendor to fold these enhancements into their product; however, sale of that product to any competitor would have to be contractually restricted for a reasonable period of time.

Project Conception

This project was conceived during the Information Technology Strategy Planning (ITSP) project. That project established a strategic information systems plan and projects to support the corporate strategic business plan developed one year earlier. The information systems plan established priorities for applications, databases, and networks (including the use of the Internet as a strategic platform). Because our members (customers) are our lifeline, a cross functional and highly integrated member services information system was identified as one of the most important projects. (It should be noted that this system would interface with another high priority system for inventory and supply chain management.)

Problem Statement

Member Services handles membership subscriptions and member orders. Subscription and order processing is, for the most part, based on a combination of manual and computerized processes that have remain largely unchanged for twenty years. Existing computer processes are based on dated batch processing that does not keep pace with the contemporary economy and industry in which we compete. Existing computer processes have been supplemented by rudimentary, user-developed PC database and spreadsheet applications that are not always fully compatible or consistent with their enterprise information system counterparts. Finally, the team conceded that most computerization was merely automating what appear to be outdated business processes. The following specific problems were discussed in a full-day meeting of the project team:

1. The constantly changing product mix has led to incompatible and often jury-rigged systems and procedures that have created numerous internal inefficiencies and customer relations' problems.

FIGURE A   *Project Charter for SoundStage Case Study*

S·O·U·N·D·S·T·A·G·E

S O U N D S T A G E   E N T E R T A I N M E N T   C L U B

2. The changing product mix creates new opportunities to create new clubs and membership options that would appeal to prospective customers; however, the current system will not support such changes. This problem is amplified by the recent Silver Screenings Video Club acquisition.

3. Directives to increase membership and sales through aggressive advertising will soon overload the current system's ability to process transactions on a timely basis. Customer shipment delays and cash flow problems are anticipated.

4. Response times to orders have already doubled during peak periods from those measures just one year past.

5. Management has suggested a "Preferred Member Program" that cannot be implemented with current data.

6. Unpaid orders have increased from 2%, only two years ago, to 4%. The current credit checking process has contributed significantly to the problem.

7. Member defaults on contracts have increased 7% in three years. It is believed that the current system inadequately enforces contracts.

8. Members have begun to complain about automatic cancellation of memberships after too brief periods of inactivity. This problem has been traced to a data integrity problem in current files.

9. Competition from other companies has led management to propose dynamic contract adjustments to retain members. The current system cannot handle this requirement.

10. Backorders are not receiving proper priority. Some backorders go for as long as three months, with many cancellations and refused deliveries. New orders frequently deplete inventory before backorders can be processed.

11. Customers have expressed dissatisfaction with the passive response order entry model, as well as current club agreements that limit flexibility of members to easily purchase products outside of a narrowly defined media or product type.

12. Existing systems look dated prompting employee complaints that those systems are not as easy to learn and use as the personal computing applications to which they have become accustomed.

13. Management is concerned that SoundStage is not exploiting the Internet as a marketing and service channel.

Initial Scope of the Project

This cross-functional project will support or impact the following business functions and external parties:

1. Marketing
2. Subscriptions
3. Sales and order entry (all sales offices)
4. Warehousing (all distribution centers)
5. Inventory control and procurement
6. Shipping and receiving (all distribution centers)
7. Accounts receivable and payable
8. Member services for all clubs
9. External parties

   a. Prospective members
   b. Members
   c. Former members
   d. Suppliers
   e. Merchandisers

It is recognized that project scope may need to be refined over the course of the project. Project scope should be defined as explicitly as possible in the first phase of the project. Any significant deviation of functionality, cost, or timetable must be reported promptly to the appropriate director. That director must promptly request and facilitate a scope change consensus meeting of the Information Systems Steering Committee. The Friedlander scope management framework will be used to adjust scope.

FIGURE A   *Continued*

S·O·U·N·D·S·T·A·G·E

S O U N D S T A G E   E N T E R T A I N M E N T   C L U B

PROJECT CHARTER–Member Services Information System   *Page 3*

Project Vision

The strategic IS plan recommended a system that will:

1. Expedite the processing of subscriptions and orders through improved data capture technology, methods, channels, and decision support. Management would like a system that extends to the Internet and World Wide Web.
2. Interface to the new bar-coding automatic identification system currently being implemented in the warehouse.
3. Reduce unpaid orders to 2% by the end of fiscal year 2002.
4. Reduce contract defaults to 5% by the end of fiscal year 2002, and 3% by the end of fiscal year 2003.
5. Support constantly changing club and agreement structures, including dynamic agreement changes during the term of an agreement.
6. Triple the order processing capacity of the unit by the end of fiscal year 2002.
7. Reduce order response time by 50% by the end of fiscal year 2002. Management has changed the definition of order response from 'order receipt-to-warehouse' to 'order-receipt-to-member-delivery'.
8. Rethink any and all underlying business processes, procedures, and policies that have any visible impact on member satisfaction and complaints.
9. Provide improved marketing analysis of subscription and promotion programs.
10. Provide improved follow-up mechanisms for orders and backorders.

Business Constraints

1. The initial version of the system must be operational in nine months. Subsequent versions should be released in six-month increments.
2. The system cannot alter any existing file or database structures in the Accounts Receivable Information System without approval of Accounting.
3. The system may be required to interface with an Enterprise Resource Planning software package that is being considered for inventory control, procurement, and warehousing.
4. As part of SoundStage's strategic goal to become ISO 9000 certified, all business processes are subject to business process redesign to improve total quality management and support continuous improvement.
5. The system must conform to the approved technology architecture approved as part of the IS strategic plan. Exceptions must be preapproved by both the Technology Architecture Committee and the Information Services Steering Committee. The system should harness the recent plan to invest in state-of-the-art desktop computing and client/server network technology.

Technology Constraints

The new system must conform to the following information technology architectural standards:

1. The current LAN architecture is client/server based on *Windows* clients running on an Ethernet and TCP/IP network using *Windows 2000* and *Windows 2000 Terminal Server* servers.
2. The current messaging architecture is based on *Outlook* clients (for e-calendar and e-mail) running on a Microsoft *Windows 2000*-based *Exchange Server*.
3. This project will require the development of one or more enterprise databases. The corporate database server standard is Microsoft *SQL Server* running on a *Windows 2000* server. Because the project may include Internet/intranet database access, the information technology architecture group has approved Microsoft *InterDev* as a candidate database access technology.
4. The project will require the development of one or more applications. The corporate application development environment must be chosen from Microsoft *Visual Basic* or Microsoft *Visual C++*. *Visual Basic* is preferred for most applications, deferring to *Visual C++* when performance becomes an issue. Because this project will be the most significant foray into electronic business and commerce undertaken at SoundStage, the information technology

---

FIGURE A   *Continued*

SOUNDSTAGE

SOUNDSTAGE ENTERTAINMENT CLUB

PROJECT CHARTER–Member Services Information System   *Page 4*

architecture group has also approved Borland *J-Builder* as a candidate Java-based application development environment.

5. Internet and intranet Web servers will be implemented using Microsoft *Internet Information Server* (IIS) running on a *Windows 2000* server.

6. Internally, all client workstations will run the *Windows 98* or *Windows 2000* desktop operating system including the *Internet Explorer* Web browser.

7. Externally, for members, any solution developed must run equally well on either the Microsoft *Internet Explorer* or AOL *Navigator* Web browsers (multiple versions) running on *Windows, Macintosh,* or *Linux* clients.

8. The project team is empowered to explore and recommend intranet and extranet technologies as appropriate to the information system requirements; however, all technologies should be approved by the information technology architecture group prior to purchase or installation.

Project Strategy

All IS development projects are subject to the following process strategies:

1. In support of the strategic goal for Information Services to achieve Level III on the Software Engineering Institute's *Capability Maturity Model*, the system must be developed in accordance with the *FAST* (Framework for Systems Techniques) development process/methodology. It is anticipated that one or more of the following *FAST* routes will be used: a combination of (a) Model-Driven Development and (b) Rapid Application Development, or (c) Commercial Off-the-Shelf System Integration.

2. Any and all model-driven documentation will be developed with the CASE tool, *System Architect 2001*.

Project Documentation and Communication

The following guidelines should be used for communications:

1. The project team will hold weekly status meetings, chaired by the project manager. All project status meetings minutes and reports will be shared with all IT directors.

2. Team members will utilize electronic mail, dialogue, and written completion criteria on a regular basis as vehicles for project communication.

3. The following directory folder shall be used to store this charter and all subsequent documentation and work-in-progress components.

   H:\information services\repository\projects\2000-001MSIS Charter\ ...

   This directory should be managed using the Intersolv *PVCS* version control software.

Project Organization and Staffing Approach

The Information Services Steering Committee is responsible for:

- Naming the project manager.

- Naming the project team upon recommendation from the project manager.

- Reviewing and approving project deliverables.

- Ensuring the project follows the management vision.

- Approving any scope, budget, and schedule changes.

- Developing tactical strategies for implementing the management vision.

FIGURE A *Concluded*

SOUNDSTAGE

MONICA

Galen has only touched the tip of the iceberg in describing the marketing business plan to you. We're looking to new markets, new marketing strategies, new membership and sales goals, even new order technologies. We want you to explore phone-based touch-tone response technology and voice recognition as part of this project. We definitely want to implement electronic commerce options using our Internet Web site. And we want to solve some of the existing system's problems while we're at it.

SANDRA

I take it this project doesn't go through the project steering committee?

GALEN

The steering committee only evaluates user-initiated system requests to assess feasibility and priority. This project comes from the strategic planning committee, which has already assessed its importance and feasibility—it was assigned top priority for new systems development.

SANDRA

Let's establish some vision for this project.

STEVEN

The way I see it, the system should provide several essential functions. First, the system should generate and process two kinds of orders: primary and secondary orders for each club.

BOB

Could you differentiate between those types of orders?

STEVEN

Sure. Primary orders are for our main products, namely audio, video, and game titles. Secondary orders are for other types of merchandise we market such as T-shirts, posters, magazines, blank tapes, supplies, and so forth. These secondary products are marketed through SoundStage, but we don't actually stock the merchandise or fill the orders. Instead we provide a

marketing and order fulfillment channel for those who do sell the merchandise. We get a percentage profit based on the merchandiser's sales through our channel. Bob, you probably get such secondary merchandise offers every month with your credit card statements.

BOB

Yes, but I didn't know how that worked. Interesting!

SANDRA

What kinds of problems exist in the current system, Steven?

STEVEN

I've got a big problem coming! I had a meeting with Rebecca Todd yesterday. I was told to expect an aggressive new marketing program over the next three years … TV, radio, newspapers, magazines, and the Web. Also, as Monica suggested, a single integrated member agreement structure will replace existing clubs' current agreement structures. Instead of tying members to an agreement based on a certain number of purchases over some time period, they will be tied to agreements based on a certain number of purchase credits over a period of time. This will allow members to purchase different types of products with various levels of credit toward fulfilling the purchase agreement.

SUSAN

I'm not sure I see the difference.

STEVEN

All right. Let's say you join the compact disc club today. You must buy six discs in the next two years. You can also purchase cassettes and videotapes, but those don't count toward fulfilling your membership agreement. Under the new approach, you will still join the club, and each compact disc you buy will establish a certain number of purchase credits toward fulfilling your membership agreement, but so will cassettes, videotapes, videodisks, and computer games—in any combination! And when you fulfill your membership agreement, you'll receive SoundStage

Dollars, which are credits that are good toward subsequent purchases.

BOB

Wow! Where do I sign up?

STEVEN

You can't! That's the problem. The current system cannot handle any of this. To give customers this new level of service, and to give Marketing the go-ahead to start advertising the service, we have to totally redesign the supporting information systems, in both Marketing and the clubs.

The second essential function is to fully integrate marketing into order entry and all the way through order fulfillment and billing. That means we must extend the system into the warehouse and purchasing functions. We have to recognize that the members are not satisfied when their order is processed. They are only satisfied when the ordered products are delivered to their homes—that means shipping gets involved too. The integration with our secondary merchandise partnerships may be complicated, but we have to deal with that angle as well.

Essential function is to provide management with faster and more reliable information to support Marketing and Member Services decision making.

SANDRA

Dick, what's the warehouse angle?

DICK

Yeah, we're not used to working so closely with you Marketing and Order Entry people. Basically, I think I'm here because of the new auto-ID system.

SUSAN

Auto-ID system?

DICK

Bar coding! We are in the middle of converting all inventory over to a common bar-coding scheme. It should eliminate order-filling errors when we finish. For some period of time, most products will have their existing product numbers and a new bar code number. Also, the acquisition of Silver Screenings requires that

SOUNDSTAGE
SOUNDSTAGE ENTERTAINMENT CLUB

all video titles will have to be renumbered to match SoundStage product numbering standards.

SANDRA

Whew! That's news to me! We'll need to gather some more facts about this bar code system. We should talk with you and your staff very soon.

STEVE

That's why I'm here. A couple of years ago I was working for Silver Screenings Video Club. I'm at least somewhat familiar with their current products and inventory schemes. We now have a good grasp of which Silver Screenings employees will transfer to SoundStage as part of the merger. I intend to assign one of the staff directly to the project team to represent their interests.

SUSAN

Can I change the subject? While I concur with the basic functions that Steven has outlined, from my perspective, the biggest problem we currently have is with the data—it's out of control. My top priority would be to get control of the data.

SANDRA

Please explain.

SUSAN

Order management requires us to bring together data from various sources. Marketing provides us with promotion and product data. Warehouse and Purchasing provide inventory data. The clubs provide data about agreements and members, and now we have three clubs with different management approaches. I'm probably missing someone, but the problem is coordinating all this data in an organized fashion. If we get control of that data, all of Steven's functions could be built around that data.

BOB

We can do that. Sandra and I can work

with the DA to design and implement a SQL Server database that consolidates all this data into a highly organized database. We'll write SQL programs to properly maintain that data and provide users with 4GLs to. . . .

SUSAN

Hold on! You're speaking a foreign language to me! DA? SQL? 4GL?

STEVE

Can I interrupt? I think we might be missing the big picture here. We now have at least three warehouses in different cities. We have regional membership offices in five cities. I'm sure that we all do things somewhat differently, but we all do a lot of the same things. Shouldn't our system communicate with each site, or duplicate itself at each site? Why reinvent the wheel at each operating location?

GALEN

I think Steve has a good point, maybe even better than he realizes. I'm a novice at this network stuff, but the Internet presumably gives us the potential of taking the store directly to members and prospective members. I'd like to think that we should try to creatively exploit Internet technology, both at and between our current operating locations and direct to our members, and even our suppliers.

BOB

We can do that too! We can LAN each operating site and create a WAN to other sites and use the Web to reach our members.

SUSAN

More technology terms! Ugh! But I do like the idea. While we're on the subject of technology, why can't you guys make our homegrown systems look and act like my PC applications? I use *Word, Excel,* and *PowerPoint,* and I really like the way they all put the commands and but-

tons in the same place. If companies like Microsoft and Lotus can do it, why not us?

SANDRA

We can, and we will. And we'll try to control our use of technical jargon and focus on business issues early in this project. Bob and I both have a tendency to speak the jargon because it is part of what we do everyday. Everybody here should feel free to stop us, just like Susan did.

It seems like we have a lot of opinions. Believe it or not, I think you have helped start to establish our project vision. To sum up, you want a system that (1) provides various functional capabilities, (2) integrates and coordinates a wide variety of data and information, (3) takes into consideration multiple operating locations including direct-to-member commerce, and (4) is as easy to learn and use as your PC applications.

Let's take a 10-minute break. When we return, we'll try to formalize some of these ideas and establish a project vision.

DISCUSSION QUESTIONS

1. Why did the different participants in this meeting have entirely different views of the same basic system?

2. Each of the different participants was concerned with different aspects of the system. Briefly organize their concerns into two or three categories.

3. Why did Bob's view of the system cause communications problems with Susan? How could Bob have better communicated with Susan and the group?

4. How do the different views of the system affect Sandra's job? How should she deal with such diverse perspectives?

"Wherever people are active and working together in an organization, they work out some sort of system, without the help of an analyst. Through these systems the people get their day-to-day work done—buying, selling, making, growing, sending, shipping, transporting, paying money, collecting money, and so on."[1] In other words, information systems are the natural by-product of people working with people.

Most of you have had at least one information systems course before coming to this course, so we'll keep this review to a minimum. The basic terminology of information systems varies slightly from author to author and course to course. It is, therefore, important that we establish our basic concepts and terminology to provide context for this book.

Most experts agree on the fundamental difference between data and information.

> **Data** are raw facts about the organization and its business transactions. Most data items have little meaning and use by themselves.

> **Information** is data that has been refined and organized by processing and purposeful intelligence. The latter, purposeful intelligence, is crucial to the definition—People provide the purpose and the intelligence that produces true information.

In other words, data are a by-product of doing business. Information is a resource created from the data to serve the management and decision-making needs of the business. Information technology has created a data and information explosion in virtually all businesses. The ability of businesses to harness and manage this data and information has become a critical success factor in most businesses.

Stated simply, information systems transform data into useful information. To serve the analysis and design focus of this book, we offer a more formal definition.

> An **information system (IS)** is an arrangement of people, data, processes, and interfaces that interact to support and improve day-to-day operations in a business as well as support the problem-solving and decision-making needs of management and users.

The role of information technology was intentionally left out of this definition. An information system exists with or without a computer. But when information technology is used, it significantly expands the power and potential of most information systems.

> **Information technology (IT)** is a contemporary term that describes the combination of computer technology (hardware and software) with telecommunications technology (data, image, and voice networks).

As we shall see throughout this chapter, various information technologies significantly influence or drive the fundamental building blocks of our information system definition: PEOPLE, DATA, PROCESSES, and INTERFACES.

Organizations are not served by a single information system, but instead by a federation of information systems that support various business functions. Figure 2.1 illustrates this idea.

> Notice that businesses have both **front-office information systems** that support business functions that reach out to customers (or constituents) as well as **back-office information systems** that support internal business operations and interact with suppliers (of materials, equipment, supplies, and services). These front- and back-office information systems feed data to management information systems and decision support systems. Contemporary

---

[1] Keith London, *The Management System: Systems Are for People* (New York: Wiley-Interscience, 1976), p. 30.

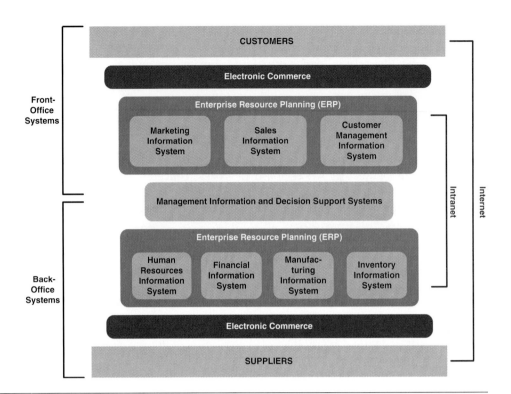

FIGURE 2.1

A Federation of Information Systems

Classes of Information System Applications

Transaction processing
Management information
Decision support
Expert systems
Office automation and work group

Transaction Processing Systems

Transaction Processing Systems

Airline reservations
Bank deposits and withdrawals
Course registration
Customer returns
Hotel check-in/check-out
Inventory procurement
Invoicing or billing
Order processing
Payment processing
Payroll
Retail point-of-sale
Supplies procurement

information systems are interfacing with customers and suppliers using electronic commerce technology over the Internet. Finally, most companies have some sort of intranet (internal to the business) to support communications between employees and the information systems.

In practice, there are several classes of information system applications (see margin). Each class serves the needs of different types of users. The tools and techniques taught in this book help you analyze, design, and build these classes of systems.

Business transactions are events that serve the mission of the business. They are the primary means by which the business interacts with its suppliers, customers, partners, employees, and government. Transactions capture and or create data about and for the business. Examples of transactions include purchases, orders, sales, reservations, registrations, vouchers, shipments, invoices, and payments.

Transaction processing systems are information system applications that capture and process data about business transactions.

Transaction processing systems can either (1) respond to business transactions (such as orders, time cards, or payments) or (2) initiate transactions (such as invoices, paychecks, or receipts), or possibly both. Also, transaction processing systems can respond to both external events (such as processing orders from customers) or internal events (such as generating production orders for the shop floor). Examples of transaction processing systems are listed in the margin.

One dimension of transaction processing system, **data maintenance,** provides for custodial updates to stored data. For example, a system must provide for the ability to add and delete CUSTOMERS and PRODUCTS, as well as to change specific facts such as CUSTOMER ADDRESS and PRODUCT PRICE.

The design of transaction processing systems tends to focus on factors such as response time, throughput (volume of transactions), accuracy, consistency, and service. Although most transaction processing systems have long been computerized, new opportunities for improvement are being driven by the trend toward business process redesign.

> **Business process redesign (BPR)** is the study, analysis, and redesign of fundamental business processes to reduce costs and/or improve value added to the business.

In many cases, existing systems automate the inefficiencies and bureaucracy of dated business practices and policies. When a business redesigns those processes and practices, the underlying transaction processing systems must also be redesigned. Thus, there are still many new opportunities to practice systems development for transaction processing systems.

Management information systems supplement transaction processing systems with management reports required to plan, monitor, and control business operations.

> A **management information system (MIS)** is an information system application that provides for management-oriented reporting. These reports are usually generated on a predetermined schedule and appear in a prearranged format.

Examples of management information systems are listed in the margin. Management information is normally produced from a shared database that stores data from many sources, including transaction processing systems.

Management information systems can present detailed information, summary information, and exception information. *Detailed information* is used for operations management as well as regulatory requirements (as imposed by the government). *Summary information* consolidates raw data to quickly indicate trends and possible problems. *Exception information* filters data to report exceptions to some rule or criteria (such as reporting only those products that are low in inventory). As long as organizations continue to recognize information as an important management resource, opportunities for MIS development will expand.

Decision support systems further extend the power of information systems.

> A **decision support system (DSS)** is an information system application that provides its users with decision-oriented information whenever a decision-making situation arises. When applied to executive managers, these systems are sometimes called **executive information systems (EIS).**

A DSS does not typically make decisions or solve problems—people do. Decision support systems are concerned with providing useful information to support the decision process. In particular, decision support systems are usually designed to support *unstructured decisions,* that is, those decision-making situations that cannot be predicted.

Does this seem impossible? Not really! DSS is based on the reality that transaction processing and management information systems have already captured the data needed to produce information for unstructured decisions. The DSS provides the decision maker with tools to access that data and analyze it to make a decision. In general, a DSS provides one or more of the following types of support to the decision maker:

- Identification of problems or decision-making opportunities (similar to exception reporting).
- Identification of possible solutions or decisions.
- Access to information needed to solve a problem or make a decision.
- Analysis of possible decisions or of variables that will affect a decision. Sometimes this is called "what if" analysis.
- Simulation of possible solutions and their likely results.

Management Information Systems

——————— ✓ ———————

Management Information Systems

Budget forecasting and analysis
Financial reporting (e.g., balance sheets, income statements, cash flow reports)
Inventory reporting
Materials requirements planning
Production scheduling
Salary analyses
Sales forecasting
Sales reporting
Schedule of classes

Decision Support Systems

Interest in decision support systems is at an all-time high. This interest is being driven by projects to develop data warehouses.

A **data warehouse** is a read-only, informational database that is populated with detailed, summary, and exception data and information generated by other transaction and management information systems. The data warehouse can then be accessed by end-users and managers with DSS tools that generate a virtually limitless variety of information in support of unstructured decisions.

DSS tools include spreadsheets (such as Microsoft *Excel*), PC-database management systems (such as Microsoft *Access*), custom reporting tools (such as Seagate Software's *Crystal Reports* and Brio Technology's *BrioQuery*), and statistical analysis programs (such as SAS Institute's *SAS*).

Expert Systems

Expert systems are an extension of the decision support system.

An **expert system** is a programmed decision-making information system that captures and reproduces the knowledge and expertise of an expert problem solver or decision maker and then simulates the "thinking" or "actions" of that expert.

Expert systems address the critical need to duplicate the expertise of experienced problem solvers, managers, professionals, and technicians. These experts often possess knowledge and expertise that cannot easily be duplicated or replaced in all organizations. Expert systems imitate the logic and reasoning of the experts within their respective fields. The following examples are real:

- A food manufacturer uses an expert system to preserve the production expertise of engineers who are nearing retirement.
- A major credit card broker uses an expert system to accelerate credit screening that requires data from multiple sites and databases.
- A plastics manufacturer uses an expert system to determine the cause of quality control problems associated with shop floor machines.

Expert systems are implemented with **artificial intelligence (AI)** technology that captures, stores, and provides access to the reasoning of the experts. Expert systems require data and information like any other information system, but they are unique in their requirement of storing rules (called *heuristics*) that simulate the reasoning of the experts who use that data and information.

Office Automation Systems

Office automation is more than word processing and spreadsheet applications.

Office automation (OA) systems support the wide range of business office activities that provide for improved work flow and communications between workers, regardless of whether or not those workers are located in the same office.

Office automation systems are concerned with getting all relevant information to those who need it. Office automation functions include word processing, electronic messages (or electronic mail), work group computing, work group scheduling, facsimile (fax) processing, imaging and electronic documents, and work flow management. We'll explore the design implications of some of these technologies in this book.

Office automation systems can be designed to support both individuals and work groups.

Personal information systems are those designed to meet the needs of a single user. They are designed to boost an individual's productivity.

Work group information systems are those designed to meet the needs of a work group. They are designed to boost the group's productivity.

Most personal information systems are developed directly by the end-user using PC office suites such as Microsoft's *Office Professional,* IBM's *Lotus SmartSuite,* or Corel's *PerfectOffice.* While developed for the individual user, such systems might be deployed to others, making them work group information systems. Also, clever end-users can often integrate their personal information systems into the other types of information systems we've discussed.

Work group applications are typically developed by departmental computing specialists using the same PC tools in conjunction with work group technology such as Microsoft's *Exchange* and *Outlook,* IBM's *Lotus Notes/Domino,* or Novell's *Group-Wise.* Through this work group technology, users can collaborate on projects. For example, the partners in a legal firm could collaboratively develop a new contract using a combination of electronic research, word processing, spreadsheets, and databases—all while working from their own offices on their own time schedules.

We've discussed several classes of information systems. In practice, these classes overlap such that you can't always differentiate one from the other. Even if you could, the various applications should ideally interoperate to complement and supplement one another. Take a few moments to study Figure 2.2. We call your attention to the following number annotations on the diagram:

Putting It All Together

❶ The first transaction process responds to an input transaction's data (e.g., an ORDER). It produces transaction information to verify the correct processing of the input transaction.

❷ The second transaction process merely produces an output transaction (e.g., an INVOICE). such a system may respond to something as simple as the passage of time (e.g., it is the END-OF-MONTH; therefore, generate all INVOICES).

❸ The first management information system simply produces reports or information (e.g., sales analysis reports) using data stored in operational databases (maintained by the aforementioned transaction processing systems).

❹ The second management information system uses business models (e.g., MRP) to produce operational management information (e.g., a production schedule).

❺ Notice that an MIS may use data from more than one operational database.

❻ Notice that snapshots of data from the operational databases populate a data warehouse. The snapshots may be taken at various time intervals, and different subsets of data may be included in various snapshots. The data in the warehouse will be organized to ensure easy access and inquiry by managers.

❼ Decision support and executive information systems applications will typically provide read-only access to the data warehouses to produce decision support and executive management information.

❽ An expert system requires a special database that stores the expertise in the form of rules and heuristics.

❾ An expert system either accepts problems as inputs (e.g., should we grant credit to a specific customer) or senses problems in the environment (e.g., is the lathe producing parts within acceptable specifications) and then responds to a problem with an appropriate solution based on the system's expertise.

❿ Personal office automation systems tend to revolve around the data and business processing needs of an individual. Such systems are typically developed by the users themselves (and run on personal computers).

⓫ Work group office automation systems are frequently message-based (e.g., e-mail-based) and are smaller-scaled solutions to departmental needs. As shown in the figure, they can access or import data from larger, transaction processing systems.

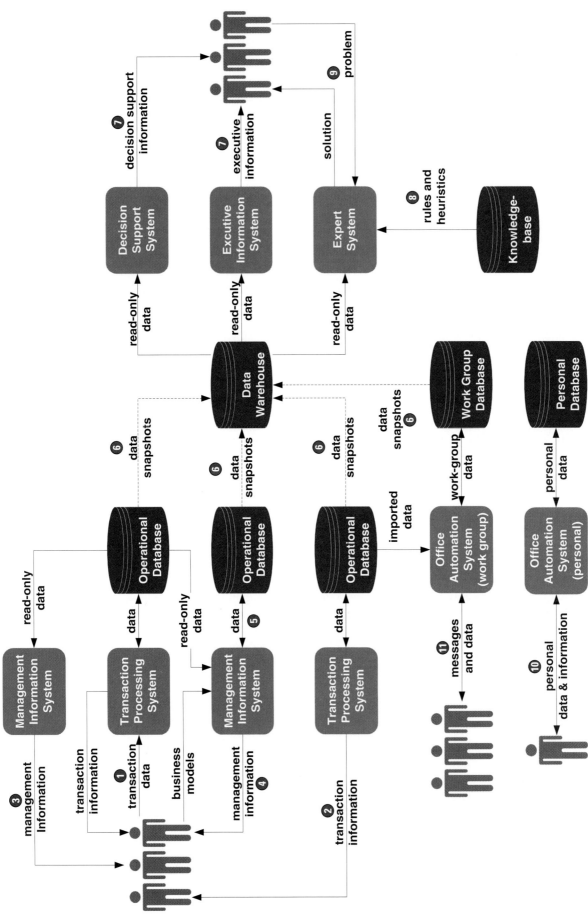

FIGURE 2.2 *Information Systems Applications*

In the average business, there will be many instances of each of these different application classes.

It has become fashionable to deal with the complexity of modern information systems using *architecture*. Information technology professionals speak of data architectures, application architectures, network architectures, software architectures, and so forth. For this book, we'll define information systems architecture as follows.

A FRAMEWORK FOR INFORMATION SYSTEMS ARCHITECTURE

> **Information systems architecture** provides a unifying framework into which various people with different perspectives can organize and view the fundamental building blocks of information systems.

Essentially, information systems architecture provides a foundation for organizing the various components of any information system you care to develop.

In this chapter, we'll build a framework for information systems architecture that was inspired by the work of John Zachman.[2] The Zachman "Framework for Information Systems Architecture" has achieved international recognition and use. The Zachman framework is a matrix (similar to the chapter map at the beginning of this chapter). The rows correspond to what Zachman calls *perspectives* of different people involved in systems development and use. The columns correspond to *focuses* on different aspects of the information system. We have adapted and extended Zachman's framework to teach (and practice) systems analysis and design.

Let's begin with system *perspectives*. Different people have different views of an information system. Managers, users, and technical specialists each view the system in different ways and in different levels of detail. We call these people *stakeholders* in the system. The stakeholders can be broadly classified into six groups:

- **System owners** pay for the system to be built and maintained. They own the system, set priorities for the system, and determine policies for its use. In some cases, system owners may also be system users.
- **System users** are the people who actually use the system to perform or support the work to be completed. System users define the business requirements and performance expectations for the system to be built.
- **System designers** are the technical specialists who design the system to meet the users' requirements. In many cases, system designers may also be system builders.
- **System builders** are the technical specialists who construct, test, and deliver the system into operation.
- **Systems analysts** facilitate the development of information systems and computer applications by bridging the communications gap that exists between nontechnical system owners and users and technical system designers and builders.
- **IT vendors and consultants** who sell hardware, software and services to businesses for incorporation into their information systems.

Each of these classes of people was described in greater detail in Chapter 1. These are *roles,* not positions or job titles. A single individual can, and frequently does, play many roles. For example, a manager may play the role of system owner, system user, and if technically literate, system designer. Many system owners and users have become sufficiently technical to suggest technical ideas and options.

---

[2] John A. Zachman, "A Framework for Information Systems Architecture,"*IBM Systems Journal* 26, no. 3 (1987), pp. 276–92.

We have identified three *focuses* for information systems development activities.

- DATA—the raw material used to create useful information.
- PROCESSES—the activities (including management) that carry out the mission of the business.
- INTERFACES—how the system interfaces with its users and other information systems.

As shown in Figure 2.3, the intersection of a perspective and a focus defines a *building block* for an information system. In the next section, we will describe all these information system building blocks.

> NOTE   Throughout this book, we have used a consistent color scheme for both the framework and the various tools that relate to, or document, the building blocks. The color scheme is based on the building blocks as follows:
>
> ■ represents something to do with DATA.
>
> ■ represents something to do with PROCESSES.
>
> ■ represents something to do with INTERFACES.

The information system building blocks do not exist in isolation. They must be carefully synchronized to avoid inconsistencies and incompatibilities within the system. For example, a database designer (a *system designer*) and a programmer (a *system builder*) have their own architectural views of the system; however, these views must be compatible if the system is going to work properly. Synchronization occurs both horizontally (across any given row) and vertically (down any given column).

In the remainder of this chapter, we'll briefly examine each focus and perspective—the building blocks of information systems.

DATA Building Blocks

When engineers design a new product, they must create a bill of materials for that product. A bill of materials says nothing about what the product is intended to do. It states only that certain raw materials and subassemblies will make up the finished product. The same analogy can be used for information systems. Data can and should be thought of as the raw material used to produce information. Consequently, we consider DATA to be one of the fundamental building blocks of an information system.

Figure 2.4 illustrates the DATA column of the framework. Notice at the bottom of the DATA column that the goal is to capture and store business data using a *database management system*. Database management system technology (such as *Access, SQL Server,* or *Oracle*) will be used to organize and store data for all information systems. Also, as you look down the DATA column, each of the stakeholders has different perspectives on the system's data. Let's examine those views and discuss their relevance to system development.

System Owners' View of DATA   The average system owner is not interested in raw data. The system owner is interested in information that adds new business knowledge.

> **Business knowledge** is the insight that is gained from timely, accurate, and relevant information. (Recall that information is a product of raw data.)

Business knowledge helps managers make intelligent decisions that enhance the organization's mission, goals, objectives, and competitiveness.

Business knowledge may initially take the form of a simple list of business entities and business rules. Examples of business entities might include CUSTOMERS, PRODUCTS, EQUIPMENT, BUILDINGS, ORDERS, and PAYMENTS. What do business entities have to do with data? Information is produced from raw data that describes these business entities. Therefore, we should quickly identify relevant business entities about which we need to capture and store data.

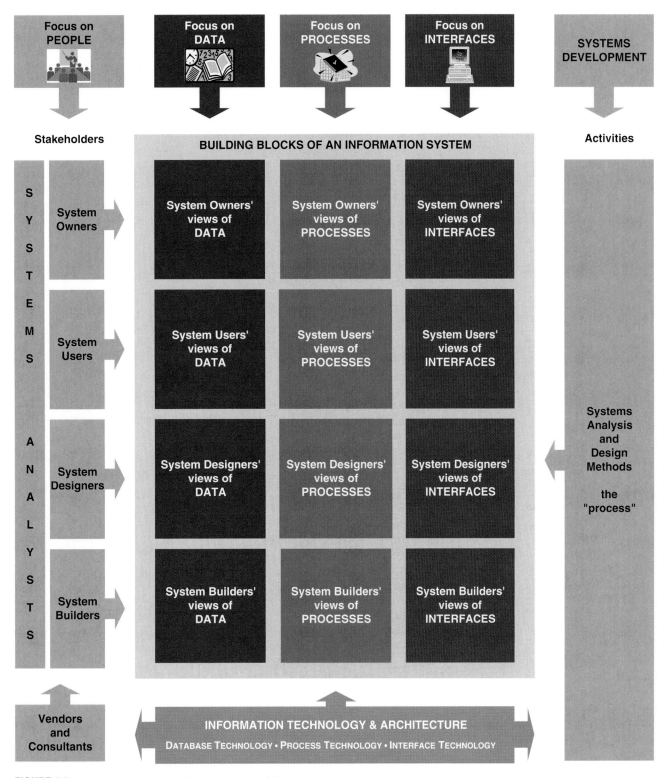

FIGURE 2.3 *Information System Perspectives and Focuses*

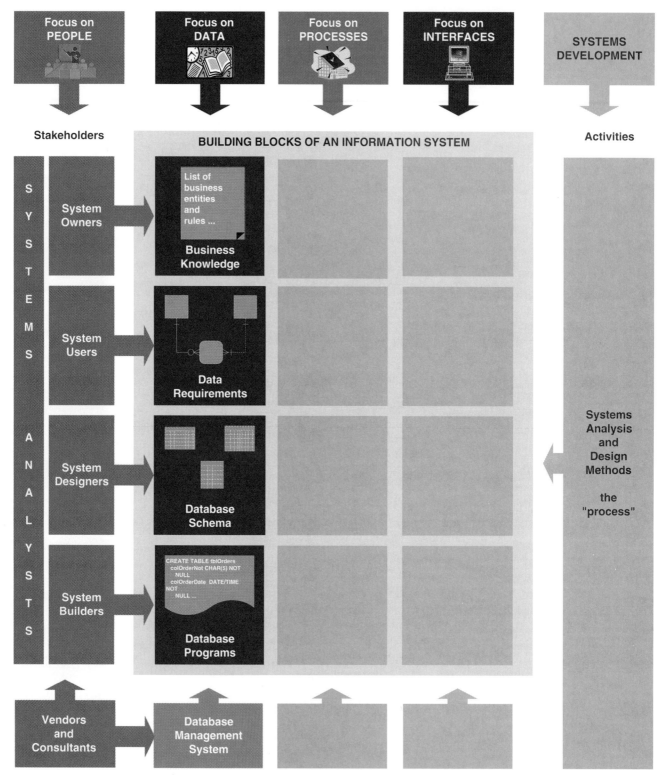

FIGURE 2.4   *A Data Focus for Information Systems*

It is also useful to understand simple business associations or rules that describe how the business entities interact. Examples of useful business rules for a sales system might include the following:

- A CUSTOMER <u>can place</u> ORDERS—an ORDER <u>must be placed by</u> a CUSTOMER.
- An ORDER <u>sells</u> PRODUCTS—a PRODUCT <u>may be sold on</u> an ORDER.

Intuitively, a system's database needs to track these business entities and rules to produce useful information (for example, "Has CUSTOMER 2846 placed any unfilled ORDERS?").

System owners are concerned with the big picture. They are generally not interested in details (such as what fields describe a CUSTOMER or ORDER). The primary role of system owners in a systems development project should be to define the scope and expectations for the project. For DATA, scope can be defined in simple terms such as the aforementioned business entities and rules. System owners define project expectations in terms of their insight into problems, opportunities, and constraints as they relate to the business entities and rules.

System Users' View of DATA   The users of an information system are the experts about the data that describe the business. As information workers, they capture, store, process, edit, and use that data every day. Unfortunately, they frequently see the data only in terms of how data are currently stored or how they think data should be stored. To them, the data are recorded on forms, stored in file cabinets, recorded in books and binders, organized into spreadsheets, or stored in computer files and databases. The challenge in systems development is to correctly identify and verify users' business data requirements.

> **Data requirements** are a representation of users' data in terms of entities, attributes, relationships, and rules. Data requirements should be expressed in a format that is independent of the technology that can or will be used to store the data.

Data requirements are an extension of the business entities and rules that were initially identified by the system owners. System users may identify additional entities and rules because of their greater familiarity with the data. More importantly, system users must specify the exact data attributes to be stored and the precise business rules for maintaining that data.

EXAMPLE   A system owner may have identified the need to store data about an entity called CUS-TOMER. System users might tell us that we need to differentiate between PROSPECTIVE CUSTOMERS, ACTIVE CUSTOMERS, and INACTIVE CUSTOMERS because they know that slightly different types of data describe each type of customer. System users can also tell us precisely what data must be stored about each type of customer. For example, an ACTIVE CUSTOMER might require such data attributes as CUSTOMER NUMBER, NAME, BILLING ADDRESS, CREDIT RATING, and CURRENT BALANCE. Finally, system users are also knowledgeable about the precise rules that govern entities and relationships. For example, they might tell us the credit rating for an ACTIVE CUSTOMER must be PREFERRED, NORMAL, or PROBATIONARY, and the default for a new customer is NORMAL. They might also specify that only an ACTIVE CUSTOMER can place an ORDER, but an ACTIVE CUSTOMER might not necessarily have any current ORDERS at any given time.

Figure 2.4 illustrates the data requirements building block using the format of a graphical picture called a data model. Systems analysts use such a tool to document and verify the system users' view of data requirements. You'll learn how to draw data models in this book.

System Designers' View of DATA   System users define the data requirements for an information system. System designers translate those data requirements into computer databases that will be made available via the information system. The system designers' view of data is constrained by the limitations of whatever database management system (DBMS) is chosen. Often, the choice has already been made and the developers must use that technology. For example, many businesses have standardized on an enterprise DBMS (such as *Oracle, DB2,* or *SQL Server*) and a work group DBMS (such as *Access* or *FoxPro*).

The system designer's view of data consists of data structures, database schemas, fields, indexes, and other technology-dependent components. Most of these technical specifications are too complex to be reasonably understood by system users. The systems analyst and/or database specialist design and document these technical views of the data. As shown in Figure 2.4, the system designers' view of data is a **database schema.** A database schema is the transformation of the data requirements (system users' view) into a set of data structures that can be implemented using the chosen DBMS. Once again, this book will teach tools and techniques for transforming user data requirements into database schemas.

System Builders' View of Data   The final view of data is relevant to the system builders. In the Data column of Figure 2.4, system builders are closest to the actual **database management system (DBMS)** technology. They must represent data in very precise and unforgiving languages. The most commonly encountered database language is *SQL (Structured Query Language).* Alternatively, many database management systems, such as *Access* and *FoxPro,* include proprietary languages or facilities for constructing a new database.

Not all information systems use database technology to store their business data. Older legacy systems were built with *flat-file* technologies such as VSAM. These flat-file data structures were constructed directly within the programming language used to write the programs that use those files. For example, in a *COBOL* program the flat-file data structures are expressed as PICTURE clauses in a DATA DIVISION. This book will not teach either database or flat-file construction languages, only place them in the context of the Data building block of information systems.

Process Building Blocks

Let's start the same way we did with the Data building block. When engineers design a new product, that product should provide some level of functionality or service. It must do something useful. Prospective customers define the desired functionality of the product, and the engineer creates a design to provide that functionality. Processes deliver the functionality of an information system. Processes perform the *work* in a system. People perform some processes. Machines, including computers, perform others.

Figure 2.5 illustrates the Process building blocks of information systems. Notice at the bottom of the Process column that an *application development environment* (ADE) such as *Visual Basic* will be used to automate selected processes. Alternatively, a commercial off-the-shelf software package might be implemented. As you look down the Process column, each stakeholder has a different view of the system's processes. Let's examine those views and discuss their relevance to system development.

System Owners' View of Processes   As usual, system owners are usually interested in the big picture—in this case, groups of high-level processes called business functions.

> **Business functions** are ongoing activities that support the business. Functions can be decomposed into other subfunctions and eventually into processes that do specific tasks.

Think of functions as groups of related processes. Typical business functions include SALES, SERVICE, MANUFACTURING, SHIPPING, RECEIVING, ACCOUNTING, and so forth. Each function is ongoing; that is, it has no starting time or stopping time.

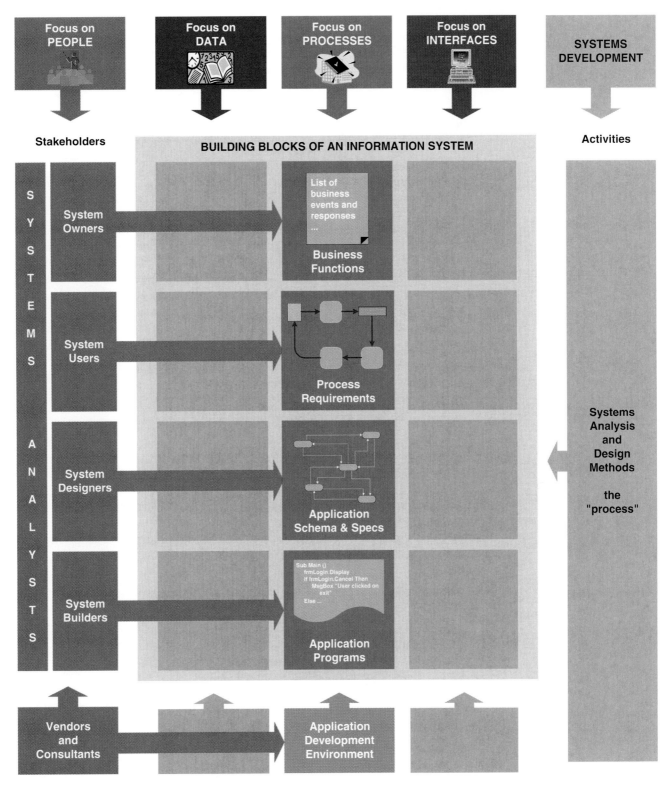

FIGURE 2.5 *A Process Focus for Information Systems*

Historically, most information systems were function-centered. The system supported one business function. An example would be a SALES INFORMATION SYSTEM that supported only the initial processing of customer orders. Today, many of these single-function information systems are being redesigned as cross-functional information systems.

> A **cross-functional information system** supports relevant business processes from several business functions without regard to traditional organizational boundaries such as divisions, departments, centers, and offices.

As a contemporary alternative to the traditional SALES INFORMATION SYSTEM, a cross-functional ORDER FULFILLMENT INFORMATION SYSTEM would also support all relevant processes subsequent to the processing of the customer order. This would include filling the order in the warehouse, shipping the products to the customer, billing the customer, and providing any necessary follow-up service to the customer—in other words, all business processes required to ensure a complete and satisfactory response to the customer order, regardless of which departments are involved.

As shown in Figure 2.5, systems development teams frequently document business functions in terms of simple lists of business events and responses to those events. Some examples of business events and responses are as follows:

- Event: CUSTOMER SUBMITS ORDER.
  Response: CUSTOMER RECEIVES ORDERED PRODUCTS.
- Event: EMPLOYEE SUBMITS PURCHASE REQUISITION FOR SUPPLIES.
  Response: EMPLOYEE RECEIVES REQUESTED SUPPLIES.
- Event: END OF MONTH.
  Response: INVOICE CUSTOMERS AGAINST ACCOUNTS.

With respect to each event and response identified, system owners would be queried about perceived problems, opportunities, goals, objectives, and constraints. The costs and benefits of developing information systems to support business functions would also be discussed. As was the case with DATA, system owners are not concerned with PROCESS details. That level of detail is identified and documented as part of the system users' view of processes.

This book will teach you how to identify and document project scope in terms of relevant business functions, business events, and responses.

System Users' View of PROCESSES   Returning again to Figure 2.5, we are ready to examine the system users' view of processes. Users are concerned with business processes.

> **Business processes** are activities that respond to business events. Business processes are the "work" performed by the system.

System users specify the business process requirements.

> **Process requirements** are a representation of the users' business processes in terms of activities, data flows, or work flow.

These process requirements must be precisely specified, especially if they are to be automated or supported by computer software. Business process requirements are frequently defined in terms of policies and procedures.

> A **policy** is a set of rules that govern a business process.

> A **procedure** is a step-by-step set of instructions and logic for accomplishing a business process.

An example of a policy is CREDIT APPROVAL, a set of rules for determining whether or not to extend credit to a customer. That credit approval policy is usually applied within the context of a specific CREDIT CHECK procedure that has established the correct steps for checking credit against the credit policy. Many of you have implemented policies and procedures in computer programming assignments.

Unfortunately, users tend to see their processes only in terms of how they are currently implemented or how they think they should be implemented. Most businesses need to rethink and redesign business processes to eliminate redundancy, increase efficiency, and streamline the entire business—all independent of any information technology to be used! The fashionable term, introduced in Chapter 1, is *business process redesign*.

Once again, the challenge in system development is to identify, express, and analyze business process requirements exclusively in business terms that can be understood by system users. Figure 2.5 illustrates the process requirements building block using the format of a graphical picture called a *process model*. Tools and techniques for process modeling and documentation of policies and procedures are taught extensively in this book.

System Designers' View of PROCESSES   As was the case with the DATA building block, the system designer's view of activities is constrained by the limitations of a specific application development environment (ADE) such as *Visual Basic*. Sometimes the analyst is able to choose that technology. But often the choices are limited by standards that specify which software and hardware technologies must be used. In either case, the designer's view of processes is technical.

Given the business processes from the system users' view, the designer must first determine which processes to automate and how to best automate those processes. Initially, the designer tends to focus on an application schema.

> An **application schema** is a model that communicates how selected business processes are, or will be, implemented using the software and hardware.

Today, many businesses purchase commercial off-the-shelf software instead of building that software in-house. In this scenario, the application schema specifies how the software package will be integrated into the enterprise.

Alternatively, when developing software in-house, these schemas may be supplemented by software specifications.

> **Software specifications** represent the technical design of business processes to be automated or supported by computer programs to be written by system builders.

Examples of software specifications include state transition diagrams, flowcharts, structure charts, or unified modeling language diagrams. You may have encountered some of these program design tools in a programming course. As was the case with DATA, some of these technical views of PROCESSES can be understood by users, but most cannot. The designers' intent is to prepare software specifications that (1) fulfill the business process requirements of system users and (2) provide sufficient detail and consistency for communicating the software design to system builders. The systems design chapters in this book teach tools and techniques for transforming business process requirements into both an application schema and software design specifications.

System Builders' View of PROCESSES   System builders represent PROCESSES using precise computer programming languages or ADEs that describe inputs, outputs, logic, and control. Examples include *COBOL, C++, Visual Basic, Powerbuilder, Smalltalk,* and *Java*. Additionally, some database management systems provide their own internal languages for programming. Examples include *Visual Basic for Applications* (in *Access*) and *PL-SQL* (in *Oracle*). All these languages are used to write application programs.

Application programs are language-based, machine-readable representations of what a software process is supposed to do, or how a software process is supposed to accomplish its task.

This book does not teach application programming. We will, however, demonstrate how some of these languages provide an excellent environment for prototyping software.

Prototyping is a technique for quickly building a functioning, but incomplete model of the information system using rapid application development tools.

Prototyping has become the design technique of choice for many system designers. Prototypes typically evolve into the final version of the system or application.

INTERFACE Building Blocks

When engineers design a new product, that product should be easy to learn and use. Today's best product engineers are obsessed with *human engineering* and *ergonomics*. Consider, for example, the evolution of videocassette recorder (VCR) programming. Engineers have greatly simplified VCR programming through innovations such as VCR Plus, a system whereby you merely punch in a number from your TV guide to automatically schedule the recording of that program—no need to select date, start time, stop time, channel, and so on. Similarly, customers also expect new products to work with their other products. For example, today's VCRs should fully integrate into your stereo system to provide a home theater experience.

Information systems have similar human interface and system interoperability expectations. Our INTERFACE building blocks provide that dimension of the system. There are two critical aspects to the information system INTERFACE.

— Information systems must provide effective and efficient interface to the system's users.
— Information systems must interface effectively and efficiently with *other* information systems—both within the business and increasingly with other businesses' information systems.

The INTERFACE building blocks of information systems are illustrated in our framework in Figure 2.6. Notice at the bottom of the INTERFACE column it utilizes *interface technology* to implement the interface. The classic example is a graphical user interface (GUI) used to implement a *Windows*-compliant application. Other interface technologies exist for integrating information systems with one another (e.g., *DCOM* and *CORBA*).

And once again, as you look down the INTERFACE column, each stakeholder has different views of the system's INTERFACES. Let's examine those views and discuss their relevance to system development.

System Owners' View of INTERFACE   The system owners' view of INTERFACE is relatively simple. Early in a systems development project the system owners need to specify:

— With which business units, employees, customers, and external businesses must the new system interface?
— Where are these business units, employees, customers, and external businesses located?
— Will the system have to interface with any other information, computer, or automated systems?

Answers to these questions help to define the interface scope of an information systems development project. Minimally, a suitable system owners' view of information system interfaces might be expressed as a simple list of business locations or systems with which the information system must interface. Again, relevant problems, opportunities, or constraints may be identified and analyzed.

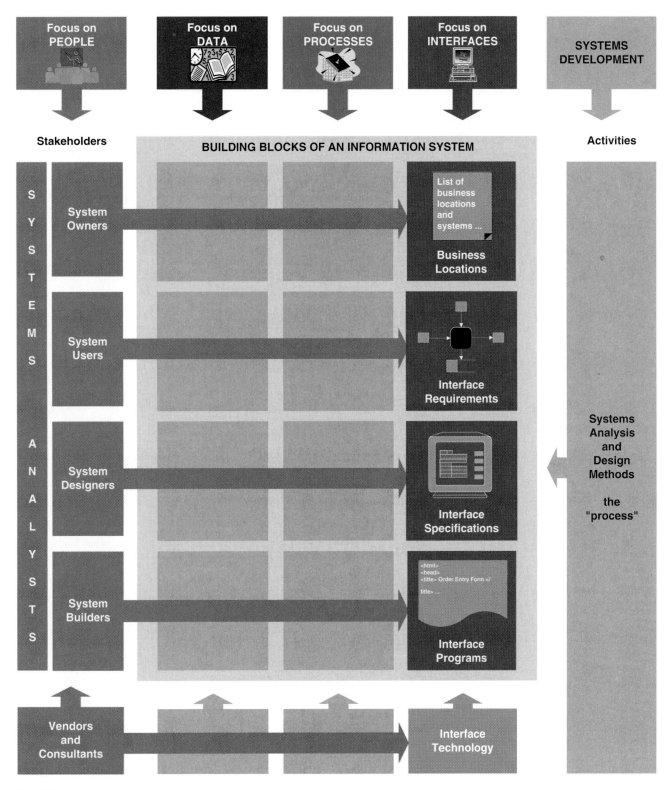

FIGURE 2.6   *An Interface Focus for Information Systems*

System Users' View of INTERFACE System users are concerned with an information system's inputs and outputs. Those inputs and outputs can take many forms; however, the interface requirements are more important than the technical format.

> **Interface requirements** are a representation of the users' inputs and outputs.

A suitable system users' view of INTERFACE might be as simple as a *context model* as shown in Figure 2.6. Such a model represents the proposed system as a single "black box" in the middle of the page (effectively hiding all the details). Inputs and outputs to that process provide a conceptual view of how the proposed system would interact with users, employees, business units, customers, and other businesses. You'll learn to draw this relatively simple context model in this book.

But the details of those inputs and outputs are equally important. System users might specify the details in the form of a list of fields (and their values) that make up the inputs or outputs. Alternatively, and because system users have become comfortable with the graphical user interface (e.g., *Windows* or Web browsers) for the system, the details might be specified in the form of prototypes. System users are increasingly demanding that their custom-built information system applications have the same "look and feel" as their favorite PC tools such as word processors and spreadsheets. This common graphical user interface makes each new application easier to learn and use.

Both list and prototype approaches to documenting the system users' view of INTERFACE will be addressed in various chapters of this book.

System Designers' View of INTERFACE System designers must be concerned with the technical design of both the user and system-to-system interfaces. We call these *interface specifications*. Let's begin with the user interface.

In some ways, it is difficult to distinguish between the system users' and system designers' views of the user interface. Given the emergence of prototyping as an interface requirements definition technique, both users and designers can be involved in input, output, and screen design. But whereas system users are interested in requirements and format, system designers have other interests such as consistency, compatibility, completeness, and user dialogues.

> A **user dialogue** (sometimes called interface navigation) describes how the user moves from window-to-window, interacting with the application programs to perform useful work.

The trend toward graphical user interfaces such as *Windows* and Web browsers has simplified life for system users, but complicated the design process for system designers. Think about it! In a typical *Windows* application, at any given time you can do many different things—type something, click the left mouse button on a menu item or tool bar icon, press the F1 key for help, maximize the current window, minimize the current window, switch to a different program, and many others.

Accordingly, the system designer views the interface in terms of various system states, events that change the system from one state to another, and responses to those events. You'll learn how to use such tools to document user dialogues in the design unit of this book.

Although not depicted in Figure 2.6, modern system designers may also design *keyless interfaces* such as bar coding, optical character recognition, pen, and voice or handwriting recognition. These alternatives reduce errors by eliminating the keyboard as a source of human error. But these interfaces, like graphical user interfaces, must be carefully designed to both exploit the underlying technology and maximize the return on what can be a sizable investment.

Finally, and as suggested earlier, system designers are also concerned with system-to-system interfaces. The existing information systems in most businesses were built with the technologies and techniques that represented best practices at the time they were developed. Some systems were built in-house. Others were purchased from software vendors or developed with consultants. As a result, integrating these heterogeneous systems can be difficult. But the need for different systems to interoperate is pervasive. Accordingly, system designers frequently spend as much or more time on system-to-system integration as they do on system development. The system designer's mission is to find or build interfaces between these systems that (1) do not create maintenance projects for the legacy systems, (2) do not compromise the superior technologies and design of the new systems, and (3) are transparent to the system users.

System Builders' View of INTERFACE   System builders construct, install, test, and implement both user and system-to-system interfaces using *interface technology* (see Figure 2.6). For user interfaces, the interface technology is frequently embedded into the application development environment (ADE) used to construct the system. For example, ADEs such as *Visual Basic, Java,* and *PowerBuilder* include all the interface technology required to construct a *Windows*- or Web browser-style GUI. Alternatively, the user interface could be constructed with a stand-alone interface technology such as HTML (used for Web-based user interfaces).

System-to-system interfaces are considerably more complex than user interfaces to construct or implement. One system-to-system interfacing technology that is currently popular is middleware.

> **Middleware** is a layer of utility software that sits between application software and systems software to transparently integrate differing technologies so that they can interoperate.

One common example of middleware is the *open database connectivity* (ODBC) tools that allow application programs to work with different database management systems without having to be rewritten. Programs written with ODBC commands can, for the most part, work with any ODBC-compliant database (which includes dozens of database management systems). Similar middleware products exist for each of the columns in our information system framework. System designers help to select and apply these products to integrate systems.

Once again, this book is not about system construction. But we present the system builder's view because the other INTERFACE views lead to the construction of the actual interfaces.

The complete framework, depicted in Figure 2.7, identifies the building blocks that must be considered when developing information systems. Throughout this textbook, we will use the framework to place systems analysis and design activities, tools, techniques, and deliverables into the context of our final, desired product—an information system. Along the way, we will make slight additions and modifications to the framework to illustrate various systems analysis and design methods and alternatives.

Before we move on, we should acknowledge an important focus that we omitted from our framework. Today's information systems are built on networks. Figure 2.8 shows a modern high-level information systems framework that demonstrates the contemporary layering of an information system's DATA, PROCESS, and INTERFACE building blocks upon a NETWORK. Today's best-designed information systems tend to separate these layers and force them to communicate across the network. This *clean layering* approach allows any one building block to the replaced with another while having little or no impact on the other building blocks. For example, the database management system technology or the user interface technology could be changed without affecting the other building blocks.

Using the Framework for Information Systems Architecture

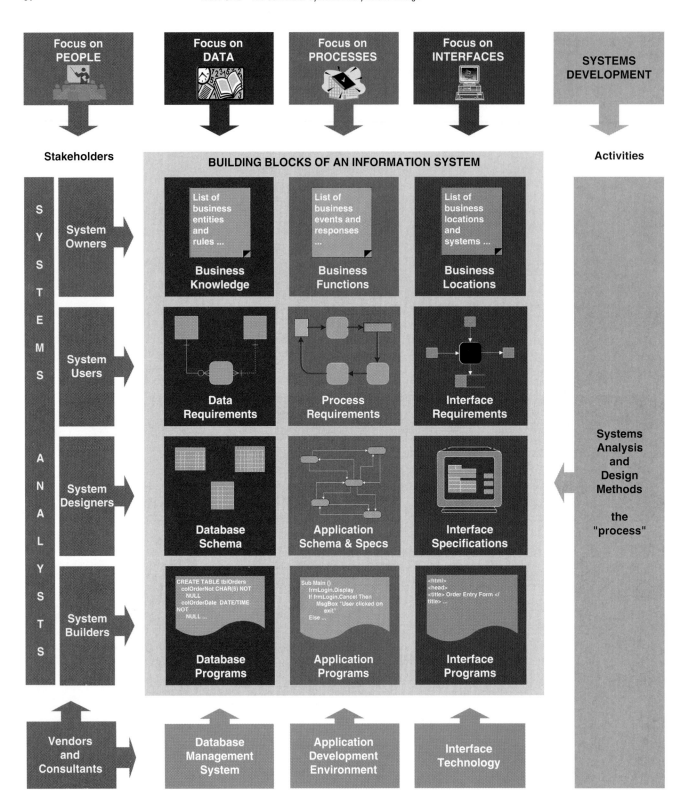

FIGURE 2.7 *Information System Building Blocks*

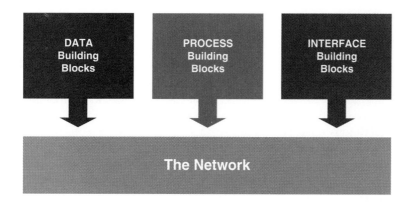

FIGURE 2.8
*The Role of the Network in
Information Systems*

John Zachman recognized the network dimension and included a technology-based COMMUNICATIONS column in his version of the framework. For the most part, the network is not designed as part of an information system development project. Instead, it already exists, having been previously designed to support multiple information systems. That's why we did not include a communications column in our framework—it is more suited to a telecommunications and network analysis and design textbook or course. But in the spirit of completeness, a COMMUNICATIONS column is depicted in the margin and summarized in the following paragraphs.

System Owners' View of COMMUNICATIONS   To the system owner, communications is not a technical or networking issue. It is merely a business reality. In its simplest form, the system owner views COMMUNICATIONS in terms of *geography*. This is illustrated in the margin, but try not to restrict your thinking to geography in the classic sense of the depicted icon, a map. True, geography includes such locations as cities, states, and countries. But it also includes locations such as campuses, sites, buildings, floors, and rooms. Also note, operating locations are not synonymous with computer centers because computer centers can network to many operating locations.

Geography has taken on a new dimension in many businesses that strategically seek to integrate their information systems into the businesses of their suppliers and customers. For example, Sears, the retailing giant, has directly linked its purchase order information system with the order processing information systems of many of its primary suppliers. Thus, Sears can avoid the delays associated with mailing or phoning orders to replenish stock. This system also gives the retailer a competitive advantage—keeping products on the shelves with minimum lead time, which reduces inventory-carrying costs and increases profits.

System Users' View of Communications   Consistent with the user views of DATA, PROCESSES, and INTERFACES, system users are the experts about the *communication requirements* between and within business locations. System users might identify some truly unique definitions of location. For example, consider a traveling sales representative as part of an order entry information system. That sales representative's location moves. He or she might be in a car, an office, the office of a customer, or working from a home office.

Each operating location's system resources can be expressed in terms of the building blocks you've already learned—DATA, PROCESSES, and INTERFACES that are needed at the location. The communications requirements are frequently expressed in terms of location-to-location data flows and the volume and frequency of those data flows. A network analyst might represent the communication requirements for information systems using a flow-like diagram as shown in the margin on page 66.

Focus on COMMUNICATIONS

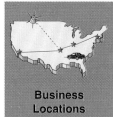

Business Locations

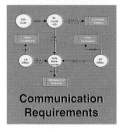

Communication Requirements

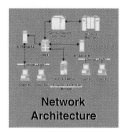

Network Architecture

Network Programs

Network Technology

A COMMUNICATIONS *Focus for Information Systems*

System Designers' View of COMMUNICATIONS   As was the case with the previous building blocks, system designers' view of COMMUNICATIONS is influenced and/or constrained by the limitations of specific telecommunications and network technologies. The emphasis shifts to specifying a network architecture that can support the communications requirements. The system designer's view is expressed in terms of physical locations for workstations, terminals, servers, and peripherals and the interconnections between those devices. The system designer, in this case a telecommunications specialist, addresses and documents these technical views of a computer network (see margin).

The technical dimension of network design is far beyond the scope of this book. But because of the trend toward distributed computing, aspiring systems analysts should pursue a conversational level of understanding of data communications. If your school does not offer such a course, we have included an excellent introductory textbook in the recommended readings for this chapter.

System Builders' View of COMMUNICATIONS   The final view of COMMUNICATIONS is relevant only to the system builders, once again, telecommunications and networking specialists. They use networking software to customize and optimize networks including node addresses, protocols, line speeds, flow controls, security, privileges, and other complex, networking parameters. This software is usually purchased and installed, but it must also be configured and tuned for performance. Examples of communications software include network operating systems such as *NetWare, Windows/NT Server,* IBM *Lotus Notes/Domino,* and *UNIX* or *Linux.*

Once again, it is not the intent of this book to teach network software and optimization. Most information system programs offer courses that can expand your understanding of telecommunications and networking.

WHERE DO YOU GO FROM HERE?

Where are we now? If you have already read Chapter 1, you learned about the participants in information system development projects with a focus on systems analysts as the key facilitators of development. If you haven't already done so, you should at least skim Chapter 1 to learn about the players in the information system development game.

In Chapter 2, you learned about the product itself—information systems—in terms of basic building blocks. We will reuse and refine the framework for information systems architecture throughout this book.

Most readers should proceed directly to Chapter 3, which introduces you to the *process* of information system development. You'll learn about information systems problem solving, methodologies, and development technology as you expand your education in the fundamentals for systems analysis and design.

SUMMARY

1. Data are raw facts about a business and its business transactions. Information is data that has been refined and organized by processing and purposeful intelligence.

2. An information system (IS) is an arrangement of people, data, processes, interfaces, and communications that interact to support and improve day-to-day operations in a business as well as support the problem-solving and decision-making needs of management and users.
   a. Information systems are enabled by information technology, a contemporary term that describes the combination of computer technology (hardware and software) with telecommunications technology (data, image, and voice networks).
   b. Front-office information systems support business functions that reach out to customers (or constituents).
   c. Back-office information systems support internal business operations as well as interact with suppliers (of materials, equipment, supplies, and services).

3. Information systems fulfill one or more of the following basic purposes:
   a. Transaction processing systems capture and process data about (or for) business transactions.
   b. Management information systems provide essential management reports required to plan, monitor, and control business operations.
   c. Decision support systems provide users, especially managers, with decision-oriented information in response to various unstructured decision and problem-solving opportunities.
      (1) Decision support systems are built around data warehouses—read-only, informational databases that are populated with detailed, summary, and exception data and information that can be accessed by end-users and managers with DSS tools that generate a virtually limitless variety of information in support of unstructured decisions.
   d. Expert systems capture and simulate the knowledge and expertise of subject matter experts so nonexperts may apply it.
   e. Office automation (OA) systems support the wide range of business office activities that provide for improved work flow and communications between workers, regardless of whether or not those workers are located in the same office.
      (1) Personal information systems are those designed to meet the needs of a single user. They are designed to boost an individual's productivity.
      (2) Work group information systems are those designed to meet the needs of a work group. They are designed to boost the group's productivity.

4. Information systems architecture provides a unifying framework into which various people with different perspectives can organize and view the fundamental building blocks of information systems. The framework used throughout this book is based on, and adapted from, the Zachman "Framework for Information Systems Architecture."
   a. The information system framework is visually presented as a matrix. The rows correspond to the perspectives of different stakeholders in the information system development process. The columns correspond to a specific focus or dimension of the information system that must be analyzed, designed, and implemented. Each cell represents one perspective and one focus.
   b. Although the cells can be studied and developed in isolation, they must be synchronized with the other cells (both across rows and down columns) to develop a successful information system.

5. The six perspectives are provided by:
   a. System owners, who pay for the system to be built and maintained. They own the system, set priorities for the system, and determine policies for its use.
   b. System users, who actually use the system to perform or support the work to be completed. System users define the business requirements and performance expectations for the system to be built.
   c. System designers, the technical specialists who design the system to meet the users' requirements. In many cases, system designers may also be system builders.
   d. System builders, the technical specialists who construct, test, and deliver the system into operation.
   e. Systems analysts, who facilitate the development of information systems and computer applications by bridging the communications gap that exists between nontechnical system owners and users and technical system designers and builders.
   f. IT vendors and consultants, who sell hardware, software, and services to businesses for incorporation into their information systems.

6. The three focuses represented in the model are:
   a. DATA—the raw material used to create useful information.
   b. PROCESSES—the activities (including management) that carry out the mission of the business.
   c. INTERFACES—how the system interfaces with its users and other information systems.

7. A fourth focus, COMMUNICATIONS, was identified in the original Zachman framework but is omitted from our framework because computer networks are not typically built in conjunction with any single information system development project. Instead, computer networks are built in separate projects to support the communication requirements for many information systems. Regardless, the COMMUNICATIONS focus can be viewed from the same perspectives as the other framework focuses.

KEY TERMS

application program, p. 60
application schema, p. 59
artificial intelligence (AI), p. 48
back-office information system, p. 45
business function, p. 56
business knowledge, p. 52
business process, p. 58
business process redesign, p. 47
cross-functional information
  system, p. 58
data, p. 45
data maintenance, p. 46
data requirement, p. 55
data warehouse, p. 48
database management system
  (DBMS), p. 56

database schema, p. 56
decision support system (DSS), p. 47
executive information system
  (EIS), p. 47
expert system, p. 48
front-office information system, p. 45
information, p. 45
information system (IS), p. 45
information systems architecture, p. 51
information technology (IT), p. 45
interface requirement, p. 62
IT vendor or consultant, p. 51
management information system
  (MIS), p. 47
middleware, p. 63
office automation (OA), p. 48

personal information system, p. 48
policy, p. 58
procedure, p. 58
process requirement, p. 58
prototyping, p. 60
software specification, p. 59
system builder, p. 51
system designer, p. 51
system owner, p. 51
system user, p. 51
systems analyst, p. 51
transaction processing system, p. 46
user dialogue, p. 62
work group information system, p. 49

REVIEW QUESTIONS

1. Define data, information, information systems, and information technology.
2. Differentiate between front- and back-office information systems.
3. What is a transaction processing system? How are transaction processing systems being affected by business process redesign?
4. Explain the role of data maintenance in a transaction processing system.
5. Differentiate between management information systems and decision support systems.
6. What role does a data warehouse play in a decision support system?
7. What is an expert system? How is it different from a decision support system?
8. What is an office automation system? Describe two types of office automation systems.
9. What is information systems architecture?

10. What is the difference between a perspective and a focus in this chapter's information system framework?
11. Name six groups of information system stakeholders.
12. List three focuses for an information system.
13. List four data perspectives.
14. List four process perspectives.
15. Differentiate between business functions and business processes.
16. Why are businesses interested in cross-functional systems?
17. Differentiate between business policies and business procedures.
18. List four interface perspectives.
19. Differentiate between user and system interfaces. With respect to user interfaces, what phenomena are driving the trend toward graphical user interfaces?
20. What is the role of the network in information systems (as it relates to the building blocks)?

PROBLEMS AND EXERCISES

1. Identify each of the following as data or information. Explain why you made the classification you did.
   a. A report that identifies, for the purchasing manager, parts that are low in stock.
   b. A customer's record in the customer database (or table).
   c. An order you place for a software package.
   d. A report your boss must modify to be able to present statistics to her boss.
   e. Your monthly credit card invoice.

2. Give an example of a completely manual information system. Describe how information technology might improve that information system.
3. In the SoundStage case study for Chapter 2, classify each attendee as system owner, user, designer, builder, analyst, or IT vendor/consultant.
4. Explain three examples of transaction processing at your college or university.
5. Explain three examples of transaction processing that might occur for a retail store.

6. For problem 3 or 4, list three examples of related management information system reports.
7. How might a decision support system evolve into an expert system.
8. An office manager has described his company's office information system in terms of word processing and spreadsheets for his staff. Explain to him why his system is actually a personal information system. Explain how it might be extended to become a true office information system.
9. Identify the type of information system application described for each of the following:
    a. A customer presents a deposit slip and cash to a bank teller.
    b. A teller gets a report from the cash register that summarizes the total cash and checks that should be in the cash drawer.
    c. Before cashing a customer's check, the teller checks the customer's account balance.
    d. The bank manager gets an end-of-day report that shows all tellers whose cash drawers don't balance with the cash register summary report.
    e. The system prints a report of all deposits and withdrawals for a given day.
    f. A chief executive officer is using his computer to obtain access to the latest stock market trends.
    g. A doctor keys in data describing symptoms of a patient; she receives a report that suggests what illness the patient is likely suffering from and a detailed explanation concerning the rationale as to why the symptoms suggest that particular illness.
    h. An employee is electronically sending a memorandum concerning a scheduled committee meeting to all committee members who are to attend.
10. The system owner's view of data is primarily concerned with entities. If the owner of a course scheduling information system is the registrar, brainstorm the entities this registrar might identify.
11. Differentiate between entities and rules. Using the course registration example in the previous exercise, give examples of each.
12. For a programming assignment you completed in a programming course, describe how each of the four process building blocks was either presented to you or developed by you.
13. Explain the difference between user interfaces and user dialogues. Why are user dialogues more difficult to design than user interfaces?

PROJECTS AND RESEARCH

1. Make an appointment to visit a systems analyst at a local information system organization. Discuss information system projects the analyst has worked on. What were some of the applications (e.g., transaction processing, management information, decision support, expert system, and office automation) being supported?
2. Obtain an organization chart for a business. (Your instructor may provide one.) Study that organization chart and try to list as many transaction processing, management information, and decision support systems that you can envision supporting the units illustrated in the chart.
3. Information systems are all around you. Consider one of your previous employers. Describe an information system that you used in terms of the four focuses and from the perspectives of a system owner, a system user, and a system designer.
4. Read John Zachman's original framework paper (see reference at end of chapter). How does the Zachman framework differ from the adaptation presented in the chapter?
5. John Zachman wrote a sequel to his historic paper on information systems architecture. In that paper, he suggested the relevance of additional focuses for information systems work. Do a literature search to find and read that paper. What new focuses did Zachman identify, and how might they be accommodated in this chapter's information systems framework?
6. Build Your Own Case Project—As part of a personalized, custom semester project, we encourage you to apply the concepts and techniques we will teach you to an information system, real or hypothetical, that is based on your current or prior work background, a serious hobby, a student organization to which you belong, or the like. We can think of no better way to master the material than to apply it to something with which you are very familiar. A team experience would better simulate the true nature of systems analysis and design; however, we realize this may not be possible because you may not be able to find another classmate who is familiar with the business, organization, or subject area you choose.
    a. For what business, organization, or subject area do you intend to design your system?
    b. What type of information system do you intend to design? Try to describe its capabilities in terms of transaction processing, management information, decision support, expertise simulation, and/or office automation.
    c. Try to describe the system in terms of the DATA, PROCESS, and INTERFACE building blocks.
    d. Does your business or organization have an existing information technology environment? Most do. If so, briefly describe that environment in terms of computers, networks, database management systems, and application development tools.

MINICASES

1. Knowledgeable University plans to support course registration and scheduling on a computer. The following client community has been designated:

   a. Curriculum deputy—one per department, estimates demand by that department's own students for each course offered by the university. This person may revise demand estimates from time to time.

   b. Department chair—determines which courses will be offered by a department, and which faculty will teach those courses.

   c. Schedule deputy—one per department, decides at what times courses will be offered, and with what enrollment limits. These parameters may change during the registration period. This is the only person who can increase or decrease enrollment limits for a course.

   d. Space deputy—allocates classroom and lecture hall space and time to departments. Also prints the schedule of classes to show students what will be offered and when.

   e. Students—submit course requests and revisions and receive schedules and fee statements.

   f. Counselors—advise students and approve all course requests and revisions. They also help students resolve time conflicts (where student has registered for two courses that meet at the same time).

   This is the cast of characters or constituents (which may be revised or supplemented by your instructor to more closely match your school). For each constituent, brainstorm and describe their likely perspectives with respect to DATA, PROCESSES, and INTERFACES.

2. Liz, an account collection manager for the bankcard office of a large bank, has a problem. Each week, she receives a listing of accounts that are past due. This report has grown from a listing of 250 accounts (two years ago) to 1,250 accounts (today). Liz has to go through the report to identify those accounts that are seriously delinquent. A seriously delinquent account is identified by several different rules, each requiring Liz to examine one or more data fields for that customer. What used to be a half-day job has become a three-days-per-week job. Even after identifying seriously delinquent accounts, Liz cannot make a final credit decision (such as a stern phone call, cutting off credit, or turning the account over to a collection agency) without accessing a three-year history of the account. Additionally, Liz needs to report what percentage of all accounts is past due, delinquent, seriously delinquent, and uncollectible. The current report doesn't give her that information. What kind of report does Liz have—detail, summary, or exception? What kind of reports does Liz need? What kind of decision support aids would be useful?

SUGGESTED READINGS

Bruce, Thomas. *Designing Quality Databases Using IDEF1X Information Models*. New York: Dorset House Publishing, 1993. This book, like much of the database industry, has embraced the Zachman framework.

Davis, Gordon B. "Knowing the Knowledge Workers: A Look at the People Who Work with Knowledge and the Technology That Will Make Them Better." *ICP Software Review*, Spring 1982, pp. 70–75. This timeless article provided our first exposure to the concept of information workers as the new majority.

Goldman, James E.; Phillip T. Rawles; and Julie R. Mariga. *Client/Server Information Systems: A Business-Oriented Approach*. New York: John Wiley & Sons, 1999. For those students who are looking for a student-oriented introduction to information technology architecture and data communications, we recommend our colleagues' book because it was written for business and information systems majors to provide both a comprehensive survey of the technology that supports today's information systems.

Intelligent Enterprise (formerly known as *Database Programming & Design*). This monthly periodical has strongly embraced the Zachman framework to unify its subject matter.

O'Brien, James. *Information Systems: The Alternate Edition*. Burr Ridge, IL: Irwin/McGraw-Hill, 1998. This is a popular information systems survey textbook targeted at the freshman/sophomore audience.

Zachman, John A. "A Framework for Information System Architecture." *IBM Systems Journal* 26, no. 3 (1987). We adapted the matrix model for information system building blocks from Mr. Zachman's conceptual framework. We first encountered John Zachman on the lecture circuit where he delivers a remarkably informative and entertaining talk on the same subject as this article. Mr. Zachman's framework has drawn professional acclaim and inspired at least one conference on his model. His framework is based on the concept that architecture means different things to different people. His framework suggests that information systems consist of three distinct "product-oriented" views—data, processes, and technology (which we renamed *communications*). We added a fourth view, interface. The Zachman framework offered six different audience-specific views for each of those product views—the ballpark and owner's views (which we renamed as owner's and user's views, respectively), the designer's and builder's views (which we combined into our designer's view), and an out-of-context view (which we called the builder's view).

**Focus on
PEOPLE**

**Focus on
DATA**

**Focus on
PROCESSES**

**Focus on
INTERFACE**

**Focus on
DEVELOPMENT**

Stakeholders

Activities

S Y S T E M S A N A L Y S T S		
	SYSTEM OWNERS →	
	SYSTEM USERS →	
	SYSTEM DESIGNERS →	
	SYSTEM BUILDERS →	

VENDORS AND CONSULTANTS

BUILDING BLOCKS OF AN INFORMATION SYSTEM

Management Expectations

The PIECES Framework

Performance ● Information ● Economics ● Control ● Efficiency ● Service

List of business entities and rules ...

**Business
Knowledge**

List of business functions and events ...

**Business
Functions**

List of business locations and systems...

**Business
Locations**

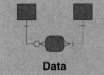

**Data
Requirements**

**Process
Requirements**

**Interface
Requirements**

**Database
Schema**

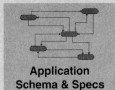 

**Application
Schema & Specs**

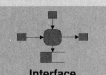

**Interface
Specifications**

```
CREATE TABLE tblOrders
  colOrderNot CHAR(5) NOT
    NULL
  colOrderDate DATE/TIME
NOT
```

**Database
Programs**

```
PROC ValidateOrder
  PERFORM ValidateCust
  REPEAT UNTIL
    NoMoreProd ...
```

**Application
Programs**

```
<html>
<head>
<title> Order Entry Form </title>
...
```

Interface Programs

INFORMATION TECHNOLOGY & ARCHITECTURE
Database Technology ● Process Technology ● Interface Technology ● Network Technology

PROJECT & PROCESS MANAGEMENT

PRELIMINARY INVESTIGATION

PROBLEM ANALYSIS

REQUIREMENTS ANALYSIS

DECISION ANALYSIS

DESIGN

CONSTRUCTION

IMPLEMENTATION

OPERATIONS AND SUPPORT

3

INFORMATION SYSTEMS DEVELOPMENT

CHAPTER PREVIEW AND OBJECTIVES

This chapter introduces the system development process used to develop information systems. System development is not a hit-or-miss process! As with any product, information systems must be carefully developed. Successful systems development is governed by fundamental, underlying principles that we will introduce in this chapter. We also introduce a representative systems development methodology as a disciplined approach to developing information systems. Although such an approach will not guarantee success, it will improve the chances of success. You will know that you understand information systems development when you can:

- Describe the motivation for a systems development process in terms of the Capability Maturity Model (CMM) for quality management.

- Differentiate between the system life cycle and a system development methodology.

- Describe eight basic principles of systems development.

- Define problems, opportunities, and directives—the triggers for systems development projects.

- Describe the PIECES framework for categorizing problems, opportunities, and directives.

- Describe the traditional, basic phases of systems development. For each phase, describe its purpose, inputs, and outputs.

- Describe cross life cycle activities that overlap all system development phases.

- Describe four basic alternative "routes" through the basic phases of systems development. Describe how routes may be combined or customized for different projects.

- Differentiate between computer-aided systems engineering (CASE), application development environments (ADEs), and process and project management technology as automated tools for systems development.

SCENE

A conference room where Sandra Shepherd has called a project planning meeting. Sitting around the table are Bob Martinez (a teammate), Terri Hitchcock (the business analyst assigned to the Member Services system project), and Gary Goldstein (the Development Center manager).

SANDRA

Good afternoon. I thought the project launch meeting went very well. I called this meeting to start planning the Member Services project. I guess some introductions are in order. I am Sandra Shepherd, senior systems analyst and project manager for this project. The permanent members of our team will be Bob Martinez, a new systems analyst here at SoundStage, and Terri Hitchcock, who is on a two-year loan to Information Services from Member Services. She will be the business analyst for the project. I'd like to introduce you both to Gary Goldstein. He manages our Development Center.

BOB

What is the Development Center?

TERRI

My question, exactly!

GARY

The Development Center was established as a result of the company's ISO 9000 quality management certification initiative. We recognized the need to establish a quality management center in information systems. Quality is a function of the process and tools we use to build information systems; therefore, we named our quality management unit a "Development Center."

The Development Center is a group of IT professionals who plan, implement, and support an information systems development environment for our systems analysts, designers, and programmers. We provide training and support for both our chosen methodology, *FAST*, and automated tools for building all information systems. Think of us as consultants to our information systems development staff.

Peg Li is the director of the Develop-

ment Center. I work for Peg and I'm primarily responsible for *FAST* methodology training, application, and improvement. You should have received my memo scheduling you for our two-day workshop to introduce your Member Services project team to *FAST*.

TERRI

What is this *FAST* thing?

GARY

Good question! Actually, *FAST* is not a "thing"—It is a standard "process" or methodology we use to develop and maintain all of our information systems. *FAST* is an acronym that stands for *Framework for the Application of Systems Techniques*.

BOB

Is it called *FAST* because it supports rapid development?

GARY

No, it is called *FAST* because it attempts to deliver the best possible information system quality in a reasonable amount of time. It does support the techniques you call rapid application development. But it also supports other techniques including structured systems analysis, information engineering, and object-oriented analysis and design . . .

TERRI

Say what?

GARY

You're right! These are strategy buzzwords. Those specific strategies are not as important as the flexibility that *FAST* gives us to choose the best development strategy for each given project. That is why we chose *FAST*. It is truly a framework for selecting our development methods, not a rigid process. It even supports the option of purchasing our information systems instead of building them in-house.

BOB

Why is *FAST* better?

GARY

As I already noted, it is more flexible. We found that to be highly desirable in our development culture. It is customizable to our standards and extensible to

emerging methods, like object-oriented. And it is online. The only documentation is a "Getting Started" booklet and a reference manual for the online service. There are also some training materials, but you receive those when you complete the various in-house courses I teach.

BOB

Online?

GARY

Yes, the methodology database is stored on the local area network, and it's available in its most current version from any developer's workstation. Simple diagrams walk the developers through the various phases, activities, roles, deliverables, and so forth. It can even automatically invoke the correct CASE and ADE tools we use to design and build our systems at the correct time during the project.

Terri, CASE and ADE tools are automated tools we use to design and build systems. You'll see them in action at the workshop.

SANDRA

Well, we need to get on with our agenda. The full team will begin the preliminary investigation phase shortly after Gary's workshop. Today we need to review the preliminary system development strategy for the purpose of preparing some project planning materials for that phase.

BOB

Such as?

SANDRA

FAST recognizes that one size does not fit all projects! The *FAST* methodology defines a series of basic project phases and deliverables; however, the methodology defines different routes through those phases and alternative deliverables based on project goals and development strategies. We need to select the appropriate route for the Member Services project.

[. . . noting nonverbal agreement]

I think we can eliminate the purchased software route, no?

SOUNDSTAGE
SOUNDSTAGE ENTERTAINMENT CLUB

TERRI

Why? Wouldn't purchasing a package get us up and running much faster?

SANDRA

It could. But we have researched some alternatives and we just cannot find any package that comes remotely close to providing us with the competitive advantage that we want from our new system. Mr. Short [the SoundStage CEO] and Mr. Kirchoff [VP of Member Services] both feel strongly that the Member Services system is where we need to build competitive advantage for SoundStage. Along with Nancy [Picard, CIO], they all feel that we can achieve that competitive advantage only through an in-house developed system that precisely meets our goals.

TERRI

That makes sense. So are there different in-house development routes?

GARY

Yes. The *FAST* methodology provides two basic routes and several variations and combinations for those routes. So far, most projects have been using the rapid application development route because the projects have been relatively small and that route results in a working application relatively quickly. But I don't think that route is necessarily best for your new project.

BOB

Why not? I would think that Terri and the user participants on the team would want a rapidly developed solution.

GARY

The scope of this project is much larger than anything we've tried with the rapid application development route. The project crosses boundaries with multiple departments, users, and managers. Also, there appears to be a directive to redesign existing business processes to significantly reduce bureaucracy and improve response time and service to members. For this type of project, *FAST* would recommend a model-driven approach.

Terri and Bob, the model-driven approach is a problem-solving strategy that forces us to thoroughly analyze our business problems and requirements in terms of pictures called system models. The process of drawing these models forces us to break a system down into manageable parts and thoroughly analyze those parts and their interrelationships. The system model serves as a communication vehicle between managers, users, designers, and builders. Essentially, it becomes the blueprint for building a new system.

SANDRA

I think I agree that a model-driven approach might be better for this type of project. But I hate to lose all of the advantages of the rapid application development.

GARY

Actually, you can merge the model-driven and rapid application development routes. That's one of the services provided by my Development Center

team. *FAST* allows us to develop a customized route that best suits any project. In your case, we might use rapid application development techniques to build initial system models. Or we could use rapid application development techniques to construct subsystems that have already been modeled. Every project is different. A strength of *FAST* is that *we* define the route and the methodology guides us through that route.

SANDRA

I vote for such a hybrid route. Let's move on . . . Terri, can you provide us with any management and user insight that may impact our plans? How about . . .

DISCUSSION QUESTIONS

1. Why would an organization follow a prescribed methodology for building information systems?

2. Some people believe that an enforced methodology for building systems stifles creativity. Why would they think that is true?

3. What are the advantages of buying an information system (including software) versus building that information system in-house?

4. Why should a methodology or strategy be adaptable to different projects?

5. How do you perceive Gary's role in this meeting and project? How does Gary's role differ from Sandra's role?

This chapter introduces the focus on information systems development (see the chapter knowledge map at the beginning of the chapter). We will examine the systems development process.

> A **systems development process** is a set of activities, methods, best practices, deliverables, and automated tools that stakeholders (Chapter 1) use to develop and maintain information systems and software.

Notice we did not say "the" process. There are as many variations on the process as there are experts. We present one such process and use it consistently throughout this book.

Why do organizations embrace standardized processes to develop information systems? Information systems are a complex product—recall the information system building blocks in Chapter 2. Perhaps this explains why as many as 70 percent

THE PROCESS OF SYSTEMS DEVELOPMENT

or more of information system development projects have failed to meet expectations, cost more than budgeted, and are delivered much later than promised. The Gartner Group suggests, "Consistent adherence to moderately rigorous methodology guidelines [a process] can provide 70 percent of [systems development] organizations with a productivity improvement of at least 30 percent within two years."[1]

Increasingly, organizations have no choice but to adopt and follow a systems development process! The U.S. government has mandated that any organization seeking to develop software or firmware for the government must adhere to certain quality management requirements. And many other organizations have aggressively committed to total quality management goals to increase competitive advantage. To realize quality and productivity improvements, many organizations have turned to frameworks such as the Capability Maturity Model.

The Capability Maturity Model

It has been shown that as an organization's standard information system development process matures, project timelines and costs decrease while productivity and quality increase. The Software Engineering Institute at Carnegie Mellon University has observed and measured this phenomenon and developed a framework to assist all organizations achieve these benefits.

> The **Capability Maturity Model (CMM)** is a framework to assess the maturity level of an organization's information systems development and management processes and products. It consists of five levels of development maturity.

The CMM has developed a wide following, both in industry and government. Software evaluation based on CMM will be used to qualify information technology contractors for all federal government projects by 2002.

The CMM framework for information systems and software is intended to help organizations improve the maturity of their systems development processes. The CMM is organized into five maturity levels (see Figure 3.1) that assess the quality of an organization's systems and software development process.

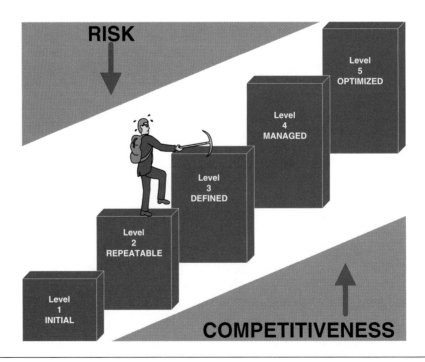

FIGURE 3.1
The Capability Maturity Model (CMM)

---

[1] Richard Hunter, "AD Project Portfolio Management," *Proceedings of the Gartner Group IT98 Symposium/Expo,* [CD-ROM].

- *Level 1—Initial:* This is sometimes called *anarchy* or *chaos*. System development projects follow no prescribed process. Each developer uses his or her own tools and methods. Success or failure is usually a function of the skill and experience of the project team. The process is unpredictable and not repeatable. A project typically encounters many crises and is frequently over budget and behind schedule. Documentation is sporadic or not consistent from one project to the next, thus creating problems for those who must maintain a system over its lifetime. Almost all organizations start at Level 1.

- *Level 2—Repeatable:* Project management processes and practices have been established to track project costs, schedules, and functionality. The focus is on project management, not systems development. A systems development process is always followed, but it may vary from project to project. Success or failure is still a function of the skill and experience of the project team; however, a concerted effort is made to repeat earlier project successes. Effective project management practices lay the foundation for standardized processes in the next level.

- *Level 3—Defined:* A standard system development process (sometimes called a methodology) has been purchased or developed, and its use has been integrated throughout the information systems/services unit of the organization. All projects use a tailored version of this software development process to develop and maintain information systems and software. As a result of using the standardized process for all projects, each project results in consistent and high-quality documentation and deliverables. The process is stable, predictable, and repeatable.

- *Level 4—Managed:* Measurable goals for quality and productivity have been established. Detailed measures of the standard system development process and product quality are routinely collected and stored in a database. There is an effort to improve individual project management based on this collected data. Thus, management seeks to become more proactive than reactive to system development problems (such as cost overruns, scope creep, schedule delays, etc.). Even when projects encounter unexpected problems or issues, the process can be adjusted based on predictable and measurable impacts.

- *Level 5—Optimized:* The standardized system development process is continuously monitored and improved based on measures and data analysis established in Level 4. This can include changing the technology and best practices used to perform activities required in the standard system development process, as well as adjusting the process itself. Lessons learned are shared across the organization, with an emphasis on eliminating inefficiencies in the systems development process while sustaining quality. In summary, the organization has institutionalized continuous systems development process improvement.

Each level is a prerequisite for the next level.

Currently, many organizations are working hard to achieve at least CMM Level 3 (sometimes driven by government mandate). A central theme to achieving Level 3 (Defined) is the use of a standard process or methodology to build or integrate systems. As shown in Table 3.1, an organization can realize significant improvements in schedule and cost by institutionalizing CMM Level 3 process improvements.[2]

As you will learn in this chapter, most systems development processes are derived from a variation of a classic system life cycle.

System Life Cycle versus System Development Methodologies

---

[2] White paper, "Rapidly Improving Process Maturity: Moving up the Capability Maturity Model Through Outsourcing," (Boston: Keane, Inc., 1997, 1998), p. 11.

TABLE 3.1	CMM Project Statistics for a 200,000 Lines-of-Code Development Project					
Organization's CMM Level	Project Duration (months)	Project Person Months	Number of Defects Shipped	Median Cost ($ millions)	Lowest Cost ($ millions)	Highest Cost ($ millions)
1	30	600	61	5.5	1.8	100+
2	18.5	143	12	1.3	.96	1.7
3	15	80	7	.728	.518	.933

Source: Master Systems, Inc.

A **system life cycle** divides the life of an information system into two stages, *systems development* and *systems operation and support*—first you build it; then you use it, keep it running, and support it. Eventually, you cycle back from operation and support to redevelopment.

Figure 3.2 illustrates the two life cycle stages and the two key events between the stages:

- When a system cycles from development to operation and support, a *conversion* must take place.
- At some point, *obsolescence* occurs and a system cycles from operation to redevelopment.

A system may be in more than one stage at the same time. For example, version one may be in *operation and support* while version two is in *development.*

So what is a systems development *methodology?* Figure 3.2 also demonstrates that a systems development methodology implements the development stage of the system life cycle.

A **systems development methodology** is a very formal and precise system development process that defines (as in CMM Level 3) a set of activities, methods, best practices, deliverables, and automated tools for system developers and project managers to use to develop and maintain most or all information systems and software.

Consistent with the goals of the CMM, methodologies ensure that a consistent, reproducible approach is applied to all projects. Methodologies reduce the risk associated with shortcuts and mistakes. Finally, methodologies produce complete and consistent documentation from one project to the next. These advantages provide one overriding benefit—as development teams and staff constantly change, the results of prior work can be easily retrieved and understood by those who follow!

Methodologies can be homegrown; however, many businesses purchase their development methodology. Examples of commercial methodologies are listed in the margin. In the case study, SoundStage has purchased a methodology. Why purchase a methodology? Most information system organizations can't afford to dedicate staff to the development of a homegrown methodology. Also, commercial methodologies are often supported by available automated tools. Finally, methodology vendors have a vested interest in keeping their methodologies current with the latest business and technology trends.

The SoundStage Entertainment Club methodology is called **FAST,** a

F ramework for the
A pplication of
S ystems
T echniques.

FAST is not a real commercial methodology. We developed it as a composite of

---

Representative Commercial Methodologies

Fusion
AD/Method
Foundation
Rational Unified Process
SUMMIT-D

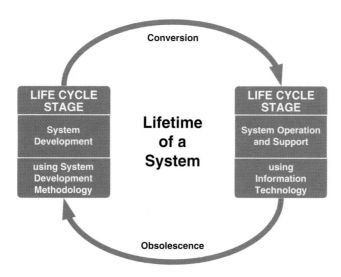

FIGURE 3.2
A Systems Life Cycle

the best methodology practices we've encountered. Like many methodologies, it is flexible enough to provide for different types of projects and strategies.

Before we study the life cycle and *FAST,* let's introduce some general principles that should underlie all systems development methodologies.[3]

Principle 1: Get the Owners and Users Involved Analysts, programmers, and other information technology specialists frequently refer to "my system." This attitude creates an "us-versus-them" conflict between technical staff and the users and management. Although analysts and programmers work hard to create technologically impressive solutions, those solutions often backfire because they don't address the real organization problems or they introduce new problems. For this reason, system owner and user involvement is necessary for successful systems development. The individuals responsible for systems development must make time for owners and users, insist on their participation, and seek agreement from them on all decisions that may affect them.

Miscommunication and misunderstandings continue to be a significant problem in systems development. However, owner and user involvement and education minimize such problems and help to win acceptance of new ideas and technological change. Because people tend to resist change, information technology is often viewed as a threat. The best way to counter that threat is through constant and thorough communication with owners and users.

Principle 2: Use a Problem-Solving Approach A methodology is a problem-solving approach to building systems. The term *problem* is used throughout this book to include real problems, opportunities for improvement, and directives from management. The classic problem-solving approach is as follows:

1. Study and understand the problem and its context.
2. Define the requirements of a suitable solution.
3. Identify candidate solutions and select the "best" solution.
4. Design and/or implement the solution.
5. Observe and evaluate the solution's impact, and refine the solution accordingly.

Systems analysts should approach all projects using a problem-solving approach.

Underlying Principles for Systems Development

---

[3] Adapted from R.I. Benjamin, *Control of the Information System Development Cycle* (New York: Wiley-Interscience, 1971).

Inexperienced problem solvers tend to eliminate or abbreviate one or more of the above steps. The result can range from (1) solving the wrong problem, to (2) incorrectly solving the problem, to (3) picking the wrong solution. A methodology's problem-solving orientation can reduce or eliminate the above risks.

Principle 3: Establish Phases and Activities   All life cycle methodologies prescribe phases and activities. The number and scope of phases and activities varies from author to author, expert to expert, and company to company. Figure 3.3 illustrates the seven phases of a representative system development methodology in the context of your information systems framework (from Chapter 2). In each phase, the stakeholders are concerned with the building blocks opposite that phase. The phases are:

- Preliminary investigation
- Problem analysis
- Requirements analysis
- Decision analysis
- Design
- Construction
- Implementation

Each phase serves a role in the problem-solving process. Some phases identify problems, while others evaluate, design, and implement solutions.

Also notice in Figure 3.3 that ongoing *project and process management* activities are included to monitor and control the phases themselves, as well as quality, costs, resources, and schedule. In addition, a system will eventually enter the operation stage of the life cycle; therefore, we included *operations and support* activities to support the final system. Both activities are depicted with a light color to reflect that they are ongoing activities.

Figure 3.4 shows that the phases tend to overlap. Also, the phases may be tailored to the special needs of any given project (e.g., deadlines, complexity, strategy, resources, and so on). In this chapter, we will describe such tailoring as "routes" through the methodology or problem-solving process.

Principle 4: Establish Standards   An organization should embrace standards for both information systems and the process used to develop those systems. In medium to large organizations, system owners, users, analysts, designers, and builders come and go. Some will be promoted; some will quit; and others will be reassigned. To promote good communication between constantly changing managers, users, and information technology professionals, you must develop standards to ensure consistent systems development.

Standards should minimally encompass the following:

- Documentation
- Quality
- Automated tools
- Information technology

These standards will be documented and embraced within the context of the chosen system development process or methodology.

The need for *documentation standards* underscores a common failure of many analysts—the failure to document as an *ongoing* activity during the life cycle. Documentation should be a working by-product of the entire systems development effort. Documentation reveals strengths and weaknesses of the system to multiple stakeholders before the system is built. It stimulates user involvement and reassures management about progress.

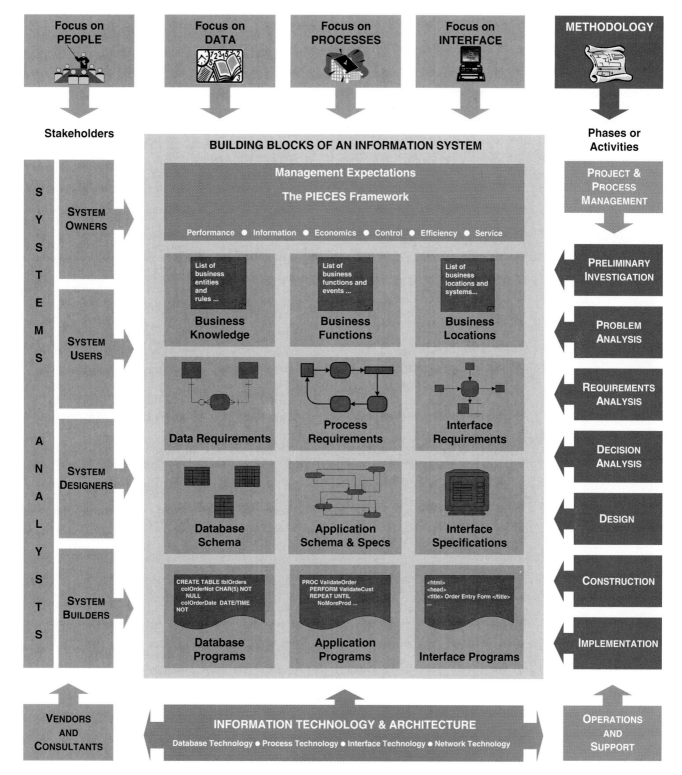

FIGURE 3.3 *Phases of a Representative Methodology*

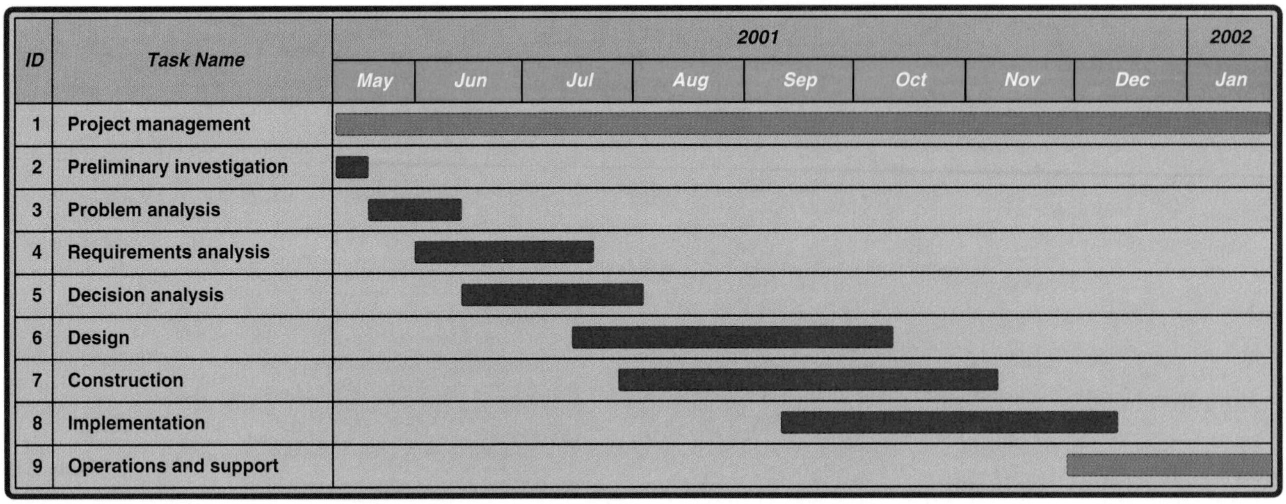

ID	Task Name	2001								2002
		May	Jun	Jul	Aug	Sep	Oct	Nov	Dec	Jan
1	Project management									
2	Preliminary investigation									
3	Problem analysis									
4	Requirements analysis									
5	Decision analysis									
6	Design									
7	Construction									
8	Implementation									
9	Operations and support									

FIGURE 3.4   *Overlap of Systems Development Phases*

Quality standards ensure that the deliverables of any phase or activity meet business and technology expectations. They minimize the likelihood of missed business problems and requirements, as well as flawed designs and program errors (bugs). Frequently, quality standards are applied to documentation produced during development, but quality standards must also be applied to the technical end products such as databases, programs, user and system interfaces, and networks.

Automated tool standards prescribe technology that will be used to develop and maintain information systems and to ensure consistency, completeness, and quality. Today's developers routinely use automated tools (such as Microsoft *Access* or *Visual Basic*) to facilitate the completion of phases and activities, produce documentation, analyze quality, and generate technical solutions. Later in this chapter, we will introduce another type of automated tool called computer-aided systems engineering (CASE).

Finally, *information technology standards* direct technology solutions and information systems to a common technology architecture or configuration. This is similar to automated tools except the focus is on the underlying technology of the finished product, the information systems themselves. For example, an organization may standardize on specific computers and peripherals, operating systems, database management systems, network topologies, user interfaces, and software architectures. The intent is to reduce effort and costs required to provide high-quality support and maintenance of the technologies themselves. Information technology standards also promote familiarity, ease of learning, and ease of use across all information systems by limiting the technology choices. Information technology standards should not inhibit the investigation or use of appropriate emerging technologies that could benefit the business.

Principle 5: Justify Systems as Capital Investments   Information systems are capital investments, just as a fleet of trucks and a new building are. Even if management fails to recognize information systems as an investment, you should not. When considering a capital investment, two issues must be addressed.

First, for any problem, there are likely to be several possible solutions. The analyst (or users) should not necessarily accept the first solution that comes to mind. The analyst who fails to look at alternatives may be doing the business a disservice. Second, after identifying alternative solutions, the systems analyst should evaluate each possible solution for feasibility, especially for cost-effectiveness and risk management.

> **Cost-effectiveness** is defined as the result obtained by striking a balance between the cost of developing and operating an information system and the benefits derived from that system.

Cost-effectiveness is measured using a technique called cost–benefit analysis, taught later in the book.

> **Risk management** is the process of identifying, evaluating, and controlling what might go wrong in a project before it becomes a threat to the successful completion of the project or implementation of the information system.

Cost–benefit analysis and risk management are important skills to be mastered.

Principle 6: Don't Be Afraid to Cancel or Revise Scope   A significant advantage of the phased approach to systems development is that it provides several opportunities to reevaluate cost-effectiveness and feasibility. There is often a temptation to continue with a project only because of the investment already made. In the long run, canceled projects are less costly than implemented disasters! This is extremely important for young analysts to remember.

Most system owners want more out of their systems than they can afford or are willing to pay for. Furthermore, the scope of most information system projects increases as the analyst learns more about the business problems and requirements as the project progresses. Unfortunately, most analysts also fail to adjust estimated costs and schedules as scope increases. As a result, the analysts frequently and needlessly accept responsibility for cost and schedule overruns.

We advocate a *creeping commitment* approach to systems development.[4] Using the creeping commitment approach, multiple feasibility checkpoints are built into any systems development methodology. At each feasibility checkpoint, all costs are considered sunk (meaning not recoverable). They are, therefore, irrelevant to the decision. Thus, the project should be reevaluated at each checkpoint to determine if it remains feasible to continue investing time, effort, and resources.

At each checkpoint, the analyst should consider the following options:

- Cancel the project if it is no longer feasible.
- Reevaluate and adjust the costs and schedule if project scope is to be increased.
- Reduce the scope if the project budget and schedule are frozen and not sufficient to cover all project objectives.

The concept of sunk costs is frequently forgotten or not used by the majority of practicing analysts, most users, and even many managers.

Principle 7: Divide and Conquer   Consider the old saying, "If you want to learn anything, you must not try to learn everything—at least not all at once." For this reason, we divide a system into subsystems and components to more easily conquer the problem and build the larger system. By dividing a larger problem (system) into more easily managed pieces (subsystems), the analyst can simplify the problem-solving process. This divide and conquer approach also complements communication and project management by allowing different pieces of the system to be delegated to different stakeholders.

---

[4] Thomas Gildersleeve, *Successful Data Processing Systems Analysis,* 2nd ed. (Englewood Cliffs, NJ: Prentice Hall, 1985), pp. 5–7.

You have learned the divide and conquer approach throughout your schooling. Since high school, you've been taught to outline a paper before you write it. Outlining is a divide and conquer approach to writing.

Principle 8: Design Systems for Growth and Change   Many systems analysts develop systems to meet only today's user requirements because of the pressure to develop the system as quickly as possible. Although this may seem to be a necessary short-term strategy, it frequently leads to long-term problems.

System scientists describe the natural and inevitable decay of all systems over time as *entropy*. As described earlier in this section, after a system is implemented it enters the operations and support stage of the life cycle. During this stage the analyst encounters the need for changes that range from correcting simple mistakes, to redesigning the system to accommodate changing technology, to making modifications to support changing user requirements. Such changes direct the analyst and programmers to rework formerly completed phases of the life cycle. Eventually, the cost of maintaining the current system exceeds the cost of developing a replacement system—the current system has reached entropy and become obsolete.

But system entropy can be managed. Today's tools and techniques make it possible to design systems that can grow and change as requirements grow and change. This book will teach you many of those tools and techniques. For now, it's more important to simply recognize that flexibility and adaptability do not happen by accident—they must be built into a system.

We have presented eight principles that should underlie any methodology. These principles are listed in the margin and can be used to evaluate any methodology, including our *FAST* methodology.

✓

Principles of System Development

Get the owners and users involved.

Use a problem-solving approach.

Establish phases and activities.

Establish standards.

Justify systems as capital investments.

Don't be afraid to cancel or revise scope.

Divide and conquer.

Design systems for growth and change.

A SYSTEMS DEVELOPMENT METHODOLOGY

In this section we'll examine a systems development process based on a hypothetical methodology called *FAST* (*Framework for the Application of Systems Techniques*). We'll begin by studying types of projects and how they get started. Then we'll introduce the basic *FAST* phases. Finally, we'll examine alternative routes through the phases for different types of projects and development strategies.

Project Identification

System owners and users initiate most projects. As shown in Figure 3.5 (see bullet ❶), the impetus for most projects is some combination of problems, opportunities, and directives.

> **Problems** are undesirable situations that prevent the organization from fully achieving its purpose, goals, and/or objectives.

For example, a business might identify that it is taking too long to fulfill customer orders. A project would be initiated to achieve more responsive and timely order fulfillment. Problems might be real, suspected, or anticipated.

> An **opportunity** is a chance to improve the organization even in the absence of specific problems. (Note: You could argue that any unexploited opportunity is a problem.)

For instance, management is usually receptive to cost-cutting ideas, even when costs are not currently considered a problem. Opportunistic improvement is the source of many of today's most interesting projects.

> A **directive** is a new requirement that's imposed by management, government, or some external influence.

For example, the Equal Employment Opportunity Commission, a government agency, may mandate that a new set of reports be produced each quarter. Similarly,

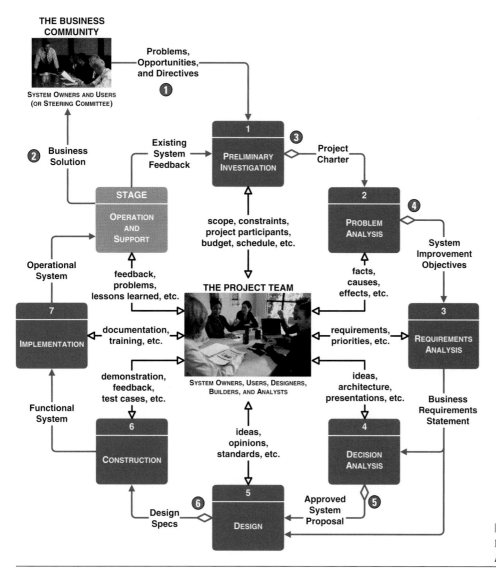

FIGURE 3.5
FAST *System Development Phases*

company management may dictate support for a new product line or policy. Some directives may be technical. For instance, systems may be strategically directed to convert from keyboard data entry to bar code technology or from conventional flat files to modern database technology.

Throughout this book, we will use the term *problem* to collectively refer to "problems, opportunities, and directives." Accordingly, *problem solving* also refers to solving problems, exploiting opportunities, and fulfilling directives.

There are far too many potential problems to list them all in this book. However, James Wetherbe developed a useful framework for classifying problems.[5] He calls it **PIECES** because the letters of each of the six categories spell the word *pieces*. The categories are:

P the need to improve *performance*.

I the need to improve *information* (and data).

E the need to improve *economics*, control costs, or increase profits.

[5] James Wetherbe and Nicholas P. Vitalari, *Systems Analysis and Design: Traditional, Best Practices,* 4th ed. (St. Paul, MN: West Publishing, 1994), pp. 196–99.

C the need to improve *control* or security.

E the need to improve *efficiency* of people and processes.

S the need to improve *service* to customers, suppliers, partners, employees, etc.

If you look again at Figure 3.3, you'll notice that we've added the PIECES framework to your information system building blocks. Figure 3.6 expands on each of the PIECES categories.

The categories of the PIECES framework overlap. Any given project is usually characterized by one or more categories, and any given problem or opportunity may have implications with respect to more than one category. PIECES is a practical framework (used in *FAST*), not just an academic exercise! We'll revisit PIECES later.

Projects can be either planned or unplanned. A planned project is the result of one of the following:

— An *information strategy plan* that has examined the business as a whole to identify those system development projects that will return the greatest strategic (long-term) value to the business.

— A *business process redesign* that has thoroughly analyzed a series of fundamental business processes to eliminate redundancy and bureaucracy and to improve efficiency and value added—now it is time to redesign the supporting information system for those business processes.

Unplanned projects can easily overwhelm the largest information services organization; therefore, they are frequently screened and prioritized by a **steering committee** of system owners and IT managers to determine which requests get approved. Those requests that are not approved are *backlogged* until resources become available (which sometimes never happens).

Project Phases

Want to learn more about the FAST *methodology? Visit the* McGraw-Hill Online Learning Center for this textbook and browse the online resource for the FAST *methodology.*

FAST, like most commercial methodologies, consists of development phases. The number of phases will vary from one methodology to another. Referring to Figure 3.5, the final output of any development methodology is the Business Solution (see bullet ❷), that solves the problems, opportunities, and directives that initiated the project. The operation and support stage lasts the lifetime of the system.

The *FAST* methodology supports both the *systems development* and *operation and support stages* of the system life cycle. The systems development stage consists of seven phases. Let's briefly discuss the phases. Throughout this discussion, we will be referring to Figure 3.3 (your information system building blocks framework) and Figure 3.5 (a methodology flowchart model). The systems development phases are screened in dark blue while the operation and support stage is screened in light blue.

Preliminary Investigation Phase   The purpose of the preliminary investigation (or survey) phase is twofold. First, it answers the question, "Is this project worth looking at?" To answer this question, you must define the perceived problems, opportunities, and directives that triggered the project and assess the risk of pursuing the project. Second, and assuming the project *is* worth looking at, the preliminary investigation phase must also establish the project charter that establishes scope, preliminary requirements and constraints, project participants, budget, and schedule.

To answer the question, "Is this project worth looking at?" the PIECES framework provides an excellent outline for discovering problems, opportunities, and directives. The goal here is not to solve the problems, just to catalog and categorize them.

The following checklist for problem, opportunity, and directive identification uses Wetherbe's PIECES framework. Note that the categories of PIECES are not mutually exclusive; some possible problems show up in multiple lists. Also, the list of possible problems is not exhaustive. The PIECES framework is equally suited to analyzing both manual and computerized systems and applications.

PERFORMANCE

A. Throughput—the amount of work performed over some period of time.

B. Response times—the average delay between a transaction or request and a response to that transaction or request.

INFORMATION (and Data)

A. Outputs
1. Lack of any information.
2. Lack of necessary information.
3. Lack of relevant information.
4. Too much information—"information overload."
5. Information that is not in a useful format.
6. Information that is not accurate.
7. Information that is difficult to produce.
8. Information that is not timely to its subsequent use.

B. Inputs
1. Data is not captured.
2. Data is not captured in time to be useful.
3. Data is not accurately captured—contains errors.
4. Data is difficult to capture.
5. Data is captured redundantly—same data captured more than once.
6. Too much data is captured.
7. Illegal data is captured.

C. Stored data
1. Data is stored redundantly in multiple files and/or databases.
2. Stored data is not accurate.
3. Data is not secure from accident or vandalism.
4. Data is not well organized.
5. Data is not flexible—not easy to meet new information needs from stored data.
6. Data is not accessible.

ECONOMICS

A. Costs
1. Costs are unknown.
2. Costs are untraceable to source.
3. Costs are too high.

B. Profits
1. New markets can be explored.
2. Current marketing can be improved.
3. Orders can be increased.

CONTROL (and Security)

A. Too little security or control
1. Input data is not adequately edited.
2. Crimes (e.g., fraud, embezzlement) are (or can be) committed against data.
3. Ethics are breached on data or information—refers to data or information getting to unauthorized people.
4. Redundantly stored data is inconsistent in different files or databases.
5. Data privacy regulations or guidelines are being (or can be) violated.
6. Processing errors are occurring (either by people, machines, or software).
7. Decision-making errors are occurring.

B. Too much control or security
1. Bureaucratic red tape slows the system.
2. Controls inconvenience customers or employees.
3. Excessive controls cause processing delays.

EFFICIENCY

A. People, machines, or computers waste time.
1. Data is redundantly input or copied.
2. Data is redundantly processed.
3. Information is redundantly generated.

B. People, machines, or computers waste materials and supplies.

C. Effort required for tasks is excessive.

D. Materials required for tasks is excessive.

SERVICE

A. The system produces inaccurate results.
B. The system produces inconsistent results.
C. The system produces unreliable results.
D. The system is not easy to learn.
E. The system is not easy to use.
F. The system is awkward to use.
G. The system is inflexible to new or exceptional situations.
H. The system is inflexible to change.
I. The system is incompatible with other systems.
J. The system is not coordinated with other systems.

FIGURE 3.6 *The PIECES Problem-Solving Framework and Checklist*

Scope definition is an important outcome of this preliminary analysis of problems. Scope defines how big the project is. Your information system building blocks (Figure 3.2) provide a useful framework to define scope. For example, scope could be defined in terms of the DATA required in the system, the business PROCESSES to be supported by the system, and the systems INTERFACES with users and other information systems.

Given the initial scope of the project, the analyst can staff the project team, estimate the budget for system development, and prepare a schedule for the remaining phases. Ultimately, this phase concludes with a "go or no-go" decision from system owners. Either the system owners agree with the proposed scope, budget, and schedule for the project, or they must either reduce scope (to reduce costs and time) or cancel the project. This feasibility checkpoint is illustrated in Figure 3.5 as a diamond (see bullet ❸).

Problem Analysis Phase   There's an old saying that suggests, "Don't try to fix it unless you understand it." With those words of wisdom, the problem analysis phase provides for a study and analysis of the existing system. There is always an existing system, regardless of whether it currently uses a computer. The problem analysis phase provides the project team with a more thorough understanding of the problems that triggered the project. The analyst frequently uncovers new problems. The problem analysis answers the question, "Will the benefits of solving these problems exceed the costs of building the system to solve these problems?"

As shown in Figure 3.5, the key input is the project charter from the preliminary investigation phase. The project team studies the existing system by collecting factual information from the system users concerning the business and the perceived problems, causes, and effects. It is also appropriate to discuss system limitations with information technology specialists who support the existing system. From all this information, the project team gains a better understanding of the existing system's problems.

Every existing system has its own terminology, history, culture, and nuances. Learning those aspects of the system is the principal activity in this phase. Depending on the complexity of the problem domain and the project schedule, the team may or may not choose to formally document the system. Your information system building blocks (Figure 3.3) provide an effective outline for studying the existing system from the perspective of system users as they see the DATA, PROCESSES, and INTERFACES.

The primary deliverable of the problem analysis phase is system improvement objectives. These objectives do not define inputs, outputs, or processes. Instead, they define the business criteria on which any new system will be evaluated. For instance, we might define an objective that states the new system must REDUCE THE TIME BETWEEN ORDER PROCESSING AND SHIPPING BY THREE DAYS, or REDUCE BAD CREDIT LOSSES BY 45 PERCENT. Think of objectives as the "grading criteria" for evaluating any new system.

In some cases, these objectives may be included in a formal detailed study report that also includes documented problems, causes, effects, and solution benefits. An oral presentation to management and users may be substituted for a formal written report.

After reviewing the findings, the system owners will either agree or disagree with the recommendations of the problem analysis phase. Thus, consistent with the creeping commitment principle, we include another "go or no-go" feasibility checkpoint at the end of the phase (see bullet ❹). The project can be:

— Canceled if the problems are deemed not worth solving.
— Approved to continue to the next phase.
— Reduced or expanded in scope (with budget and schedule modifications) and then approved to continue to the next phase.

Requirements Analysis Phase   Given management approval to continue, now you can design a new system, right? Not yet! What capabilities should the new system provide for its users? What data must be captured and stored? What performance level is expected? Careful! This requires decisions about *what* the system must do, not *how* it should do those things. The next phase of the methodology is to define and prioritize *business* requirements. This is called the requirements analysis phase.

The analyst approaches the users to find out what they need or want out of the new system. This is perhaps the most important phase of the methodology. Errors and omissions in requirements analysis result in user dissatisfaction with the final system and costly modifications.

Essentially, the purpose of requirements analysis is to identify the DATA, PROCESS, and INTERFACE requirements (Figure 3.3) for the users of a new system. Most importantly, the purpose is to specify these requirements without prematurely expressing computer alternatives and technology details; at this point, keep analysis at the business level!

As shown in Figure 3.5, the requirements analysis phase is triggered by the approved system improvement objectives from the problem analysis phase. From the system users, the team collects and discusses requirements and priorities. This information is collected through interviews, questionnaires, and facilitated meetings. The challenge to the team is to validate those requirements. The new system objectives from the problem analysis phase provide some validation.

The deliverable for the requirements analysis phase is a business requirements statement. This could be a very small or very large document depending on the methodology or system complexity.

You should never skip the requirements analysis phase. One common complaint about new systems and applications is that they don't satisfy the users' needs. This is usually because information systems specialists found it difficult to separate *what* the user needed from *how* the new system would work. They became so preoccupied with the technical solution that they failed to consider the users' essential requirements.

Decision Analysis Phase   Given the business requirements statement, there are usually numerous alternative ways to design a new information system to fulfill those requirements. Some of the pertinent questions include the following:

— How much of the system should be computerized?
— Should we purchase software or build it ourselves (called the make-versus-buy decision)?
— Should we design the system for an internal network, or should we design a Web-based solution?
— What emerging information technologies might be useful for this application?

The questions are answered in the decision analysis phase of the methodology. The purpose of this phase is to identify candidate solutions, analyze those candidate solutions for feasibility, and recommend a candidate system as the target solution to be designed.

As shown in Figure 3.5, some level of validated business requirements triggers the decision analysis phase. The project team solicits ideas and opinions from a diverse audience, possibly including IT vendor presentations. Many modern organizations have information technology and architecture standards that constrain the number of candidate solutions that might be considered and analyzed. The existence of such standards is illustrated at the bottom of your information system building blocks model (Figure 3.3).

After identifying candidate solutions, each candidate is evaluated by the following criteria:

— **Technical feasibility**—Is the solution technically practical? Does the staff have the technical expertise to design and build this solution?
— **Operational feasibility**—Will the solution fulfill the users' requirements? To what degree? How will the solution change the users' work environment? How do the users feel about such a solution?

- **Economic feasibility**—Is the solution cost-effective (as defined earlier in the chapter)?
- **Schedule feasibility**—Can the solution be designed and implemented within an acceptable time period?
- **Risk feasibility**—What's the probability of a successful implementation using the technology and approach?

The project team is usually looking for the *most* feasible solution—the solution that offers the best combination of technical, operational, economic, schedule, and risk feasibilities.

The key deliverable of the decision analysis phase is an approved system proposal. This proposal may be written or verbal. Several outcomes are possible from the configuration phase. The "go or no-go" feasibility checkpoint (see bullet **❺**) may result in one of the following options:

- Approve and fund the system proposal for design and construction (possibly including an increased budget and timetable if scope has significantly expanded).
- Approve or fund one of the alternative candidate solutions.
- Reject all the candidate solutions and either cancel the project or send it back for new recommendations.
- Approve a reduced-scope version of the proposed solution.

Design Phase   Given the approved system proposal from the decision analysis phase, you can finally design the target system. The purpose of the design phase is to transform the business requirements statement from the requirements analysis phase into design specifications for construction. In other words, the design phase addresses *how* technology will be used in the new system. Design requires soliciting ideas and opinions from users, vendors, and IT specialists. Also, design requires adherence to internal technical design standards that ensure completeness, usability, reliability, performance, and quality.

In the information system framework (Figure 3.3), the design phase is concerned with technology-based views of the system's DATA, PROCESSES, and INTERFACES. You may have been exposed to such technical specifications in either programming or database courses. Design specifications can take many forms including written documentation or working computer-generated prototypes of the new system.

No new information system exists in isolation from other existing information systems in an organization. Consequently, system design always includes design for integration. The new system must be integrated with other information systems as well as into the business itself.

Notice in Figure 3.5 that we have included one final "go or no-go" feasibility checkpoint for the project (see bullet **❻**). A project is rarely canceled after the design phase, unless it is hopelessly over budget or behind schedule. But scope could be decreased to produce a minimum acceptable product in a specified time frame. Or the schedule could be extended to build a more complete solution in multiple versions. The project plan (schedule and budget) would need to be adjusted to reflect these decisions.

Modern methodologies show a trend toward merging the design and construction phases.

Construction Phase   Given some level of design specifications, we can construct and test system components for that design. Eventually, we will have built the functional system. The purpose of the construction phase is twofold: (1) to build and test a system that fulfills business requirements and design specifications, and (2) to implement the interfaces between the new system and existing systems. The construction phase may also involve installation of purchased or leased software.

The information system framework (Figure 3.3) identifies the relevant building blocks and activities for the construction phase. The project team must construct the database, application programs, and user and system interfaces. Some of these components may already exist (subject to enhancement).

You may already have some experience with part of this activity—application programming. Programs can be written in many different languages, but the current trend is toward the use of visual and object-oriented programming languages such as *Visual Basic, C++,* or *Java.* As components are constructed, they are typically demonstrated to users to solicit feedback.

One important aspect of construction is conducting tests of both individual system components as well as the overall system. Once tested, a system (or version of a system) is ready for implementation.

Implementation Phase   New systems usually represent a departure from the way business is currently done; therefore, the analyst must provide for a smooth transition from the old system to the new system and help users cope with normal start-up problems. Thus, the implementation phase delivers the production system into operation.

The functional system from the construction phase is the key input to the implementation phase (see Figure 3.5). Analysts must also train users, write various manuals, load files and databases, and perform final testing. System users provide continuous feedback as new problems and issues arise. In the information system framework (Figure 3.3) the implementation phase considers the same building blocks as the construction phase.

To provide a smooth transition to the new system, a conversion plan should be prepared. This plan may call for an abrupt switch where the old system is terminated and replaced by the new system on a specific date. Alternatively, the old and new systems may run in parallel until the new system has been deemed acceptable to replace the old system.

The implementation phase also involves training individuals that will use the final system and developing documentation to aid system users. The implementation phase usually includes an audit to gauge the success of the completed project. This activity represents a TQM effort that will contribute to the success of future systems projects.

The deliverable of the implementation phase (and the project) is the operational system that will enter the operation and support stage of the life cycle.

Operation and Support Stage   Once the system is operating, it delivers the business solution to the user community. It will still require ongoing system support for the remainder of its life.

> **System support** is the ongoing technical support for users, as well as the maintenance required to fix any errors, omissions, or new requirements that may arise.

System support consists of the following ongoing activities:

— *Assisting users.* Regardless of how well the users have been trained and how good the end-user documentation is, users will eventually require additional assistance—unanticipated problems arise, new users are added, and so forth.

— *Fixing software defects (bugs).* Software defects are errors that slipped through the software testing.

— *Recovering the system.* A system failure may result in a program "crash" and/or loss of data. Human error or a hardware or software failure may have caused this. The systems analyst or technical support specialists may then be called on to recover the system—that is, to restore a system's files and databases and to restart the system.

— *Adapting the system to new requirements.* New requirements may include new business problems, new business requirements, new technical problems, or new technology requirements.

Eventually, we expect the user feedback and problems to indicate that it is time to start over and reinvent the system. In other words, the system has reached entropy and a new project to create a new system should be initiated.

Cross Life Cycle Activities

System development also involves a number of cross life cycle activities.

> **Cross life cycle activities** are activities that overlap many or all phases of the methodology.

These activities, listed in the margin, are not explicitly depicted in Figure 3.5. Let's briefly examine each of these activities.

Fact-Finding   There are many occasions for fact-finding during the project.

> **Fact-finding** is the formal process of using research, interviews, meetings, questionnaires, sampling, and other techniques to collect information about systems, requirements, and preferences. It is also called *information gathering* or *data collection*.

Fact-finding is most crucial to the preliminary investigation, problem analysis, and requirements analysis phases of a project. During these phases the project team learns about a business's vocabulary, problems, opportunities, constraints, requirements, and priorities. Fact-finding is also used during the decision analysis, design, construction, and implementation phases but to a lesser extent. During these latter phases the project team researches technical alternatives and solicits feedback on technical designs, standards, and working components.

Documentation and Presentations   Communication skills are essential to the successful completion of a project. Two forms of communication that are common to systems development projects are documentation and presentation.

> **Documentation** is the activity of recording facts and specifications for a system for current and future reference.

> **Presentation** is the activity of communicating findings, recommendations, and documentation for review by interested users and managers. Presentations may be either written or verbal.

Documentation and presentation opportunities span all the phases. In Figure 3.7, the blue arrows represent various instances of presentations or deliverables of a phase. The purple arrows represent documentation, knowledge, and other artifacts of systems development that are stored in a repository.

> A **repository** is a database where system developers store all documentation, knowledge, and products for one or more information systems or projects.

Feasibility Analysis   Consistent with our creeping commitment approach to systems development, feasibility analysis is a cross life cycle activity.

> **Feasibility** is a measure of how beneficial the development of an information system would be to an organization.

> **Feasibility analysis** is the activity by which feasibility is measured and assessed.

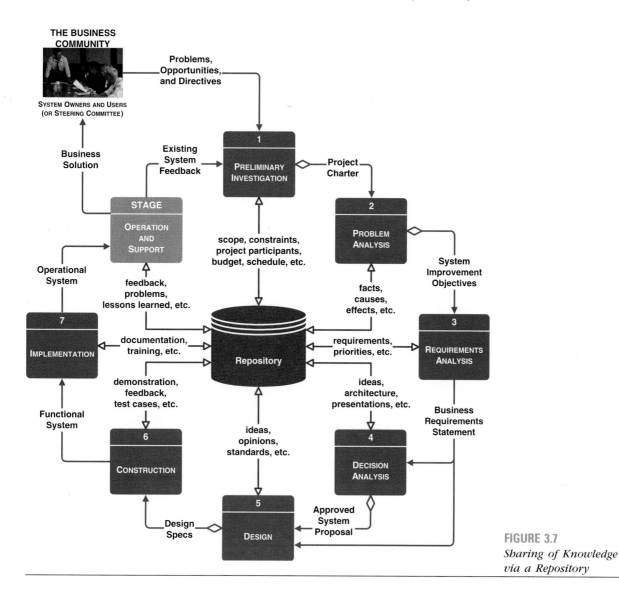

FIGURE 3.7
*Sharing of Knowledge
via a Repository*

Different measures of feasibility are applicable in different phases of the methodology. These measures include technical, operational, economic, schedule, and risk feasibility.

Process and Project Management   Recall that the CMM considers systems development to be a *process* that must be managed on a project-by-project basis. For this reason and others, process and project management are ongoing, cross life cycle activities.

Process management is an ongoing activity that documents, manages the use of, and improves an organization's chosen methodology (the "process") for systems development. Process management is concerned with the activities, deliverables, and quality standards to be applied to *all* projects.

Project management is the activity of defining, planning, directing, monitoring, and controlling a project to develop an acceptable system within the allotted time and budget.

Process management defines the methodology to be used on every project. That process must be documented and improved based on lessons learned from using the methodology. Project management is concerned with administering a single instance of the methodology as applied to a single project.

Failures and limited successes of systems development projects far outnumber very successful projects. One reason for this is that many systems analysts are unfamiliar with or undisciplined in the application of tools and techniques of systems development. But most failures are attributed to poor leadership and management. This mismanagement results in unfulfilled or unidentified requirements, cost overruns, and late delivery.

ALTERNATIVE ROUTES AND METHODS

There are many routes to any given destination. You could take the interstates, highways, or back roads. You could fly! The decision on which route is best depends on your goals and priorities. Do you want to get there fast, or do you want to see the sights? How much are you willing to spend? Are you comfortable with the means of travel? Just as you would pick your route and means to a travel destination, you can and should pick a route and means for a systems development destination.

So far, we've described a basic set of phases that comprise our *FAST* methodology. At one time, a "one size fits all" methodology was common for most projects. But today, a variety of types of projects, technologies, and development strategies exist—one size can no longer fit all projects! Like many contemporary methodologies, *FAST* provides alternative routes through the phases to accommodate different types of projects, technology goals, developer skills, and development strategies. The selection of a route usually occurs during the preliminary investigation phase, which is common to all routes.

In this section, we will describe each *FAST* route. These routes are not exclusive. Any given project may use a variation on, or a combination of, the routes described below.

Model-Driven Development Route

One of the oldest and most commonly used approaches to analyzing and designing information systems is based on modeling.

> **Modeling** is the act of drawing one or more graphical representations (or pictures) of a system. Modeling is a communication technique based on the old saying, "a picture is worth a thousand words."

In the *FAST* methodology, system models are used to illustrate and communicate the DATA, PROCESS, or INTERFACE building blocks of information systems. This approach is called model-driven development.

> **Model-driven development (MDD)** techniques emphasize the drawing of models to help visualize and analyze problems, define business requirements, and design information systems.

Model-driven development is illustrated in Figure 3.8. The model-driven approach usually does not skip or consolidate the basic phases we described earlier. The model-driven route takes on the appearance of a waterfall, suggesting that phases must generally be completed in sequence. As shown in the figure, it is possible to back up to correct mistakes or omissions; however, such rework is often difficult, time consuming, and costly. We call your attention to the following notes that correspond to the numbered bullets in Figure 3.8.

❶ The model-driven approach places a premium on planning because projects tend to be large and rework tends to be costly. During preliminary investigation, simple system models can be useful to visualize project scope. Defining scope is critical to estimating time and costs to complete the model-driven route.

❷ Some system modeling techniques call for extensive modeling of the current system to identify problems and opportunities for improvement. This is especially

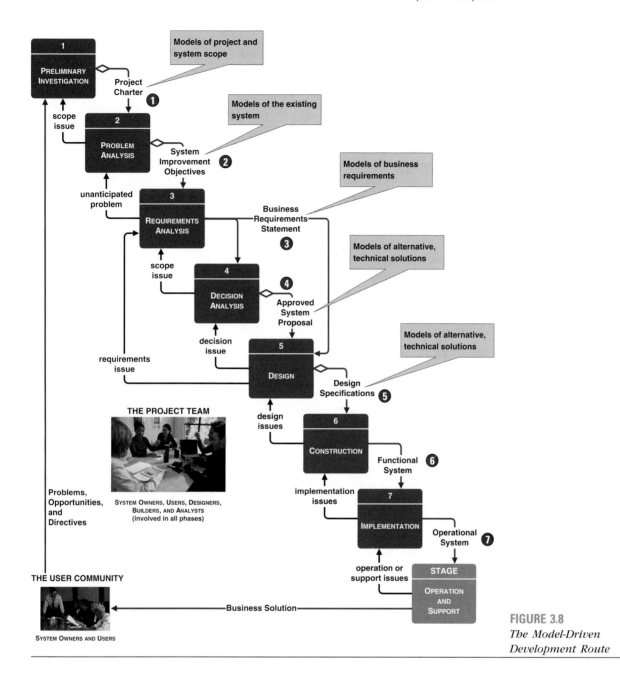

FIGURE 3.8
The Model-Driven Development Route

true if the underlying business processes are believed to be inefficient or bureaucratic. This may also prove helpful if the system is very large and complex.

❸ Most model-driven techniques require analysts to document business requirements with "logical" system models. Logical models show only "what" a system must be or must do. They are implementation *in*dependent; that is, they depict the system independent of any possible technical implementation.

❹ Many model-driven techniques require analysts to document alternative technical solutions with "physical" system models. Physical models show not only what a system is or does, but also how the system is physically and technically implemented. They are implementation *de*pendent because they reflect technology choices and the limitations of those technology choices.

⑤ Detailed physical models are a common product of many system design techniques. Examples include database schemas, structure charts, and flowcharts. They serve as a blueprint for construction of the new system.

⑥ Construction translates the physical system models into software.

⑦ During implementation, system models can be included in training and user manuals for the new system.

Essentially system models are produced as a portion of the deliverables for each phase. Once implemented, the models serve as documentation for any changes that might be needed during the operation and support stage of the life cycle. The model-driven approach offers several advantages:

— It minimizes planning overhead because all the phases are planned up front. (This does not mean the project cannot become infeasible and get canceled.)

— Requirements analysis tends to be more thorough and better documented in the model-driven approach.

— Alternative technical solutions tend to be more thoroughly analyzed in the model-driven approach.

— System designs tend to be more sound, stable, adaptable, and flexible because they are model-based and more thoroughly analyzed *before* they are built.

— The approach is effective for systems that are well understood but so complex that they require large project teams to complete.

— The approach works well when fulfilling user expectations and quality are more important than cost and schedule.

There are also several disadvantages of model-driven development. Most often cited is the long duration of projects; it takes time to collect the facts, draw the models, and validate those models. This is especially true if users are uncertain or imprecise about their system requirements. The models can only be as good as the users' understanding of those requirements. Second, pictures are not software; some argue that this reduces the users' role in a project to passive participation. Most users don't get excited about pictures. Instead, they want to see working software and they gauge project progress by the existence of software (or its absence). Finally, the model-driven approach is considered by some to be inflexible; users must fully specify requirements before design; design must fully document technical specifications before construction; and so forth. Some view such rigidity as impractical. Regardless, model-driven approaches remain popular.

There are several different model-driven techniques. They differ primarily in terms of the types of models that they require the analyst to draw and validate. Let's briefly examine three common model-driven development techniques that will be taught in this book. Please note that we are only introducing the techniques here, not the models. We'll teach the models later in the "how to" chapters.

Structured Analysis and Design   Structured analysis and design were two of the first formal system modeling techniques.

> **Structured analysis** is a PROCESS-centered technique that is used to model business requirements for a system. Structured analysis introduced a modeling tool called the *data flow diagram,* used to illustrate business process requirements.

> **Structured design** is a PROCESS-centered technique that transforms the structured analysis models into good software design models. Structured design introduced a modeling tool called *structure charts,* used to illustrate software structure to fulfill business requirements.

Recall that the information system building blocks included several possible focuses: DATA, PROCESSES, and INTERFACES. Structured analysis and design tends to focus primarily on the PROCESS building blocks. Thus, data flow diagrams and structure charts focus on business and software processes in the system. (The techniques have evolved to also consider the DATA building blocks as a secondary emphasis.)

Data flow diagrams and structure charts contributed significantly to reducing the communication gap that often exists between the nontechnical system owners and users and the technical system designers and builders. For this reason, structured analysis and design are still among the most widely practiced modeling techniques. Modern structured analysis and design are taught in this book.

Information Engineering   Many organizations have chosen information engineering modeling techniques.

> **Information engineering (IE)** is a DATA-centered, but PROCESS-sensitive technique used to model business requirements and design systems that fulfill those requirements. Information engineering emphasized a modeling tool called *entity relationship diagrams* to model business requirements.

With respect to the information system building blocks, information engineering tends to focus primarily on the DATA building blocks. Thus, entity relationship diagrams focus on data required in the system. Information engineering *subsequently* used data flow diagrams and structure charts to model the system's PROCESSES that capture, store, and use that DATA.

Technically, information engineering is more ambitious than we have described here. At the risk of oversimplification, information engineering attempts to use its techniques on the organization as a whole—modeling the organization's entire data requirements, as opposed to a single information system in the organization. Regardless, the DATA-centered technique in information engineering remains a practical option for developing individual information systems. The underlying modeling techniques for information engineering are also taught in this book.

Object-Oriented Analysis and Design   Object-oriented analysis and design are the new kids on the block. The concepts behind this exciting new strategy (and technology) are covered extensively in Part VI of the book, but a simplified introduction is appropriate here.

For the past 30 years, techniques such as structured analysis and design and information engineering deliberately separated concerns of DATA from those of PROCESSES. In other words, data and process models are separate and distinct. But these concerns must be synchronized. Object techniques are an attempt to eliminate the separation of concerns about DATA and PROCESS.

> **Object-oriented analysis and design (OOAD)** attempts to merge the data and process concerns into singular constructs called objects. Object-oriented analysis and design introduced object diagrams that document a system in terms of its objects and their interactions.

Business objects might correspond to real things of importance in the business such as customers and the orders they place for products. Each object consists of *both* the data that describes the object and the processes that can create, read, update, and delete that object.

With respect to the information system building blocks, object-oriented analysis and design significantly changes the paradigm. The DATA and PROCESS columns are essentially merged into a single OBJECT column. The models then focus on identifying objects, building objects, and assembling appropriate objects into useful information systems.

Object models are also extendable to the INTERFACE building blocks of the information system framework. Most contemporary computer user interfaces are already based on object technology. For example, both the Microsoft *Windows* and Netscape *Navigator* interfaces use standard objects such as windows, frames, drop-down menus, radio buttons, check boxes, scroll bars, and the like. Consider a *window* object. It can be characterized in terms of data such as height, width, color, and so on. And it has methods such as minimizing, maximizing, resizing, and so forth. Object programming technologies such as *C++, Java, Smalltalk,* and *Visual Basic* are used to construct and assemble such INTERFACE objects.

Rapid Application Development Route

In response to the faster pace of the economy, rapid application development has become a popular route for accelerating systems development.

Rapid application development (RAD) techniques emphasize extensive user involvement in the rapid and evolutionary construction of working prototypes of a system to accelerate the system development process. RAD is sometimes called a *spiral approach* because you repeatedly spiral through the phases to construct a system in various degrees of completeness and complexity.

The basic ideas of RAD are:

— To more actively involve system users in the analysis, design, and construction activities.
— To organize systems development into a series of focused, intense workshops jointly involving system owners, users, analysts, designers, and builders.
— To accelerate the requirements analysis and design phases through an iterative construction approach.
— To reduce the amount of time until the users begin to see a working system.

RAD uses prototypes to accelerate requirements analysis and system design.

A **prototype** is a smaller-scale, representative or *working* model of the users' requirements or a proposed design for an information system.

The basic principle behind prototyping is that users know what they want when they see it working. In RAD, a prototype eventually evolves into the final information system. Rapid application development is illustrated in Figure 3.9. The following notes correspond to the numbered bullets:

❶ All projects should be scoped and planned. This phase differs from model-driven development in that you rarely plan in as much detail. The RAD route (or any other route) is selected in the preliminary investigation phase.

❷ The problem analysis, requirements analysis, and decision analysis phases of the basic methodology (from Figure 3.5) are consolidated into a *single* accelerated analysis phase. The requirements statement may include some system models, but not nearly in the same level of detail and rigidity as in the model-driven approach.

❸ After analysis, the RAD approach iterates through a "prototyping loop" (phases 3–6) until the prototype is considered a "candidate" system for implementation.

❹ Design may use models, but not to the extent of a pure model-driven approach. The final design evolves from the subsequent working prototypes.

❺ The construction phase is depicted with a larger symbol to indicate the greater time spent constructing the working prototypes.

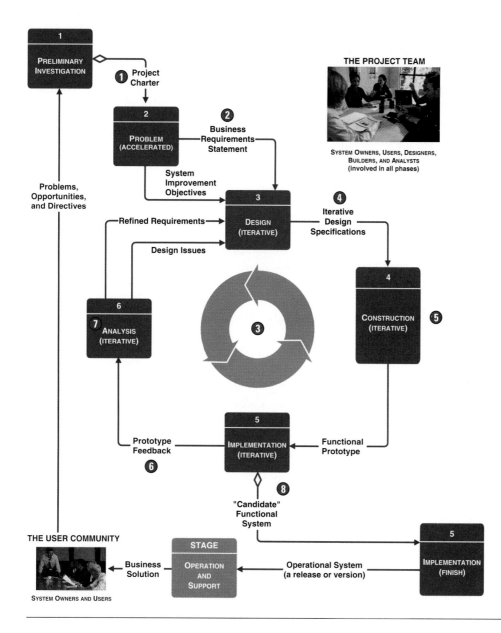

THE PROJECT TEAM

SYSTEM OWNERS, USERS, DESIGNERS, BUILDERS, AND ANALYSTS (involved in all phases)

FIGURE 3.9
The Rapid Application Development Route

6 Each prototype is initially implemented only to the extent that the users are given the opportunity to "experience" working with that prototype. The expectation is that users will clarify requirements, identify new requirements, and provide feedback on design (e.g., ease of learning, ease of use) for the next iteration through the design-by-prototyping loop.

7 Analysis is revisited based on user feedback to the prototype. This analysis tends to focus on revising requirements and identifying user concerns with the design. Notice, analysis cycles back to iterative design and the prototyping loop repeats itself.

8 Eventually, a prototype will be deemed worthy of implementation. This version of the system is placed into operation. The next version of the system may continue through the design-by-prototyping loop.

Although not a rigid requirement for RAD, the duration of the prototyping loop can be limited using a technique called timeboxing.

A **timebox** is a nonextendable period of time, usually 60 to 120 days, by which a candidate system must be placed into operation.

Advocates of timeboxing argue that management and user enthusiasm for a project can be enhanced and sustained because a working version of the system is implemented on a regular basis.

The RAD approach offers several advantages:

— It is useful for projects in which user requirements are uncertain or imprecise.

— It encourages active user and management participation (as opposed to a passive reaction to nonworking system models). This increases end-user enthusiasm for the project.

— Projects have higher visibility and support because of the extensive user involvement throughout the process.

— Users and management see working, software-based solutions more rapidly than in model-driven development.

— Errors and omissions tend to be detected earlier in prototypes than in system models.

— Testing and training is a natural by-product of the underlying prototyping approach.

— The iterative approach is a more "natural" fit because change is an expected factor during development.

— It reduces risk because you test the technical solution iteratively instead of making a wholesale commitment to any solution.

RAD is not without its critics. Some argue that RAD encourages a "code, implement, and repair" mentality that increases lifetime costs required to operate, support, and maintain the system. RAD prototypes can easily solve the wrong problems because problem analysis is abbreviated or ignored. A RAD-based prototype may discourage analysts from considering other, more worthy technical alternatives. Finally, the RAD emphasis on speed can hurt quality because of ill-advised shortcuts through the methodology.

RAD is most popular for smaller system projects. For this reason, we will teach both prototyping and RAD techniques in appropriate chapters of this book.

Commercial Off-the-Shelf Package Software Route

Sometimes it makes more sense to buy an information system solution than to build one in-house. Today, many organizations undertake in-house development of a system only if they cannot purchase a system that fulfills most of their requirements and expectations. Some organizations choose not to build any software solution that does not give them competitive advantage. Accordingly, our methodology includes a commercial off-the-shelf software route.

Commercial off-the-shelf (COTS) software is a software package or solution that is purchased to support one or more business functions and information systems.

The ultimate COTS software solution is enterprise resource planning.

An **enterprise resource planning (ERP)** software product is a fully integrated collection of information systems that span most basic business functions required by a major corporation. These systems include accounting and financials, human resources, sales and procurement, inventory management, production planning and control, and so on.

Examples of ERP solutions include *SAP, PeopleSoft,* and *Oracle Applications.* These ERP solutions provide all the core information system functions for an entire business. For many organizations, these core applications are similar from one business to the next; therefore, it makes sense to buy them instead of building them in-house. At the same time, an ERP implementation and integration represents the single largest information system project ever undertaken by that organization. It can cost tens of millions of dollars and require an army of managers, users, analysts, technical specialists, programmers, and consultants. The *FAST* methodology's route for COTS software is not really intended for ERP projects. Most ERP vendors provide their own COTS methodology (and consulting partners) to help their customers implement such a massive software solution.

Instead, our *FAST* methodology provides a COTS route for other types of information system solutions that may be purchased and implemented by a business. For example, an organization might purchase a COTS solution for a single business function such as accounting, human resources, or procurement. That COTS solution must be selected, installed, customized, and integrated into the business and its other existing information systems. The COTS route helps you do this.

The basic ideas of our COTS software route are:

— Packaged software solutions must be carefully selected to fulfill business needs—"You get what you ask and pay for."

— Packaged software solutions are not only costly to purchase, but they also can be costly to implement. The COTS route can actually be more expensive to implement than an in-house, model-driven or RAD development route.

— Software packages must usually be customized for, and integrated into, the business. Additionally, software packages usually require the redesign of existing business processes to adapt to the software.

— Software packages rarely fulfill all business requirements to the user's complete satisfaction. Thus, some level of in-house systems development is necessary to meet those unfulfilled requirements.

The COTS route is illustrated in Figure 3.10. The following notes correspond to the numbered bullets:

1 Again all projects should be scoped and planned. The COTS route (or any other route) is selected in the preliminary investigation phase.

2 Problem analysis usually includes some initial market research to identify available COTS solutions and the criteria used to evaluate such applications software.

3 After defining requirements, the requirements must be communicated to candidate technology vendors. The business requirements are formatted as a request for proposals and sent to vendors for their response.

4 Vendors submit proposals for their software. These proposals are evaluated against appropriate business and technical criteria. A contract and order is negotiated with the winning vendor for the software, and possibly for services necessary to install and maintain the software.

5 The vendor provides the software and services for installation and implementation of the basic software. This may include customization of the software.

6 Purchased software must be integrated into both the business and its other information systems. Some integration requirements can be identified as part of the business requirements statement. Other integration issues are not discovered until the software has been installed and customized. Decisions need to be made regarding what business processes need to be redesigned and what complementary in-house software needs to be developed.

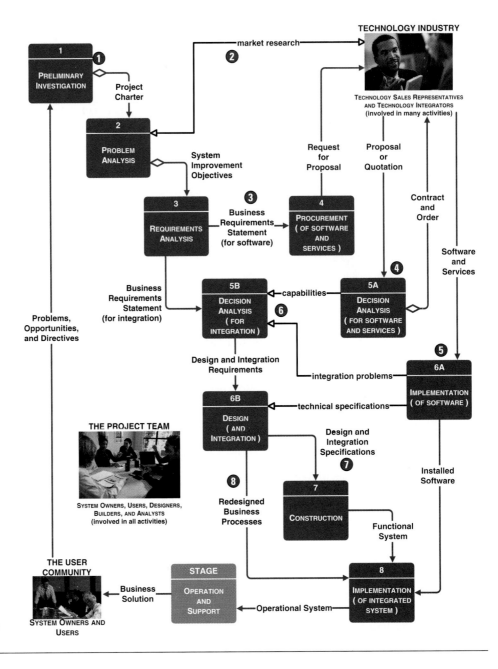

FIGURE 3.10
The Commercial Off-the-Shelf Software Route

❼ Design specifications include both those for integrating the COTS software with other existing information systems as well as those for complementary software to be built in-house to fulfill requirements not addressed by the COTS solution.

❽ Business processes are redesigned so that they correctly interoperate with the installed software.

The COTS approach offers several advantages:

— New systems can be implemented more quickly because extensive programming is not required. (This is not always true.)

— Many businesses cannot afford the staffing and expertise to develop in-house solutions.

— COTS vendors spread their development costs across all customers who purchase their software. Thus, they can invest in continuous improvements in features, capabilities, and usability that individual businesses cannot always afford.

— The COTS vendor assumes responsibility for significant system improvements and error corrections. For example, COTS vendors are typically responsible for solving systemic problems such as year 2000 and euro compatibility.

— Many business functions are more similar than dissimilar for all businesses in a given industry. For example, business functions for organizations in the health care industry are more alike than different. It does not make good business sense for each organization to "reinvent the wheel."

As with the other routes, COTS has disadvantages. A successful COTS implementation is dependent on the long-term success and viability of the COTS vendor—if the vendor goes out of business you can lose your technical support and future improvements. A purchased system rarely reflects the ideal solution that the business might achieve with an in-house developed system that could be customized to the precise expectations of management and the users. Also, there is almost always at least some resistance to changing business processes to adapt to the software. Some users will have to give up or assume new responsibilities. And some people may resent changes they perceive to be technology-driven instead of business-driven. Finally, if you customize the purchased software, future upgrades may have to be repeated and will be costly and tedious.

Regardless, the trend toward purchased software and solutions cannot be ignored. Many businesses require that a COTS alternative be considered before starting any type of in-house development project. Some experts estimate that by the year 2005 businesses will purchase 75 percent of their new information system applications. For this reason, we will teach COTS tools and techniques in appropriate chapters of this book.

The *FAST* routes are not mutually exclusive. Any given project may elect to or be required to use a combination of more than one route or a variation on a route to complete a project. The route to be used is always selected during the preliminary investigation phase. A new hybrid route, ideally suited to the project, might be created during this phase. Or perhaps a previously developed hybrid path could be used or adapted for the project. Let's briefly examine a few commonly used hybrids illustrated in Figure 3.11. Remember that there are many more possibilities than we can briefly describe here.

Hybrid Approaches

Rapid Architected Development Route   A rapid architected development route is shown in Figure 3.11(a). In this hybrid, you start the project as if you were using the RAD approach, but then you switch to a model-driven approach. First you build a prototype and use the prototype to build the system models. For example, it is not uncommon for developers to use Microsoft *Access* to prototype a database and application. Such prototypes can be converted into data and object models for a more thorough and rigorous analysis of requirements. The models would then be used to construct the final system using technologies other than Microsoft *Access* (such as *Visual Basic* and *Oracle*). This approach offers the RAD advantages of user involvement and prototypes, as well as the model-driven advantages of rigorous specification and quality checks.

Multiple Implementation Route   A multiple implementation route is depicted in Figure 3.11(b). In this hybrid, you follow the model-driven approach through the decision analysis phase. During decision analysis, you divide the system into subprojects that deal with different subsystems. Each subproject follows its own model-driven, waterfall route to include design, construction, and implementation. The subsystems eventually must be integrated into a single, cohesive solution.

Staged Implementation Route   A staged implementation route is illustrated in Figure 3.11(c). In this hybrid, you once again follow the model-driven approach through

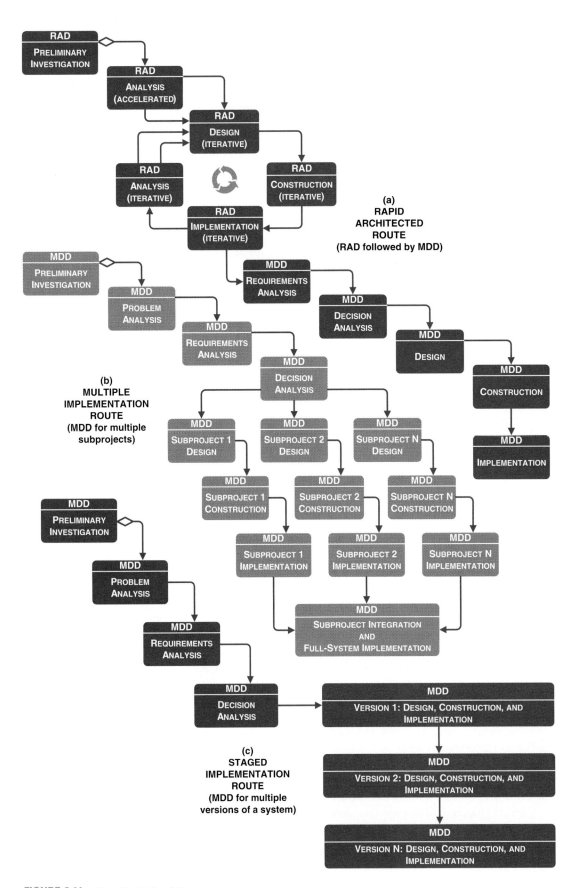

FIGURE 3.11   *Sample Hybrid Routes*

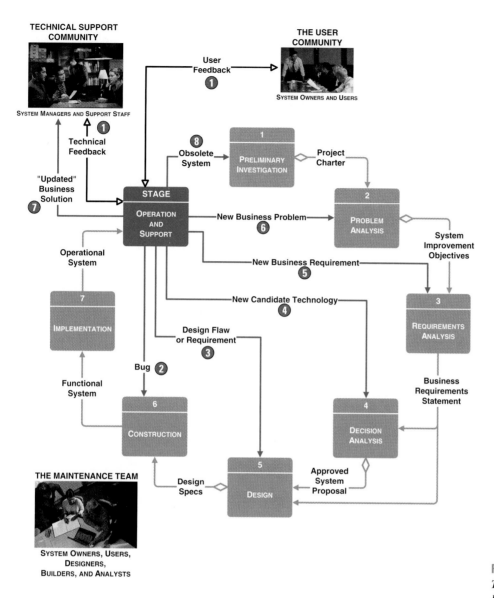

FIGURE 3.12
The Maintenance and Reengineering Route

the decision analysis phase. During decision analysis, you define a series of staged design, construction, and implementation phases that result in a series of operational versions of the desired system. In other words, you design, construct, and implement version 1.0 to provide the minimal, usable system capabilities and features. Each subsequent version adds to those capabilities and features.

All routes ultimately result in placing a new system into operation and support. The maintenance and reengineering route is intended to guide projects through the operation and support stage of the life cycle. Figure 3.12 depicts the maintenance and reengineering route of the *FAST* methodology. As illustrated in the figure, maintenance and reengineering projects are merely smaller-scale versions of the *FAST* methodology as applied to operations and support goals. We call your attention to the following numbered bullets on the figure:

The Maintenance and Reengineering Route

❶ Maintenance and reengineering projects are triggered by some combination of user and technical feedback. Such feedback may identify new problems, opportunities, or constraints.

As demonstrated in the figure, the starting point for addressing any new problems, opportunities, and constraints differs based on the nature and complexity of the request.

❷ The simplest fixes are software errors or bugs. Such a project begins in the construction phase to correct the error and proceeds immediately to implementation.

❸ Sometimes a design flaw becomes apparent after implementation. For example, users may frequently make the same mistake due to a confusing screen design. Alternatively, a new design constraint such as year 2000 compatibility might be introduced. For this type of reengineering project, the design phase would need to be revisited, followed by the construction and implementation phases.

❹ Technical circumstances might dictate that the system be upgraded to new technology or versions of technology. For example, an organization may standardize on the newest version of a particular database management system such as *SQL Server* or *Oracle*. For this type of reengineering project, the decision analysis phase may need to be revisited to determine the risk and feasibility of converting an existing, operational database to the new version. As appropriate, such a project would subsequently complete the (re)design, construction, and implementation phases.

❺ Businesses constantly change; therefore, business requirements for a system also change. One common trigger for a reengineering project is a new or revised business requirement. Given the requirement, the requirements analysis phase must be revisited with a focus on impact of the new requirement on the existing system. Based on requirements analysis, the project would then proceed to the decision analysis, design, construction, and implementation phases.

By now, you've noticed that maintenance and reengineering projects initiate in different phases of the basic methodology. You might be concerned that repeating these phases would require excessive time and effort. Keep in mind, however, that the scope of maintenance and reengineering projects is much smaller than the original project that created the operational system. Thus, the work required in each phase is much less than you might have expected.

❻ Again, as businesses change, significant new problems, opportunities, and constraints can be encountered. In this type of project, work begins with the problem analysis phases and proceeds as necessary to the subsequent phases.

❼ The final result of any type of maintenance or reengineering project is an updated business solution. This may include revised programs, databases, interfaces, or business processes. This updated system remains in operation.

❽ As described earlier in the chapter, we expect a system to eventually reach entropy. The business and/or technical problems and requirements have become so troublesome as to warrant starting over.

That completes our introduction to the systems development methodology and routes. Now we will introduce the role of automated tools that support contemporary system development.

AUTOMATED TOOLS AND TECHNOLOGY

You may be familiar with the old fable of the shoemaker whose own children had no shoes. That situation is not unlike the one faced by some system developers. For years we've been applying information technology to solve our users' business problems; however, we've been slow to apply that same technology to our own problem—developing information systems. In the not-too-distant past, the principal tools of the systems analyst were paper, pencil, and flow-chart template.

Today, entire suites of automated tools have been developed, marketed, and installed to assist system developers. While system development methodologies do not always require automated tools, most methodologies benefit from such technology. Some of the most commonly cited benefits include:

— Improved productivity—through automation of tasks.

— Improved quality—because automated tools check for completeness, consistency, and contradictions.

— Better and more consistent documentation—because the tools make it easier to create and assemble consistent, high-quality documentation.

— Reduced lifetime maintenance—because of the aforementioned system quality improvements combined with better documentation.

— Methodologies that really work—through rule enforcement and built-in expertise.

Chances are that your future employer is using or will be using this technology to develop systems. We will demonstrate the use of various automated tools throughout this textbook. There are three classes of automated tools for developers: computer-aided systems engineering, application development environments, and project and process managers. Let's briefly examine each of these classes of automated tools.

CASE—Computer-Aided Systems Engineering

System developers have long aspired to transform information systems and software development into engineering-like disciplines. The terms systems engineering and software engineering are based on a vision that systems and software development can and should be performed with engineering-like precision and rigor. Such precision and rigor are consistent with the model-driven approach to system development. To help systems analysts better perform system modeling, the industry developed automated tools called computer-aided systems engineering.

> **Computer-aided systems engineering (CASE)** tools are software programs that automate or support the drawing and analysis of system models and provide for the translation of system models into application programs.

Think of CASE as software that is used to design and construct other software. This is very similar to the computer-aided design (CAD) technology used by most contemporary engineers to design other products such as vehicles, structures, machines, and so forth. Representative CASE products are listed in the margin. Most CASE products run on personal computers as depicted in Figure 3.13. Many of these tool vendors have abandoned the term *CASE* in their marketing strategies and now simply refer to the technology as modeling tools.

CASE Tools

Oracle's *Designer 2000*
Platinum's *Erwin*
Rational's *ROSE*
Popkin's *System Architect 2001*
Sterling's *COOL* product family
Visible Systems' *Visible Analyst*
Visio's *Visio Enterprise*

CASE Repositories At the center of any true CASE tool's architecture is a database called a repository. The repository concept was introduced earlier (see Figure 3.7).

> A **CASE repository** is a system developers' database. It is a place where developers can store system models, detailed descriptions and specifications, and other products of systems development. Synonyms include *dictionary* and *encyclopedia*.

Around the CASE repository is a collection of tools or facilities to create system models and documentation.

FIGURE 3.13   *Using Automated Tools for Systems Development*

CASE Facilities   To use the repository, the CASE tools provide some of the following facilities, illustrated in Figure 3.14.

— *Diagramming tools* are used to draw the system models required or recommended in most system development methodologies. Usually, the shapes on one system model can be linked to other system models and to detailed descriptions (see next item below).

— *Description tools* are used to record, delete, edit, and output detailed documentation and specifications. The descriptions can be associated with shapes appearing on system models that were drawn with the diagramming tool.

— *Prototyping tools* can be used to construct system components including inputs and outputs. These inputs and outputs can be associated with both the aforementioned system models and the descriptions.

— *Quality management tools* analyze system models, descriptions and specifications, and prototypes for completeness, consistency, and conformance to accepted rules of the methodologies.

— *Documentation tools* are used to assemble, organize, and report on system models, descriptions and specifications, and prototypes that can be reviewed by system owners, users, designers, and builders.

— *Design and code generator tools* automatically generate database designs and application programs or significant portions of those programs.

Forward and Reverse Engineering   As previous stated, CASE technology automates system modeling. Today's CASE tools provide two distinct ways to develop system models—forward and reverse engineering.

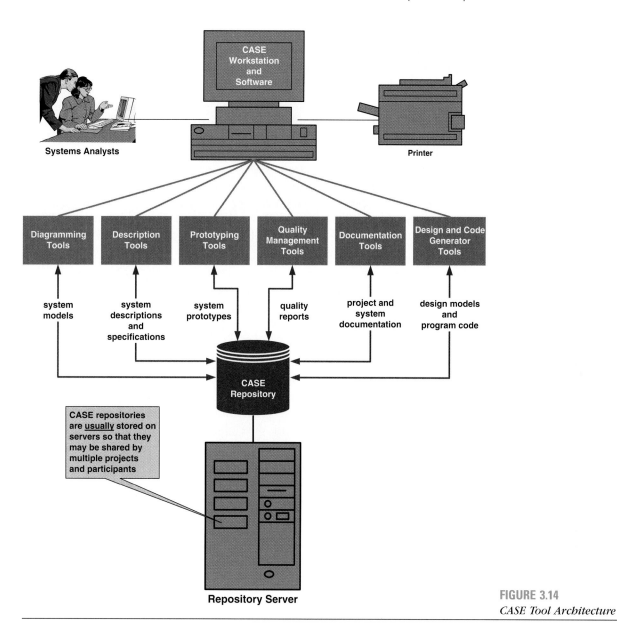

FIGURE 3.14
CASE Tool Architecture

Forward engineering requires the systems analyst to draw system models, either from scratch or from templates. The resulting models are subsequently transformed into program code.

Reverse engineering allows a CASE tool to read existing program code and transform that code into a representative system model that can be edited and refined by the systems analyst.

Think of reverse engineering as allowing you to generate a flowchart from an existing program.

CASE tools that allow for bidirectional, forward, and reverse engineering are said to provide for "round-trip engineering." For example, you reverse engineer a poorly designed system into a system model, edit and improve that model, and then forward engineer the new model into an improved system.

The emphasis on speed and quality in software development has resulted in RAD approaches. The potential for RAD has been amplified by the transformation of programming language compilers into complete application development environments.

Application Development Environments

ADE Tools

Allaire's *Cold Fusion* (for Web appli-
cation development)
IBM's *Visual Age* product family
  (*C++, Smalltalk, Java,* and others).
InPrise's *Delphi* and *J Builder* (*Java*)
Microsoft's *Visual Studio* (*Visual
  Basic, C++,* and *Java*)
Microsoft *Access* (SQL and *Visual
  Basic for Applications*)
Oracle's *Designer 2000*
Sybase's *PowerBuilder*
Symantec's *Visual Cafe* (*Java*)

Application development environments (ADE) are integrated software development tools that provide all the facilities necessary to develop new application software with maximum speed and quality. A common synonym is *integrated development environment (IDE)*.

ADEs make prototyping and RAD programming possible; however, ADEs are not exclusive to RAD techniques. Most programming language compilers are now integrated with a full ADE. Examples are listed in the margin (with the core programming language in parentheses). You may be familiar with one or two of the most popular ADEs used in college courses such as Microsoft *Visual Basic* and Microsoft *Access*. (The latter is more than just a database management system!)

Application development environments provide a number of productivity and quality management facilities. The ADE vendor provides some of these facilities. Third-party vendors provide many other facilities that can integrate into the ADE.

— *Programming languages or interpreters* are the heart of an ADE. Powerful debugging features and assistance are usually provided to help programmers quickly identify and solve programming problems.

— *Interface construction tools* help programmers quickly build the user interfaces using a component library.

— *Middleware* is software that helps programmers integrate the software being developed with various databases and computer networks.

— *Testing tools* are used to build and execute test scripts that can consistently and thoroughly test software.

— *Version control tools* help multiple programmer teams manage multiple versions of a program, both during development and after implementation.

— *Help authoring tools* are used to write online help systems, user manuals, and online training.

— *Repository links* permit the ADE to integrate with CASE tool products as well as other ADEs and development tools.

Process and Project Managers

A third class of automated tools helps us manage the systems development methodology and projects that use the methodology. While CASE tools and ADEs are intended to support analysis, design, and construction of new information systems and software, process and project management tools are intended to support cross life cycle activities.

A **process manager** is an automated tool that helps to document and manage a methodology and routes, its deliverables, and quality management standards.

Client/Server Connection, Ltd.'s *CS/10000* is an example of an automated process management tool. *CS/10000* acts as an adviser regarding system development activities, cost estimation, alternative technical solutions, and many other facets of systems development.

Project manager is an automated tool to help plan system development activities (preferably using the approved methodology), estimate and assign resources (including people and costs), schedule activities and resources, monitor progress against schedule and budget, control and modify schedule and resources, and report project progress.

Microsoft's *Project* and Applied Business Technology's *Project Manager Workbench* are examples of automated project management tools.

Because project management is the subject of the next chapter, you'll learn more about these automated tools in that chapter.

WHERE DO YOU GO FROM HERE?

This chapter, along with the first two, completes the minimum context for studying systems analysis and design. We have described that context in terms of the three Ps—participants (with special emphasis on stakeholders), product (the information system), and process (the system development methodology). Armed with this understanding, you are now ready to study systems analysis and/or design methods.

For many of you Chapter 4, "Project Management," will provide a more complete context for studying systems analysis and design. It builds on Chapter 3 by emphasizing a variety of management issues related to the system development methodology. These include methodology management, resource management, management of expectations, change management, and others.

For those of you who skip the project management chapter, your next assignment will depend on whether your course is:

- A comprehensive survey of systems development.
- A systems analysis-only course.
- A systems design-only course.
- A methodology "route" specific course (e.g., RAD, COTS, or a specific model-driven route such as object-oriented).

For either a system development survey course or a systems analysis-only course, we recommend you continue to Chapter 5, "Systems Analysis." In that chapter you will study the preliminary investigation, problem analysis, requirements analysis, and decision analysis phases of the FAST methodology in greater depth.

If you are in a system design-only course, you might still want to quickly review Chapter 5 for context. Then continue your study into Chapter 9, "System Design." In Chapter 9, you will learn more about the different strategies for design and construction of information systems. You will more closely examine the design and construction phases of the *FAST* methodology.

If you are in a route-specific course, we still recommend a quick overview of both the traditional systems analysis and system design chapters. Your instructor will subsequently direct you to chapters that are appropriate for the desired methodology route.

SUMMARY

1. A systems development process is a set of activities, methods, best practices, deliverables, and automated tools that stakeholders use to develop and maintain information systems and software.

2. The Capability Maturity Model (CMM) is a framework to assess the maturity level of an organization's information system development and management processes and products. It consists of five levels of maturity as measured by a set of guidelines called the key process areas.

   a. Level 1—Initial: This is sometimes called anarchy or chaos.

   b. Level 2—Repeatable: Project management processes and practices are established to track project costs, schedules, and functionality.

   c. Level 3—Defined: A standard systems development process (sometimes called a methodology) is purchased or developed and integrated throughout the information systems/services unit of the organization.

   d. Level 4—Managed: Measurable goals for quality and productivity are established.

   e. Level 5—Optimizing: The standardized systems development process is continuously monitored and improved based on measures and data analysis established in Level 4.

3. A system life cycle divides the life of an information system into two stages, *systems development* and *systems operation and support*—first you build it; then you use it. A systems development methodology is a very formal and precise system development process that defines a set of activities, methods, best practices, deliverables, and automated tools that system developers and project managers use to develop and maintain information systems and software.

4. The following principles should underlie all systems development methodologies:

   a. Get the owners and users involved.

   b. Use a problem-solving approach.

    c. Establish phases and activities.

    d. Establish standards.

    e. Justify systems as capital investments.

    f. Don't be afraid to cancel or revise scope.

    g. Divide and conquer.

    h. Design systems for growth and change.

5. System development projects are triggered by problems, opportunities, and directives.

    a. A problem is an undesirable situation that prevents the organization from fully achieving its purpose, goals, and/or objectives.

    b. An opportunity is a chance to improve the organization even in the absence of specific problems.

    c. A directive is a new requirement that's imposed by management, government, or some external influence.

6. Wetherbe's PIECES framework is useful for categorizing problems, opportunities, and directives. The letters of the PIECES acronym correspond to Performance, Information, Economics, Control, Efficiency, and Service.

7. Traditional, basic systems development phases include:

    a. Preliminary investigation.

    b. Problem analysis.

    c. Requirements analysis.

    d. Decision analysis.

    e. Design.

    f. Construction.

    g. Implementation.

8. Cross life cycle activities are activities that overlap many or all phases of the methodology. They may include:

    a. Fact-finding, the formal process of using research, interviews, meetings, questionnaires, sampling, and other techniques to collect information about systems, requirements, and preferences.

    b. Documentation, the activity of recording facts and specifications for a system for current and future reference. Documentation is frequently stored in a repository, a database where system developers store all documentation, knowledge, and products for one or more information systems or projects.

    c. Presentation, the activity of communicating findings, recommendations, and documentation for review by interested users and managers. Presentations may be either written or verbal.

    d. Feasibility analysis, the activity by which feasibility, a measure of how beneficial the development of an information system would be to an organization, is measured and assessed.

    e. Process management, the ongoing activity that documents, manages the use of, and improves an organization's chosen methodology (the "process") for systems development.

    f. Project management, the activity of defining, planning, directing, monitoring, and controlling a project to develop an acceptable system within the allotted time and budget.

9. There are different routes through the basic systems development phases. An appropriate route is selected during the preliminary investigation phase. Typical routes include:

    a. Model-driven development techniques emphasize the drawing of models to help visualize and analyze problems, define business requirements, and design information systems. Alternative model-driven strategies include:

        (1) Structured analysis and design.

        (2) Information engineering.

        (3) Object-oriented analysis and design.

    b. Rapid application development (RAD) techniques emphasize extensive user involvement in the rapid and evolutionary construction of working prototypes of a system to accelerate the system development process.

    c. Commercial off-the-shelf (COTS) techniques focus on the purchase and integration of a software package or solution to support one or more business functions and information systems.

    d. The maintenance and reengineering route is intended to guide projects through the operation and support stage of the life cycle.

10. Automated tools support all systems development phases.

    a. Computer-aided systems engineering (CASE) tools are software programs that automate or support the drawing and analysis of system models and provide for the translation of system models into application programs.

        (1) A CASE repository is a system developers' database. It is a place where developers can store system models, detailed descriptions and specifications, and other products of system development.

        (2) Forward engineering requires the systems analyst to draw system models, either from scratch or from templates. The resulting models are subsequently transformed into program code.

        (3) Reverse engineering allows a CASE tool to read existing program code and transform that code into a representative system model that can be edited and refined by the systems analyst.

    b. Application development environments (ADEs) are integrated software development tools that provide all the facilities necessary to develop new application software with maximum speed and quality.

    c. Process management tools help us document and manage a methodology and routes, its deliverables, and quality management standards.

    d. Project management tools help us plan system development activities (preferably using the approved methodology), estimate and assign resources (including people and costs), schedule activities and resources, monitor progress against schedule and budget, control and modify schedule and resources, and report project progress.

KEY TERMS

application development environment (ADE), p. 110

Capability Maturity Model (CMM), p. 76

CASE repository, p. 107

commercial off-the-shelf (COTS), p. 100

computer-aided systems engineering (CASE), p. 107

cost-effectiveness, p. 83

cross life cycle activity, p. 92

directive, p. 84

documentation, p. 92

economic feasibility, p. 90

enterprise resource planning (ERP), p. 100

fact-finding, p. 92

feasibility, p. 92

feasibility analysis, p. 92

forward engineering, p. 109

information engineering (IE), p. 97

model-driven development (MDD), p. 94

modeling, p. 94

object-oriented analysis and design (OOAD), p. 97

operational feasibility, p. 89

opportunity, p. 84

PIECES, p. 85

presentation, p. 92

problem, p. 84

process management, p. 93

process manager, p. 110

project management, p. 93

prototype, p. 98

rapid application development (RAD), p. 98

repository, p. 92

reverse engineering, p. 109

risk feasibility, p. 90

risk management, p. 83

schedule feasibility, p. 90

steering committee, p. 86

structured analysis, p. 96

structured design, p. 96

systems development methodology, p. 78

systems development process, p. 75

system life cycle, p. 78

system support, p. 91

technical feasibility, p. 89

timebox, p. 100

REVIEW QUESTIONS

1. What is a systems development process?
2. What is the Capability Maturity Model (CMM)? Why is it getting so much attention?
3. Describe the five levels of the Capability Maturity Model.
4. What is the difference between the *system life cycle* and a *systems development methodology?*
5. Why do companies use methodologies? Why do many choose to purchase a methodology?
6. What are the eight fundamental principles of systems development? Explain what you would do to incorporate those principles into a systems development process.
7. Describe four types of information systems development standards.
8. What are cost-effectiveness and risk management and how do they impact systems development?
9. What is the creeping commitment approach?
10. What is entropy and how does it impact systems development?
11. List the three triggers for a systems development project.
12. Differentiate among problems, opportunities, and directives.
13. Name the six problem classes of the PIECES framework.
14. Differentiate between the development stage and operation and support stage of the life cycle.
15. Identify and briefly describe the seven basic phases that are common to most modern systems development methodologies.

16. Describe each phase of the FAST life cycle in terms of purpose, inputs, and outputs.
17. Describe five types of feasibility.
18. Name four cross life cycle activities.
19. What is a repository?
20. Differentiate between process management and project management.
21. Why should a methodology have alternative "routes?"
22. What is a model? What is model-driven development?
23. List three model-driven development techniques.
24. What is rapid application development (RAD)?
25. What is timeboxing? Why is it popular?
26. What is commercial off-the-shelf (COTS) software? Why does COTS software solution require a route?
27. List three common hybrid routes.
28. What are the benefits of using automated tools in systems development?
29. List three classes of automated tools for systems development.
30. Differentiate between computer-aided systems engineering and application development environments as automated tools for systems development.
31. What is the role of a CASE repository?
32. What is the difference between forward and reverse engineering as it relates to CASE?

PROBLEMS AND EXERCISES

1. Two years ago, United Express, Inc. purchased a systems development methodology. At a meeting with other CEOs, United's president proudly proclaimed that United's information systems unit had achieved its goal of CMM Level 3 compliance. Later, the CIO of United cautioned the CEO that the methodology was only being used on a few moderately sized projects. Is United's information systems unit CMM Level 3 compliant? Why or why not?

2. Why has the Capability Maturity Model (CMM) reenergized the commercial systems development methodology industry?

3. Using the PIECES framework, evaluate your local course registration system. Do you see any problems or opportunities? What would be some possible directives that might require the system to change? (Alternative: Substitute any system with which you are familiar.)

4. Assume you are given a programming assignment that requires you to modify a computer program. Explain the problem-solving approach you would use. How is this approach similar to the phased approach of the methodology presented in this chapter?

5. A systems analyst is considering three alternative technical solutions for a new information system. All three solutions fulfill the users' business requirements. One solution was determined to be the most compatible with the developers' technical expertise. A second solution was determined to be the fastest solution to implement. The third solution is the most cost-effective solution. Is there a feasible solution? If so, how would the analyst make the final decision?

6. Which phases of the *FAST* methodology presented in this chapter do the following tasks characterize?
   a. The analyst demonstrates a prototype of a new Microsoft *Windows* user interface for reserving a company vehicle.
   b. The analyst observes the order entry clerks to determine how a customer's order is processed.
   c. The analyst specifies the structure of a database to support production scheduling.
   d. The analyst teaches the plant manager how to generate a new predefined report using the new microcomputer.
   e. A plant supervisor describes the content of a new procurement report that would simplify the tracking of purchase orders.
   f. The claims adjuster describes to an analyst the results of lost customer business due to delays in the claims processing system.
   g. The analyst is preparing an initial schedule and budget for the Student Admissions Contact System that was recently approved by the steering committee.
   h. The analyst is installing the microcomputer and database management system needed to run the petty-cash management system.
   i. The analyst is reviewing the company's organizational chart to identify who becomes involved in payroll authorizations and vacation/sick leave approvals.
   j. The analyst is comparing the pros and cons of two software packages that may fulfill management's need for facilities maintenance and renewal.
   k. An analyst is testing the latest version of a computer program that will more quickly identify materials that are out of stock, based on projected production of the products that use the materials.

7. Explain why the maintenance and reengineering route of the *FAST* methodology is a small-scale version of the full methodology.

8. Management has approached you to develop a new system. The project will last seven months. Management wants a budget next week. You will not be allowed to deviate from that budget. Explain why you shouldn't overcommit to early estimates. Defend the creeping commitment approach as it applies to cost estimating. What would you do if management insisted on the up-front estimate with no adjustments?

9. You have a user who has a history of impatience—encouraging shortcuts through the systems development process and then blaming the analysts for systems that fail to fulfill expectations. By now, you should understand the phased approach to systems development. For each phase, compile a list of possible consequences to use when the user suggests a shortcut through or around that activity.

10. Our company does not have a methodology but realizes it needs one. The president has suggested that we start a project to build a methodology. Explain to her the advantages of purchasing a methodology instead.

11. You are in a meeting with two other systems analysts. One is arguing that the business should adopt a model-driven development strategy. The other is arguing that a rapid application development strategy should be used instead. Explain to them how the two approaches might complement one another.

12. Differentiate among structured analysis and design, information engineering, and object-oriented analysis and design in terms of the information systems building blocks.

13. In a staff meeting, a middle-level manager says, "I suggest we dump our methodology and replace it with CASE technology." Respond to this manager.

14. Given the seven basic phases of the book's systems development methodology, explain how fact-finding might be used in each phase. (Alternative: Substitute any of the cross life cycle activities.)

15. Some people believe that automated tools will eventually replace programmers or systems analysts. What do you think? Why or why not?

16. When describing the phased approach that you plan to follow in developing a new information system, your client

asks why your approach is missing feasibility analysis and project management phases. How would you respond?

17. Timeboxing is usually associated with the rapid application development route. Explain how you might use that technique in the model-driven development route.

18. Golden Gateway University is considering building versus buying its new procurement management information system. The director of procurement is strongly biased toward purchasing one of the many new Web-based procurement software packages. "Purchasing a package makes more sense than building one in-house. We can have the system up and running almost immediately. Furthermore, we would not have to do any design or programming, which always takes an excessive amount of time. Don't you agree?" Respond to the director.

PROJECTS AND RESEARCH

1. For your development center's distribution, prepare a quick-reference guide (one-page maximum) to advise project teams on when to select either model-driven development or rapid application development as the preferred systems development route for a given project.

2. Make an appointment to visit a systems analyst at a local information system installation. Discuss the analyst's current project. What problems, opportunities, and directives triggered the project? How do they relate to the PIECES framework?

3. Visit a local information system installation. Compare the methodology with the one in this chapter. Evaluate the company's methodology with respect to the eight systems development principles. (Alternative: Substitute the system development methodology used in another systems analysis and design book, possibly assigned by your instructor.)

4. Read the mini-book *The One Minute Methodology* by Ken Orr (see Suggested Readings). Write a paper that describes what the author was trying to teach, and relate it to the subject presented in this chapter.

5. Search the Web for information about commercial or organizational systems development methodologies. Compare and contrast at least one methodology with the simple methodology described in this chapter.

6. Visit a local information system installation. What automated tools are being used for systems development? Classify each as CASE, ADE, or project management technology. Pick an automated tool being used and describe its capabilities as compared to the typical capabilities described in this chapter. Did you discover any new capabilities?

7. Search the Web for the CASE marketplace. (Be careful. The industry's terminology is constantly changing.) What are some products and their vendors? On which operating systems do they run? How are they similar and different? What do they cost?

8. Research the current trade literature about CASE experiences. What do the writers like about their CASE tools? What are their major complaints? How would you rate their overall satisfaction? Why? What are some future directions of the CASE tool industry?

MINICASES

1. Jeannine Strothers, investment manager, has submitted numerous requests for a new investment tracking system. She needs to make quick decisions regarding possible investments and divestments. One hour can cost her thousands of dollars in profits for her company.

   She has finally given up on Information Systems for not giving her requests high enough priority to get service. She goes to a computer store and buys a microcomputer along with spreadsheet, database, and word processing software. The computer store salesperson suggests she (1) build a database of her investments and options, (2) subscribe to a computer investment database (accessed via a modem in the microcomputer), (3) feed data from her database and the bulletin board into the spreadsheet, (4) play "what if" investment games on the spreadsheet, and then (5) update the database to reflect her final decisions. The word processor could draw data from the database for form letters and mailing lists.

   After discussing her plans with Jeff, a systems analyst at another company, he suggests she take a systems analysis and design course before beginning to use the spreadsheet and database. The local computer store says she doesn't need any systems analysis and design training to develop systems using the spreadsheet and database programs. The store personnel's reasoning is that spreadsheets and database tools are not programming languages; therefore, she shouldn't need analysis and design to build systems with them. Is the computer store correct? Why or why not? Can you persuade Jeannine to take the systems analysis and design course? What would your arguments be?

2. Jeannine Strothers, the impatient manager in the earlier minicase, did not take Jeff's advice. She built the new system, but she can't get top management to allow her to use it. And she's run into a number of other problems.

   First, the financial comptroller has been reevaluating company investment strategies and policies. Jeannine wasn't aware of that. The new system does not account for the new policies being considered.

   Her own staff has rejected the investment and divestment recommendations generated by the system. She used Information Systems' existing file structure to design those orders, only to find out that her clerks had abandoned those files two years ago because they didn't include the data necessary to fully analyze alternative investments. Her staff is also critical of the design, saying that minor mistakes send them off into the "twilight zone" with no easy way to recover.

   The computer link to the investment database service has been useless. The data received and its format are not compatible with systems requirements. Although other database services are available, the current database has been prepaid for two years. Based on her recent experience, Jeannine is now skeptical of such services.

   Some of her subordinate managers are insisting on graphic reports. She's not sure how to convert the data of either package to a graphic format (assuming it is possible).

   To top off her problems, she isn't sure that her existing local database structure can be modified to meet new requirements without having to rewrite all the programs. And her boss is not sure he wants to invest the money in a consultant to fix the problems.

   Jeannine's analyst friend Jeff is not very sympathetic to her problems: "Jeannine, I don't have any quick answers for you. You've taken too many shortcuts through the system development process. When we do a system, we go through a carefully thought-out process. We thoroughly study the problem, define needs, evaluate options, design the system and its interfaces, and only then do we begin programming."

   Jeannine replies, "Wait a minute. I only bought a microcomputer. It's not the same as your complex computer networks. I didn't see the need for going through the ritual you guys use for network applications. Besides, I didn't have the time to do all those steps."

   Jeff's parting words are philosophical: "You didn't have the time to do it right. Where will you find the time to do it over?"

   What principles did Jeannine violate? Why do so many people today fall prey to the belief that the system development process for a PC application is somehow different than for larger applications? What conclusions can you draw from Chapter 2 (the information system building blocks chapter) that might help Jeannine learn from her mistakes? What would you recommend to Jeannine if she were to decide to reapproach her manager with a plan to salvage the system?

3. Evaluate the following scenarios using the PIECES framework. Do not be concerned that you are unfamiliar with the application. That isn't unusual for a systems analyst. Use the PIECES framework to brainstorm potential problems or opportunities you would ask the user about.

   a. The staff benefits and payroll counselor is having some problems. Her job is to counsel employees on their benefit options. The company has just negotiated a new medical insurance package that requires employees to choose from among several health maintenance organizations (HMOs) and preferred provider plans (PPPs). The HMOs and PPPs vary according to employee classifications, contributions, deductibles, beneficiaries, services covered, and service providers permitted. The intent was to provide the most flexible benefits possible for employees, to minimize costs to the company, and to control costs to the insurance underwriters (which would affect subsequent premiums charged back to the company).

      The counselor will be called on to help employees select the best plan for themselves. She currently responds manually to such requests. But the current options are more straightforward than those under the new plan. She can explain the options, what they do and do not cover, what they cost and may cost, and the pros and cons. However, current employees' distrust of the new plan suggests she will need to provide more specific suggestions and responses to employees.

      She may have to work up scenarios—possibly worst-case scenarios—for many employees. The scenarios will have to be personalized for each employee's income, marital and family status, current health risks, and so on. In working up a few sample scenarios, she discovered, first, that it takes one full day to get salary and personnel data from the Information Systems department. Second, employee data is stored in many files that are not always properly updated. When conflicting data becomes apparent, she can't continue her projections until that conflict has been resolved. Third, the computations are complex. It often takes one full day or more to create investment and/or retirement scenarios for a single employee. Fourth, there are some concerns that projections are being provided to unauthorized individuals, such as former spouses or nonimmediate relatives. Finally, the complexity of the variations in the calculations (there are a lot of "If this, do that" calculations) results in frequent errors, many of which probably go undetected.

   b. The manager of a tool and die shop needs help with job processing and control. Jobs are currently processed by hand. First, a job number is established. Next, the job supervisor estimates time and materials for the job. This is a time-consuming process, and delays are common. Then, the job is scheduled for a specific day and estimated time.

On the day the job is to be worked on, materials orders have to be issued. If materials aren't available, the order has to be rescheduled.

Time cards are completed in the shop when workers fulfill the work order. These time cards are used to charge back time to the customer. Time cards are processed by hand, and the final calculations are entered on the work order. The work orders are checked for accuracy and sent to the IS department, where accounting records are updated and the customer is billed.

The problem is that the customer frequently calls to inquire about costs already incurred on a work order, but it's not possible to respond because IS sends a report of all work orders only once a month. Also, management has no idea of how good initial estimates are or how much work is being done on any tool or machine, or by any worker.

c. State University's Development Office raises funds for improving instructional facilities and laboratories at the university. It has uncovered a sensitive problem: The data is out of control.

The Development Office keeps considerable redundant data on past gifts and givers, as well as prospective benefactors. This results in multiple contacts for the same donor—and people don't like to be asked to donate to a single university over and over!

To further complicate matters, the faculty and administrators in most departments conduct their own fund-raising and development campaigns, again resulting in duplication of contact lists.

Contacts with possible benefactors are not well coordinated. Some prospective givers are contacted too often, and others are overlooked. It is currently impossible to generate lists of prospective givers based on specific criteria (e.g., prior history and socioeconomic level), even though data on hundreds of criteria have been collected and stored. Gift histories are nonexistent, which makes it impossible to establish contribution patterns that would help various fund-raising campaigns.

4. Century Tool and Die, Inc., is a major manufacturer of industrial tools and machines located in Newark, New Jersey. Larry is the assistant Accounts Receivable (AR) manager. Valerie is a systems analyst for the Information Systems department. Valerie and Larry are meeting to discuss their current project—improving the recently implemented Accounts Receivable information system. As they start to discuss the project, they are interrupted by Robert Washington, the executive vice president of Finance, and Gene Burnett, the AR manager. Larry suddenly looks very nervous. And for good reason—he had suggested the new system, and it has not turned out as promised. Gene's support had been lukewarm, at best. Robert initiates the conversation.

"We've got big problems," said Robert. "This new AR system is a disaster. It has cost the company more than $625,000, not to mention lost customer goodwill and pending legal costs. I can't afford this when the board of directors is complaining about declining return on investment. I want some answers. What happened?"

Gene responds, "I was never really in favor of this project. Why did we need this new computer system?"

Larry gets defensive. "Look, we were experiencing cash flow problems on our accounts. The existing system was too slow to identify delinquencies and incapable of efficiently following up on those accounts. I was told to solve the problem. A manual system would be inefficient and error-prone. Therefore, I suggested an improved computer-based system."

Gene replies, "I'm not against the computer. I approved the original computer-based system. And I realized that a new system might be needed. It's just that you and Valerie decided to redesign the system without considering alternatives—just like that! In my opinion, you should always analyze options. And let's suppose that a new computerized system was our best option. Why did we have to build the system from scratch? There are good AR software packages available for purchase. I . . ."

Sensing a confrontation, Robert interrupts, "Gene has a point, Larry. Still, the system you proposed was defended as feasible. And yet it failed! Valerie, as lead analyst, you proposed the new system, correct?"

Nervously, Valerie responds, "Yes, with Larry's help."

Robert continues, "And you wrote this feasibility report early in the project. Let's see. You proposed replacing the current batch AR system with an online system using a database management package."

Valerie replies, "Strictly speaking, the database management package wasn't needed. We could have used the existing VSAM files."

With a troubling look, Robert says, "The report says you needed it. I paid $15,000 to get you that package!"

Valerie answers, "Bill, our database administrator, made that recommendation. The AR system was to be the pilot database project."

Robert continues. "You also proposed using a network of microcomputers as a front end to the mainframe computer?"

"Yes," answers Valerie. "Larry felt a mainframe-based system would take too long to design and implement. With microcomputers, we could just start writing the necessary transaction programs and then transfer the data to the database on the mainframe computer."

Robert looked puzzled. "I'm a former engineer. It seems to me that some sort of design work should have been done no matter what size computer you used . . . *[brief pause]* In any case, the bottom line in this report is that your projected benefits outweighed the lifetime costs. You projected a 22 percent annual return on investment. Where did you get that number?"

Valerie answers, "I met with Larry four times—about six hours total, I'd say—and Larry explained the problems, described the requirements, made suggestions, and then projected the costs, benefits, and rate of return."

Robert replies, "But that return hasn't been realized, has it? Why not?" After a long pause, Robert continues, "Valerie, what happened after this proposal was approved?"

"We spent the next nine months building the system."

Robert responds, "And how much did you have to do with that, Larry?"

"Not a lot, sir. Valerie occasionally popped into my office to clarify requirements. She showed me sample reports, files, and screens. Obviously, she was making progress, and I had no reason to believe that the project was off schedule."

Robert continues his investigation. "Were you on schedule, Valerie?"

"I don't believe so, Mr. Washington. My team and I were having some problems with certain business aspects of the system. Larry was unfamiliar with those aspects and he had to go to the account clerks and the accountants for answers."

Visibly irritated, Robert asks, "I don't get it! Why didn't you go to the clerks and accountants?"

Although the question was directed to Valerie, Gene replies, "I can answer that. I designated Larry as Valerie's contact. I didn't want her team wasting my people's time—they have jobs to do!"

Robert is silent for a moment, but then he responds, "Something about that bothers me, Gene. In any case, when this project got seriously behind, Larry, why didn't you consider canceling it, or at least reassessing the feasibility?"

"We did!" answers Larry. "Gene expressed concern about progress about seven months into the project. We called a meeting with Valerie. At that meeting, we learned that the new database system wasn't working properly. We also found that we needed more memory and storage on the microcomputers. And to top it off, Valerie and her staff seemed to have little understanding of the business nature of our problems and needs."

This time, Valerie gets defensive. "As I already pointed out, I wasn't permitted contact with the users during the first seven months. Besides, we were asked by Larry to start programming as quickly as possible so that we would be able to show evidence of progress."

Looking at Larry, Robert asks, "Larry, I'm no computer professional; however, my gut instinct suggests that some design or prototyping should have been done first."

Valerie responds for Larry, "Yes, but that would have required end-user participation, which Larry and Gene would not permit."

Larry interrupts. "As I was saying, we considered canceling the project. But I pointed out that $150,000 had already been spent. It would be stupid to cancel a project at that point. I did reassess the feasibility and concluded that the project could be completed in four more months for another $50,000."

Robert responds, "Nobody asked me if I wanted to spend that extra money."

Larry answers, "We realize that the system hasn't worked out as well as we had hoped. We are trying to redesign . . ."

Robert interrupts, "As well as you hoped? That's an understatement! Let me read you some excerpts from Gene's last monthly report. Customer accounts have mysteriously disappeared, deleted without explanation. Later, we discovered that data-entry clerks didn't know that the F2 key deletes a record. Also, customers have been legally credited for payments that were never made! Customers have been double billed in some cases! Reports generated by the system are late, inaccurate, and inadequate. Cash flow has been decreased by 35 percent! My sales manager claims that some customers are taking their business elsewhere. And the legal department says we may be sued by two customers and that it will be impossible to collect on those accounts where customers received credit for nonpayments. You tell me, what would you do if you were in my shoes?"

a. Evaluate this project against the eight principles of systems development.

b. What would you do if you were in Mr. Washington's shoes? How would you react to Larry's performance? Valerie's? Gene's?

c. What did Valerie do wrong? Was she in control of her own destiny on this project? Why or why not?

d. What did Larry or Gene do wrong? Can either be held responsible for the failure of a computer project when they have limited computer literacy or experience?

e. What was wrong with the feasibility report? Did Valerie and Larry meet often enough? Was the input to that report sufficient? Did the team commit to a solution too early? Did programming begin too soon? Why or why not?

f. Why were Valerie and her staff uncomfortable with the business problem and needs?

g. Should the project have been canceled? What about the $150,000 investment that had already been made?

h. If you were Valerie or Larry, what would you have done differently?

5. Assume that you are a newly hired CASE champion for a major company. You are to make a presentation to management proposing that large sums of money be spent to buy a methodology and CASE tools for your company. Research the market and prepare a spreadsheet that details projected costs and benefits. State all your assumptions. Don't forget training and support.

SUGGESTED READINGS

Application Development Strategies (monthly newsletter). Arlington, MA: Cutter Information Corporation. This is our favorite theme-oriented periodical that follows system development strategies, methodologies, CASE and ADE, and other relevant trends. Each issue focuses on a single theme.

Application Development Trends (monthly periodical). Framingham, MA: Software Productivity Group, Inc. This is our favorite periodical for keeping up with the latest trends in methodology and automated tools. Each month's issue features several articles on different topics and products.

Benjamin, R. I. *Control of the Information System Development Cycle.* New York: Wiley-Interscience, 1971. Benjamin's 16 axioms for managing the systems development process inspired our adapted principles to guide successful systems development.

DeMarco, Tom. *Structured Analysis and System Specification.* Englewood Cliffs, NJ: Prentice Hall, 1978. This is the classic book on the structured systems analysis, a process-centered, model-driven methodology.

Gane, Chris. *Rapid Systems Development.* Englewood Cliffs, NJ: Prentice Hall, 1989. This book presents a nice overview of RAD that combines model-driven development and prototyping in the correct balance.

Gildersleeve, Thomas. *Successful Data Processing Systems Analysis,* 2nd ed. Englewood Cliffs, NJ: Prentice Hall, 1985. We are indebted to Gildersleeve for the creeping commitment approach.

Harmon, Paul, and Mark Watson. *Understanding UML: The Developer's Guide.* San Francisco: Morgan Kaufmann Publishers, Inc., 1998. This book describes an emerging standard for object-oriented analysis and design, a contemporary model-driven methodology.

Martin, James. *Information Engineering: Volumes 1–3.* Englewood Cliffs, NJ: Prentice Hall, 1989 (Vol. 1), 1990 (Vols. 2, 3). These works present information engineering, data-centered, model-driven methodology.

McConnell, Steve. *Rapid Development.* Redmond, WA: Microsoft Press. Chapter 7 of this excellent reference book provides what may be the definitive summary of system development life cycle and methodology variations that we call "routes" in our book.

Orr, Ken. *The One Minute Methodology.* New York: Dorsett House Publishing, 1990. Must reading for those interested in exploring the need for methodology. This very short book can be read in one sitting. It follows the story of an analyst's quest for the development silver bullet, "the one minute methodology."

Paulk, Mark C.; Charles V. Weber; Bill Curtis; and Mary Beth Chrissis. *The Capability Maturity Model: Guidelines for Improving the Software Process.* Reading, MA: Addison-Wesley Publishing Company, 1995. This book fully describes version 1.1 of the Capability Maturity Model. Version 2.0 was under development at the time we were writing this chapter.

Wetherbe, James, and Nicholas P. Vitalari. *Systems Analysis and Design: Best Practices,* 4th ed. St. Paul, MN: West, 1994. We are indebted to Wetherbe for the PIECES framework.

Zachman, John A. "A Framework for Information System Architecture." *IBM Systems Journal* 26, no. 3 (1987). This article presents a popular conceptual framework for information systems survey and the development of an information architecture.

Focus on PEOPLE

Focus on DATA

Focus on PROCESSES

Focus on INTERFACE

Focus on DEVELOPMENT

Stakeholders

Activities

SYSTEMS ANALYSTS

- SYSTEM OWNERS
- SYSTEM USERS
- SYSTEM DESIGNERS
- SYSTEM BUILDERS

BUILDING BLOCKS OF AN INFORMATION SYSTEM

Management Expectations

The PIECES Framework

Performance • Information • Economics • Control • Efficiency • Service

List of business entities and rules ...

Business Knowledge

List of business functions and events ...

Business Functions

List of business locations and systems ...

Business Locations

Data Requirements

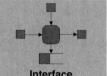

Process Requirements

Interface Requirements

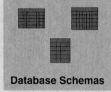

Database Schemas

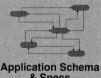

Application Schema & Specs

Interface Specifications

```
CREATE TABLE tblOrders
colOrderNot CHAR(5) NOT
NULL
colOrderDate  DATE/TIME NOT
NULL ...
```

Database Programs

```
PROC ValidateOrder
 PERFORM ValidateCust
 REPEAT UNTIL
  NoMoreProd ...
```

Application Programs

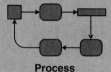

```
<html>
<head>
<title> Order Entry Form </
title> ...
```

Interface Programs

- PROJECT & PROCESS MANAGEMENT
- PRELIMINARY INVESTIGATION
- PROBLEM ANALYSIS
- REQUIREMENTS ANALYSIS
- DECISION ANALYSIS
- DESIGN
- CONSTRUCTION
- IMPLEMENTATION

VENDORS AND CONSULTANTS

INFORMATION TECHNOLOGY & ARCHITECTURE
Database Technology • Process Technology • Interface Technology • Network Technology

OPERATIONS AND SUPPORT

4

PROJECT MANAGEMENT

CHAPTER PREVIEW AND OBJECTIVES

Project management skills are greatly in demand in the information technology community. Project management is a natural extension of the previous chapter's introduction to system development. Project management is deserving of entire books; however, this chapter provides a process-centric survey of key project management tools and techniques as they apply to systems analysis and design. You will know that you understand the basics of project management when you can:

— Define the terms *project* and *project management,* and differentiate between project and process management.

— Describe the causes of failed information systems and technology projects.

— Describe the basic competencies required of project managers.

— Describe the basic functions of project management.

— Differentiate between PERT and Gantt charts as project management tools.

— Describe the role of project management software as it relates to project management tools.

— Describe eight activities in project management.

— Define *joint project planning* and its role in project management.

— Define *scope* and a write a *statement of work* to document scope.

— Use a *work breakdown structure* to decompose a project into tasks.

— Estimate tasks' durations, and specify intertask dependencies on a PERT chart.

— Assign resources to a project and produce a project schedule with a Gantt chart.

— Assign people to tasks and direct the team effort.

— Use critical path analysis to adjust schedule and resource allocations in response to schedule and budget deviations.

— Manage user expectations of a project and adjust project scope.

SCENE

Bob, Sandra, and Terri Hitchcock (the business analyst assigned to the project) are in Sandra's office for a meeting.

BOB

I'm eager to get started here! I've been reading some documentation on the *FAST* methodology. I assume this is the first meeting for the preliminary investigation phase?

SANDRA

I appreciate your enthusiasm, Bob, but we need to first develop a project plan.

BOB

I thought the *FAST* methodology *is* the project plan?

SANDRA

Actually, no. The *FAST* methodology provides the basic system development process for all SoundStage projects. But the process is essentially a template that must be customized for each individual project. We rarely delete tasks from a *FAST* route, but we frequently add tasks or split tasks, and we must estimate the time required for each task and assign people and other resources to each task. That's the purpose of this meeting.

BOB

You're talking about project management.

SANDRA

Yes, specifically, I'm talking about the project planning steps of project management. This meeting will not actually produce a project plan. What we want to do today is to plan for the project planning meeting.

TERRI

Plan for a plan?

SANDRA

Yes. We call our planning method joint project planning or JPP. It is a meeting, or series of meetings, that results in a consensus agreement to a project goal, schedule, and budget.

[pause]

Project planning is a painful but essential activity. Many projects fail because of an incomplete or poorly developed plan. There was no agreement on what the scope of the project is and isn't. For that reason, the schedule and budget can get quickly out of control, and project stakeholders get frustrated, and everyone starts blaming one another. We've learned—the hard way, I'm afraid—that it is important to at least try to identify project scope, assumptions, and constraints before committing to any schedule or budget. Indeed, the scope drives the schedule and budget.

TERRI

I assume that's why I'm here—as a representative of the user community?

SANDRA

Yes, but today your role is to help us identify other user and management community stakeholders. It is unlikely that any one user or manager can fully represent all stakeholders' scope interests.

BOB

But shouldn't we have other Information Services stakeholders here as well? After all, I assume that programmers, database designers, network people, and other technical specialists will need to be committed to the project as well.

SANDRA

Absolutely! That's why our goal today is to plan for the plan. More precisely, we will be planning the joint project planning meetings. And one of the deliverables we need to produce from this meeting is a list of the stakeholders for this project. We want to get them involved in the JPP. Everyone must agree to the scope; everyone must agree to commit necessary resources; and everyone must agree to the schedule and budget as directed to the agreed-on scope.

BOB

So who needs to be involved in our JPP?

SANDRA

Typically, we like to have a facilitator, someone who will not be involved in the project but who can facilitate the JPP to the goal. This is going to sound weird, but I have had great success using a friend of mine in Human Resources as the facilitator. She has some IT background, but not enough that she interjects her own opinions into the process. She has proven her ability to lead a discussion to a project plan in the past.

TERRI

You're talking about Jackie Roland, right? I once participated in sessions she facilitated to change our business process for staff performance evaluations. She was terrific!

SANDRA

Yes, I was thinking Jackie, if she's available. Regardless, I want an outside facilitator. Of course, I'll be in the session since I must manage the resulting plan. And Bob, I'd like you to be my scribe. You'll take notes and help me translate the planning data into Microsoft *Project,* our project management software.

BOB

I can do that. I actually learned a little about project management in my systems analysis and design course. And I completed an independent study project with my instructor to learn more about Microsoft *Project.* I had always hoped that my knowledge of *Project* would lead to some project management opportunities.

SANDRA

Some day, Bob, but project management is a lot more than just Microsoft *Project. Project* is just a tool. Project management is a process that starts at the beginning of a project, extends through a project, and doesn't culminate until after the project is completed.

Let's get back to our JPP participants list. We typically invite a representative from our Centers for Excellence to serve as a methodology, tools, and technology consultant to the process to be used to develop each new system. Of course, Terri, you need to participate in this process. As a business analyst, you are going to serve as my liaison back to the user community.

S O U N D S T A G E

S O U N D S T A G E   E N T E R T A I N M E N T   C L U B

But I also need to know who from the community needs to be involved in scope definition.

TERRI

Let's see. Galen Kirchoff is our executive sponsor. He's paying for all this, so he should obviously be involved. Our key management sponsor is Sharon Stone, and she should be involved. Maybe I should also ask Steven, Susan, or Monica. And maybe we should ask Dick. The warehousing operations are likely to be impacted by the project. In fact, he is management sponsor of the bar-coding project. Their club requirements will drive this project.

SANDRA

Let me know who can make the commitments that we are seeking.

Ok, Bob. Let's see how much of that *FAST* documentation you remember. Who do you think we need from Information Services?

BOB

Obviously, Peg Li, the director of the Development Center, should be involved. She will have to commit programmer and contractor resources to this project. And I haven't learned everyone's name yet, but wouldn't the data administrator need to be involved? After all,

we will need resources for database design and implementation, no?

SANDRA

Good. Go on.

BOB

Well, we've talked about extending the new member services system to the Internet. I'd say that the telecommunications administrator should be represented. Other than you and I, that should be it.

SANDRA

Almost! We typically get someone from Computer Operations to participate. I'm impressed, Bob. Both of you should be aware that I've already chatted with Nancy Picard. Nancy has committed representatives from each of the areas Bob suggested to the JPP. Those individuals will need to commit specific people to the project plan.

TERRI

So what do we need to do now?

SANDRA

We need to prepare an agenda for the JPP. Obviously, we need to invite the people we've been discussing to participate in the JPP and get their commitment to those meetings. We need to arrange for a room. I like to go off-site;

perhaps to a local hotel conference room. We would need to arrange for a continental breakfast, catered lunch, and refreshments. And most of all, we need to establish an agenda of deliverables to be produced at this meeting.

DISCUSSION QUESTIONS

1. Without a well-prepared project plan, what could go wrong in a project and lead to missed schedules and budget overruns?

2. What is the difference between a methodology, such as *FAST,* and project management?

3. What is the difference between Microsoft *Project* and project management? Why is this distinction important?

4. What would be the advantages and disadvantages of having a noninformation technology specialist or involved user serving as facilitator of the JPP? If there were no experienced facilitators outside of Information Services, who else might serve as a facilitator?

5. Prepare an announcement memo and agenda for the JPP. Be sure to include deliverables. (You are encouraged to "cheat" by reading the chapter!)

Most of you are familiar with Murphy's Law, which suggests, "If anything can go wrong, it will." Murphy has motivated numerous pearls of wisdom about projects, machines, people, and why things go wrong. This chapter will teach you strategies, tools, and techniques for project management as applied to information systems projects.

The demand for project managers in the information systems community is strong. Typically, IS project managers come from the ranks of experienced IS developers such as systems analysts. While it is unlikely that your first job responsibilities will include project management, you should immediately become aware of project management processes, tools, and techniques. Eventually you will combine this knowledge with development experience plus your own observation of project managers to form the basis for your own career opportunities in project management.

Before we can define *project management,* we should first define *project.* There are as many definitions as there are authors, but we like the definition put forth by Wysocki, Beck, and Crane:

WHAT IS PROJECT MANAGEMENT?

> "A **project** is a [temporary] <u>sequence</u> of <u>unique</u>, <u>complex</u>, and <u>connected</u> <u>activities</u> having <u>one goal</u> or purpose and that must be completed by a <u>specific time</u>, <u>within budget</u>, and <u>according to specification</u>."[1]

The keywords are underlined. As applied to information system development, we note the following:

— A system development process or methodology, such as *FAST*, defines a <u>sequence of activities</u>, mandatory and optional.

— Every system development project is <u>unique</u>; that is, it is different from every other system development project that preceded it.

— The activities that comprise system development are relatively <u>complex</u>. They require the skills that you are learning in this book, and they require that you be able to adapt concepts and skills to changing conditions and unanticipated events.

— By now, you've already learned that the phases and activities that make up a system development methodology are generally <u>sequential</u>. While some tasks may overlap, many tasks are dependent on the completion of other tasks.

— The development of an information system represents a <u>goal</u>. Several objectives may need to be met to achieve that goal.

— Although many information system development projects do not have absolute deadlines or <u>specified times</u> (there are exceptions), they are notoriously completed later than originally projected. This is becoming less acceptable to upper management given the organizationwide pressures to reduce cycle times for products and business processes.

— Few information systems are completed <u>within budget</u>. Again, upper management is increasingly rejecting this tendency.

— Information systems must satisfy the business, user, and management expectations <u>according to specifications</u> (which we call *requirements* throughout this book).

For any systems development project, effective *project management* is necessary to ensure that the project meets the deadline, is developed within an acceptable budget, and fulfills customer expectations and specifications.

> **Project management** is the process of scoping, planning, staffing, organizing, directing, and controlling the development of an acceptable system at a minimum cost within a specified time frame.

You learned in Chapter 3 that project management is a cross life cycle activity because it overlaps all phases of any systems development methodology.

Corporate rightsizing has changed the structure and culture of most organizations, and hence, project management. More flexible and temporary interdepartmental teams that are given greater responsibility and authority for the success of organizations have replaced rigid hierarchical command structures and permanent teams. Contemporary system development methodologies depend on having teams that include both technical and nontechnical users, managers, and information technologists all directed to the project goal. These dynamic teams require leadership and project management.

Organizations take different approaches to project management. One approach is to appoint a project manager from the ranks of the team (once it has been formed). Alternatively, many organizations believe that successful project managers apply a unique body of knowledge and skills that must be learned. These orga-

---

[1] Robert K. Wysocki; Robert Beck, Jr.; and David B. Crane, *Effective Project Management: How to Plan, Manage, and Deliver Projects on Time and within Budget* (New York: John Wiley & Sons, 1995), p. 38.

nizations tend to hire and/or develop professional project managers who are then assigned to one or more projects.

The prerequisite for good project management is a well-defined system development process. In Chapter 3, we introduced the Capability Maturity Model (CMM) as a framework for quality management that is based on a sound systems development process. It is important to differentiate between project and process management. Project management was defined above.

> **Process management** is an ongoing activity that documents, manages the use of, and improves an organization's chosen methodology (the "process") for systems development. Process management is concerned with the activities, deliverables, and quality standards to be applied to *all* projects.

In other words, the scope of process management is all projects; whereas, the scope of project management is a single project. Chapter 3 introduced the process and process management. This chapter will focus on project management.

What causes a project to succeed or fail? Chapter 3 introduced eight basic principles of systems development that are critical success factors for all projects. See Chapter 3 for a review of those principles. From a project management perspective, a project is considered a success if:

The Causes of Failed Projects

- The resulting information system is acceptable to the customer.
- The system was delivered "on time."
- The system was delivered "within budget."
- The system development process had a minimal impact on ongoing business operations.

But not all projects are successful. Failures and limited successes far outnumber successful information systems. Project mismanagement can undermine the best application of the systems analysis and design methods taught in this book. We can develop an appreciation for the importance of project management by studying the mistakes of some project managers. Let's examine some project mismanagement problems and consequences.

- *Failure to establish upper-management commitment to the project*—Sometimes commitment changes during a project.
- *Lack of organization's commitment to the system development methodology*—Many system development methodologies do little more than collect dust.
- *Taking shortcuts through or around the system development methodology*—Project teams often take shortcuts for one or more of the following reasons:
  - The project gets behind schedule and the team wants to catch up.
  - The project is over budget and the team wants to make up costs by skipping steps.
  - The team is not trained or skilled in some of the methodology's activities and requirements, so it skips them.
- *Poor expectations management*—All users and managers have expectations of the project. Over time, these expectations may change. This can lead to two undesirable situations.
  - **Scope creep** is the unexpected growth of user expectations and business requirements for an information system as the project progresses. Unfortunately, the schedule and budget can be adversely affected by such changes.
  - **Feature creep** is the uncontrolled addition of technical features to a system under development without regard to schedule and budget.
- *Premature commitment to a fixed budget and schedule*—You can rarely make accurate estimates of project costs and schedule before completing a detailed

problem analysis or requirements analysis. Premature estimates are inconsistent with the creeping commitment approach introduced in Chapter 3.

- *Poor estimating techniques*—Many systems analysts estimate by making a best-calculated estimate (jokingly referred to as a "guesstimate") and then doubling that number. This is not a scientific approach.

- *Overoptimism*—Systems analysts and project managers tend to be optimists. As project schedules slip, they respond, "So we've lost a day or two! It's no big deal. We can make it up later." They fail to recognize that certain tasks are dependent on other tasks. Because of these dependencies, a schedule slip in one phase or activity will cause corresponding slips in many other phases and activities, thus contributing to cost overruns.

- *The mythical man-month* (Brooks, 1975)—As the project gets behind schedule, project leaders frequently try to solve the problem by assigning more people to the team. It doesn't work! There is no linear relationship between time and number of personnel. The addition of personnel usually creates more communication problems, causing the project to get even further behind schedule.

- *Inadequate people management skills*—Managers tend to be thrust into management, not prepared for management responsibilities. This problem is easy to identify. No one seems to be in charge; customers don't know the status of the project; teams don't meet regularly to discuss and monitor progress; team members aren't communicating with one another; and the project is always said to be "95 percent complete."

- *Failure to adapt to business change*—If the project's importance changes during the project, or if the management or the business reorganizes, projects should be reassessed for compatibility with those changes and their importance to the business.

- *Insufficient resources*—This could be due to poor estimating or to other priorities, or it could be that the staff resources assigned to a project do not possess the necessary skills or experience.

- *Failure to "manage to the plan"*—Various factors may cause the project manager to become sidetracked from the original project plan.

This brings us to *the* major cause of project failure—*Most project managers were not educated or trained to be project managers!* Just as good programmers don't always go on to become good systems analysts, good systems analysts don't automatically perform well as project managers. To be a good project manager, you should be educated and skilled in the "art of project management."

The Project Management Body of Knowledge

The Project Management Institute was created as a professional society to guide the development and certification of *professional* project managers. The institute created the **"Project Management Body of Knowledge" (PMBOK)** for the education and certification of professional project managers. This chapter's content was greatly influenced by the PMBOK.

Project Manager Competencies   Good project managers possess a core set of competencies. Table 4.1 summarizes these competencies. Some of these competencies can be taught, in courses, books, and professional workshops; however, some come only with professional experience in the field. First, you usually cannot manage a process you have never used. Second, you cannot manage a project without understanding the business and culture that provides a context for the project. The competencies summarized in Table 4.1 can be studied in greater detail in the Wysocki et al. book referenced at the end of this chapter.

TABLE 4.1	*Project Manager Competencies*	
Competency	**Explanation**	**How to Obtain?**
	Business Achievement Competencies	
Business awareness	Ties every systems project to the mission, vision, and goals of the organization.	Business experience
Business partner orientation	Keeps managers and users involved throughout a systems project.	Business experience
Commitment to quality	Ensures that every systems project contributes to the quality expectation of the organization as a whole.	Business experience
	Problem-Solving Competencies	
Initiative	Demonstrates creativity, calculated risks, and persistence necessary to get the job done.	Business experience
Information gathering	Skillfully obtains the factual information necessary to analyze, design, and implement the information system.	Chapter 6 in this book plus business experience.
Analytical thinking	Can assess and select appropriate system development processes and use project management tools to plan, schedule, and budget for system development.	This chapter
	Can solve problems through the analytical approach of decomposing systems into their parts and then reassembling the parts into improved systems.	Chapters 7–9 and Module A in this book plus business experience
Conceptual thinking	Understands systems theory and applies it to systems analysis and design of information systems.	Chapters 2 and 5–9 in this book
	Influence Competencies	
Interpersonal awareness	Understands, recognizes, and reacts to interpersonal motivations and behaviors.	Can be learned in courses but requires business experience
Organizational awareness	Understands the politics of the organization and how to use them in a project.	Business experience
Anticipation of impact	Understands implications of project decisions and manages expectations and risk.	Introduced in this chapter but requires business experience
Resourceful use of influence	Skillfully obtains cooperation and consensus of managers, users, and technologists to solutions.	Business experience
	People Management Competencies	
Motivating others	Coaches and directs individuals to overcome differences and achieve project goals as a team.	Business experience
Communication skills	Communicates effectively, both orally and in writing, in the context of meetings, presentations, memos, and reports.	Can be learned in courses but usually requires business experience
Developing others	Ensures that project team members receive sufficient training, assignments, supervision, and performance feedback required to complete projects.	Business experience
Monitoring and controlling	Develops the project plan, schedule, and budget and continuously monitors progress and makes adjustments when necessary.	Tools and techniques taught in this chapter, but requires project experience
	Self-Management Competencies	
Self-confidence	Consistently makes and defends decisions with a strong personal confidence in the process and/or facts.	Business experience
Stress management	Works effectively under pressure or adversity.	Business experience
Concern for credibility	Consistently and honestly delivers on promises and solutions. Maintains technical or business currency in the field as appropriate.	Business experience
Flexibility	Capable of adjusting process, management style, or decision making based on situations and unanticipated problems.	Business experience

Source: Adapted from Robert K. Wysocki; Robert Beck, Jr.; and David B. Crane, *Effective Project Management: How to Plan, Manage, and Deliver Projects on Time and within Budget* (New York: John Wiley & Sons, 1995).

Project Management Functions   The project manager is not necessarily a senior analyst who happens to be in charge. A project manager must apply a set of skills different from those applied by the analyst. The basic functions of a manager have been studied and refined by management theorists for many years. These functions include scoping, planning, staffing, organizing, scheduling, directing, controlling, and closing.

- **Scoping**—Scope defines the boundaries of the project. If you cannot scope project expectations and constraints, you can't plan activities, estimate costs, or manage expectations.

- **Planning**—Planning identifies the required tasks to complete the project. This is based on the manager's understanding of the project goal and the methodology used to achieve the goal.

- **Estimating**—Each task that is required to complete the project must be estimated. How much time will be required? How many people will be needed? What skills will be needed? What tasks must be completed before other tasks are started? Can some of the tasks overlap? These are all estimating issues. Some of these issues can be resolved with the project modeling tools that will be discussed later in this chapter.

- **Scheduling**—Given the project plan, the project manager is responsible for scheduling all project activities. The project schedule should be developed with an understanding of the required tasks, task duration, and task prerequisites.

- **Organizing**—Members of the project team should understand their own individual roles and responsibilities as well as their reporting relationship to the project manager.

- **Directing**—Once the project has begun, the project manager directs the team's activities. Every project manager must demonstrate people management skills to coordinate, delegate, motivate, advise, appraise, and reward team members.

- **Controlling**—Perhaps the manager's most difficult and important function is controlling the project. Few plans will be executed without problems and delays. The project manager must monitor and report progress against goals, schedule, and costs and make appropriate adjustments when necessary.

- **Closing**—Good project managers always assess successes and failures at the conclusion of a project. They learn from their mistakes and plan for continuous improvement of the systems development process.

All the above functions are dependent on interpersonal *communication* among the project manager, the team, and other managers.

Project Management Tools and Techniques—PERT and Gantt Charts   The PMBOK also includes tools and techniques to support project managers. Two such tools are PERT and Gantt charts.

PERT, which stands for *Project Evaluation and Review Technique,* was developed in the late 1950s to plan and control large weapons development projects for the U.S. Navy.

> A **PERT chart** is a graphical network model that depicts a project's tasks and the relationships between those tasks.

A sample PERT chart is illustrated in Figure 4.1. PERT was developed to make clear the *interdependence* between project tasks before those tasks are scheduled. The boxes represent project tasks (we used phases from Chapter 3). (The content of the boxes can be adjusted to show various project attributes such as schedule and actual start and finish times.) The arrows indicate that one task is dependent on the start or completion of another task.

The Gantt chart, first conceived by Henry L. Gantt in 1917, is the most commonly used project scheduling and progress evaluation tool.

> A **Gantt chart** is a simple horizontal bar chart that depicts project tasks against a calendar. Each bar represents a named project task. The tasks are listed vertically in the left-hand column. The horizontal axis is a calendar timeline.

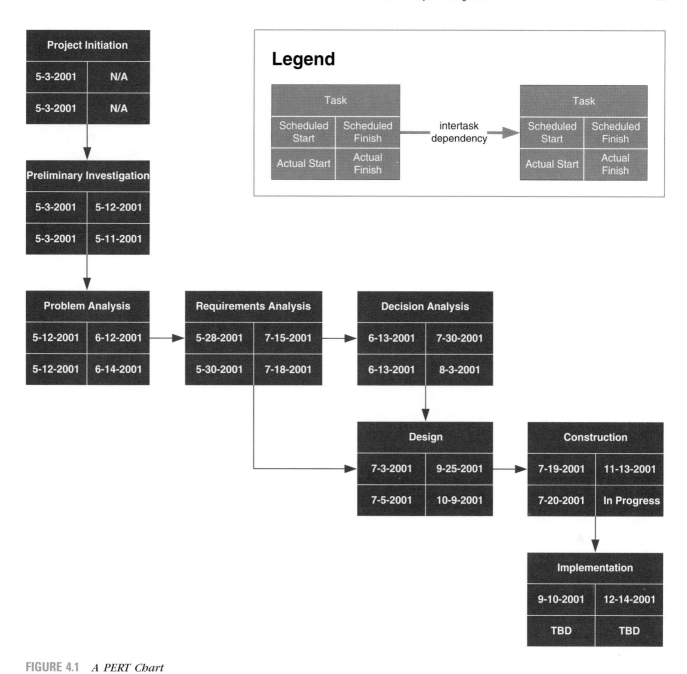

FIGURE 4.1   *A PERT Chart*

Figure 4.2 illustrates a phase-level Gantt chart, once again based on Chapter 3. We used the same project that was illustrated in Figure 4.1.

Gantt charts offer the advantage of clearly showing *overlapping* tasks, that is, tasks that can be performed at the same time. The bars can be shaded to clearly indicate percentage completion and project progress. The figure demonstrates which phases are ahead and behind schedule at a glance. The popularity of Gantt charts stems from their simplicity—they are easy to learn, read, prepare, and use.

Gantt and PERT charts are not mutually exclusive. Gantt charts are more effective when you are seeking to communicate schedule. PERT charts are more effective when you want to study the relationships between tasks.

Project Management Software   Project management software is routinely used to help project managers plan projects, develop schedules, develop budgets, monitor

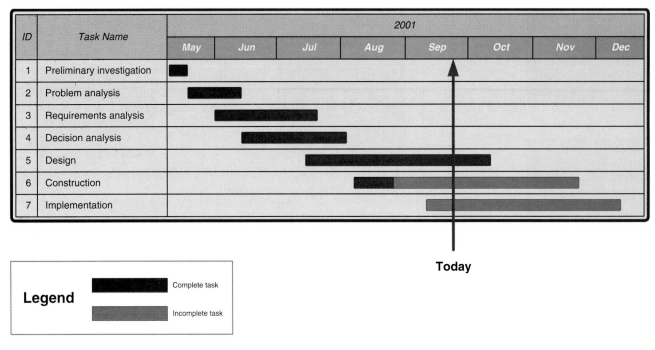

ID	Task Name	2001							
		May	Jun	Jul	Aug	Sep	Oct	Nov	Dec
1	Preliminary investigation								
2	Problem analysis								
3	Requirements analysis								
4	Decision analysis								
5	Design								
6	Construction								
7	Implementation								

Today

Legend   Complete task

Incomplete task

FIGURE 4.2   *A Gantt Chart*

---

progress and costs, generate reports, and effect change. Representative automated project management tools are listed in the margin.

We will teach you project modeling and management techniques in the context of project management software. We used Microsoft *Project* because that tool is frequently available to students and institutions at special academic prices through their college bookstore. Microsoft *Project,* like most project management software tools, supports both PERT and Gantt charts.

Figure 4.3(a) illustrates one possible Microsoft *Project* Gantt chart for the Sound-Stage Member Services project. We call your attention to the following numbered bullets.

❶ The black bars are *summary tasks* that represent project *phases* that are further decomposed into other tasks.

❷ The red bars indicate tasks have been determined to be "critical" to the schedule, meaning that any extension to the duration of those tasks will delay other tasks and the project as a whole. We'll talk more about critical tasks later.

❸ The blue bars indicate tasks that are not critical to the schedule, meaning they have some slack time during which delays will not affect other tasks and the project as a whole.

❹ The red arrows indicate prerequisites between two critical tasks.

(The blue arrows indicate prerequisites between two noncritical tasks.)

❺ The teal diamonds indicate milestones—events that have no duration. They signify the end of some significant task or deliverable.

Figure 4.3(b) shows a Microsoft *Project* PERT chart based on the same project plan that was illustrated in the Gantt chart. The contents of each cell in the task rectangles are highly customizable in Microsoft *Project.* (The rectangles on this MS *Project* PERT chart had to be painstakingly rearranged to produce a readable PERT chart with no overlapping lines.)

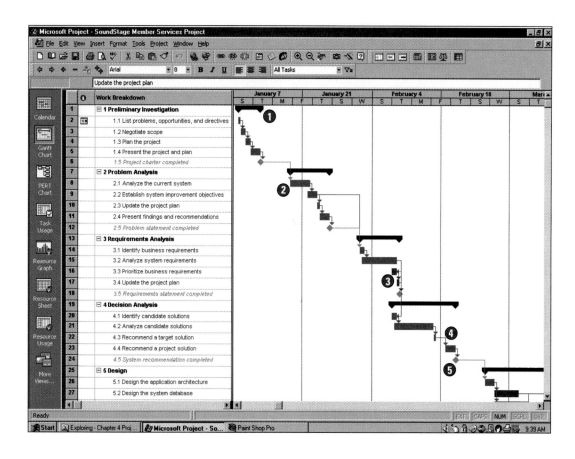

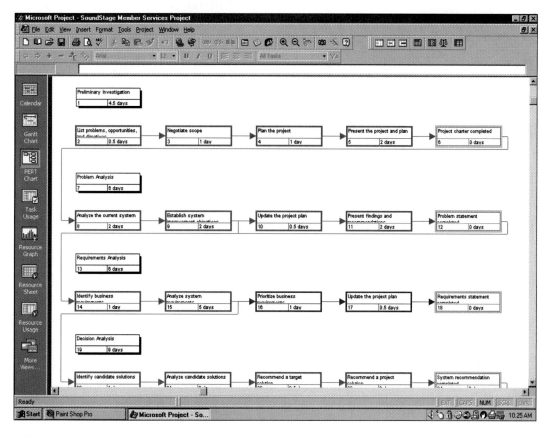

FIGURE 4.3   *Microsoft* Project *Gantt and PERT Charts*

THE PROJECT MANAGEMENT LIFE CYCLE

Recall from Chapter 3 that the Capability Maturity Model defines a framework for assessing the quality of an organization's information systems development activities. CMM Level 1 is defined as "initial" and characterized as the lack of any consistent project or process management function. The first stage of maturity improvement is to implement a consistent project management function—called CMM Level 2. In this section we introduce a project management life cycle representative of CMM Level 2 maturity.

Figure 4.4 illustrates a project management process or life cycle. Recall that project management is a cross life cycle activity; that is, project management activ-

FIGURE 4.4   *A Project Management Life Cycle*

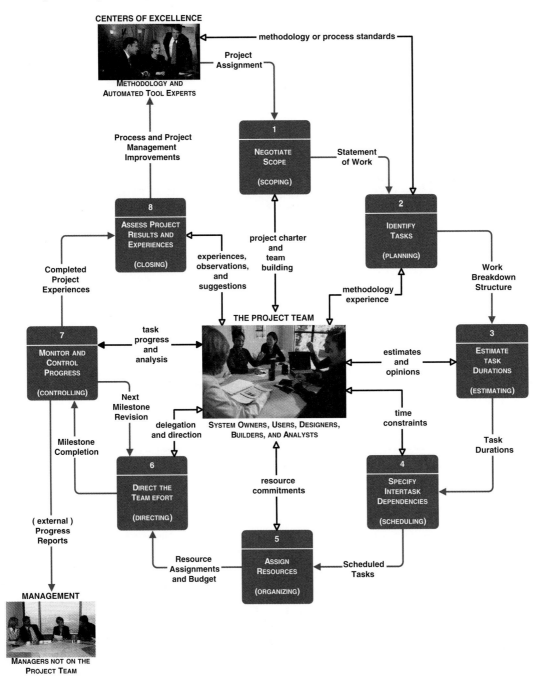

ities overlap all the system development phases that were introduced in Chapter 3. The illustrated project management activities correspond to classic management functions: scoping, planning, estimating, scheduling, organizing, directing, controlling, and closing.

The project management process shown in Figure 4.4 incorporates a joint project planning (JPP) technique.[2]

> **Joint project planning (JPP)** is a strategy wherein all stakeholders in a project (meaning system owners, users, analysts, designers, and builders) participate in a one-to-three-day project management workshop, the result of which is consensus agreement on project scope, schedule, resources, and budget. (Subsequent workshops or meetings may be required to adjust scope, budget, and schedule.)

Notice that in JPP, the project team is *actively* involved in all inputs and deliverables of all project management activities.

In the following subsections, we will review each of the illustrated project management activities and teach you how to use appropriate project management tools and techniques.

Perhaps the most important prerequisite to effective project management occurs at the beginning. All parties must agree to the project scope before any attempt is made to identify and schedule tasks or to assign resources (people) to those tasks. Scope defines the expectations of a project, and expectations ultimately determine satisfaction and degrees of success. Accordingly, the negotiation of project scope is a necessary activity in the project management life cycle. What is scope?

> **Scope** defines the boundaries of a project—What part of the business is to be studied, analyzed, designed, constructed, implemented, and ultimately improved?

Scope also defines those aspects of a system that are considered outside of the project. The answers to five basic questions influence the negotiation of project scope:

- Product—What do you want?
- Quality—How good do you want it to be?
- Time—When do you want it?
- Cost—How much are you willing to pay for it?
- Resources—What resources are you willing or able to bring to the table?

Negotiation of the above factors is a give-and-take activity that includes much iteration. The deliverable is an agreed on statement of work.

> A **statement of work** is a narrative description of the work to be performed as part of a project. Common synonyms include *scope statement, project definition, project overview,* and *document of understanding.*

In consulting engagements, the statement of work has become a commonly used contract between the consultant and client. But the approach works equally well for internal system development projects to establish a contract between business management and the project manager and team. According to Keane, Inc., a leading project management consulting firm,

> The statement of work affirms that the project manager understands who is really in charge of the effort, who is controlling the purse strings, what is the formal and informal organization within which the project will be developed, who are the "kings and queens" that have interest, and other similar but mainly nontechnical issues. It estab-

Activity 1—Negotiate Scope

Visit the McGraw-Hill Online Learning Center for this textbook and browse the online resource for both samples of a statement of work for the SoundStage case study as well as a Microsoft Word template and instructions you can use to write your own statement of work document.

---

[2] Wysocki, Beck, and Crane, *Effective Project Management: How to Plan, Manage, and Deliver Projects on Time and within Budget,* p. 38.

STATEMENT OF WORK

I.   Purpose
II.   Background
        A. Problem, opportunity, or directive statement
        B. History leading to project request
        C. Project goal and objectives
        D. Product description
III.   Scope
        (notice the use of the information system building blocks)
        A. Stakeholders
        B. Data
        C. Processes
        D. Locations
IV.   Project Approach
        A. Route
        B. Deliverables
V.   Managerial Approach
        A. Team-building considerations
        B. Manager and experience
        C. Training requirements
        D. Meeting schedules
        E. Reporting methods and frequency
        F. Conflict management
        G. Scope management
VI.   Constraints
        A. Start date
        B. Deadlines
        C. Budget
        D. Technology
VII.   Ballpark Estimates
        A. Schedule
        B. Budget
VIII.   Conditions of Satisfaction
        A. Success criteria
        B. Assumptions
        C. Risks
IX.   Appendixes

FIGURE 4.5
An Outline for a Statement of Work

lishes a firm business relationship between the project manager and both the customer and the extended project team.[3]

An outline for a typical statement of work document is shown in Figure 4.5. The size of the document will vary in different organizations. It may be as small as one to two pages or it may run several pages.

Activity 2—Identify Tasks

Given the project scope, the next activity is to identify project tasks. Tasks identify the work to be done. Typically, this work is defined in a top-down, outline manner. In Chapter 3, you learned about system development routes and their phases. But phases are too large and complex to plan and schedule a project. We need to decompose phases into activities and tasks until each task represents a manageable amount of work that can be planned, scheduled, and assigned. Some

---

[3] Updated and revised by Donald H. Plumber, *Productivity Management: Keane's Project Management Approach for Systems Development* (Boston: Keane, Inc., 1995), p. 5.

experts advocate decomposing tasks until the tasks represent an amount of work that can be completed in two weeks or less.

Ultimately, the project manager will determine the level of detail in the outline; however, most system development methodologies decompose phases for you—into suggested activities and tasks. These activities and tasks are not necessarily carved in stone; that is, most methodologies allow for some addition, deletion, and changing of activities and tasks based on the unique nature of each project. One popular tool used to identify and document project activities and tasks is a work breakdown structure.

A **work breakdown structure (WBS)** is a hierarchical decomposition of the project into phases, activities, and tasks.

Work breakdown structures can be drawn using top-down hierarchy charts similar to organization charts (Figure 4.6). But in Microsoft *Project,* a WBS is depicted using a simple outline style, indentation of activities and tasks on the Gantt chart "view" of the project. Microsoft *Project* also offers a military numbering scheme to represent hierarchical decomposition of a project as follows:

1 Phase 1 of the project
   1.1 Activity 1 of Phase 1
      1.1.1 Task 1 of Activity 1 in Phase 1
      1.1.2 Task 2 of Activity 1 in Phase 1
   1.2 Activity 2 of Phase 1 . . .
2 Phase 2 of the project . . .

Visit the McGraw-Hill Online Learning Center for this textbook and browse the online resource for both samples of a work breakdown structure for the SoundStage case study as well as Microsoft Project *templates and instructions you can use to document your own work breakdown structures based on various routes through the* FAST *methodology.*

FIGURE 4.6
A Graphical Work Breakdown Structure

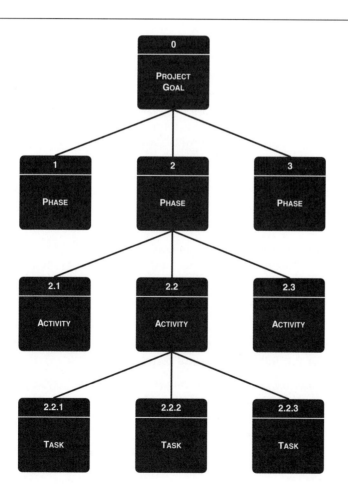

If you reexamine Figure 4.3(a), you will notice that Microsoft *Project* provides a column for the WBS in the Gantt chart. Also notice its use of the indentation and numbering to differentiate between tasks and subtasks.

We may want to include special tasks in a WBS called milestones.

Milestones are events that signify the accomplishment or completion of major deliverables during a project.

In information systems projects, an example might be the completion of all the tasks associated with producing a major project deliverable such as a requirements statement (see Chapter 3). It might be useful to distinguish milestones from other tasks in a WBS by using special formatting, such as italics.

Activity 3—Estimate Task Durations

Given a work breakdown structure with a suitable level of detail, the project manager must estimate duration for each task. Duration of any task is a random variable subject to factors such as the size of the team, number of users, availability of users, aptitudes of users, complexity of the business system, information technology architecture, experience of team personnel, time committed to other projects, and experience with other projects.

Most system development methodologies not only define tasks but also provide baseline estimates for task duration. The project manager must adjust these baselines into reasonable estimates for each unique project.

In Microsoft *Project,* all phases, activities, and tasks of a methodology are simply called *tasks.* The work breakdown structure then consists of both summary and primitive tasks. A summary task is one that consists of other tasks (such as phases and activities). A primitive task is one that does not consist of any other tasks. It is these primitive tasks for which we must estimate duration. (Like most project management software, Microsoft *Project* will automatically calculate the duration of all summary tasks based on the estimated durations of their component primitive tasks.)

For those primitive tasks that are not milestones, we must estimate duration. When estimating task duration, it is important to understand the concept of *elapsed time.* Elapsed time takes into consideration two important people factors:

- Efficiency—No worker performs at 100 percent efficiency. Most people take coffee breaks, lunch breaks, restroom breaks, and time to read their e-mail, check their calendars, participate in nonproject work, and even engage in idle conversation. Experts differ on just how productive the average worker is, but one commonly used figure is 75 percent.
- Interruptions—People experience phone calls, visitors, and other unplanned interruptions that increase the time required for project work. This is variable for different workers. Interruptions can consume as little as 10 percent of your day or as much as 50 percent.

Why is this important? Given a task that could be completed in 10 hours with 100 percent efficiency and no interruptions, and assuming a worker efficiency of 75 percent and 15 percent interruptions, the true estimate for the task would be

$$10 \text{ hours} \div 0.75 \text{ efficiency} < 13.3 \text{ hours} \div (1.00 - 0.15 \text{ interruptions})$$
$$< 15.7 \text{ hours}$$

There are many techniques for estimating task duration. For the sake of demonstration, we offer the following classical technique.

1. *Estimate the minimum amount of time it would take to perform the task.* We'll call this the **optimistic duration (OD).** The optimistic duration assumes that even the most likely interruptions or delays, such as occasional employee illnesses, will *not* happen.
2. *Estimate the maximum amount of time it would take to perform the task.* We'll call this the **pessimistic duration (PD).** The pessimistic duration assumes

that nearly anything that can go wrong will go wrong. All possible interruptions or delays, such as labor strikes, illnesses, training, inaccurate specification of requirements, equipment delivery delays, and underestimation of the systems complexity, are assumed to be inevitable.

3. Estimate the **expected duration (ED)** *that will be needed to perform the task.* Don't just take the median of the optimistic and pessimistic durations. Attempt to identify interruptions or delays that are most likely to occur, such as occasional employee illnesses, inexperienced personnel, and occasional training.

4. Calculate the **most likely duration (D)** as follows:

$$D = \frac{(1 \times OD) + (4 \times ED) + (1 \times PD)}{6}$$

where 1, 4, and 1 are default weights used to calculated a weighted average of the three estimates.

Developing OD, PD, and ED estimates can be tricky and require experience. Several techniques are used in estimating. Three of the most common techniques are:

— *Decomposition*—a simple technique wherein a project is decomposed into small, manageable pieces that can be estimated based on historical data of past projects and similarly complex pieces.

— *COCOMO (pronounced like Kokomo)*—a model-based technique wherein standard parameters based on prior projects are applied to the new project to estimate duration of a project and its tasks.

— *Function points*—another model-based technique wherein the "end product" of a project is measured based on number and complexity of inputs, outputs, files, and queries. The number of function points is then compared to projects that had a similar number of function points to estimate duration.

Some automated project management tools, such as *CS/10000* and *Cost•Xpert,* provide expert system technology that makes these estimates for you based on your answers to specific questions. See the references for information about an evaluation copy of *Cost•Xpert.*

Milestones (as defined in the previous subsection) have no duration. They simply happen. In Microsoft *Project,* milestones are designated by setting the duration to *zero.* (In the Gantt chart, those zero duration tasks change from bars to diamonds.)

Given the duration estimates for all tasks, we can now begin to develop a project schedule. The project schedule depends not only on task durations but also on intertask dependencies. In other words, the start or completion of individual tasks may depend on the start or completion of other tasks. There are four types of intertask dependencies:

— *Finish-to-start* (FS)—The finish of one task triggers the start of another task.

— *Start-to-start* (SS)—The start of one task triggers the start of another task.

— *Finish-to-finish* (FF)—Two tasks must finish at the same time.

— *Start-to-finish* (SF)—The start of one task signifies the finish of another task.

Intertask dependencies can be established and depicted in both Gantt and PERT charts. Figure 4.7 illustrates how to enter intertask dependencies in the Gantt chart view in Microsoft *Project.* We call your attention to the following annotated bullets:

❶ Intertask dependencies may be entered in the Gantt chart view in the *Predecessors* column by entering the dependent tasks' row numbers. Note that a task can have zero, one, or many predecessors.

Microsoft Project *supports the weighted average technique, which is called PERT analysis. Its use is optional. First, open the* PERT Weights *dialogue box to adjust the weights based on your experience and history—for example, if historical data suggests your projects tend to come closer to the optimistic or pessimistic estimates. Next, open the* PERT Entry Form. *You will see all the tasks with columns for D, OD, ED, and PD. Enter OD, ED, and PD for every task using the* PERT Entry Form. *Finally, you must use the* PERT Calculator *icon to determine a most likely duration (D) for each task. That most likely duration (D) will be carried forward to the Gantt and PERT charts.*

Activity 4—Specify Intertask Dependencies

② Intertask dependencies may also be entered (or modified) by opening the *Task Information* dialogue box for a given task. In this dialogue box, you can specify type of dependency using the drop-down list as shown.

③ The type of dependency can be entered in the *Task Information* dialogue box for any given dependent task.

④ Intertask dependencies are graphically illustrated in the Gantt chart as arrows between the bars that represent each task. Arrows may begin or terminate on the left side (to indicate a "start" dependency) or right side (to indicate a "finish" dependency).

Milestones (as defined earlier) almost always have several predecessors to signify those tasks that must be completed before you can say that the milestone has been achieved.

Given the start date for a project, the tasks to be completed, the task durations, and the intertask dependencies, the project can now be scheduled. There are two approaches to scheduling:

Forward scheduling establishes a project start date and then schedules forward from that date. Based on the planned duration of required tasks, their interdependencies, and the allocation of resources to complete those tasks, a projected project completion date is calculated.

Reverse scheduling establishes a project deadline and then schedules backward from that date. Tasks, their duration, interdependencies, and resources must be considered to ensure that the project can be completed by the deadline.

FIGURE 4.7 *Entering Intertask Dependencies in Microsoft* Project

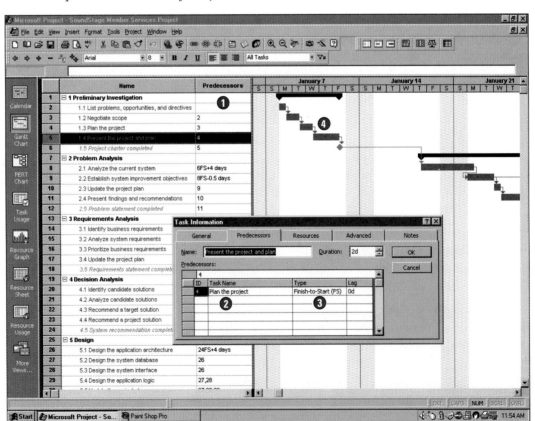

Each task can be given its own start and finish date. Like most project management tools, Microsoft *Project* actually builds the schedule for you as you enter the task durations and intertask dependencies (predecessors). On the Gantt chart, the task bars are expanded to reflect duration and shifted left and right to reflect start and end dates. Microsoft *Project* can also produce a traditional calendar view of the final schedule as shown in Figure 4.8.

The previous steps resulted in "a" schedule, but not "the" schedule! We have yet to consider the allocation of resources to the project. Resources include the following:

Activity 5—Assign Resources

— *People*—inclusive of all the system owners, users, analysts, designers, builders, external agents, and clerical help that will be involved in the project in any way.

— *Services*—a service such as a quality review that may be charged on a per use basis.

— *Facilities and equipment*—inclusive of all rooms and technology that will be needed to complete the project.

— *Supplies and materials*—everything from pencils, paper, notebooks, toner cartridges, and so on.

— *Money*—a translation of all of the above into the language of accounting— budgeted dollars!

The availability of resources, especially people and facilities, can significantly alter the project schedule.

FIGURE 4.8   *The Project Schedule in Calendar View*

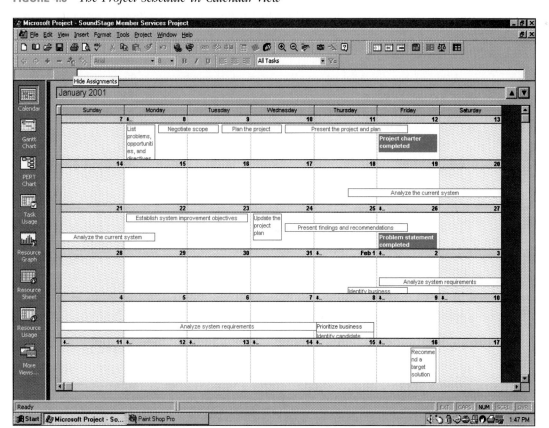

✓

Representative Roles in a Project

Auditor
Business analyst
Business subject matter expert
Database administrator
Executive sponsor
Information systems manager
JAD facilitator
JAD scribe
Management sponsor
Network administrator
Programmer
Project manager
System modeler

Most system development methodologies identify *people* resources required for each task in the form of roles. A role is not the same as a job title. Think of a role as a "hat" that someone wears because they possess a certain skill(s). Any given individual may be capable of wearing many hats (thus playing many roles). Also, many people may possess the skills required to play a given role. The project manager's task is to either assign specific people to fill roles or to gain commitments from management to provide people to fill roles. Representative roles from the *FAST* methodology are listed in the margin.

In Microsoft *Project,* roles and assignments are specified in the *Resource Sheet* view as shown in see Figure 4.9(a). Predefined roles and resources may be available in the chosen methodology and route templates.

❶ The project manager enters the names or titles of people [roles] in the *resource name* column. Resources may also include specific services, facilities, equipment, supplies, materials, and so forth.

❷ Notice that *Project* provides a column for establishing what percentage of a resource will be allocated to the project. For example, a database administrator might be allocated one-quarter time (25 percent) to a project. Allocations greater than 100 percent indicate a need for more than one person to fill a given role in the project. For example, you would specify the need for the equivalent of 2½ full-time programmers by setting *Max. Units* to 250 percent for that resource.

❸ *Project* also allows the project manager to estimate the cost of each resource. These costs can be estimated based on company history, consulting contracts, or internal cost accounting standards. Notice that both standard and overtime costs can be estimated. These costs are usually based on standards to protect information about anyone's actual salary.

❹ Each resource has a *calendar* that considers the standard workweek and holidays, as well as individual vacations and other commitments.

Given the resources, they now can be specifically assigned to tasks as shown in Figure 4.9(b). As resources are assigned to the tasks, the project manager would specify the units of that resource that will be required to complete each assigned task. (This may be a *percentage* of a person's time needed for that task.)

As these resources are formally assigned, the schedule will be adjusted (which happens automatically in tools such as *Project*). If you enter the cost of resources, tools such as Microsoft *Project* will also automatically calculate and maintain a budget based on the resources and schedule.

Assigning People to Tasks   Recruiting the "right" team members can make or break a project. The following are guidelines on selecting and recruiting the team.

— *Recruit talented, highly motivated people.* Highly skilled and motivated team members are more likely to overcome project obstacles unaided and are more likely to meet project deadlines and produce quality work.

— *Select the best task for each person.* All workers have strengths and weaknesses. Effective project managers learn to exploit the strengths of team members and avoid assigning tasks to team members not skilled in those tasks.

— *Promote team harmony.* Select team members who will complement and work well with each other.

— *Plan for the future.* Include junior personnel with potential to be mentored by project leaders. Junior personnel might not be as productive as the seasoned veterans, but you will need them and have to rely on them on future projects.

— *Keep the team size small.* By limiting the team size, communication overhead and difficulties will be reduced. A two-person team has only 1

(a)

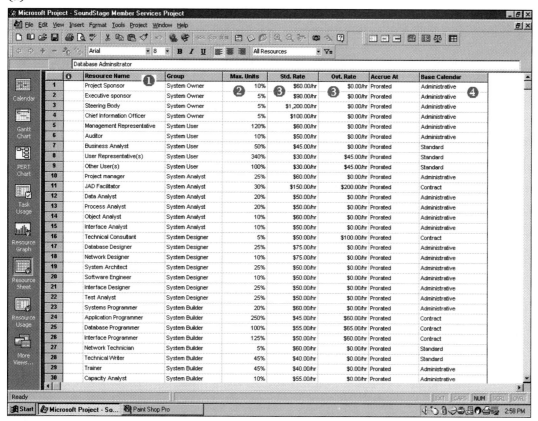

(b)

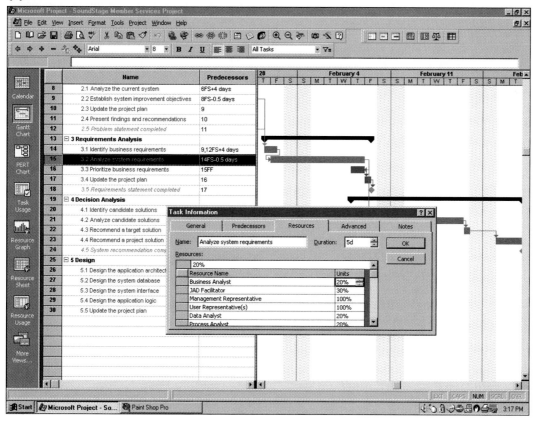

FIGURE 4.9 *Defining and Assigning Project Resources*

communication path. A four-person team has 6 communication paths and a 50-person team has at least 1,200 communication paths. The more communication paths there are, the greater the probability that there will be increased communication problems. By the same token the teams should be large enough to provide adequate backup and coverage in key skills if a team member is lost.

Resource Leveling   So far, we have identified tasks, task durations, and intertask dependencies and assigned resources to each task to produce the project schedule. It is common to overallocate resources when assigning resources to tasks. By overallocate we mean that at certain times in a project, we have assigned more resources than we have available. For example, during a specific period in the project (day, week, etc.), we may have assigned a specific person(s) to work on multiple tasks that add up to more hours than the person(s) has available to work during that period. This renders the overall schedule infeasible because the overallocated resource cannot reasonably complete all assigned tasks according to schedule. To correct this problem, project managers must use a technique called resource leveling.

> **Resource leveling** is a strategy used to correct resource overallocations by some combination of *delaying* or *splitting tasks*.

Let's briefly explain both approaches.

Delaying tasks is based on the concepts of *critical path* and *slack time*. When it comes to the project schedule, some tasks are more sensitive to schedule delays than others. For this reason, project managers must become aware of the critical path for a project.

> The **critical path** for a project is that sequence of dependent tasks that have the largest sum of *most likely durations*. The critical path determines the earliest possible completion date of the project. (We previously described how to estimate *most likely duration* for a task.)

These critical path tasks have no slack time available—thus, any delay in completion of any of the tasks on the critical path will cause an overall delay in the completion of the entire project. The opposite of a critical task is one that has some slack time.

> The **slack time** available for any noncritical task is the amount of delay that can be tolerated between the starting time and completion time of a task without causing a delay in the completion date of the entire project.

Tasks that have slack time can get behind schedule by an amount less than or equal to the slack time without having any impact on the project's final completion date. The availability of slack time in certain tasks gives us the opportunity to delay the start of the tasks to level resources while not affecting the project completion date. Of course, it may be necessary to delay a critical path task to level resources, unless you can split the task.

Splitting tasks involves breaking a task into multiple tasks to assign alternate resources to the tasks. Thus, a single task for which a resource was overallocated is now apportioned to two or more resources that are (presumably) not overallocated. Splitting tasks requires identifying and assigning new resources such as analysts, contractors, or consultants.

Resource leveling can be tedious to perform manually. For each resource, the project manager needs to know the total time available to the project for the resource, all task assignments made to the resource, and the sum of all durations of those task assignments over various time periods. All project management software tools, such as Microsoft *Project,* automatically determine critical paths and slack times. This enables those same software tools to track resource allocations and automatically perform resource leveling. It is extraordinarily rare for any modern project manager to manually level resource assignments.

Resource leveling will be an ongoing activity because the schedule and resource assignments are likely to change over the course of a project.

Schedule and Budget Given a schedule based on leveled resources and given the cost of each resource (e.g., cost per hour of a systems analyst or database administrator) the project manager can produce a printed (or Web-based) document that communicates the project plan to all concerned parties. Project management tools will provide multiple views of a project such as calendars, Gantt chart, PERT chart, resource and resource leveling reports, and budget reports. All that remains is to direct resources to the completion of project tasks and deliverables.

Communication The statement of work, timetable for major deliverables, and overall project schedule should be communicated to all parties interested in the project. This communication should also include a plan for reporting progress, both orally and in writing, the frequency of those communications, and a contact person and method for parties to submit feedback and suggestions. A corporate intranet can be an effective way to keep everyone informed of project progress and issues.

All the preceding project management activities led to a master plan for the project. It's now time to execute that plan. There are several dimensions to directing the team effort. Tom Demarco states in his book *The Deadline: A Novel About Project Management,* the hardest job in management is people.

Few new project managers are skilled supervisors of people. Most learn supervision through their own experiences as subordinates—things they liked and disliked about those who supervised them. Yet there are many references for learning about the best practices in supervision. This topic could easily require an entire chapter. In the margin checklist, we provide a classic list of project supervision recommendations from *The People Side of Systems* by Keith London.

As noted by McLeod and Smith, "Individuals brought together in a systems development team do not form a close-knit unit immediately." They explain that teams go through stages of team development as shown in Figure 4.10.

We also direct you to a couple of the shortest and most valuable books ever written on the subject of supervision. You could easily read both books overnight!

The One Minute Manager by Kenneth Blanchard and Spencer Johnson is a classic, fun, and indispensable aid to anyone managing people for the first time. In just over 100 pages, the authors share the simple secrets of managing people and achieving success through the actions of subordinates. The book highlights three basic secrets:

— One-minute goal setting.
— One-minute praisings.
— One-minute reprimands.

This book should be part of every college graduate's personal library!

Most young and many experienced managers have difficulty with the subtle arts of delegation and accountability. Worse still, they let subordinates reverse-delegate tasks back to the manager. This leads to poor time management and manager frustration. In *The One Minute Manager Meets the Monkey,* Kenneth Blanchard teams with William Oncken and Hal Burrows to help managers overcome this problem.

The solution is based on Oncken's classic principle of "The care and feeding of monkeys." Monkeys are "problems" that managers delegate to their subordi-

Activity 6—Direct the Team Effort

——— ✓ ———

10 Hints for Project Leadership

Be consistent
Provide support
Don't make promises you can't keep
Praise in public, criticize in private
Be aware of morale danger points
Set realistic deadlines
Set perceivable targets
Explain and show, rather than do
Don't rely just on [status reports]
Encourage a good team spirit

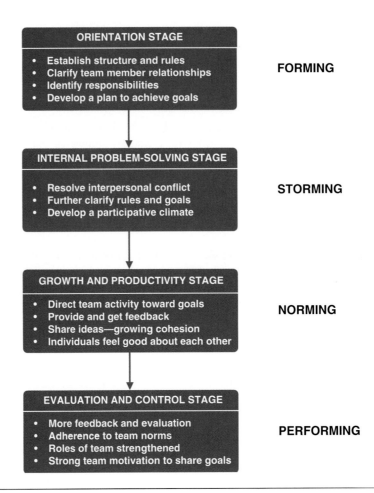

FIGURE 4.10

Stages of Team Maturity

Source: Adapted from Graham McLeod and Derek Smith, *Managing Information Technology Projects* (Cambridge, MA: Course Technology, 1996).

nates who, in turn, attempt to reverse-delegate back to the manager. In this 125-page book the authors teach managers how to keep the monkeys on the subordinates' backs. Doing so increases the manager's available work time, accelerates task accomplishment by subordinates, and teaches subordinates how to take responsibility and solve their own problems.

Activity 7—Monitor and Control Progress

While executing the project, the project manager must control the project, that is, monitor its progress against the scope, schedule, and budget. The manager must report progress and, when necessary, adjust scope, schedule, and resources.

Progress Reporting   One of the project manager's responsibilities is to report progress to superiors. Progress reporting should be frequent enough to establish accountability and control, but not so frequent as to become a burden and impediment to real project progress. For example, Keane, Inc., a consulting firm, recommends that progress reports or meetings occur every two weeks—consistent with the firm's project planning strategy that decomposes projects into tasks that produce deliverables that require no more than 80 work hours.

Project progress reports can be verbal or written. Figure 4.11 illustrates a template for a written progress report. Project progress reports (or presentations) should be honest and accurate, even if the news is less than good. Project progress reports should report successes but should clearly identify problems and concerns such that they can be addressed before they escalate unto major issues or catastrophes.

As tasks are completed, progress can be recorded in Microsoft *Project* (see Figure 4.12). We call your attention to the following Gantt progress items:

❶ All the tasks in the preliminary investigation phase are complete as indicated by the yellow lines that run the full length of each task bar. Notice that because all these tasks are complete, they are no longer critical—the bars have changed from red to blue.

❷ In the problem analysis phase, only the first task, "Analyze the current system," is 100 percent complete.

❸ Notice that the "Establish system improvement objectives" task bar has a partial yellow line running 60 percent of its length. This indicates the task is about 60 percent complete. The task bar is still red because any delay in completing the task will still threaten the project completion date.

❹ All remaining tasks shown in the displayed chart have not been started. Actual progress will be recorded when the task is started, in process, or completed.

❺ Progress for any given task is recorded in the task information dialogue box for that task. In this example, the project manager is recording 10 percent completion of the named task.

Visit the McGraw-Hill Online Learning Center for this textbook and browse the online resource for both samples of a project status report for the SoundStage case study as well as Microsoft Word and PowerPoint templates and instructions you can use to document your own project progress based on various routes through the FAST methodology.

FIGURE 4.11
Outline for a Progress Report

PROJECT PROGRESS REPORT

I.   Cover page
    A. Project name or identification
    B. Project manager
    C. Date of report
II.   Summary of progress
    A. Schedule analysis
    B. Budget analysis
    C. Scope analysis
        (describe any changes that may have an impact on future progress)
    D. Process analysis
        (describe any problems encountered with strategy or methodology)
    E. Gantt progress chart(s)
III. Activity analysis
    A. Tasks completed since last report
    B. Current tasks and deliverables
    C. Short-term future tasks and deliverables
IV.   Previous problems and issues
    A. Action item and status
    B. New or revised action items
        1. Recommendation
        2. Assignment of responsibility
        3. Deadline
V.   New problems and issues
    A. Problems
        (actual or anticipated)
    B. Issues
        (actual or anticipated)
    C. Possible solutions
        1. Recommendation
        2. Assignment of responsibility
        3. Deadline
VI.   Attachments
    (include relevant printouts from project management software)

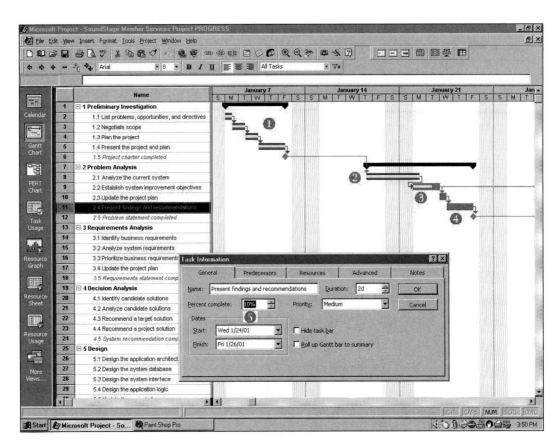

FIGURE 4.12   *Progress Reporting on a Gantt Chart*

Microsoft *Project* also provides a number of preconfigured and customizable reports that can present useful project status information.

Change Management   It is not uncommon for scope to grow out of control even when a properly completed statement of work was agreed on early in the planning process. We refer to scope growth as "change." As noted by Keane, Inc., "Change is frequently a point of contention between the customer and the information systems organization, because they disagree on whether a particular function is a change or a part of the initial agreement." The inevitability of scope change necessitates that we have a formal strategy and process to deal with change and its impact on schedule and budget. **Change management** is intended to protect the project manager and team from being held accountable for schedule and budget overruns that were driven by changes in scope.

Changes can be the result of various events and factors including:

— An omission in defining initial scope (as documented in the statement of work).

— A misunderstanding of the initial scope (the desired product is more complicated than originally communicated or perceived).

— An external event such as government regulations that create new requirements.

— Organizational changes, such as mergers, acquisitions, and partnerships, that create new business problems and opportunities (not to mention "players").

— Availability of better technology.

- Shifts in planned technology that force unexpected and significant changes to the business organization, culture, and/or processes.
- The users or management simply wanting the system to do more than they originally requested or agreed to.
- Management reducing the funding for the project or imposing an earlier deadline.

A change management system includes a collection of procedures to document a change request and define the steps necessary to consider the change based on the expected impact of the change. Most change management systems require that a change request form be initiated by one or more project stakeholders (e.g., system owners, users, analysts, designers, or builders). Ideally, these change requests are considered by a *change control board* (CCB) that is responsible for approving or rejecting all change requests. The CCB's composition typically includes members of the project team as well as outsiders who may have an interest or stake in the project. The CCB's decision should be based on impact analysis.

Feasibility impact analysis should assess the importance of the change to the business, the impact of the change on the project schedule, and the impact of the change on the project budget and long-term operating costs.

Ultimately, change management boils down to managing the expectations of the customer. In the next section, we introduce a simple but conceptually sound framework for managing expectations and their impact on project schedule and budget.

Expectations Management Experienced project managers often complain that managing system owners' and users' expectations of a project is more difficult than managing cost, schedule, people, or quality. In this section we introduce a simple tool that we'll call an *expectations management matrix* that can help project managers deal with this problem. We first learned about this tool from Dr. Phil Friedlander, a consultant and trainer then with McDonnell Douglas. He attributes the matrix to "folklore" but also credits Jerry Gordon of Majer, and Ron Leflour, a project management educator/trainer. Dr. Friedlander's paper is listed in the Suggested Readings for this chapter. We have slightly adapted the tool for this presentation.

Every project has goals and constraints when it comes to cost, schedule, scope, and quality. In an ideal world, you could optimize each of these parameters. Management often has that expectation. Reality, however, suggests that you can't optimize them all—you must strike a balance that is both feasible and acceptable to management. That is the purpose of the expectations management matrix.

An **expectations management matrix** is a rule-driven tool for helping management understand the dynamics and impact of changing project parameters such as cost, schedule, scope, and quality.

The basic matrix, shown in Figure 4.13, consists of three rows and three columns (plus headings). The rows correspond to the measures of success in any project: cost, schedule, and scope and/or quality. The columns correspond to priorities: first, second, and third. To establish expectations, we assign names to the priorities as follows:

- *Maximize or minimize*—The measure of success that is determined to be the most important for a given project.
- *Constrain*—The second most important of the three measures of success in a project.
- *Accept*—The least important of the three measures in a project.

Most managers would ideally like to give equal priority to all three measures;

PRIORITIES → ↓MEASURES OF SUCCESS	Max or Min	Constrain	Accept
Cost			
Schedule			
Scope and/or Quality			

FIGURE 4.13
*A Management
Expectations Matrix*

experience suggests that the three measures tend to balance themselves naturally. For example, if you increase scope or quality requirements, it will take more time and/or money. If you try to get any job done faster, you generally have to reduce scope or quality requirements, or pay more money to compensate. The management expectations matrix helps (or forces) management to understand this through three simple rules:

— For any project, you must record three Xs within the nine available cells.
— No row may contain more than one X. In other words, a single measure of success must have one and only one priority.
— No column may contain more than one X. In other words, there must be a first, second, and third priority.

Let's illustrate the tool using Dr. Friedlander's own example. In 1961 President John F. Kennedy established a major project—land a man on the moon and return him safely before the end of the decade. Figure 4.14 shows the realistic expectations of the project. Let's walk through the example.

— The system owner (the public) had both scope and quality expectations. The scope (or requirement) was to successfully land a man on the moon. The quality measure was to return the man (or men) safely. Because the public would expect no less from the new space program, this had to be made the first priority. In other words, we had to maximize safety and minimize risk as a first priority. Hence, we record the X in column 1, row 3.
— At the time of the project's inception, the Soviet Union was ahead in the race to space. This was a matter of national pride; therefore, the second priority was to get the job done by the end of the decade. We call this the project constraint—there is no need to rush the deadline, but we don't want to miss the deadline. Thus, we record the second X in column 2, row 2.
— By default, the third priority had to be cost (estimated at $20 billion in 1961). By making cost the third priority, we are not stating that cost will not be controlled. We are merely stating that we may have to accept cost overruns to achieve the scope and quality requirement by the constrained deadline.

History records that we achieved the scope and quality requirement, and did so in 1969. The project actually cost well in excess of $30 billion, a 50 percent cost overrun. Did that make the project a failure? On the contrary, most people perceived the project a grand success. The government managed the public's expectations of the project in realizing that maximum safety and minimum risk, plus meeting the deadline (beating the Soviets!), was an acceptable trade-off for the cost overrun! The government brilliantly managed public opinion. Systems development project managers can learn a valuable lesson from this balancing act.

At the beginning of any project, the project manager should consider introducing the system owner to the expectations matrix concept and should work with the system owner to complete the matrix. For most projects, it would be difficult to record all the scope and quality requirements in the matrix. Instead, they would be listed in the statement of work. The estimated costs and deadlines could be recorded directly in the matrix.

The project manager doesn't establish the priorities; he or she merely enforces the rules of the matrix. This sounds easy, but it rarely is! Many managers are unwilling to be pinned down on the priorities—"Shouldn't we be able to maximize everything?" These managers need to be educated about the reason for the priorities. We always try to maximize all three measures of success because it makes us look that much better. But we need to know the priorities if we cannot maximize all three measures. This helps us make intelligent compromises instead of merely guessing right or wrong.

What if your system owner refuses to prioritize? The tool is less useful then, except as a mechanism for documenting your concerns before they become disasters. A system owner who refuses to set priorities is a manager who may be setting the project manager up for a no-win performance review. And as Dr. Friedlander points out, "Those who do not 'believe' the principles [of the matrix] will eventually 'know' the truth. You do not have to believe in gravity, but you will hit the ground just as hard as the person who does."

Let's assume you have a management expectations matrix that conforms to the aforementioned rules. How does this help you manage expectations? During the course of the average systems development project, priorities are not stable. Various factors such as the economy, government, and company politics can change the priorities. Budgets may become more or less constrained. Deadlines may become more or less important. Quality can become more important. And, most frequently, requirements increase. As already noted, these changing factors affect all the measures in some way. The trick is to manage expectations despite the ever-changing project parameters.

The technique is relatively straightforward. Whenever the "max/min measure"

PRIORITIES → / ↓MEASURES OF SUCCESS	Max or Min	Constrain	Accept
Cost • **$20 billion (estimated)**			**X**
Schedule • **Dec 31, 1969 (deadline)**		**X**	
Scope and/or Quality • **Land a man on the moon** • **Get him back safely**	**X**		

FIGURE 4.14

Management of Expectations for the Lunar Landing Project

or the "constrain measure" begins to slip, you have a potential expectations management problem. For example, suppose you are faced with the following priorities (see Figure 4.15):

— Explicit requirements and quality expectations were established at the start of a project and given the highest priority.

— An absolute maximum budget was established for the project.

— You agreed to shoot for the desired deadline, but the system owner(s) accepted the reality that if something must slip, it should be schedule.

Now suppose that during systems analysis, significant and unanticipated business problems are identified. The analysis of these problems has placed the project behind schedule. Furthermore, solving the new business problems substantially expands the user requirements for the new system. How do you, as project manager, react? First, don't overreact to the schedule slippage—schedule slippage was the "accept" priority in the matrix. The scope increase (in the form of several new requirements) is the more significant problem because the added requirements will increase the cost of the project. Cost is the constrained measure of success. As it stands, we have an expectations problem. It is time to review the matrix with the system owner.

First, the system owner needs to be made aware of which measure or measures are in jeopardy and why. Then together, the project manager and system owner can discuss courses of action. Several courses of action are possible:

— The resources (cost and/or schedule) can be reallocated. Perhaps the system owner can find more money somewhere. All priorities would remain the same (noting, of course, the revised deadline based on schedule slippages already encountered during systems analysis).

— The budget might be increased, but it would be offset by additional planned schedule slippages. For instance, by extending the project into a new fiscal year, additional money might be allocated without taking any money from existing projects or uses. This solution is shown in Figure 4.16.

— The user requirements (or quality) might be reduced through prioritizing those requirements and deferring some number of those requirements until Version 2 of the system. This alternative would be appropriate if the budget cannot be increased.

— Finally, measurement priorities can be changed.

FIGURE 4.15 *A Typical Initial Expectations Matrix*

PRIORITIES → ↓MEASURES OF SUCCESS	Max or Min	Constrain	Accept
Cost		X	
Schedule			X
Scope and/or Quality	X		

PRIORITIES → ↓MEASURES OF SUCCESS	Max or Min	Constrain	Accept
Cost • **Adjusted budget**		**X+** Increase budget	
Schedule • **Adjusted deadline**			**X-** Extend deadline
Scope and/or Quality • **Adjusted scope**	**X+** Accept expanded requirements		

FIGURE 4.16
*Adjusting Expectations
(a sample)*

Only the system owner may initiate priority changes. For example, the system owner may agree that the expanded requirements are worth the additional cost. He or she allocates sufficient funds to cover the requirements but migrates priorities such that minimizing cost becomes the highest priority (see Figure 4.17, Step 1). But now the matrix violates a rule—there are two *X's* in column 1. To compensate, we must migrate the scope and/or quality criterion to another column, in this case, the constrain column (see Figure 4.17, Step 2). Expectations have been adjusted. In effect, the system owner is freezing growth of requirements and still accepting schedule slippage.

There are three final comments about priority changes. First, priorities may change more than once during a project. Expectations can be managed through any number of changes as long as the matrix is balanced (meaning it conforms to our rules). Second, expectation management can be achieved through any combination of priority changes and resource adjustments. Finally, system owners can initiate priority changes even if the project is on schedule. For example, government regulation might force an uncompromising deadline on an existing project. That would suddenly migrate our "accept" schedule slippages to "max constrain." The other *X's* would have to be migrated to rebalance the matrix.

The expectations management matrix is a simple tool, but sometimes the simple tools are the most effective!

Schedule Adjustments—Critical Path Analysis   When it comes to the project schedule, some tasks are more sensitive to schedule delays than others. For this reason, project managers must become aware of the critical path and slack times for a project. Let's review the definitions of critical path and slack time.

> The **critical path** for a project is that sequence of dependent tasks that have the largest sum of *most likely durations*. The critical path determines the earliest completion date of the project. (Recall that we previously described how to estimate *most likely duration* for a task.)

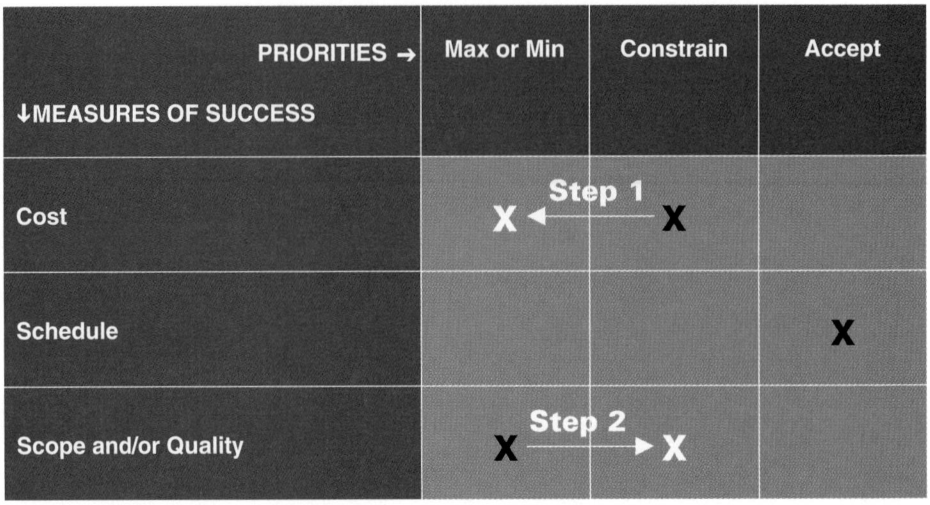

PRIORITIES → ↓MEASURES OF SUCCESS	Max or Min	Constrain	Accept
Cost	X ← **Step 1** X		
Schedule			X
Scope and/or Quality	X **Step 2** → X		

FIGURE 4.17
Changing Priorities

These tasks have no slack time available; thus, any delay in completion of any of the tasks on the critical path will cause a delay in the completion of the entire project.

The opposite of a critical task is one that has some slack time.

> The **slack time** available for any noncritical task is the amount of delay that can be tolerated between the starting time and completion time of a task without causing a delay in the completion date of the entire project.

Tasks that have slack time can get behind schedule by an amount less than or equal to the slack time without having any impact on the project's final completion date.

Understanding the critical path and slack time in a project are indispensable to the project manager. Knowledge of such project factors influences the people management decisions to be made by the project manager. Emphasis can and should be placed on the critical path tasks, and if necessary, resources might be temporarily diverted from tasks with slack time to help get critical tasks back on schedule.

The critical path and slack time for a project can be depicted on both Gantt and PERT charts; however, PERT charts are generally preferred because they more clearly depict intertask dependencies that define the critical path. Most project management software, including Microsoft *Project,* automatically calculates and highlights the critical path based on intertask dependencies combined with durations. It is useful, however, to understand how the critical path and slack times are calculated.

Consider the following hypothetical example. A project consists of eight primitive tasks shown in Figure 4.18. The most likely duration (in days) for each task is recorded. There are four distinct sequences of tasks in a project. They are:

Path 1:	A → B → C → D → I
Path 2:	A → B → C → E → I
Path 3:	A → B → C → F → G → I
Path 4:	A → B → C → F → H → I

The total of most likely duration times for each path is calculated as follows:

Path 1:	3 + 2 + 2 + 7 + 5 = 19
Path 2:	3 + 2 + 2 + 6 + 5 = 18
Path 3:	3 + 2 + 2 + 3 + 2 + 5 = 17
Path 4:	3 + 2 + 2 + 3 + 1 + 5 = 16

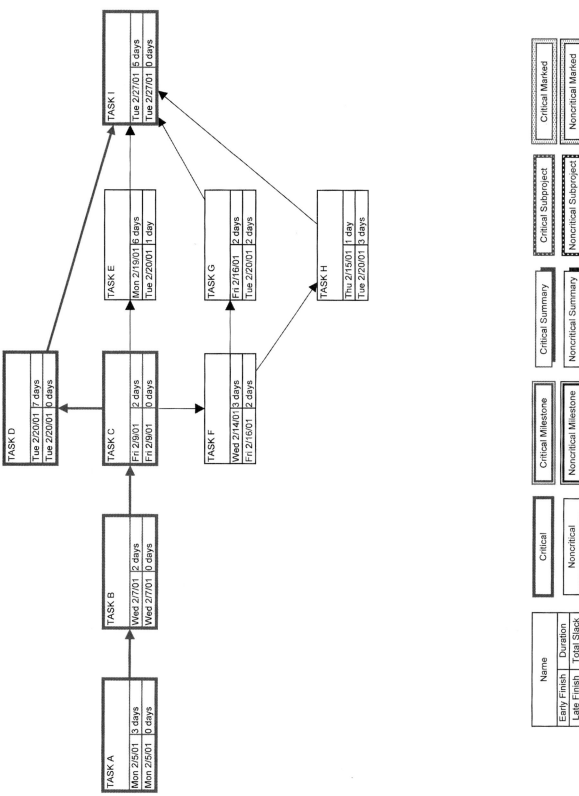

FIGURE 4.18 *Critical Path Analysis*

WHERE DO YOU GO FROM HERE?

Where you go from here depends on where you are coming from and where you want to go. If you are reading through the chapters sequentially, you should probably move on to **Chapter 5, Systems Analysis,** to expand on your understanding of systems analysis tasks, tools, and techniques. Alternatively, if you are enrolled in a system design-focused course, you might skip ahead to either **Chapter 9, Feasibility and the System Proposal** (which marks the end of systems analysis), or to **Chapter 10, Systems Design** (which provides an in-depth look at the activities of system design, prototyping, and rapid application development).

Some instructors have deferred this project management chapter to the end of your course. If so, you may be interested in expanding your knowledge of project management tools, techniques, and methods. Some schools offer a project management course. If not, you may find that your systems analysis and design instructor might supervise you to complete an independent study course on the subject. If so, we direct you to two specific references at the end of this chapter as possible texts: (1) the Wysocki et al. book is well organized around the Project Management Body of Knowledge that we presented in our chapter, and (2) the McLeod and Smith book is especially comprehensive in its coverage of project management dimensions that we could not cover fully in our chapter.

In this example, path 1 is the *critical path* at 19 days. (Note: You can have multiple critical paths if they have the same total duration.)

In this example, tasks E, F, and G are not on the critical path; they each have some slack time. For example, task E is included in a path that has one day less duration than the critical path; therefore, task E can get behind by as much as one day without adversely affecting the project completion date. Similarly, tasks F and G can *combine* for a maximum slack of two days without delaying the entire schedule.

In Figure 4.18, the critical path is shown in red. The tasks that have slack capacity are shown in black. Similarly, project management software also uses color to differentiate critical path tasks in a Gantt or PERT chart.

Activity 8—Assess Project Results and Experiences

Learn from your mistakes! Embrace continuous process improvement! This final activity involves soliciting feedback from project team members (including customers) concerning their project experiences and suggestions aimed at improving the project and process management of the organization. Project review(s) should be conducted to answer the following fundamental questions:

- Did the final product meet or exceed user expectations?
- Did the project come in on schedule?
- Did the project come in under budget?

The answers to these questions should be followed up with the basic question "why or why not?" Subsequently, and based on the responses to the above questions, changes should be made to improve the system development and project management methods that will be used on future projects. Suggestions for improvements are communicated to the "Centers for Excellence," which can modify standards and processes, as well as share useful ideas and experiences with other project teams that may solicit their help or expertise. Project assessments often contribute improvements to specific project deliverables (milestones), processes or tasks that created the deliverables, and the overall management of the project.

SUMMARY

1. A project is a (temporary) sequence of unique, complex, and connected activities having one goal or purpose and that must be completed by a specific time, within budget, and according to specification.

2. Project management is the process of scoping, planning, staffing, organizing, directing, and controlling the development of an acceptable system at a minimum cost within a specified time frame.

3. Process management is an ongoing activity that documents, manages the use of, and improves an organization's chosen methodology (the "process") for systems development. Process management is concerned with the activities, deliverables, and quality standards to be applied to *all* projects.

4. From a project management perspective, a project is considered a success if the resulting information system is acceptable to the customer, the system was delivered "on time" and "within budget," and the system development process had a minimal impact on ongoing business operations.

5. The Project Management Institute has created the "Project Management Body of Knowledge" (PMBOK) for the education and certification of professional project managers. It addresses:
   a. Project manager competencies.
   b. Project management functions, which include:
      i)   Scoping   v)   Organizing
      ii)   Planning   vi)   Directing
      iii) Estimating   vii) Controlling
      iv)   Scheduling   viii) Closing
   c. Tools and techniques such as:
      i) PERT charts, graphical network models that depict a project's tasks and the relationships between those tasks.
      ii) Gantt charts, simple horizontal bar charts that depict project tasks against a calendar.
   d. Project management software.

6. Project management is a cross life cycle activity; that is, project management tasks that overlap all the system development phases. A project management process is essential to achieving CMM Level 2 maturity.

7. Joint project planning (JPP) is a strategy wherein all stakeholders in a project participate in a one-to-three-day project management workshop, the result of which is consensus agreement on project scope, schedule, resources, and budget.

8. The tasks of project management include:
   a. Negotiate scope. Scope defines the boundaries of a project and is included in the statement of work, a narrative description of the work to be performed as part of a project.
   b. Identify tasks. A work breakdown structure (WBS) is a hierarchical decomposition of the project into its tasks and subtasks. Some tasks represent the completion of milestones or the completion of major deliverables during a project.
   c. Estimate task durations. There are many techniques and tools for estimating task durations.
   d. Specify intertask dependencies. The start or completion of individual tasks may be dependent on the start or completion of other tasks. These dependencies impact the completion of any project.
      i) Forward and reverse scheduling are two strategies for generating the preliminary schedule. The former works forward from a project start date, while the latter works backward from a project deadline.
   e. Assign resources. The following resources may impact a project schedule: People, services, facilities and equipment, supplies and materials, and money.
      i) Such resources must be assigned to tasks to develop a schedule.
      ii) Resource leveling is a strategy used to correct resource overallocations by some combination of *delaying* or *splitting tasks*. Resource leveling requires knowledge of:
         (1) The critical path—that sequence of dependent tasks that have the largest sum of *most likely durations*. The critical path determines the earliest possible completion date of the project.
         (2) Slack time—the amount of delay that can be tolerated between the starting time and completion time of a task without causing a delay in the completion date of the entire project.
   f. Direct the team effort. One of the most important dimensions of directing the team effort is the supervision of people.
   g. Monitor and control progress. During the project, the project manager must monitor project progress against the scope, schedule, and budget, and when necessary, make adjustments to scope, schedule, and resources.
      i)   Progress reporting is an essential control process that uses communication to keep a project within scope, on time, and within budget.
      ii) A complete project plan provides mechanisms and a process to manage requests for changes to scope. This is called change management.
      iii) Change management frequently requires a project manager to manage the expectations of user management and users themselves. An expectations management matrix is a rule-driven tool for helping management understand the dynamics and impact of changing project parameters such as cost, schedule, scope, and quality.
      iv) Schedule adjustments are required when a project's scope changes, or when other factors drive schedule or budget out of the projected range.
   h. Assess project results and experiences. This final activity involves soliciting feedback from project team members (including customers) concerning their project experiences and suggestions aimed at improving the project and process management of the organization.

KEY TERMS

change management, p. 146
closing, p. 128
controlling, p. 128
critical path, pp. 142,151
directing, 128
estimating, p. 128
expectations management matrix, p. 147
expected duration (ED), p. 137
feature creep, p. 125
forward scheduling, p. 138
Gantt chart, p. 128
joint project planning (JPP), p. 133

milestone, p. 136
most likely duration (D), p. 137
optimistic duration (OD), p. 136
organizing, p. 128
PERT chart, p. 128
pessimistic duration (PD), p. 136
planning, p. 128
process management, p. 125
project, p. 124
project management, p. 124
project management body of
   knowledge (PMBOK), p. 126

resource leveling, p. 142
reverse scheduling, p. 138
scheduling, p. 128
scope, p. 133
scope creep, p. 125
scoping, p. 128
slack time, pp. 142, 152
statement of work, p. 133
work breakdown structure (WBS),
   p. 135

REVIEW QUESTIONS

1. What is project management? What are the characteristics that define a project?
2. Differentiate between project and process management.
3. List four criteria for project success.
4. List at least five project mismanagement problems and their consequences.
5. What are scope creep and feature creep? How do they impact project success?
6. What is PMBOK and how has it contributed to the practice of project management?
7. Define eight functions performed by every project manager.
8. Compare and contrast Gantt and PERT charts.
9. What is joint project planning?
10. What role does a statement of work play in scope definition? Why is it important to project management?
11. What is a work breakdown structure? Why is it important to project planning?

12. Differentiate between a task and a milestone. Why don't milestones have duration?
13. List two factors that impact task duration and our ability to estimate task duration accurately.
14. List four types of intertask dependencies. Why are they important to project planning?
15. List five types of resources that can be assigned to a project and managed.
16. Differentiate between forward and reverse scheduling. When might each be more appropriate to a given project?
17. What is resource leveling and when must a project manager do it? Name two techniques for resource leveling.
18. What is the critical path for a project? What is slack time? How are they related?
19. What is change management? Why is it important?
20. What is expectations management and how does it impact project management?

PROBLEMS AND EXERCISES

1. Speculate as to the advantages and disadvantages of appointing project managers from the ranks of senior systems analysts versus appointing them from a pool of full-time project managers. What do you think of the idea of appointing project managers from the ranks of the user community?
2. Write a job advertisement for a professional project manager. Prepare a set of interview questions for the applicants.
3. How does a project management life cycle (or process) contribute to an organization's ability to achieve CMM Level 2 certification?
4. Explain to a nontechnical manager why process and project management are separate but complementary. Spec-

ulate as to the problems that might be caused by standardizing and applying either one without the other.
5. Based on a project (school or work) that you have mismanaged or seen to be mismanaged, what were some causes of the mismanagement that resulted in cost overruns, late delivery, or poor quality? Explain how these problems that result from mismanaged projects are related.
6. Give some examples that differentiate between scope and feature creep.
7. Which three of the project management activities do you think will be most difficult? Why? What can you do to make them easier by the time you will have to perform them?

8. You work for Taylor County Information Services. You have been assigned to a new project for the county sheriff's department. You have asked for a meeting to define a statement of work. The sheriff does not understand the need for such a document because in Taylor County all projects are initiated on a standard request for system services form. He asks why that form can't serve as the statement of work? Write a memo to explain to the sheriff how a more formal statement of work will benefit both his office and yours.

9. Your financial aid office must respond annually to changing federal regulations. Financial aid applications must be sent out annually by February 15 to allow for sufficient processing time. Therefore, each year the information systems unit must be ready to process new applications by March 1st. What type of project scheduling strategy must be used and why?

10. For a project that will last this entire academic term (at your school), define basic project calendars for faculty, staff, and/or students.

11. Why might different managers, users, analysts, and technical specialists have different calendars for the same project?

12. Using system development, give examples of summary and primitive tasks. Also give examples of milestones.

13. How can expectations mismanagement lead to *perceived* project failure? How can this be avoided?

14. Using the default formula presented in this chapter, calculate the most likely duration for the following tasks:

Task	Optimistic Duration	Pessimistic Duration	Expected Duration	Most Likely Duration?
A	3	6	4	?
B	1	3	2	?
C	4	7	6	?
D	2	5	3	?
E	3	9	6	?
F	3	3	4	?

15. Draw a PERT chart for the following tasks. The most likely durations have already been calculated for you. Using next Monday as the project start date, determine the most likely date by which the project can be completed. Show the critical path for the project. Construct a Gantt chart for the project.

Task	Duration	Predecessors	Critical Task?
A. Preliminary investigation	1	None	?
B. Problem analysis	2	A	?
C. Data requirements analysis	3	B	?
D. Process requirements analysis	6	C	?
E. Database design	2	C	?
F. Interface design	3	C, D, E	?
G. Process and program design	4	C, E, F	?
H. Programming and unit testing	7	E, F, G	?
I. System testing	2	H	?
J. Installation	2	I	?

16. Something has gone wrong in the above project. Interface design was based on the assumption of a Windows user interface, but now management has decided the system should also support a Web browser interface. You estimate that the Web screen will require an additional two days to design, prototype, and test. What does this do to the schedule and projected project completion date?

17. For each of the tasks listed below, determine the earliest and latest completion time, the slack time, and whether or not the task is on the critical path. Draw a PERT chart and highlight the critical path.

Task	Predecessors	Duration	Earliest Completion	Latest Completion	Critical Path?
A	None	1	?	?	?
B	None	3	?	?	?
C	A	3	?	?	?
D	B, C	5	?	?	?
E	C	1	?	?	?
F	D	4	?	?	?
G	E, F	3	?	?	?
H	G	2	?	?	?
I	G	2	?	?	?
J	H, I	5	?	?	?
K	J	6	?	?	?

18. Choose any one critical plus one noncritical task other than J or K from above. Double the duration and determine the overall schedule impact. What choices would you have to get the project back on schedule?

19. Consider your degree curriculum. Courses should be tasks. Degrees should be milestones.
    a. Prepare a work breakdown structure for the "project" of earning a degree(s).
    b. Prepare a PERT chart to reflect intertask dependencies and most likely durations. If possible, try to use class time instead of calendar time.
    c. Determine the critical path. Record slack time for each course in the task boxes.
    d. When do you hope to receive your degree? Based on that deadline, use reverse scheduling to determine if your deadline is feasible. Prepare a Gantt chart and record your progress as of today's date (including the *partial* completion of your current schedule of classes).

20. Select a programming assignment from one of your past or current courses.
    a. Convert the assignment into a statement of work.
    b. Prepare a work breakdown structure for the "project."
    c. Prepare a PERT chart to reflect intertask dependencies and estimate most likely durations.
    d. Determine the critical path. Record slack time for each course in the task boxes.
    e. Prepare a Gantt chart for the project.

21. At the beginning of a project, management decided that highest priority should be placed on meeting an absolute deadline for a project. Second highest priority is to be placed on living within the allocated project budget. Draw the expectations management matrix for this project.

22. During the project initiated in Exercise 21, things started going wrong. Creeping requirements set in. Both the deadline and budget are in jeopardy. Using the expectations management matrix, identify alternatives for adjusting the project. Who should make the decision?

PROJECTS AND RESEARCH

1. Research the project management certification process defined by the Project Management Institute. What are the minimum qualifications for certification? What must you do to get certified as a project manager? How can you best prepare for certification?

2. Research project management software in your local library. Present a report to management that identifies important selection criteria for selecting a project management software package.

3. Complete a tutorial for Microsoft *Project* (or its equivalent). Evaluate the package. Analyze the package's strengths and weaknesses.

4. Make an appointment to visit an information systems project manager (or systems analyst with project management experience). What techniques does he or she use to plan and control projects? Why? Is project management software used? If so, what does the project manager like and dislike about that software?

5. If your course requires a team project, volunteer to serve as project team manager to gain academic experience in project management. Try to apply the techniques taught in this book. This experience might be arranged as either extra credit or a separate independent study course.

6. If your systems course requires you to complete a real or simulated development project, prepare a management expectations matrix with your client (or your instructor acting as your client). During the semester, review the matrix with your client (or instructor) and make appropriate adjustments. At the end of the project, be prepared to defend your management of expectations as well as your progress. (Note: Your instructor may assign a subjective grade to your management of expectations. That may not seem fair; however, it is realistic. Project success is as much perceived as it is real.)

7. Discuss the project manager competencies described in Table 4.1 with a real project manager. Determine the relative importance of each competency as seen by that manager and the rationale. Do you agree or disagree?

8. The PERT and Gantt charts in this chapter were based on the "classic" or "standard" route in the *FAST* system development methodology. Chapter 3 presented alternative routes such as model-driven development, rapid application development, and commercial off-the-shelf software. Select one of these alternative routes and adjust the Gantt and PERT charts for that route.

MINICASES

1. Fun & Games, Inc. is a successful developer and manufacturer of board, electronic, and computer games. The company is headquartered in Cleveland, Ohio. Jan Lampert, Applications Development manager, has requested a meeting with Steven Beltman, systems analyst and project manager for a new distribution project recently placed into production.

   "Steven, I want to discuss the distribution project your team completed last month. Now that the system has been operational for a few weeks, we need to evaluate the performance of you and your team. Frankly, Steven, I'm a little disappointed."

   "Me too! I don't know what happened! We used the standard methodology and tools, but we still had problems."

   "You still have some, Steven. The production system isn't exactly getting rave reviews from either users or managers."

   Steven replies, " I know."

   Jan continues, "Well, I've talked to several of the analysts, programmers, and end-users on the project, and I've drawn a few conclusions. Obviously, the end-users are less than satisfied with the system. You took some shortcuts in the methodology, didn't you?"

   "We had to, Jan! We got behind schedule. We didn't have time to follow the methodology to the letter."

   Jan explains, "But now we have to do major parts of the system over. If you didn't have time to do it right, where will you find time to do it over? You see, Steven,

systems development is more than tools, techniques, and methodologies. It's also a management process. In addition to your missing the boat on end-user requirements, I note two other problems. And both of them are management problems. The system was over budget and late. The projected budget of $35,000 was exceeded by 42 percent. The project was delivered 13 weeks behind schedule. Most of the delays and cost overruns occurred during programming. The programmers tell me that the delays were caused by rework of analysis and design specifications. Is this true?"

   Steven answers, "Yes, for the most part."

   Jan continues, "Once again, those delays were probably caused by the shortcuts taken earlier. The shortcuts you took during analysis and design were intended to get you back on schedule. Instead, they got you further behind schedule when you got into the programming phase."

   "Not all the problems were due to shortcuts," says Steven. "The users' expectations of the system changed over the course of the project."

   "What do you mean?" asks Jan.

   Steven answers, "The initial list of general requirements was one-page long. Many of those requirements were expanded and supplemented by the users during the analysis and design phases."

   Jan interrupts, "The old 'creeping requirements syndrome.' How did you manage that problem?"

   Steven replies, "Manage it? Aren't we supposed to simply give in? If they want it, you give it to them."

"Yes," answers Jan, "but were the implications of the creeping requirements discussed with project's management sponsor?"

Steven answers, "Not really! I don't recall any schedule or budget adjustments. We should explain that to them now."

"An excuse?" inquires Jan.

Steven replies, " I guess that's not such a good idea. But the project grew. How would you have dealt with the schedule slippage during analysis?"

Jan answers, " If I were you, I would have reevaluated the scope of the project when I first saw it changing. In this case, either project scope should have been reduced or project resources—schedule and budget—should have been increased. [pause] Don't be so glum! We all make mistakes. I had this very conversation with my boss seven years ago. You're going to be a good project manager. That's why I've decided to send you to this project management course and workshop."

a. What did Steven do wrong? How would you have done it differently?

b. Should Jan share any fault for the problems encountered in this project?

c. Why would it be a mistake to use creeping requirements as an excuse for the project mismanagement?

SUGGESTED READINGS

Blanchard, Kenneth, and Spencer Johnson. *The One Minute Manager.* New York: Berkley Publishing Group, 1981, 1982. Arguably, this is one of the best people management books ever written. Available in most bookstores, it can be read overnight and used for discussion material for the lighter side of project management (or any kind of management). This is must reading for all college students with management aspirations.

Blanchard, Kenneth; William Oncken, Jr.; and Hal Burrows. *The One Minute Manager Meets the Monkey.* New York: Simon & Schuster, 1988. A sequel to *The One Minute Manager,* this book effectively looks at the topic of delegation and time management. The monkey refers to Oncken's classic article, ``Managing Management Time: Who's Got the Monkey?'' as printed in the *Harvard Business Review* in 1974. The book teaches managers how to achieve results by helping their staff (their monkeys) solve their own problems.

Brooks, Fred. *The Mythical Man-Month.* Reading, MA.: Addison-Wesley, 1975. A classic set of essays on software engineering (also known as systems analysis, design, and implementation). Emphasis is on managing complex projects.

Catapult, Inc. *Microsoft Project 98: Step by Step.* Redmond, WA: Microsoft Press, 1997. An update for *Project 2000* is expected.

Demarco, Tom. *The Deadline: A Novel About Project Management.* New York: Dorset House Publishing, 1997. This would be an excellent companion to a project management text, especially for a graduate-level course. It demonstrates the "good, bad, and ugly" of project management, told as a story.

Duncan, William R., Director, and Standards Committee. *A Guide to the Project Management Body of Knowledge.* Upper Darby, PA: Project Management Institute, 1996. This is a concise overview of the generally accepted project management body of knowledge and practices used for certification of project managers.

Friedlander, Phillip. ``Ensuring Software Project Success with Project Buyers,'' *Software Engineering Tools, Techniques,* *and Practices 2,* no. 6 (March/April 1992), pp. 26–29. We adapted our expectations management matrix from Dr. Friedlander's work.

Kernzer, Harold. *Project Management: A Systems Approach to Planning, Scheduling, and Controlling,* 4th ed. New York: Van Nostrand Reinhold, 1989. Many experts consider this book to be the definitive work in the field of project management. Dr. Kernzer's seminars and courses on the subject are renowned.

London, Keith. *The People Side of Systems.* New York: McGraw-Hill, 1976. This is a timeless classic about various people aspects of systems work. Chapter 8, "Handling a Project Team," does an excellent job of teaching the leadership aspects of project management.

McLeod, Graham, and Derek Smith. *Managing Information Technology Projects.* Cambridge, MA: Course Technology, 1996. If you are looking for a good academic book for a course or independent study project to expand your knowledge of IT project and process management, this is it! The book provides a comprehensive treatment of virtually all dimensions of IT project and process management.

Roetzheim, William H., and Reyna A. Beasley. *Software Project Cost & Schedule Estimating: Best Practices.* Upper Saddle River, NJ: Prentice Hall, 1998. This is one of the more complete books on the subject of estimating techniques. Better still, the book includes evaluation copies of *Cost•Xpert, Risk•Xpert,* and *Strategy•Xpert* (™ of Marotz, Inc.), software tools for estimating.

Wysocki, Robert K., Robert Beck, Jr., and David B. Crane. *Effective Project Management: How to Plan, Manage, and Deliver Projects on Time and within Budget.* New York: John Wiley & Sons, 1995. Buy this book! This is our new benchmark for introducing project management. It is easy to read and worth its weight in gold. We were surprised how compatible the book is with past editions of our book, and our project management directions continue to be influenced by this work.

SYSTEMS ANALYSIS METHODS

The five chapters in Part Two introduce you to systems analysis activities and methods. Chapter 5, Systems Analysis, provides the context for all the subsequent chapters by introducing the activities of *systems analysis*. Systems analysis is the most critical phase of a project. During systems analysis we learn about the existing business system, come to understand its problems, define objectives for improvement, and define the detailed business requirements that must be fulfilled by *any* subsequent technical solution. Clearly, any subsequent system design and implementation of a new system depends on the quality of the preceding systems analysis. Systems analysis is often shortchanged in a project because (1) many analysts are not skilled in the concepts and logical modeling techniques to be used, and (2) many analysts

do not understand the significant impact of those shortcuts. Chapter 5 introduces you to systems analysis and its overall importance in a project. Subsequent chapters teach you specific systems analysis skills with an emphasis on logical system modeling.

Chapter 6, Requirements Discovery, teaches various fact-finding techniques and strategies used to solicit user requirements for a new system. You will learn how to use Use Cases to document facts that are obtained through these methods.

In Chapter 7, Data Modeling and Analysis, we teach you *data modeling,* a technique for organizing and documenting the stored data requirements for a system. You will learn to draw entity relationship diagrams as a tool for structuring business data that will eventually be designed as a database. These models

will capture the business associations and rules that must govern the data.

Chapter 8, Process Modeling, introduces *process modeling*. It explains how data flow diagrams can be used to depict the essential business processes in a system, the flow of data through a system, and policies and procedures to be implemented by processes. If you've done any programming, you recognize the importance of understanding the business processes for which you are trying to write the programs.

Chapter 9, Feasibility and the System Proposal, teaches you how to brainstorm possible system solutions, analyze those solutions for feasibility, select the best overall solution, and then present your recommendation in the form of a written and oral proposal to management.

Focus on PEOPLE

Focus on DATA

Focus on PROCESSES

Focus on INTERFACES

Focus on DEVELOPMENT

Stakeholders

Activities

S Y S T E M S   A N A L Y S T S

SYSTEM OWNERS

SYSTEM USERS

SYSTEM DESIGNERS

SYSTEM BUILDERS

VENDORS AND CONSULTANTS

BUILDING BLOCKS OF AN INFORMATION SYSTEM

Management Expectations

The PIECES Framework

Performance ● Information ● Economics ● Control ● Efficiency ● Service

List of business entities and rules ...

Business Knowledge

List of business functions and events ...

Business Functions

List of business locations and systems...

Business Locations

Data Requirements

Process Requirements

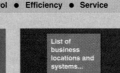

Interface Requirements

Database Schema

Application Schema & Specs

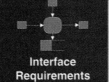

Interface Specifications

```
CREATE TABLE tblOrders
  colOrderNot CHAR(5) NOT
  NULL
  colOrderDate  DATE/TIME
  NOT
```

Database Programs

```
PROC ValidateOrder
  PERFORM ValidateCust
  REPEAT UNTIL
    NoMoreProd ...
```

Application Programs

```
<html>
<head>
<title> Order Entry Form </title>
...
```

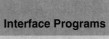

Interface Programs

PROJECT & PROCESS MANAGEMENT

PRELIMINARY INVESTIGATION

PROBLEM ANALYSIS

REQUIREMENTS ANALYSIS

DECISION ANALYSIS

DESIGN

CONSTRUCTION

IMPLEMENTATION

OPERATIONS AND SUPPORT

INFORMATION TECHNOLOGY & ARCHITECTURE

Database Technology ● Process Technology ● Interface Technology ● Network Technology

5

SYSTEMS ANALYSIS

CHAPTER PREVIEW AND OBJECTIVES

In this chapter you will learn more about the systems analysis phases in a systems development project: the preliminary investigation, problem analysis, requirements analysis, and decision analysis phases. The first three phases are collectively referred to as systems analysis. The latter phase provides transition between systems analysis and systems design. You will know that you understand the process of systems analysis when you can:

— Define systems analysis and relate the term to the preliminary investigation, problem analysis, requirements analysis, and decision analysis phases of the systems development methodology.

— Describe a number of systems analysis approaches for solving business system problems.

— Describe the preliminary investigation, problem analysis, requirements analysis, and decision analysis phases in terms of your information system building blocks.

— Describe the preliminary investigation, problem analysis, requirements analysis, and decision analysis phases in terms of purpose, participants, inputs, outputs, techniques, and steps.

— Identify those chapters and modules in this textbook that can help you learn specific systems analysis tools and techniques.

Although some of the tools and techniques of systems analysis are introduced in this chapter, it is *not* the intent of this chapter to teach those tools and techniques. This chapter teaches only the *process* of systems analysis. The tools and techniques will be taught in the subsequent five chapters.

SOUNDSTAGE

SOUNDSTAGE ENTERTAINMENT CLUB

SCENE

The SoundStage project has been launched and a preliminary investigation has been completed (Chapter 2). This episode begins shortly after the project's executive sponsor (the person who pays for the system to be built) has approved the initial feasibility assessment and scope and has authorized a detailed systems analysis. We join Sandra, Bob, and Terri Hitchcock (the business analyst assigned to this project) as they plot strategy in Sandra's office.

SANDRA

Good morning! I asked you here so Bob and I could review our strategy for systems analysis. Hopefully, this won't take long.

TERRI

As you know, I was just appointed to this business analyst position, and I'm not that familiar with your methods and terminology yet. I'm kind of looking forward to working for the next year or so in Information System Services. I have to admit, I avoided most of the computer courses when I was in school. To be honest, I thought you were all nerds—no offense!

BOB

None taken. Can't live with us; can't live without us, no?

TERRI

I guess so! Bob, your e-mail message said we are starting the systems analysis phase?

BOB

Actually, we've already started. The preliminary investigation was part of the systems analysis, but it was intentionally completed in two days—quick and dirty, so to speak.

SANDRA

But in our methodology, systems analysis consists of four phases. The ultimate goal of systems analysis is to produce a system proposal to improve the member services information system. That proposal will trigger the design, construction, and implementation of the proposed system.

TERRI

That's funny! I thought the system proposal was already done. Didn't I just read a memo or report that said we are going to produce a Web-based information system to streamline member services and order processing? That sounds like a proposal to me.

SANDRA

Not entirely! It is true that most projects begin with someone's vision for a new and improved system. The document to which you referred came from a strategic planning project. It outlined a number of perceived problems, some opportunities and challenges, and established a vision for a Web-based member services system. But the analysis that led to that conclusion was abbreviated out of necessity. That planning project was looking at SoundStage as a whole. Clearly, the executive committee and team were aware of the emergence of the Internet, and they were appropriately looking at the competitive threats that such technology might pose for SoundStage. To their credit, we are already seeing the development of some of those threats. And that is why this project was given the highest priority of all the projects in the strategic plan.

TERRI

So why is that report not our system proposal?

BOB

Good question, Terri. And Sandra's overview provided some hints to the answer. The initial report identified some *perceived* problems and opportunities. No doubt, most of those problems and opportunities are real. But your member services staff had only limited input to the report because the study was conducted at a very high level of detail. And the initial report described one solution, albeit a good one. But extending the Internet to our customers is only one possible solution. What about members who are not on the Internet, or who are still wary of placing orders over the Internet? And what about the connec-

tions to our existing systems? Do we replace those systems or interface to them?

TERRI

So you are saying we have not looked at alternatives, or the implications of alternatives?

BOB

That's right.

TERRI

And I can confirm that most of us in Member Services had limited user input to the strategic plan. For the most part, managers were involved. Several staff members have complained about that. They felt that the full context of the problems and opportunities and their impact on existing operations and member services were not fully understood and taken into consideration.

SANDRA

Precisely! But that is not a criticism of the strategic planning project team. They were charged with establishing a high-level vision that would sustain and grow our market leadership in light of a changing economy and new technological opportunities and threats. Now, we have been charged with implementing a small but critical piece of their vision. Therefore . . .

TERRI

. . . we need to develop a better understanding of where we are today. Sorry, I didn't mean to interrupt.

SANDRA

Not a problem, and you are correct. That understanding is especially important for Bob, myself, and other information technology specialists who will become involved. You know your business, but as systems analysts, Bob and I need to develop some comfort level with your business environment, your vocabulary, your problems, and your constraints.

TERRI

And to do that, you will be getting many more of our member services staff involved than was true in the strategic planning project?

SOUNDSTAGE

SOUNDSTAGE ENTERTAINMENT CLUB

BOB

Yes. In fact, it will be extremely important that we have contact with everyone who might have a stake in this project ... at least at some level. I anticipate that we will have many group meetings. During this phase of systems analysis, we also seek to develop a more thorough understanding of your business problems, their causes, and their ultimate effects.

TERRI

And I can tell you that the strategic planning project only revealed the tip of the iceberg when it comes to problems.

SANDRA

There you go! That's why we need to study the current system as part of our systems analysis.

TERRI

Part? There's more.

BOB

Yes, our methodology requires that we also define the business requirements for solving those problems and exploiting your opportunities.

TERRI

Such as the Web solution!

SANDRA

No. The Web represents a technology solution, one of many possible technology solutions we could develop. When we talk of defining your business requirements, we are interested in those requirements that *any* technological solution would have to fulfill.

TERRI

How do you do that in systems analysis?

BOB

Any number of ways! We can draw pictures that represent your business requirements, or we could build prototypes of some of your requirements.

TERRI

Prototypes sounds kind of contradictory! On the one hand, you are saying that you want to focus on business requirements independent of technology. But then you say you might build prototypes to learn about requirements. But aren't prototypes a technological solution?

SANDRA

Yes, but they are not necessarily *the* solution. If we decide to build prototypes, we'll probably use an easy technology to rapidly build those prototypes. We won't be committing to the prototyping technology. In fact, it probably *won't* be the technology used in the final system. Our sole purpose of prototyping will be to discover your business requirements.

TERRI

I see ... kind of! I'm sure I'll understand it better once we get into it. But when are we ever going to get back to that system proposal that you said was the ultimate product of systems analysis?

BOB

Once we understand your current system, and we have defined business requirements for your new system, then we can identify possible technological solutions and analyze them for feasibility. That feasibility analysis becomes the basis for our system proposal.

TERRI

Actually, it all seems very logical when you spell it out that way.

SANDRA

So let's get started with our plans to collect some factual information about the current member services system.

BOB

I've already collected some of the existing system documentation. And also found a correspondence file with several memos that predated the start of this project. The first thing I noticed is ...

DISCUSSION QUESTIONS

1. What role does systems analysis appear to play in a project?

2. How would you characterize the focus of the early phases of this project? Why are technical concerns being avoided?

3. What kind of pictures would you draw to illustrate an existing or proposed information system?

4. What kinds of questions would facilitate discussion of problems and their underlying causes and effects?

5. How can a *technology*-based approach such as prototyping help systems analysts better understand *business* requirements independent of technology solutions?

6. Can you think of any alternatives to a pure Web-based member services information system? Why might it be appropriate to consider other alternatives?

In Chapter 3 you learned about the systems development process. In that chapter we briefly examined each phase. In this chapter we take a closer look at those phases that are collectively referred to as systems analysis. Let's begin with the formal requirements analysis of systems analysis.

WHAT IS SYSTEMS ANALYSIS?

> **Systems analysis** is a problem-solving technique that decomposes a system into its component pieces for the purpose of studying how well those component parts work and interact to accomplish their purpose.

Presumably, we do a systems analysis in order to subsequently perform a systems design.

Systems design (also called *systems synthesis*) is a complementary problem-solving technique (to systems analysis) that reassembles a system's component pieces back into a complete system—it is hoped an improved system. This may involve adding, deleting, and changing pieces relative to the original system.

This chapter will focus on systems analysis. Chapter 10 will do the same for systems design.

Systems analysis is a term that collectively describes the early phases of systems development. Figure 5.1 uses color to identify the systems analysis phases in the context of the full classic route for our *FAST* methodology (from Chapter 3). There has never been a universally accepted definition of information systems analysis. In fact, there has never been universal agreement on when systems analysis ends and when systems design begins. But for purposes of this textbook, we offer the following definition.

Information **systems analysis** is defined as those development phases in a project that primarily focus on the <u>business</u> problem, independent of any technology that can or will be used to implement a solution to that problem.

FIGURE 5.1
The Context of Systems Analysis

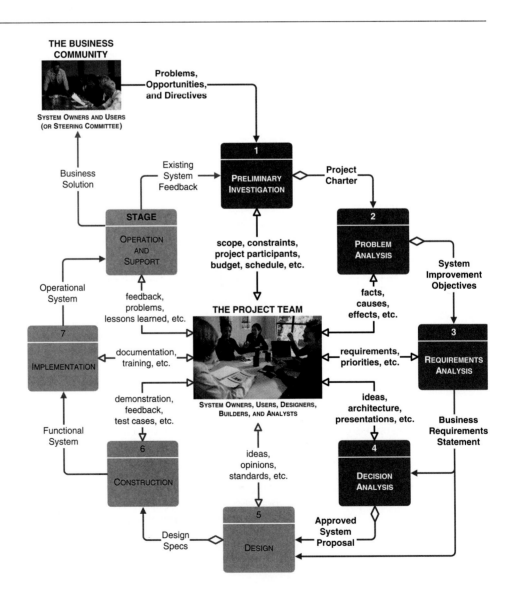

Notice that emphasis is placed on <u>business</u> issues, not technical or implementation concerns.

Systems analysis is driven by the business concerns of system owners and system users. Hence, it addresses the DATA, PROCESS, and INTERFACE building blocks from system owners' and system users' perspectives. The systems analysts serve as facilitators of systems analysis. This context was illustrated in the chapter map that preceded the objectives for this chapter.

The documentation and deliverables produced by systems analysis tasks are typically stored in a repository.

> A **repository** is a location (or set of locations) where systems analysts, systems designers, and system builders keep the documentation associated with one or more systems or projects.

A repository may be created for a single project or shared by all projects and systems. A repository is normally implemented as some combination of the following:

- A network directory of word processing, spreadsheet, and other computer-generated files that contain project correspondence, reports, and data.
- One or more CASE tool dictionaries or encyclopedias (as discussed in Chapter 3).
- Printed documentation (such as that stored in binders and system libraries).
- An *intra*net website interface to the above components (useful for communication).

Hereafter, we will refer to these components collectively as "the repository."

This chapter examines each of our four systems analysis phases in greater detail. But first, let's examine some overall strategies for systems analysis.

SYSTEMS ANALYSIS APPROACHES

Fundamentally, systems analysis is about *problem solving*. There are many approaches to problem solving; therefore, there are many approaches to systems analysis. Some of the more popular systems analysis approaches include *structured analysis, information engineering, discovery prototyping,* and *object-oriented analysis*. These approaches are often viewed as *competing* alternatives. In reality, certain combinations can and should actually complement one another.

Let's briefly examine these approaches. <u>The intent here is to develop a high-level understanding only. Subsequent chapters in this unit will teach you the underlying techniques.</u>

(Recall from Chapter 3 that methodology "routes" are sometimes defined for these approaches.)

Model-Driven Analysis Approaches

Structured analysis, information engineering, and object-oriented analysis are examples of model-driven approaches.

> **Model-driven analysis** emphasizes the drawing of pictorial system models to document and validate both existing and/or proposed systems. Ultimately, the system model becomes the blueprint for designing and constructing an improved system.

> A **model** is a representation of either reality or vision. Just as "a picture is worth a thousand words," most models use pictures to represent the reality or vision.

Examples of models with which you may already be familiar include flowcharts, structure or hierarchy charts, and organization charts.

Today, model-driven approaches are almost always enhanced by the use of automated tools. Some analysts draw system models with general-purpose graphics software such as *Visio Professional* or *Corel Flow.* Other analysts and organizations require the use of repository-based CASE or modeling tools such as *System Architect, Visio Enterprise, Visible Analyst,* or *Rational ROSE.* CASE tools offer the advantage of consistency and completeness analysis as well as rule-based error checking.

Let's briefly examine today's three most popular model-driven analysis approaches. Model-driven analysis approaches are featured in the model-driven methodologies and routes that were introduced in Chapter 3.

Structured Analysis   *Structured analysis* was one of the first formal approaches for systems analysis of information systems. In its various dialects, it is still one of the most widely practiced approaches.

> **Structured analysis** is a model-driven, PROCESS-centered technique used to either analyze an existing system, define business requirements for a new system, or both. The models are pictures that illustrate the system's component pieces: processes and their associated inputs, outputs, and files.

By PROCESS-centered, we mean the emphasis in this technique is on the PROCESS building blocks in your information system framework. Over the years, the technique has evolved to also model the DATA and INTERFACE building blocks as a secondary emphasis. But PROCESSES have always been the primary focus of the system models produced using structured analysis.

Structured analysis is simple in concept. Systems analysts draw a series of process models called **data flow diagrams (DFD)** (Figure 5.2) that depict the existing and/or proposed processes in a system along with their inputs, outputs, and files. Ultimately, these process models serve as blueprints for both business

FIGURE 5.2

A Simple Process Model (also called a data flow diagram)

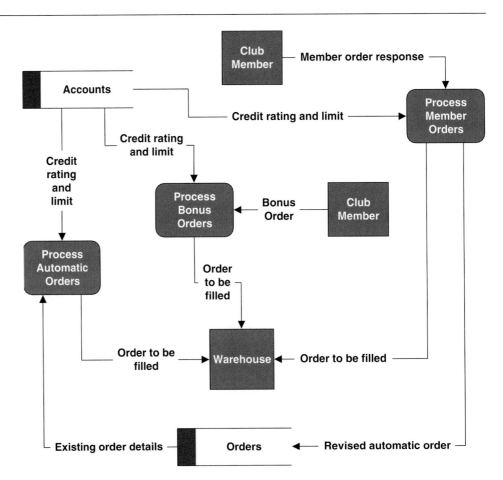

processes to be implemented and computer programs to be purchased or constructed.

Today, process modeling is enjoying a revival thanks to the renewed emphasis on *business process redesign*. Through business process redesign, organizations seek to study the fundamental business processes to increase throughput and efficiency and to reduce waste and costs. Process models, in various forms including data flow diagrams, help organizations to visualize their processes and ultimately eliminate or reduce bureaucracy. Thus, process models can return value to an organization even if the goal is not to design or purchase software to automate those processes!

Information Engineering   Many organizations have evolved from a structured analysis approach to an *information engineering* approach.

> **Information engineering (IE)** is a model-driven and DATA-centered, but PROCESS-sensitive technique to plan, analyze, and design information systems. IE models are pictures that illustrate and synchronize the system's data and processes.

The data models in IE are called entity relationship diagrams (see Figure 5.3), borrowed from the classic toolkit of database designers. The process models in IE use the same data flow diagrams that were invented for structured analysis. IE's contribution was to define an approach for integrating and synchronizing the data and process models.

Information engineering is said to be a DATA-centered paradigm because it emphasizes the study and requirements analysis of DATA requirements <u>before</u> those of the PROCESS and INTERFACE requirements. This is based on the belief that data is a corporate resource that should be planned and managed. Accordingly, systems analysts draw **entity relationship diagrams (ERDs)** to model the system's raw data before they draw the *data flow diagrams* that illustrate how that data will be captured, stored, used, and maintained.

Both information engineering and structured analysis attempt to synchronize data and process models. The two approaches differ only in which model is drawn first. Information engineering draws the data models first, while structured analysis draws the process models first. The need for both models and their synchronization should make sense to you because we know, from Chapter 2, that all information systems include both DATA and PROCESS building blocks.

Object-Oriented Analysis   Object-oriented analysis is the new kid on the block. The concepts behind this exciting new strategy (and technology) are covered extensively in Part Five, Module A, but a simplified introduction is appropriate here.

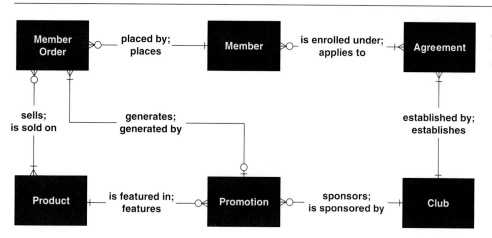

FIGURE 5.3
A Simple Data Model (also called an entity relationship diagram)

For the past 30 years, most system development approaches have deliberately separated concerns of DATA from those of PROCESSES. Most systems analysis techniques (such as structured analysis and information engineering) also separated these concerns. Although most systems analysis methods have made significant attempts to synchronize data and process models, the results have been only partially successful.

Object technologies have since emerged to eliminate this artificial separation of concerns about data and processes. Instead, specific data and the processes that create, read, update and delete that data are integrated into constructs called an OBJECT. The only way to create, read, update, or delete the object's data (called *properties*) is through one of its embedded processes (called *methods*). You may already be familiar with some programming languages based on object technology—*Visual Basic, C++, Powerbuilder, Object Pascal* (as in *Delphi*), *Smalltalk,* and *Java*—these are the languages of choice for today's new information systems!

The trend toward using object technology to build new applications created the need for a complementary systems analysis approach, object-oriented analysis.

> **Object-oriented analysis (OOA)** is a model-driven technique that integrates DATA and PROCESS concerns into constructs called OBJECTS. OOA models are pictures that illustrate the system's objects from various perspectives such as structure and behavior.

OOA has become so popular that a modeling standard has evolved around it. The *Unified Modeling Language* (or UML) provides a graphical syntax for an entire series of **object models,** such as the model illustrated in Figure 5.4. The UML defines several different types of diagrams that collectively model an information system or application in terms of objects.

FIGURE 5.4

An Object Model (using the Unified Modeling Language standard)

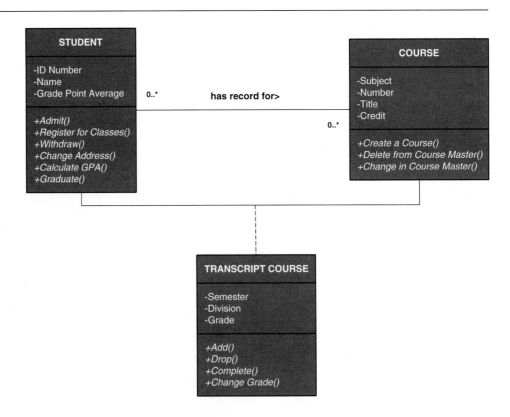

Discovery prototyping and rapid architected development are examples of accelerated analysis approaches.

> **Accelerated analysis** approaches emphasize the construction of prototypes to more rapidly identify business and user requirements for a new system.

> A **prototype** is a small-scale, incomplete, but <u>working</u> sample of a desired system. Prototypes cater to the "I'll know what I want when I see it" way of thinking that is characteristic of many users and managers.

By incomplete, we mean that a prototype will not include the error checking, input data validation, security, and processing completeness of a finished application. Nor will it be as polished or offer the user help as in a final system. But it can quickly identify the most crucial of business-level requirements. Sometimes the prototypes can evolve into the completed information systems and applications.

Accelerated analysis approaches emphasize the INTERFACE building blocks in your information system framework by constructing sample forms and reports. At the same time, prototypes also address the DATA and PROCESS building blocks.

These approaches are common in rapid application development (RAD) methodologies and routes (introduced in Chapter 3). RAD approaches require automated tools. While some repository-based CASE tools include very simple RAD facilities, most analysts use true RAD programming environments such as *Access, Visual Basic,* or *Delphi.*

Let's briefly examine two currently popular accelerated analysis approaches.

Discovery Prototyping   Discovery prototyping uses rapid development technology to help users discover their business requirements.

> **Discovery prototyping** (sometimes called *requirements prototyping*) is used to identify the users' business requirements by having them react to a quick-and-dirty implementation of those requirements.

For example, it is very common for systems analysts to use a simple development tool such as Microsoft *Access* to rapidly create a simple database, user input forms, and sample reports to solicit user responses as to whether the database, forms, and reports truly represent business requirements.

In discovery prototyping, we try to discourage users from worrying about the final "look and feel" of the prototypes—that can be changed during system design! Therein lies the primary criticism of prototyping—Microsoft's wizards can quickly generate some very elegant and visually appealing prototypes. This can encourage a premature focus on and commitment to design. Users can also be misled to believe: (1) that the completed system can be built just as rapidly, or (2) that tools such as *Access* can be used to build the final system. First, while tools such as *Access* can accelerate systems development, their use in discovery prototyping is fast only because we omit most of the detailed database and application programming required for a complete and secure application. Second, these tools typically cannot support the database sizes, number of users, and network traffic that is required of most enterprise applications.

Still, discovery prototyping is a preferred and recommended approach. Unfortunately, some systems analysts and developers are using discovery prototyping to replace model-driven design, only to learn what true engineers have known for years—you cannot prototype without some amount of more formal design. Enter rapid architecture analysis.

Rapid Architecture Analysis   Rapid architecture analysis is an accelerated analysis approach that also builds system models.

> **Rapid architecture analysis** attempts to derive system models (as described earlier in this section) from existing systems or discovery prototypes.

Rapid architecture analysis is made possible by reverse engineering technology that is included in many automated tools such as CASE and programming languages (again, as introduced in Chapter 3).

> **Reverse engineering** technology reads the program code for an existing database, application program, and/or user interface and automatically generates the equivalent system model.

The resulting system model can then be edited and improved by systems analysts and users to provide a blueprint for a new and improved system. Rapid architecture analysis is a blending of model-driven and accelerated analysis approaches. There are two different techniques to apply rapid architecture analysis:

— Most systems have already been automated to some degree and exist as legacy information systems. Many CASE tools can read the underlying database structures and/or application programs and reverse engineer them into various system models. Those models serve as a point of departure for defining model-driven user requirements analysis.

— If prototypes have been built in tools such as Microsoft *Access* or *Visual Basic,* those prototypes can sometimes be reverse engineered into their equivalent system models. The system models usually better lend themselves to analyzing the users' requirements for consistency, completeness, stability, scalability, and flexibility to future change. Also, the system models can frequently be forward engineered by the same CASE tools and ADEs into databases and application templates or skeletons that will use more robust enterprise-level database and programming technology.

Both techniques address the previous issue that engineers rarely prototype in the total absence of a more formal design and, at the same time, preserve the advantages of accelerating the systems analysis phases.

Requirements Discovery Methods

Both model-driven and accelerated systems analysis approaches attempt to express user requirements for a new system, either as models or prototypes. But both approaches are dependent on the more subtle need to identify and manage those requirements. Furthermore, the requirements for systems are dependent on the analysts' ability to discover the problems and opportunities that exist in the current system; thus, analysts must become skilled in identifying problems, opportunities, and requirements! Consequently, all approaches to systems analysis require some form of requirements discovery.

> **Requirements discovery** includes those techniques to be used by systems analysts to identify or extract system problems and solution requirements from the user community.

Let's briefly survey a couple of common requirements discovery approaches.

Fact-Finding Techniques   Fact-finding techniques are an essential skill for all systems analysts.

> **Fact-finding** (or **information gathering**) is a classical set of techniques used to collect information about system problems, opportunities, solution requirements, and priorities.

The fact-finding techniques covered in this book (in the next chapter) include:

— Sampling of existing documentation, reports, forms, files, databases, and memos.

— Research of relevant literature, benchmarking others' solutions, and site visits.

— Observation of the current system in action and the work environment.

- Questionnaires and surveys of the management and user community.
- Interviews of appropriate managers, users, and technical staff.

Joint Requirements Planning   The classic fact-finding techniques listed above are invaluable; however, they can be time-consuming in their classic forms. Alternatively, requirements discovery and management can be significantly accelerated using joint requirements planning techniques.

> **Joint requirements planning (JRP)** techniques use facilitated workshops to bring together all the system owners, system users, systems analysts, and some systems designers and builders to jointly perform systems analysis. JRP is generally considered a part of *joint application development* (or *JAD),* a more comprehensive application of the techniques to the entire system development process.

A JRP-trained or -certified analyst usually plays the role of *facilitator* for a workshop that will typically run from three to five full working days. This workshop can replace weeks or months of classical fact-finding and follow-up meetings.

JRP provides a working environment in which to accelerate all systems analysis tasks and deliverables. It promotes enhanced system owner and system user participation in systems analysis. But it also requires a facilitator with superior mediation and negotiation skills to ensure that all parties receive appropriate opportunities to contribute to the system's development.

JRP is typically used in conjunction with the model-driven analysis approaches we described earlier and is typically incorporated in rapid application development (RAD) methodologies and routes (introduced in Chapter 3).

One interesting contemporary application of systems analysis methods is business process redesign.

Business Process Redesign Methods

> **Business process redesign** (or BPR—also called **business process reengineering**) is the application of systems analysis methods to the goal of dramatically changing and improving the fundamental business processes of an organization, independent of information technology.

The interest in BPR was driven by the discovery that most current information systems and applications merely automated existing and inefficient business processes. Automated bureaucracy is still bureaucracy; automation does not necessarily contribute value to the business and it may actually subtract value from the business. Introduced in Chapter 1, BPR is one of many types of projects triggered by the trends we call *total quality management* (TQM) and *continuous process improvement* (CPI).

Some BPR projects focus on all business processes, regardless of their automation. Each business process is thoroughly studied and analyzed for bottlenecks, value returned, and opportunities for elimination or streamlining. Once the business processes have been redesigned, most BPR projects conclude by examining how information technology might best be applied to the improved business processes. This may create new information system and application development projects to implement or support the new business processes.

BPR is also applied within the context of information system development projects. It is not uncommon for projects to include a study of existing business processes to identify problems, bureaucracy, and inefficiencies that may be addressed in requirements for new and improved information systems and computer applications.

BPR has also become common in IS projects that will be based on the purchase and integration of commercial off-the-shelf (COTS) software. The purchase of COTS software usually requires a business to adapt its business processes to fit the software. An analysis of existing business processes during systems analysis is usually part of such projects.

FAST Systems Analysis Strategies

Like most commercial methodologies, our hypothetical *FAST* methodology does not impose a single approach on systems analysts. Instead, it integrates all the popular approaches introduced in the preceding paragraphs. The SoundStage case study will demonstrate these methods in the context of a typical first assignment for a systems analyst. The systems analysis techniques will be applied within the framework of:

— Your information system building blocks (from Chapter 2).
— The *FAST* phases (from Chapter 3).
— *FAST* tasks that implement a phase (described in this chapter).

Given this context for studying systems analysis, we can now explore the systems analysis phases and tasks.

THE PRELIMINARY INVESTIGATION PHASE

Recall from Chapter 3 that the **preliminary investigation phase** is the first phase of the classic systems development process. In other methodologies this might be called the *initial study phase, survey phase,* or *planning phase.* The preliminary investigation answers the question, "Is this project worth looking at?" To answer this question, the preliminary investigation must define the scope of the project and the <u>perceived</u> problems, opportunities, and directives that triggered the project. Assuming the project *is* deemed worth considering, the survey phase must also establish the project plan in terms of scope, development strategy or "route," schedule, resource requirements, and budget.[1]

The context for the preliminary investigation phase is shaded in red in Figure 5.5. Notice that the preliminary investigation phase is concerned primarily with the system owners' view of the existing system. System owners tend to be concerned with the big picture, not details. Furthermore, they determine whether resources will be committed to the project.

Figure 5.6 is the first of four task diagrams we will introduce in this chapter to take a closer look at each systems analysis phase. A *task diagram* shows the work (= tasks) that should be performed to complete a phase. Our task diagrams do not mandate any specific methodology, but we will describe in the accompanying paragraphs those approaches, tools, and techniques you might want to consider for each task. Figure 5.6 shows the tasks required for the preliminary investigation phase. These task diagrams are only templates. The project team and project manager may expand on or alter these templates to reflect the unique needs of any given project.

As shown in Figure 5.6, the final deliverable for the preliminary investigation phase is completion of a **project charter.** (Such major deliverables are indicated in each task diagram in all capital letters and a larger font size.) A project charter defines the project scope, plan, methodology, standards, and so on. Completion of the project charter represents the first milestone in a project.

The preliminary investigation phase is intended to be quick. The entire phase should not exceed two or three days for most projects. The preliminary investigation typically includes the following tasks:

1.1 List problems, opportunities, and directives.
1.2 Negotiate preliminary scope.
1.3 Assess project worth.
1.4 Plan the project.
1.5 Present the project and plan.

Let's now examine each of these tasks in greater detail.

---

[1] If your course or reading has already included Chapter 4, you should recognize these planning elements to be part of project management. Chapter 4 surveyed and demonstrated the process used by project managers to develop a project plan.

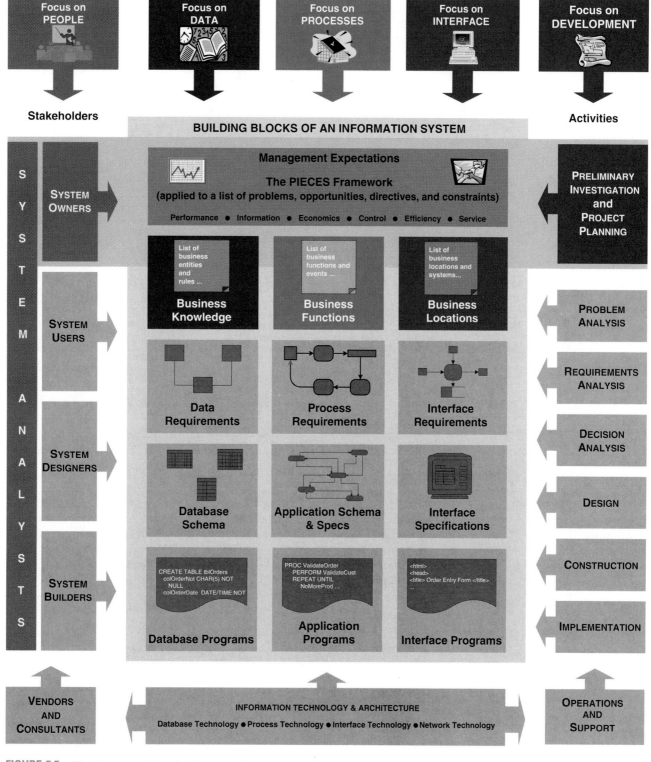

FIGURE 5.5   *The Context of the Preliminary Investigation Phase of Systems Analysis*

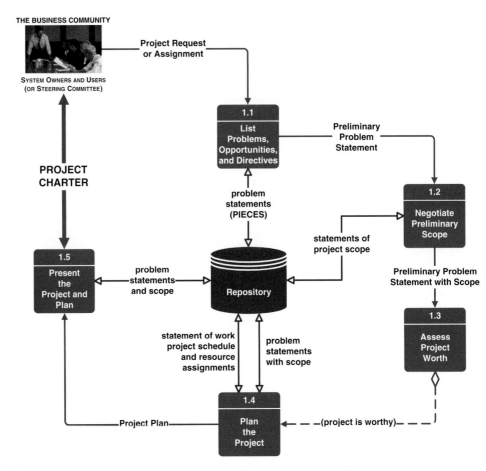

FIGURE 5.6
Tasks for the Preliminary Investigation Phase of Systems Analysis

Task 1.1—List Problems, Opportunities, and Directives

One important task in the preliminary investigation phase is to establish an initial baseline of the problems, opportunities, and/or directives that triggered the project. Each problem, opportunity, and directive is assessed with respect to urgency, visibility, tangible benefits, and priority. Any additional, detailed analysis is not relevant at this stage. It may, however, be useful to list any perceived constraints (limits) on the project such as deadlines, maximum budget, or general technology.

A senior systems analyst or project manager usually leads this task. Most of the other participants are broadly classified as system owners. This includes the executive sponsor, the highest-level manager who will pay for and support the project. It also includes managers of all organizational units that may be affected by the system and possibly information systems managers. System users, system designers, and system builders are not typically involved in this task.

As shown in Figure 5.6, a project request or assignment triggers the task. This trigger may take one of several alternative forms. It may be as simple as a memorandum of authority from an information systems steering body. Or it may be a memorandum from a business team or unit requesting systems development. Some organizations require that all project requests be submitted on a standard request for service form such as Figure 5.7.

Also shown in Figure 5.6 is the key deliverable of this task, the preliminary problem statement, consisting of the problems, opportunities, and directives that were identified. These problem statements are also stored in the repository for later use in the project. Figure 5.8 is a sample document that summarizes problems, opportunities, and directives in terms of:

— *Urgency.* In what time frame must/should the problem be solved or the

Visit the McGraw-Hill Online Learning Center for this textbook and browse the on-line resource for a Microsoft Word template you can use to write your own preliminary problem statements. The first half of the template deals with problem statements.

SoundStage Entertainment Club
Information System Services
Phone: 494-0666 Fax: 494-0999
Internet: http://www.soundstage.com
Intranet: http://www.soundstage.com/iss

REQUEST FOR INFORMATION SYSTEM SERVICES

DATE OF REQUEST	SERVICE REQUESTED FOR DEPARTMENT(S)
January 10, 2001	Member Services, Warehouse, Shipping

SUBMITTED BY (key user contact)
Name	Sarah Hartman
Title	Business Analyst, Member Services
Office	B035
Phone	494-0867

EXECUTIVE SPONSOR (funding authority)
Name	Galen Kirkhoff
Title	Vice President, Member Services
Office	G242
Phone	494-1242

TYPE OF SERVICE REQUESTED:
- ☐ Information Strategy Planning
- ☒ Business Process Analysis and Redesign
- ☒ New Application Development
- ☐ Other (please specify) _____
- ☐ Existing Application Enhancement
- ☐ Existing Application Maintenance (problem fix)
- ☐ Not Sure

BRIEF STATEMENT OF PROBLEM, OPPORTUNITY, OR DIRECTIVE (attach additional documentation as necessary)
The information strategy planning group has targeted member services, marketing, and order fulfillment (inclusive of shipping) for business process redesign and integrated application development. Currently serviced by separate information systems, these areas are not well integrated to maximize efficient order services to our members. The current systems are not adaptable to our rapidly changing products and services. In some cases, separate systems exist for similar products and services. Some of these systems were inherited through mergers that expanded our products and services. There also exist several marketing opportunities to increase our presence to our members. One example includes Internet commerce services. Finally, the automatic identification system being developed for the warehouse must fully interoperate with member services.

BRIEF STATEMENT OF EXPECTED SOLUTION
We envision completely new and streamlined business processes that minimize the response time to member orders for products and services. An order shall not be considered fulfilled until it has been received by the member. The new system should provide for expanded club and member flexibility and adaptability of basic business products and services.
   We envision a system that extends to the desktop computers of both employees and members, with appropriate shared services provided across the network, consistent with the ISS distributed architecture. This is consistent with strategic plans to retire the AS/400 central computer and replace it with servers.

ACTION (ISS Office Use Only)

☐ **Feasibility assessment approved**

☒ **Feasibility assessment waived**

Assigned to Sandra Shepherd _____

Approved Budget $ 450,000 _____

Start Date ASAP   **Deadline** ASAP

☐ **Request delayed**   **Backlogged until date:** _____

☐ **Request rejected**   **Reason:** _____

Authorized Signatures:

Rebecca J. Todd
Chair, ISS Executive Steering Body

Galen Kirkhoff
Project Executive Sponsor

FORM ISS-100-RFSS (Last revised December, 1999)

FIGURE 5.7
A Request for Systems Services

opportunity or directive realized? A rating scale could be developed to consistently answer this question.

- *Visibility.* To what degree would a solution or new system be visible to customers and/or executive management? Again, a rating scale could be developed for the answers.

- *Benefits.* Approximately how much would a solution or new system increase annual revenues or reduce annual costs. This is often a guess, but if all participants are involved in that guess, it should prove sufficiently conservative.

- *Priority.* Based on the above answers, what are the consensus priorities for each problem, opportunity, or directive. If budget or schedule becomes a problem, these priorities will help to adjust project scope.

PROBLEM STATEMENTS

PROJECT:	Member Services Information System	PROJECT MANAGER:	Sandra Shepherd
CREATED BY:	Sandra Shepherd	LAST UPDATED BY:	Robert Martinez
DATE CREATED:	January 15, 2001	DATE LAST UPDATED:	January 17, 2001

Brief Statements of Problem, Opportunity, or Directive	Urgency	Visibility	Annual Benefits	Priority or Rank	Proposed Solution
1. Order response time as measured from time of order receipt to time of customer delivery has increased to an average of 15 days.	ASAP	High	$175,000	2	New development
2. The recent acquisitions of Private Screenings Video Club and GameScreen will further stress the throughput requirements for the current system.	6 months	Med	75,000	2	New development
3. Currently, three different order entry systems service the audio, video, and game divisions. Each system is designed to interface with a different warehousing system; therefore, the intent to merge inventory into a single warehouse has been delayed.	6 months	Med	515,000	2	New development
4. There is a general lack of access to management and decision-making information. This will become exasperated by the acquisition of two additional order processing systems (from Private Screenings and GameScreen)	12 months	Low	15,000	3	After new system is developed, provide users with easy-to-learn and use reporting tools.
5. There currently exists data inconsistencies in the member and order files.	3 months	High	35,000	1	Quick fix; then new development
6. The Private Screenings and GameScreen file systems are incompatible with the SoundStage equivalents. Business data problems include data inconsistencies and lack of input edit controls.	6 months	Med	unknown	2	New development. Additional quantification of benefit might increase urgency.
7. There is an opportunity to open order systems to the Internet, but security and control is an issue.	12 months	Low	unknown	4	Future version of newly developed system
8. The current order entry system is incompatible with the forthcoming automatic identification (bar coding) system being developed for the warehouse.	3 months	High	65,000	1	Quick fix; then new development

FIGURE 5.8 *Sample Problem Statements*

— *Possible solutions* (OPT). At this early stage of the project, possible solutions are best expressed in simple terms such as *(a)* leave well enough alone, *(b)* a quick fix, *(c)* a simple-to-moderate enhancement of the existing system, *(d)* redesign the existing system, or *(e)* design a new system. The participants listed for this task are well-suited to an appropriately high-level discussion of these options.

The PIECES framework that was introduced in Chapter 3 can be used to categorize problems, opportunities, directives, and constraints.

The primary techniques used to complete this task include fact-finding and meetings with system owners. These techniques are taught in Chapter 6.

Task 1.2—Negotiate Preliminary Scope

Scope defines the boundary of the project—those aspects of the business that will and will not be included. Scope can change during the project; however, the initial project plan must establish the preliminary scope. Then if the scope changes significantly, all parties involved will have a better appreciation for why the budget and schedule have also changed. This task can occur in parallel with the prior task.

Once again, a senior systems analyst or project manager usually leads this task. Most of the other participants are broadly classified as system owners. This includes the executive sponsor, managers of all organizational units that may be impacted by the system, and possibly information systems managers. System users, system designers, and system builders are not typically involved in this task.

As shown in Figure 5.6, this task uses the preliminary problem statement produced by the previous task. Those problems, opportunities, and directives form the basis for defining scope. The statements of project scope are added to the repository for later use. These statements are also formally documented as the task deliverable, preliminary problem statement with scope.

Scope can be defined easily within the context of your information system building blocks. For example, a project's scope can be described in terms of:

— What types of DATA describe the system being studied? For example, a sales information system may require data about such things as CUSTOMERS, ORDERS, PRODUCTS, and SALES REPRESENTATIVES.

— What business PROCESSES are included in the system being studied? For example, a sales information system may include business processes for CATALOG MANAGEMENT, CUSTOMER MANAGEMENT, ORDER ENTRY, ORDER FULFILLMENT, ORDER MANAGEMENT, and CUSTOMER RELATIONSHIP MANAGEMENT.

— How must the system INTERFACE with users, locations, and other systems? For example, potential interfaces for a sales information system might include CUSTOMERS, SALES REPRESENTATIVES, SALES CLERKS AND MANAGERS, REGIONAL SALES OFFICES, and the ACCOUNTS RECEIVABLE and INVENTORY CONTROL INFORMATION SYSTEMS.

Notice that each statement of scope can be described as a simple list. We don't necessarily define the items in the list. Nor are we very concerned with precise requirements analysis. And we definitely are not concerned with any time-consuming steps such as modeling or prototyping.

Once again, the primary techniques used to complete this task are fact-finding and meetings. Many analysts prefer to combine this task with both the previous and next task and accomplish them within a single meeting.

Visit the McGraw-Hill Online Learning Center for this textbook and browse the on-line resource for a Microsoft Word template you can use to write your own preliminary problem statements with scope. The second half of the template deals with scope.

Task 1.3—Assess Project Worth

This is where we answer the question, "Is this project worth looking at?" At this early stage of the project, the question may actually boil down to a best guess—Will solving the problems, exploiting the opportunities, or fulfilling the directives return enough value to offset the costs that we will incur to develop this system? It is impossible to do a thorough feasibility analysis based on the limited facts we've collected to date.

Again, a senior systems analyst or project manager usually leads this task. But the system owners including the executive sponsor, the business unit managers, and the information systems managers should make the decision.

As shown in Figure 5.6, the completed preliminary problem statement with scope triggers the task. This provides the level of information required for this preliminary assessment of worth. There is no physical deliverable other than the go or no-go decision. There are actually several alternative decisions. The project can be approved or cancelled, and project scope can be renegotiated (increased or decreased!). Obviously, the remaining tasks in the preliminary investigation phase are necessary only if the project has been deemed worthy and approved to continue.

Task 1.4—Plan the Project

If the project has been deemed worthy to continue, we can now plan the project in depth. The initial project plan should consist of at least the following:

— A preliminary master plan that includes schedule and resource assignments for the entire project. This plan will be updated at the end of each phase of the project. This is sometimes called a *baseline plan.*

— A detailed plan and schedule for completing the next phase of the project (the problem analysis phase).

The task is the responsibility of the *project manager.* Most project managers find it useful to include as much of the project team, including system owners, users, designers, and builders, as possible. Chapter 4 used the term *joint project planning* to describe the team approach to building a project plan.

As shown in Figure 5.6, this task is triggered by the go or no-go decision to continue the project. This decision represents a consensus agreement on the project's scope, problems, opportunities, directives, and worthiness. (This worthiness must still be presented and approved.) The problem statements with scope are the key input (from the repository). The deliverable of this task is the project plan. The statement of work (see Chapter 4) and project schedule and resource assignments are also added to the repository for continuous monitoring and, as appropriate, updating. The schedule and resources are typically maintained in the repository as a project management software file.

The techniques used to create a project plan were covered in Chapter 4. Today, these techniques are supported by project management software such as Microsoft *Project*. Chapter 4 also discussed the detailed steps to complete the plan.

Task 1.5—Present the Project and Plan

In most organizations, there are more potential projects than resources to staff and fund those projects. Unless our project has been predetermined to be of the highest priority (by some sort of prior tactical or strategic planning process), then it must be presented and defended to a steering body for approval.

> A **steering body** is a committee of executive business and system managers that studies and prioritizes competing project proposals to determine which projects will return the most value to the organization and thus should be approved for continued system development. It is also called a *steering committee*.

Any steering body should consist mainly of noninformation systems professionals or managers. Many organizations designate vice presidents to serve on a steering body. Other organizations assign the direct reports of vice presidents to the steering body. And some organizations utilize two steering bodies, one for vice presidents and one for their direct reports. Information systems managers serve on the steering body only to answer questions and to communicate priorities back to developers and project managers.

Regardless of whether or not a project requires steering committee approval, it is equally important to formally launch the project and communicate the project, goals, and schedule to the entire business community. Opening the lines of communication is an important capstone to the preliminary investigation. For this reason, we advocate conducting a *project kickoff event* and creating an *intranet project website*. The project kickoff meeting is open to the entire business community, not just the business units affected and the project team. The intranet project website establishes a community portal to all nonsensitive news and documentation concerning the project.

Ideally, the executive sponsor should jointly facilitate the task with the chosen project manager. The visibility of the executive sponsor establishes instant credibility and priority to all who participate in the kickoff meeting. Other kickoff meeting participants should include the entire project team, including assigned system owners, users, analysts, designers, and builders. And ideally, the kickoff meeting should be open to any and all interested staff from the business community. This builds community awareness and consensus while reducing both the volume and consequences of rumor and misinformation. For the intranet component, a Webmaster or Web author should be assigned to the project team.

As shown in Figure 5.6, this task is triggered by the completion of the project plan. The problem statements and scope are available from the repository. The deliverable is the project charter, which is usually a document. It includes various elements that define the project in terms of participants, problems,

Visit the McGraw-Hill Online Learning Center for this textbook and browse the on-line resource for Microsoft Word and Microsoft PowerPoint templates you can use to create your own project charters.

opportunities, and directives; scope; methodology; statement of work to be completed; deliverables; quality standards; schedule; and budget. The project charter should be added to the project website for all to see. Elements of the project charter may also be reformatted as slides and handouts (using software such as Microsoft *PowerPoint*) for inclusion in the project kickoff event.

Effective interpersonal and communication skills are the keys to this task. These include principles of persuasion, selling change, business writing, and public speaking.

The participants in the preliminary investigation phase might decide the project is not worth proposing. It is also possible the steering body may decide that other projects are more important. Or the executive sponsor might not endorse the project. In each of these instances, the project is terminated. Little time and effort have been expended. On the other hand, with the blessing of all system owners and steering committee, the project can now proceed to the problem analysis phase.

Visit the McGraw-Hill Online Learning Center for this textbook and browse the on-line resource for a supplemental reading module called Interpersonal Skills for Systems Analysts. This module described techniques that can enhance oral and written communications skills specifically employed by systems analysts at various times during a project.

THE PROBLEM ANALYSIS PHASE

The statement "Don't try to fix it unless you understand it" aptly describes the **problem analysis phase** of systems analysis. There is always a current or existing system, regardless of the degree to which it is automated with information technology. The problem analysis phase provides the analyst with a more thorough understanding of the problems, opportunities, and/or directives that triggered the project. This phase answers the questions, "Are the problems really worth solving?" and "Is a new system really worth building?" In other methodologies, the problem analysis phase may be known as the *study phase, study of the current system, detailed investigation phase,* or *feasibility analysis phase.*

You can rarely skip the problem analysis phase. You almost always need <u>some</u> level of understanding of the current system. But there may be reasons to accelerate the problem analysis phase. First, if the project was triggered by a strategic or tactical plan, the worthiness of the project is probably not in doubt—the problem analysis phase would be reduced to understanding the current system, not analyzing it. Second, a project may be initiated by a directive (such as compliance with a governmental directive and deadline). Again, project worthiness is not in doubt. Finally, some methodologies and organizations deliberately consolidate the problem analysis and requirements analysis phases to accelerate systems analysis.

The goal of the problem analysis phase is to study and understand the problem domain well enough to thoroughly analyze its problems, opportunities, and constraints. Some methodologies encourage a very detailed understanding of the current system and document that system in painstaking detail using system models such as data flow diagrams. Today, except when business processes must be redesigned, the effort required and the value added by such detailed modeling is questioned and usually bypassed. Thus, the current version of our hypothetical *FAST* methodology encourages only enough system modeling to refine our understanding of project scope and problem statement and to define a common vocabulary for the system.

The context for the problem analysis phase is shaded in tan in Figure 5.9. Notice that the problem analysis phase is concerned primarily with the system owners' and users' views of the existing system. We build on the lists created in the preliminary investigation phase to analyze the DATA, PROCESS, and INTERFACE building blocks of the existing system. Also notice that we imply minimal system modeling. We may still use the PIECES framework to analyze each building block for problems, causes, and effects.

Figure 5.10 is the task diagram for the problem analysis phase. The final deliverable and milestone of this phase is to produce system improvement objectives that address problems, opportunities, and directives.

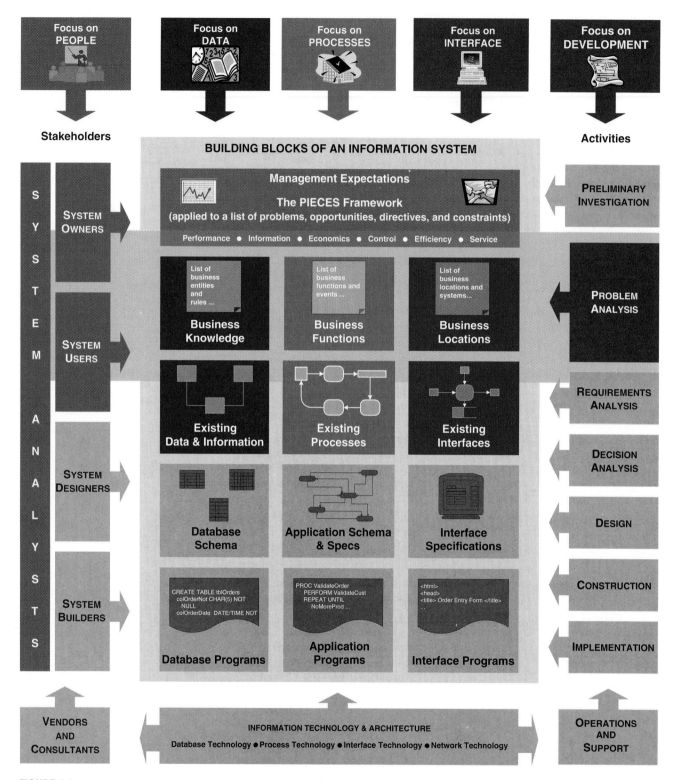

FIGURE 5.9 *The Context of the Problem Analysis Phase of Systems Analysis*

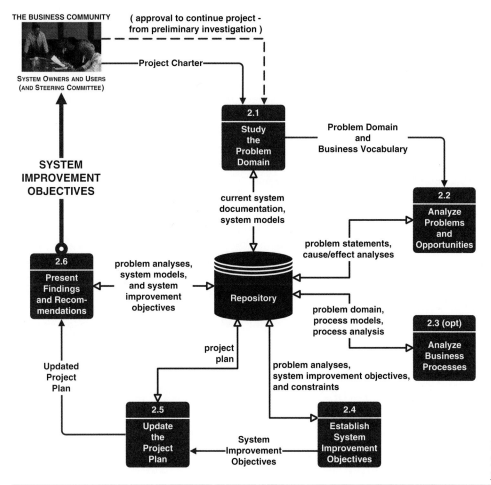

FIGURE 5.10

Tasks for the Problem Analysis Phase of Systems Analysis

Depending on the size of the system, its complexity, and the degree to which project worthiness is already known, the illustrated tasks may consume one to six weeks. Most of these tasks can be accelerated by JRP-like sessions. The problem analysis phase typically includes the following tasks:

2.1 Study the problem domain.

2.2 Analyze problems and opportunities.

2.3 (optional) Analyze business processes.

2.4 Establish system improvement objectives.

2.5 Update the project plan.

2.6 Present findings and recommendations.

Let's now examine each of these tasks in greater detail.

During the problem analysis phase, the *team* initially attempts to learn about the current system. Each system owner, user, and analyst brings a different level of understanding to the system—different detail, different vocabulary, different perceptions, and different opinions. A well-conducted study can prove revealing to all parties, including the system's own management and users. It is important to study the problem domain, that domain in which the business problems, opportunities, directives, and constraints exist.

This task will be led by the project manager, but facilitated by the lead systems analyst. It is not uncommon for one individual to play both roles (as Sandra does in the SoundStage case). Other systems analysts may also be involved as they

Task 2.1—Study the Problem Domain

conduct interviews, scribe for meetings, and document findings. A comprehensive study should include representative system owners and users from all business units that will be supported or affected by the system and project. It is extraordinarily important that enough users be included to encompass the full scope of the system being studied. In some organizations, one or more experienced users are "lent" to the project full-time as *business analysts;* however, it is rare that any one user can fully represent the interests of all users. Business analysts can, however, serve as facilitators to get the right people involved and sustain effective communication back to the business units and management. System designers and builders are rarely involved in this task unless they are interviewed to determine any technical limitations of the current system.

In Figure 5.10, this task is triggered by approval to continue the project from the preliminary investigation phase. (The dashed line indicates this approval is an event or trigger, not a data or information flow.) The approval comes from the system owners or steering committee. The key informational input is the project charter and any current system documentation that may exist in the repository and program libraries for the current system. Current system documentation doesn't always exist. And when it does exist, it must be carefully checked for currency—most such documentation is notoriously out of date because analysts and programmers are not always diligent about updating that documentation as changes occur throughout the lifetime of a system.

The deliverables of this task are an understanding of the problem domain and business vocabulary. Your understanding of the existing problem domain should be documented so that it can be verified that you truly understand it. There are several ways to document the problem domain. Drawing system models of the current system can help, but they can lead to a phenomenon called "analysis paralysis" in which the desire to produce perfect models becomes counterproductive to the schedule. Another approach might be to use your information system building blocks as a framework to list and define the system domain.

- DATA—List all the things about which the system currently stores data (in files, databases, forms, etc.). Define each thing in business terms. For example, "An ORDER is a business transaction in which a customer requests to purchase products."

  Additionally, we could list all the reports currently produced by the current system and describe their purpose or use. For example, "The open orders report describes all orders that have not been filled within one week of their approval to be filled. The report is used to initiate customer relationship management through personal contact."

- PROCESSES—Define each business event for which a business response (process) is currently implemented. For example, "A customer places a new order" or "A customer requests changes to a previously placed order" or "A customer cancels an order."

- INTERFACES—Define all the locations that the current system serves and all the users at each of those locations. For example, "The system is currently used at regional sales offices in San Diego, Dallas, St. Louis, Indianapolis, Atlanta, and Manhattan. Each regional sales office has a sales manager, assistant sales manager, administrative assistant, and 5 to 10 sales clerks, all of whom use the current system. Each region is also home to 5 to 30 sales representatives who are on the road most days, but who upload orders and other transactions each evening."

  Another facet of interfaces is system interfaces—that is, interfaces that currently exist between the current information system and other information systems and computer applications. These can be quickly listed and described by the information systems staff.

Ultimately, the organization's system development methodology and project plan will determine what types and level of documentation are expected.

The business vocabulary deliverable is too often shortchanged. Understanding the business vocabulary of a system is an excellent way of understanding the system itself. It bridges the communication gap that often exists or develops between business and technology experts.

If you elect to draw system models during this task, we suggest if you want to learn *anything,* you must not try to learn *everything*—at least not all in this task. To avoid analysis paralysis we suggest that the following system models may be appropriate:

- DATA—A one-page data model is very useful for establishing business vocabulary and rules. Data modeling is taught in Chapter 7.

- PROCESSES—A one- or two-page functional decomposition diagram should prove sufficient to get a feel for the current system processing. Decomposition modeling is taught in Chapter 8.

- INTERFACES—A one-page context diagram or use-case diagrams are very useful for illustrating the system's inputs and outputs with other organizations, business units, and systems. Context diagrams are taught in Chapter 8. Use-case diagrams are taught in Part Five, Module A.

Several other techniques and skills are useful to develop an understanding of an existing system. Fact-finding techniques (taught in the next chapter) are critical to learning about any existing system. Also, joint requirements planning, or JRP techniques, (also taught in the next chapter) can accelerate this task. Finally, the ability to clearly communicate back to users what you've learned about a system is equally crucial.

Visit the McGraw-Hill Online Learning Center for this textbook and browse the on-line resource for a Microsoft Word template that can be used to document the scope lists for this task.

Task 2.2—Analyze Problems and Opportunities

In addition to learning about the current system, the project team must work with system owners and system users to *analyze problems and opportunities.* Problems and opportunities were identified earlier in the preliminary investigation phase, but those initial problems may be only symptoms of other problems not as well known or understood by the users. Besides, we haven't yet really analyzed any of those problems in the classical sense.

True problem analysis is a difficult skill to master, especially for inexperienced systems analysts. Experience suggests that most new systems analysts (and many system owners and users) try to solve problems without truly analyzing them. They might state a problem like this: "We need to . . ." or "We want to . . ." In doing so, they are stating the problem in terms of a solution. More effective problem solvers have learned to truly analyze the problem before stating any possible solution. They analyze each perceived problem for causes and effects.

> **Cause-and-effect analysis** is a technique in which problems are studied to determine their causes and effects.

An effect may be a symptom of a different, more deeply rooted or basic problem. That problem must also be analyzed for causes and effects until the causes and effects do not yield symptoms of other problems. Cause-effect analysis leads to true understanding of problems and can lead to not so obvious, but more creative and valuable solutions.

Systems analysts facilitate this task; however, all systems owners and users should actively participate in the cause-effect analysis. They are the problem domain experts. System designers and builders are not usually involved in this process unless they are asked to analyze technical problems that may exist in the current system.

PROBLEMS, OPPORTUNITIES, OBJECTIVES, AND CONSTRAINTS MATRIX

Project:	Member Services Information System	Project Manager:	Sandra Shepherd
Created by:	Robert Martinez	Last Updated by:	Robert Martinez
Date Created:	January 21, 2001	Date Last Updated:	January 31, 2001

CAUSE-AND-EFFECT ANALYSIS		SYSTEM IMPROVEMENT OBJECTIVES	
Problem or Opportunity	**Causes and Effects**	**System Objective**	**System Constraint**
1. Order response time is unacceptable.	1. Throughput has increased while number of order clerks was downsized. Time to process a single order has remained relatively constant. 2. System is too keyboard dependent. Many of the same values are keyed for most orders. Net result is (with the current system) each order takes longer to process than is ideal. 3. Data editing is performed by the AS/400. As that computer has approached its capacity, order edit responses have slowed. Because order clerks are trying to work faster to keep up with the volume, the number of errors has increased. 4. Warehouse picking tickets for orders were never designed to maximize the efficiency of order fillers. As warehouse operations grew, order filling delays were inevitable.	1. Decrease the time required to process a single order by 30%. 2. Eliminate keyboard data entry for as much as 50% of all orders. 3. For remaining orders, reduce as many keystrokes as possible by replacing keystrokes with point-and-click objects on the computer display screen. 4. Move data editing from a shared computer to the desktop. 5. Replace existing picking tickets with a paperless communication system between member services and the warehouse.	1. There will be no increase in the order processing workforce. 2. Any system developed must be compatible with the existing *Windows 95* desktop standard. 3. New system must be compatible with the already approved automatic identification system (for bar coding).

Page 1 of 5

FIGURE 5.11   *A Sample Cause-and-Effect Analysis*

Visit the McGraw-Hill Online Learning Center for this textbook and browse the on-line resource for a Microsoft Word template you can use to write your own cause-effect analysis. The second half of the template deals with scope.

As shown in Figure 5.10, the team's understanding of the system domain and business vocabulary triggers this task. This understanding of the problem domain is crucial because the team should not attempt to analyze problems unless they understand the domain in which those problems occur. The other informational input to this task is the initial problem statements (from the preliminary investigation phase). The deliverables of this task are the updated problem statements and the cause-effect analyses for each problem and opportunity. Figure 5.11 illustrates one way to document the cause-and-effect analysis.

Once again, fact-finding and JRP techniques are crucial to this task. These techniques as well as cause-effect analysis are taught in the next chapter.

Task 2.3—Analyze Business Processes

This task is appropriate only to *business process redesign* (BPR) projects or system development projects that build on or require significant business process redesign. In such a project, the team is asked to examine its business processes in much greater detail to measure the value added or subtracted by each process as it relates to the total organization. Business process analysis can be politically charged. System owners and users alike can become very defensive about their existing business processes. The analysts involved must keep the focus on the processes, not the people who perform them, and constantly remind everyone that the goal is to identify opportunities for fundamental business change that will benefit the business and everyone in the business.

Either a systems analyst(s) or business analyst(s) facilitates the task. Ideally, the analysts should be experienced, trained, or certified in BPR methods. The only other participants should be appropriate system owners and users. Business process analysis should avoid any temptation to focus on information technology solutions until well after the business processes have been redesigned for maximum efficiency. Some analysts find it useful to assume the existence of "perfect people" and "perfect technology" that can make anything "possible." They ask, "If the world were perfect, would we need this process?"

As depicted in Figure 5.10, a business process analysis task is dependent only on some problem domain knowledge (from Task 2.1). The deliverables of this task are business "as is" process models and process analyses. The process models can look very much like data flow diagrams (Figure 5.2) except they are significantly annotated to show: (1) the volume of data flowing through the processes, (2) the response times of each process, and (3) any delays or bottlenecks that occur in the system. The process analysis data provides additional information such as: *(a)* the cost of each process, *(b)* the value added by each process, and *(c)* the consequences of eliminating or streamlining the process. Based on the "as is" models and their analysis, the team develops "to be" models that redesign the business processes to eliminate redundancy and bureaucracy and to increase efficiency and service.

Several techniques are applicable to this task. Once again, fact-finding techniques and facilitated team meetings (Chapter 6) are invaluable. Also, process modeling techniques (Chapter 8) are critical to BPR success.

Given our understanding of the current system's scope, problems, and opportunities, we can now *establish system improvement objectives.* This task establishes the criteria against which any improvements to the system will be measured and identifies any constraints that may limit flexibility in achieving those improvements. The criteria for success should be measured in terms of objectives.

Task 2.4—Establish System Improvement Objectives

> An **objective** is a measure of success. It is something that you expect to achieve, if given sufficient resources.

Objectives represent the first attempt to establish expectations for any new system. In addition to objectives, we must also identify any known constraints.

> A **constraint** is something that will limit your flexibility in defining a solution to your objectives. Essentially, constraints cannot be changed.

Deadlines and budgets are examples of constraints.

The systems analysts facilitate this task. Other participants include the same system owners and users who have participated in other tasks in this problem analysis phase. Again, we are not yet concerned with technology; therefore, system designers and builders are not involved.

This task is triggered by the problem analyses completed in Tasks 2.2 and 2.3. For each verified and significant problem, the analysts and users should define specific system improvement objectives. They should also identify any constraints that may limit or prevent achieving the system improvement objectives.

System improvement objectives should be precise, measurable statements of business performance that define the expectations for the new system. Some examples might include the following:

- Reduce the number of uncollectible customer accounts by 50 percent within the next year.
- Increase by 25 percent the number of loan applications that can be processed during an eight-hour shift.
- Decrease by 50 percent the time required to reschedule a production lot when a workstation malfunctions.

The following is an example of a poor objective: *Create a delinquent accounts report.* This is a poor objective because it states only a requirement, not an actual objective. Now, let's reword that objective: *Reduce credit losses by 20 percent through earlier identification of delinquent accounts.* This gives us more flexibility.

The delinquent accounts report would work, but a customer delinquency inquiry might provide an even better way to achieve the same objective.

System improvement objectives may be tempered by identifiable constraints. Constraints fall into four categories as listed below (with examples):

— *Schedule:* The new system must be operational by April 15.

— *Cost:* The new system cannot cost more than $350,000.

— *Technology:* The new system must be on-line, or all new systems must use the DB2 database management system.

— *Policy:* The new system must use double-declining balance inventory techniques.

The last two columns of Figure 5.11 document typical system improvement objectives and constraints.

Task 2.5—Update the Project Plan

Recall that project scope is a moving target. Based on our initial understanding and estimates from the preliminary investigation phase, scope may have grown or diminished in size and complexity. (Growth is much more common!) Now that we're approaching the completion of the problem analysis phase, we should reevaluate project scope and *update the project plan* accordingly.

The project manager in conjunction with system owners and the entire project team facilitate this task. The systems analysts and system owners are the key individuals in this task. The analysts and owners should consider the possibility that not all objectives may be met by the new system. The new system may be larger than expected, and they may have to reduce the scope to meet a deadline. In this case the system owner will rank the objectives in order of importance. Then, if the scope must be reduced, the higher-priority objectives will tell the analyst what's most important.

As shown in Figure 5.10, this task is triggered by completion of the system improvement objectives. The initial project plan is another key input and the updated project plan is the key output. The updated plan should now include a detailed plan for the requirements analysis phase that should follow. The techniques and steps for updating the project plan were taught in Chapter 4, Project Management.

Task 2.6—Present Findings and Recommendations

As with the preliminary investigation phase, the problem analysis phase concludes with a communication task. We must *present findings and recommendations* to the business community.

The project manager and executive sponsor should jointly facilitate this task. Other meeting participants should include the entire project team, including assigned system owners, users, analysts, designers, and builders. As usual, the meeting should be open to any and all interested staff from the business community. Also, if an intranet website was established for the project, it should have been maintained throughout the problem analysis phases to ensure continuous communication of project progress.

This task is triggered by the completion of the updated project plan. Informational inputs include the problem analyses, any system models, the system improvement objectives, and any other documentation that was produced during the problem analysis phase. Appropriate elements are combined into the **system improvement objectives.** The format may be a report, a verbal presentation, or an inspection by an auditor or peer group (called a walkthrough). An outline for a written report is shown in Figure 5.12.

Interpersonal and communications skills are essential to this task. Systems analysts should be able to write a formal business report and make a business presentation without getting into technical issues or alternatives.

This concludes the problem analysis phase. One of the following decisions must be made after the conclusion of this phase

— Authorize the project to continue, as is, to the requirements analysis phase.

Visit the McGraw-Hill Online Learning Center for this textbook and browse the on-line resource for Microsoft Word and PowerPoint templates you can use to present your own problem analysis findings and recommendations to an appropriate audience.

FIGURE 5.12	*An Outline for a System Improvement Objectives and Recommendations Report*

Analysis of the Current _____ System

I. Executive summary (approximately 2 pages)
   A. Summary of recommendation
   B. Summary of problems, opportunities, and directives
   C. Brief statement of system improvement objectives
   D. Brief explanation of report contents
II. Background information (approximately 2 pages)
   A. List of interviews and facilitated group meetings conducted
   B. List of other sources of information that were exploited
   C. Description of analytical techniques used
III. Overview of the current system (approximately 5 pages)
   A. Strategic implications (if the project is part of or impacts an existing information systems strategic plan)
   B. Models of the current system
      1. Interface model (showing project scope)
      2. Data model (showing project scope)
      3. Geographic models (showing project scope)
      4. Process model (showing functional decomposition only)
IV. Analysis of the current system (approximately 5–10 pages)
   A. Performance problems, opportunities, and cause-effect analysis
   B. Information problems, opportunities, and cause-effect analysis
   C. Economic problems, opportunities, and cause-effect analysis
   D. Control problems, opportunities, and cause-effect analysis
   E. Efficiency problems, opportunities, and cause-effect analysis
   F. Service problems, opportunities, and cause-effect analysis
V. Detailed recommendations (approximately 5–10 pages)
   A. System improvement objectives and priorities
   B. Constraints
   C. Project plan
      1. Scope reassessment and refinement
      2. Revised master plan
      3. Detailed plan for the definition phase
VI. Appendixes
   A. Any detailed system models
   B. (other documents as appropriate)

— Adjust the scope, cost, and/or schedule for the project and then continue to the requirements analysis phase.

— Cancel the project due to either (1) lack of resources to further develop the system, (2) realization that the problems and opportunities are simply not as important as anticipated, or (3) realization that the benefits of the new system are not likely to exceed the costs.

With some level of approval of the system owners, the project can now proceed to the requirements analysis phase.

THE REQUIREMENTS ANALYSIS PHASE

Many inexperienced analysts make a critical mistake after completing the problem analysis phase. They begin looking at alternative solutions, particularly technical solutions. One of the most frequently cited errors in new information systems is illustrated in the statement, "Sure the system works, and it is technically impressive, but it just doesn't do what we needed it to do." The **requirements analysis phase** defines the business requirements for a new system.

The key in the requirements analysis phase is *what*, not *how!* Analysts are frequently so preoccupied with the *technical* solution that they inadequately define

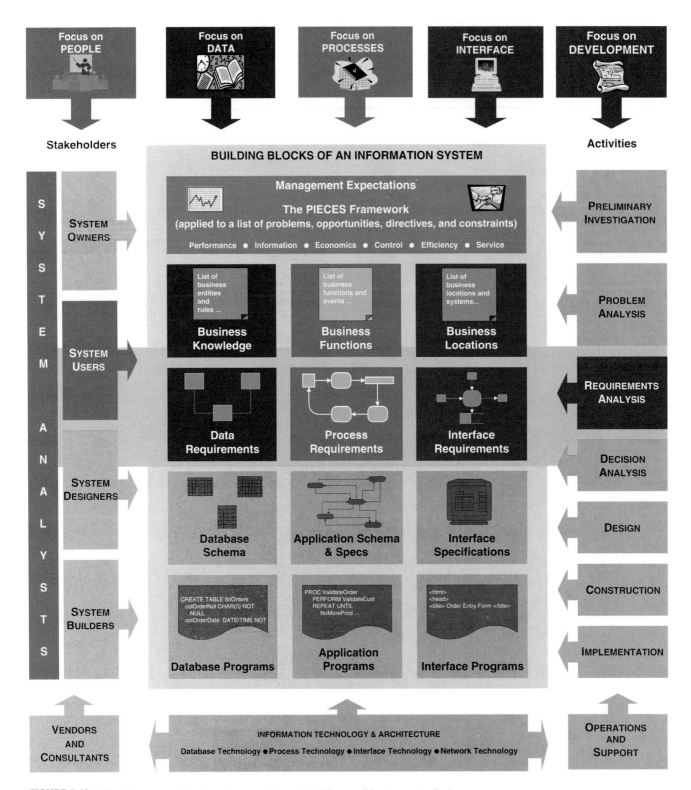

FIGURE 5.13   *The Context of the Requirements Analysis Phase of Systems Analysis*

the *business* requirements for that solution. The requirements analysis phase answers the question, "<u>What</u> do the users need and want from a new system?" This phase is critical to the success of any new information system! In different methodologies the requirements analysis phase might be called the *definition phase* or *logical design phase.*

You can never skip the requirements analysis phase, but some methodologies integrate the problem analysis and requirements analysis phases into a single phase. New systems will always be evaluated on whether or not they fulfilled business objectives and requirements, regardless of how impressive or complex the technological solution might be!

Once again, your information systems building blocks (Figure 5.13) can serve as a useful framework for documenting the information systems requirements. Notice that we are still concerned with the system users' perspectives. Also notice that for most of the building blocks, it is possible to draw system models to document the requirements for a new and improved system. Alternatively, prototypes could be built to discover requirements.

Figure 5.14 illustrates the typical tasks of the requirements analysis phase. The final deliverable and milestone is to produce a **business requirements statement** that will fulfill the system improvement objectives identified in the previous phase. One of the first things you may notice in this task diagram is that most of the tasks are not sequential as in previous task diagrams. Instead, many of these tasks occur in parallel as the team works toward completing the requirements statement.

The requirements analysis phase typically includes the following tasks:

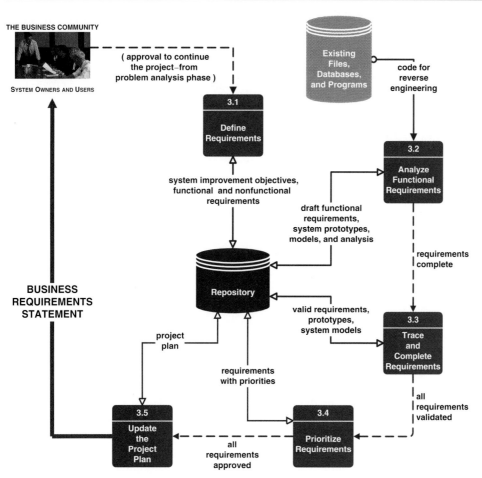

FIGURE 5.14

Tasks for the Requirements Analysis Phase of Systems Analysis

3.1 Define requirements.

3.2 Analyze functional requirements.

3.3 Trace and complete requirements.

3.4 Prioritize requirements.

3.5 Update the project plan.

Let's now examine each of these tasks in greater detail.

Task 3.1—Define Requirements

The initial task of the requirements analysis phase is to *identify requirements*. While this may seem to be an easy or trivial task, it is often the source of many errors, omissions, and conflicts. The foundation for this task was established in the problem analysis phase when we identified system improvement objectives. Minimally, this task translates those objectives into an outline of functional and nonfunctional requirements that will be needed to meet the objectives.

> A **functional requirement** is a description of activities and services a system must provide.

Functional requirements are frequently identified in terms of inputs, outputs, processes, and stored data that are needed to satisfy the system improvement objectives.

> A **nonfunctional requirement** is a description of other features, characteristics, and constraints that define a satisfactory system.

Examples of nonfunctional requirements include performance (throughput and response time); ease of learning and use; budgets, costs, and cost savings; timetables and deadline; documentation and training needs; quality management; and security and internal auditing controls.

Rarely will this definition task identify *all* the functional or nonfunctional business requirements. But the outline will frame your thinking as you proceed to later tasks that will add new requirements and details to the outline. Thus, neither completeness nor perfection is a goal of this task.

Systems analysts facilitate the task. They also document the results. Obviously, system users are the primary source of business requirements. Some system owners may also participate in this task because they played a role in framing the system improvement objectives that will guide the task. System designers and builders should not be involved because they tend to prematurely redirect the focus to the technology and technical solutions.

As shown in Figure 5.14, this task (and phase) is triggered by the approval to continue the project from the problem analysis phase. The key input is the system improvement objectives from the problem analysis phase (via the repository). Of course, all relevant information from the problem analysis phase is available from the repository for reference as needed.

The only deliverables of this task are the draft functional and nonfunctional requirements. Various formats can work. In its simplest format, the outline could be divided into four logical sections: the original list of system improvement objectives and for each objective a sublist of *(a)* inputs, *(b)* processes, *(c)* outputs, and *(d)* stored data needed to fulfill the objective. Many CASE tools, such as *System Architect* provide facilities for documenting raw, narrative requirements to its dictionary, encyclopedia, or repository. This can be useful as a completeness check against the system models or prototypes that will be developed in later tasks to fulfill these requirements.

The PIECES framework that was earlier used to identify problems, opportunities, and constraints can also be used as a framework for defining draft requirements.

Several techniques are applicable to this task. Joint requirement planning (JRP) is the preferred technique for rapidly outlining business requirements. Alternatively, the analysts could use other fact-finding methods such as surveys and interviews. Both JRP and fact-finding are taught in the next chapter.

Identifying draft requirements is only the tip of the iceberg. We must *analyze functional requirements* such that they can be verified and communicated to both business and technical audiences. Business users must understand requirements so they can prioritize the needs and justify the expenditures for any technical solution. Technicians must understand functional requirements so they can transform them into appropriate technical solutions. There are two approaches to functional requirements documentation and validation, *system modeling* and *prototyping*.

Task 3.2—Analyze Functional Requirements

Recall that system models are pictures (similar to flowcharts) that express requirements or designs. The best systems analysts can develop logical system models that provide no hint of how the system will or might be implemented. This is called *logical design*.

> **Logical system models** depict <u>what</u> a system is or <u>what</u> a system must do— *not* how the system will be implemented. Because logical models depict the *essential* requirements of a system, they are sometimes called *essential system models*. (*The opposite of logical system models are physical system models that are used to depict how a system will implement the logical system requirements.*)

Logical models express business requirements—sometimes referred to as the **logical design**—as opposed to the technical solution, which is called the *physical design*. In theory, by focusing on the logical design of the system, the project team will:

- Appropriately separate business concerns from technical solutions.
- Be more likely to conceive and consider new and different ways to improve business processes.
- Be more likely to consider different, alternative technical solutions (when the time comes for physical design).

An alternative or complementary approach to system modeling of business requirements is to build discovery prototypes. Prototyping is typically used in the requirements analysis phase to establish user interface requirements (inputs and outputs). Although discovery prototyping is optional, it is frequently applied to systems development projects, especially when the users are having difficulty stating or visualizing their business requirements. The philosophy is that the users will recognize their requirements when they see them.

Regardless of the approach taken, the intent is to *analyze* the requirements. Requirements are analyzed for accuracy, urgency, consistency, flexibility, and feasibility, to name a few criteria. Recall that the classical definition of analysis is to separate a system into its component parts for the purpose of studying how those components interact to accomplish the objectives of the system. System modeling (and to a lesser degree prototyping) fits this definition perfectly. System modeling is almost always based on rules for completeness, correctness, and consistency of the components that make up the models. You will soon be studying these rules in our system modeling chapters.

Systems analysts facilitate this analysis task. They also document and analyze the results. As usual, system users are the primary source of factual input to the task. System owners tend to be less involved due to the higher level of detail. System designers should not be involved because they tend to prematurely redirect the focus to the technology and technical solutions. System builders should only be involved to the extent of any work they may contribute in the form of building prototypes.

Figure 5.14 demonstrates that this task is not dependent on the completion of any prior task. As draft functional requirements become known or approved, the analysts and users can immediately begin to respond by constructing system models and prototypes. As described earlier in this chapter, it may be possible to *reverse engineer* some system models directly from existing databases and program libraries. Also, prototypes can often be reverse engineered into system models. If you use model-driven techniques, you will probably use some combination of the following models that correspond to your information system building blocks:

- DATA—All systems capture and store data. *Data models* such as entity relationship diagrams are used to document and analyze the detailed data requirements for new systems. These data models eventually serve as the starting point for designing databases.

- PROCESSES—All systems perform work. *Process models* such as data flow diagrams are used to model the work flow through systems; the detailed content of all inputs, outputs, and files; and the detailed logic of all processes. These process models serve as a starting point for designing computer applications and programs.

- INTERFACES—No system exists in isolation from people, other systems, or other organizations. *Interface models,* such as context diagrams (Figure 5.15) and use-case diagrams (Part Five, Module A) depict the external inputs and outputs to and from the system, and their sources and destinations. These interface models serve as the basis for designing user and system interfaces.

As was suggested earlier in this chapter, many systems analysts are now experimenting with *object models* as an alternative to data and process models.

FIGURE 5.15

A Simple Interface Model (also called a context diagram)

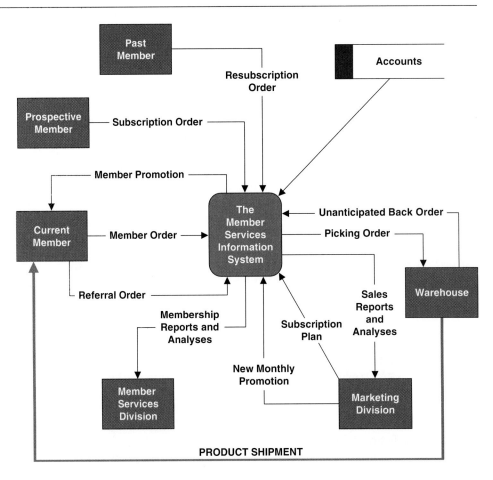

The following techniques are essential for analyzing requirements.

- Data, process, and interface modeling techniques, taught extensively in Chapters 7 and 8, are necessary to graphically visualize requirements so that they <u>can</u> be analyzed. Emerging object modeling techniques are taught in the appendixes.

- Most systems analysts use either diagramming software, such as *Visio Professional,* or CASE software, such as *System Architect,* to construct and maintain system models and their underlying detailed documentation. CASE tools offer the advantage of analysis; they can check for completeness, consistency, and correctness in system models. We used *Visio Professional* to draw most of the line art in this book, and we used *System Architect* to demonstrate the advantages and capabilities of a full-blown CASE tool. Various diagramming and CASE tool software bundling options are described in the preface.

- Prototyping techniques are demonstrated throughout the book; however, it is beyond this book's scope to teach prototyping techniques. Fortunately, most schools are teaching prototyping technology within the context of their programming and database curricula. Look for courses that use development tools such as Microsoft *Access* and *Visual Basic.* While these courses typically teach such technologies in the context of complete application development, you will quickly realize how the basic techniques can be used to rapidly construct prototypes or even throwaway prototypes.

- You can't build models without facts. Fact-finding techniques are taught in the next chapter.

Task 3.3—Trace and Complete Requirements

Given the completed set of requirements and analysis based on some combination of system models and prototypes, it is recommended that system owners and users be given a final opportunity to validate requirements. Presumably, each requirement has been verified in some form of model or prototype. This new task is more of a completeness check. In Task 3.1, we established an outline of draft requirements, usually in narrative or outline form. Now we check to ensure that those draft requirements have been adequately defined in some form, either model or prototype. Some analysts call this *requirements tracing,* that is, tracing each system model and/or prototype back to the functional requirement that it fulfills and ensuring that all functional requirements are included in our system models and prototypes.

When tracing requirements, it may be necessary to revisit and make changes and additions to system models and prototypes. Many CASE tools offer facilities for linking raw requirements with the system models that fulfill those requirements.

In addition to tracing the functional requirements for completeness, the project team needs to associate nonfunctional requirements with the functional requirements. For example, a nonfunctional security requirement may need to be associated with various system models, prototypes, and/or detailed specifications to ensure that these nonfunctional aspects of the target system will be fulfilled in the system design and construction phases.

The project manager and systems analysts facilitate the task. System owners and users should obviously play the key role of ensuring completeness of the requirements specification.

Figure 5.14 shows the inputs and outputs that we have described for this task: the valid requirements (both functional and nonfunctional), final system models (which may be updated and supplemented by the task), and final prototypes (which also may be updated and supplemented by this task).

Task 3.4—Prioritize Requirements

We stated earlier that the success of a systems development project can be measured in terms of the <u>degree</u> to which business requirements are met. But not all requirements are created equal. If a project gets behind schedule or over budget, it may be useful to recognize which requirements are more important than others. Thus, given the validated requirements, system owners and users should *prioritize business requirements.*

Prioritization of business requirements also enables a popular technique called timeboxing.

> **Timeboxing** is a technique that delivers information systems functionality and requirements through versioning. The development team selects the smallest subset of the system that, if fully implemented, will return immediate value to the system owners and users. That subset is developed, ideally with a time frame of six to nine months or less. Subsequently, value-added versions of the system are developed in similar time frames.

Timeboxing requires that the priorities be clearly understood.

Systems analysts facilitate the prioritization task. System owners and users establish the actual priorities. System designers and builders are not involved in the task. The task is triggered by the validated requirements, an event or milestone. You cannot adequately prioritize an incomplete set of requirements. The deliverable of this task is the requirements with priorities. Priorities can be classified according to their relative importance.

— A *mandatory requirement* is one that must be fulfilled by the minimal system, version 1.0! The system is useless without it. There is a temptation to label too many requirements as mandatory. A mandatory requirement cannot be ranked because it is essential to any solution. In fact, if an alleged mandatory requirement can be ranked, it is actually a desirable requirement.

— A *desirable requirement* is one that is not essential to version 1.0. It may still be essential to the vision of some future version. Desirable requirements can and should be ranked. Using version numbers as the ranking scheme is an effective way to communicate and categorize desirable requirements.

Task 3.5—Update the Project Plan

Here again, recall that project scope is a moving target. Now that we've identified the business system requirements, we should step back and redefine our understanding of the project scope and update our project plan accordingly. The team must consider the possibility that the new system may be larger than originally expected. If so, the team must adjust the schedule, budget, or scope accordingly. We should also secure approval to continue the project into the next phase. (Work may have already started on the design phases; however, the decisions still require review.)

The project manager in conjunction with system owners and the entire project team facilitate this task. As usual, the project manager and system owners are the key individuals in this task. They should consider the possibility that the requirements now exceed the original vision that was established for the project and new system. They may have to reduce the scope to meet a deadline or increase the budget to get the job done.

As shown in Figure 5.16, this task is triggered by completion of the approved requirements, including priorities. The up-to-date project plan is the other key input and it is updated in the repository as appropriate. The completed business requirements statement completes the task, as well as the requirements analysis phase. The tools, techniques, and steps for maintenance of the project plan were covered in Chapter 4, Project Management.

A consolidation of all requirements, system models, prototypes, and supporting documentation comprises the business requirements statement. All elements

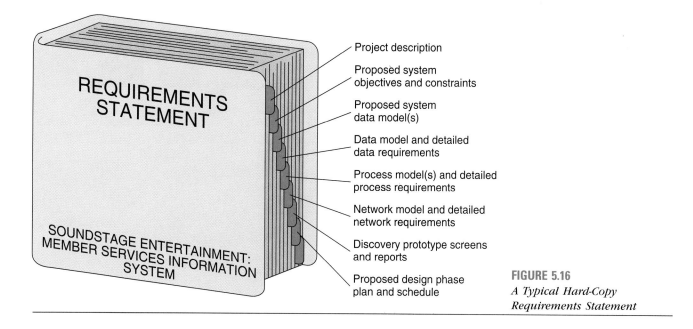

Project description

Proposed system objectives and constraints

Proposed system data model(s)

Data model and detailed data requirements

Process model(s) and detailed process requirements

Network model and detailed network requirements

Discovery prototype screens and reports

Proposed design phase plan and schedule

FIGURE 5.16
A Typical Hard-Copy Requirements Statement

of the requirements statement are stored in the repository, but most systems analysts find it useful to keep a printed copy of that documentation for reference and reporting. This hard-copy requirements statement might be organized as shown in Figure 5.16. It is not recommended that typical users be exposed to this rather imposing document. The contents of the requirements statement are best presented to system owners and users in subsets (for example, only the data models and their associated documentation).

Ongoing Requirements Management

The requirements analysis phase is now complete. Or is it? It was once popular to freeze the business requirements before beginning the system design and construction phases. But today's economy has become increasingly fast paced. Businesses are measured on their ability to quickly adapt to constantly changing requirements and opportunities. Information systems can be no less responsive than the business itself. Thus, requirements analysis really never ends. While we quietly transition to the remaining phases of our project, there remains an ongoing need to continuously manage requirements through the course of the project and the lifetime of the system.

Requirements management defines a process for system owners, users, analysts, designers, and builders to submit proposed changes to requirements for a system. The process specifies how changes are to be requested and documented, how they will be logged and tracked, when and how they will be assessed for priority, and how they will be eventually be satisfied (if they are ever satisfied).

THE DECISION ANALYSIS PHASE

Given the business requirements for an improved information system, we can finally address how the new system—including computer-based alternatives—might operate. The **decision analysis phase** identifies candidate solutions, analyzes those candidate solutions, and recommends a target system that will be designed, constructed, and implemented. Chances are that someone has already championed a vision for a technical solution. But alternative solutions, perhaps better ones, nearly always exist. During the decision analysis phase, it is imperative that you identify options, analyze those options, and then sell the best solution based on the analysis.

Once again, your information systems building blocks (Figure 5.17) can serve as a useful framework for the decision analysis phase. Information technology

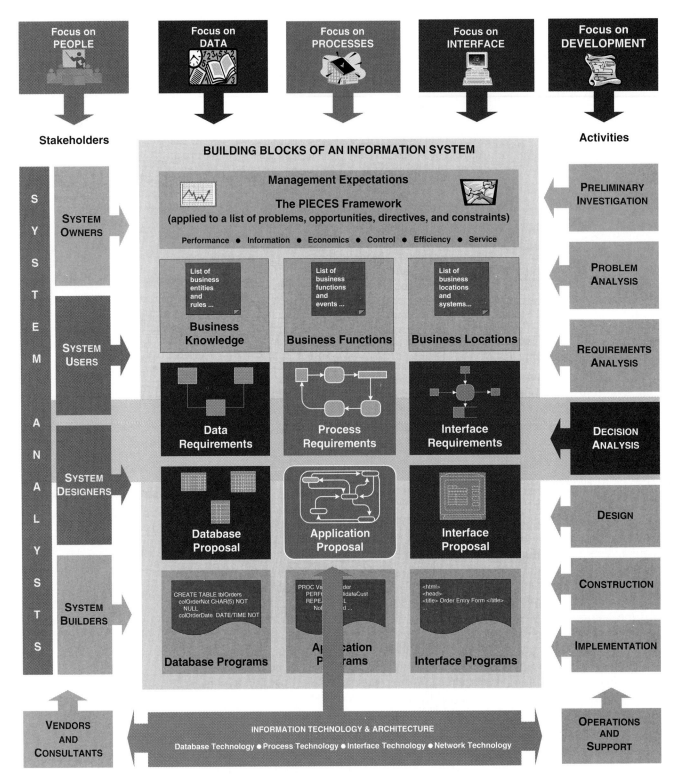

Focus on PEOPLE

Focus on DATA

Focus on PROCESSES

Focus on INTERFACE

Focus on DEVELOPMENT

Stakeholders

Activities

S
Y
S
T
E
M

A
N
A
L
Y
S
T
S

SYSTEM OWNERS

SYSTEM USERS

SYSTEM DESIGNERS

SYSTEM BUILDERS

VENDORS AND CONSULTANTS

BUILDING BLOCKS OF AN INFORMATION SYSTEM

Management Expectations

The PIECES Framework
(applied to a list of problems, opportunities, directives, and constraints)

Performance ● Information ● Economics ● Control ● Efficiency ● Service

List of business entities and rules ...

List of business functions and events ...

List of business locations and systems...

Business Knowledge

Business Functions

Business Locations

Data Requirements

Process Requirements

Interface Requirements

Database Proposal

Application Proposal

Interface Proposal

CREATE TABLE tblOrders
colOrderNot CHAR(5) NOT NULL
colOrderDate DATE/TIME NOT

PROC Ve...der
PERF...lidateCust
REPE...L
No...d ...

<html>
<head>
<title> Order Entry Form </title>
...

Database Programs

Application Programs

Interface Programs

INFORMATION TECHNOLOGY & ARCHITECTURE
Database Technology ● Process Technology ● Interface Technology ● Network Technology

PRELIMINARY INVESTIGATION

PROBLEM ANALYSIS

REQUIREMENTS ANALYSIS

DECISION ANALYSIS

DESIGN

CONSTRUCTION

IMPLEMENTATION

OPERATIONS AND SUPPORT

FIGURE 5.17   *The Context of the Decision Analysis Phase of Systems Analysis*

and architecture now begin to influence the decisions we must make. In some cases, we must work within standards. In other cases, we can look to apply different or emerging technology. You should also notice that our perspective is in transition—from those of the system users to those of the system designers. Again, this reflects our transition from pure business concerns to technology. But we are not yet designing. The building blocks indicate our goal to develop a proposal that will fulfill requirements.

Figure 5.18 illustrates the typical tasks of the decision analysis phase. The final deliverable and milestone is to produce a **system proposal** that will fulfill the business requirements identified in the previous phase. The decision analysis phase typically includes the following tasks:

4.1 Identify candidate solutions.

4.2 Analyze candidate solutions.

4.3 Compare candidate solutions.

4.4 Update the project plan.

4.5 Recommend a solution.

Let's now examine each of these tasks in greater detail.

Given the business requirements established in the definition phase of systems analysis, we must first identify alternative candidate solutions. Some candidate solutions will be posed by design ideas and opinions from system owners and users. Others may come from various sources including: systems analysts, systems designers, technical consultants, and other IS professionals. And some technical choices

Task 4.1—Identify Candidate Solutions

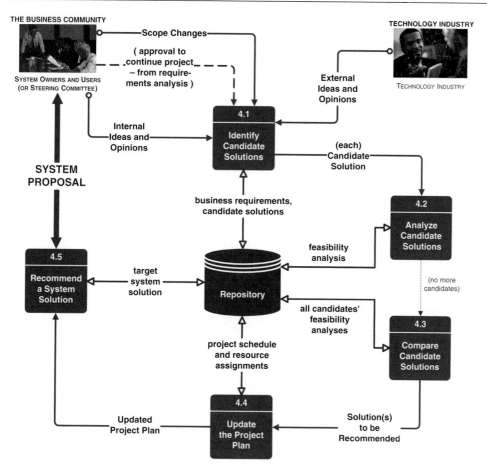

FIGURE 5.18

Tasks for the Decision Analysis Phase of Systems Analysis

may be limited by a predefined, approved technology architecture. It is not the intent of this task to evaluate the candidates. Rather, it simply defines possible candidate solutions to be considered.

The systems analysts facilitate this task. System owners and users are not normally directly involved in this task, but they may contribute ideas and opinions that start the task. For example, an owner or user may have read an article or perhaps learned how some competitor or acquaintance's similar system was implemented. In either case, it is politically sound to consider the ideas. System designers and builders such as database administrators, network administrators, technology architects, and programmers are also sources of ideas and opinions.

As shown in Figure 5.18, this task is formally triggered by the approval to continue the project from the requirements analysis phase. In reality, ideas and opinions have been generated and captured since the preliminary investigation phase—it is human nature to suggest solutions throughout any problem-solving process. In addition to the project team, both internal and external sources can generate ideas and opinions. Each idea generated is considered to be a candidate solution to the business requirements.

The amount of information describing the characteristics of any one candidate solution may become overwhelming. A candidate matrix, such as Figure 5.19, is a useful tool for effectively capturing, organizing, and comparing the characteristics for different candidate solutions.

As has been the case throughout this chapter, fact-finding and group facilitation techniques such as JRP are the main techniques used to research candidate system solutions. Fact-finding and group facilitation techniques are taught in the next chapter. Also, Chapter 9, Feasibility and the Systems Proposal, will teach you how to generate candidate system solutions and document them in the matrix.

Visit the McGraw-Hill Online Learning Center for this textbook and browse the on-line resource for a Microsoft Word template you can use to create your own candidate systems matrix.

Task 4.2—Analyze Candidate Solutions

Each candidate system solution must be analyzed for feasibility. This can occur as each candidate is identified or after all candidates have been identified. Feasibility analysis should not be limited to costs and benefits. Most analysts evaluate solutions against at least four criteria:

— *Technical feasibility.* Is the solution technically practical? Does the staff have the technical expertise to design and build this solution?

— *Operational feasibility.* Will the solution fulfill the user's requirements? To what degree? How will the solution change the user's work environment? How do users feel about such a solution?

— *Economic feasibility.* Is the solution cost-effective?

— *Schedule feasibility.* Can the solution be designed and implemented within an acceptable time period?

When completing this task, the analysts and users must take care not to compare the candidates. The feasibility analysis is performed on each individual candidate without regard to the feasibility of other candidates. This approach discourages the analyst and users from prematurely making a decision concerning which candidate is the best.

Again, the systems analysts facilitate the task. Usually systems owners and users analyze operational, economic, and schedule feasibility. Systems designers and builders usually contribute to those analyses and play the critical role in analyzing technical feasibility.

Figure 5.18 shows that the task is triggered by the completion of each candidate solution; however, it is acceptable to delay the task until all candidate solutions have been identified. Input to the actual feasibility analyses comes from the various team participants; however, it is not uncommon for external experts (and

Characteristics	Candidate 1	Candidate 2	Candidate 3	Candidate ...
Portion of System Computerized Brief description of that portion of the system that would be computerized in this candidate.	COTS package Platinum Plus from Entertainment Software Solutions would be purchased and customized to satisfy Member Services required functionality.	Member Services and warehouse operations in relation to order fulfillment.	Same as candidate 2.	
Benefits Brief description of the business benefits that would be realized for this candidate.	This solution can be implemented quickly because it's a purchased solution.	Fully supports user required business processes for SoundStage Inc. Plus more efficient interaction with member accounts.	Same as candidate 2.	
Servers and Workstations A description of the servers and workstations needed to support this candidate.	Technically architecture dictates Pentium Pro, MS Windows NT class servers and Pentium, MS Windows NT 4.0 workstations (clients).	Same as candidate 1.	Same as candidate 1.	
Software Tools Needed Software tools needed to design and build the candidate (e.g., database management system, emulators, operating systems, languages, etc.). Not generally applicable if applications software packages are to be purchased.	MS Visual C++ and MS Access for customization of package to provide report writing and integration.	MS Visual Basic 5.0 System Architect 3.1 Internet Explorer	MS Visual Basic 5.0 System Architect 3.1 Internet Explorer	
Application Software A description of the software to be purchased, built, accessed, or some combination of these techniques.	Package Solution	Custom Solution	Same as candidate 2.	
Method of Data Processing Generally some combination of: on-line, batch, deferred batch, remote batch, and real-time.	Client/Server	Same as candidate 1.	Same as candidate 1.	
Output Devices and Implications A description of output devices that would be used, special output requirements (e.g., network, preprinted forms, etc.), and output considerations (e.g., timing constraints).	(2) HP4MV department laser printers (2) HP5SI LAN laser printers	(2) HP4MV department laser printers (2) HP5SI LAN laser printers (1) PRINTRONIX bar-code printer (includes software & drivers) Web pages must be designed to VGA resolution. All internal screens will be designed for SVGA resolution.	Same as candidate 2.	
Input Devices and Implications A description of input methods to be used, input devices (e.g., keyboard, mouse, etc.), special input requirements (e.g., new or revised forms from which data would be input), and input considerations (e.g., timing of actual inputs).	Keyboard & mouse	Apple "Quick Take" digital camera and software (15) PSC Quickscan laser bar-code scanners (1) HP Scanjet 4C Flatbed Scanner Keyboard & mouse	Same as candidate 2.	
Storage Devices and Implications Brief description of what data would be stored, what data would be accessed from existing stores, what storage media would be used, how much storage capacity would be needed, and how data would be organized.	MS SQL Server DBMS with 100GB arrayed capability.	Same as candidate 1.	Same as candidate 1.	

FIGURE 5.19   *A Candidate Systems Matrix*

influences) to also provide data. Each candidate's feasibility analysis for each candidate is saved in the repository for later comparison to other candidates.

Fact-finding techniques again play a role in this systems analysis task. But the ability to perform a feasibility analysis on a candidate system solution is essential. That technique is taught in Chapter 9, Feasibility and the System Proposal.

Task 4.3—Compare Candidate Solutions

Visit the McGraw-Hill Online Learning Center for this textbook and browse the on-line resource for a Microsoft Word template you can use to create your own feasibility matrix.

Once the feasibility analysis has been completed for each candidate solution, we can now compare the candidates and select one or more solutions to recommend to the system owners and users. First, any infeasible candidates are usually eliminated from further consideration. Since we are looking for the most feasible solution of those remaining, we will identify and recommend the candidate that offers the "best overall" combination of technical, operational, economic, and schedule feasibilities. Rarely is one candidate the best in all the criteria.

Once again, the systems analysts facilitate the task. System designers and builders should be available to answer any technical feasibility questions. But ultimately, the system owners and users should be empowered to drive the final analysis and recommendation.

In Figure 5.18, this task is triggered by the completion of the feasibility analysis of all candidate solutions (no more candidate solutions). The input is all of the candidates' feasibility analyses. Once again, a matrix can be used to communicate the large volume of information about candidate solutions. The feasibility matrix in Figure 5.20 allows for a side-by-side comparison of the different feasibility analyses for a number of candidates.

The deliverable of this task is the solution(s) to be recommended. If more than one solution is recommended, priorities should be established.

Again, feasibility analysis techniques (and the matrix) will be taught in Chapter 9, Feasibility and the System Proposal.

Task 4.4—Update the Project Plan

A recurring theme exists throughout this chapter. We are continually updating our project plan as we learn more about a system, its problems, its requirements, and its solutions. We are adjusting scope accordingly. Thus, based on our recommended solution(s), we should once again reevaluate project scope and *update the project plan* accordingly.

The project manager in conjunction with system owners and the entire project team facilitate this task. The systems analysts and system owners are the key individuals in this task. But because we are transitioning into technical system design, we need to begin involving our system designers and builders in the project plan updates.

As shown in Figure 5.18, this task is triggered by completion of the solution(s) to be recommended. The latest project schedule and resource assignments must be reviewed and updated. The updated project plan is the key output. The updated plan should now include a detailed plan for the system design phase that will follow. The techniques and steps for updating the project plan were taught in Chapter 4, Project Management.

Task 4.5—Recommend a Solution

As with the preliminary investigation and problem analysis phases, the decision analysis phase concludes with a communication task. We must *recommend a system solution* to the business community.

The project manager and executive sponsor should jointly facilitate this task. Other meeting participants should include the entire project team, including assigned system owners, users, analysts, designers, and builders. And as usual, the meeting should be open to any and all interested staff from the business community. Also, if an intranet website was established for the project, it should have been maintained throughout the problem analysis phases to ensure continuous communication of project progress.

This task is triggered by the completion of the updated project plan. The target system solution (from Task 4.3) is reformatted for presentation as a system proposal. The format may be a report, a verbal presentation, or an inspection by an auditor or peer group (called a walkthrough). An outline for a written report is shown in Figure 5.21.

Visit the McGraw-Hill Online Learning Center for this textbook and browse the on-line resource for both Microsoft Word and PowerPoint templates you can use to create your own system proposal.

Interpersonal and communications skills are essential to this task. Soft skills such as salesmanship and persuasion become important. (Many schools offer speech and communications courses on these subjects.) Systems analysts should be able to write a formal business report and make a business presentation without getting into technical issues or alternatives.

This concludes the decision analysis phase. And it also concludes our coverage of systems analysis.

THE NEXT GENERATION OF SYSTEMS ANALYSIS

Predicting the future of systems analysis is not easy, but we'll make an attempt. CASE technology will continue to improve, making it easier to model system requirements. First, CASE tools will include object modeling to support emerging object-oriented analysis techniques. While some CASE tools will be purely object-

FIGURE 5.20   *A Feasibility Analysis Matrix*

Feasibility Criteria	Weight	Candidate 1	Candidate 2	Candidate 3	Candidate ...
Operational Feasibility **Functionality**. A description of to what degree the candidate would benefit the organization and how well the system would work. **Political**. A description of how well received this solution would be from both user management, user, and organization perspective.	30%	Only supports Member Services requirements and current business processes would have to be modified to take advantage of software functionality Score: 60	Fully supports user required functionality. Score: 100	Same as candidate 2. Score: 100	
Technical Feasibility **Technology**. An assessment of the maturity, availability (or ability to acquire), and desirability of the computer technology needed to support this candidate. **Expertise**. An assessment of the technical expertise needed to develop, operate, and maintain the candidate system.	30%	Current production release of Platinum Plus package is version 1.0 and has only been on the market for 6 weeks. Maturity of product is a risk and company charges an additional monthly fee for technical support. Required to hire or train C++ expertise to perform modifications for integration requirements. Score: 50	Although current technical staff has only Powerbuilder experience, the senior analysts who saw the MS Visual Basic demonstration and presentation have agreed the transition will be simple and finding experienced VB programmers will be easier than finding Powerbuilder programmers and at a much cheaper cost. MS Visual Basic 5.0 is a mature technology based on version number. Score: 95	Although current technical staff is comfortable with Powerbuilder, management is concerned with recent acquisition of Powerbuilder by Sybase Inc. MS SQL Server is a current company standard and competes with SYBASE in the Client/Server DBMS market. Because of this we have no guarantee future versions of Powerbuilder will "play well" with our current version SQL Server. Score: 60	
Economic Feasibility Cost to develop: Payback period (discounted): Net present value: Detailed calculations:	30%	 Approximately $350,000. Approximately 4.5 years. Approximately $210,000. See Attachment A. Score: 60	 Approximately $418,040. Approximately 3.5 years. Approximately $306,748. See Attachment A. Score: 85	 Approximately $400,000. Approximately 3.3 years. Approximately $325,500. See Attachment A. Score: 90	
Schedule Feasibility An assessment of how long the solution will take to design and implement.	10%	Less than 3 months. Score: 95	9–12 months Score: 80	9 months Score: 85	
Ranking	**100%**	**60.5**	**92**	**83.5**	

FIGURE 5.21   *An Outline for a Typical System Proposal*

I.   Introduction
    A. Purpose of the report
    B. Background of the project leading to this report
    C. Scope of the project
    D. Structure of the report
II. Tools and techniques used
    A. Solution generated
    B. Feasibility analysis (cost–benefit)
III. Information systems requirements
IV. Alternative solutions and feasibility analysis
V.   Recommendations
VI. Appendixes

---

oriented, we believe that the demand for other types of modeling support (e.g., data modeling for databases and process modeling for BPR), will place a premium on comprehensive CASE tools that can support many types of models. Second, the reverse engineering technology in CASE tools will continue to improve our ability to more quickly generate first-draft system models from existing databases and application programs.

In the meantime, RAD technology will continue to enable accelerated analysis approaches such as prototyping. We also expect the trend for RAD and CASE tools to interoperate through reverse and forward engineering to further simplify both system modeling and discovery prototyping.

Object-oriented analysis will eventually replace structured analysis and information engineering as the best practice for systems analysis. This change may not occur as rapidly as object purists would like, but it will occur all too rapidly for a generation of systems analysts who were skilled in the older methods. There is a grand opportunity for talented young analysts to lead the transition; however, career opportunities will remain strong for analysts who know data modeling, which will continue to be used for database design. Also, the process modeling renaissance will continue as BPR projects continue to proliferate.

We also predict our systems proposals will continue to get more interesting. As the Internet, e-commerce, and e-business become increasingly pervasive in our economy, systems analysts will be proposing new alternatives to old problems. There will be a fundamental change in business and information systems to use these new technologies.

One thing will *not* change! We will continue to need systems analysts who understand how to fundamentally investigate and analyze business problems and define the logical business requirements as a preface to system design. But we will all have to do that with increased speed and accuracy to meet the accelerated systems development schedules required in tomorrow's faster-paced economy.

W H E R E   D O   Y O U   G O   F R O M   H E R E ?

This chapter provided a detailed overview of the systems analysis phases of a project. You are now ready to learn some of the skills introduced in this chapter. For most students, this would be the ideal time to study fact-finding techniques that were identified as critical to almost every phase and task that was described in this chapter. Chapter 6 teaches these skills.

Or you could jump directly into the system modeling chapters. The sequencing of the system modeling chapters is flexible; however, we prefer and recommend Chapter 7, Data Modeling and Analysis, be studied first. All information systems include databases, and data modeling is an essential skill for database development. Also, it is easier to synchronize early data models with later process models than vice versa. Your instructor may prefer that you first study Chapter 8, Process Modeling. Advanced courses may jump straight to the appendices. Module A teaches *object modeling* using the emerging object-oriented analysis techniques.

If you do jump straight to a system modeling chapter from this chapter, make a commitment to return to Chapter 6 to study the fact-finding techniques. Regardless of how well you master system modeling, that modeling skill depends on your ability to discover and collect the business facts that underlie the models.

For those of you who have already completed a systems analysis course, this chapter was probably scheduled as only a review or context for systems design. For you, we suggest that you merely review the system modeling chapters and proceed directly to Chapter 10, Systems Design. That chapter will pick up where this chapter left off.

Do you want to know still more about the *process* of systems analysis? The On-line Learning Center for this book includes several more details and templates for the tasks described in this book. You can drill on the tasks included in our diagrams or examine those details in various Microsoft *Project* templates.

SUMMARY

1. Formally, systems analysis is the dissection of a system into its component pieces. As a problem-solving phase, it precedes systems design. With respect to information systems development, systems analysis is the preliminary investigation of a proposed project, the study and problem analysis of the existing system, the requirements analysis of business requirements for the new system, and the decision analysis for alternative solutions to fulfill the requirements.

2. The results of systems analysis are stored in a repository for use in later phases and projects.

3. There are several popular or emerging strategies for systems analysis. These techniques can be used in combination with one another.

   a. Model-driven analysis techniques emphasize the drawing of pictorial system models that represent either a current reality or a target vision of the system.

      1) Structured analysis is a technique that focuses on modeling processes.

      2) Information engineering is a technique that focuses on modeling data.

      3) Object-oriented analysis is a technique that focuses on modeling objects that encapsulate the concerns of data and processes that act on that data.

   b. Accelerated analysis approaches emphasize the construction of working models of a system in an effort to

accelerate systems analysis.

      1) Discovery prototyping is a technique that focuses on building small-scale, functional subsystems to discover requirements.

      2) Rapid architecture analysis attempts to automatically generate system models from either prototypes or existing systems. The automatic generation of models requires reverse engineering technology.

   c. Both model-driven and accelerated system analysis approaches are dependent on requirements discovery techniques to identify or extract problems and requirements from system owners and users.

      1) Fact-finding is the formal process of using research, interviews, questionnaires, sampling, and other techniques to collect information.

      2) Joint requirements planning (JRP) techniques use facilitated workshops to bring together all interested parties and accelerate fact-finding.

   d. Business process redesign focuses on simplifying and streamlining fundamental business processes before applying information technology to those processes.

4. Each phase of systems analysis (preliminary investigation, problem analysis, requirements analysis, and decision analysis) can be understood in the context of the information system building blocks: DATA, PROCESSES, and INTERFACES.

5. The preliminary investigation phase determines the worthiness of the project and creates a plan to complete those projects deemed worthy of a detailed study and analysis. To accomplish the preliminary investigation phase, the systems analyst will work with the system owners and users to: *(a)* list problems, opportunities, and directives; *(b)* negotiate preliminary scope; *(c)* assess project worth; *(d)* plan the project, and *(e)* present the project to the business community. The deliverable for the preliminary investigation phase is a project charter that must be approved by system owners and/or a decision-making body, commonly referred to as the steering committee.

6. The purpose of the problem analysis phase is to answer the questions: Are the problems really worth solving? and Is a new system really worth building? To answer these questions, the problem analysis phase thoroughly analyzes the alleged problems and opportunities first identified in the preliminary investigation phase. To complete the problem analysis phase, the analyst will continue to work with the system owner, system users, and other IS management and staff. The systems analyst and appropriate participants will: *(a)* study the problem domain; *(b)* thoroughly analyze problems and opportunities, *(c)* optionally, analyze business processes; *(d)* establish system improvement objectives and constraints; *(e)* update the project plan; and *(f)* present the findings and recommendations. The deliverables for the problem analysis phase are the system improvement objectives.

7. The requirements analysis phase identifies what the new system is to do without the consideration of technology; in other words, define the business requirements for a new system. As in the preliminary investigation and problem analysis phases, the analyst actively works with system users and owners as well as other IS professionals. To complete the requirements analysis phase, the analyst and appropriate participants will: *(a)* define requirements; *(b)* analyze functional requirements using system modeling and/or discovery prototyping; *(c)* trace and complete the requirements statement; *(d)* prioritize the requirements; and *(e)* update the project plan and scope. The deliverable of the requirements analysis phase is the business requirements statement. Because requirements are a moving target with no finalization, requirements analysis also includes an ongoing task to manage changes to the requirements.

8. The purpose of the decision analysis phase is to transition the project from business concerns to technical solutions by identifying, analyzing, and recommending a technical system solution. To complete the decision analysis phase, the analyst and appropriate participants will: *(a)* define candidate solutions; *(b)* analyze candidate solutions for feasibility (technical, operational, economic, and schedule feasibility); *(c)* compare feasible candidate solutions to select one or more recommended solutions; *(d)* update the project plan based on the recommended solution; and *(e)* present and defend the target solution. The deliverable of the decision analysis phase is the system proposal.

KEY TERMS

accelerated analysis, p. 171
business process redesign (BPR), p. 173
business process reengineering, p. 173
business requirements statement, p. 191
cause-and-effect analysis, p. 185
constraint, p. 187
data flow diagram (DFD), p. 168
decision analysis phase (of systems analysis), p. 197
discovery prototyping, p. 171
entity relationship diagram (ERD), p. 169
fact-finding, p. 172
functional requirement, p. 192
information engineering (IE), p. 169

information gathering, p. 172
joint requirements planning (JRP), p. 173
logical design, p. 193
logical system models, p. 193
model, p. 167
model-driven analysis, p. 167
nonfunctional requirement, p. 192
object models, p. 170
objective, p. 187
object-oriented analysis (OOA), p. 170
preliminary investigation phase (of systems analysis), p. 174
problem analysis phase (of systems analysis), p. 181

project charter, p. 174
prototype, p. 171
rapid architecture analysis, p. 171
repository, p. 167
requirements analysis phase (of systems analysis), p. 189
requirements discovery, p. 172
reverse engineering, p. 172
steering body, p. 180
structured analysis, p. 168
system improvement objectives, p. 188
system proposal, p. 199
systems analysis, pp. 165, 166
systems design, p. 166
timeboxing, p. 196

REVIEW QUESTIONS

1. What is the difference between systems analysis and systems design? How does the focus of information systems analysis differ from that of information systems design?
2. Whose concerns are addressed by systems analysis?
3. What role does a repository play in systems analysis?
4. Differentiate between model-driven analysis and accelerated analysis approaches? When can they be used in a complementary fashion?
5. What is the difference between structured analysis and information engineering?
6. What is object-oriented analysis? How is it similar to, and different from, modern structured analysis and information engineering?
7. What is discovery prototyping? Is it an alternative to model-driven development? Why or why not?
8. What is rapid architecture analysis and how does it differ from model-driven analysis?
9. What is reverse engineering and what role can it play in systems analysis?
10. Describe two approaches to requirements discovery? Why is requirements discovery not an alternative to model-driven and accelerated analysis approaches?
11. What is business process redesign? What is the role of systems analysis in business process redesign?
12. List and describe the purpose of the four phases of systems analysis.
13. What important question is addressed during the preliminary investigation phase? How might the information systems building blocks be used to identify the general level of understanding required for the phase?
14. List the tasks required to complete the preliminary investigation phase.
15. What important question is addressed during the problem analysis phase? How might the information systems building blocks be used to analyze problems and opportunities during the problem analysis phase?
16. List the tasks required to complete the problem analysis phase.
17. What is cause-and-effect analysis? What is the risk of not performing it?
18. Name two situations when business process redesign might be performed in the context of a traditional information systems development project.
19. Differentiate between an objective and a constraint.
20. What important question is addressed during the requirements analysis phase? How might the information systems building blocks be used to identify requirements?
21. List the tasks required to complete the requirements analysis phase.
22. Differentiate between logical and physical system models.
23. Differentiate between logical and physical design.
24. What is timeboxing? How is it related to requirements prioritizations?
25. What important question is addressed during the decision analysis phase? How might the information systems building blocks be used to identify requirements?
26. List the tasks of the decision analysis phase.
27. Briefly describe four types of feasibility criteria.

PROBLEMS AND EXERCISES

1. How do the classical definitions of systems analysis and systems design relate to systems modeling? How about discovery prototyping?
2. What modeling approaches do structured analysis and information engineering have in common? Explain the fundamental differences they take in using those models.
3. What role does CASE play in systems modeling? What is the relationship between CASE, reverse engineering, and system modeling?
4. How does object-oriented analysis address the problem of data and process model synchronization that has proven problematic in methods such as structured analysis and information engineering.
5. Joseph has long believed that developing system models results in a better understanding between analysts and users regarding system requirements, both business and technical. At the same time, he appreciates the speed with which prototypes define user requirements when compared to model-driven approaches. He longs for a way to integrate the best features and value of the two approaches. Make a recommendation.
6. Reverse engineering has one limitation that can render its models as valueless. What is that limitation?
7. In practice, classical fact-finding techniques must co-exist with contemporary facilitated group techniques. Why is this likely true?
8. Mary has facilitated several "student services" business process redesign projects in the registrar's office. These projects were not directed by information systems development and were focused primarily on improving manual processes by eliminating bureaucracy and increasing efficiency. Now, the registrar is sponsoring a project to replace 13 legacy computer applications with an integrated software package that will be purchased instead of built in-house. How will business process redesign come into play in this new project? How will it differ from the BPR projects that Mary had previously facilitated?
9. When might a preliminary investigation phase be skipped and why?
10. What are the possible final outcomes of the preliminary investigation phase?
11. Differentiate between the preliminary investigation and

problem analysis phases of the systems analysis process.

12. What is the difference between systems analysis and logical design?

13. What is the value of system modeling during the systems analysis phases? What types of models might be developed during the preliminary investigation, problem analysis, and requirements analysis phases? Do you see any value of drawing systems models during the decision analysis phase?

14. Why is the vocabulary of a business problem so important during systems analysis? When is that vocabulary established?

15. Why must problems be analyzed for causes and effects before establishing system improvement objectives?

16. Define well-formulated educational objectives for your academic career. Describe at least two constraints that could limit your ability to achieve your objectives.

17. Why is a program flowchart not a <u>logical</u> system model.

18. "We have used the PIECES framework to categorize our system requirements. For each requirement, we have indicated whether it is mandatory or desirable. The set of all mandatory requirements have been ranked, as has the set of all desirable requirements." What is wrong with this scheme.

19. Why does requirements analysis never really end for a project? How can this reality be accommodated?

20. Why did several systems analysis phases conclude with a presentation task when appropriate system owners and users were included in virtually all the systems analysis tasks?

21. Several systems analysis phases included a task to "update the project plan." Why is it necessary to continually monitor and update the project plan?

PROJECTS AND RESEARCH

1. How might the PIECES framework introduced in Chapter 3 be used in the problem analysis phase of systems analysis? Use the PIECES framework to evaluate your local course registration system. Do you see problems or opportunities? (Alternative: Substitute any system with which you are familiar.)

2. How might the PIECES framework be used in requirements analysis? Demonstrate by way of example.

3. Interview a systems analyst or system manager in the information services unit of a local organization. How do they initiate projects? What type of report or document do they produce to justify a project in the early stages of development?

4. Interview a systems analyst in a local organization. What techniques do they use to ensure that the problems stated by their users are worthy of solution? Do you quantify that worth? If so, how? If not, why not?

5. Interview a system manager in a local organization. What strategy does the manager use to verify and document business requirements for a system? Which of the strategies described in this book most closely correspond to the strate-

gies? How does the manager teach these strategies to his analysts? How do users react to these strategies? How does the manager enforce the use of his strategies? What new techniques is he exploring? (Ask about object-oriented analysis, BPR, information engineering, and JRP.)

6. Research the subject of "getting the requirements correct." While strategies such as information engineering and object-oriented analysis provide mechanisms for documenting requirements, their value in discovering requirements has often been criticized. Find out why, and offer your perspectives on how the popular techniques might be improved.

7. If you have read and studied Chapter 4, develop PERT and Gantt chart templates that include all the systems analysis tasks described in this chapter.

8. Repeat problem 7 for a specific *FAST* development route (e.g., model-driven RAD, or COTS).

9. Research model-driven methodologies and report on which methodologies are currently popular. (Note: Just because a methodology loses popularity does not mean analysts have abandoned its preferred models.)

MINICASES

1. Colleens Financial Services is a nationwide financial services company headquartered in Omaha, Nebraska. Senior systems analyst Fred McNamara is meeting with Ken Borelli, the MIS manager at Colleens Financial Services. Fred has just completed an evaluation of a new software package, a fourth-generation programming language (4GL). In addition to evaluating the 4GL, Fred has been asked to learn about a popular systems development approach called prototyping that uses 4GLs to build working models of systems. Fred is meeting with Ken to give

him his assessment of the 4GL product and prototyping as an alternative approach to systems development.

Ken starts the conversation. "So what do you think about that new software product? Is it worthy of being called a 4GL?"

Fred replies, "Without a doubt! Third-generation programming languages like COBOL can't compare to it! I think this product can do wonders for the systems staff. It is very user friendly, and it provides a number of facilities that assist in developing a complete system."

"What kind of facilities does it provide?" asks Ken.

Fred answers, "To give you some idea, I used a facility to develop a database containing actual data, another facility to produce a relatively complex printed report against that database, and other facilities to develop menus and other input and output screens all in a fraction of the time that would have been required with COBOL."

Ken says, "You said it was very user friendly. Does that mean you didn't have to spend a lot of time referencing manuals?"

"I hardly used the manuals," answers Fred. "The facilities simply led me through a series of questions or prompts. All I had to do was answer the questions. The 4GL generated the program code, which I subsequently executed. The productivity implications are tremendous!"

Ken responds enthusiastically, "Sounds like the product is a good investment. And what about prototyping? Do you think prototyping is something we should consider doing as an alternative to our current approach to developing systems?"

"Well," answers Fred, "prototyping certainly takes advantage of tools such as 4GLs. The strategy is very simple. It emphasizes the development of a working model of the target system, instead of traditional paper specifications. You begin building the model by first defining the database requirements for the new or desired system. Identifying the database requirements is relatively simple, since the data for most systems already exists, either in computer files or manual forms. Afterward, you then build and load a database using the 4GL. Once you have the database built, the rest is easy. You can use various facilities to quickly generate the menus, reports, and input and output screens. The analyst does not have to worry about whether the working model is totally complete or accurate. The end-user is encouraged to review the model and provide the analyst with feedback. If the model, say a report, is not acceptable, the analyst simply makes requested changes and reviews it again with the end-user at some later time. This repetitive process and active end-user participation are considered essential and to be encouraged."

"Now that's a new one!" exclaims Ken. "I certainly agree with the idea of encouraging end-user participation. But this attitude of encouraging or expecting to keep redoing work would be difficult to adjust to."

Fred responds, "I'm sure some of us old-timers will have some difficulty adjusting to this type of thinking."

Ken pauses momentarily and then says, "It sounds like this new 4GL and prototyping should be pursued further. Both 4GL and prototyping seem to offer some productivity gains. I've been looking for a way to get the end-users more involved in the systems development process, and I think this prototyping approach is the answer. There is one other thing: I suspect my staff's morale would be improved. I believe my staff would be motivated by this new 4GL product and by prototyping's emphasis on building a model as a basis for performing systems development."

Fred nods, "I agree! I can't wait to start my next project. I've got just the project picked out. I received a request for a new employee benefits system from the Personnel Department. I thought I'd use the 4GL in conjunction with the prototyping approach to complete this project. I won't be needing any programmers since I'll be developing the system myself while I'm working with the end-users. I've already drafted a memo asking Personnel to provide me with some sample records, forms, reports, and other materials that will help me identify their data storage requirements. From those samples, I'll be able to build a database for the new system. Then I'll start meeting with the end-users to define and implement screens and reports."

Ken suddenly looks concerned. "Hold your horses! I do have some concerns about this approach to systems development. Neither the tool nor the approach justifies a departure from the systems life cycle concept we follow here at Colleens. And that's exactly what you're proposing. You're proposing to select a project, define some basic requirements, and jump right into the design and construction of a new system. That I won't have. I want you to reconsider things."

Ken reaches for a systems development standards manual on his bookshelf. He turns to a figure that depicts the systems life cycle phases and continues, "Notice that our systems development life cycle includes several systems analysis phases, including a survey or preliminary investigation phase. Do you fully understand why we require this first phase?"

Fred answers, "Sure, that's where we perform a very quick study of the proposed project request. We try to gain a quick understanding of the size, scope, and complexity of the project."

"You're half right," replies Ken. "But why must we complete the phase? I'll tell you why! Because we receive numerous project requests from our end-users! We have a limited number of resources. We can't take on all the requests. It is the purpose of this phase to address the seriousness of the problems and to prioritize the project request against other requests."

Fred pauses, and then he replies, "I see what you're saying. I sort of jumped the gun by picking this Personnel project without considering other project requests that might be given higher priority."

"Good!" says Ken. "Now let's consider the second phase, the detailed problem analysis phase. If the project is to be pursued further, we then conduct a detailed study or investigation of the current system. We want to gain an understanding of the causes and effects of all problems and to appreciate the benefits that might be derived from any existing opportunities. We don't want to bypass this phase. This phase ensures that any new system we propose solves the problems encountered in the current system."

Fred, after glancing at his notes, says, "That brings us to the requirements analysis phase, right?"

"Right," answers Ken. "For all practical purposes, this is where you were proposing to begin your project. You were going to define data storage requirements for a new employee benefits system. I assume you would also attempt to identify other requirements. What particularly bothered me was that I didn't get an impression that you intended to study the database requirements to ensure data reliability and flexibility of form. And what about completeness checks? What were you going to do to ensure that processes were sufficient to ensure data will be properly maintained?"

Fred looks frustrated. "I'm beginning to lose my confidence in this prototyping approach. I was about to make some big mistakes by trying to bypass several important problem-solving phases and tasks."

Ken smiles and replies, "Listen, there's no reason to write off prototyping. So long as we base our prototypes on some sound design principles, I think we can still achieve all the benefits that you described earlier. I have another concern, though. We will eventually want to look at alternative solutions such as manual versus computer-based systems, on-line versus batch systems, and the like. These options should be evaluated for technical, operational, and economical feasibility. It seems to me that this task should precede extensive prototyping. We don't want to prematurely commit to less feasible or infeasible solutions."

Fred says, "I understand. It looks like prototyping has potential, especially in the requirements analysis phase of analysis. And I suspect that prototyping will greatly accelerate our systems design and implementation phases."

a. Do you think Fred did a thorough evaluation of the fourth-generation software product? What benefits do you believe can be derived from using such a tool?

b. Did Fred view prototyping as an alternative to the traditional systems development life cycle? If so, how should he have viewed it?

c. What systems analysis phases would have been skipped by the prototyping approach Fred proposed to follow? What do you think would have been the results of the employee benefits project if Fred had approached the project in the manner he originally envisioned?

2. A company is considering awarding your consulting firm a contract to develop a new and improved system. But, at the beginning, the company wants to commit only to systems analysis. The firm is concerned about your ability to understand its problems and needs. And most of all, it wants to see what kind of computer-based solutions you propose before it contracts you to design and implement a new system. Write a letter of proposal that will address the company's concerns.

3. You have been a systems analyst with your current employer for the past five years. During this time, the projects that you were assigned came directly from your IS manager. All the projects were originally submitted to the IS manager on formal request forms by various users within the company. But now things are different! Several months ago, the company hired a consulting firm to help it develop a strategic plan for the business. How might the resulting strategic plan affect the IS manager in determining which future projects will receive commitment of IS resources? Explain how a systems plan might affect the phases and tasks you perform during systems analysis.

4. You have recently completed the preliminary investigation phase for an assigned systems project. You have become concerned with your ability to complete the study of the current system and the requirements analysis of new system requirements within a reasonable time frame. The project is particularly challenging given that it involves numerous users, having differing vocabularies and system perspectives, who are located across several departments. What strategy would you use to reduce the amount of time required to complete the study and requirements analysis phases? How would this strategy deal with the issues presented by the diverse user community?

SUGGESTED READINGS

Application Development Trends (monthly periodical). Natick, MA: Software Productivity Group, a ULLO International company. This is our favorite systems development periodical that follows systems analysis and design strategies, methodologies, CASE, and other relevant trends. Visit the website at www.adtmag.com.

Coad, Peter, and Edward Yourdon. *Object-Oriented Analysis,* 2nd ed. Englewood Cliffs, NJ: Yourdon Press, 1991. This book is a great way to expose yourself to objects and the relationship of object methods to everything that preceded them.

Gane, Chris. *Rapid Systems Development.* Englewood Cliffs, NJ: Prentice Hall, 1989. This book presents a nice overview of RAD that combines model-driven development and prototyping in the correct balance.

Gause, Donald C., and Gerald M. Weinberg. *Are Your Lights On? How to Figure Out What the Problem REALLY Is.* New York: Dorsett House Publishing, 1990. Here's a title that should really get you thinking, and the entire book addresses one of the least published aspects of systems analysis; namely, problem solving.

Hammer, Mike. "Reengineering Work: Don't Automate, Obliterate." *Harvard Business Review,* July–August 1990, pp. 104–11. Dr. Hammer is a noted expert on business process redesign. This seminal paper examines some classic cases where the technique dramatically added value to businesses.

Wetherbe, James. *Systems Analysis and Design: Best Practices,* 4th ed. St. Paul, MN: West Publishing, 1994. We are indebted to Dr. Wetherbe for the PIECES framework.

Wood, Jane, and Denise Silver. *Joint Application Design: How to Design Quality Systems in 40% Less Time.* New York: John Wiley & Sons, 1989. This book provides an excellent in-depth presentation of joint application development (JAD).

Yourdon, Edward. *Modern Structured Analysis.* Englewood Cliffs, NJ: Yourdon Press, 1989. This update to the classic DeMarco text on the same subject defines the current state-of-the-practice for the structured analysis approach.

Zachman, John A. "A Framework for Information System Architecture." *IBM Systems Journal* 26, no. 3 (1987). This article presents a popular conceptual framework for information systems survey and the development of an information architecture.

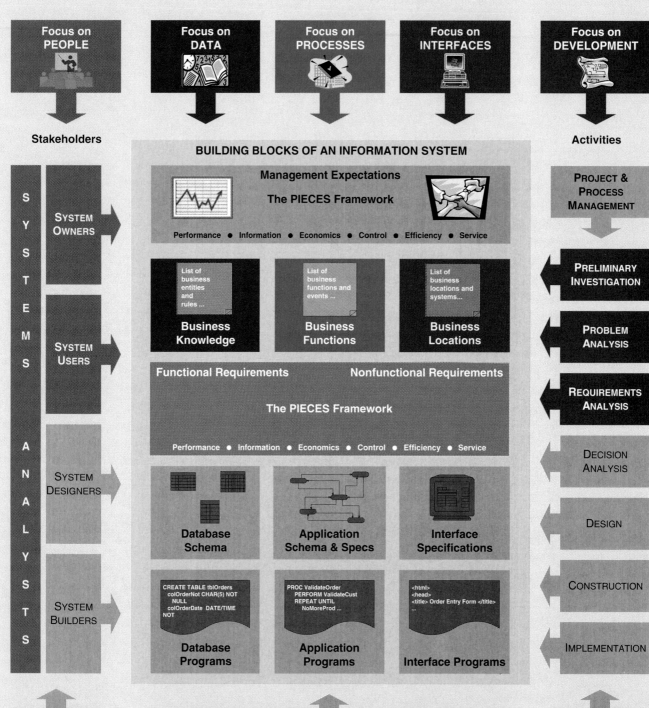

Focus on PEOPLE

Focus on DATA

Focus on PROCESSES

Focus on INTERFACES

Focus on DEVELOPMENT

Stakeholders

Activities

BUILDING BLOCKS OF AN INFORMATION SYSTEM

SYSTEMS ANALYSTS

SYSTEM OWNERS

SYSTEM USERS

SYSTEM DESIGNERS

SYSTEM BUILDERS

VENDORS AND CONSULTANTS

Management Expectations

The PIECES Framework

Performance ● Information ● Economics ● Control ● Efficiency ● Service

List of business entities and rules ...

Business Knowledge

List of business functions and events ...

Business Functions

List of business locations and systems...

Business Locations

Functional Requirements   **Nonfunctional Requirements**

The PIECES Framework

Performance ● Information ● Economics ● Control ● Efficiency ● Service

Database Schema

Application Schema & Specs

Interface Specifications

CREATE TABLE tblOrders colOrderNot CHAR(5) NOT NULL colOrderDate DATE/TIME NOT

Database Programs

PROC ValidateOrder PERFORM ValidateCust REPEAT UNTIL NoMoreProd ...

Application Programs

```
<html>
<head>
<title> Order Entry Form </title>
...
```

Interface Programs

PROJECT & PROCESS MANAGEMENT

PRELIMINARY INVESTIGATION

PROBLEM ANALYSIS

REQUIREMENTS ANALYSIS

DECISION ANALYSIS

DESIGN

CONSTRUCTION

IMPLEMENTATION

OPERATIONS AND SUPPORT

INFORMATION TECHNOLOGY & ARCHITECTURE
Database Technology ● Process Technology ● Interface Technology ● Network Technology

6

REQUIREMENTS DISCOVERY

CHAPTER PREVIEW AND OBJECTIVES

In this chapter you will learn about the tools and techniques necessary to discover and analyze requirements. You will learn how to use various fact-finding techniques to gather information about the system's problems, opportunities, and directives. Once you have the information, you will learn how to write requirements and prepare a document to be presented to the project stakeholders. You will know that you understand requirements discovery tools and techniques when you can:

— Define system requirements and differentiate between functional and nonfunctional requirements.

— Understand the activity of problem analysis and be able to create an Ishikawa (fishbone) diagram to aid in problem solving.

— Understand the concept of requirements management.

— Identify seven fact-finding techniques and characterize the advantages and disadvantages of each.

— Understand six guidelines for effective listening.

— Understand what body language and proxemics are and why a systems analyst should care.

— Characterize the typical participants in a JRP session and describe their roles.

— Complete the planning process for a JRP session, including selecting and equipping the location, selecting the participants, and preparing an agenda.

— Describe several benefits of using JRP as a fact-finding technique.

— Describe a fact-finding strategy that will make the most of your time with end-users.

— Describe various techniques to document and analyze requirements.

— Understand use cases and be able to document them.

SOUNDSTAGE

SOUNDSTAGE ENTERTAINMENT CLUB

SCENE

The SoundStage project is in the midst of the systems analysis phases. The development team has decided to hold group work sessions called joint requirements planning (JRP) sessions in order to decrease the time necessary to gather requirements information. This episode begins as Sandra, Bob, Terri, and Galen Kirchoff (executive sponsor) are planning the sessions. We join Sandra, Bob, Terri, and Galen in Galen's office.

SANDRA

Good morning, Galen! I'm glad you could take the time from your busy schedule to meet with us today.

GALEN

No problem at all, Sandra. I'm sure you are all aware of my commitment to this project and the visibility it has to the entire organization. I am willing to help in anyway I can to ensure its successful implementation.

SANDRA

Thank you, Galen. The reason I called for this meeting is to continue our planning for the joint requirements planning or JRP sessions.

GALEN

When are the sessions scheduled?

SANDRA

We have tentatively scheduled them to start four weeks from today, provided that we can reserve the location we want. Bob, what did you find out about reserving the rooms?

BOB

I contacted both the Marriott and the Hilton downtown. They both have the space and the facilities we need, and they are available during the time period we requested. The Marriott is more expensive, but it is a little more convenient in terms of parking and area restaurants.

TERRI

Food shouldn't be an issue because we will have a continental breakfast, snacks, and lunch catered by the hotel. Is cost going to be an issue?

GALEN

It's not if we can justify it. Refresh my memory. Why are we holding these off-site again?

SANDRA

We have many issues to cover and we need all the participants' undivided attention. We need to avoid any interruptions and distractions if the sessions are to be productive. If we were to hold the sessions at our office, it would be too tempting for people to go back to their desk to check their e-mail or phone messages and risk being detained to work on a problem. We are on a tight, aggressive schedule and we cannot afford people to be tardy or absent. Which reminds me, we need to make sure to tell all the session participants to leave their pagers and cell phones in their car or at home. I don't want them going off in the middle of a meeting. I will arrange for the hotel to take messages to allow people to contact us in case of an emergency.

GALEN

Terri, go ahead and make the final arrangements for what we need with the Marriott. If you have any problems with the purchase order let me know.

SANDRA

I have also contracted with an outside agency to provide an experienced JRP facilitator to lead the sessions.

GALEN

Sandra, I was told you were trained to be a facilitator. Why aren't you fulfilling that role?

SANDRA

Good question. A facilitator has to be impartial to any decisions that will be addressed and must not report to any-body attending the session. This helps to alleviate any pressure or intimidation the facilitator might experience. Also, our employees may talk more openly to an outsider. Plus in my experience an outsider can better control a meeting. They can reprimand people if they get out of line and not fear the repercussions. Since I'm closely associated with the project, it wouldn't be wise for me to facilitate the sessions.

GALEN

Is there a potential that we could have problems at these sessions?

SANDRA

I don't think so, but we are addressing some tough issues such as the reasons for so many member contract defaults. Because of the sensitivity of the problem and because it spans many departments, people may get emotional when we start identifying causes and try to outline a solution. A good facilitator can channel all that emotional energy into positive outcomes.

GALEN

I see. I will do my best to make sure everyone checks their egos at the door so we can avoid any such conflicts. Let's go over the list of participants . . .

DISCUSSION QUESTIONS

1. Do you think the cost of conducting joint requirements planning sessions can be justified? How would you do it?

2. What would be the benefit of an organization having its own trained JRP facilitators? Would the size of the organization make a difference?

3. What measures can the executive sponsor take to ensure the JRP sessions are held in an orderly manner and that all attendees can participate without fear of repercussions? Does the executive sponsor need to attend all sessions?

In Chapter 3 we discussed the several phases of system development. Each phase is important and necessary to effectively design, construct, and ultimately implement a system to meet the users' (stakeholders') needs. But to develop such a system, we first must be able to correctly identity, analyze, and understand what the users' requirements are, or what the user wants the system to do.

> **Requirements discovery** includes those techniques to be used by systems analysts to identify or extract system problems and solution requirements from the user community.

Requirements discovery for a system depends on the analysts' ability to first discover and then analyze problems and opportunities that exist in the current system.

> **Problem analysis** is the activity of identifying the problem, understanding the problem (including causes and effects), and understanding any constraints that may limit the solution.

Let's examine the concepts of system requirements and the process of discovering and documenting those requirements. What are system requirements?

> A **system requirement** (also called a *business requirement*) is a description of the needs and desires for an information system. A requirement may describe functions, features (attributes), and constraints.

System requirements define the services the system is to provide and prescribe constraints for its operation. In documenting the system requirements for a new information system, an analyst will likely identify dozens of unique requirements. As depicted in the chapter map, to simplify the presentation of requirements and to make them more readable, understandable, and traceable, requirements are often categorized as *functional* versus *nonfunctional*.

> A **functional requirement** is a function or feature that must be included in an information system to satisfy the business need and be acceptable to the users.

A functional requirement is an action of the system and usually is written using an action verb phrase. For example, the following are all examples of functional requirements.

- Process a checking account deposit.
- Calculate the GPA for a student.
- Capture the account holder identification information.

Many practitioners like to preface the above phrases with the words, "*The system should . . .*" This is a simple testing mechanism that you can use to verify if the proposed requirement is a functional requirement. If the resulting sentence makes sense, then it can be determined to be a functional requirement. Using this test we can rephrase the previous statements above to verify that they are indeed functional requirements.

- <u>The system should</u> process a checking account deposit.
- <u>The system should</u> calculate the GPA for a student.
- <u>The system should</u> capture the account holder identification information.

Let's now examine some statements that are examples of requirements that are not functional.

- Be user friendly.
- Be able to accommodate users at different physical locations.

Since these phrases do not include a strong action verb phrase they are not functional requirement; they are considered nonfunctional requirements.

A **nonfunctional requirement** is a description of the features, characteristics, and attributes of the system as well as any constraints that may limit the boundaries of the proposed solution.

There are many classifications of nonfunctional requirements. The PIECES framework, (see Figure 6.1) introduced in Chapter 3 provides an excellent tool for classifying nonfunctional requirements. Classifying the various types of requirements enables like requirements to be grouped for reporting, tracking, and validation purposes, plus it aids in identifying possible overlooked requirements.

Essentially, the purpose of requirements discovery and management is to correctly identify the DATA, PROCESS, and INTERFACE functional and nonfunctional requirements for the users of a new system. These correspond to our IS building blocks, which we introduced in Chapter 2. Most importantly, the objective is to

FIGURE 6.1	*Types of Nonfunctional Requirements*
Requirement Type	**Explanation**
Performance	Performance requirements represent the performance the system is required to exhibit to meet the needs of users. — What is the acceptable throughput rate? — What is the acceptable response time?
Information	Information requirements represent the information that is pertinent to the users in terms of content, timeliness, accuracy, and format. — What are the necessary inputs and outputs? When must they happen? — Where is the required data to be stored? — How current must the information be? — What are the interfaces to external systems?
Economy	Economy requirements represent the need for the system to reduce costs or increase profits. — What are the areas of the system where costs must be reduced? — How much should costs be reduced or profits be increased? — What are the budgetary limits? — What is the timetable for development?
Control (and Security)	Control requirements represent the environment in which the system must operate, as well as the type and degree of security that must be provided. — Must access to the system or information be controlled? — What are the privacy requirements? — Does the criticality of the data necessitate the need for special handling (backups, off-site storage, etc.) of the data?
Efficiency	Efficiency requirements represent the system's ability to produce outputs with minimal waste. — Are there duplicate steps in the process that must be eliminated? — Are there ways to reduce waste in the way the system uses its resources?
Service	Service requirements represent needs in order for the system to be reliable, flexible, and expandable. — Who will use the system and where are they located? — Will there be different types of users? — What are the appropriate human factors? — What training devices and training materials are to be included in the system? — What training devices and training materials are to be developed and maintained separately from the system, such as stand-alone computer-based training (CBT) programs or databases? — What are the reliability/availability requirements? — How should the system be packaged and distributed? — What documentation is required?

specify these requirements without prematurely expressing computer alternatives and technology details; at this point, keep the analysis at the business level.

Why is correctly identifying requirements important? Figure 6.2. depicts three possible solutions to an ambiguous requirement based on three different interpretations. Which one is correct? Too often members of the development team make requirements assumptions only to learn when they deliver the final product that it doesn't meet the users' expectations. Such a "misunderstanding" of the users' needs has been a significant problem of system development for years. When requirements are wrong, any one of the following could happen:

- The system may cost more than projected.
- The system may be delivered later than promised.
- The system may not meet the users' expectations and that dissatisfaction may cause them not to use it.
- Once in production, the costs of maintaining and enhancing the system may be excessively high.
- The system may be unreliable and prone to errors and downtime.
- The reputation of the IT staff on the team is tarnished because any failure, regardless of who is at fault, will be perceived as a mistake by the team.

The impact in terms of cost can be staggering, as shown in the following table by Barry W. Boehm, a noted expert in information technology economics.[1] He studied several software development projects to determine the costs for errors in requirements that weren't discovered until later in the development process.

Relative Cost to Fix an Error

Phase in Which Found	Cost Ratio
Requirements	1
Design	3–6
Coding	10
Development Testing	15–40
Acceptance Testing	30–70
Operation	40–1,000

Based on these findings, an erroneous requirement that goes undetected and unfixed until the operation phase may cost 1,000 times more than if it were

---

Requirement:
Create a means to transport a single individual from home to place of work.

FIGURE 6.2
Example of an Ambiguous Requirements Statement

Management Interpretation	I T Interpretation	User Interpretation

---

[1] Donald C. Gause and Gerald M. Weinberg, *Exploring Requirements: Quality Before Design* (New York: Dorset House Publishing, 1989), pp. 17–18.

detected and fixed in the requirements phase! Therefore, the goal of the systems analyst is to define system requirements that meet the following criteria:

- Consistent—the requirements are not conflicting or ambiguous.
- Complete—the requirements describe all possible system inputs and responses.
- Feasible—the requirements can be satisfied based on the available resources and constraints (feasibility analysis is covered in Chapter 9).
- Required—the requirements are truly needed and fulfill the purpose of the system.
- Accurate—the requirements are stated correctly.
- Traceable—the requirements directly map to the functions and features of the system.
- Verifiable—the requirements are defined so they can be demonstrated during testing.

This can be a time-consuming, difficult, and frustrating process that often leads organizations and individuals to take shortcuts to save time and money. But this shortsightedness often leads to the problems mentioned before. Now that we understand our goal, let's look at the process.

THE PROCESS OF REQUIREMENTS DISCOVERY

Requirements discovery consists of the following activities:

- Problem discovery and analysis.
- Requirements discovery.
- Documenting and analyzing requirements.
- Requirements management.

Let's now examine each of these activities in detail.

Problem Discovery and Analysis

As previously stated, requirements solve problems. For systems analysts to be successful, they must be skilled in problem analysis. To fully understand problem analysis, let's use the following example. A mother takes her young child to the doctor because she is ill. The first thing the doctor tries to do is identify the problem. The child has an earache, a fever, and a runny nose. Are these the problems? The mother has been giving the child pain medicine to ease the pain, but the child has not gotten better. The mother is treating the symptoms and not the real problem. Fortunately, the doctor is trained to analyze further. After examining the child the doctor concludes the child has an ear infection, which is the root cause of the child's symptoms. Now that the problem has been identified and analyzed, it is time for the doctor to recommend a cure (solution). Normally an antibiotic is prescribed to cure an ear infection, but the doctor first needs to determine if there are any constraints on the medicine. How old is and what is the weight of the child? Is the child allergic to anything? Can she swallow pills? Once these limitations are known, a prescription can be generated. Systems analysts use the same problem-solving process as a doctor, but instead of diagnosing medical problems they diagnosis system problems.

One common mistake inexperienced systems analysts make when trying to analyze problems is identifying a symptom as a problem. As a result they may design and implement a solution that doesn't solve the real problem or may cause new problems. Development teams use an Ishikawa diagram to identify, analyze, and solve problems.

The **Ishikawa diagram** is a graphical tool used to identify, explore, and depict problems and the causes and effects of those problems. It is often referred to as a **cause-and-effect diagram** or a **fishbone diagram** (because it resembles the skeleton of a fish).

The fishbone diagram is the brainchild of Kaoru Ishikawa, who pioneered quality management processes in the Kawasaki shipyards of Japan and became a founding father of modern management.

The basic concept of the fishbone diagram is that the name of the problem of interest is entered at the right of the diagram (or the *fish's head*). The possible causes of the problem are drawn as *bones* off the *main backbone*. Typically, these bones are labeled the four basic categories—materials, machines, manpower, and methods (the Four Ms). Other names can be used to suit the problem at hand. Alternative or additional categories include places, procedures, policies, and people (the Four Ps) or surroundings, suppliers, systems, and skills (the Four Ss).

The key is to have three to six main categories that encompass all possible areas of causes. Brainstorming techniques (defined later in the chapter) are commonly performed to add causes to the main *bones*. When the fishbone is complete, it depicts all the possibilities about what could be the root cause for the designated problem. The development team can then use the diagram to agree on the most likely causes of the problem and how they should be acted upon. Figure 6.3 is an example of a fishbone diagram depicting the SoundStage problem of member defaults on contracts. The diagram places the problem to be solved in the box at the far right. The five areas that have been identified as categories of causes (people-members, methods, contracts, materials, and policies) are listed in boxes above and below the *fish's skeleton* connected by arrows (*bones*) pointing to the fish's backbone. The actual causes of the problem for each category are depicted as arrows pointing to the category arrow (*bone*).

Requirements Discovery

Given an understanding of the problems, you can now start to define requirements. For today's systems analysts to be successful in defining system requirements they must be skilled in effective methods for gathering information—fact-finding. Fact-finding is used across the entire development cycle, but it is extremely critical in the requirements analysis phase.

> **Fact-finding** is the formal process of using research, interviews, questionnaires, sampling, and other techniques to collect information about problems, requirements, and preferences. It is also called *information gathering*.

FIGURE 6.3 *SoundStage Fishbone Diagram*

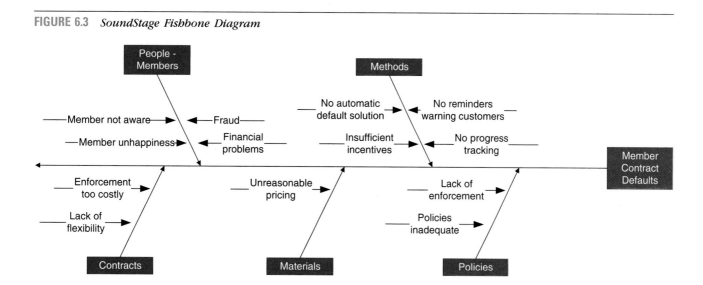

Tools, such as data, process, and object models, will eventually be used to document facts, and conclusions will be drawn from facts. But if you can't collect the facts, you can't use those tools. Fact-finding skills must be learned and practiced!

Facts are in the domain of the business application and its end-users. Therefore, the analyst must collect those facts to effectively apply the documentation tools and techniques. During systems analysis phases, the analyst learns about the vocabulary, problems, opportunities, constraints, requirements, and priorities of a business and a system.

What types of facts must be collected? Fortunately, we have a framework to help us determine what facts need to be collected, no matter what project we are working on. Throughout the system development process, we are looking at an existing or target information system. In Chapter 2 we saw that any information system can be examined in terms of three building blocks: DATA, PROCESSES, and INTERFACES. Those three building blocks were depicted using our matrix model. The facts that describe any information system also correspond nicely with the building blocks of that matrix model.

Fact-Finding Ethics   During your fact-finding exercises you often come across or analyze sensitive information. It could be a file of an aerospace company's pricing structure for a contract bid or employee profile data that includes salaries, performance history, medical history, and career plans. The analyst must protect the security and privacy of any facts or data. Many people and organizations in this highly competitive atmosphere are looking for an edge to get ahead. Careless systems analysts that leave sensitive documents in view on their desks or publicly discuss sensitive data could cause great harm to the organization or to individuals. If the data would fall into the wrong people's hands, the systems analyst may lose respect, credibility, or the confidence of users and management. In some cases they would be responsible for the invasion of a person's privacy and could be liable!

Most corporations make every effort to ensure they conduct business in an ethical manner because the laws require them to. Many corporations require their employees to attend annual training seminars on company ethics and reinforce the learning by displaying banners or signs that contain the company's code of conduct and ethics statements throughout the workplace in highly visible locations. Companies may distribute copies of their ethics policies to all employees or they may post policies on the Web, making them easily accessible to employees. Ethics policies document expected and required behavior. Violations of these policies could lead to disciplinary action or even termination. Ethics play a crucial role in fact-finding.

What Fact-Finding Methods Are Available?   Now that we have a framework for our fact-finding activities, we can introduce seven common fact-finding techniques:

- Sampling of existing documentation, forms, and databases.
- Research and site visits.
- Observation of the work environment.
- Questionnaires.
- Interviews.
- Prototyping.
- Joint requirements planning

An understanding of each of these techniques could become essential to your success. An analyst usually applies several of these techniques during a single systems project. To be able to select the most suitable technique for use in any given situation, you will have to learn the advantages and disadvantages of each. They are discussed in detail later in the chapter.

When performing fact-finding activities, the systems analyst must assemble or document the information (or *draft requirements*) in an organized, understandable, and meaningful way. These initial documents will provide direction for the modeling techniques the systems analyst will use to analyze the requirements to determine <u>what</u> are the correct requirements for the project. Once those have been identified the systems analyst formalizes the requirements in a document that will be reviewed and approved by the users.

Documenting the Draft Requirements   Systems analysts use various tools to document their initial findings in draft form. They write *use cases* to describe the system functions from the perspective of external users and in the manner and terminology in which they understand. *Decision tables* are used to document an organization's complex business policies and decision-making rules, and *requirements tables* are used to document each specific requirement. Each of these tools is examined in more detail later in the chapter.

Analyzing Requirements   Fact-finding activities usually produce requirements that conflict with each other. Requirements are solicited from many different sources, and each person has his or her own opinions and desires for the functionality and features of the new system. The goal of requirements analysis is to discover and resolve the problems with the requirements and reach agreement on any modifications to satisfy the stakeholders. The process is concerned with the initial requirements gathered from the stakeholders. These requirements are usually incomplete and documented in an informal way in instruments such as use cases, tables, and reports. The focus is to reach agreement on the stakeholder's needs; in other words, it should answer the question "do we have the right system requirements for the project?" Inevitably these draft requirements contain many problems such as:

- Missing requirements.
- Conflicting requirements.
- Infeasible requirements.
- Overlapping requirements.
- Ambiguous requirements.

These types of requirements problems are very common in many of the requirement documents written today and can be extremely costly to fix later in the development cycle if left unresolved. System modeling techniques such as process, data, and object modeling are excellent tools for the systems analyst to use for analyzing requirements in order to eliminate mistakes. These techniques are covered in depth in Chapters 7, 8, and Module A.

It was also previously mentioned that stakeholders should agree on the resulting system requirements, necessitating negotiation among stakeholders during analysis. If multiple stakeholders submit requirements that are in conflict with each other or if the proposed requirements are too ambitious, the stakeholders must negotiate, often under the guidance of the systems analyst, to agree on any modifications or simplifications to the system requirements. They also must agree on the criticality and priority of the requirements. This is crucial to ensure the success of the development effort.

The fact-finding and requirements analysis activities are very closely associated and are often interleaved. If requirements discovered during the fact-finding process are found to be a problem, the analyst may perform analysis activities on those select items in order to resolve those problems before continuing to elicit additional system needs and desires.

This chapter has focused primarily on the business side of requirements, but additional technical requirements exist that are physical in nature. Examples of technical

**Documenting and
Analyzing Requirements**

requirements include specifying a required software package or a hardware plat-form. These types of requirements will be discussed in depth in Chapter 9.

Formalizing Requirements   System requirements are usually documented in a for-mal way to communicate the requirements to the key stakeholders. This docu-ment serves as the contract between the system owners and the development team on what is going to be provided in terms of a new system. Thus, it may go through many revisions and reviews before everyone agrees and authorizes its contents. There is no standard name or format for this document; names used include requirements statement, requirements specification, requirements defini-tion, and functional specification, and the format is usually tailored to the orga-nization's needs. Those companies that provide information systems and software to the U.S. government are required to use the format and naming conventions specified in the government's published standards document MIL-STD-498.[2] Many organizations have created their own standards adapted from MIL-STD-498 because of its thoroughness and because many people are familiar with it. In this book we will use the term *requirements definition document.* Figure 6.4 provides a sample outline of one. This document will be consolidated with other project information to form the requirements statement, which is the final deliverable of the requirements analysis phase. A requirements definition document should con-sist of the following.

— Functions and services the system should provide.
— Nonfunctional requirements including the system's features, characteristics, and attributes.
— Constraints that restrict the development of the system or under which the system must operate.
— Information about other systems the system must interface with.

The requirements definition document is probably the most widely read and ref-erenced document of all the project documentation. System owners and users use

---

FIGURE 6.4	*Sample Requirements Definition Outline*

Requirements Definition Report

1. Introduction
   1.1. Purpose
   1.2. Background
   1.3. Scope
   1.4. Definitions, Acronyms, and Abbreviations
   1.5. References
2. General Project Description
   2.1. System Objectives
3. Requirements and Constraints
   3.1. Functional Requirements
   3.2. Nonfunctional Requirements
4. Conclusion
   4.1. Outstanding Issues
Appendix (optional)

---

[2] MIL-STD-498 is a standard that merges DOD-STD-2167A and DOD-STD-7935A to define a set of activities and documentation suitable for the development of both weapon systems and automated information systems.

it to specify their requirements and any changes that may arise. Managers use it to prepare project plans and estimates, and developers use it to understand what is required and to develop tests to validate the system. Taking the time to write the requirements correctly, concisely, and clearly will not only save time from a schedule point of view, but also save costs and reduce the risk of costly requirements errors. Performing requirements validation will help us achieve that goal.

> **Requirements validation** checks the requirements definition document for accuracy, completeness, consistency, and conformance to standards.

Requirements validation is performed on a final draft of the requirements definition document after all input has been solicited from the system owners and users. The purpose of this activity is for the systems analyst to ensure the requirements are written correctly. Examples of errors the systems analyst might find are:

- System models that contain errors.
- Typographical or grammar errors.
- Requirements that conflict with each other.
- Ambiguous or poorly worded requirements.
- Lack of conformance to quality standards required for the document.

Requirements Management

Over the lifetime of the project it is very common for new requirements to emerge and existing requirements to change once a requirements definition document has been approved. Some studies have shown that as much as 50 percent or more of the requirements will change before the system is put into production. Obviously, this can be a major headache for the development team. To help alleviate the many problems this can cause it is necessary to perform requirements management.

> **Requirements management** is the process of managing change to the requirements.

Requirements management encompasses the policies, procedures, and processes that govern how a change to a requirement is handled. It specifies how a change request should be submitted, how it is analyzed for impact to scope, schedule, and cost; how it's approved or rejected; and how the change is implemented if approved. If you desire additional information about this subject, many resources are available, and some are listed at the end of this chapter.

REQUIREMENTS DISCOVERY METHODS

Systems analysts need an organized method of collecting facts. They especially need to develop a detective-like mentality to be able to discern relevant facts. In this section we present popular, alternative fact-finding techniques.

Sampling of Existing Documentation, Forms, and Files

When you are studying an existing system, you can develop a good feel for the system by studying existing documentation, forms, and files. A good analyst always gets facts first from existing documentation rather than from people.

Collecting Facts from Existing Documentation   The first document the analyst should seek out is the organization chart. Next, the analyst may want to trace the history that led to the project. To accomplish this, the analyst may want to collect and review documents that describe the problem. These include:

- Interoffice memoranda, studies, minutes, suggestion box notes, customer complaints, and reports that document the problem area.
- Accounting records, performance reviews, work measurement reviews, and other scheduled operating reports.
- Information systems project requests—past and present.

In addition to documents that describe the problem, there are usually documents that describe the business function being studied or designed. These documents may include:

— The company's mission statement and strategic plan.
— Formal objectives for the organization subunits being studied.
— Policy manuals that may place constraints on any proposed system.
— Standard operating procedures (SOPs), job outlines, or task instructions for specific day-to-day operations.
— Completed forms that represent actual transactions at various points in the processing cycle.
— Samples of manual and computerized databases.
— Samples of manual and computerized screens and reports.

Also, don't forget to check for documentation of previous system studies and designs performed by systems analysts and consultants. This documentation may include:

— Various types of flowcharts and diagrams.
— Project dictionaries or repositories.
— Design documentation, such as inputs, outputs, and databases.
— Program documentation.
— Computer operations manuals and training manuals.

All documentation collected should be analyzed to determine the information's currency. Don't discard outdated documentation. Just keep in mind that additional fact-finding will be needed to verify or update the facts collected. As you review existing documents, take notes, draw pictures, and use systems analysis and design tools to model what you are learning or proposing for the system.

Document and File Sampling Techniques   Because it would be impractical to study every occurrence of every form or record in a file or database, systems analysts normally use sampling techniques to get a large enough cross section to determine what <u>can</u> happen in the system.

Sampling is the process of collecting a representative sample of documents, forms, and records.

The key word is *representative*—the systems analyst seeks to sample enough forms to represent the full nature and complexity of the data. Experienced analysts avoid the pitfalls of sampling blank forms—they tell little about how the form is used, not used, or misused. When studying documents or records from a database table, you should study enough samples to identify all the possible processing conditions and exceptions. You use statistical sampling techniques to determine if the sample size is large enough to be representative.

The size of the sample depends on how representative you want the sample to be. There are many sampling issues and factors, which is a good reason to take an introductory statistics course. One simple and reliable formula for determining sample size is

$$\text{Sample size} = 0.25 \times (\text{Certainty factor/Acceptable error})^2$$

The certainty factor depends on how certain you want to be that the data sampled will not include variations not in the sample. The certainty factor is calculated from tables (available in many industrial engineering texts). A partial example of the table is given here.

Desired Certainty	Certainty Factor
95%	1.960
90	1.645
80	1.281

Suppose you want 90 percent certainty that a sample of invoices will contain no unsampled variations. Your sample size, SS, is calculated as follows:

$$SS = 0.25(1.645/0.10)^2 = 68$$

We need to sample 68 invoices to get the desired accuracy. If we want a higher level of certainty, we would have to sample a larger number of invoices.

Now suppose we know from experience that 1 in every 10 invoices varies from the norm. Based on this knowledge we can alter the above formula by replacing the heuristic .25 with $p(1-p)$.

$$SS = p(1-p)(1.645/0.10)^2$$

Where p is the proportion of invoices with variances.

By using this formula, we can reduce the number of samples required to get the desired accuracy.

$$SS = .10(1-.10)(1.645/0.10)^2 = 25$$

How do we choose our 25 invoices? Two commonly used sampling techniques are randomization and stratification.

Randomization is a sampling technique characterized as having no predetermined pattern or plan for selecting sample data.

Therefore, we just randomly choose 25 invoices based on the sample size calculated above.

Stratification is a systematic sampling technique that attempts to reduce the variance of the estimates by spreading out the sampling—for example, choosing documents or records by formula—and by avoiding very high or low estimates.

For computerized files, stratification sampling can be executed by writing a simple program. For instance, suppose our invoices were in a database that had a volume of approximately 250,000 invoices. Recall that our sample size needs to include 25 invoices. We will simply write a program that prints every 10,000th record (250,000/25). For manual files and documents, we could execute a similar scheme.

Research and Site Visits

A second fact-finding technique is to thoroughly research the problem domain. Most problems are not unique. Others have solved them before us. Organizations often contact or perform site visits at companies they know have experienced similar problems. If these companies are willing to share, valuable information can be obtained that may save tremendous time and cost in the development process. Memberships in professional societies such as the Association for Information Technology Professionals (AITP) or the Association for Information Systems (AIS) among others can provide a network of useful contacts.

Computer trade journals and reference books are also a good source of information. They can provide you with information on how others have solved similar problems, plus you can learn whether software packages exist to solve your problem. And now with recent advances in cyberspace, you don't even have to leave your desk to do it. Exploring the Internet via your personal computer can provide you with immeasurable amounts of information.

Observation of the Work Environment

Observation is an effective data-collection technique for obtaining an understanding of a system.

> **Observation** is a fact-finding technique wherein the systems analyst either participates in or watches a person perform activities to learn about the system.

This technique is often used when the validity of data collected through other methods is in question or when the complexity of certain aspects of the system prevents a clear explanation by the end-users.

Collecting Facts by Observing People at Work   Even with a well-conceived observation plan, the systems analyst is not assured that fact-finding will be successful. The following story, which appears in a book by Gerald M. Weinberg, *Rethinking Systems Analysis and Design,* gives an entertaining yet excellent example of some pitfalls of observation.[3]

The Railroad Paradox

About 30 miles from Gotham City lay the commuter community of Suburbantown. Each morning, thousands of Suburbanites took the Central Railroad to work in Gotham City. Each evening, Central Railroad returned them to their waiting spouses, children, and dogs.

Suburbantown was a wealthy suburb, and many of the spouses liked to leave the children and dogs and spend an evening in Gotham City with their mates. They preferred to precede their evening of dinner and theater with browsing among Gotham City's lush markets. But there was a problem. To allow time for proper shopping, a Suburbanite would have to depart for Gotham City at 2:30 or 3:00 in the afternoon. At that hour, no Central Railroad train stopped in Suburbantown.

Some Suburbanites noted that a Central train did pass through their station at 2:30, but did not stop. They decided to petition the railroad, asking that the train be scheduled to stop at Suburbantown. They readily found supporters in their door-to-door canvass. When the petition was mailed, it contained 253 signatures. About three weeks later, the petition committee received the following letter from the Central Railroad:

Dear Committee

Thank you for your continuing interest in Central Railroad operations. We take seriously our commitment to providing responsive service to all the people living along our routes, and greatly appreciate feedback on all aspects of our business. In response to your petition, our customer service representative visited the Suburbantown station on three separate days, each time at 2:30 in the afternoon. Although he observed with great care, *on none of the three occasions were there any passengers waiting for a southbound train.*

We can only conclude that there is no real demand for a southbound stop at 2:30, and must therefore regretfully decline your petition.

Yours sincerely,

Customer Service Agent

Central Railroad

What are the lessons learned from this story? For one, use the appropriate fact-finding technique for the problem at hand. Observation was an incorrect choice. Why would anyone be waiting for a 2:30 train when everyone knew the train

---

[3] Gerald M. Weinberg, *Rethinking Systems Analysis and Design,* pp. 23–24. Copyright © 1988, 1982 by Gerald M. Weinberg. Reprinted by permission of Dorset House Publishing, 353 W. 12th St., New York, NY 10014 (212-620-4053/1-800-DH-BOOKS/www.dorsethouse.com). All rights reserved.

didn't stop? A second lesson to be learned is to verify your fact-finding results with the user. Based on the user feedback, you may discover that you need to try other fact-finding techniques to gather additional information. Never jump to conclusions!

Observation Advantages and Disadvantages   Observation can be a very useful and beneficial fact-finding technique provided you have the ability to observe thoroughly and accurately. The pros and cons of observation include the following:

Advantages

- Data gathered by observation can be highly reliable. Sometimes observations are conducted to check the validity of data obtained directly from individuals.
- The systems analyst is able to see exactly what is being done. Complex tasks are sometimes difficult to clearly explain in words. Through observation, the systems analyst can identify tasks that have been missed or inaccurately described by other fact-finding techniques. Also, the analyst can obtain data describing the physical environment of the task (e.g., physical layout, traffic, lighting, noise level).
- Observation is relatively inexpensive compared with other fact-finding techniques. Other techniques usually require substantially more employee release time and copying expenses.
- Observation allows the systems analyst to do work measurements.

Disadvantages

- Because people usually feel uncomfortable when being watched, they may unwittingly perform differently when being observed. The famous Hawthorne experiment proved that the act of observation can alter behavior.
- The work being observed may not involve the level of difficulty or volume normally experienced during that time period.
- Some systems activities may take place at odd times, causing a scheduling inconvenience for the systems analyst.
- The tasks being observed are subject to various types of interruptions.
- Some tasks may not always be performed in the manner in which they are observed by the systems analyst. For example, the systems analyst might have observed how a company filled several customer orders. However, the procedures observed may have been those steps used to fill a number of regular customer orders. If any of those orders had been special orders (e.g., an order for goods not normally kept in stock), the systems analyst would have observed a different set of procedures being executed.
- If people have been performing tasks in a manner that violates standard operating procedures, they may temporarily perform their jobs correctly while you are observing them. In other words, people may let you see what they want you to see.

Guidelines for Observation   How does the systems analyst obtain facts through observation? Does one simply arrive at the observation site and begin recording everything that's viewed? Of course not. Much preparation should occur first. The analyst must determine how data will actually be captured. Will special forms on which to quickly record data be necessary? Will the individuals being observed be bothered by having someone watch and record their actions? When are the low, normal, and peak periods of operations for the task to be observed? The systems analyst must identify the ideal time to observe a particular aspect of the system.

Observation should first be conducted when the workload is normal. Afterward, observations can be made during peak periods to gather information for measuring the effects caused by the increased volume. The systems analyst might also obtain samples of documents or forms that will be used by those being observed. A great deal of planning and preparation must be done.

The sampling techniques discussed earlier are also useful for observation. In this case, the technique is called work sampling.

> **Work sampling** is a fact-finding technique that involves a large number of observations taken at random intervals.

This technique is less threatening to the people being observed because the observation period is not continuous. When using work sampling, you need to predefine the operations of the job to be observed. Then calculate a sample size as you did for document and file sampling. Make that many random observations, being careful to observe activities at different times of the day. By counting the number of occurrences of each operation during the observations, you will get a feel for how employees spend their days.

With proper planning completed, the actual observation can be done. Effective observation is difficult to carry out. Experience is the best teacher; however, the following guidelines may help you develop your observation skills:

— Determine the who, what, where, when, why, and how of the observation.
— Obtain permission from appropriate supervisors or managers.
— Inform those who will be observed of the purpose of the observation.
— Keep a low profile.
— Take notes during or immediately following the observation.
— Review observation notes with appropriate individuals.
— Don't interrupt the individuals at work.
— Don't focus heavily on trivial activities.
— Don't make assumptions.

Living the System   In this type of observation the systems analyst actively performs the role of the user for a short time. This is one of the most effective ways to learn about problems and requirements of the system. By filling the user's shoes, a systems analyst quickly gains an appreciation for what the user experiences and what she has to do to perform the job. This type of role-playing gives the systems analyst a firsthand education on the business processes and functions, as well as the problems and challenges associated with them.

Questionnaires

Another fact-finding technique is to conduct surveys through questionnaires.

> **Questionnaires** are special-purpose documents that allow the analyst to collect information and opinions from respondents.

The document can be mass-produced and distributed to respondents, who can then complete the questionnaire on their own time. Questionnaires allow the analyst to collect facts from a large number of people while maintaining uniform responses. When dealing with a large audience, no other fact-finding technique can tabulate the same facts as efficiently.

Collecting Facts by Using Questionnaires   The use of questionnaires has been heavily criticized and is often avoided by systems analysts. Many systems analysts claim that the responses lack reliable and useful information. But questionnaires can be an effective method for gathering facts, and many of these criticisms can be attributed to the inappropriate or ineffective use of questionnaires by systems analysts. Before using questionnaires, you should first understand the pros and cons associated with their use.

Advantages

- Most questionnaires can be answered quickly. People can complete and return questionnaires at their convenience.

- Questionnaires provide a relatively inexpensive means for gathering data from a large number of individuals.

- Questionnaires allow individuals to maintain anonymity. Therefore, individuals are more likely to provide the real facts, rather than telling you what they think their boss would want them to.

- Responses can be tabulated and analyzed quickly.

Disadvantages

- The number of respondents is often low.

- There's no guarantee that an individual will answer or expand on all the questions.

- Questionnaires tend to be inflexible. There's no opportunity for the systems analyst to obtain voluntary information from individuals or to reword questions that may have been misinterpreted.

- It's not possible for the systems analyst to observe and analyze the respondent's body language.

- There is no immediate opportunity to clarify a vague or incomplete answer to any question.

- Good questionnaires are difficult to prepare.

Types of Questionnaires   There are two formats for questionnaires, free-format and fixed-format.

Free-format questionnaires offer the respondent greater latitude in the answer. A question is asked, and the respondent records the answer in the space provided after the question.

Here are two examples of free-format questions:

- What reports do you currently receive and how are they used?

- Are there any problems with these reports (e.g., are they inaccurate, is there insufficient information, or are they difficult to read and/or use)? If so, please explain.

Such responses may be difficult to tabulate. Also, the respondents' answers may not match the questions asked. To ensure good responses in free-format questionnaires, the analyst should phrase the questions in simple sentences and not use words, such as *good,* that can be interpreted differently by different respondents. The analyst should also ask questions that can be answered with three or fewer sentences. Otherwise, the questionnaire may take up more time than the respondent is willing to sacrifice.

The second type of questionnaire is fixed-format.

Fixed-format questionnaires contain questions that require selection of predefined responses.

Given any question, the respondent must choose from the available answers. This makes the results much easier to tabulate, but the respondent cannot provide additional information that might prove valuable. There are three types of fixed-format questions.

1. For **multiple-choice questions,** the respondent is given several answers. The respondent should be told if more than one answer can be selected. Some multiple-choice questions allow for very brief free-format responses when

none of the standard answers apply. Examples of multiple-choice, fixed-format questions are:

Do you feel that backorders occur too frequently?

☐ YES   ☐ NO

Is the current accounts receivable report that you receive useful?

☐ YES   ☐ NO

If no, please explain.

2. For **rating questions,** the respondent is given a statement and asked to use supplied responses to state an opinion. To prevent built-in bias, there should be an equal number of positive and negative ratings. The following is an example of a rating fixed-format question:

The implementation of quantity discounts would cause an increase in customer orders.

☐ Strongly agree

☐ Agree

☐ No opinion

☐ Disagree

☐ Strongly disagree

3. For **ranking questions,** the respondent is given several possible answers, which are to be ranked in order of preference or experience. An example of a ranking fixed-format question is:

Rank the following transactions according to the amount of time you spend processing them:

_____% new customer orders

_____% order cancellations

_____% order modifications

_____% payments

Developing a Questionnaire   Good questionnaires are "designed." If you write your questionnaires without designing them first, you reduce your chances of success. The following procedure is effective:

1. Determine what facts and opinions must be collected and from whom you should get them. If the number of people is large, consider using a smaller, randomly selected group of respondents.

2. Based on the needed facts and opinions, determine whether free- or fixed-format questions will produce the best answers. A combination format that permits optional free-format clarification of fixed-format responses is often used.

3. Write the questions. Examine them for construction errors and possible misinterpretations. Make sure that the questions don't offer your personal bias or opinions. Edit the questions.

4. Test the questions on a small sample of respondents. If your respondents had problems with them or if the answers were not useful, edit the questions.

5. Duplicate and distribute the questionnaire.

Interviews

The personal interview is generally recognized as the most important and most often used fact-finding technique.

> **Interviews** are a fact-finding technique whereby the systems analysts collect information from individuals through face-to-face interaction.

Interviewing can be used to achieve any or all of the following goals: find facts, verify facts, clarify facts, generate enthusiasm, get the end-user involved, identify requirements, and solicit ideas and opinions. There are two roles assumed in an interview. The systems analyst is the **interviewer,** responsible for organizing and conducting the interview. The system user or system owner is the **interviewee,** who is asked to respond to a series of questions.

There may be one or more interviewers and/or interviewees. In other words, interviews may be conducted one-on-one or many-to-many. Unfortunately, many systems analysts are poor interviewers. In this section you will learn how to conduct proper interviews.

Collecting Facts by Interviewing Users   The most important element of an information system is people. No other fact-finding technique places as much emphasis on people as interviews, but people have different values, priorities, opinions, motivations, and personalities. Therefore, to use the interviewing technique, you must possess good human relations skills for dealing effectively with different types of people. Like other fact-finding techniques, interviewing isn't the best method for all situations. Interviewing has its advantages and disadvantages, which should be weighed against those of other fact-finding techniques.

Advantages

— Interviews give the analyst an opportunity to motivate the interviewee to respond freely and openly to questions. By establishing rapport, the systems analyst is able to give the interviewee a feeling of actively contributing to the systems project.

— Interviews allow the systems analyst to probe for more feedback from the interviewee.

— Interviews permit the systems analyst to adapt or reword questions for each individual.

— Interviews give the analyst an opportunity to observe the interviewee's nonverbal communication. A good systems analyst may be able to obtain information by observing the interviewee's body movements and facial expressions as well as by listening to verbal replies to questions.

Disadvantages

— Interviewing is a very time-consuming, and therefore costly, fact-finding approach.

— Success of interviews is highly dependent on the systems analyst's human relations skills.

— Interviewing may be impractical due to the location of interviewees.

Interview Types and Techniques   There are two types of interviews, unstructured and structured.

Unstructured interviews are conducted with only a general goal or subject in mind and with few, if any, specific questions. The interviewer counts on the interviewee to provide a framework and direct the conversation.

This type of interview frequently gets off the track, and the analyst must be prepared to redirect the interview back to the main goal or subject. For this reason, unstructured interviews don't usually work well for systems analysis and design.

In **structured interviews** the interviewer has a specific set of questions to ask of the interviewee.

Depending on the interviewee's responses, the interviewer will direct additional questions to obtain clarification or amplification. Some of these questions may be planned and others spontaneous. **Open-ended questions** allow the interviewee to respond in any way that seems appropriate. An example of an open-ended question is "Why are you dissatisfied with the report of uncollectible accounts?" **Closed-ended questions** restrict answers to either specific choices or short, direct responses. An example of such a question might be "Are you receiving the report of uncollectible accounts on time?" or "Does the report of uncollectible accounts contain accurate information?" Realistically, most questions fall between the two extremes.

How to Conduct an Interview

Your success as a systems analyst is at least partially dependent on your ability to interview. A successful interview involves selecting appropriate individuals to interview, preparing extensively for the interview, conducting the interview properly, and following up on the interview. Here we examine each of these steps in more detail. Let's assume that you've identified the need for an interview and you have determined exactly what kinds of facts and opinions you need.

Select Interviewees   You should interview the end-users of the information system you are studying. A formal organizational chart will help you identify these individuals and their responsibilities. You should attempt to learn as much as possible about each individual before the interview. Attempt to learn their strengths, fears, biases, and motivations. The interview can then be geared to take the characteristics of the individual into account.

Always make an appointment with the interviewee. Never just drop in. Limit the appointment to somewhere between a half hour and an hour. The higher the management level of the interviewee, the less time you should schedule. If the interviewee is a clerical, service, or blue-collar worker, get the supervisor's permission before scheduling the interview. Be certain that the location you want for the interview will be available during the time the interview is scheduled. Never conduct an interview in the presence of your officemates or the interviewee's peers.

Prepare for the Interview   Preparation is the key to a successful interview. An interviewee can easily detect an unprepared interviewer and may resent the lack of preparation because it wastes valuable time. When the appointment is made, the interviewee should be notified about the subject of the interview. To ensure that all pertinent aspects of the subject are covered, the analyst should prepare an interview guide.

> An **interview guide** is a list of specific questions the interviewer will ask the interviewee.

The interview guide may also contain follow-up questions that will be asked only if the answers to other questions warrant the additional answers. A sample interview guide is presented in Figure 6.5. The agenda is carefully laid out with the specific time allocated to each question. Time should also be reserved for follow-up questions and redirecting the interview. Questions should be carefully chosen and phrased. Most questions begin with the standard who, what, when, where, why, and how much type of wording. Avoid the following types of questions:

— *Loaded questions,* such as "Do we have to have both of these columns on the report?" The question conveys the interviewer's personal opinion on the issue.

— *Leading questions,* such as "You're not going to use this OPERATOR CODE, are you?" The question leads the interviewee to respond, "No, of course not," regardless of actual opinion.

— *Biased questions,* such as "How many codes do we need for FOOD CLASSIFICATION in the INVENTORY FILE? I think 20 ought to cover it." Why bias the interviewee's answer with your own?

FIGURE 6.5	*Sample Interview Guide*	

Interviewee: Jeff Bentley, Accounts Receivable Manager
Date: Tuesday, March 23, 2000
Time: 1:30 P.M.
Place: Room 223, Admin. Bldg.
Subject: Current Credit-Checking Policy

Time Allocated	Interviewer Question or Objective	Interviewee Response
1 to 2 min.	**Objective** Open the interview: — Introduce ourselves. — Thank Mr. Bentley for his valuable time. — State the purpose of the interview—to obtain an understanding of the existing credit-checking policies.	
5 min.	**Question 1** What conditions determine whether a customer's order is approved for credit? **Follow-up**	
5 min.	**Question 2** What are the possible decisions or actions that might be taken once these conditions have been evaluated? **Follow-up**	
3 min.	**Question 3** How are customers notified when credit is not approved for their order? **Follow-up**	
1 min.	**Question 4** After a new order is approved for credit and placed in the file containing orders that can be filled, a customer might request that a modification be made to the order. Would the order have to go through credit approval again if the new total order cost exceeds the original cost? **Follow-up**	
1 min.	**Question 5** Who are the individuals that perform the credit checks? **Follow-up**	
1 to 3 min.	**Question 6** May I have permission to talk to those individuals to learn specifically how they carry out the credit-checking process? **Follow-up** If so: When would be an appropriate time to meet with each of them?	
1 min.	**Objective** Conclude the interview: — Thank Mr. Bentley for his cooperation and assure him that he will be receiving a copy of what transpired during the interview.	
21 minutes	Time allotted for base questions and objectives	
9 minutes	Time allotted for follow-up questions and redirection	
30 minutes	Total time allotted for interview (1:30 p.m. to 2:00 p.m.)	

General Comments and Notes:

Additional guidelines for questions are provided below. You should especially avoid threatening or critical questions. The purpose of the interview is to investigate, not to evaluate or criticize.

Interview Question Guidelines

- Use clear and concise language.
- Don't include your opinion as part of the question.
- Avoid long or complex questions.
- Avoid threatening questions.
- Don't use "you" when you mean a group of people.

Conduct the Interview   The actual interview can be characterized as consisting of three phases: the opening, body, and conclusion. The **interview opening** is intended to influence or motivate the interviewee to participate and communicate by establishing an ideal environment. When establishing an environment of mutual trust and respect, you should identify the purpose and length of the interview and explain how the gathered data will be used. Here are three ways to effectively begin an interview:

- Summarize the apparent problem, and explain how the problem was discovered.
- Offer an incentive or reward for participation.
- Ask the interviewee for advice or assistance.

The **interview body** represents the most time-consuming phase. During this phase, you obtain the interviewee's responses to your list of questions. Listen closely and observe the interviewee. Take notes concerning both verbal and nonverbal responses from the interviewee. It's very important for you to keep the interview on track. Anticipate the need to adapt the interview to the interviewee. Often questions can be bypassed if they have been answered earlier in part of an answer to another question, or they can be deleted if determined to be irrelevant, based on what you've already learned during the interview. Finally, probe for more facts when necessary.

During the **interview conclusion,** you should express your appreciation and provide answers to any questions posed by the interviewee. The conclusion is very important for maintaining rapport and trust with the interviewee.

The importance of human relations skills in interviewing cannot be overemphasized. These skills must be exercised throughout the interview. Here is a set of rules that should be followed during an interview.

Do	**Avoid**
- Be courteous.	- Continuing an interview unnecessarily.
- Listen carefully.	- Assuming an answer is finished or leading nowhere.
- Maintain control.	
- Probe.	- Revealing verbal and nonverbal clues.
- Observe mannerisms and nonverbal communication.	- Using jargon.
	- Revealing your personal biases.
- Be patient.	- Talking instead of listening.
- Keep interviewee at ease.	- Assuming anything about the topic and the interviewee.
- Maintain self-control.	- Tape recording—a sign of poor listening skills.

Follow Up on the Interview   To help maintain good rapport and trust with interviewees, you should send them a memo that summarizes the interview. This memo should remind the interviewees of their contributions to the project and allow them the opportunity to clarify any misinterpretations that you may have derived during the interview. In addition, the interviewees should be given the opportunity to offer additional information they may have failed to bring out during the interview.

Listening   When most people talk about communication skills they think of speaking and writing. The skill of listening rarely gets mentioned, but it may be the most important skill during the interviewing process. To conduct a successful interview you must distinguish between hearing and listening. "To hear is to recognize that someone is speaking, to listen is to understand what the speaker wants to communicate."[4]

We have been conditioned most of our lives not to listen. We ignore our quarreling brothers and sisters while we enjoy our favorite music CD, or we study by blocking out distractions such as noisy roommates. We have learned not to listen, but we can also learn how to listen effectively.

When working with users to solve their problems, getting the users to communicate may be difficult. The following guidelines can open the lines of communication.

- *Approach the session with a positive attitude.* Approaching the project or person with a negative attitude is fighting a losing battle. You have a job to do! Make the best of it and look at it as a fun, pleasurable experience.

- *Set the other person at ease.* Presenting a cheerful attitude can help the person relax. Start by talking about the person's interests or hobbies. Showing an interest in his personal life can serve as an icebreaker.

- *Let them know you are listening.* Always maintain eye contact when listening and use a response such as a head nod or "uh-huh" to indicate that you acknowledge what the other person is saying. Have good posture and even sit on the edge of your seat and lean forward. This will tell the speaker that you are really interested in what she is saying.

- *Ask questions.* To make sure you clearly understand what the person is saying or to clarify a point, ask a question. This will show that you are listening and will also give the other person the opportunity to expand on the answer.

- *Don't assume anything.* One of the worst things you can do is to get in a hurry and be impatient with the speaker. For example, you assume you know what the other person is going to say so you cut in and finish the sentence, possibly missing what the person was going to say and probably irritating the speaker. Or you interrupt or stop the speaker because you may have already heard that information and you believe it is not applicable to what you are doing, thus risking missing a valuable piece of information. Don't assume anything! Art Linkletter learned this lesson on his popular television show, "House Party," when he asked a child a philosophical question:

  On my show I once had a child tell me he wanted to be an airline pilot. I asked him what he'd do if all the engines stopped out over the Pacific Ocean. He said, "First I would tell everyone to fasten their seatbelts, and then I'd find my parachute and jump out."

  While the audience rocked with laughter, I kept my attention on the young man to see if he was being a smart alec. The tears that sprang into his eyes alerted me

---

[4] Thomas R. Gildersleeve, *Successful Data Processing Systems Analysis* (Englewood Cliffs, NJ: Prentice Hall, 1978), p. 93.

to his chagrin more than anything he could have said, so I asked him why he'd do such a thing. His answer revealed the sound logic of a child: "I'm going for gas . . . I'm coming back!"[5]

— *Take notes* Taking notes serves two purposes. First, by jotting down brief notes while the other person is speaking, you give him the impression that what he has to say is important enough that you want to write it down. Second, it helps you remember the major points of the meeting later.

Body Language and Proxemics   What is body language, and why should a systems analyst care about it during the interviewing process?

> **Body language** is all the nonverbal information that we all communicate and are usually unaware of.

Research has determined a startling fact—of a person's total feelings, only 7 percent are communicated verbally (in words), 38 percent are communicated by the tone of voice used, and 55 percent are communicated by facial and body expressions. If you listen only to someone's words, you are missing most of what she has to say!

For this discussion, we will focus on just three aspects of body language: facial disclosure, eye contact, and posture. Facial disclosure means you can sometimes understand how people feel by watching the expressions on their faces. Many common emotions have easily recognizable facial expressions associated with them. However, the face is one of the most controlled parts of the body. Some people who are aware that their expressions often reveal what they are thinking are very good at controlling these expressions.

Another form of nonverbal communication is eye contact. Eye contact is the least controlled part of the face. Have you ever spoken to someone who will not look directly at you? How did it make you feel? A continual lack of eye contact may indicate uncertainty. A normal glance is usually from three to five seconds in length; however, direct eye contact time should increase with distance. As an analyst, you need to be careful not to use excessive eye contact with threatened users so that you won't further intimidate them. Direct eye contact can cause strong feelings, either positive or negative, in other people. If eyes are "the window to the soul," be sure to search for any information they may provide.

Posture is the least controlled aspect of the body. As such, body posture holds a wealth of information for the astute analyst. Members of a group who are in agreement tend to display the same posture. A good analyst will watch the audience for changes in posture that could indicate anxiety, disagreement, or boredom. An analyst should normally maintain an "open" body position signaling approachability, acceptance, and receptiveness. In special circumstances, the analyst may choose to use a confrontation angle of head-on or at a 90-degree angle to another person to establish control and dominance.

In addition to the information communicated by body language, individuals also communicate via proxemics.

> **Proxemics** is the relationship between people and the space around them. Proxemics is a factor in communications that can be controlled by the knowledgeable analyst.

People tend to be very territorial about their space. Observe where your classmates sit in a course that does not have assigned seats. Or the next time you are talking with someone, deliberately move much closer or farther away and see what happens. A good analyst is aware of four spatial zones:

---

[5] Donald Walton, *Are You Communicating? You Can't Manage Without It* (New York: McGraw-Hill Cos., 1989), p. 31.

- Intimate zone—closer than 1.5 feet.
- Personal zone—from 1.5 feet to 4 feet.
- Social zone—from 4 feet to 12 feet.
- Public zone—beyond 12 feet.

Certain types of communications occur only in some of these zones. For example, an analyst conducts most interviews with system users in the personal zone. But the analyst may need to move back to the social zone if the user displays any signs (body language) of being uncomfortable. Sometimes increasing eye contact can make up for a long distance that can't be changed. Many people use the fringes of the social zone as a "respect" distance.

We have examined some of the informal ways that people communicate their feelings and reactions. A good analyst will use all the information available, not just the written or verbal communications of others.

Another type of fact-finding technique is prototyping. Prototyping was introduced in Chapter 3 for use in rapid application development (RAD). The concept behind prototyping was building a small working model of the users' requirements or a proposed design for an information system. This type of prototyping is usually a design technique, but the approach can be applied earlier in the systems development life cycle to perform fact-finding and requirements analysis. This is performed by building discovery prototypes.

Discovery Prototyping

> **Discovery prototyping** is the act of building a small-scale, representative or working model of the users' requirements to discover or verify those requirements.

Discovery prototyping is frequently applied to systems development projects, especially when the development team is having problems defining system requirements. The philosophy is that users will recognize their requirements when they see them. The prototype should be developed quickly so it can be used during the development process. Usually, only the areas where the requirements are not clearly understood are prototyped. This means that a lot of desired functionality may be left out and quality assurance may be ignored. Also, nonfunctional requirements such as performance and reliability may be less stringent than they would be for the final product. Frequently, alternate technologies other than the ones used for the final software will be used to build the discovery prototypes. In these cases the prototypes are most likely discarded when the system has been finished. This "throwaway" approach is primarily used to gather information and develop ideas for the system concept. Many areas of a proposed system may not be clearly understood or some features may be a technical challenge for the developers. Creating discovery prototypes enables the developers as well as the users to better understand and refine the issues involved with developing the system. This technique minimizes the risk of a system being delivered that doesn't meet user needs or one that can't fulfill technical requirements.

Discovery prototyping has its advantages and disadvantages, which should be weighed against those of other fact-finding techniques for every fact-finding situation.

Advantages
- Allows users and developers to experiment with the software and develop an understanding of how the system might work.
- Aids in determining the feasibility and usefulness of the system before high development costs are incurred.
- Serves as a training mechanism for users.

— Aids in building system test plans and scenarios to be used last in the system testing process.

— May minimize the time spent for fact-finding and help define more stable and reliable requirements.

Disadvantages

— Developers may need to be trained in the prototyping approach.

— Users may develop unrealistic expectations based on the performance, reliability, and features of the prototype. Prototypes can only simulate system functionality and are incomplete in nature. Care must be taken to educate the users of this fact and not to mislead them.

— Doing prototyping may extend the development schedule and increase the development costs.

Joint Requirements Planning (JRP)

Separate interviews of end-users and management have been the classic fact-finding technique practiced during systems development. However, many analysts and organizations have discovered the great flaw of interviewing—separate interviews often lead to conflicting facts, opinions, and priorities, not to mention significant time and effort being expended. For these reasons, many organizations are using the group work session as a substitute for interviews. One example of the group work session approach is joint requirements planning. This and similar techniques generally require extensive training to work as intended. However, they can significantly decrease the time spent on fact-finding in one or more phases of the life cycle.

> **Joint requirements planning (JRP)** is a process whereby highly structured group meetings are conducted to analyze problems and define requirements. JRP is a subset of a more comprehensive *joint application development* or *JAD* technique that encompasses the entire systems development process.

JRP (and JAD) techniques are becoming increasingly common in systems planning and systems analysis to obtain group consensus on problems, objectives, and requirements.

In this section, you will learn about the participants of a JRP session and their roles. We will also learn how to plan and conduct a JRP session, the tools and techniques that are used during a JRP session, and the benefits to be achieved through JRP.

JRP Participants   Joint requirements planning sessions include a variety of participants and roles. Each participant is expected to attend and actively participate for the entire JRP session. The individuals involved in a typical JRP session include the sponsor, the facilitator, the users and managers, the scribe, and the IT staff.

— *Sponsor.* Any successful JRP session requires a single person, called the **JRP sponsor,** to serve as its *champion.* This person is normally an individual who is in top management (not IT or IS management) and has authority that spans the different departments and users who are to be involved in the systems project. The sponsor gives full support to the systems project by encouraging designated users to willingly and actively participate in the JRP sessions. Recalling the "creeping commitment" approach to system development, it is the sponsor who usually makes final decisions regarding the go or no-go direction of the project.

The sponsor plays a visible role during a JRP session by "kicking off" the meeting with introductions of the participants. Often, the sponsor will also make closing remarks for the session. The sponsor also works closely with the JRP leader to plan the session by helping identify individuals from the user community who should attend and determining the time and location for the JRP session.

— *Facilitator.* JRP sessions also involve a single individual who plays the role of the leader or facilitator. The **JRP facilitator** is usually responsible for leading all sessions that are held for a systems project. This individual is someone who: has excellent communication skills, possesses the ability to negotiate and resolve group conflicts, has a good knowledge of the business, has strong organizational skills, is impartial to decisions that will be addressed, and does not report to any of the JRP session participants.

It sometimes is difficult to find an individual within the company who possesses all these traits. Thus, companies often must provide extensive JRP training or hire an expert from outside the organization to fill this role. Many systems analysts are trained to become JRP and JAD facilitators.

The role of the JRP facilitator is to plan the JRP session, conduct the session, and follow through on the results. During the session, the facilitator is responsible for leading the discussion, encouraging the attendees to actively participate, resolving conflicts that may arise, and ensuring the goals and objectives of the meeting are fulfilled. It is the JRP facilitator's responsibility to establish the ground rules that will be followed during the meeting and ensure that the participants abide by these rules.

— *Users and Managers.* Joint requirements planning includes a number of participants from the user and management sectors of an organization who are given release time from their day-to-day jobs to devote themselves to the JRP sessions. These participants are normally chosen by the project sponsor, who must be careful to ensure that each person has the business knowledge to contribute during the fact-finding sessions. The project sponsor must exercise authority and encouragement to ensure that these individuals will be committed to actively participate.

A typical JRP session may involve anywhere from a relatively small number of user and management people to a dozen or more. The role of the users during a JRP session is to effectively communicate business rules and requirements, review design prototypes, and make acceptance decisions. The role of the managers during a JRP session is to approve project objectives, establish project priorities, approve schedules and costs, and approve identified training needs and implementation plans. Overall, both users and managers are depended on to ensure that their critical success factors are being addressed.

— *Scribe(s).* A JRP session also includes one or more **scribes** who are responsible for keeping records pertaining to everything discussed in the meeting. These records are published and disseminated to the attendees immediately after the meeting in order to maintain the momentum that has been established by the JRP session and its members. This need to quickly publish the records is reflected in more scribes using CASE tools to capture many facts (documented using data and process models) that are communicated during a JRP session. Thus, it is advantageous for the scribe to possess strong knowledge of systems analysis and design and be skilled with using CASE tools. Systems analysts frequently play this role.

— *IT Staff.* A JRP session may also include a number of IT personnel who primarily listen and take notes regarding issues and requirements voiced by the users and managers. Normally, IT personnel do not speak unless invited to do so. Rather, any questions or concerns they have are usually directed to the JRP facilitator immediately after or before the JRP session. It is the JRP facilitator who initiates and facilitates discussion of issues by users and managers.

The IT staff in the JRP session usually consists of members of the project team. These members may work closely with the scribe to develop models and other documentation related to facts communicated during the meeting. Specialists may also be called on to gain information regarding special technical issues and concerns that may arise. When the situation warrants, the JRP facilitator may prompt the IT professional to address the technical issue.

How to Plan JRP Sessions   Most JRP sessions span three to five days, but they occasionally last up to two weeks. The success of any JRP session depends on proper planning and effectively carrying out that plan. Some preparation is necessary well before the JRP session can be performed. Before planning a JRP session, the analyst must work closely with the executive sponsor to determine the scope of the project that is to be addressed through JRP sessions. It is also important to determine the high-level requirements and expectations of each JRP session. This normally involves interviewing selected individuals who are responsible for departments or functions that are to be addressed by the project. Finally, before planning the JRP session, the analyst must ensure that the executive sponsor is willing to commit people, time, and other resources to the session.

Planning for a JRP session involves three steps: selecting a location for the JRP session, selecting JRP participants, and preparing an agenda for the JRP session. Let's examine each of these planning steps in detail.

1. *Selecting a Location for JRP Sessions.* When possible, JRP sessions should be conducted away from the company workplace. Most local hotels or universities have facilities designed to host group meetings. By holding the JRP session at an off-site location, the attendees can concentrate on the issues and activities related to the JRP session and avoid interruptions and distractions that would occur at their regular workplace. Regardless of the location of the JRP session, all attendees should be required to attend and be prohibited from returning to their regular workplace.

   A JRP session typically requires several rooms. A conference room is required in which the entire group can meet to address JRP issues. Also, if the JRP session includes many people, several breakout rooms may be needed for small groups to focus on specific issues.

   The conference or main meeting room should comfortably hold all the attendees. The room should be fully equipped with tables, chairs, and other items that meet the needs of all attendees. Figure 6.6 depicts a typical room layout for a JRP session. Typical visual aids for a JRP room should include white board or blackboard, one or more flipcharts, and overhead projectors.

   The room should also include computer equipment needed by scribes to record facts and issues communicated during the session. The computer should include software packages to support the various types of records or documentation to be captured and later published by the scribe. Such software may include CASE tool, word processing, spreadsheet, presentation package, prototyping software (i.e., 4GL), printer; copier (or quick access), and computer projection capability. Computer equipment (except that used for prototyping) should be located at the rear of the room so it doesn't distract participants. Personal interaction of the participants should be the focus of the session, not technology.

   Finally, the room should include notepads and pencils for users, managers, and other attendees. Attendees should also be provided with name tags and place cards. Plan on also providing snacks and drinks to make the attendees as comfortable as possible. Creature comforts are very important because JRP sessions are very intensive and often run the entire day.

2. *Selecting JRP Participants.* As mentioned earlier, the analyst, executive sponsor, and managers establish the needs and expectations for a JRP session. It is also their responsibility to select the JRP leader. Ideally, an

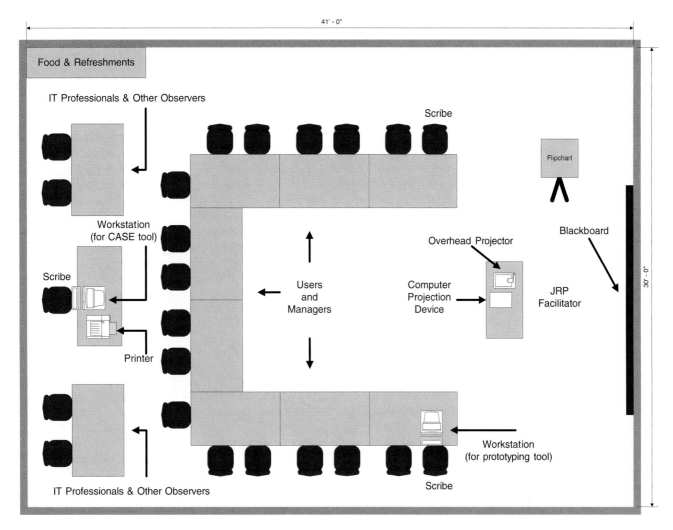

FIGURE 6.6   *Typical Room Layout for JRP Session*

experienced JRP facilitator may be available in-house. If not, an individual may be selected to obtain the extensive training needed to conduct JRP sessions. Many companies opt to hire a qualified person from outside the organization, which eliminates bias on the part of the facilitator.

One or more scribes must also be selected. Because scribes must possess technical skills (word processing, CASE, data and process modeling, etc.), they are usually selected from among the organization's IT professionals. Sometimes the duties of the scribe are shared—one scribe may be responsible for taking notes and meeting presentation materials and another may be focused on documenting technical requirements and issues such as developing data models, process models, and prototypes.

In addition to scribes, other IT professionals must be selected to be involved in the JRP session. Once again, the role of these IT professionals is primarily to listen and learn about the users' and manager' needs. Usually all IT individuals assigned to the project team are involved in the JRP session. Other IT specialists may also be assigned to address specific technical issues pertaining to the project.

Finally, the analyst and managers must select individuals from the user community. While all managers will participate, clearly not all users can attend and participate. The number of attendees must be kept at a level that provides a conducive environment for each attendee to be able to actively participate. Thus, only those users who are able to clearly articulate facts and opinions will be invited. These should be key individuals who are knowledgeable about their business area. Unfortunately, managers are often very dependent on these individuals to run their business area and are often hesitant to release them from their duties. Thus, the analyst must ensure that management is committed to the JRP project and willing to not only permit but also require these key individuals to participate.

3. *Preparing a JRP Session Agenda.* Preparation is the key to a successful JRP session. The JRP facilitator must prepare documentation to brief the participants about the scope and objectives of the sessions. In addition, an agenda for each JRP session should be prepared and distributed before each session. The agenda dictates issues to be discussed during the session and the amount of time allotted to each item.

The agenda should contain three parts: the opening, body, and conclusion. The opening is intended to communicate the expectations of the session, to communicate the ground rules, and to influence or motivate the attendees to participate. The body is intended to detail the topics or issues to be addressed in the JRP session. Finally, the conclusion is intended to allow time to summarize the day's session and to remind the attendees of unresolved issues (to be carried forward).

How to Conduct a JRP Session   The JRP session begins with opening remarks, introductions, and a brief overview of the agenda and objectives for the session. The JRP facilitator will direct the session by following the prepared script. To successfully conduct the session, the facilitator should follow these guidelines:

— Do not unreasonably deviate from the agenda.
— Stay on schedule (agenda topics are allotted specific times).
— Ensure that the scribe is able to take notes (this may mean having the users and managers restate their points more slowly or clearly).
— Avoid the use of technical jargon.
— Apply conflict resolution skills.
— Allow for ample breaks.
— Encourage group consensus.
— Encourage user and management participation without allowing individuals to dominate the session.
— Make sure that attendees abide by the established ground rules for the session.

One goal of a JRP session may be to generate possible ideas to solve a problem. One approach to this goal is called brainstorming.

Brainstorming is a technique for generating ideas during group meetings. Participants are encouraged to generate as many ideas as possible in a short time without any analysis until all the ideas have been exhausted.

Contrary to what you might believe, brainstorming is a formal technique that requires discipline. These guidelines should be followed to ensure effective brainstorming:

1. Isolate the appropriate people in a place that will be free from distractions and interruptions.
2. Make sure that everyone understands the purpose of the meeting (to generate ideas to solve the problem) and focuses on the problem(s).
3. Appoint one person to record ideas. This person should use a flipchart, chalkboard, or overhead projector that can be viewed by the entire group.

4. Remind everyone of the brainstorming rules:

   a. Be spontaneous. Call out ideas as fast as they occur.

   b. Absolutely no criticism, analysis, or evaluation of any kind is permitted while the ideas are being generated. Any idea may be useful, if only to spark another idea.

   c. Emphasize quantity of ideas, not necessarily quality.

5. Within a specified time period, team members call out their ideas as quickly as they can think of them.

6. After the group has run out of ideas and all ideas have been recorded, then and only then should the ideas be analyzed and evaluated.

7. Refine, combine, and improve the ideas that were generated earlier.

With a little practice and attention to these rules, brainstorming can be a very effective technique for generating ideas to solve problems.

As mentioned earlier, the success of a JRP session is highly dependent on planning and the skills of the JRP facilitator and scribes. These skills only get better through training and experience. JRP sessions are usually concluded with an evaluation questionnaire for the participants to complete, in the hope that the responses will help ensure the likelihood of future JRP successes.

The end product of a JRP session is typically a formal written document. This document is essential in confirming the specifications agreed on during the session to all participants. The content and organization of the specification depend on the objectives of the JRP session. The analyst may provide a different set of specifications to different participants based on their role—for example, a manager may receive more of a summary version of the document provided to the user participants (especially in those cases in which the system owners had minimal actual involvement in the JRP session).

Benefits of JRP   Joint requirements planning offers many benefits as an alternative fact-finding and development approach. More companies are beginning to realize its advantages and are incorporating JRP into their existing methodologies. An effectively conducted JRP session offers the following benefits:

- JRP actively involves users and management in the development project (encouraging them to take ownership in the project).

- JRP reduces the amount of time required to develop systems. This is achieved by replacing traditional, time-consuming one-on-one interviewing of each user and manager with group meetings. The group meetings aid in obtaining consensus among the users and managers as well as resolving conflicting information and requirements.

- When JRP incorporates prototyping as a means for confirming requirements and obtaining design approvals, the benefits of prototyping are realized.

Achieving a successful JRP session depends on the JRP facilitator and his or her ability to plan and facilitate the JRP session.

An analyst needs an organized method for collecting facts. An inexperienced analyst will frequently jump right into interviews. "Go to the people. That's where the real facts are!" Wrong! This attitude fails to recognize an important fact of life: People must complete their day-to-day jobs! Your job is not their main responsibility. Your demand on their time is their money lost. Now you may be thinking, "But I thought you've been saying that the system is for people and that direct end-user involvement in systems development is essential! Aren't you contradicting yourselves?"

Not at all! Time is money. To waste your end-users' time is to waste your company's money. To make the most of the time that you spend with end-users, don't

A FACT-FINDING STRATEGY

jump right into interviews. Instead, first collect all the facts you can by using other methods. Consider the following step-by-step strategy:

1. Learn all you can from existing documents, forms, reports, and files. You'll be surprised how much of the system becomes clear without any people contact.

2. If appropriate, observe the system in action. Agree not to ask questions. Just watch and take notes or draw pictures. Make sure the workers know that you're not evaluating individuals. Otherwise, they may perform in a more efficient manner than normal.

3. Given all the facts that you've already collected, design and distribute questionnaires to clear up things you don't fully understand. This is also a good time to solicit opinions on problems and limitations. Questionnaires require your end-users to give up some of their time, but they choose when to best make that sacrifice.

4. Conduct your interviews (or group work sessions). Because you have already collected most of the pertinent facts by low-user-contact methods, you can use the interviews to verify and clarify the most difficult issues and problems. (Alternatively, consider using JRP techniques to replace or complement interviews).

5. (Optional). Build discovery prototypes for any functional requirements that are not understood or if requirements need to be validated.

6. Follow up. Use appropriate fact-finding techniques to verify facts (usually interviews or observation).

The strategy is not sacred. Although a fact-finding strategy should be developed for every pertinent phase of systems development, every project is unique. Sometimes observation and questionnaires may be inappropriate. But the idea should always be to collect as many facts as possible before using interviews.

DOCUMENTING REQUIREMENTS METHODS

Over the life of the project, requirements data will be read and reread by the development team and users many times. Good systems analysts use techniques and tools that allow them to clearly and concisely specify system requirements.

Use Cases

A technique that has its roots in object modeling but has gained popularity in nonobject development is documenting use cases.

> A **use case** is a behaviorally related sequence of steps (a scenario), both automated and manual, for the purpose of completing a single business task.

Use cases describe the system functions from the perspective of external users and in the manner and terminology they understand. To accurately and thoroughly accomplish this demands a high level of user involvement.

Use cases are the results of decomposing the scope of system functionality into many smaller statements of system functionality. The creation of use cases has proven to be an excellent technique to better understand and document system requirements. A use case itself is not considered a functional requirement, but the story (scenario) the use case tells consists of one or more requirements. Use cases are initiated or triggered by external users called actors.

> An **actor** represents anything that needs to interact with the system to exchange information. An actor is a user, a role, which could be an external system as well as a person.

An actor initiates system activity, a use case, for the purpose of completing some business task. An actor represents a role fulfilled by a user interacting with the system and is not meant to portray a single individual or job title. Let's use the example of a college student enrolling for the fall semester's courses. The actor would be the *student* and the business event, or use case, would be *enrolling in course.* In many information systems there are business events triggered by the calendar or the time on a clock. Consider the following examples:

— The billing system for a credit card company automatically generates its bills on the fifth day of the month (billing date).
— A bank reconciles its check transactions every day at 5:00 P.M.
— On a nightly basis a report is automatically generated listing which courses have been closed to enrollment (no open seats available) and which courses are still open.

These events are example of temporal events.

A **temporal event** is a system event that is triggered by time.

Who would be the actor? All the example events listed above were performed (or triggered) automatically—when it became a certain date or time. We say the actor of a temporal event is time.

Use cases are constructed during requirements analysis, but they can be used in the entire system development process. During analysis, the use cases are used to model functionality of the proposed system and are the starting point for identifying the data entities or objects of the system. During design and construction, use cases aid the developers in programming. Because use cases contain an enormous amount of system functionality detail, they will be a constant resource for validating and testing the development of the system design.

Using use cases for requirements discovery provides the following benefits:

— Facilitates user involvement.
— Gives a view of the desired system's functionality from an external person's viewpoint.
— Creates an effective tool for validating requirements.
— Supplies an effective communication tool.

A typical information system may consist of dozens of use cases. During requirements analysis we strive to identify and document only the most critical, complex, and important ones because of time and cost considerations. In this section we will examine how to identify and document a use case.

How to Document a Use Case

Identifying Actors and Use Cases   Potential actors and use cases can be identified by working with the users and identifying the primary inputs and outputs of the system and the external parties that submit and receive them. The primary inputs that trigger business events within the organization are considered use cases, and the external parties that provide those inputs are considered actors. Inputs that are the result of system requests are not considered inputs—such as a credit card company responding to an authorization request. Use cases are named using the name of the input preceded by an action verb. For example, if *purchase order* was the input a good name for the use case would be *submit purchase order.* Use cases that are temporal events are usually identified as a result of analyzing the key outputs of the system. For example, any output that is generated on the basis of time or a date, such as monthly or annual reports, is considered a use case and the actor is time. Figure 6.7 is a partial listing of the actors and use cases that have been identified for the SoundStage Member Services System.

Documenting High-Level Use Cases   At this point it is useful to create high-level use cases. Each one tersely describes the events identified giving the name of the use case, the actors associated with it, and a brief description of the use case. Figure 6.8 is a high-level use case description for the Member Services System's *Submit New Member Order* use case. It is wise to start with high-level use cases to quickly obtain an understanding of the events and magnitude of the system.

FIGURE 6.7		*Partial List of Actors and Use Cases for SoundStage Member Services System*	
Actor	**Input**	**Use Case Name**	**Use Case Description**
Potential member	New subscription	Submit New Subscription	Joins the club by subscribing ("Take any 12 CDs for one penny and agree to buy 4 more at regular prices within two years.")
Member	Change of address	Submit Change of Address	Changes address (including e-mail and privacy code)
Member	New member order	Place New Member Order	Places order
Marketing	New member subscription program	Submit New Member Subscription Program	Establishes a new membership subscription plan to entice new members
Marketing	New promotion	Submit New Promotion	Initiates a promotion (Note: A promotion features specific titles, usually new, the company is trying to sell at a special price. These promotions are integrated into a catalog sent, or communicated, to all members.)
Time	N/A	Generate Quarterly Promotion Analysis	Prints quarterly promotion analysis report

Author: S. Shepherd   Date: 03/01/2000

Use Case Name:	New Member Order
Actor(s):	Member
Description:	This use case describes the process of a member submitting an order for SoundStage products. On completion, the member will be sent a notification that the order was accepted.

FIGURE 6.8

An Example of a High-Level Use Case

Documenting the Use Case Course of Events   For each high-level use case identified, we must now expand it to include the use case's typical course of events and its alternate courses. A use case's typical course of events is a step-by-step description starting with the actor initiating the use case and going until the end of the business event. In this section we include only the major steps that happen the majority of the time (its typical course). The alternate course documents the exceptions or the conditional branching of the use case. Figure 6.9 represents a requirements use case narrative for the Member Services System's *Submit New Member Order* use case. Notice that it includes the following items:

❶ A reference to the requirement(s) in which it can be traced to.

❷ A typical event course describing the use case's major steps, from beginning to end of this interaction with the actor.

❸ Alternate courses describing exceptions to the typical course of events.

❹ Precondition describing the state the system is in before the use case is executed.

❺ Postcondition describing the state the system is in after the use case is executed.

❻ An assumptions section, which includes any nonbehavioral issues, such as performance or security, that is associated with the use case but is difficult to model within the use case's course of events.

Author: <u>S. Shepherd</u>   Date: <u>10/05/2000</u>

Use Case Name:	Submit New Member Order	
Actor(s):	Member	
Description:	This use case describes the process of a member submitting an order for SoundStage products. On completion, the member will be sent a notification that the order was accepted.	
References:	MSS-1.0 ①	
Typical Course of Events: ②	**Actor Action** **Step 1:** This use case is initiated when a member submits an order to be processed. **Step 7:** This use case concludes when the member receives the order confirmation notice.	**System Response** **Step 2:** The member's personal information such as address is validated against what is currently recorded in member services. **Step 3:** The member's credit status is checked with Accounts Receivable to make sure no payments are outstanding. **Step 4:** For each product being ordered, validate the product number and then check the availability in inventory and record the ordered product information. **Step 5:** Create a picking ticket for the member order containing all ordered products that are available and route it to the warehouse for processing. **Step 6:** Generate an order confirmation notice indicating the status of the order and send it to the member.
Alternate Courses: ③	**Step 2:** If the club member has indicated an address or telephone number change on the promotion order, update the club member's record with the new information. **Step 3:** If Accounts Receivable returns a credit status that the customer is in arrears, send an order rejection notice to the member. **Step 4:** If the product number is not valid, send a notification to the member requesting them to submit a valid product number. If the product being ordered is not available, record the ordered product information and mark as "back-ordered."	
Pre-condition: ④	Orders can only be submitted by members.	
Post-condition: ⑤	Member order has been recorded and the picking ticket has been routed to the warehouse.	
Assumptions: ⑥	None at this time.	

FIGURE 6.9   *An Example of a Requirements Use Case*

In this chapter we have presented use cases as technique for documenting and verifying requirements. Additional coverage of use cases and use case modeling can be found in the object-oriented analysis and design chapters found later in this book.

Decision Tables

Many organizations have complex policies and decision-making rules that drive their business processes. Decision tables are a popular tool in expressing and analyzing these requirements.

A **decision table** is a tabular form of presentation that specifies a set of conditions and their corresponding actions.

An in-depth discussion of decision tables is deferred until Chapter 8 because of their close association with process modeling.

Requirements Tables

In some organizations requirements are written in paragraph form, but in others requirements are expressed in a table format such as the one presented in Figure 6.10. Tables facilitate requirements traceability.

Requirements traceability is the ability to trace a system function or feature back to the requirement that mandates it.

FIGURE 6.10	An Example of a Requirements Table Format
Requirement	**Explanation**
Requirement number:	Indicate a unique number or identifier of the requirement
Requirement title:	Assign short phrase indicating nature of the requirement
Requirement text:	Provide a textual statement of the requirement
Requirement type:	Indicate the requirement type
Requirement details and constraints:	Functional characteristics or dimensions
Revision date and revision number:	Indicate the acceptance date and revision number of current (accepted/baselined) version
Criticality:	Must, want, or optional

FIGURE 6.11	Partial List of Member Services System Requirements

Requirement	**Explanation**
Requirement number:	MSS-1.0
Requirement title:	Process New Member Order
Requirement text:	The system should be able to process new member orders. Within this process it should be able to validate member demographic information, verify creditworthiness, inquire and modify inventory levels based on quantity of product ordered, initiate back-order process in the event of insufficient inventory to fulfill order, and send an order confirmation once the order has been placed.
Requirement type:	Functional
Requirement details and constraints:	Member credit status will be obtained from the Accounts Receivable system. A picking ticket, containing the available ordered items, must be generated and routed to the warehouse.
Revision date and revision number:	Version 1.0
Criticality:	Must

Requirement	**Explanation**
Requirement number:	MSS-14.0
Requirement title:	One Hour Order Confirmation Notice
Requirement text:	An e-mail notice must be generated and sent to the member, within one hour from the time the member placed the order.
Requirement type:	Performance
Requirement details and constraints:	The member's e-mail address must be stored on the system within the member's profile. The one-hour constraint applies only to the sending of the notification and not when it's received by the member. Related requirement(s): MSS-1.0
Revision date and revision number:	Version 1.0
Criticality:	Must

Requirement	**Explanation**
Requirement number:	MSS-16.0
Requirement title:	Member Account Security
Requirement text:	Members can access and view only information regarding their own accounts.
Requirement type:	Control
Requirement details and constraints:	Member accounts will be password and ID protected. The Member Services System will generate IDs and passwords and send them to the members. In the event the member forgets his ID or password, a new set will need to be generated.
Revision date and revision number:	Version 1.0
Criticality:	Must

Requirements tables can be very valuable if a requirement changes and you need to evaluate what impact that change may have on other requirements. By documenting requirements in this manner, related and dependent requirements can be grouped together or queried to make the systems analysts job easier. This style of documentation also facilitates the prioritization of requirements by being able to specify the criticality of the requirement. It facilitates the tracking of changes of requirements, and it provides a list to check to make sure all needs were met when users are ready to accept the finished system. Figure 6.11 lists examples of requirements tables of the SoundStage Member Services System.

Many CASE tools such as Popkin's *System Architect 2001* also provide facilities to document requirements. Figure 6.12 is an example of this.

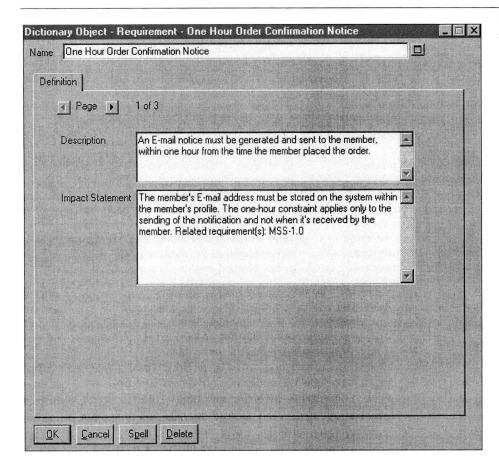

FIGURE 6.12   *System Architect Requirement Example*

W H E R E   D O   Y O U   G O   F R O M   H E R E ?

This chapter provided you with the necessary tools and techniques to discover requirements. Most methodologies require some level of system modeling to analyze requirements. Accordingly, system modeling techniques are the next logical topic to learn. Most of you will proceed directly to Chapter 7, Data Modeling and Analysis. All information systems include databases, and data modeling is an essential skill for database development. Also, it is easier to synchronize early data models with later process models than vice versa. Your instructor may prefer that you first study Chapter 8, Process Modeling. Process modeling is an effective way to analyze and document functional system requirements. Advanced courses may elect to jump straight to Part Five. Module A, Object Modeling, teaches emerging object modeling techniques.

SUMMARY

1. System requirements define the services the system is to provide and prescribe constraints for its operation. In documenting the system requirements for a new information system, an analyst will likely identify dozens of unique requirements. To simplify the presentation of the requirements and to make them more readable, understandable, and traceable, requirements are often categorized as *functional* versus *nonfunctional*.

2. For systems analysts to be successful, they must be skilled in problem analysis. This is the ability to identify problems, understand the problems (including causes and effects), and understand any constraints that limit the solution. A popular tool used by development teams to identify, analyze, and solve problems is an Ishikawa diagram.

3. Effective fact-finding techniques are crucial to the application of systems analysis and design methods during systems projects. Fact-finding is performed during all phases of the systems development life cycle. To support development activities, the analyst must collect facts about end-users, the business, data and information resources, and information systems components.

4. Conducting business in an ethical manner is a required practice, and analysts need to be more aware of the implications of not being ethical.

5. Fact-finding activities usually produce requirements that conflict with each other. This is because requirements are solicited from many different sources and each person has his own opinions and desires for the functionality and features of the new system. Systems analysts perform requirements analysis to discover and resolve the problems with the requirements and reach agreement on any modifications to satisfy the stakeholders.

6. System requirements are usually documented in a *formal* way to communicate the requirements to the key stakeholders of the system. This document usually serves as the contract between the system owners and the development team on what is going to be provided in terms of a new system.

7. Over the lifetime of the project it is very common for new requirements to emerge and existing requirements to change once a requirements definition document has been approved. Requirements management encompasses the policies, procedures, and processes that govern how a change is handled.

8. There are seven common fact-finding techniques.
   a. The sampling of existing documents and files can provide many facts and details with little or no direct personal communication being necessary. The analyst should collect historical documents, business operations manuals and forms, and information systems documents.
   b. Research is an often-overlooked technique based on the study of other similar applications. It now has become more convenient with the Internet and World Wide Web. Site visits are a special form of research.
   c. Observation is a fact-finding technique in which the analyst studies people doing their jobs.
   d. Questionnaires are used to collect similar facts from a larger number of individuals.
   e. Interviews are the most popular but the most time-consuming fact-finding technique. When interviewing, the analyst meets individually with people to gather information.
      1) When most people talk about communication skills they think of speaking and writing, but the skill of listening may be the most important, especially during the interviewing process.
      2) Research has determined that of a person's total feelings, only 7 percent are communicated verbally (in words), 38 percent are communicated by the tone of voice used, and 55 percent are communicated by facial and body expressions. If you only listen to someone's words, you are missing most of what they have to say! Experienced systems analysts pay close attention to body language and proxemics.
   f. Discovery prototyping is frequently applied to systems development projects, especially when the development team is having problems defining system requirements. The philosophy is that the users will recognize their requirements when they see them. It is important that the prototype be developed quickly so it can be used during the development process.
   g. Many analysts find flaws with interviewing—separate interviews often lead to conflicting facts, opinions, and priorities. The end result is numerous follow-up interviews and/or group meetings. For this reason, many organizations use the group work session known as a joint requirements planning as a substitute for interviews.
      1) Joint requirements planning sessions include a variety of participants and roles. Each participant is expected to attend and actively participate for the entire duration of the JRP session.
      2) An effective JRP session involves extensive planning. Planning for a JRP session involves three steps: selecting a location, selecting JRP participants, and preparing an agenda.

9. Because "time is money," it is wise and practical for the systems analyst to use a fact-finding strategy to maximize the value of time spent with the end-users.

10. Systems analysts should assemble or document the information they have gathered in an understandable and meaningful way. Good systems analysts use techniques and tools that allow them to clearly specify the system requirements.
    a. Use cases describe the system functions from the perspective of external users and in a manner and terminology they understand.
    b. Many organizations have complex policies and decision-making rules that drive their business processes. Decision tables are a popular tool in expressing and analyzing these requirements.
    c. Many organizations document their requirements in tabular format called a requirements table.

KEY TERMS

actor, p. 244
body language, p. 236
brainstorming, p. 242
cause-and-effect diagram, p. 219
closed-ended questions, p. 232
decision table, p. 247
discovery prototyping, p. 237
fact-finding, p. 219
fishbone diagram, p. 219
fixed-format questionnaires, p. 229
free-format questionnaires, p. 229
functional requirement, p. 215
interview body, p. 234
interview conclusion, p. 234
interviewee, p. 231
interviewer, p. 231

interview guide, p. 232
interview opening, p. 234
interviews, p. 230
Ishikawa diagram, p. 218
joint requirements planning (JRP),
   p. 238
JRP facilitator, p. 239
JRP sponsor, p. 238
multiple-choice questions, p. 229
nonfunctional requirement, p. 216
observation, p. 226
open-ended questions, p. 232
problem analysis, p. 215
proxemics, p. 236
questionnaires, p. 228
randomization, p. 225

ranking questions, p. 230
rating questions, p. 230
requirements discovery, p. 215
requirements management, p. 223
requirements traceability, p. 247
requirements validation, p. 223
sampling, p. 224
scribe, p. 239
stratification, p. 225
structured interviews, p. 231
system requirement, p. 215
temporal event, p. 245
unstructured interviews, p. 231
use case, p. 244
work sampling, p. 228

REVIEW QUESTIONS

1. Compare functional requirements and nonfunctional requirements.
2. What is an Ishikawa diagram? What is its purpose?
3. What is fact-finding?
4. What should be included in a requirements definition document?
5. What is requirements validation?
6. What are the seven common fact-finding techniques?
7. What question does analyzing the requirements hope to answer?
8. When collecting facts from existing documentation, what is the first document the analyst should seek out? What should the analyst do next?
9. What is sampling? What are two commonly used sampling techniques? How are they different?
10. What is work sampling?
11. What are four advantages of using questionnaires? What are four disadvantages?
12. Identify and briefly describe the two types of questionnaires.
13. Which fact-finding technique is generally recognized as the most important and most often used?
14. What is body language? Give an example of using body language.
15. List the four spatial zones and the characteristics of each.
16. Define discovery prototyping. List three advantages and three disadvantages of discovery prototyping.
17. What is joint requirements planning?
18. Who are the participants in a JRP session?
19. Who is a JRP sponsor and what is the sponsor's role?
20. Who is a JRP facilitator and what is the facilitator's role?
21. What characteristics are important for a JRP facilitator?
22. What is a scribe and what is their role in JRP?
23. What steps are involved in planning a JRP session?
24. Where should a JRP session be located and why?
25. What are some of the resources that should be made available at the JRP location?
26. What important guidelines should a JRP facilitator adhere to in running a JRP session?
27. What are the three rules of brainstorming?
28. What benefits can be derived through a successful JRP project?
29. Why should a systems analyst use a fact-finding strategy when working with an end-user?
30. Define the term *use case*. Give an example.
31. Define the term *actor*. Give an example.
32. Define the term *temporal event*. Who is the actor that initiates a temporal event?
33. What are the different text sections of a use case? Define each one.
34. What is requirements traceability?

PROBLEMS AND EXERCISES

1. Explain how a requirements error, if left undetected, costs more to fix as the project progresses through the life cycle.
2. List the common problems often found with draft requirements. What can the analyst do to eliminate these mistakes?
3. Explain why it is necessary to formally document requirements. How does this benefit the users? The development team?
4. Explain the concept of requirements management. Why is it necessary to have requirements management?
5. Explain how the information systems building blocks can serve as a framework in determining what facts need to be collected during systems development.
6. Explain how an organization chart can aid in planning for fact-finding. What are some potential drawbacks to using an existing organizational chart?
7. A systems analyst wants to study documents stored in a large metal file cabinet. The cabinet contains several hundred records describing product warranty claims. The analyst wishes to study a sample of the records in the file and to be 95 percent certain (certainty factor = 1.960) that the data from which the sample is taken will not include variations not in the sample. How many sample records should the analyst retrieve to get this desired accuracy?
8. For the sample size in Exercise 7, explain two specific strategies for selecting the samples.
9. Describe how you would use form and/or file sampling in each phase of the systems development life cycle. If you think sampling would be inappropriate for any of these phases, explain why.
10. Repeat Exercise 9 for the technique of observation.
11. Make a list of things that might affect your work performance when you are being observed performing your job. What could an observer do to eliminate these concerns or problems?
12. Repeat Exercise 9 for the questionnaire technique.
13. Give two examples of free-format questions and two examples of each of the following types of fixed-format questions:
   a. Multiple choice.
   b. Rating.
   c. Ranking.
14. Repeat Exercise 9 for the interviewing technique.
15. Explain the difference between a structured and an unstructured interview. When is each type of interview appropriately used?
16. Explain the meaning of the statements, "To hear is to recognize that someone is speaking. To listen is to understand what the speaker wants to communicate."
17. List and explain six guidelines for effective listening.
18. List three aspects of body language. Why should a systems analyst care about body language?
19. Prepare a sample interview guide to use in obtaining facts from your academic adviser describing the course registration policies and procedures.
20. Explain how JRP may significantly decrease the amount of systems development time.
21. Why might an organization choose to use a JRP facilitator from outside the company?
22. Explain why IT personnel should primarily take a back-seat role in a JRP session?
23. Research the topic of conflict resolution. Describe one or more approaches that the leader may use to resolve conflicts.
24. Write a memo that would defend your recommendations to conducting a JRP session at an off-site location.
25. List the benefits of using use cases.
26. Compare a use case typical course of events versus a use case alternate course.
27. Prepare a use case complete with typical and alternate courses documenting the event of you purchasing a can of soda from a soda machine.

PROJECTS AND RESEARCH

1. Mr. Art Pang is the Accounts Receivables manager. You have been assigned to do a study of Mr. Pang's current billing system, and you need to solicit facts from his subordinates. Mr. Pang has expressed his concern that, although he wishes to support you in your fact-finding efforts, his people are extremely busy and must get their jobs done. Write a memo to Mr. Pang describing a fact-finding strategy that you could follow to maximize your fact-finding while minimizing the release time required for his subordinates.

2. Group decision support systems (GDSS) are popular for dealing with conflict resolutions. Research this topic and write a brief overview of its principles and how it is applied to JRP-like situations.
3. Research available software GDSS products. What would the resource implications be for using a GDSS product in a JRP session?

MINICASES

1. Working with your teammates, research any problem that you are aware of with your school's registration system, financial-aid system, or the like. Use brainstorming techniques to identify the problem's causes and effects and construct an Ishikawa diagram.

2. Schedule an interview with the business analyst responsible for your school's registration system. Based on the information she provides perform the following:
   a. Identify the actors and the use cases they initiate.
   b. Complete a use case description for the event of a student enrolling for a course. For a student dropping a course.

SUGGESTED READINGS

Andrews, D. C., and N. S. Leventhal. *Fusion Integrating IE, CASE and JAD: A Handbook for Reengineering the Systems Organization.* Englewood Cliffs, NJ: Prentice Hall, 1993.

Berdie, Douglas R., and John F. Anderson. *Questionnaires: Design and Use.* Metuchen, NJ: Scarecrow Press, 1974. A practical guide to the construction of questionnaires. Particularly useful because of its short length and illustrative examples.

Davis, William S. *Systems Analysis and Design.* Reading, MA: Addison-Wesley, 1983. Provides useful pointers for preparing and conducting interviews.

Dejoie, Roy; George Fowler; and David Paradice. *Ethical Issues in Information Systems.* Boston: Boyd and Fraser, 1991. Focuses on the impact of computer technology on ethical decision making in today's business organizations.

Fitzgerald, Jerry; Ardra F. Fitzgerald; and Warren D. Stallings, Jr. *Fundamentals of Systems Analysis.* 2nd ed. New York: Wiley, 1981. A useful survey text for the systems analyst. Chapter 6, "Understanding the Existing System," does a good job of presenting fact-finding techniques in the study phase.

Gane, C. *Rapid Systems Development.* New York: Rapid Systems Development, Inc., 1987. This book provides a good discussion on how to lead a group meeting/interview.

Gause, Donald C., and Gerald M. Weinberg. *Exploring Requirements: Quality Before Design.* New York: Dorset House Publishing, 1989. An excellent book describing the techniques of requirements discovery.

Gildersleeve, Thomas R. *Successful Data Processing System Analysis.* Englewood Cliffs, NJ: Prentice Hall, 1978. Chapter 4, "Interviewing in Systems Work," provides a comprehensive look at interviewing specifically for the systems analyst. A thorough sample interview is scripted and analyzed in this chapter.

Jacobson, Ivar; Magnus Christerson; Patrik Jonsson; and Gunnar Overgaard. *Object-Oriented Software Engineering–A Use Case Driven Approach.* Wokingham, England: Addison Wesley, 1992. This book presents detailed coverage of how to identify and document use cases.

London, Keith R. *The People Side of Systems.* New York: McGraw-Hill, 1976. Chapter 5, "Investigation versus Inquisition," provides a very good people-oriented look at fact-finding, with considerable emphasis on interviewing.

Lord, Kenniston W., Jr., and James B. Steiner. *CDP Review Manual: A Data Processing Handbook.* 2nd ed. New York: Van Nostrand Reinhold, 1978. Chapter 8, "Systems Analysis and Design," provides a more comprehensive comparison of the merits and demerits of each fact-finding technique. This material is intended to prepare data processors for the Certificate in Data Processing examinations, one of which covers systems analysis and design.

Miller, Irwin, and John F. Freund. *Probability and Statistics For Engineers.* Englewood Cliffs, NJ: Prentice Hall, 1965. Introductory college textbook on probability and statistics.

Mitchell, Ian; Norman Parrington; Peter Dunne; and John Moses. "Practical Prototyping, Part One." *Object Currents,* May 1996. First of a three-part series of articles that explores prototyping and how you can benefit from it. Prototyping is an integral part of JRP.

Salvendy, G., ed. *Handbook of Industrial Engineering.* New York: Wiley, 1974. A comprehensive handbook for industrial engineers; systems analysts are, in a way, a type of industrial engineer. Excellent coverage on sampling and work measurement.

Stewart, Charles J., and William B. Cash, Jr. *Interviewing: Principles and Practices.* 2nd ed. Dubuque, IA: Brown, 1978. Popular college textbook that provides broad exposure to interviewing techniques, many of which are applicable to systems analysis and design.

Walton, Donald. *Are You Communicating? You Can't Manage Without It.* New York: McGraw-Hill, 1989. This book is an easy-to-use guidebook on the process of communications and a must for anyone who must work with people and influence them.

Weinberg, Gerald M. *Rethinking Systems Analysis and Design.* Boston: Little, Brown and Company, 1982. A book created to stimulate a new way of thinking.

Wood, Jane, and Denise Silver. *Joint Application Design.* New York: John Wiley & Sons, 1989. This book provides a comprehensive overview of IBM's joint application design technique.

Focus on
PEOPLE

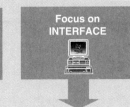
Focus on
DATA

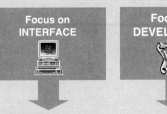
Focus on
PROCESSES

Focus on
INTERFACE

Focus on
DEVELOPMENT

Stakeholders

Activities

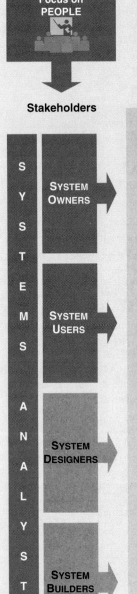

S
Y
S
T
E
M
S

A
N
A
L
Y
S
T
S

SYSTEM OWNERS

SYSTEM USERS

SYSTEM DESIGNERS

SYSTEM BUILDERS

VENDORS AND CONSULTANTS

PROJECT & PROCESS MANAGEMENT

PRELIMINARY INVESTIGATION

PROBLEM ANALYSIS

REQUIREMENTS ANALYSIS

DECISION ANALYSIS

DESIGN

CONSTRUCTION

IMPLEMENTATION

OPERATIONS AND SUPPORT

BUILDING BLOCKS OF AN INFORMATION SYSTEM

Management Expectations

The PIECES Framework

Performance ● Information ● Economics ● Control ● Efficiency ● Service

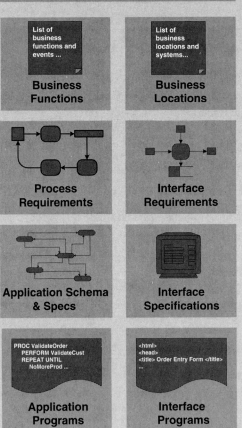

List of business entities and rules ...

Business Knowledge

List of business functions and events ...

Business Functions

List of business locations and systems...

Business Locations

Data Requirements

Process Requirements

Interface Requirements

Database Schema

Application Schema & Specs

Interface Specifications

CREATE TABLE tblOrders
  colOrderNot CHAR(5) NOT
  NULL
  colOrderDate DATE/TIME
NOT

Database Programs

PROC ValidateOrder
  PERFORM ValidateCust
  REPEAT UNTIL
    NoMoreProd ...

Application Programs

<html>
<head>
<title> Order Entry Form </title>
...

Interface Programs

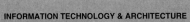
INFORMATION TECHNOLOGY & ARCHITECTURE
Database Technology ● Process Technology ● Interface Technology ● Network Technology

7

DATA MODELING AND ANALYSIS

CHAPTER PREVIEW AND OBJECTIVES

This is the first of two graphical systems modeling chapters. In this chapter you will learn how to use a popular data modeling tool, *entity relationship diagrams,* to document the data that must be captured and stored by a system, independently of showing how that data is or will be used—that is, independently of specific inputs, outputs, and processing. You will also learn about a data analysis technique called *normalization* that is used to ensure that a data model is a "good" data model. You will know data modeling and data analysis as systems analysis tools and techniques when you can:

— Define systems modeling and differentiate between logical and physical system models.

— Define data modeling and explain its benefits.

— Recognize and understand the basic concepts and constructs of a data model.

— Read and interpret an entity relationship data model.

— Explain when data models are constructed during a project and where the models are stored.

— Discover entities and relationships.

— Construct an entity relationship context diagram.

— Discover or invent keys for entities and construct a key-based diagram.

— Construct a fully attributed entity relationship diagram and describe all data structures and attributes to the repository or encyclopedia.

— Normalize a logical data model to remove impurities that can make a database unstable, inflexible, and nonscalable.

— Describe a useful tool for mapping data requirements to business operating locations.

SOUNDSTAGE

SOUNDSTAGE ENTERTAINMENT CLUB

SCENE

We begin this episode soon after the project executive sponsor has approved the initial system improvement objectives. It's time to model the new system requirements, beginning with data. Sandra is facilitating a meeting with the following staff:

- David Hensley, representing Legal Services.
- Ann Martinelli, representing Member Services.
- Sally Hoover, representing Member Services.
- Joe Bosley, representing Marketing.
- Antonio Scarpachi, representing the warehouse.
- Bob Martinez, assigned to sketch the data model.
- Sarah Hartman, assigned to take detailed notes.

SANDRA

The purpose of this meeting is to discover the business data that your new system needs to capture and store. I'd like to keep this discussion as nontechnical as possible. Try not to think about files and databases. Instead, I'd like you to focus exclusively on the data and explain to us how that data describes your business.

[pause, noting consensus]

In an earlier meeting, you identified the following things or subjects about which we agreed that the system must capture and store data: agreements, members, member orders, products, and promotions. Have we missed anything?

ANTONIO

I'm not sure. What about merchandise?

SANDRA

Is that another name for product?

ANTONIO

Not exactly. I'm referring to general merchandise like posters and T-shirts. They do not fall into the same category as products like CDs, videos, and games.

ANN

I don't agree, Antonio. Merchandise is simply another type of product. At least that's the way we treat it in Marketing.

SANDRA

OK, are there other types of products? For example, should we treat audio, video, and game products as separate types?

ANN

I would! They are described by different pieces of data. For example, we need data such as category, media, screen aspect ratio, and motion picture rating for a video. On the other hand, we need a different category and media and a content advisory code for an audiotape or disk.

SANDRA

Different category and media?

ANN

Of course! The categories for a video include science fiction and adventure. But the categories for audio are entirely different, such as popular and country/western. Media for videos include VHS tapes and laserdiscs, while media for audio titles include cassettes and CDs.

ANTONIO

You're absolutely right, but not all the data is different.

SANDRA

What do you mean?

ANTONIO

For instance, every audio, video, and game title will have a universal product code in our bar-coding scheme. Come to think of it, so will the general merchandise I spoke of earlier. Let's see, I'm sure that audio, video, and games share some unique data . . . Yes, they all have a title, catalog description, and copyright date.

SALLY

And let's not forget system objectives! We want every audio, video, and game title to count toward fulfilling any membership agreement; therefore, they

must each be assigned some sort of credit value that counts toward fulfilling an agreement. That value will be based on the price of the item . . .

JOE

Or how badly we want to get rid of the inventory!

[laughter]

SANDRA

Bob, I think we have a generalization hierarchy here.

JOE

Say what?

SANDRA

Sorry, it's a technical term that Bob and I have to worry about. Let's get back to this discussion. Are there any data that describe all products, regardless of all these types of products?

ANTONIO

I already mentioned the universal product codes, but there are some other common data. Because the UPCs are not yet implemented, each product has an existing product number that we can't get rid of. For every product, we need to know the quantity in stock.

ANN

Almost every product has a manufacturer's suggested retail price and a club default retail price. At any given time, a product could have a special retail price. I'd also like to have some historical data about the sales of each product.

SANDRA

Such as?

ANN

Oh, number of units sold this month; this year; lifetime.

SANDRA

Anything else?

[pause]

Let's get back to this product type thing. So we now know that a product is either general merchandise, an audio title, a video title, or a game title—and that each of those types has its own unique data.

ANTONIO

No, I said that the three types of titles have some common data just like all products have some common data.

SANDRA

You're right. I heard that. Let's try again. A product is either general merchandise or a title. Some data describes all products. Some data is unique to either merchandise or titles. Subsequently, a title is either an audio, video, or game title; each of which has some unique data requirements. Let's break one of those down. What attributes describe a game title?

SALLY

I can answer that one. A game title is described by all the general product and title data we've been discussing, but it is also described by its manufacturer, category, platform, media type, number of players, and a parent advisory code.

SANDRA

I assume this category is different from that for audio and video titles?

SALLY

Yes, it represents game categories such as sports and fantasy.

SANDRA

What is a platform?

SALLY

That is the type of computer the game

runs on. Media type tells us whether the game comes on a cartridge or disk.

SANDRA

Now that brings up an interesting question. Many titles, including audio, video, and game, are distributed on several media. For example, I can get the Beatles Anthology on either CD or cassette . . .

DAVID

Or DVD and possibly others.

SANDRA

That leads to my question. Is the same title on different media considered different products?

ANTONIO

Absolutely! They have different product numbers, and they will have different UPCs. That's how we avoid sending the wrong format to a member.

SANDRA

OK. Can a title be an audio, video, and game title?

DAVID

I'm not sure I understand your question.

SANDRA

I own the audio soundtrack to *Jurassic Park,* a copy of the videotape, and a game based on the movie. Are those three titles?

DAVID

Yes, they are three different titles. And I suspect that each title is offered on several different media; therefore, it could

be said that we have a dozen or so *Jurassic Park* products.

SARAH

Are you sure you want to build a database? Sounds complicated!

[We can exit the meeting now.]

DISCUSSION QUESTIONS

1. Should the different types of products each have their own file (or database table) or should they be consolidated? What are the business implications of that decision? Each of the different participants was concerned with different aspects of the system. Briefly organize their concerns into two or three categories.

2. How would you deal with the issue of category and media having different values for different types of products?

3. Why can't SoundStage simply replace the current product number scheme with the planned universal product code scheme for identifying its products?

4. The group spent a lot of time identifying characteristics of different subject areas. What else would you need to know about each characteristic (e.g., game category) to properly store that data?

Systems models play an important role in systems development. As a systems analyst or user, you will constantly deal with unstructured problems. One way to structure such problems is to draw models.

A **model** is a representation of reality.

Models can be built for existing systems as a way to better understand those systems or for proposed systems as a way to document business requirements or technical designs. An important concept, in both this chapter and the next, is the distinction between logical and physical models.

Logical models show *what* a system is or does. They are implementation *in*dependent; that is, they depict the system independent of any technical implementation. As such, logical models illustrate the *essence* of the

AN INTRODUCTION TO SYSTEMS MODELING

system. Popular synonyms include **essential model,** *conceptual model,* and *business model.*

Physical models show not only *what* a system is or does, but also *how* the system is physically and technically implemented. They are implementation *de*pendent because they reflect technology choices and the limitations of those technology choices. Synonyms include *implementation model* and *technical model.*

Let's say we need a system that can store students' GRADE POINT AVERAGES. For any given student, the GRADE POINT AVERAGE must be between 0.00 and 4.00 where 0.00 reflects an average grade of F and 4.00 reflects an average grade of A. We further know that a new student does not have a grade point average until he receives his first grade report. Every statement in this paragraph reflects a business fact. It does not matter which technology we use to implement these facts; the facts are constant—they are logical. Thus, GRADE POINT AVERAGE is a *logical* attribute with a *logical* domain of 0.00–4.00 and a *logical* default value of null (meaning nothing).

Now, let's get physical. First we have to decide how we are going to store the above attribute. Let's say we will use a table column in a Microsoft *Access* database. We need a column name. Let's assume our company's programming standards require the following name: ColGradePointAvg. That is the *physical* field name that will be used to implement the logical attribute GRADE POINT AVERAGE. (It is not important that you understand why.) Additionally, we could define the following *physical* properties for the database column ColGrade-PointAvg:

Data type	NUMBER
Field size	SINGLE (as in "single precision")
Decimal places	2
Default value	NULL (as in "none")
Validation rule	BETWEEN 0 AND 4 INCLUSIVE
Required?	NO

Systems analysts have long recognized the importance of separating business and technical concerns (or the logical versus the physical concerns). That is why they use logical system models to depict business requirements and physical system models to depict technical designs. Systems analysis activities tend to focus on the logical system models for the following reasons:

— Logical models remove biases that are the result of the way the current system is implemented or the way any one person thinks the system might be implemented. Thus, we overcome the "we've always done it that way" syndrome. Consequently, logical models encourage creativity.

— Logical models reduce the risk of missing business requirements because we are too preoccupied with technical details. Such errors are almost always much more costly to correct after the system is implemented. By separating what the system must do from how the system will do it, we can better analyze the requirements for completeness, accuracy, and consistency.

— Logical models allow us to communicate with end-users in nontechnical or less technical languages. Thus, we don't lose "business" requirements in the technical jargon of the computing discipline.

This chapter will present data modeling as a technique for defining business requirements for a database.

Data modeling is a technique for organizing and documenting a system's DATA. Data modeling is sometimes called database modeling because a data model is eventually implemented as a database. It is also sometimes called *information modeling.*

Figure 7.1 is an example of a simple data model called an *entity-relationship diagram,* or ERD. This diagram makes the following business assertions:

— We need to store data about CUSTOMERS, ORDERS, and INVENTORY PRODUCTS.

— The value of CUSTOMER NUMBER <u>uniquely</u> identifies one and only one CUSTOMER. The value of ORDER NUMBER <u>uniquely</u> identifies one and only one ORDER. The value of PRODUCT NUMBER <u>uniquely</u> identifies one and only one INVENTORY PRODUCT.

— For a CUSTOMER we need to know the CUSTOMER NAME, SHIPPING ADDRESS, BILLING ADDRESS, and BALANCE DUE. For an ORDER we need to know ORDER DATE and ORDER TOTAL COST. For an INVENTORY PRODUCT we need to know PRODUCT NAME, PRODUCT UNIT OF MEASURE, and PRODUCT UNIT PRICE.

— A CUSTOMER has placed zero, one, or more current or recent ORDERS.

— An ORDER is placed by exactly one CUSTOMER. The value of CUSTOMER NUMBER (as recorded in ORDER) identifies that CUSTOMER.

— An ORDER sold one or more ORDERED PRODUCTS. Thus, an ORDER must contain at least one ORDERED PRODUCT.

— An INVENTORY PRODUCT may have been sold as zero, one, or more ORDERED PRODUCTS.

— An ORDERED PRODUCT identifies a single INVENTORY PRODUCT on a single ORDER. The ORDER NUMBER (for an ORDERED PRODUCT) identifies the ORDER, and the PRODUCT NUMBER (for an ORDERED PRODUCT) identifies the INVENTORY PRODUCT. Together, they identify one and only one ORDERED PRODUCT.

— For each ORDERED PRODUCT we need to know QUANTITY ORDERED and UNIT PRICE AT TIME OF ORDER.

After you study this chapter, you will be able to read data models and construct them.

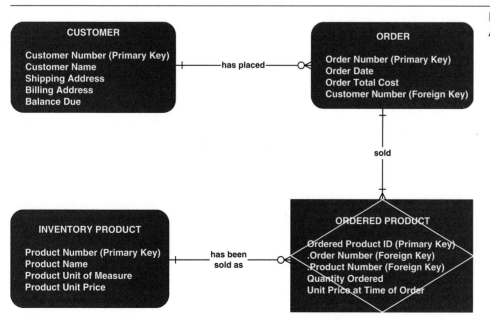

FIGURE 7.1 *An Entity Relationship Data Model*

SYSTEM CONCEPTS FOR DATA MODELING

There are several notations for data modeling. The actual model is frequently called an **entity relationship diagram** (ERD) because it depicts data in terms of the entities and relationships described by the data. There are several notations for ERDs. Most are named after their inventor (e.g., Chen, Martin, Bachman, Merise) or after a published standard (e.g., IDEF1X). These data modeling "languages" generally support the same fundamental concepts and constructs. We have adopted the Martin (information engineering) notation because of its widespread use and CASE tool support.

Let's explore some basic concepts that underlie all data models.

Entities

All systems contain data—usually lots of data! Data describes "things." Consider a school system. A school system includes data that describes things such as STUDENTS, TEACHERS, COURSES, and CLASSROOMS. For any of these things, it is not difficult to imagine some of the data that describes any given instance of the thing. For example, the data that describes a particular student might include NAME, ADDRESS, PHONE NUMBER, DATE OF BIRTH, GENDER, RACE, MAJOR, and GRADE POINT AVERAGE, to name a few.

We need a concept to abstractly represent all instances of a group of similar things. We call this concept an entity.

> An **entity** is something about which the business needs to store data. Formal synonyms include *entity type* and *entity class.*

In system modeling, we find it useful to assign each abstract concept to a shape. In this book, an entity will be drawn as a rectangle with rounded corners (see margin). This shape represents all instances of the named entity. For example, the entity STUDENT represents all students in the system.

We can refine our definition of entity to identify specific classes of entities.

An Entity

> An **entity** is a class of persons, places, objects, events, or concepts about which we need to capture and store data.

Each entity is distinguishable from the other entities. Categories of entities (and examples) include:

Persons: AGENCY, CONTRACTOR, CUSTOMER, DEPARTMENT, DIVISION, EMPLOYEE, INSTRUCTOR, STUDENT, SUPPLIER. Notice that a person entity class can represent individuals, groups, or organizations.

Places: SALES REGION, BUILDING, ROOM, BRANCH OFFICE, CAMPUS.

Objects: BOOK, MACHINE, PART, PRODUCT, RAW MATERIAL, SOFTWARE LICENSE, SOFTWARE PACKAGE, TOOL, VEHICLE MODEL, VEHICLE. An object entity can represent actual objects (such as a specific software license) or specifications for a type of object (such as specifications for different software packages).

Events: APPLICATION, AWARD, CANCELLATION, CLASS, FLIGHT, INVOICE, ORDER, REGISTRATION, RENEWAL, REQUISITION, RESERVATION, SALE, TRIP.

Concepts: ACCOUNT, BLOCK OF TIME, BOND, COURSE, FUND, QUALIFICATION, STOCK.

It is important to distinguish between an entity and its instances.

> An **entity instance** is a single occurrence of an entity.

For example, the entity STUDENT may have multiple instances: Mary, Joe, Mark, Susan, Cheryl, and so forth. In data modeling, we do not concern ourselves with individual students because we recognize that each student is described by similar pieces of data.

If an entity is something about which we want to store data, then we need to identify what specific pieces of data we want to store about each instance of a given entity. We call these pieces of data *attributes*.

Attributes

> An **attribute** is a descriptive property or characteristic of an entity. Synonyms include *element, property,* and *field.*

As noted at the beginning of this section, each instance of the entity STUDENT might be described by the following attributes: NAME, ADDRESS, PHONE NUMBER, DATE OF BIRTH, GENDER, RACE, MAJOR, GRADE POINT AVERAGE, and others.

We can now extend our graphical abstraction of the entity to include attributes by recording those attributes inside the entity shape along with the name (see margin).

Some attributes can be logically grouped into superattributes called compound attributes.

> A **compound attribute** is one that actually consists of other attributes. Synonyms in different data modeling languages are numerous: *concatenated attribute, composite attribute,* and *data structure.*

For example, a student's NAME is actually a compound attribute that consists of LAST NAME, FIRST NAME, and MIDDLE INITIAL. In the margin, we demonstrate one <u>possible</u> notation for compound attributes. Notice that a period is placed at the beginning of each primitive attribute that is included in the composite attribute.

STUDENT

Name
.Last Name
.First Name
.Middle Initial
Address
.Street Address
.City
.State or Province
.Country
.Postal Code
Phone Number
.Area Code
.Exchange Number
.Number Within Exchange
Date of Birth
Gender
Race
Major
Grade Point Average

Attributes and Compound Attributes

Domains   When analyzing a system, we should define those values for an attribute that are legitimate or that make business sense. The values for each attribute are defined in terms of three properties: data type, domain, and default.

> The **data type** for an attribute defines what type of data can be stored in that attribute.

Data typing should be familiar to those of you who have written computer programs; declaring types for variables is common to most programming languages. For purposes of systems analysis and business requirements definition, it is useful to declare logical (nontechnical) data types for our business attributes. For the sake of argument, we will use the logical data types shown in Table 7.1.

An attribute's data type constrains its domain.

> The **domain** of an attribute defines what values an attribute can legitimately take on.

TABLE 7.1	*Representative Logical Data Types for Attributes*
Logical Data Type	**Logical Business Meaning**
NUMBER	Any number, real or integer.
TEXT	A string of characters, inclusive of numbers. When numbers are included in a TEXT attribute, it means we do not expect to perform arithmetic or comparisons with those numbers.
MEMO	Same as TEXT but of an indeterminate size. Some business systems require the ability to attach potentially lengthy notes to a given database record.
DATE	Any date in any format.
TIME	Any time in any format.
YES/NO	An attribute that can only assume one of these two values.
VALUE SET	A finite set of values. In most cases, a coding scheme would be established (e.g., FR = freshman, SO = sophomore, JR = junior, SR = senior, etc.)
IMAGE	Any picture or image.

Eventually, system designers must use technology to enforce the business domains of all attributes. Table 7.2 demonstrates how logical domains might be expressed for each data type.

Finally, every attribute should have a logical default value.

The **default value** for an attribute is the value that will be recorded if not specified by the user.

Table 7.3 shows possible default values for an attribute. Notice that NOT NULL is a way to specify that each instance of the attribute must <u>have</u> a value; while, NULL is a way to specify that some instances of the attribute may be optional, or not have a value.

Identification   An entity has many instances, perhaps thousands or millions. There exists a need to uniquely identify each instance based on the data value of one or more attributes. Thus, every entity must have an identifier or key.

A **key** is an attribute, or a group of attributes, that assumes a unique value for each entity instance. It is sometimes called an *identifier*.

For example, each instance of the entity STUDENT might be uniquely identified by the key STUDENT NUMBER attribute. No two students can have the same STUDENT NUMBER. Sometimes more than one attribute is required to uniquely identify an instance of an entity.

A group of attributes that uniquely identifies an instance of an entity is called a **concatenated key.** Synonyms include *composite key* and *compound key*.

For example, each TAPE entity instance in a video store might be uniquely identified by the concatenation of TITLE NUMBER plus COPY NUMBER. TITLE NUMBER by

TABLE 7.2	*Representative Logical Domains for Logical Data Types*	
Data Type	**Domain**	**Examples**
NUMBER	For integers, specify the range: {minimum−maximum} For real numbers, specify the range and precision: {minimum.precision−maximum.precision}	{10−99} {1.000−799.999}
TEXT	TEXT (maximum size of attribute) *Actual values are usually infinite; however, users may specify certain narrative restrictions.*	TEXT (30)
MEMO	*Not applicable. There are no logical restrictions on size or content.*	*Not applicable.*
DATE	Variation on the MMDDYYYY format. To accommodate the year 2000, do not abbreviate year to YY. Formatting characters are rarely stored; therefore, do not include hyphens or slashes.	MMDDYYYY MMYYYY YYYY
TIME	For AM/PM times: HHMMT or For military times: HHMM	HHMMT HHMM
YES/NO	{YES, NO}	{YES, NO} {ON, OFF}
VALUE SET	{value#1, value#2, . . . , value#n} or {table of codes and meanings}	{FRESHMAN, SOPHOMORE, JUNIOR, SENIOR} {FR = FRESHMAN SO = SOPHOMORE JR = JUNIOR SR = SENIOR}
IMAGE	*Not applicable; however, any known characteristics of the images will eventually prove useful to designers.*	*Not applicable.*

TABLE 7.3	*Permissible Default Values for Attributes*	
Default Value	**Interpretation**	**Examples**
A legal value from the domain (as described above)	For an instance of the attribute, if the user does not specify a value, then use this value.	0 1.00 FR
NONE or NULL	For an instance of the attribute, if the user does not specify a value, then leave it blank.	NONE NULL
REQUIRED or NOT NULL	For an instance of the attribute, require the user to enter a legal value from the domain. (This is used when no value in the domain is common enough to be a default, but some value must be entered.)	REQUIRED NOT NULL

itself would be inadequate because the store may own many copies of a single title. COPY NUMBER by itself would also be inadequate since we presumably have a *copy #1* for every title we own. We need both pieces of data to identify a specific tape (e.g., *copy #7* of *Star Wars: The Phantom Menace*). In this book, we will give a name to the group as well as the individual attributes. For example, the concatenated key for TAPE would be recorded as follows:

TAPE ID

.TITLE NUMBER

.COPY NUMBER

Frequently, an entity may have more than one key. For example, the entity EMPLOYEE may be uniquely identified by SOCIAL SECURITY NUMBER, or company-assigned EMPLOYEE NUMBER, or E-MAIL ADDRESS. Each of these attributes is called a candidate key.

> A **candidate key** is a "candidate to become the primary key" of instances of an entity. It is sometimes called a *candidate identifier*. (A candidate key may be a single attribute or a concatenated key.)

> A **primary key** is that candidate key that will most commonly be used to uniquely identify a single entity instance.

The default for a primary key is always NOT NULL because if the key has no value it cannot serve its purpose to identify an instance of an entity.

> Any candidate key that is not selected to become the primary key is called an **alternate key.** A common synonym is *secondary key*.

In the margin, we demonstrate our notation for primary and alternate keys. All candidate keys must be either primary or alternate; therefore, we do not use a separate notation for candidate keys.

Sometimes, it is also necessary to identify a subset of an entity's instances as opposed to a single instance. For example, we may require a simple way to identify all male students and all female students.

> A **subsetting criteria** is an attribute (or concatenated attribute) whose finite values divide all entity instances into useful subsets. This is sometimes referred to as an *inversion entry*.

In our STUDENT entity, the attribute GENDER divides the instances of STUDENT into two subsets: male students and female students. In general, subsetting criteria are useful only when an attribute has a finite (meaning limited) number of legitimate values. For example, GRADE POINT AVERAGE would not be a good subsetting criteria because there are 999 possible values between 0.00 and 4.00 for that attribute. The margin art demonstrates a notation for subsetting criteria.

STUDENT

Student Number (Primary Key)
Social Security Number
  (Alternate Key)
Name
.Last Name
.First Name
.Middle Initial
Address
.Street Address
.City
.State or Province
.Country
.Postal Code
Phone Number
.Area Code
.Exchange Number
.Number Within Exchange
Date of Birth
Gender (Subsetting Criteria 1)
Race (Subsetting Criteria 2)
Major (Subsetting Criteria 3)
Grade Point Average

Keys and Subsetting Criteria

Relationships

Conceptually, entities and attributes do not exist in isolation. The things they represent interact with and impact one another to support the business mission. Thus, we introduce the concept of a relationship.

> A **relationship** is a natural business association that exists between one or more entities. The relationship may represent an event that links the entities or merely a logical affinity that exists between the entities.

Consider, for example, the entities STUDENT and CURRICULUM. We can make the following business assertions that link students and courses:

— A current STUDENT IS ENROLLED IN one or more CURRICULA.

— A CURRICULUM IS BEING STUDIED BY zero, one, or more STUDENTS.

The underlined verb phrases define business relationships that exist between the two entities.

We can graphically illustrate this association between STUDENT and CURRICULUM as shown in Figure 7.2. The connecting line represents a relationship. Verb phrases describe the relationship. Notice that all relationships are implicitly bidirectional, meaning they can be interpreted in both directions (as suggested by the above business assertions). Data modeling methods may differ in their naming of relationships—some include both verb phrases and others include a single verb phrase.

Cardinality   Figure 7.2 also shows the complexity or *degree* of each relationship. For example, in the above business assertions, we must also answer the following questions:

— Must there exist an instance of STUDENT for each instance of CURRICULUM? No! Must there exist an instance of CURRICULUM for each instance of STUDENT? Yes!

— How many instances of CURRICULUM can exist for each instance of STUDENT? Many! How many instances of STUDENT can exist for each instance of CURRICULUM? Many!

We call this concept cardinality.

> **Cardinality** defines the minimum and maximum number of occurrences of one entity that may be related to a single occurrence of the other entity. Because all relationships are bidirectional, cardinality must be defined in both directions for every relationship.

A popular graphical notation for cardinality is shown in Figure 7.3. Sample cardinality symbols were demonstrated in Figure 7.2.

Conceptually, cardinality tells us the following rules about the data we want to store:

— When we insert a STUDENT instance in the database, we <u>must</u> link (associate) that STUDENT to at least one instance of CURRICULUM. In business terms, "a student cannot be admitted without declaring a major." (Most schools would include an instance of CURRICULUM called "undecided" or "undeclared.")

— A STUDENT <u>can</u> study more than one CURRICULUM, and we must be able to store data that indicates all CURRICULA for a given STUDENT.

FIGURE 7.2
A Relationship (Many-to-Many)

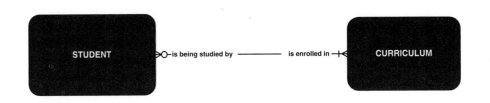

CARDINALITY INTERPRETATION	MINIMUM INSTANCES	MAXIMUM INSTANCES	GRAPHIC NOTATION
Exactly one (one and only one)	1	1	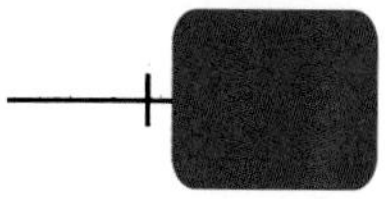 — or 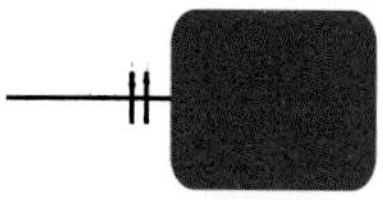 —
Zero or one	0	1	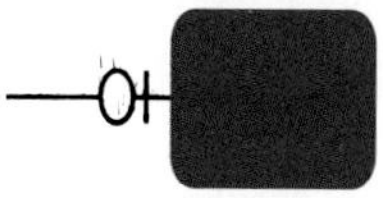
One or more	1	many (>1)	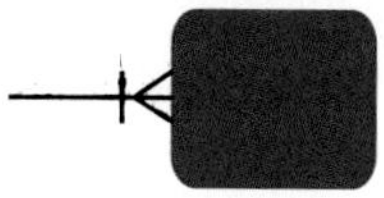
Zero, one, or more	0	many (>1)	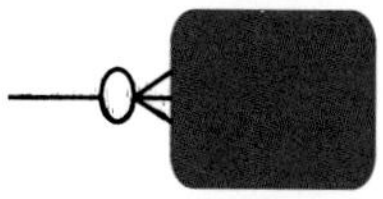
More than one	>1	>1	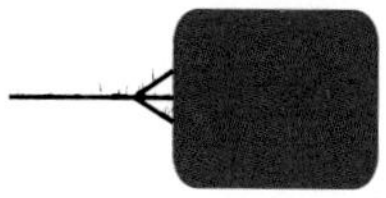

FIGURE 7.3
Cardinality Notations

- We must insert a CURRICULUM before we can link (associate) STUDENTS to that CURRICULUM. That is why a CURRICULUM can have zero students—no students have yet been admitted to that CURRICULUM.
- Once a CURRICULUM has been inserted into the database, we can link (associate) many STUDENTS with that CURRICULUM.

Degree   Another measure of the complexity of a data relationship is its degree.

The **degree** of a relationship is the number of entities that participate in the relationship.

All the relationships we've explored so far are *binary* (degree = 2). In other words, two different entities participated in the relationship.

Relationships may also exist between different instances of the same entity. We call this a **recursive relationship** (degree = 1). For example, in your school a course may be a prerequisite for other courses. Similarly, a course may have several other courses as its prerequisite. Figure 7.4 demonstrates this many-to-many recursive relationship.

Relationships can also exist between more than two different entities. These are sometimes called N-ary relationships. An example of a *3-ary* or *ternary relationship* is shown in Figure 7.5. An N-ary relationship is illustrated with a new entity construct called an associative entity.

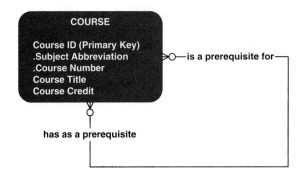

FIGURE 7.4
A Recursive Relationship

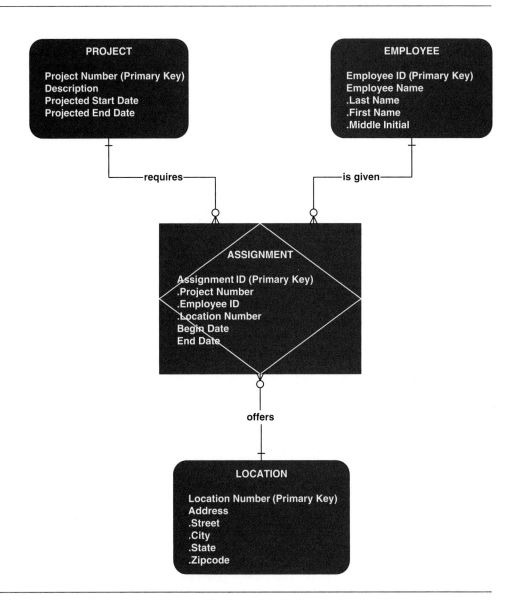

FIGURE 7.5
A Ternary Relationship

An **associative entity** is an entity that inherits its primary key from more than one other entity (called *parents*). Each part of that concatenated key points to one and only one instance of each of the connecting entities.

In Figure 7.5 the associative entity ASSIGNMENT (notice the unique shape) matches an EMPLOYEE, a LOCATION, and a PROJECT. For each instance of ASSIGNMENT, the key indicates which EMPLOYEE ID, which LOCATION NUMBER, and which PROJECT NUMBER are combined to form that assignment.

Also as shown in Figure 7.5, an associative entity can be described by its own nonkey attributes. In addition to the primary key, an ASSIGNMENT is described by the attributes BEGIN DATE and END DATE. If you think about it, none of these attributes describes an EMPLOYEE, LOCATION, or PROJECT—they describe a single instance of the relationship between an instance of each of those entities.

Foreign Keys   A relationship implies that instances of one entity are related to instances of another entity. We should be able to identify those instances for any given entity. The ability to identify specific related entity instances involves establishing foreign keys.

> A **foreign key** is a primary key of one entity that is contributed to (duplicated in) another entity to identify instances of a relationship. A foreign key (always in a child entity) always matches the primary key (in a parent entity).

In Figure 7.6(a), we demonstrate the concept of foreign keys with our simple data model. Notice that the <u>maximum</u> cardinality for DEPARTMENT is "one," whereas the maximum cardinality for CURRICULUM is "many." In this case, DEPARTMENT is called the *parent* entity and CURRICULUM is the *child* entity. The primary key is always contributed by the parent to the child as a foreign key. Thus, an instance of CURRICULUM now has a foreign key DEPARTMENT NAME whose value points to the instance of DEPARTMENT that offers that curriculum. (Foreign keys are never contributed from child to parent.)

In our example, the relationship between CURRICULUM and DEPARTMENT is referred to as a nonidentifying relationship.

> **Nonidentifying relationships** are those in which each of the participating entities has its own independent primary key. In other words, none of the primary key attributes is shared.

The entities CURRICULUM and DEPARTMENT are also referred to as **strong** or independent entities because neither depends on any other entity for its identification. But sometimes a foreign key may participate as part of the primary key of the child entity. For example, in Figure 7.6(b) the parent entity BUILDING contributes its key to the entity ROOM. Thus, BUILDING NAME serves as a foreign key to relate a ROOM and BUILDING and in conjunction with ROOM ID to uniquely identify a given instance of ROOM. In those situations the relationship between the parent and child entity is referred to as an identifying relationship.

> **Identifying relationships** are those in which the parent entity contributes its primary key to become part of the primary key of the child entity.

The child entity of any identifying relationship is frequently referred to as a **weak entity** because its identification is dependent on the parent entity's existence.

Most popular CASE tools and data modeling methods use different notations to distinguish between identifying and nonidentifying relationships and between strong and weak entities. In Figure 7.7, we use a *dashed line* notation to represent the nonidentifying relationship between PASSENGER and SEAT ASSIGNMENT. Because part of the primary key of SEAT ASSIGNMENT is the foreign key FLIGHT NUMBER from the parent entity FLIGHT, the relationship is an identifying relationship and is represented using a *solid line*. Finally, seat assignment is a weak entity because it receives the primary key of flight to compose its own primary key. A weak entity is represented using a symbol composed of a *rounded rectangle within a rounded rectangle*.

NOTE   To reinforce the above concepts of identifying and nonidentifying relationships and strong versus weak entities and to be consistent with most popular data modeling methods and most widely used CASE tools, the authors use the above modeling notations on all subsequent data modeling examples presented in the book.

What if you cannot differentiate between parent and child? For example, in Figure 7.8(a) on p. 270 we see that a CURRICULUM enrolls zero, one, or more STUDENTS. At the same time, we see that a STUDENT is enrolled in one or more CURRICULA. The maximum cardinality on both sides is "many." So, which is the parent and which is the child? You can't tell! This is called a nonspecific relationship.

FIGURE 7.6   *Foreign Keys*

(a)

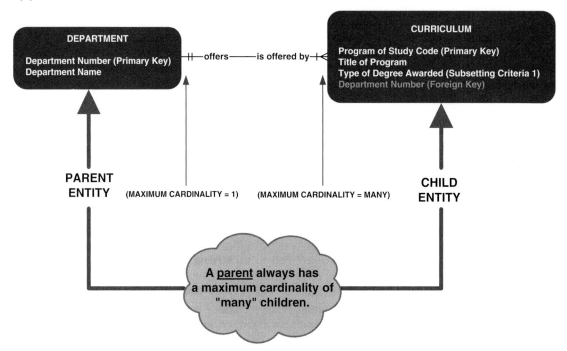

(b)

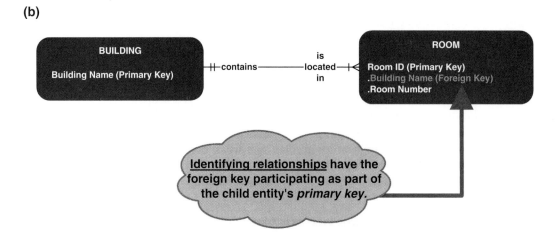

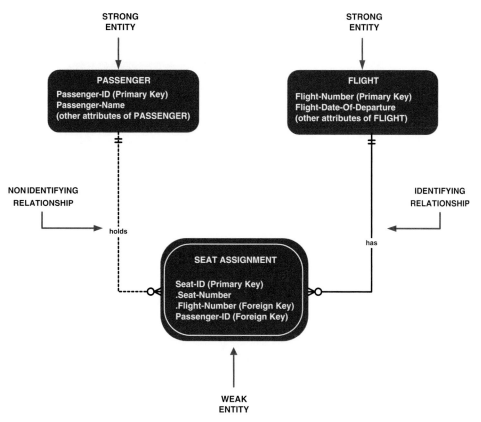

STRONG
ENTITY

STRONG
ENTITY

PASSENGER

Passenger-ID (Primary Key)
Passenger-Name
(other attributes of PASSENGER)

FLIGHT

Flight-Number (Primary Key)
Flight-Date-Of-Departure
(other attributes of FLIGHT)

NONIDENTIFYING
RELATIONSHIP

IDENTIFYING
RELATIONSHIP

holds

has

SEAT ASSIGNMENT

Seat-ID (Primary Key)
.Seat-Number
.Flight-Number (Foreign Key)
Passenger-ID (Foreign Key)

WEAK
ENTITY

FIGURE 7.7
*Notations for Weak Entity and
Nonidentifying Relationship*

A **nonspecific relationship** (or **many-to-many relationship**) is one in which many instances of one entity are associated with many instances of another entity.

Such relationships are suitable only for preliminary data models and should be resolved as quickly as possible.

Many nonspecific relationships can be resolved into a pair of one-to-many relationships. As illustrated in Figure 7.8(b), each entity becomes a parent. A new, *associative entity* is introduced as the child of each parent. In Figure 7.8(b), each instance of MAJOR represents <u>one</u> STUDENT's enrollment in <u>one</u> CURRICULUM. If a student is pursuing two majors, that student will have two instances of the entity MAJOR.

Study Figure 7.8(b) carefully. For associative entities, the cardinality from child to parent is always <u>one and only one.</u> That makes sense because an instance of MAJOR must correspond to one and only one STUDENT and one and only one CURRICULUM. The cardinality from parent to child depends on the business rule. In our example, a STUDENT must declare <u>one or more</u> MAJORS. Conversely, a CURRICULUM is being studied by zero, one, or more MAJORS—perhaps it is new and no one has been admitted to it yet. An associative entity can also be described by its own nonkey attributes (such as DATE ENROLLED and CURRENT CANDIDATE FOR DEGREE?). Finally, associative entities inherit the primary keys of the parents; thus, all associative entities are weak entities.

Not all nonspecific relationships can and should be automatically resolved as described above. Occasionally nonspecific relationships result from the failure of the analyst to identify the existence of other entities. For example, examine the relationship between CUSTOMER and PRODUCT in Figure 7.9(a) on p. 271. Recognize that the relationship "orders" between CUSTOMER and PRODUCT suggests an event

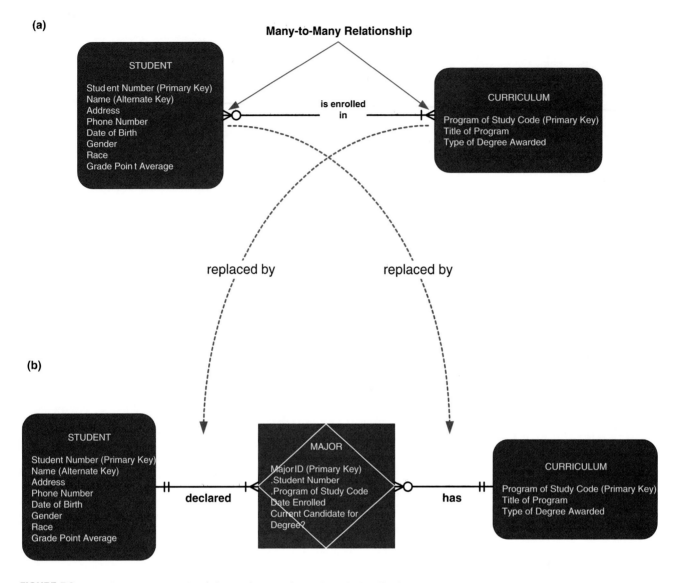

FIGURE 7.8   *Resolving Nonspecific Relationships with an Associative Entity*

about which a user might want to store data! That event represents an event entity called ORDER depicted in Figure 7.9(b). In reality, CUSTOMER and PRODUCT are not a natural and direct relationship as was depicted in Figure 7.9(a). Rather, they are related indirectly, by way of an ORDER. Thus, our many-to-many relationship was replaced by separate relationships between CUSTOMER, ORDER, and PRODUCT. Notice that the relationship between ORDER and PRODUCT is a many-to-many relationship. That relationship would need to be resolved by replacing it with an associative entity and two one-to-many relationships as is illustrated in Figure 7.9(c).

Finally, some nonspecific relationships can be resolved by introducing separate relationships. Notice the many-to-many relationship between TRANSFER and BANK ACCOUNT shown in Figure 7.10(a) on p. 272. While it is true that a TRANSFER transaction involves many BANK ACCOUNTS and a BANK ACCOUNT may be involved in many TRANSFER transactions, we must be careful! Data modeling notations can sometimes mislead us. Technically, a single TRANSFER transaction involves two BANK ACCOUNTS. When we know the specific maximum number of occurrences of a many-to-many relationship, it often suggests that our original relationship is weak or too general. Notice in Figure 7.10(b) that our relationship "involves" was

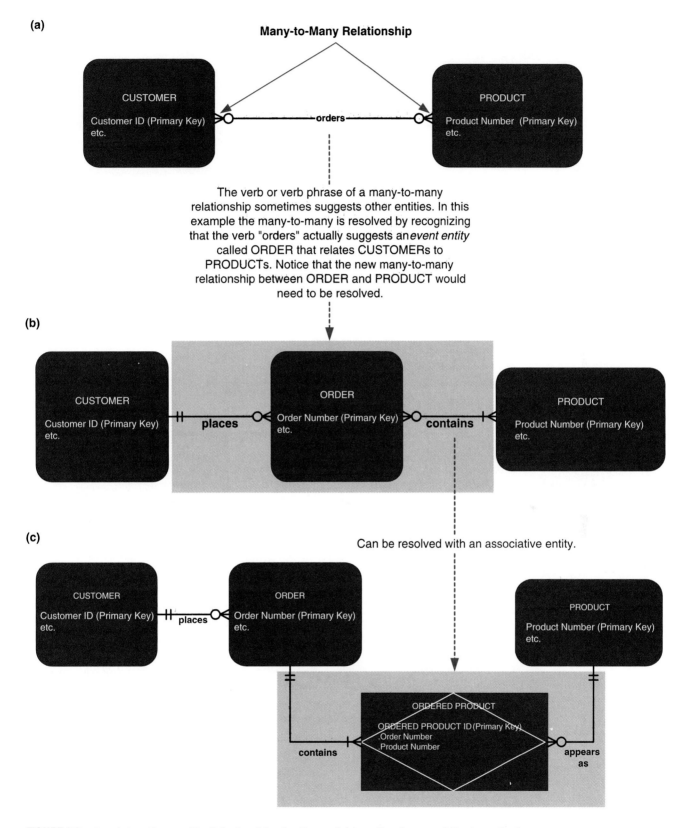

FIGURE 7.9 *Resolving Nonspecific Relationships by Recognizing a Fundamental Business Entity*

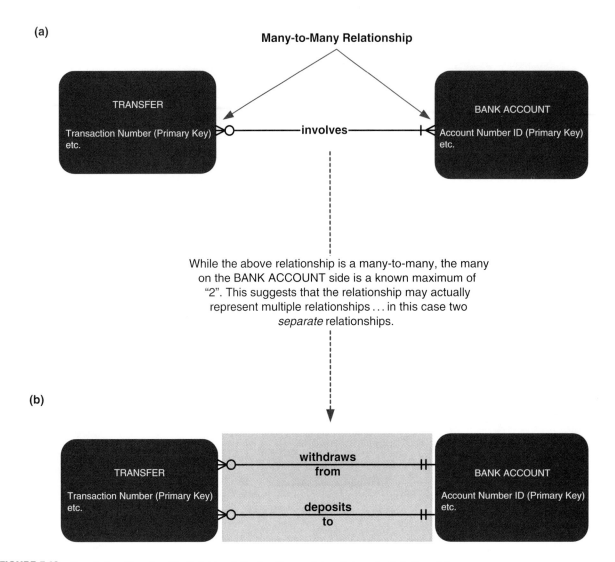

(a)

Many-to-Many Relationship

TRANSFER

Transaction Number (Primary Key)
etc.

─────involves─────

BANK ACCOUNT

Account Number ID (Primary Key)
etc.

While the above relationship is a many-to-many, the many
on the BANK ACCOUNT side is a known maximum of
"2". This suggests that the relationship may actually
represent multiple relationships ... in this case two
separate relationships.

(b)

TRANSFER

Transaction Number (Primary Key)
etc.

**withdraws
from**

**deposits
to**

BANK ACCOUNT

Account Number ID (Primary Key)
etc.

FIGURE 7.10   *Resolving Nonspecific Relationships by Recognizing Separate Relationships*

replaced by two separate one-to-many relationships that more accurately describe the business relationships between a TRANSFER and BANK ACCOUNTS.

Generalization   Most people associate the concept of generalization with modern object-oriented techniques. In reality, the concepts have been applied by data modelers for many years. Generalization is an approach that seeks to discover and exploit the commonalities between entities.

> **Generalization** is a technique wherein the attributes that are common to several types of an entity are grouped into their own entity, called a *supertype*.

Consider, for example a typical school. A school enrolls STUDENTS and employs EMPLOYEES (in a university, a person could be both). There are several attributes that are common to both entities; for example, NAME, GENDER, RACE, MARITAL STATUS, and possibly even a key based on SOCIAL SECURITY NUMBER. We could consolidate these common attributes into an entity supertype called PERSON.

> An entity **supertype** is an entity whose instances store attributes that are common to one or more entity subtypes.

The entity supertype will have one or more *one-to-one* relationships to entity *subtypes*. These relationships are sometimes called "is a" relationships (or "was a," or "could be a") because each instance of the supertype "is also an" instance of one or more subtypes.

> An entity **subtype** is an entity whose instances inherit some common attributes from an entity supertype and then add other attributes that are unique to an instance of the subtype.

In our example, "a PERSON is an employee, or a student, or both." The top half of Figure 7.11 illustrates this generalization ❶ as a hierarchy. Notice that the subtypes STUDENT and EMPLOYEE have inherited attributes from PERSON, as well as adding their own. (Unfortunately, most CASE tools do not migrate the inherited attributes.)

Extending the metaphor, an entity can be both a supertype and subtype. Returning to Figure 7.11 we see that a STUDENT (which was a subtype of PERSON) has its own subtypes. In the diagram, we see that a STUDENT is ❷ either a PROSPECT, or a CURRENT STUDENT, or a FORMER STUDENT (having left for any reason other than graduation), and ❸ a STUDENT could be an ALUMNUS. These additional subtypes inherit all the attributes from STUDENT as well as those from PERSON. Finally, notice that all subtypes are "weak" entities.

Through inheritance, the concept of generalization in data models permits us to reduce the number of attributes through the careful sharing of common attributes. The subtypes not only inherit the attributes, but also the data types, domains, and defaults of those attributes. This can greatly enhance the consistency with which we treat attributes that apply to many different entities (e.g., dates, names, addresses, currency, etc.).

In addition to inheriting attributes, subtypes also inherit relationships to other entities. For instance, all EMPLOYEES and STUDENTS inherit the relationship ❹ between PERSON and ADDRESS. But only EMPLOYEES inherit the relationship ❺ with CONTRACTS. And only an ALUMNUS can be related to ❻ an AWARDED DEGREE.

THE PROCESS OF LOGICAL DATA MODELING

Now that you understand the basic concepts of data models, we can examine the process of data modeling. When do you do it? How many data models may be drawn? What technology exists to support the process?

Data modeling may be performed during various types of projects and in multiple phases of projects. Data models are progressive; there is no such thing as the "final" data model for a business or application. Instead, a data model should be considered a living document that will change in response to a changing business. Data models should ideally be stored in a repository so they can be retrieved, expanded, and edited over time. Let's examine how data modeling may come into play during systems planning and analysis.

Strategic Data Modeling

Many organizations select application development projects based on strategic information systems plans. Strategic planning is a separate project. This project produces an information systems strategy plan that defines an overall vision and architecture for information systems. Almost always, this architecture includes an **enterprise data model.** Information engineering is a methodology that embraces this approach.

An enterprise data model typically identifies only the most fundamental of entities. The entities are typically defined (as in a dictionary), but they are not described in terms of keys or attributes. The enterprise data model may or may not include relationships (depending on the planning methodology's standards and the level of detail desired by executive management). If relationships are included, many of them will be nonspecific (a concept introduced earlier in the chapter).

How does an enterprise data model affect subsequent applications development? Part of the information strategy plan identifies application development projects and prioritizes them according to whatever criteria management deems appropriate.

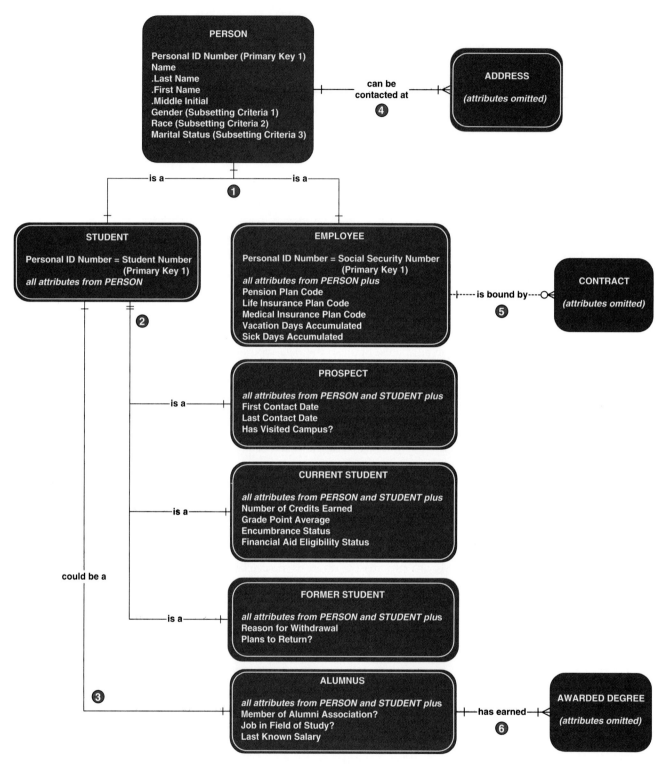

FIGURE 7.11 *A Generalization Hierarchy*

As those projects are started, the appropriate subsets of the information systems architecture, including a subset of the enterprise data model, are provided to the applications development team as a point of departure.

The enterprise data model is usually stored in a corporate repository. When the application development project is started, the subset of the enterprise data model (as well as the other models) is exported from the corporate repository into a project repository. Once the project team completes systems analysis and design, the expanded and refined data models are imported back into the corporate repository.

In systems analysis and in this chapter, we will focus on *logical* data modeling as a part of systems analysis. The data model for a single information system is usually called an **application data model.**

Data models are rarely constructed during the preliminary investigation phase of systems analysis. The short duration of that phase makes them impractical. If an enterprise data model exists, the subset of that model that is applicable to the project might be retrieved and reviewed as part of the phase requirement to establish context. Alternatively, the project team could identify a simple list of entities, the things about which team members think the system will have to capture and store data.

Unfortunately, data modeling is rarely associated with the problem analysis phase of systems analysis. Some analysts prefer to draw process models (Chapter 8) to document the current system, but many analysts report that data models are far superior for the following reasons:

- Data models help analysts to quickly identify business vocabulary more completely than process models.
- Data models are almost always built more quickly than process models.
- A complete data model can fit on a single sheet of paper. Process models often require dozens of sheets of paper.
- Process modelers frequently and too easily get hung up on unnecessary detail.
- Data models for existing and proposed systems are far more similar than process models for existing and proposed systems. Consequently, there is less work to throw away as you move into later phases.

We agree! A problem analysis phase model includes only entities and relationships, but no attributes—called a **context data model.** The intent is to refine our understanding of scope, not to get into details about the entities and business rules. Many relationships may be nonspecific.

Many automated tools provide the ability to read existing system files and databases and translate them into "physical" data models. These physical data models can then be transformed into their equivalent "logical" data model. This translation capability benefits both the problem analysis and the requirements analysis phases.

The requirements analysis results in a logical data model that is developed in stages as follows:

1. We begin by constructing the **context data model** to establish the project scope. If a context data model was already developed during problem analysis, that model may be revised to reflect new requirements and project scope.

2. Next, a **key-based data model** will be drawn. This model will eliminate nonspecific relationships, add associative entities, and include primary and alternate keys. The key-based model will also include precise cardinalities and any generalization hierarchies.

3. Next, a **fully attributed data model** will be constructed. The fully attributed model includes all remaining descriptive attributes and subsetting criteria. Each attribute is defined in the repository with data types, domains, and defaults (in what is sometimes called a **fully described data model**).

4. The completed data model is analyzed for adaptability and flexibility through a process called **normalization.** The final analyzed model is referred to as a **normalized data model.**

Data Modeling during Systems Analysis

This data requirements model requires a team effort that includes systems analysts, users and managers, and data analysts. A data administrator often sets standards for and approves all data models.

Ultimately, during the decision analysis phase, the data model will be used to make implementation decisions—the best way to implement those requirements with database technology. In practice, this decision may have already been standardized as part of a database architecture. For example, SoundStage has already standardized on two database management systems: Microsoft *Access* for personal and work-group databases, and Microsoft *SQL Server* for enterprise databases.

Finally, data models cannot be constructed without appropriate facts and information as supplied by the user community. These facts can be collected by a number of techniques such as sampling of existing forms and files, research of similar systems, surveys of users and management, and interviews of users and management. The fastest method of collecting facts and information and simultaneously constructing and verifying the data models is joint requirements planning. JRP uses a carefully facilitated group meeting to collect the facts, build the models, and verify the models—usually in one or two full-day sessions. Fact-finding and information-gathering techniques were fully explored in Chapter 6. Table 7.4 summarizes some questions that may be useful for fact-finding and information-gathering as it pertains to data modeling.

Looking Ahead to Systems Design

During system design, the logical data model will be transformed into a physical data model (called a *database schema*) for the chosen database management system. This model will reflect the technical capabilities and limitations of that data-

TABLE 7.4	*JRP and Interview Questions for Data Modeling*
Purpose	**Candidate Questions**
Discover the system entities	What are the subjects of the business? In other words, what types of persons, organizations, organizational units, places, things, materials, or events are used in or interact with this system about which data must be captured or maintained? How many instances of each subject exist?
Discover the entity keys	What unique characteristic (or characteristics) distinguishes an instance of each subject from other instances of the same subject? Are there any plans to change this identification scheme in the future?
Discover entity subsetting criteria	Are there any characteristics of a subject that divide all instances of the subject into useful subsets? Are there any subsets of the above subjects for which you have no convenient way to group instances?
Discover attributes and domains	What characteristics describe each subject? For each of these characteristics: (1) what type of data is stored? (2) who is responsible for defining legitimate values for the data? (3) what are the legitimate values for the data? (4) is a value required? and (5) is there any default value that should be assigned if you don't specify otherwise?
Discover security and control needs	Are there any restrictions on who can see or use the data? Who is allowed to create the data? Who is allowed to update the data? Who is allowed to delete the data?
Discover data timing needs	How often does the data change? Over what period of time is the data of value to the business? How long should we keep the data? Do you need historical data or trends? If a characteristic changes, must you know the former values?
Discover generalization hierarchies	Are all instances of each subject the same? That is, are there special types of each subject that are described or handled differently? Can any of the data be consolidated for sharing?
Discover relationships and degrees	What events occur that imply associations between subjects? What business activities or transactions involve handling or changing data about several different subjects of the same or a different type?
Discover cardinalities	Is each business activity or event handled the same way or are there special circumstances? Can an event occur with only some of the associated subjects, or must all the subjects be involved?

base technology, as well as the performance tuning requirements suggested by the database administrator. Any further discussion of database design is deferred until Chapter 12.

Data models are stored in a repository. In a sense, the data model is **metadata**—that is, data about the business's data. Computer-aided systems engineering (CASE) technology, introduced in Chapter 3, provides the repository for storing the data model and its detailed descriptions. Most CASE products support computer-assisted data modeling and database design. Some CASE products (such as Logic Works *ERwin*) only support data modeling and database design. CASE takes the drudgery out of drawing and maintaining these models and their underlying details.

Using a CASE product, you can easily create professional, readable data models without the use of paper, pencil, erasers, and templates. The models can be easily modified to reflect corrections and changes suggested by end-users—you don't have to start over! Also, most CASE products provide powerful analytical tools that can check your models for mechanical errors, completeness, and consistency. Some CASE products can even help you analyze the data model for consistency, completeness, and flexibility. The potential time and quality savings are substantial.

As mentioned earlier, some CASE tools support reverse engineering of existing file and database structures into data models. The resulting data models represent "physical" data models that can be revised and reengineered into a new file or database, or they may be translated into their equivalent "logical" model. The logical data model could then be edited and forward engineered into a revised physical data model, and subsequently a file or database implementation.

CASE tools do have their limitations. Not all data modeling conventions are supported by all CASE products. And different CASE tools adopt slightly different notations for the same data modeling methods. Therefore, it is very likely that any given CASE product may force a company to adapt its methodology's data modeling symbols or approach so that it is workable within the limitations of the CASE tool.

All the SoundStage data models in the next section of this chapter were created with Popkin Systems and Software's CASE tool, *System Architect 2001*. For the case study, we provide you the printouts exactly as they came off our printers. We did not add color. The only modifications by the artist were the bullets that call your attention to specific items of interest on the printouts. All of the entities, attributes, and relationships on the SoundStage data models were automatically cataloged into *System Architect*'s project repository (which it calls an encyclopedia). Figure 7.12 illustrates some of *System Architect*'s screens as used for data modeling.

Automated Tools for Data Modeling

HOW TO CONSTRUCT DATA MODELS

You now know enough about data models to read and interpret them. But as a systems analyst or knowledgeable end-user, you must learn how to construct them. We will use the SoundStage Entertainment Club project to teach you how to construct data models.

> NOTE   This example teaches you to draw the data model from scratch. In reality, you should always look for an existing data model. If such models exist, the data management or data administration group usually maintains them. Alternatively, you could reverse engineer a data model from an existing database.

Entity Discovery

The first task in data modeling is relatively easy. You need to discover those fundamental entities in the system that are or might be described by data. You should not restrict your thinking to entities about which the end-users know they want to store data. There are several techniques that may be used to identify entities.

— During interviews or JRP sessions with system owners and users, pay attention to key words in their discussion. For example, during an interview

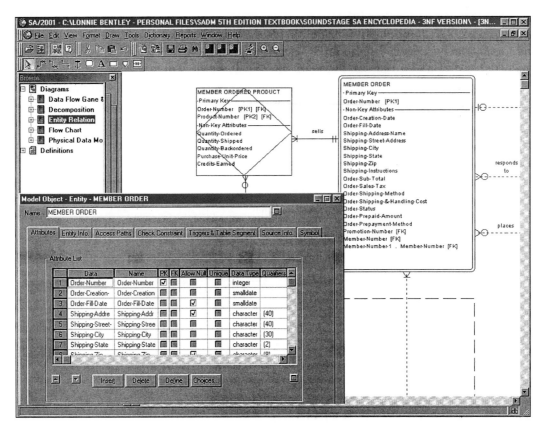

FIGURE 7.12   *Screen Capture of* System Architect *CASE Tool*

with an individual discussing SoundStage's business environment and activities, a user may state, "We have to keep track of all our <u>members</u> and their bound <u>agreements.</u>" Notice that the key words in this statement are MEMBERS and AGREEMENTS. Both are entities!

— During interviews or JRP sessions, specifically ask system owners and users to identify things about which they would like to capture, store, and produce information. Those things often represent entities that should be depicted on the data model.

— Another technique for identifying entities is to study existing forms, files, and reports. Some forms identify event entities. Examples include ORDERS, REQUISITIONS, PAYMENTS, DEPOSITS, and so forth. But most of these same forms also contain data that describes other entities. Consider a registration form used in your school's course enrollment system. A REGISTRATION is itself an event entity. But the average registration form also contains data that describes other entities, such as STUDENT (a person), COURSES (which are concepts), INSTRUCTORS (other persons), ADVISOR (yet another person), DIVISIONS (another concept), and so forth. Studying the computerized registration system's computer files, databases, or outputs could also discover these same entities.

— Technology may also help you identify entities. Some CASE tools can reverse engineer existing files and databases into <u>physical</u> data models. The analyst must usually clean up the resulting model by replacing physical names, codes, and comments with their logical, business-friendly equivalents.

While these techniques may prove useful in identifying entities, they occasionally play tricks on you. A simple, quick quality check can eliminate false entities.

Ask your user to specify the number of instances of each entity. A true entity has multiple instances—dozens, hundreds, thousands, or more! If not, the entity does not exist.

As entities are discovered, give them simple, meaningful, business-oriented names. Entities should be named with nouns that describe the person, event, place, object, or thing about which we want to store data. Try not to abbreviate or use acronyms. Names should be singular so as to distinguish the logical concept of the entity from the actual instances of the entity. Names may include appropriate adjectives or clauses to better describe the entity—for instance, an externally generated CUSTOMER ORDER must be distinguished from an internally generated STOCK ORDER.

For each entity, define it in business terms. Don't define the entity in technical terms, and don't define it as "data about . . ." Try this! Use an English dictionary to create a draft definition, and then customize it for the business at hand. Your entity names and definitions should establish an initial glossary of business terminology that will serve both you and future analysts and users for years.

Our SoundStage management and users initially identified the entities listed in Table 7.5. Notice how the definitions contribute to establishing the vocabulary of the system.

The next task in data modeling is to construct the context data model. The context data model should include the fundamental business entities that were previously discovered as well as their natural relationships.

The Context Data Model

Relationships should be named with verb phrases that, when combined with the entity names, form simple business sentences or assertions. Some CASE tools, such as *System Architect,* let you name the relationships in both directions. Otherwise, always name the relationship from parent to child.

We have completed this task in Figure 7.13. This figure represents a data model created in *System Architect.* Once we begin mapping attributes, new entities and relationships may surface. The numbers below reference those same numbers in Figure 7.13. The ERD communicates the following:

❶ An AGREEMENT <u>binds</u> one or more MEMBERs. While relationships may be named in only one direction (parent-to-child), the other direction is implicit. For example, it is implicit that a MEMBER <u>is bound to</u> one and only one AGREEMENT.

TABLE 7.5	*Fundamental Entities for the SoundStage Project*
Entity Name	**Business Definition**
AGREEMENT	A contract whereby a member agrees to purchase a certain number of products within a certain time. After fulfilling that agreement, the member becomes eligible for bonus credits that are redeemable for free or discounted products.
MEMBER	An active member of one or more clubs.
	Note: A target system objective is to reenroll inactive members as opposed to deleting them.
MEMBER ORDER	An order generated for a member as part of a monthly promotion, or an order initiated by a member.
	Note: The current system only supports orders generated from promotions; however, customer-initiated orders have been given a high priority as an added option in the proposed system.
TRANSACTION	A business event to which the Member Services System must respond.
PRODUCT	An inventoried product available for promotion and sale to members.
	Note: System improvement objectives include (1) compatibility with new bar code system being developed for the warehouse, and (2) adaptability to a rapidly changing mix of products.
PROMOTION	A monthly or quarterly event whereby special product offerings are made available to members.

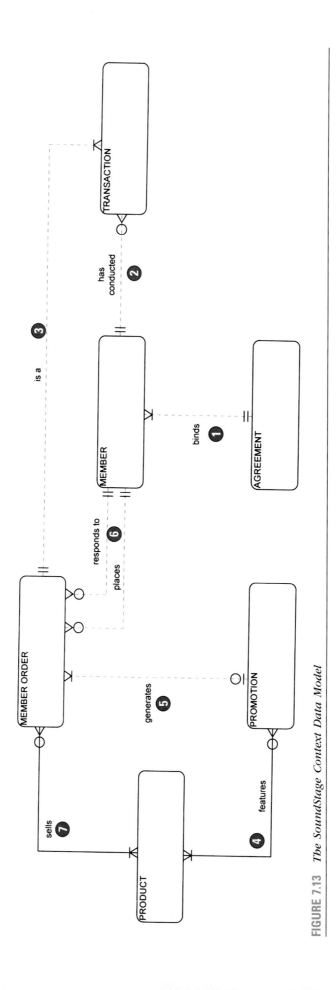

FIGURE 7.13 *The SoundStage Context Data Model*

❷ A MEMBER <u>has conducted</u> zero, one, or more TRANSACTIONS. Implicitly, a given TRANSACTION was <u>conducted by</u> one and only one MEMBER.

❸ A MEMBER ORDER <u>is a</u> TRANSACTION. In fact, a given MEMBER ORDER may correspond to many TRANSACTIONS (for example, a new member order, a cancelled member order, a changed member order, etc.) But a given TRANSACTION may or may not <u>represent</u> a MEMBER ORDER.

❹ A PROMOTION <u>features</u> one or more PRODUCTS. Implicitly, a PRODUCT <u>is featured in</u> zero, one, or more PROMOTIONS. For example, a CD that appeals to both country/western and light rock audiences might be featured in the promotion for both. Since products greatly outnumber promotions, most products are never featured in a promotion.

❺ A PROMOTION <u>generates</u> many MEMBER ORDERS. Implicitly, a MEMBER ORDER <u>is generated for</u> zero or one PROMOTION. Why zero? In the new system, a member will be able to initiate his or her own order.

❻ It is permissible for more than one relationship to exist between the same two entities if the separate relationships communicate different business events or associations. Thus, a MEMBER <u>responds to</u> zero, one, or more MEMBER ORDERS. This relationship supports the promotion-generated orders. A MEMBER <u>places</u> zero, one, or more MEMBER ORDERS. This relationship supports member-initiated orders. In both cases, a MEMBER ORDER <u>is placed by</u> (<u>is responded to by</u>) exactly one MEMBER.

Although we didn't need it for this double relationship, some CASE tools (including *System Architect*) provide a symbol for recording Boolean relationships (such as AND, OR). Thus, for any two relationships, a Boolean symbol could be used to establish that instances of the relationships must be mutually exclusive (= OR) or mutually contingent (= AND).

❼ A MEMBER ORDER <u>sells</u> one or more PRODUCTS. Implicitly, a PRODUCT is <u>sold on</u> zero, one, or more MEMBER ORDERS.

If you read each of the preceding items carefully, you probably learned a great deal about the SoundStage system. Data models have become increasingly popular as a tool for describing the business context for system projects.

The next task is to identify the keys of each entity. The following guidelines are suggested for keys:[1]

The Key-Based Data Model

1. The value of a key should not change over the lifetime of each entity instance. For example, NAME would be a poor key since a person's last name could change by marriage or divorce.
2. The value of a key cannot be null.
3. Controls must be installed to ensure that the value of a key is valid. This can be accomplished by precisely defining the domain and using the database management system's validation controls to enforce that domain.
4. Some experts (Bruce) suggest you avoid **intelligent keys.** An intelligent key is a business code whose structure communicates data about an entity instance (such as its classification, size, or other properties). A code is a group of characters and/or digits that identifies and describes something in the business system. They argue that because those characteristics can change, it violates rule number 1 above.

   We disagree. Business codes can return value to the organization because

---

[1] Adapted from Thomas A. Bruce, *Designing Quality Databases with IDEF1X Information Models.* Copyright © 1992 by Thomas A. Bruce. Reprinted by permission of Dorset House Publishing, 353 W. 12th St., New York, NY 10014 (212-620-4053/1-800-DH-BOOKS/www.dorsethouse.com). All rights reserved.

they can be quickly processed by humans without the assistance of a computer.

a. There are several types of codes. They can be combined to form effective means for entity instance identification.

   (1) **Serial codes** assign sequentially generated numbers to entity instances. Many database management systems can generate and constrain serial codes to a business's requirements.

   (2) **Block codes** are similar to serial codes except that block numbers are divided into groups that have some business meaning. For instance, a satellite television provider might assign 100–199 as PAY PER VIEW channels, 200–299 as CABLE channels, 300–399 to SPORT channels, 400–499 to ADULT PROGRAMMING channels, 500–599 to MUSIC-ONLY channels, 600–699 to INTERACTIVE GAMING channels, 700–799 to INTERNET channels, 800–899 to PREMIUM CABLE channels, and 900–999 to PREMIUM MOVIE AND EVENT channels.

   (3) **Alphabetic codes** use finite combinations of letters (and possibly numbers) to describe entity instances. For example, each STATE has a unique two-character alphabetic code. Alphabetic codes must usually be combined with serial or block codes to uniquely identify instances of most entities.

   (4) In **significant position codes,** each digit or group of digits describes a measurable or identifiable characteristic of the entity instance. Significant digit codes are frequently used to code inventory items. The codes you see on tires and lightbulbs are examples of significant position codes. They tell us about characteristics such as tire size and wattage, respectively.

   (5) **Hierarchical codes** provide a top-down interpretation for an entity instance. Every item coded is factored into groups, subgroups, and so forth. For instance, we could code employee positions as follows:
- First digit identifies classification (e.g., clerical, faculty, etc.).
- Second and third digits indicate level within classification.
- Fourth and fifth digits indicate calendar of employment.

b. The following guidelines are suggested when creating a business coding scheme:

   (1) Codes should be expandable to accommodate growth.

   (2) The full code must result in a unique value for each entity instance.

   (3) Codes should be large enough to describe the distinguishing characteristics, but small enough to be interpreted by people *without a computer.*

   (4) Codes should be convenient. A new instance should be easy to create.

5. Consider inventing a surrogate key instead to substitute for large concatenated keys of independent entities. This suggestion is not practical for associative entities because each part of the concatenated key is a foreign key that must precisely match its parent entity's primary key.

Figure 7.14 is the key-based data model for the SoundStage project. Notice that the primary key is specified for each entity.

❶ Many entities have a simple, single-attribute primary key.

❷ We resolved the nonspecific relationship between MEMBER ORDER and PRODUCT by introducing the associative entity MEMBER ORDERED PRODUCT. Each associative entity instance represents one product on one member order. The parent entities contributed their own primary keys to comprise the associative entity's concatenated key. *System Architect* places a "PK1" next to ORDER NUMBER to indicate that it is "part one" of the concatenated primary key and a "PK2" beside PRODUCT NUMBER to indicate that it is "part two" of the

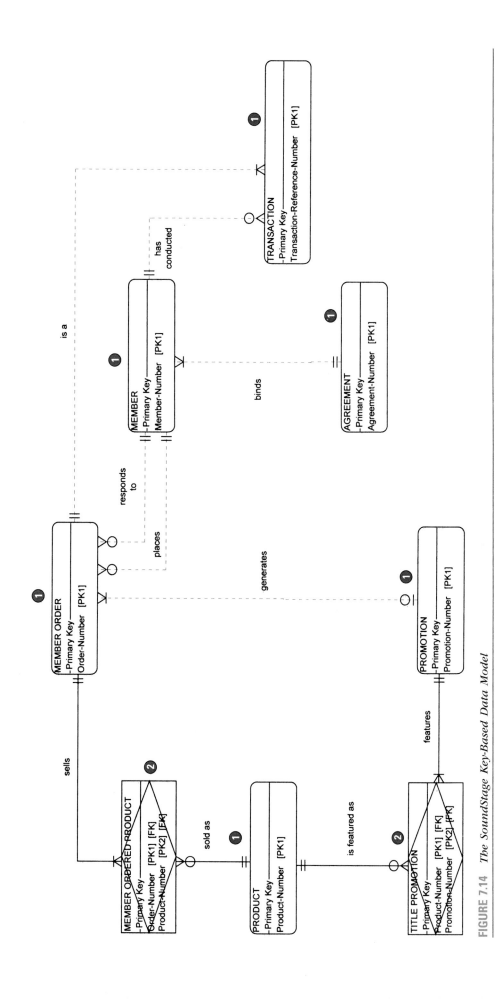

FIGURE 7.14 *The SoundStage Key-Based Data Model*

concatenated key. Also notice that each attribute in that concatenated key, by itself, is a foreign key that points back to the correct parent entity instance.

Likewise, the nonspecific relationship between PRODUCT and PROMOTION was resolved using an associative entity, TITLE PROMOTION, that also inherits the keys of the parent entities.

When developing this model, look out for a couple of things. If you cannot define keys for an entity, it may be that the entity doesn't really exist—that is, multiple occurrences of the so-called entity do not exist. Thus, assigning keys is a good quality check before fully attributing the data model. Also, if two or more entities have identical keys, they are in all likelihood the same entity.

Generalized Hierarchies

At this time, it would be useful to identify any generalization hierarchies in the business domain. The SoundStage project at the beginning of this chapter identified at least one supertype/subtype structure. Subsequent discussions did uncover a generalization hierarchy. Thus, our key-based model was revised as shown in Figure 7.15. We had to lay out the model somewhat differently because of the hierarchy; however, the relationships and keys that were previously defined have been retained. We call your attention to the following:

❶ The SoundStage CASE tool automatically draws a dashed box around a generalization hierarchy.

❷ The subtypes inherit the keys of the supertypes.

❸ We disconnected PROMOTION from PRODUCT as it was shown earlier and reconnected it to the subtype TITLE. This was done to accurately assert the business rule that MERCHANDISE is never featured on a PROMOTION—only TITLES.

The Fully Attributed Data Model

It may seem like a trivial task to identify the remaining data attributes; however, analysts not familiar with data modeling frequently encounter problems. To accomplish this task, you must have a thorough understanding of the data attributes for the system. These facts can be discovered using top-down approaches (such as brainstorming) or bottom-up approaches (such as form and file sampling). If an enterprise data model exists, some (perhaps many) of the attributes may have already been identified and recorded in a repository.

The following guidelines are offered for attribution.

— Many organizations have naming standards and approved abbreviations. The data administrator usually maintains such standards.

— Choose attribute names carefully. Many attributes share common base names such as NAME, ADDRESS, DATE. Unless the attributes can be generalized into a supertype, it is best to give each variation a unique name such as:

CUSTOMER NAME	CUSTOMER ADDRESS	ORDER DATE
SUPPLIER NAME	SUPPLIER ADDRESS	INVOICE DATE
EMPLOYEE NAME	EMPLOYEE ADDRESS	FLIGHT DATE

Also, remember that a project does not live in isolation from other projects, past or future. Names must be distinguishable across projects.

Some organizations maintain reusable, global templates for these common base attributes. This promotes consistent data types, domains, and defaults across all applications.

— Physical attribute names on existing forms and reports are frequently abbreviated to save space. Logical attribute names should be clearer—for example, translate the order form's attribute COD into its logical equivalent, AMOUNT TO COLLECT ON DELIVERY; translate QTY into QUANTITY ORDERED, and so forth.

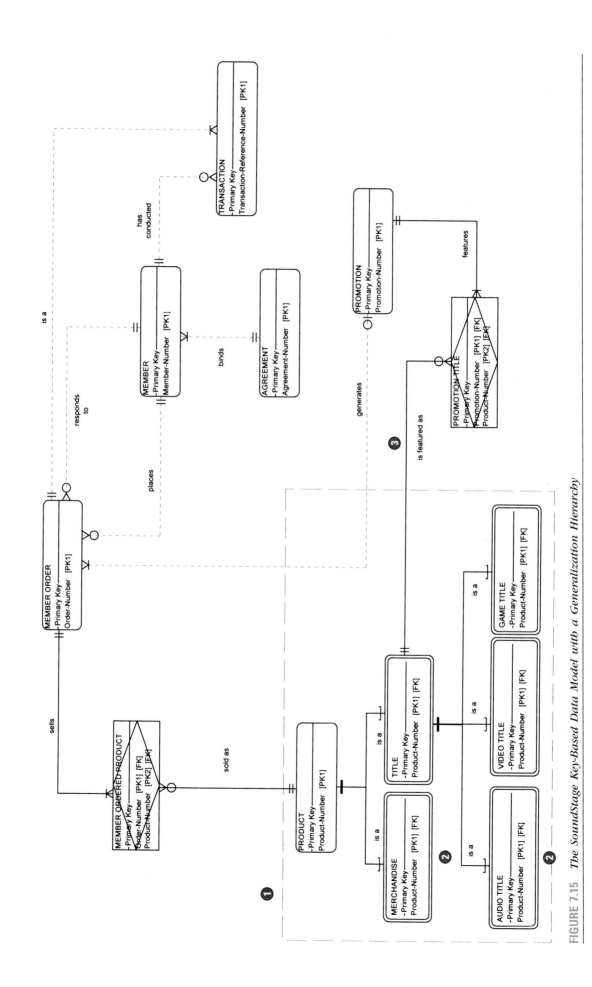

FIGURE 7.15 *The SoundStage Key-Based Data Model with a Generalization Hierarchy*

- Many attributes take on only YES or NO values. Try naming these attributes as questions. For example, the attribute name CANDIDATE FOR A DEGREE? suggests the values are YES and NO.

  Each attribute should be mapped to only one entity. If an attribute truly describes different entities, it is probably several different attributes. Give each a unique name.

- Foreign keys are the exception to the nonredundancy rule—they identify associated instances of related entities.

- An attribute's domain should not be based on logic. For example, in the SoundStage case we learned the values of MEDIA were dependent on the type of product. If the product type is a video, the media could be VHS tape, 8mm tape, laserdisc, or DVD. If the product type is audio, the media could be cassette tape, CD, or MD. The best solution would be to assign separate attributes to each domain: AUDIO MEDIA and VIDEO MEDIA.

Figure 7.16 provides the mapping of data attributes to entities for the definition phase of our SoundStage systems project. While the fully attributed model identifies all the attributes to be captured and stored in our future database, the descriptions for those attributes are incomplete; they require domains. Most CASE tools provide extensive facilities for describing the data types, domains, and defaults for all attributes to the repository. Additionally, each attribute should be defined for future reference.

ANALYZING THE DATA MODEL

While a data model effectively communicates database requirements, it does not necessarily represent a *good* database design. It may contain structural characteristics that reduce flexibility and expansion or create unnecessary redundancy. Therefore, we must prepare our fully attributed data model for database design and implementation.

This section will discuss the characteristics of a *quality* data model—one that will allow us to develop an ideal database structure. We'll also present the process used to analyze data model quality and make necessary modifications before database design.

What Is a Good Data Model?

What makes a data model good? We suggest the following criteria:

- *A good data model is simple.* As a general rule, the data attributes that describe any given entity should describe only that entity. Consider, for example, the following entity definition:

  COURSE REGISTRATION =   COURSE REGISTRATION NUMBER (PRIMARY KEY) +
                                         COURSE REGISTRATION DATE +
                                         STUDENT ID NUMBER (A FOREIGN KEY) +
                                         STUDENT NAME +
                                         STUDENT MAJOR +
                                           One or more COURSE NUMBERS

  Do STUDENT NAME and STUDENT MAJOR really describe an instance of course registration? Or do they describe a different entity, say STUDENT? The same argument could be applied to STUDENT ID NUMBER, but on further inspection, that attribute is needed to "point" to the corresponding instance of the STUDENT entity. Another aspect of simplicity is stated as follows: Each attribute of an entity instance can have only one value. Looking again at the previous example, COURSE NUMBER can have as many values for one COURSE REGISTRATION as the student elects.

- *A good data model is essentially nonredundant.* This means that each data attribute, <u>other than foreign keys</u>, describes at most one entity. In the prior

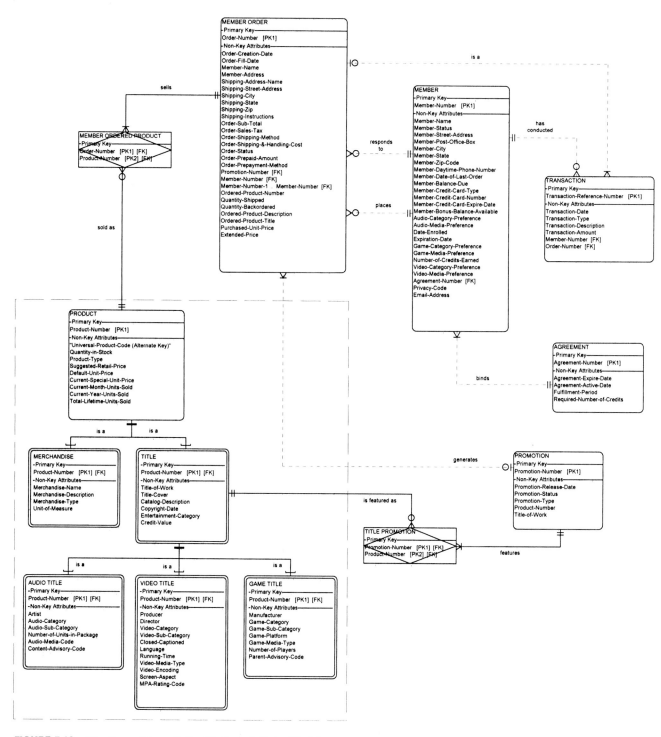

FIGURE 7.16   *The SoundStage Fully Attributed Data Model*

example, it is not difficult to imagine that STUDENT NAME and STUDENT MAJOR might also describe a STUDENT entity. We should choose. Based on the previous bullet, the logical choice would be the STUDENT entity. There may also exist subtle redundancies in a data model. For example, the same attribute might be recorded more than once under different names (synonyms).

— *A good data model should be flexible and adaptable to future needs*. In the absence of this criteria, we would tend to design databases to fulfill only today's business requirements. Then, when a new requirement becomes known, we can't easily change the databases without rewriting many or all of the programs that used those databases. While we can't change the reality that most projects are application-driven, we can make our data models as application-independent as possible to encourage database structures that can be extended or modified without impact to current programs.

So how do we achieve the above goals? How can you design a database that can adapt to future requirements that you cannot predict? The answer lies in data analysis.

Data Analysis

The technique used to improve a data model in preparation for database design is called data analysis.

> **Data analysis** is a process that prepares a data model for implementation as a simple, nonredundant, flexible, and adaptable database. The specific technique is called normalization.

> **Normalization** is a data analysis technique that organizes data attributes such that they are grouped to form nonredundant, stable, flexible, and adaptive entities.

Normalization is a three-step technique that places the data model into first normal form, second normal form, and third normal form.[2] Don't get hung up on the terminology—it's easier than it sounds. For now, let's establish an initial understanding of these three formats.

— Simply stated, an entity is in **first normal form (1NF)** if there are no attributes that can have more than one value for a single instance of the entity. Any attributes that can have multiple values actually describe a separate entity, possibly an entity and relationship.

— An entity is in **second normal form (2NF)** <u>if it is already in 1NF and</u> if the values of all nonprimary key attributes are dependent on the full primary key—not just part of it. Any nonkey attributes that are dependent on only part of the primary key should be moved to any entity where that partial key is actually the full key. This may require creating a new entity and relationship on the model.

— An entity is in **third normal form (3NF)** <u>if it is already in 2NF and</u> if the values of its nonprimary key attributes are not dependent on any other nonprimary key attributes. Any nonkey attributes that are dependent on other nonkey attributes must be moved or deleted. Again, new entities and relationships may have to be added to the data model.

Normalization Example

There are numerous approaches to normalization. We have chosen to present a nontheoretical and nonmathematical approach. We'll leave the theory, relational algebra, and detailed implications to the database courses and textbooks.

As usual, we'll use the SoundStage case study to demonstrate the steps. Let's begin by referring to the fully attributed data model that was developed earlier (see Figure 7.16). Is it a normalized data model? No. Let's identify the problems and walk through the steps of normalizing our data model.

First Normal Form   The first step in data analysis is to place each entity into 1NF. In Figure 7.16, which entities are not in 1NF?

You should find two—MEMBER ORDER and PROMOTION. Each contains a *repeating group,* that is, a group of attributes that can have multiple values for a single

---

[2] Database experts have identified additional normal forms. Third normal form removes most data anomalies. We leave a discussion of advanced normal forms to database textbooks and courses.

instance of the entity {denoted by the brackets}. These attributes repeat many times "as a group." Consider, for example, the entity MEMBER ORDER. A single MEMBER ORDER may contain many products, therefore, the attributes ORDERED PRODUCT NUMBER, ORDERED PRODUCT DESCRIPTION, ORDERED PRODUCT TITLE, QUANTITY ORDERED, QUANTITY SHIPPED, QUANTITY BACKORDERED, PURCHASED UNIT PRICE, and EXTENDED PRICE attributes may (and probably do) repeat for each instance of MEMBER ORDER.

Similarly, since a PROMOTION may feature more than one PRODUCT TITLE, the PRODUCT NUMBER and TITLE OF WORK attributes may repeat. How do we fix these anomalies in our model?

Figures 7.17 and 7.18 demonstrate how to place these two entities into 1NF. The original entity is depicted on the left side of the page. The 1NF entities are on the right side of the page. Each figure shows how normalization changed the data model and attribute assignments. For your convenience, the attributes that are affected are in boldface type and in small capital letters.

In Figure 7.17, we first removed the attributes that can have more than one value for an instance of the MEMBER ORDER entity. That alone places MEMBER ORDER in 1NF. But what do we do with the removed attributes? We can't remove them entirely from the model—they are part of the business requirements! Therefore, we moved the entire group of attributes to a new entity, MEMBER ORDERED PRODUCT. Each instance of these attributes describes one PRODUCT on a single MEMBER ORDER. Thus, if a specific order contains five PRODUCTS, there will be five instances of the MEMBER ORDERED PRODUCT entity. Each entity instance has only one value for each attribute; therefore, the new entity is also in first normal form.

Another example of 1NF is shown in Figure 7.18 for the PROMOTION entity. As before, we moved the repeating attributes to a different entity, TITLE PROMOTION.

All other entities are already in 1NF because they do not contain any repeating groups.

Second Normal Form   The next step of data analysis is to place the entities into 2NF. Recall that it is required that you have already placed all entities into 1NF. Also recall that 2NF looks for an attribute whose value is determined by only part of the primary key—not the entire concatenated key. Accordingly, entities that have a single attribute primary key are already in 2NF. That takes care of PRODUCT (and its subtypes), MEMBER ORDER, MEMBER, PROMOTION, AGREEMENT, and TRANSACTION. Thus, we need to check only those entities that have a concatenated key—MEMBER ORDERED PRODUCT and TITLE PROMOTION.

First, let's check the MEMBER ORDERED PRODUCT entity. Most of the attributes are dependent on the full primary key. For example, QUANTITY ORDERED makes no sense unless you have *both* an ORDER NUMBER and a PRODUCT NUMBER. Think about it! By itself, ORDER NUMBER is inadequate since the order could have as many quantities ordered as there are products on the order. Similarly, by itself, PRODUCT NUMBER is inadequate since the same product could appear on many orders. Thus, QUANTITY ORDERED requires both parts of the key and is dependent on the full key. The same could be said of QUANTITY SHIPPED, QUANTITY BACKORDERED, PURCHASE UNIT PRICE, and EXTENDED PRICE.

But what about ORDERED PRODUCT DESCRIPTION and ORDERED PRODUCT TITLE? Do we really need ORDER NUMBER to determine a value for either? No! Instead, the values of these attributes are dependent only on the value of PRODUCT NUMBER. Thus, the attributes are *not* dependent on the full key; we have uncovered a *partial dependency* anomaly that must be fixed. How do we fix this type of normalization error?

Refer to Figure 7.19 on p. 272. To fix the problem, we simply move the non-key attributes, ORDERED PRODUCT DESCRIPTION and ORDERED PRODUCT TITLE, to an entity that only has PRODUCT NUMBER as its key. If necessary, we would have to create this entity, but the PRODUCT entity with that key already exists. But we have

MEMBER ORDER (unnormalized)

Order-Number (Primary Key)
Order-Creation-Date
Order-Fill-Date
Member-Number (Foreign Key)
Member-Number-2 (Foreign Key)
Member-Name
Member-Address
Shipping-Address
Shipping Instructions
Promotion-Number (Foreign Key)
1 { ORDERED-PRODUCT-NUMBER } N
0 { ORDERED-PRODUCT-DESCRIPTION } N
0 { ORDERED-PRODUCT-TITLE } N
1 { QUANTITY-ORDERED } N
1 { QUANTITY-SHIPPED } N
1 { QUANTITY-BACKORDERED } N
1 { PURCHASED-UNIT-PRICE } N
1 { EXTENDED-PRICE } N
Order-Sub-Total-Cost
Order-Sales-Tax
Ship-Via-Method
Shipping-Charge
Order-Status
Prepaid-Amount
Prepaid-Method

CORRECTION

MEMBER ORDER (1NF)

Order-Number (Primary Key)
Order-Creation-Date
Order-Automatic-Fill-Date
Member-Number (Foreign Key)
Member-Number-2 (Foreign Key)
Member-Name
Member-Address
Shipping-Address
Shipping Instructions
Promotion-Number (Foreign Key)
Order-Sub-Total-Cost
Order-Sales-Tax
Ship-Via-Method
Shipping-Charge
Order-Status
Prepaid-Amount
Prepaid-Method

sells

MEMBER ORDERED PRODUCT (1NF)

Order-Number (Primary Key 1 and Foreign Key)
PRODUCT-NUMBER (PRIMARY KEY 2 AND FOREIGN KEY)
ORDERED-PRODUCT-DESCRIPTION
ORDERED-PRODUCT-TITLE
QUANTITY-ORDERED
QUANTITY-SHIPPED
QUANTITY-BACKORDERED
PURCHASED-UNIT-PRICE
EXTENDED-PRICE

sold
as

PRODUCT (1NF)

Product-Number (Primary Key)
Universal-Product-Code (Alternate Key)
Quantity-in-Stock
Product-Type
Suggested-Retail-Price
Default-Unit-Price
Current-Special-Unit-Price
Current-Month-Units-Sold
Current-Year-Units-Sold
Total-Lifetime-Units-Sold

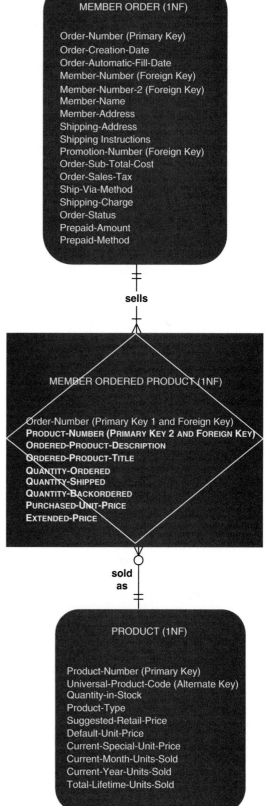

FIGURE 7.17 *First Normal Form (1NF)*

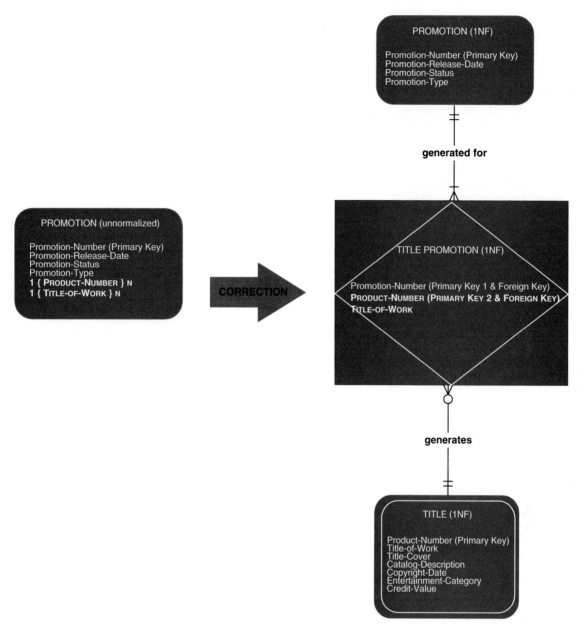

FIGURE 7.18   *First Normal Form (1NF)*

to be careful because PRODUCT is a supertype. Upon inspection of the subtypes, we discover that the attributes are already in the MERCHANDISE and TITLE entities, albeit under a synonym. Thus, we didn't actually have to move the attributes from the MEMBER ORDERED PRODUCT entity; we just deleted them as redundant attributes.

Next, let's examine the TITLE PROMOTION entity. The concatenated key is the combination of PROMOTION NUMBER and PRODUCT NUMBER. TITLE OF WORK is dependent on the PRODUCT NUMBER portion of the concatenated key. Thus, TITLE OF WORK is removed from TITLE PROMOTION (see Figure 7.20). Notice that TITLE OF WORK already existed in the entity TITLE, which has a product number as its primary key.

Third Normal Form   We can further simplify our entities by placing them into 3NF. <u>Entities are required to be in 2NF before beginning 3NF analysis.</u> Third normal form analysis looks for two types of problems, *derived data* and *transitive*

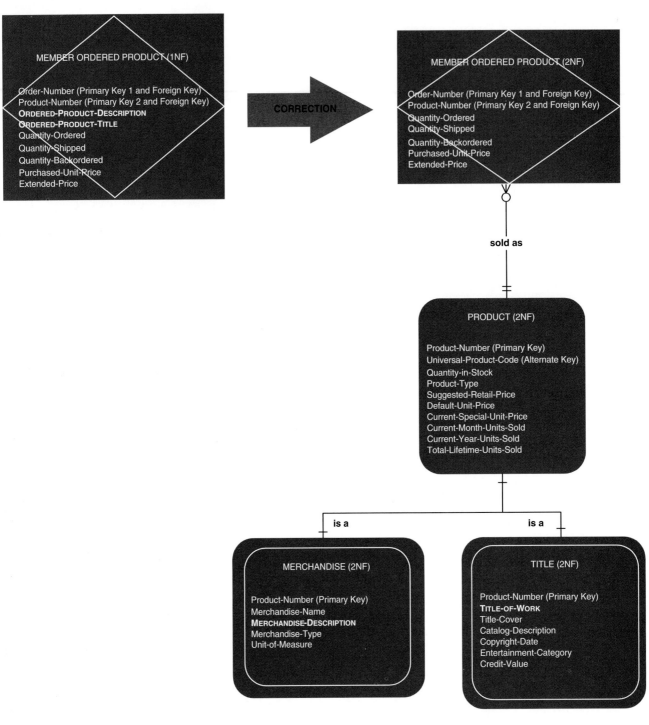

FIGURE 7.19   *Second Normal Form (2NF)*

dependencies. In both cases, the fundamental error is that nonkey attributes are dependent on other nonkey attributes.

The first type of 3NF analysis is easy—examine each entity for derived attributes.

Derived attributes are those whose values can either be calculated from other attributes or derived through logic from the values of other attributes.

If you think about it, storing a derived attribute makes little sense. First, it wastes disk storage space. Second, it complicates what should be simple updates. Why?

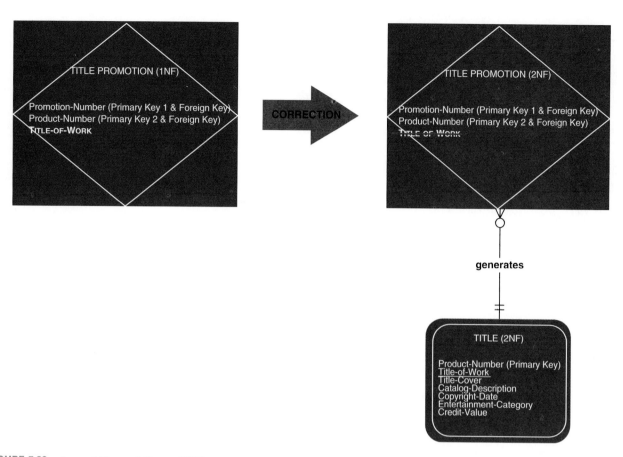

FIGURE 7.20   *Second Normal Form (2NF)*

Every time you change the base attributes, you must remember to reperform the calculation and also change its result.

For example, look at the MEMBER ORDERED PRODUCT entity in Figure 7.21. The attribute EXTENDED PRICE is calculated by multiplying QUANTITY ORDERED by PURCHASED UNIT PRICE. Thus, EXTENDED PRICE (a nonkey attribute) is not dependent on the primary key as much as it is dependent on the nonkey attributes, QUANTITY ORDERED and PURCHASED UNIT PRICE. Thus, we correct the entity by deleting EXTENDED PRICE.

Sounds simple, right? Well, not always! There is disagreement on how far you take this rule. Some experts argue that the rule should be applied only within a single entity. Thus, these experts would not delete a derived attribute if the attributes required for the derivation were assigned to *different* entities. We agree based on the argument that a derived attribute that involves multiple entities presents a greater danger for data inconsistency caused by updating an attribute in one entity and forgetting to subsequently update the derived attribute in another entity.

Another form of 3NF analysis checks *for transitive dependencies.* A transitive dependency exists when a nonkey attribute is dependent on another nonkey attribute (other than by derivation). This error usually indicates that an undiscovered entity is still embedded within the problem entity. Such a condition, if not corrected, can cause future flexibility and adaptability problems if a new requirement eventually requires us to implement that undiscovered entity as a separate database table.

Transitive analysis is performed only on those entities that do not have a concatenated key. In our example, this includes PRODUCT, MEMBER ORDER, PROMOTION, AGREEMENT, MEMBER, and TRANSACTION. For the entity PRODUCT, all the nonkey attrib-

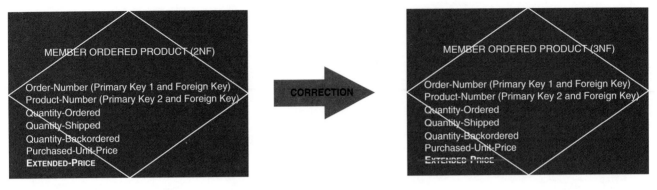

FIGURE 7.21   *Third Normal Form (3NF)*

utes are dependent on the primary key, and only the primary key. Thus, PROD-UCT is already in third normal form. A similar analysis of PROMOTION, AGREEMENT, and TRANSACTION reveals that they are also in third normal form.

But look at the entity MEMBER ORDER in Figure 7.22. In particular, examine the attributes MEMBER NAME and MEMBER ADDRESS. Are these attributes dependent on the primary key, ORDER NUMBER? No! The primary key ORDER NUMBER in no way determines the value of MEMBER NAME and MEMBER ADDRESS. On the other hand, the values of MEMBER NAME and MEMBER ADDRESS are dependent on the value of another non*primary* key in the entity, MEMBER NUMBER.

How do we fix this problem? MEMBER NAME and MEMBER ADDRESS need to be moved from the MEMBER ORDER entity to an entity whose primary key is just MEM-BER NUMBER. If necessary, we would create that entity, but in our case we already have a MEMBER entity with the required primary key. As it turns out, we don't need to really move the problem attributes because they are already assigned to the MEMBER entity. We did, however, notice that MEMBER ADDRESS was a synonym for MEMBER STREET ADDRESS. We elected to keep the latter term in MEMBER.

Several normal forms beyond 3NF exist. Each successive normal form makes the data model simpler, less redundant, and more flexible. However, systems analysts (and most database experts) rarely take data models beyond 3NF. Consequently, we will leave further discussion of normalization to database textbooks.

The first few times you normalize a data model, the process will appear slow and tedious. However, with time and practice, it becomes quick and routine. Many experienced modelers significantly reduce the modeling time and effort by doing normalization during attribution (they are able to do normalization at the time they are developing the fully attributed data model). It may help to always remember the following ditty (source unknown), which nicely summarizes first, second, and third normal forms:

> An entity is said to be in third normal form if every nonprimary key attribute is dependent on the primary key, the whole primary key, and nothing but the primary key.

Simplification by Inspection   Normalization is a fairly mechanical process. But it is dependent on naming consistencies in the original data model (before normalization). When several analysts work on a common application, it is not unusual to create problems that won't be taken care of by normalization. These problems are best solved through simplification by inspection, a process wherein a data entity in 3NF is further simplified by such efforts as addressing subtle data redundancy.

The final, normalized data model is presented in Figure 7.23 on p. 296.

CASE Support for Normalization   Many CASE tools claim to support normalization concepts. They read the data model and attempt to isolate possible normalization

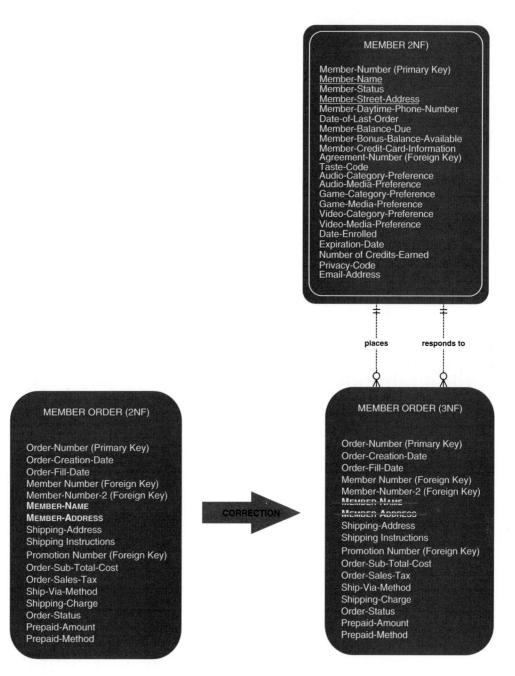

FIGURE 7.22   *Third Normal Form (3NF)*

errors. On close examination, most CASE tools can normalize only to first normal form. They accomplish this in one of two ways. They look for many-to-many relationships and resolve those relationships into associative entities. Or they look for attributes specifically described as having multiple values for a single entity instance. (One could argue that the analyst should have recognized that as a 1NF error and not described the attributes as such.)

It is exceedingly difficult for a CASE tool to identify second and third normal form errors. That would require the CASE tool to have the intelligence to recognize partial and transitive dependencies. In reality, such dependencies can be discovered only through less-than-routine examination by the systems analysts or database experts.

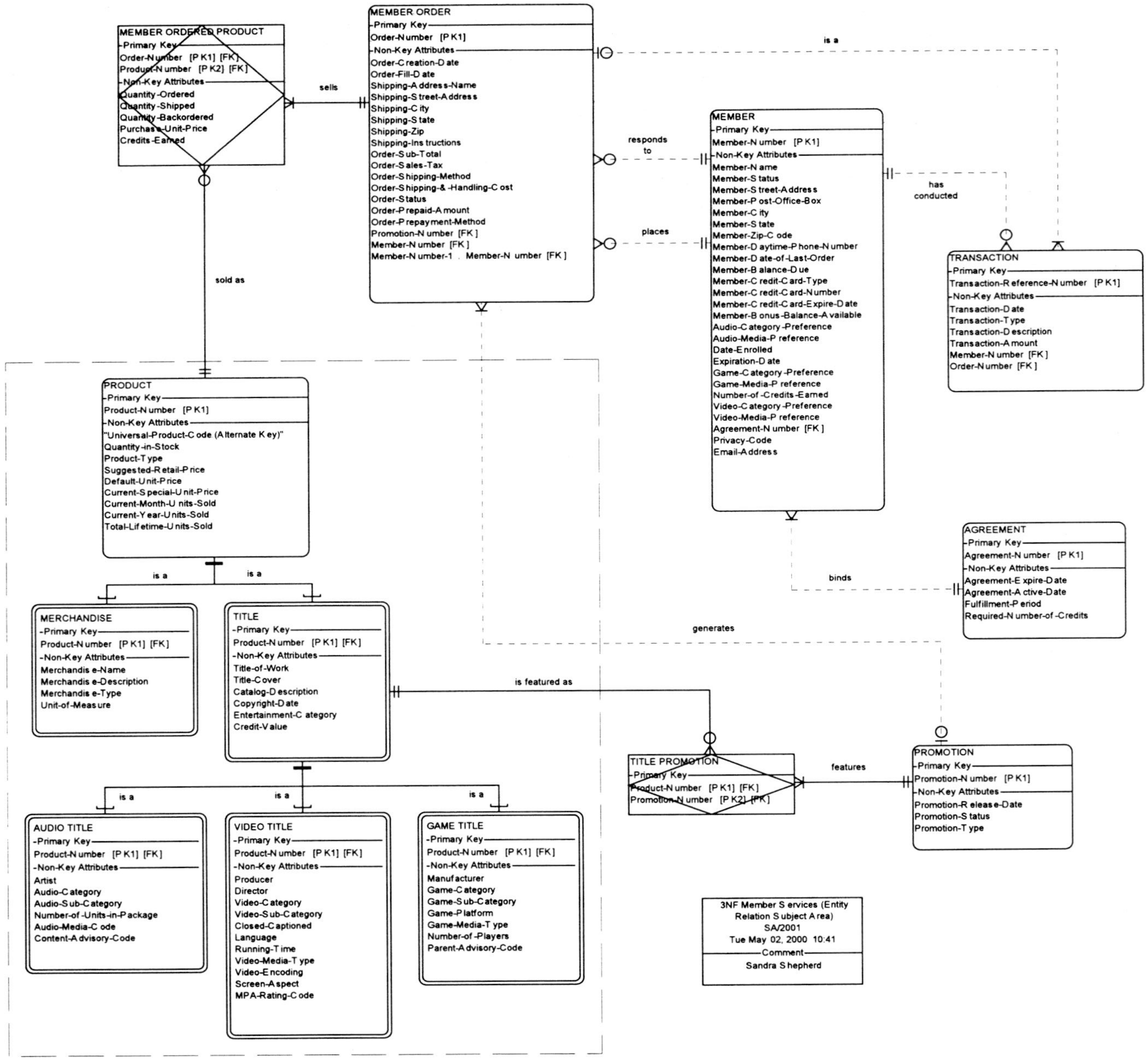

FIGURE 7.23 *SoundStage Logical Data Model in Third Normal Form (3NF)*

While a logical data model is effective for describing what data is to be stored for a new system, it does not communicate those requirements on a business operating location basis. We need to identify what data and access rights are needed at which locations. Specifically, we might ask the following business questions:

MAPPING DATA REQUIREMENTS TO LOCATIONS

- Which subset of the entities and attributes are needed to perform the work at each location?
- What level of access is required?
- Can the location *create* instances of the entity?
- Can the location *read* instances of the entity?
- Can the location *delete* instances of the entity?
- Can the location *update* existing instances of the entity?

Systems analysts have found it useful to define these logical requirements in the form of a *data-to-location-CRUD matrix*.

A **data-to-location-CRUD matrix** is a table in which the rows indicate entities (and possible attributes); the columns indicate locations; and the cells (the intersection of rows and columns) document level of access where C = create, R = read, U = update or modify, and D = delete or deactivate.

Figure 7.24 illustrates a typical data-to-location-CRUD matrix. The decision to include or not include attributes is based on whether locations need to be restricted as to which attributes they can access. Figure 7.24 also demonstrates the ability to

---

FIGURE 7.24   *Data-to-Location-CRUD Matrix*

Data-to-Location-CRUD Matrix

Entity . Attribute	Customers	Kansas City	. Marketing	. Advertising	. Warehouse	. Sales	. A/R	Boston	. Sales	. Warehouse	San Francisco	. Sales	San Diego	. Warehouse
Customer	INDV					ALL	ALL		SS	SS		SS		SS
.Customer Number	R				R	CRUD	R		CRUD	R		CRUD		R
.Customer Name	RU				R	CRUD	R		CRUD	R		CRUD		R
.Customer Address	RU				R	CRUD	R		CRUD	R		CRUD		R
.Customer Credit Rating	X					R	RU		R			R		
.Customer Balance Due	R					R	RU		R			R		
Order	INDV		ALL		SS	ALL			SS	SS		SS		SS
.Order Number	SRD		R	CRUD	R	CRUD	R		CRUD	R		CRUD		R
.Order Date	SRD		R	CRUD	R	CRUD	R		CRUD	R		CRUD		R
.Order Amount	SRD		R	CRUD		CRUD	R		CRUD	R		CRUD		R
Ordered Product	INDV		ALL		SS	ALL			SS	SS		SS		SS
.Quantity Ordered	SUD		R	CRUD	R	CRUD	R		CRUD			CRUD		
.Ordered Item Unit Price	SUD		R	CRUD		CRUD	R		CRUD			CRUD		
Product	ALL		ALL	ALL	ALL	ALL			ALL	ALL		ALL		ALL
.Product Number	R		CRUD	R	R	R			R	R		R		R
.Product Name	R		CRUD	R	R	R			R	R		R		R
.Product Description	R		CRUD	RU	R	R			R	R		R		R
.Product Unit of Measure	R		CRUD	R	R	R			R	R		R		R
.Product Current Unit Price	R		CRUD	R		R			R	R		R		R
.Product Quantity on Hand	X				RU	R			R	RU		R		RU

INDV = individual	**ALL** = ALL	**SS** = subset	**X** = no access
S = submit C = create	R = read	U = update	D = delete

document that a location requires access only to a subset (designated SS) of entity instances. For example, each sales office might need access only to those customers in its region.

In some methodologies and CASE tools, you can define *views* of the data model for each location. A view includes only the entities and attributes to be accessible for a single location. If views are defined, they must also be kept in sync with the master data model. (Most CASE tools do this automatically.)

WHERE DO YOU GO FROM HERE?

Most of you will proceed directly to Chapter 8, Process Modeling. Whereas data modeling was concerned with data independently from how that data is captured and used (data at rest), process modeling shows how the data will be captured and used (data in motion). At your instructor's discretion, some of you may jump to Module A, Object Modeling. Object modeling has many parallels with data modeling. An object includes attributes, but it also includes all the processes that can act on and use those methods.

If you want to immediately learn how to implement data models as databases, you should skim or read Chapter 12, Database Design. In that chapter, the logical data models are transformed into physical database schemas. With CASE tools, the code to create the database can be generated automatically.

SUMMARY

1. A model is a representation of reality. Models can be built for existing systems as a way to better understand those systems or for proposed systems as a way to document business requirements or technical designs.
   a. Logical models document the business requirements to show what a system is or does. They are implementation independent; that is, they depict the system independent of any technical implementation. As such, logical models illustrate the essence of the system.
   b. Physical models show not only *what* a system is or does, but also *how* the system is physically and technically implemented. They are implementation *de*pendent because they reflect technology choices and the limitations of those technology choices.
2. Data modeling is a technique for organizing and documenting a system's DATA. Data modeling is sometimes called database modeling because a data model is usually implemented as a database.
3. There are several notations for data modeling. The actual model is frequently called an entity relationship diagram (ERD) because it depicts data in terms of the entities and relationships described by the data.
4. An entity is something about which the business needs to store data. Classes of entities include: persons, places, objects, events, or concepts.
5. An entity instance is a single occurrence of an entity class.
6. Pieces of data we want to store about each instance of a given entity are called an attribute. An attribute is a descriptive property or characteristic of the entity. Some attributes can be logically grouped into superattributes called compound attributes.

7. When analyzing a system, we should define those values for an attribute that are legitimate or that make business sense. The values for each attribute are defined in terms of three properties: data type, domain, and default:
   a. The data type defines what class of data can be stored in that attribute.
   b. The domain of an attribute defines what values an attribute can legitimately take on.
   c. The default value for an attribute is the value that will be recorded if not specified by the user.
8. Every entity must have an identifier or key. A key is an attribute, or a group of attributes, that assumes a unique value for each entity instance.
   a. A group of attributes that uniquely identifies an instance of an entity is called a concatenated key.
   b. A candidate key is a "candidate to become the primary identifier" of instances of an entity.
   c. A primary key is the candidate key that will most commonly be used to uniquely identify a single entity instance.
   d. Any candidate key that is not selected to become the primary key is called an alternate key.
   e. Sometimes, it is also necessary to identify a subset of entity instances as opposed to a single instance. A subsetting criteria is an attribute (or concatenated attribute) whose finite values divide all entity instances into useful subsets.
9. A relationship is a natural business association that exists between one or more entities. The relationship may represent an event that links the entities or merely a logical affinity that exists between the entities. All relationships

are implicitly bidirectional, meaning they can be interpreted in both directions.

10. Cardinality defines the minimum and maximum number of occurrences of one entity for a single occurrence of the related entity. Because all relationships are bidirectional, cardinality must be defined in both directions for every relationship.

11. The degree of a relationship is the number of entity classes that participate in the relationship. Not all relationships are binary. Some relationships may be recursive relationships wherein the relationship exists between different instances of the same entity. Relationships can also exist between more than two different entities, as in the case of a 3-ary or ternary relationship.

12. An associative entity is an entity that inherits its primary key from more than one other entity (parents). Each part of that concatenated key points to one and only one instance of each of the connecting entities.

13. A foreign key is a primary key of one entity that is contributed to (duplicated in) another entity to identify instances of a relationship. A foreign key (always in a child entity) always matches the primary key (in a parent entity).

14. Nonidentifying relationships are those in which each of the participating entities has its own independent primary key, of which none of the primary key attributes is shared. The entities in a nonidentifying relationship are referred to as *strong* or independent entities because neither depends on any other entity for its identification. Identifying relationships are those in which the parent entity contributes its primary key to become part of the primary key of the child entity. The child entity of any identifying relationship is referred to as a *weak* entity because its identification is dependent on the existence of the parent entity's existence.

15. A nonspecific relationship (or many-to-many relationship) is one in which many instances of one entity are associated with many instances of another entity. Such relationships are suitable only for preliminary data models and should be resolved as quickly as possible.

16. Generalization is an approach that seeks to discover and exploit the commonalities between entities. It is a technique wherein the attributes are grouped to form entity supertypes and subtypes.
    a. An entity supertype is an entity whose instances store attributes that are common to one or more entity subtypes.
    b. An entity subtype is an entity whose instances inherit some common attributes from an entity supertype and then add other attributes that are unique to an instance of the subtype.

17. A logical data model is developed in the following stages:
    a. Entities are discovered and defined.
    b. A context data model is built. A context data model contains only business entities and relationships identified by the system owners and users.
    c. A key-based data model is built. The key-based model eliminates nonspecific relationships and adds associative entities. All entities in the model are given keys.
    d. A fully attributed model is built. This model shows all the attributes to be stored in the system.
    e. A fully described model is built. Each attribute is defined in the dictionary and described in terms of properties such as domain and security.
    f. The completed data model is then analyzed for adaptability and flexibility through a process called *normalization*. The final analyzed model is referred to as a third normal form data model.

18. A logical data model does not communicate data requirements on a business operating location basis. Systems analysts have found it useful to define these requirements in the form of a data-to-location-CRUD matrix.

KEY TERMS

alphabetic codes, p. 282
alternate key, p. 263
application data model, p. 275
associative entity, p. 266
attribute, p. 261
block codes, p. 282
candidate key, p. 263
cardinality, p. 264
compound attribute, p. 261
concatenated key, p. 262
context data model, p. 275
data analysis, p. 288
data modeling, p. 259
data-to-location-CRUD matrix, p. 297
data type, p. 261
default value, p. 262
degree, p. 265
derived attributes, p. 292
domain, p. 261

enterprise data model, p. 273
entity, p. 260
entity instance, p. 260
entity relationship diagram, p. 260
essential model, p. 258
first normal form (1NF), p. 288
foreign key, p. 267
fully attributed data model, p. 275
fully described data model, p. 275
generalization, p. 272
hierarchical codes, p. 282
identifying relationship, p. 267
intelligent keys, p. 281
key, p. 262
key-based data model, p. 275
logical model, p. 257
many-to-many relationship, p. 269
metadata, p. 277
model, p. 257

nonidentifying relationships, p. 267
nonspecific relationship, p. 269
normalization, pp. 275, 288
normalized data model, p. 275
physical model, p. 258
primary key, p. 263
recursive relationship, p. 265
relationship, p. 264
second normal form (2NF), p. 288
serial codes, p. 282
significant position codes, p. 282
strong entity, p. 267
subsetting criteria, p. 263
subtype, p. 273
supertype, p. 272
third normal form (3NF), p. 288
weak entity, p. 267

REVIEW QUESTIONS

1. Differentiate between logical and physical models. Give three reasons why logical models are superior for structuring business requirements.
2. What is data modeling? What is the actual data model that is created called?
3. What is an entity? What are the five categories of entities?
4. Differentiate between entities and entity instances. Why don't we try to model instances?
5. What are attributes? What are compound attributes? Give an example (not from the chapter) of each.
6. What are the three aspects of domain description for attributes?
7. Differentiate between candidate keys, primary keys, and alternate keys. Can each of these be a concatenated key?
8. What is a subsetting criteria? Why is it important?
9. What is a relationship? Why are relationships important to identify and describe? What is a nonspecific relationship?
10. Differentiate between cardinality and degree.
11. What is a recursive relationship? Give an example other than the one provided in this chapter.
12. What is an associative entity? What role does it play in ternary relationships? What role does it play in resolving nonspecific relationships?

13. What is the difference between an identifying and a nonidentifying relationship?
14. What is a "weak" entity?
15. What role does a foreign key play in implementing a relationship?
16. Differentiate between supertype and subtype entities.
17. Briefly describe three possible ways a nonspecific relationship might be resolved.
18. What is generalization, and what is its value?
19. Differentiate between an enterprise and application data model.
20. Explain the tasks used to construct an application data model.
21. Differentiate between data modeling and data analysis.
22. Give three characteristics of a good data model.
23. List and briefly describe the three steps of normalization.
24. What is a "derived attribute?" Give examples.
25. What is a transitive dependency? During which normal form are transitive dependencies resolved? Give an example of a transitive dependency.
26. Explain simplification by inspection.
27. What does the acronym CRUD stand for?
28. How does a data-to-location-CRUD matrix supplement a data model?

PROBLEMS AND EXERCISES

1. A database designer has complained about plans to construct a logical data model. He believes we should just design the database with the database management system. Give three reasons why requirements should be specified in an implementation-independent fashion.
2. Given the following data model, interpret it as a series of declarative sentences.

3. Why do some systems analysts believe that data modeling is the most important aspect of business requirements modeling?
4. Most data entities correspond to persons, objects, events, or locations in the business environment. Give three examples of each data entity class.
5. Explain why a concatenated key of NAME and ADDRESS would not be a good key.
6. What entities are described on your class schedule form or your school's course registration form? Draw a context data model to support academic scheduling at your school.
7. Give two examples of each of the following data relationship complexities: one-to-one (1:1), one-to-many (1:M or M:1), and many-to-many (M:M). Draw an ERD for each of

your examples. Be sure to label data entities using nouns and label data relationships using verbs. Annotate the graph to communicate the relationship complexity.
8. Resolve the nonspecific relationship in the previous exercise.
9. All vehicles in the state of _____ must be licensed. Some data is common to all vehicles, but certain types of vehicles require their own data. Construct a generalization hierarchy to represent this scenario.
10. Explain the origins of the primary key for an associative entity. Next, explain how the parts of the primary key serve as foreign keys.
11. Given the following entity relationship diagram (ERD) on the next page, answer the following questions:
    a. Which entity or entities, are "weak" entities?
    b. What is the primary key of SHIPPED PRODUCT?
    c. What is the primary key of PRODUCT TEST?
    d. What do you know about the key of SAMPLE?
    e. Is the relationship between SAMPLE and QA TEST an identifying or nonidentifying relationship?

For Problem 11

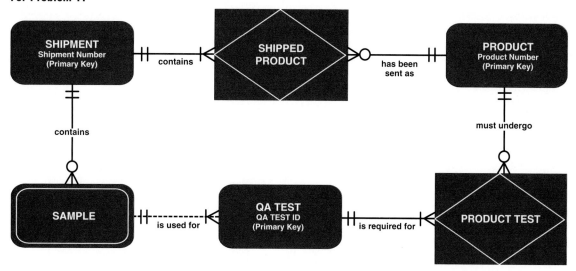

12. For the final SoundStage data model (3NF), construct sample tables that contain enough instances of data to demonstrate every possible cardinality shown. Invent your own data.

13. Create an example of a ternary relationship other than those demonstrated in this book. Why is a ternary relationship different from three binary relationships?

14. Should all nonspecific relationships be resolved by automatically introducing an associative entity? If not, explain why an alternative solution may be chosen.

15. Draw a simple key-based data model that includes at least one identifying relation and one nonidentifying relation.

16. Draw a simple key-based data model that depicts a "weak" entity other than an associative entity.

17. Transform the following entities into 3NF entities. Draw the original and final data models, and state any reasonable assumptions.

   AIRCRAFT, which is described by the following:

   > AIRCRAFT ID NUMBER (primary key), AIRCRAFT CODE, AIRCRAFT DESCRIPTION, and NUMBER OF SEATS.

   FLIGHT, which is described by the following:

   > FLIGHT NUMBER (primary key), DEPARTURE CITY, one or more ARRIVAL CITY(s), MEAL CODE, and one or more FLIGHT DATES and CURRENT SEAT PRICES.

   PASSENGER, which is described by the following:

   > PASSENGER NUMBER (primary key), PASSENGER NAME, FREQUENT FLYER NUMBER, one or more of the following: FLIGHT NUMBER, SEAT NUMBER, QUOTED SEAT PRICE, AMOUNT PAID, and BALANCE DUE.

18. What is the difference between a data entity in first normal form (1NF) and second normal form (2NF)? Give an example of an entity in 1NF and show its conversion to 2NF.

19. What is the difference between a data entity in second normal form (2NF) and third normal form (3NF)? Give an example of an entity in 2NF and show its conversion to 3NF.

20. During the preliminary investigation and problem analysis phases, an analyst collected numerous samples, including documents, forms, and reports. Explain how these samples will prove useful for data modeling.

PROJECTS AND RESEARCH

1. Obtain three sample business forms from a business, your school, or your instructor. What entities are described by the fields on the forms? Draw a context data model and an attribute of that model using a combination of the forms and brainstorming. Write a paragraph or two explaining how you would present the data model to a group of end-users who are not familiar with computer concepts.

2. Using the chapter map at the beginning of the chapter, develop a complete set of interview questions for the context data model you constructed in Problem 1.

3. Given the following narrative description of entities and their relationships, prepare a draft entity relationship diagram (ERD). Be sure to state any reasonable assumptions that you are making.

   Burger World Distribution Center serves as a supplier to 45 Burger World franchises. You are involved with a project to build a database system for distribution. Each franchise submits a day-by-day projection of sales for each of Burger World's menu products (the products listed on the menu at each restaurant) for the coming month. All menu products require ingredients and/or packaging items. Based on projected sales for the store, the system must generate a day-by-day ingredients need and then collapse those needs into one-per-week purchase requisitions and shipments.

4. As part of a semester project, identify a client with a small application development need. Departmental applications are ideal! Construct a complete data model for that client.

5. Obtain a data model developed in an information systems shop. (You can also find complete data models in many database textbooks.) If the model uses a different graphical notation, convert it to the notation in this book. (That should prove to you that different notations are more alike than they are different.) Thoroughly critique the data model. Make improvements and explain why you believe they are improvements.

6. Given the sample form that follows, prepare a list of entities and their associated data attributes as determined from the document. Then, completely normalize the entities to 3NF and draw a hypothetical ERD. Your instructor should be the final interpreter for the form.

For Project 6

| PURCHASING REQUISITION Form 12 Rev.1988 | INSTRUCTIONS — INCLUDE IN EACH REQUISITION ONLY SUCH ARTICLES AS MAY BE PURCHASED FROM ONE FIRM. IF SPECIAL HANDLING IS DESIRED, NOTE. SEE REVERSE SIDE FOR SPECIAL COMMENTS BY REQUESTOR. |

DEPARTMENT COMPLETES UNSHADED AREA		PURCHASING COMPLETES SHADED AREA	ORDER NO.

DEPT. OR FUNCTION: Computer Information Systems
COMMITMENT NO.
COMMODITY CODE ORDER TYPE

M F C	RES CODE.	ACCOUNT NUMBER			DEPT. REFERENCE	AMOUNT		FUND EXPIRATION DATE
		FUND	CENTER DEPT.— PROJ.	OBJECT				
1				5-6207		8,736.00		
2				5-6106		399.00		
3				5-6107		84.00		

SHIP TO STAFF MEMBER: Jonathan Doe
DEPT. 242
BUILDING & ROOM: Administration

ORDER DATE
FOLLOW UP
PRICING METHOD
☐ RQ #
☐ 1 Phone/Verbal Quote
☐ 2 Agreement/Contract
☐ 3 Price List on File
☐ 4 Repair Negotiation
☐ 5 None of the above
BUYER

REQUISITIONER'S PHONE NO. 555-4545
MATERIAL WILL BE USED FOR
VENDOR SUGGESTED: IBM Main Street Somewhere, IN 47906
VENDOR NAME
VENDOR NUMBER

FOB: ☐ 1 DESTINATION ☐ 2 DESTINATION PREPAY & ADD ☐ 3 SHIPPING POINT ☐ 4 SHIPPING POINT FREIGHT ALLOWED ☐ 5 SEE BELOW
VIA
TERMS

ITEM #	ITEM DESCRIPTION	MFC	QUANTITY	UNIT	UNIT PRICE	EXTENDED PRICE	DELIVER ON	EST.	COMM.
	IBM PS/2 Model 70 86 8570-121	1	1		7,995.00	4,797.00			
	IBM PS/2 2-8 MB Memory Module Expansion Option #5211	1	1		1,695.00	1,017.00			
	IBM PS/2 2MB Memory Module Kit #5213	1	3		1,395.00	2,511.00			
	IBM 8513 PS/2 Color Display	1	1		685.00	411.00			
	IBM 8770 PS/2 Mouse	2	1		95.00	57.00			
	IBM PC Network Adapter II/A #150122	2	1		570.00	342.00			
	IBM DOS 3.3	3	1		120.00	84.00			

REQUESTED — HEAD OF DEPT. *Thomas J. Mathien* DATE 6-3-99
APPROVED — FOR THE COMPTROLLER DATE
BYPASS APPROVAL REQUESTED ☐
PURCHASING APPROVALS PA AD DIR

RECOMMENDED — DEAN OR ADMINISTRATOR DATE
APPROVED — FOR THE EXECUTIVE VICE PRESIDENT AND TREASURER DATE
APPROVAL SIGNATURE/DATE
OCGBA PREAUDIT
BY: DATE:

MINICASES

1. The MIS department of our business wants to build a database to track all our hardware and software. We own workstations, network servers, and peripherals. The department wants to keep track of software packages as well as the licenses for those packages. Some software licenses are for single machines. We can install them on network servers, but we can only permit as many network users as we own licenses. We also own network licenses. A single network license authorizes a specific number of users. Nonnetwork licenses may be installed on either workstations or servers. Network licenses may be installed only on servers. We want to keep track of where software licenses are installed. Some licenses may not be installed anywhere at any given time. We must also be able to prove the legality of any software we have installed. Each license must be traced to either a purchase order, a gift, or a loan. We may also have certain software on order. We order packages, but we receive licenses. Construct the data model and attribute it through brainstorming.

2. Most students have bank accounts. Construct a data model that shows the relationships that exist among customers, different types of accounts (e.g., checking, savings, loan, funds), and transactions (e.g., deposits, withdrawals, payments, ATMs). Attribute your model such that it could be used to produce a consolidated bank statement.

3. To schedule classes, your school needs to know about courses that can be offered, instructors and their availabil-

ity, equipment requirements for courses, and rooms (and their equipment). From the courses that can be scheduled, they select the courses that will be scheduled. For each of those courses, they schedule one or more classes (sometimes called sections or divisions). The schedulers must assign classes to instructors, rooms, and time slots. The schedulers are constrained by the reality that (1) some courses cannot conflict because many students take them during the same term, (2) instructors cannot be in two places at the same time, and (3) rooms cannot be double-booked. Construct a data model to help the schedulers. (Caution: This problem requires some thought. Depending on your instructor's course policies, it might help to work in groups.) Clearly state any reasonable assumptions.

4. Given the data attributes and entities in the opposite column, indicate which attributes could be identifiers for each of the entities. You may have to combine attributes or even add some attributes that are not listed. Map all the attributes to their appropriate entity. Remember, each attribute should describe one and only one entity. Draw a rough draft entity relationship diagram.

5. Sunset Valley Distributors recently completed a major conversion project. Several months ago, Sunset decided to move into the database era. Many of its computer-based files had become unreliable, difficult to maintain, and too inflexible to be used to fulfill many end-user reporting and inquiry requests. A DBMS seemed to be the obvious solution. Two systems analysts were primarily responsible for the conversion project, which took several months. The systems analysts had decided to simply implement each of the computer-based files as a separate table in their relational database. Once the conversion was completed, the same problems that existed with the file-based system reappeared in the database system. Reports contained inaccurate data, report and inquiry requests could not easily be obtained, and data maintenance was still difficult. A consultant was hired to investigate the problems. The consultant acknowledged that many of the problems resulted because the analysts failed to do data modeling. Explain the importance of doing data modeling when designing databases.

For Minicase 4

Green Acres Real Estate System

Entities:

Seller	Buyer	Listing
House	Offer	Property
Closing	Showing	Room

Attributes:

Seller name	House style
Square foot size	Closing location
Seller address	Listing price
Number of bathrooms	Room type
Garage size	Property size
Showing date	Showing time
Garage location	Room size
Buyer name	Elementary school zone
Basement size	Buyer phone number
House heating method	Closing date
Property description	Offer amount
Offer date	Sales terms

SUGGESTED READINGS

Bruce, Thomas A. *Designing Quality Databases with IDEF1X Information Models*. New York: Dorset House Publishing, 1992. We actually use this book as a textbook in our database analysis and design course. IDEF1X is a rich, standardized syntax for data modeling (which Bruce calls information modeling). The graphical language looks different, but it communicates the same system concepts presented in our book. The language is supported by at least two CASE tools: Logic Works' *ERwin* and Popkin's *System Architect*. The book includes two case studies.

Hay, David C. *Data Model Patterns: Conventions of Thought*. New York: Dorset House Publishing, 1996. What a novel book! This book starts with the premise that most business data models are derivatives of some basic, repeatable patterns. CASE vendors, how about including these patterns in CASE tools as reusable templates?

Martin, James, and Clive Finkelstein. *Information Engineering*. 3 volumes. New York: Savant Institute, 1981. Information engineering is a formal, database, and fourth-generation language-oriented methodology. The graphical data modeling language of information engineering is virtually identical to ours. Data modeling is covered in Volumes I and II.

Reingruber, Michael, and William Gregory. *The Data Modeling Handbook*. New York: John Wiley & Sons, 1994. This is an excellent book on data modeling and is particularly helpful in ensuring the quality and accuracy of data models.

Schlaer, Sally, and Stephen J. Mellor. *Object-Oriented Systems Analysis: Modeling the World in Data*. Englewood Cliffs, NJ: Yourdon Press, 1988. Forget the title! "Object-oriented" means something different than when this book was written, but it is still one of the easiest to read books on the subject of data modeling.

Teorey, Toby J. *Database Modeling & Design: The Fundamental Principles*. 2nd ed. San Francisco: Morgan Kaufman Publishers, 1994. This book is somewhat more conceptual than the others in the list, but it provides useful insights into the practice of data modeling.

Focus on PEOPLE

Focus on DATA

Focus on PROCESSES

Focus on INTERFACES

Focus on DEVELOPMENT

Stakeholders

Activities

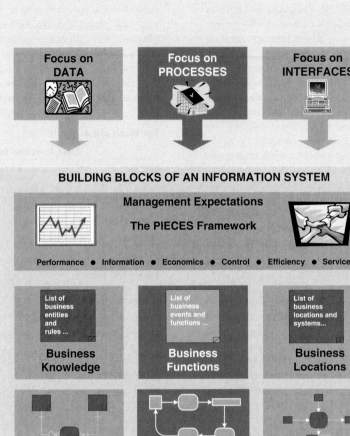

BUILDING BLOCKS OF AN INFORMATION SYSTEM

Management Expectations

The PIECES Framework

Performance ● Information ● Economics ● Control ● Efficiency ● Service

SYSTEM OWNERS

List of business entities and rules ...

Business Knowledge

List of business events and functions ...

Business Functions

List of business locations and systems...

Business Locations

PROJECT & PROCESS MANAGEMENT

PRELIMINARY INVESTIGATION

SYSTEM USERS

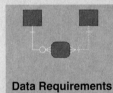

Data Requirements

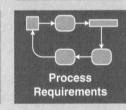

Process Requirements

Interface Requirements

PROBLEM ANALYSIS

REQUIREMENTS ANALYSIS

SYSTEM DESIGNERS

Database Schema

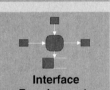

Application Schema & Specs

Interface Specifications

DECISION ANALYSIS

DESIGN

SYSTEM BUILDERS

```
CREATE TABLE tblOrders
    colOrderNot CHAR(5) NOT
    NULL
    colOrderDate DATE/TIME
NOT
```
Database Programs

```
PROC ValidateOrder
    PERFORM ValidateCust
    REPEAT UNTIL
        NoMoreProd ...
```
Application Programs

```
<html>
<head>
<title> Order Entry Form </title>
...
```
Interface Programs

CONSTRUCTION

IMPLEMENTATION

VENDORS AND CONSULTANTS

INFORMATION TECHNOLOGY the "enabler"

OPERATIONS AND SUPPORT

8

PROCESS MODELING

CHAPTER PREVIEW AND OBJECTIVES

This is the second of two systems modeling chapters. In this chapter you will learn how to draw **data flow diagrams,** a popular process model that documents a system's processes and their data flows. You will know process modeling as a systems analysis tool when you can:

— Define systems modeling and differentiate between logical and physical system models.

— Define process modeling and explain its benefits.

— Recognize and understand the basic concepts and constructs of a process model.

— Read and interpret a data flow diagram.

— Explain when to construct process models and where to store them.

— Construct a context diagram to illustrate a system's interfaces with its environment.

— Identify use cases, external and temporal business events for a system.

— Perform event partitioning and organize events in a functional decomposition diagram.

— Draw event diagrams and then merge those event diagrams into system diagrams.

— Draw primitive data flow diagrams, and describe the elementary data flows and processes in terms of data structures and procedural logic (Structured English and decision tables), respectively.

— Document the distribution of processes to locations.

— Synchronize data and process models using a CRUD matrix.

SOUNDSTAGE

SOUNDSTAGE ENTERTAINMENT CLUB

SCENE

We begin this episode shortly after the project executive sponsor has approved the initial system improvement objectives (*FAST* definition phase). It's time to model the new system requirements, beginning with data. Sandra is facilitating a meeting with the following staff:

- David Hensley, representing Legal Services.
- Ann Martinelli, representing Member Services.
- Sally Hoover, representing Member Services.
- Joe Bosley, representing Marketing.
- Antonio Scarpachi, representing Warehouse.
- Bob Martinez, assigned to sketch the data model.
- Sarah Hartman, assigned to take detailed notes.

SANDRA

Well, I thought the data model session went well. You each have a copy of the model in your packet.

[You have it in Chapter 7.]

The purpose of this facilitated session is to discover all the business events to which your new system must provide a response. As usual, I'd like to keep this discussion as nontechnical as possible. Let's try not to think about computer programs, especially those of you who have written programs. Instead, I'd like you to focus exclusively on the business events, the inputs that trigger the events, and all possible responses or outputs from the events. Any questions?

DAVID

Just the obvious one! It may help to understand why we are interested in events today, as well as your definition of an event. Also, it would help me understand the relationship between event and the data model that we constructed in the last session.

SANDRA

Good questions. I'd like to answer the last question first. The data model iden-

tified the things about which the system must capture and store data. We called them *entities.* But, as you may have noticed, I carefully steered our discussions away from when and how that data would be captured, stored, or even used. Today we want to identify the processes that will do that. Those processes will eventually become programs that our staff will have to write. But ultimately, those programs are business processes that respond to everyday events in your environment. So today, we will try to discover the events to make sure that we design a system that responds to each and every one of them.

ANTONIO

Makes sense to me. Now what do you consider an event to be?

SANDRA

An event is just something that happens.

ANTONIO

An input?

SANDRA

Not always. We recognize many, if not most, events because they show up as a transaction or input. But some events don't have an actual input. For example, the passage of time can trigger some events. For example, the last day of the month triggers various reporting events.

ANN

So far in this project, everything seems like common sense. I really appreciate the way we seem to keep focusing on business issues and requirements. I assume we will get technical sooner or later.

BOB

It's coming. But we don't want to write programs that do not fulfill your day-to-day needs or that fail to recognize less common events that are nonetheless important. This afternoon we are going to form breakout groups to construct simple business pictures of each event. Tomorrow we'll review those pictures as a group and make any changes. Then Sandra, Sarah, and I will combine

those pictures into a picture of your overall system. That picture will show all of the *work* that must be performed as part of the system.

DAVID

Let's do it!

SANDRA

You'll notice that I have written the names of all the data entities from the data model on the blackboard. These are the things about which you decided the system must capture and store data. The stored attributes of each entity are documented in your packet for reference.

[Noting that everyone had found the data model, Sandra continued . . .]

One way to do this is to ask ourselves, "For each entity, what business events might cause us to create a new instance of the entity, delete or deactivate an instance of the entity, or update an instance of the entity?" Those are events. But the big question is, "Where do we start?"

SALLY

Do you mean with which entity?

SANDRA

Yes.

JOE

Well why don't we simply ask ourselves which entity has to exist first. We can't do business without *agreements.* *Members* establish memberships using the agreements. Only after all this happens can we worry about things like *products, promotions,* and *orders.*

SANDRA

So you suggest we start with events that affect agreements?

JOE

Yes.

BOB

You know, this strategy does make some sense. Several methodologies advocate the study of an entity's life history to discover essential processes. Let's give it a try. Is there a format to these events?

SANDRA

I'd suggest simple sentences such as "customer places a new order" or "customer cancels an order" or "time to invoice customers"—sentences that describe not only the event but also who or what triggers the event.

JOE

As the marketing representative, I should take the lead here. Try these on for size. One—Marketing establishes a new agreement. Two—Marketing deletes an agreement.

ANN

Do you like the word *delete?* An agreement doesn't just go away, does it?

JOE

Technically, you're right. First we decertify the agreement so that new members cannot join. Later, once members have been relocated or canceled, we delete the agreement.

DAVID

Do we ever make changes to an agreement between the time it is created and deleted?

JOE

Well, we may have to correct errors.

BOB

OK. Let me read the events back to you. One—Marketing establishes a new agreement. Two—Marketing decertifies an agreement. Three—Marketing relocates members to new agreement. Four—Marketing cancels members that don't accept a new agreement. Five—Marketing deletes an agreement.

ANN

Wait a minute. Marketing doesn't mess with memberships. We do that. Change "Marketing" to "Membership" in the sentences.

JOE

Sorry, Ann's correct . . . What's next, members?

SANDRA

Yes, but I'd like to spend a few more minutes with each of these events first. I'd like to identify the inputs and outputs for each of these events . . .

[The meeting goes on.]

DISCUSSION QUESTIONS

1. How can the study of data help to identify processes?

2. Given the information that Sandra and Bob are trying to collect, what type of picture are they trying to draw? Why?

3. How will event discovery benefit system design and programming?

4. What types of events might be missed if the group focuses exclusively on data entities and how they are created, updated, and deleted?

5. Brainstorm some events that would create, update, and delete members.

In Chapter 5 you were introduced to systems analysis activities that called for drawing system models. System models play an important role in system development. As a systems analyst or user, you will constantly deal with unstructured problems. One way to structure such problems is to draw models.

> A **model** is a representation of reality. Just as a picture is worth a thousand words, most system models are pictorial representations of reality.

Models can be built for existing systems as a way to better understand those systems or for proposed systems as a way to document business requirements or technical designs. An important concept is the distinction between logical and physical models.

> **Logical models** show *what* a system is or does. They are implementation-*in*dependent; that is, they depict the system independent of any technical implementation. As such, logical models illustrate the *essence* of the system. Popular synonyms include *essential model, conceptual model,* and *business model.*

> **Physical models** show not only *what* a system is or does, but also *how* the system is physically and technically implemented. They are implementation-*de*pendent because they reflect technology choices and the limitations of those technology choices. Synonyms include *implementation model* and *technical model.*

Systems analysts have long recognized the value of separating business and technical concerns. That is why they use logical system models to depict business

AN INTRODUCTION TO SYSTEMS MODELING

requirements and physical system models to depict technical designs. Systems analysis activities tend to focus on the logical system models for the following reasons:

— Logical models remove biases that are the result of the way the current system is implemented or the way that any one person thinks the system might be implemented. Thus, we overcome the "we've always done it that way" syndrome. Consequently, logical models encourage creativity.

— Logical models reduce the risk of missing business requirements because we are too preoccupied with technical details. Such errors can be costly to correct after the system is implemented. By separating what the system must do from how the system will do it, we can better analyze the requirements for completeness, accuracy, and consistency.

— Logical models allow us to communicate with end-users in nontechnical or less technical languages. Thus, we don't lose requirements in the technical jargon of the computing discipline.

In this chapter we will focus exclusively on *logical* process modeling during systems analysis.

> **Process modeling** is a technique for organizing and documenting the structure and flow of data through a system's PROCESSES and/or the logic, policies, and procedures to be implemented by a system's PROCESSES.

In the context of your information system building blocks (see the chapter map at the beginning of the chapter), *logical* process models are used to document an information system's PROCESS focus from the perspective of the system owners and system users (the intersection of the PROCESS column with the system owner and system user rows). Also notice that one special type of process model, called a *context diagram,* illustrates the INTERFACE focus from the perspective of the system owners and users. Theoretically, it is possible to recover data flow diagrams by reverse engineering existing application programs. In practice, this technology is not as mature as reverse engineering for data models.

Process modeling originated in classical software engineering methods; therefore, you may have encountered various types of process models such as program structure charts, logic flowcharts, or decision tables in an application programming course. In this chapter, we'll focus on a systems analysis process model, *data flow diagrams* (DFDs).

> A **data flow diagram (DFD)** is a tool that depicts the flow of data through a system and the work or processing performed by that system. Synonyms include *bubble chart, transformation graph,* and **process model.**

We'll also introduce a DFD planning tool called *decomposition diagrams.* Finally, we'll also study *context diagrams,* a process-like model that actually illustrates a system's interfaces to the business and outside world, including other information systems.[1]

A simple data flow diagram is illustrated in Figure 8.1. In the design phase, some of these business processes might be implemented as computer software (either built in-house or purchased from a software vendor). If you examine this data flow diagram, you should find it easy to read, even before you complete this chapter—that has always been the advantage of DFDs. There are only three symbols and one connection:

— The rounded rectangles represent *processes* or work to be done. Notice that they are illustrated in the PROCESS color from your information system framework.

---

[1] In classical structured analysis, context diagrams are considered to be another type of process model. But in object-oriented analysis, they illustrate scope and interfaces. In this edition, we have chosen the latter definition.

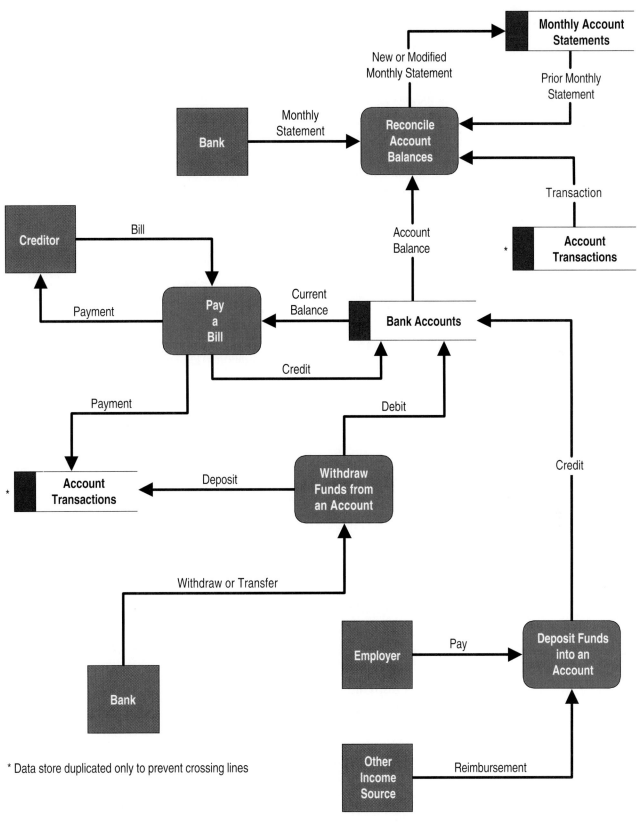

FIGURE 8.1   *A Simple Data Flow Diagram*

— The squares represent *external agents*—the boundary of the system. Notice that they are illustrated in the INTERFACE color from your information system framework.

— The open-ended boxes represent *data stores,* sometimes called files or databases. If you have already read Chapter 7, these data stores correspond to all instances of a single entity in a data model. Accordingly, they have been illustrated with the DATA color from your information systems framework.

— The arrows represent *data flows,* or inputs and outputs, to and from the processes.

Don't confuse data flow diagrams with flowcharts! Program design frequently involves the use of flowcharts. But data flow diagrams are very different! Let's summarize the differences.

— Processes on a data flow diagram can operate in parallel. Thus, several processes might be executing or working simultaneously. This is consistent with the way businesses work. On the other hand, processes on flowcharts can execute only one at a time.

— Data flow diagrams show the flow of data through the system. Their arrows represent paths down which data can flow. Looping and branching are not typically shown. On the other hand, flowcharts show the sequence of processes or operations in an algorithm or program. Their arrows represent pointers to the next process or operation. This may include looping and branching.

— Data flow diagrams can show processes that have dramatically different timing. For example, a single DFD might include processes that happen hourly, daily, weekly, yearly, and on-demand. This doesn't happen in flowcharts.

Data flow diagrams have been popular for more than 20 years, but the interest in DFDs has been recently renewed because of their applicability in **business process redesign (BPR).** As businesses have come to realize that most data processing systems have merely automated outdated, inefficient, and bureaucratic business processes, there is renewed interest in streamlining those business processes. This is accomplished by first modeling those business processes for the purpose of analyzing, redesigning, and/or improving them. Subsequently, information technology can be applied to the improved business processes in creative ways that maximize the value returned to the business. We'll revisit this trend at the end of the chapter.

There are several competing symbol sets for DFDs. Most are named after their inventors (e.g., DeMarco/Yourdon, Gane/Sarson) or after a published standard (e.g., IDEF0, SSADM). Some analysts will argue semantics, but these data flow diagramming "languages" generally support the same fundamental concepts and constructs. We have adopted the Gane and Sarson (structured analysis) notation because of its popularity and CASE tool support.

SYSTEM CONCEPTS FOR PROCESS MODELING

This chapter teaches a *technique* of systems analysis. Most systems analysis techniques are strongly rooted in *systems thinking.*

> **Systems thinking** is the application of formal systems theory and concepts to systems problem solving.

Systems theory and concepts help us understand the way systems are organized and how they work. Techniques teach us how to apply the theory and concepts to build useful real-world systems. If you understand the underlying concepts, you can better adapt the techniques to ever-changing problems and conditions. Therein lies your true opportunity for competitive advantage and security in today's business world.

Let's explore some of the basic concepts that underlie all process models.

Process Concepts

Recall from Chapter 2 that a fundamental building block of information systems is PROCESSES. All information systems include processes—usually lots of them! Information system processes respond to business events and conditions and transform

DATA (another building block) into useful information. We need a way to model processes and understand their interactions with the system's environment, other systems, and other processes.

A System *Is* a Process   The word *system* is a common one that is used to describe almost any orderly arrangement of ideas or constructs. People speak of educational systems, computer systems, management systems, business systems, and, of course, information systems. In the oldest and simplest of all system models, a system *is* a process.

A common theme in systems analysis is the use of models to view or present a system. As shown in Figure 8.2, the simplest process model of a system is based on inputs, outputs, and the system itself—viewed as a process. The process symbol defines the boundary of the system. The system is inside the boundary; the environment is outside that boundary. The system exchanges inputs and outputs with its environment. Because the environment is always changing, well-designed systems have a feedback and control loop to allow the system to adapt itself to changing conditions.

Consider a business as a system. It operates within an environment that includes customers, suppliers, competitors, other industries, and the government. Its inputs include materials, services, new employees, new equipment, facilities, money, and orders (to name but a few). Its outputs include products and/or services, waste materials, retired equipment, former employees, and money (payments). It monitors its environment to make necessary changes to its product line, services, operating procedures, competitors, and the economy.

A rounded rectangle (the Gane and Sarson notation) is used throughout this chapter to represent a process (see margin). Different process modeling notations use a circle (the DeMarco/Yourdon notation) or a rectangle (the SSADM/IDEF0 notation). The choice is often dependent on your methodology and CASE tool features. But what is a process?

A **process** is work performed on, or in response to, incoming data flows or conditions. A synonym is *transform*.

**Gane & Sarson shape;
used throughout this book**

DeMarco/Yourdon shape

SSADM/IDEF0 shape

Process Symbols

FIGURE 8.2
The Classical Process Model of a System

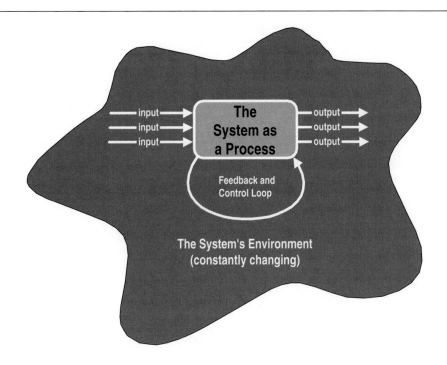

Although processes can be performed by people, departments, robots, machines, or computers, we once again want to focus on *what* work or action is being performed (the *logical* process), not on who or what is doing that work or activity (the *physical* process). For instance, in Figure 8.1 we included the logical process WITHDRAW FUNDS FROM AN ACCOUNT. We did not indicate how this would be done. Intuitively, we can think of several physical implementations such as using an ATM, a bank's drive-through service, or actually going inside the bank.

Process Decomposition   A complex system is usually too difficult to fully understand when viewed as a whole (meaning, *as a single process*). Therefore, in systems analysis we separate a system into its component subsystems, which are decomposed into smaller subsystems, until we have identified manageable subsets of the overall system (see Figure 8.3). We call this technique *decomposition*.

> **Decomposition** is the act of breaking a system into its component subsystems, processes, and subprocesses. Each level of *abstraction* reveals more or less detail (as desired) about the overall system or a subset of that system.

You have already applied decomposition in various ways. Most of you have *outlined* a term paper—this is a form of decomposition. Many of you have partitioned a medium-to-large-sized computer program into subprograms that could be developed and tested independently before they are integrated. This is also decomposition.

In systems analysis, decomposition allows you to partition a system into logical subsystems of processes for improved communication, analysis, and design. A diagram similar to Figure 8.3 can be a little difficult to construct when dealing with all but the smallest of systems. Figure 8.4 demonstrates an alternative layout that is supported by many CASE tools and development methodologies. It is called a *decomposition diagram*. We'll use it extensively in this chapter.

FIGURE 8.3

A System Consists of Many Subsystems and Processes

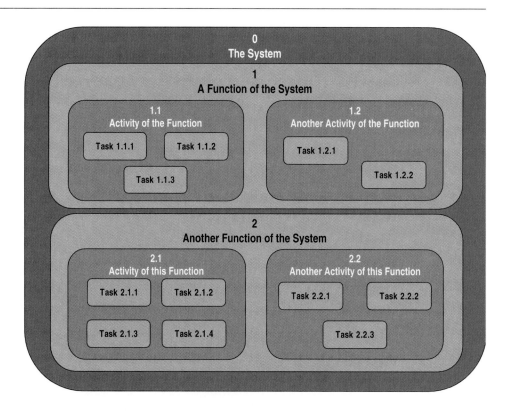

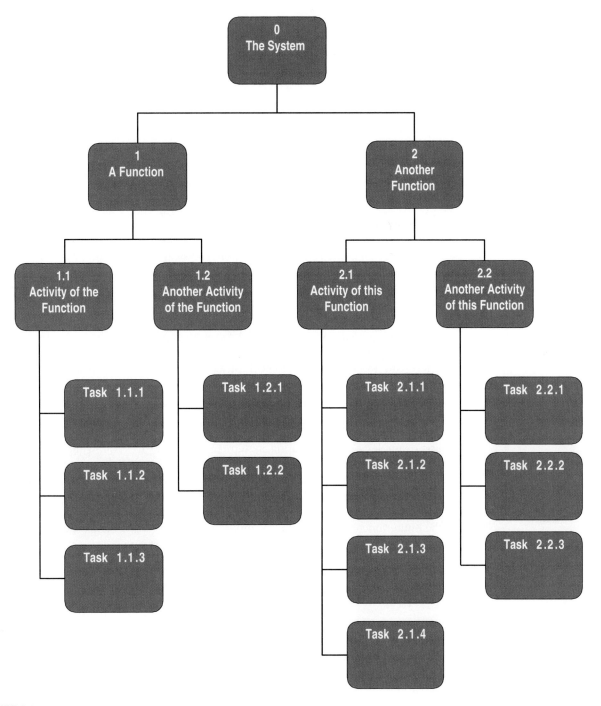

FIGURE 8.4   *A Decomposition Diagram (for Figure 8.3)*

A **decomposition diagram,** also called a hierarchy chart, shows the top-down functional decomposition and structure of a system.

A decomposition diagram is essentially a planning tool for more detailed process models, namely, data flow diagrams. The following rules apply:

- Each process in a decomposition diagram is either a *parent process,* a *child process* (of a parent), or both.
- A parent *must* have two or more children—a single child does not make sense because that would not reveal any additional detail about the system.

— In most decomposition diagramming standards, a child may have only one parent.

— Finally, a child of one parent may be the parent of its own children.

The upper and lower halves of the decomposition diagram in Figure 8.4 demonstrate two styles for laying out the processes and connections. You may use either or both as necessary to present an uncluttered model. Some models may require multiple pages for maximum clarity.

The connections on a decomposition diagram do not contain arrowheads because the diagram is meant to show *structure,* not *flow.* Also, the connections are not named. Implicitly they all have the same name—CONSISTS OF—since the sum of the child processes for a parent process *equals* the parent process.

Logical Processes and Conventions   *Logical* processes are work or actions that <u>must</u> be performed no matter <u>how</u> you implement the system. Each logical process is (or will be) implemented as one or more physical processes that may include work performed by people, work performed by robots or machines, or work performed by computer software. It doesn't matter which implementation is used, however, because logical processes should only indicate <u>that</u> there is work that must be done.

Naming conventions for logical processes depend on where the process is in the decomposition diagram/data flow diagram and the type of process depicted. There are three types of logical processes: *functions, events,* and *elementary processes.*

> A **function** is a set of related and <u>ongoing</u> activities of the business. A function has no start or end; it just continuously performs its work as needed.

For example, a manufacturing system may include the following functions (subsystems): PRODUCTION PLANNING, PRODUCTION SCHEDULING, MATERIALS MANAGEMENT, PRODUCTION CONTROL, QUALITY MANAGEMENT, and INVENTORY CONTROL. Each of these functions may consist of dozens or hundreds of more discrete processes to support specific activities and tasks. Functions are named with nouns that describe the entire function. Additional examples are: ORDER ENTRY, ORDER MANAGEMENT, SALES REPORTING, CUSTOMER RELATIONS, and RETURNS AND REFUNDS.

> An **event** is a logical unit of work that must be completed as a whole. An event is triggered by a discrete input and is completed when the process has responded with appropriate outputs. Events are sometimes called *transactions.*

Functions consist of processes that respond to events. For example, the MATERIALS MANAGEMENT function may respond to the following events: TEST MATERIAL QUALITY, STOCK NEW MATERIALS, DISPOSE OF DAMAGED MATERIALS, DISPOSE OF SPOILED MATERIALS, REQUISITION MATERIALS FOR PRODUCTION, RETURN UNUSED MATERIALS FROM PRODUCTION, ORDER NEW MATERIALS, and so on. Each of these events has a trigger and response that can be defined by its inputs and outputs.

Using *modern* structured analysis techniques such as those advocated by McMenamin, Palmer, Yourdon, and the Robertsons (see the suggested readings at the end of the chapter), system functions are ultimately decomposed into business events. Each business event is represented by a single process that will respond to that event. Event process names tend to be very general. We will adopt the convention of naming event processes as follows: PROCESS_____, RESPOND TO _____, or GENERATE_____, where the blank would be the name of the event (or its corresponding input). Sample event process names are: PROCESS CUSTOMER ORDER, PROCESS CUSTOMER ORDER CHANGE, PROCESS CUSTOMER CHANGE OF ADDRESS, RESPOND TO CUSTOMER COMPLAINT, RESPOND TO ORDER INQUIRY, RESPOND TO PRODUCT PRICE CHECK, GENERATE BACK-ORDER REPORT, GENERATE CUSTOMER ACCOUNT STATEMENT, and GENERATE INVOICE.

An event process can be further decomposed into elementary processes that illustrate in detail how the system must respond to an event.

> **Elementary processes** are discrete, detailed activities or tasks required to complete the response to an event. In other words, they are the lowest level of detail depicted in a process model. A common synonym is **primitive process.**

Elementary processes should be named with a strong action verb followed by an object clause that describes what the work is performed on (or for). Examples of elementary process names are: VALIDATE CUSTOMER IDENTIFICATION, VALIDATE ORDERED PRODUCT NUMBER, CHECK PRODUCT AVAILABILITY, CALCULATE ORDER COST, CHECK CUSTOMER CREDIT, SORT BACK ORDERS, GET CUSTOMER ADDRESS, UPDATE CUSTOMER ADDRESS, ADD NEW CUSTOMER, and DELETE CUSTOMER.

Logical process models omit any processes that do nothing more than move or route data, thus leaving the data unchanged. Physical business systems frequently implement such processes, but they are not essential; in fact, they are increasingly considered unnecessary bureaucracy. Thus, you should omit any process that corresponds to a secretary or clerk receiving and simply forwarding a variety of documents to their next processing location. In the end, you should be left only with logical processes that:

— *Perform computations* (calculate grade point average).
— *Make decisions* (determine availability of ordered products).
— *Sort, filter, or otherwise summarize data* (identify overdue invoices).
— *Organize data into useful information* (generate a report or answer a question).
— *Trigger other processes* (turn on the furnace or instruct a robot).
— *Use stored data* (create, read, update, or delete a record).

Be careful to avoid three common mechanical errors with processes (illustrated in Figure 8.5):

— Process 3.1.2 has inputs but no outputs. We call this a **black hole** because data enters the process and then disappears. In most cases, the modeler simply forgot the output.
— Process 3.1.3 has outputs but no input. Unless you are David Copperfield, this is a **miracle!** In this case, the input flows were likely forgotten.
— In Process 3.1.1 the inputs are insufficient to produce the output. We call this a **gray hole.** There are several possible causes including: (1) a misnamed process, (2) misnamed inputs and/or outputs, or (3) incomplete facts. Gray holes are the most common errors—and the most embarrassing. Once handed to a programmer, the input data flows to a process (to be implemented as a program) must be sufficient to produce the output data flows.

Process Logic   Decomposition diagrams and data flow diagrams will prove very effective tools for identifying processes, but they are not good at showing the logic inside those processes. Eventually, we will need to specify detailed *instructions* for the elementary processes on a data flow diagram. Consider, for example, an elementary process named CHECK CUSTOMER CREDIT. By itself, the named process is insufficient to explain the logic needed to CHECK CUSTOMER CREDIT. We need an effective way to model the logic of an elementary process. Ideally, our logic model should be equally effective for communicating with users (who must verify the business accuracy of the logic) and programmers (who may have to implement the business logic in a programming language).

We can rule out flowcharts. While they do model process logic, most end-users tend to be extremely inhibited by them. The same would be true of pseudocode

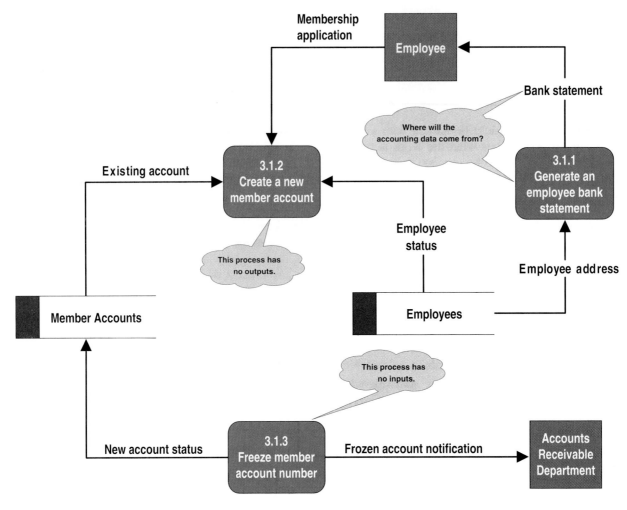

FIGURE 8.5   *Common Errors on Data Flow Diagrams*

and other popular programming logic tools. We can also rule out natural English. It is too often imprecise and frequently subject to interpretation (and misinterpretation). Figure 8.6 summarizes some common problems encountered by those who attempt to use natural English as a procedural language.

To address this problem, we require a tool that marries some of the advantages of natural English with some of the rigor of programming logic tools.

> **Structured English** is a language and syntax, based on the relative strengths of structured programming and natural English, for specifying the underlying logic of elementary processes on process models (such as *data flow diagrams*).

An example of Structured English is shown in Figure 8.7. (The numbers and letters at the beginning of each statement are optional. Some end-users like them because they further remove the programming "look and feel" from the specification.)

For you programmers out there, Structured English is *not* pseudocode. It does not concern itself with declarations, initialization, linking, and such technical issues. It does, however, borrow some of the logical constructs of *structured programming* to overcome the lack of structure and precision in the English language. Think of it as the marriage of natural English language with the syntax of structured programming.

FIGURE 8.6   *Problems with Natural English as a Procedure Specification Language*

- Many of us do not write well, and we also tend not to question our writing abilities.
- Many of us are too educated! It's often difficult for a highly educated person to communicate with an audience that may not have had the same educational opportunities. For example, the average college graduate (including most analysts) has a working vocabulary of 10,000 to 20,000 words; on the other hand, the average noncollege graduate has a working vocabulary of around 5,000 words.
- Some of us write everything like it was a program. If business procedures required such precision, we'd write everything in a programming language.
- Too often, we allow the jargon and acronyms of computing to dominate our language.
- English statements frequently have an excessive or confusing scope. How would you carry out this procedure: "If customers walk in the door and they do not want to withdraw money from their account or deposit money to their account or make a loan payment, send them to the trust department." Does this mean that the only time you should not send the customer to the trust department is when he or she wishes to do all three of the transactions? Or does it mean that if a customer does not wish to perform at least one of the three transactions, that customer should not be sent to the trust department?
- We overuse compound sentences. Consider the following procedure: "Remove the screws that hold the outlet cover to the wall. Remove the outlet cover. Disconnect each wire from the plug, but first make sure the power to the outlet has been turned off." An unwary person might try to disconnect the wires before turning off the power!
- Too many words have multiple definitions.
- Too many statements use imprecise adjectives. For example, a loan officer asks a teacher to certify that a student is in good academic standing. What is good?
- Conditional instructions can be imprecise. For example, if we state that "all applicants under the age of 19 must secure parental permission," do we mean less than 19, or less than or equal to 19?
- Compound conditions tend to show up in natural English. For example, if credit approval is a function of several conditions—credit rating, credit ceiling, annual dollar sales for the customer in question—then different combinations of these factors can result in different decisions. As the number of conditions and possible combinations increases, the procedure becomes more and more tedious and difficult to write.

Source: Adapted from Leslie Matthies, *The New Playscript Procedure* (Stamford, CT: Office Publications, Inc., 1977).

---

1. For each CUSTOMER NUMBER in the data store CUSTOMERS:
   a. For each LOAN in the data store LOANS that matches the above CUSTOMER NUMBER:
      1) Keep a running total of NUMBER OF LOANS for the CUSTOMER NUMBER.
      2) Keep a running total of ORIGINAL LOAN PRINCIPAL for the CUSTOMER NUMBER.
      3) Keep a running total of CURRENT LOAN BALANCE for the CUSTOMER NUMBER.
      4) Keep a running total of AMOUNTS PAST DUE for the CUSTOMER NUMBER.
   b. If the TOTAL AMOUNTS PAST DUE for the CUSTOMER NUMBER is greater than 100.00 then
         1) Write the CUSTOMER NUMBER and data in the data flow LOANS AT RISK.
      Else
         1) Exclude the CUSTOMER NUMBER and data from the data flow LOANS AT RISK.

FIGURE 8.7
Using Structured English to Document an Elementary Process

The overall structure of a Structured English specification is built using the fundamental constructs that have governed structured programming for nearly three decades. These constructs (summarized in Figure 8.8) are:

- A *sequence* of simple, declarative sentences—one after another. Compound sentences are discouraged because they frequently create ambiguity. Each sentence uses strong, action verbs such as GET, FIND, RECORD, CREATE, READ, UPDATE, DELETE, CALCULATE, WRITE, SORT, MERGE, or anything else recognizable or understandable to users. A formula may be included as part of a sentence (e.g., CALCULATE GROSS PAY = HOURS WORKED × HOURLY WAGE).

- A *conditional* or *decision structure* indicates that a process must perform different steps under well-specified conditions. There are two variations (and a departure) on this construct.

Note that some procedure writers suggest that IF, CASE, REPEAT, DO, *and* FOR *constructs be terminated with an* END *statement. We show these in red in Figure 8.8; however, we don't recommend them because they make the procedure specification look too much like a computer program.*

Structured English Procedural Structures

Construct	Sample Template
Sequence of steps – unconditionally perform a sequence of steps.	[Step 1] [Step 2] … [Step n]

Simple condition steps – if the specified condition is true, then perform the first set of steps. Otherwise, perform the second set of steps.
 	
Use this construct if the condition has only two possible values.
 	
(Note: The second set of conditions is optional.)
	If [truth condition]
	    then
	        [sequence of steps or other conditional steps]
	else
	        [sequence of steps or other conditional steps]
	~~End If~~

Complex condition steps – test the value of the condition and perform the appropriate set of steps.
 	
Use this construct if the condition has more than two values.
	Do the following based on [condition]:
	    Case 1: If [condition] = [value] then
	        [sequence of steps or other conditional steps]
	Case 2: If [condition] = [value] then
	        [sequence of steps or other conditional steps]
	    …
	    Case n: If [condition] = [value] then
	        [sequence of steps or other conditional steps]
	~~End Case~~

Multiple conditions – test the value of multiple conditions to determine the correct set of steps. Use a decision table instead of nested if-then-else Structured English constructs to simplify the presentation of complex logic that involves combinations of conditions. *A decision table is a tabular presentation of complex logic in which rows represent conditions and possible actions, and columns indicate which combinations of conditions result in specific actions.*	<table><tr><td>DECISION TABLE</td><td>Rule</td><td>Rule</td><td>Rule</td><td>Rule</td></tr><tr><td>[Condition]</td><td>value</td><td>value</td><td>value</td><td>value</td></tr><tr><td>[Condition]</td><td>value</td><td>value</td><td>value</td><td>value</td></tr><tr><td>[Condition]</td><td>value</td><td>value</td><td>value</td><td>value</td></tr><tr><td>[Sequence of steps or conditional steps]</td><td>X</td><td></td><td></td><td></td></tr><tr><td>[Sequence of steps or conditional steps]</td><td></td><td>X</td><td>X</td><td></td></tr><tr><td>[Sequence of steps or conditional steps]</td><td></td><td></td><td></td><td>X</td></tr></table> Although it isn't a Structured English construct, a decision table can be named, and referenced within a Structured English procedure.
One-to-many iteration – repeat the set of steps until the condition is false. Use this construct if the set of steps must be performed <u>at least</u> once, regardless of the condition's initial value.	**Repeat the following until** [truth condition]: [sequence of steps or conditional steps] ~~End Repeat~~

Zero-to-many iteration – repeat the set of steps until the condition is false.
 	
Use this construct if the set of steps is conditional based on the condition's initial value.
	Do While [truth condition]:
	    [sequence of steps or conditional steps]
	~~End Do~~
	                    - OR -
	For [truth condition]:
	    [sequence of steps or conditional steps]
	~~End For~~

FIGURE 8.8   *Structured English Constructs*

- The IF-THEN-ELSE construct specifies that one set of steps should be taken if a specified condition is true, but a different set of steps should be specified if the specified condition is false. The steps to be taken are typically a *sequence* of one or more sentences as described above.
- The CASE construct is used when there are more than two sets of steps to choose from. Once again, these steps usually consist of the aforementioned *sequential* statements. The case construct is an elegant substitute for an *IF-THEN-ELSE IF-THEN-ELSE IF-THEN* . . . construct (which is very convoluted to the average user).
- For logic based on multiple conditions and combinations of conditions (which programmers call a *nested IF*), *decision tables* are a far more elegant logic modeling tool. Decision tables will be introduced shortly.

- An *iteration* or *repetition* structure specifies that a set of steps should be repeated based on some stated condition. There are two variations on this construct.
  - The DO-WHILE construct indicates that certain steps are repeated zero, one, or more times based on the value of the stated condition. Note that these steps may not execute at all if the condition is not true when the condition is first tested.
  - The REPEAT-UNTIL construct indicates that certain steps are repeated one or more times based on the value of the stated condition. Note that a REPEAT-UNTIL set of steps must execute at least once, unlike the DO-WHILE set of actions.

Additionally, Structured English places the following restrictions on process logic:

- Only strong, imperative verbs may be used.
- Only names that have been defined in the project dictionary may be used. These names may include those of data flows, data stores, entities (from data models; see Chapter 5), attributes (the specified data fields or properties contained in a data flow, data store, or entity), and domains (the specified legal values for attributes).
- Formulas should be stated clearly using appropriate mathematical notations. In short, you can use whatever notation is recognizable to the users. Make sure each operand in a formula is either input to the process in a data flow or a defined constant.
- Undefined adjectives and adverbs (the word *good,* for instance) are not permitted unless clearly defined in the project dictionary as legal values for data attributes.
- Blocking and indentation are used to set off the beginning and ending of constructs and to enhance readability. (Some authors and models encourage the use of special verbs such as ENDIF, ENDCASE, ENDDO, and ENDREPEAT to terminate constructs. We dislike this practice because it gives the Structured English too much of a pseudocode or programming look and feel.)
- When in doubt, user readability should always take priority over programmer preferences.

Structured English should be precise enough to clearly specify the required business procedure to a programmer or user. But it should not be so inflexible that you spend hours arguing over syntax.

Many processes are governed by complex combinations of conditions that are not easily expressed with Structured English. This is most commonly encountered in business policies.

A **policy** is a set of rules that governs some process in the business.

In most firms, policies are the basis for decision making. For instance, a credit card company must bill cardholders according to various policies that adhere to restrictions imposed by state and federal governments (maximum interest rates and minimum payments, for instance). Policies consist of *rules* that can often be translated into computer programs if the users and systems analysts can accurately convey those rules to the computer programmer.

Fortunately, there are ways to formalize the specification of policies and other complex combinations of conditions. One such logic modeling tool is a decision table.

A **decision table** is a tabular form of presentation that specifies a set of conditions and their corresponding actions.

Decision tables, unfortunately, don't get enough respect! People who are unfamiliar with them tend to avoid them. Even the CASE tools ignore them. But decision tables are very useful for specifying complex policies and decision-making rules. Figure 8.9 illustrates the three components of a simple decision table.

— Condition stubs (the upper rows) describe the conditions or factors that will affect the decision or policy.

— Action stubs (the lower rows) describe, in the form of statements, the possible policy actions or decisions.

— Rules (the columns) describe which actions are to be taken under a specific combination of conditions.

The figure depicts a check-cashing policy that appears on the back of a check-cashing card for a grocery store. This same policy has been documented with a

---

FIGURE 8.9 *A Sample Decision Table*

A SIMPLE POLICY STATEMENT

CHECK CASHING IDENTIFICATION CARD

A customer with check cashing privileges is entitled to cash personal checks of up to $75.00 and payroll checks from companies pre-approved by *LMART*. This card is issued in accordance with the terms and conditions of the application and is subject to change without notice. This card is the property of *LMART* and shall be forfeited upon request of *LMART*.

SIGNATURE *Charles C. Parker, Jr.*
EXPIRES **May 31, 2001**

THE EQUIVALENT POLICY DECISION TABLE

Conditions and Actions	Rule 1	Rule 2	Rule 3	Rule 4
C1: Type of check	personal	payroll	personal	payroll
C2: Check amount less than or equal to $75.00	yes	doesn't matter	no	doesn't matter
C3: Company accredited by *LMART*	doesn't matter	yes	doesn't matter	no
A1: Cash the check	X	X		
A2: Don't cash the check			X	X

Condition Stubs

Action Stubs

Rules

decision table. Three conditions affect the check-cashing decision: the type of check, whether the amount of the check exceeds the maximum limit, and whether the company that issued the check is accredited by the store. The actions (decisions) are either to cash the check or to refuse to cash the check. Notice that each combination of conditions defines a rule that results in an action, denoted by an **x**.

One final logic modeling comment is in order. Both decision tables and Structured English can describe a single elementary process. For example, a legitimate statement in a Structured English specification might read DETERMINE WHETHER OR NOT TO CASH THE CHECK USING THE DECISION TABLE, LMART CHECK CASHING POLICY.

Data Flows

Processes respond to inputs and generate outputs. Thus, at a minimum, all processes have at least one input and one output *data flow*. Data flows are the communications between processes and the system's environment. Let's examine some of the basic concepts and conventions of data flows.

Data in Motion   A data flow is *data in motion*. The flow of data between a system and its environment or between two processes inside a system is *communication*. Let's study this form of communication.

> A **data flow** represents an input of data to a process or the output of data (or information) from a process. A data flow is also used to represent the creation, reading, deletion, or updating of data in a file or database (called a *data store* on the DFD).

Think of a data flow as a highway down which packets of known composition travel. The name implies what type of data may travel down that highway. This highway is depicted as a solid line with arrow (see margin).

The *packet* concept is critical. Data that should travel together should be shown as a single data flow, no matter how many *physical* documents might be included. The packet concept is illustrated in Figure 8.10, which shows the correct and incorrect ways to show a logical data flow packet.

The *known composition* concept is equally important. A data flow is composed of either actual data attributes (also called *data structures*—more about them later) or other data flows.

> A **composite data flow** is a data flow that consists of other data flows. They are used to combine similar data flows on high-level data flow diagrams to make those diagrams easier to read.

For example, in Figure 8.11(a), a high-level DFD consolidates all types of orders into a composite data flow called ORDER. In Figure 8.11(b), a more detailed data

Data flow name ⟶

Data Flow Symbol

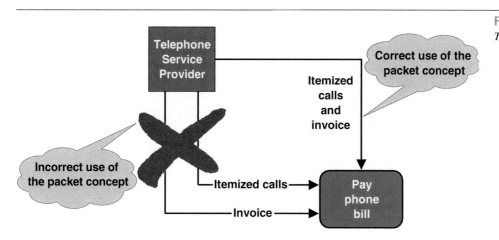

FIGURE 8.10
The Data Flow Packet Concept

(a) High-Level DFD

(b) More Detailed DFD

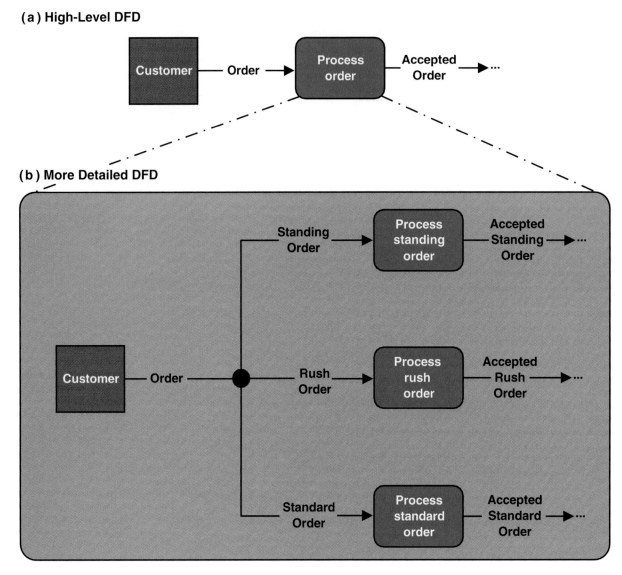

FIGURE 8.11   *Composite and Elementary Data Flows*

flow diagram shows specific types of orders: STANDING ORDER, RUSH ORDER, and STANDARD ORDER. These different orders require somewhat different processing. (The small, black circle is called a **junction.** It indicates that any given ORDER is an instance of only one of the order types.)

Another common use of composite data flows is to consolidate all reports and inquiry responses into one or two composite flows. There are two reasons for this. First, these outputs can be quite numerous. Second, many modern systems provide extensive user-defined reports and inquiries that cannot be predicted before the system's implementation and use.

Before we exit this introduction to *data* flows, we should acknowledge that some data flow diagramming methods also recognize *nondata* flows called control flows.

A **control flow** represents a condition or nondata event that triggers a process. Think of it as a condition to be monitored while the system works. When the system realizes that the condition meets some predetermined state, the process to which it is input is started.

The classic information system example is *time.* For example, a report generation process may be triggered by the temporal event END-OF-MONTH. In real-time systems, control flows often represent real-time conditions such as TEMPERATURE and ALTITUDE. In most methodologies that distinguish between data and control flows, the control flow is depicted as a dashed line with arrow (see margin).

Typically, information systems analysts have dealt mostly with data flows; however, as information systems become more integrated with real-time systems (such as manufacturing processes and computer-integrated manufacturing), the need to distinguish the concept of control flows becomes necessary.

– – – Control flow name – – →

Control Flow Symbol

***Logical* Data Flows and Conventions**   While we recognize that data flows can be implemented a number of ways (e.g., telephone calls, business forms, bar codes, memos, reports, computer screens, and computer-to-computer communications), we are interested only in *logical* data flows. Thus, we are only interested that the flow is needed (not how we will implement that flow). Data flow names should discourage premature commitment to any possible implementation.

Data flow names should be descriptive nouns and noun phrases that are singular, as opposed to plural (ORDER—not ORDERS). We do not want to imply that occurrences of the flow must be implemented as a *physical* batch.

Data flow names also should be unique. Use adjectives and adverbs to help to describe how processing has changed a data flow. For example, if an input to a process is named ORDER, the output should not be named ORDER. It might be named VALID ORDER, APPROVED ORDER, ORDER WITH VALID PRODUCTS, ORDER WITH APPROVED CREDIT, or any other more descriptive name that reflects what the process did to the original order.

Logical data flows to and from data stores require special naming considerations (see Figure 8.12). (Data store names are plural, and the numbered bullets match the note to the figure.)

- Only the *net* data flow is shown. Intuitively, you may realize that you have to get a record to update it or delete it. But unless data is needed for some other purpose (e.g., a calculation or decision), the "read" action is not shown. This keeps the diagram uncluttered.

❶ A data flow from a data store to a process indicates that data is to be "read" for some specific purpose. The data flow name should clearly indicate what data is to be read. This is shown in Figure 8.12.

❷ A data flow from a process to a data store indicates that data is to be created, deleted, or updated in/from that data store. Again, as shown in Figure 8.12 these data flows should be clearly named to reflect the specific action performed (such as NEW CUSTOMER, CUSTOMER TO BE DELETED, or UPDATED ORDER ADDRESS).

Notice that the names suggest the classic actions that can be performed on a file, namely: CREATE, READ, UPDATE, and DELETE (CRUD). In a real DFD, we would not actually record these action names on the diagram.

No data flow should ever go unnamed. Unnamed data flows are frequently the result of flowchart thinking (e.g., step 1, step 2, etc.). If you can't give the data flow a reasonable name, it probably does not exist!

Consistent with our goal of *logical* modeling, data flow names should describe the data flow without describing how the flow is or could be implemented. Suppose, for example, that end-users explain their system as follows: *"We fill out Form 23 in triplicate and send it to . . ."* The logical name for the "Form 23" data flow might be COURSE REQUEST. This logical name eliminates physical, implementation biases—the idea that we must use a *paper form,* and the notion that we must use carbon copies. Ultimately, this will free us to consider other physical

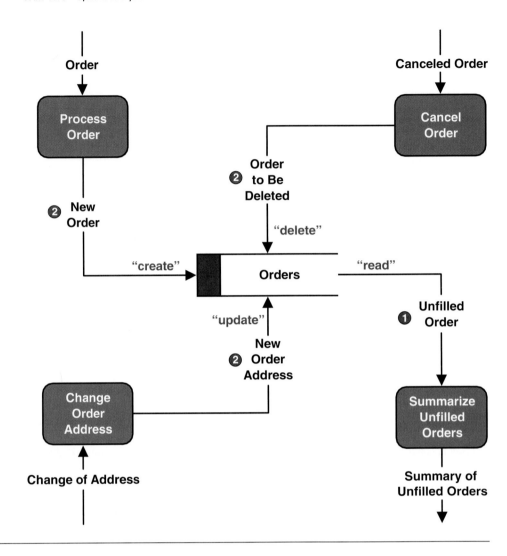

FIGURE 8.12
*Data Flows to and from
Data Stores*

alternatives such as Touch-Tone phone responses, on-line registration screens, or even e-business Internet pages!

Finally, all data flows must begin and/or end at a process because data flows are the inputs and outputs of a process. Consequently, all the data flows on the left side of Figure 8.13 are illegal. The corrected diagrams are shown on the right side.

Data Flow Conservation   For many years we have tried to improve business processes by automating them. It hasn't always worked or worked well because the business processes were designed to process data flows in a precomputing era. Consider the average business form. It is common to see the form divided into sections that are designed for different audiences. The first recipient completes his part of the form; the next recipient completes her part; and so forth. At certain points in this processing sequence, a copy of the form might even be detached and sent to another recipient who initializes a new multiple-part form that requires transcribing much of the same data from the initial form. In our own university, we've seen examples where poor form design requires the same data to be typed a dozen times!

Now, if the flow of current data is computerized based on the current business forms and processes, the resulting computer programs will merely automate these inefficiencies. This is precisely what has happened in most businesses! Today, a new emphasis on *business process redesign* encourages management, users, and systems analysts to identify and eliminate these inefficiencies <u>before</u> designing any new information system. We can support this trend in *logical* data flow diagrams by practicing data conservation.

> **Data conservation,** sometimes called "starving the processes," requires that a data flow contain only the data that is truly needed by the receiving process.

By ensuring that processes receive only as much data as they really need, we simplify the interface between those processes. To practice data conservation, we must precisely define the data composition of each (noncomposite) data flow. Data composition is expressed in the form of *data structures*.

Data Structures   Ultimately, a data flow contains data items called attributes.

> A **data attribute** is the smallest piece of data that has meaning to the end-users and the business. (This definition also applies to *attributes* as they were presented in Chapter 7.)

FIGURE 8.13   *Illegal Data Flows*

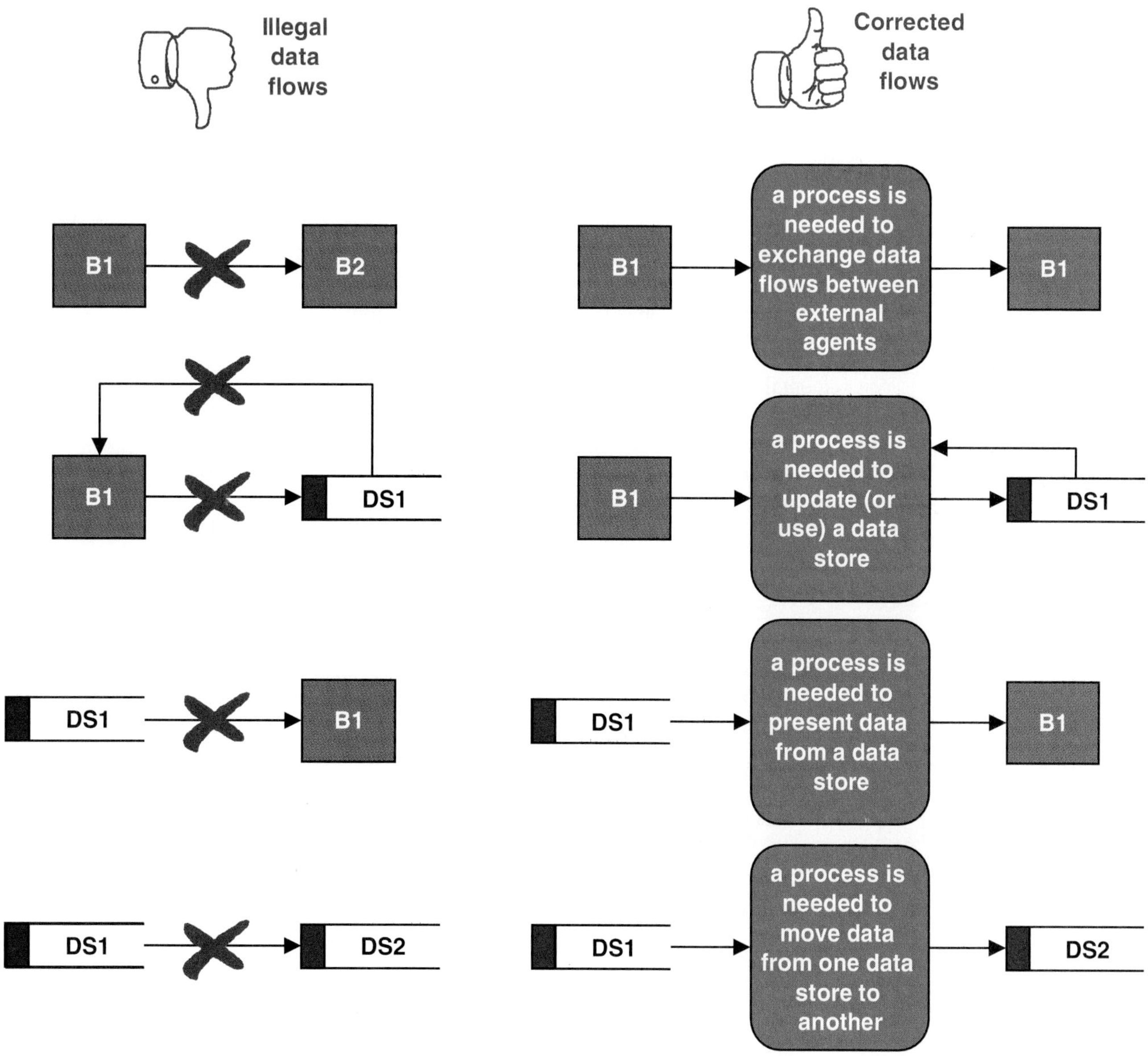

Sample attributes for the data flow ORDER might include ORDER NUMBER, ORDER DATE, CUSTOMER NUMBER, SHIPPING ADDRESS (which consists of attributes such as STREET ADDRESS, CITY, and ZIP CODE), ORDERED PRODUCT NUMBERS, QUANTITY(ies) ORDERED, and so on. Notice that some attributes occur once for each instance of ORDER, while others may occur several times for a single instance of ORDER.

The data attributes that comprise a data flow are organized into data structures.

Data structures are specific arrangements of data attributes that define the organization of a single instance of a data flow.

Data flows can be described in terms of the following types of data structures:

- A *sequence* or group of data attributes that occur one after another.
- The *selection* of one or more attributes from a set of attributes.
- The *repetition* of one or more attributes.

The most common data structure notation is a Boolean algebraic notation that is required by many CASE tools. Other CASE tools and methodologies support proprietary, but essentially equivalent notations. A sample data structure for the data flow ORDER is presented in Figure 8.14. This algebraic notation uses the following symbols:

=	Means "consists of" or "is composed of."
+	Means "and" and designates *sequence*.
[. . .]	Means "only one of the attributes within the brackets may be present"—designates *selection*. The attributes in the brackets are separated by commas.
{. . .}	Means that the attributes in the braces may occur many times for one instance of the data flow—designates *repetition*. The attributes inside the braces are separated by commas.
(. . .)	Means the attribute(s) in the parentheses are optional—no value— for some instances of the data flow.

In our experience, all data flows can be described in terms of these fundamental constructs. Figure 8.15 on p. 328 demonstrates each of the fundamental constructs using examples. Returning to Figure 8.14, notice that the constructs are combined to describe the data content of the data flow.

The importance of defining the data structures for every data flow should be apparent—you are defining the business data requirements for each input and output! These requirements must be determined before any process could be implemented as a computer program. This standard notation provides a simple but effective means for communicating between end-users and programmers.

Domains   An attribute is a piece of data. When analyzing a system, it makes sense that we should define those values for an attribute that are legitimate, or that make sense. The values for each attribute are defined in terms of two properties: data type and domain.

The **data type** for an attribute defines what class of data can be stored in that attribute.

The **domain** of an attribute defines what values an attribute can legitimately take on.

The concepts of data type and domain were introduced in Chapter 7. See that discussion and Tables 7.1 and 7.2 for a more complete description of data type and domain.

DATA STRUCTURE	ENGLISH INTERPRETATION

ORDER =
    ORDER NUMBER +
    ORDER DATE +
    [PERSONAL CUSTOMER NUMBER,
      CORPORATE ACCOUNT NUMBER] +
    SHIPPING ADDRESS = ADDRESS +
    (BILLING ADDRESS = ADDRESS) +
    1 { PRODUCT NUMBER +
      PRODUCT DESCRIPTION +
      QUANTITY ORDERED +
      PRODUCT PRICE +
      PRODUCT PRICE SOURCE +
      EXTENDED PRICE } N +
    SUM OF EXTENDED PRICES +
    PREPAID AMOUNT +
    (CREDIT CARD NUMBER + EXPIRATION DATE)
    (QUOTE NUMBER)
 	
ADDRESS =
    (POST OFFICE BOX NUMBER) +
    STREET ADDRESS +
    CITY +
    [STATE, MUNICIPALITY] +
    (COUNTRY) +
    POSTAL CODE
	An instance of ORDER consists of:
	    ORDER NUMBER and
	    ORDER DATE and
	    Either PERSONAL CUSTOMER NUMBER
	      or CORPORATE ACCOUNT NUMBER
	    and SHIPPING ADDRESS (which is equivalent to
	      ADDRESS)
	    and optionally: BILLING ADDRESS (which is
	      equivalent to ADDRESS)
	    and one or more instances of:
	      PRODUCT NUMBER and
	      PRODUCT DESCRIPTION and
	      QUANTITY ORDERED and
	      PRODUCT PRICE and
	      PRODUCT PRICE SOURCE and
	      EXTENDED PRICE
	    and SUM OF EXTENDED PRICES and
	    PREPAID AMOUNT and
	    optionally: both CREDIT CARD NUMBER and
	      EXPIRATION DATE
	    and optionally: QUOTE NUMBER
	
	An instance of ADDRESS consists of:
	    optionally: POST OFFICE BOX NUMBER and
	    STREET ADDRESS and
	    CITY and
	    Either STATE or MUNICIPALITY
	    and optionally: COUNTRY
	    and POSTAL CODE

FIGURE 8.14   *A Data Structure for a Data Flow*

Divergent and Convergent Flows   It is sometimes useful to depict diverging or converging data flows on a data flow diagram.

A **diverging data flow** is one that splits into multiple data flows.

Diverging data flows indicate that all or parts of a single data flow are routed to different destinations.[2]

A **converging data flow** is the merger of multiple data flows into a single data flow.

Converging data flows indicate that data flows from different sources can (must) come together as a single packet for subsequent processing.

Diverging and converging data flows are depicted as shown in Figure 8.16 on p. 329. Notice that we do not include a process to "route" the flows. The flows

---

[2] Some experts suggest that diverging data flows should be used only when all data in the flow is routed to all destinations. We prefer the classic DeMarco definition that allows all *or* parts of the flow to be routed to different processes.

Data Structure	Format by Example (relevant portion is boldfaced)	English Interpretation (relevant portion is boldfaced)
Sequence of Attributes – The sequence data structure indicates one or more attributes that may (or must) be included in a data flow.	WAGE AND TAX STATEMENT = **TAXPAYER IDENTIFICATION NUMBER +** **TAXPAYER NAME +** **TAXPAYER ADDRESS +** **WAGES, TIPS, AND COMPENSATION +** **FEDERAL TAX WITHHELD + ...**	An instance of WAGE AND TAX STATEMENT consists of: **TAXPAYER IDENTIFICATION NUMBER and** **TAXPAYER NAME and** **TAXPAYER ADDRESS and** **WAGES, TIPS, AND COMPENSATION and** **FEDERAL TAX WITHHELD and ...**
Selection of Attributes – The selection data structure allows you to show situations where different sets of attributes describe different instances of the data flow.	ORDER = (**PERSONAL CUSTOMER NUMBER,** **CORPORATE ACCOUNT NUMBER**) + ORDER DATE + ...	An instance of ORDER consists of: **Either PERSONAL CUSTOMER NUMBER or** **CORPORATE ACCOUNT NUMBER; and** ORDER DATE and ...
Repetition of Attributes – The repetition data structure is used to set off a data attribute or group of data attributes that may (or must) repeat themselves a specified number of times for a single instance of the data flow. The minimum number of repetitions is usually *zero* or *one*. The maximum number of repetitions may be specified as "n" meaning "many" where the actual number of instances varies for each instance of the data flow.	CLAIM = POLICY NUMBER + POLICYHOLDER NAME + POLICYHOLDER ADDRESS + 0 { **DEPENDENT NAME +** **DEPENDENT'S RELATIONSHIP** } N + 1 { **EXPENSE DESCRIPTION +** **SERVICE PROVIDER +** **EXPENSE AMOUNT** } N	An instance of CLAIM consists of: POLICY NUMBER and POLICYHOLDER NAME and POLICYHOLDER ADDRESS and **zero or more instances of:** **DEPENDENT NAME and** **DEPENDENT'S RELATIONSHIP** and **one or more instances of:** **EXPENSE DESCRIPTION and** **SERVICE PROVIDER and** **EXPENSE ACCOUNT**
Optional Attributes – The optional notation indicates that an attribute, or group of attributes in a <u>sequence or selection data structure</u> may not be included in all instances of a data flow. *Note: For the repetition data structure, a minimum of "zero" is the same as making the entire repeating group "optional."*	CLAIM = POLICY NUMBER + POLICYHOLDER NAME + POLICYHOLDER ADDRESS + (**SPOUSE NAME +** **DATE OF BIRTH**) + ...	An instance of CLAIM consists of: POLICY NUMBER and POLICYHOLDER NAME and POLICYHOLDER ADDRESS and **optionally, SPOUSE NAME and** **DATE OF BIRTH** and ...
Reusable Attributes – For groups of attributes that are contained in many data flows, it is desirable to create a separate data structure that can be reused in other data structures.	DATE = MONTH + DAY + YEAR	Then, the reusable structures can be included in other data flow structures as follows: ORDER = ORDER NUMBER ... + DATE INVOICE = INVOICE NUMBER ... + DATE PAYMENT = CUSTOMER NUMBER ... + DATE

FIGURE 8.15 *Data Structure Constructs*

simply diverge from or converge to a common flow. The following notations, not supported by all CASE tools, are used in this book.

❶ The small square *junction* means "and." This means that each time the process is performed, it must input (or output) all the diverging or converging data flows. *(Some DFD notations simply place a + between the data flows.)*

❷ The small black *junction* means "exclusive or." This means that each time the process is performed, it must input (or output) only one of the diverging or converging data flows. *(Some DFD notations simply place an * between the data flows.)*

❸ In the absence of one diverging or converging data flows, the reader should assume an "inclusive or." This means that each time the process is performed, it may input any *or* all of the depicted data flows.

With the above rules, the most complex of business process and data flow combinations can be depicted.

External Agents

All information systems respond to events and conditions in the system's environment. The environment of an information system includes *external agents* that form the boundary of the system and define places where the system interfaces with its environment.

An **external agent** defines a person, organization unit, other system, or other organization that lies outside the scope of the project but that interacts with the system being studied. External agents provide the net inputs into a system and receive net outputs from a system. Common synonyms include **external entity** (not to be confused with *data entity* as introduced in Chapter 7).

The term *external* means "external to the system being analyzed or designed." In practice, an external agent may actually be outside of the business (such as government agencies, customers, suppliers, and contractors), or it may be inside

---

FIGURE 8.16   *Diverging and Converging Data Flows*

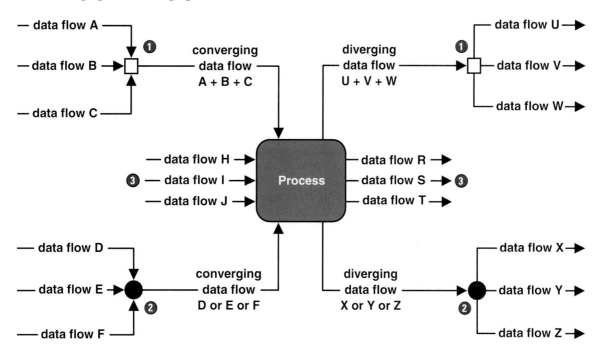

Gane and Sarson shape

Demarco/Yourdon shape

External Agent Symbols

the business but outside of the project and system scope (such as other departments, other business functions, and other internal information systems). An external agent is represented by a square on the data flow diagram. The DeMarco/Yourdon equivalent is a rectangle (see margin).

It is important to recognize that work and activities are occurring inside the external agent, but that work and those activities are said to be "out of scope" and not subject to change. Thus, the data flows between your system and these boundaries should not cause substantive change to the work or activities performed by the external agents.

To be sure, the external agents of an information system are rarely fixed. As project scope and goals change, the scope of an information system can either grow or shrink. If the system scope grows, it can consume some of the original external agents—in other words, what was once considered outside of the system is now considered inside the system (*as new processes*).

Similarly, if the system scope shrinks (because of budget or schedule constraints), processes that were once considered to be inside the system may become external agents.

External agents on a logical data flow diagram may include people, business units, other internal systems with which your system must interact, and external organizations. Their inclusion on the logical DFD means that your system interacts with these agents. They are almost always one of the following:

— An office, department, division, or individual within your company that provides net inputs to that system, receives net outputs from that system, or both.

— An organization, agency, or individual that is outside your company but that provides net inputs to, or receives net outputs from, your system. Examples include CUSTOMERS, SUPPLIERS, CONTRACTORS, BANKS, and GOVERNMENT AGENCY(ies).

— Another business or information system—possibly, though not necessarily, computer-based—that is separate from your system but with which your system must interface. Very few information systems do not interface with other information systems. It is becoming common to interface information systems with those of other businesses.

— One of your system's end-users or managers. In this case, the user or manager is either a net source of data to be input to your system and/or a net destination of outputs to be produced by your system.

External agents should be named with descriptive, singular nouns, such as REGISTRAR, SUPPLIER, MANUFACTURING SYSTEM, or FINANCIAL INFORMATION SYSTEM. External agents represent fixed, *physical* systems; therefore, they can have very physical names or acronyms—even on a <u>logical</u> DFD! For example, an external agent representing our school's financial management information system would be called FMIS. If an external agent describes an individual, we recommend job titles or role names instead of proper names (for example, use ACCOUNT CLERK, not *Mary Jacobs*).

To avoid crossing data flow lines on a DFD, it is permissible to duplicate external agents on DFDs. But as a general rule, external agents should be located on the perimeters of the page, consistent with their definition as a system boundary.

Data Stores

Most information systems capture data for later use. The data is kept in a data store, the last symbol on a data flow diagram. It is represented by the open-end box (see margin on page 331).

A **data store** is an "inventory" of data. Synonyms include *file* and *database* (although those terms are too implementation-oriented for essential process modeling).

If data flows are *data in motion,* think of data stores as *data at rest.*

Ideally, essential data stores should describe "things" about which the business wants to store data. These things include:

Persons: AGENCY, CONTRACTOR, CUSTOMER, DEPARTMENT, DIVISION, EMPLOYEE, INSTRUCTOR, OFFICE, STUDENT, SUPPLIER. Notice that a person entity can represent either individuals, groups, or organizations.

Places: SALES REGION, BUILDING, ROOM, BRANCH OFFICE, CAMPUS.

Objects: BOOK, MACHINE, PART, PRODUCT, RAW MATERIAL, SOFTWARE LICENSE, SOFTWARE PACKAGE, TOOL, VEHICLE MODEL, VEHICLE. An object entity can represent actual objects (such as SOFTWARE LICENSE) or specifications for a type of object (such as SOFTWARE PACKAGE).

Events: APPLICATION, AWARD, CANCELLATION, CLASS, FLIGHT, INVOICE, ORDER, REGISTRATION, RENEWAL, REQUISITION, RESERVATION, SALE, TRIP.

Concepts: ACCOUNT, BLOCK OF TIME, BOND, COURSE, FUND, QUALIFICATION, STOCK.

Data Store

Gane and Sarson shape

Data Store

DeMarco/Yourdon shape

Data Store Symbols

NOTE   If the above list looks familiar, it should! A data store represents *all occurrences* of a data entity—defined in Chapter 7 as something about which we want to store data. As such, the data store represents the synchronization of a system's process model with its data model.

If you do data modeling before process modeling, identification of most data stores is simplified by the following rule:

There should be one data store for each data entity on your entity relationship diagram. (We even include associative and weak entity data stores on our models.)

If, on the other hand, you do process modeling before data modeling, data store discovery tends to be more arbitrary. In that case, our best recommendation is to identify existing implementations of files or data stores (e.g., computer files and databases, file cabinets, record books, catalogs, etc.) and then rename them to reflect the logical "things" about which they store data. Consistent with information engineering strategies, we recommend that data models precede the process models.

Generally, data stores should be named as the plural of the corresponding data model entity. Thus, if the data model includes an entity named CUSTOMER, the process models will include a data store named CUSTOMERS. This makes sense because the data store, by definition, stores all instances of the entity. Avoid physical terms such as *file, database, file cabinet, file folder,* and the like.

As was the case with boundaries, it is permissible to duplicate data stores on a DFD to avoid crossing data flow lines. Duplication should be minimized.

Now that you understand the basic concepts of process models, we can examine building a process model. When do you do it? How many process models may be drawn? What technology exists to support the development of process models?

THE PROCESS OF LOGICAL PROCESS MODELING

Strategic Systems Planning

Many organizations select application development projects based on strategic information system plans. Strategic planning is a separate project that produces an information systems strategy plan that defines an overall vision and architecture for information systems. This architecture frequently includes an **enterprise process model.** (The plan usually has other architectural components that are not important to this discussion.)

An enterprise process model typically identifies only business areas and functions. Events and detailed processes are rarely examined. Business areas and functions are identified and mapped to other enterprise models such as the enterprise data model (Chapter 7). Business areas and functions are subsequently prioritized into application development projects. Priorities are usually based on which business areas, functions, and supporting applications will return the most value to the business as a whole.

An enterprise process model is stored in a corporate repository. Subsequently, as application development projects are started, subsets of the enterprise process model are exported to the project teams to serve as a starting point for building more detailed process models (including data flow diagrams). Once the project team completes systems analysis and design, the expanded and refined process models are returned to the corporate repository.

Process Modeling for Business Process Redesign

A supplemental reading module about BPR can be found in the McGraw-Hill Online Learning Center website for this book.

Business process redesign (BPR) has been discussed several times in this book and chapter. Recall that BPR projects analyze business processes and then redesign them to eliminate inefficiencies and bureaucracies before any (re)application of information technology. To redesign business processes, we must first study the existing processes. Process models play an integral role in BPR.

Each BPR methodology recommends its own process model notations and documentation. Most of the models are a cross between data flow diagrams and flowcharts. The diagrams tend to be very *physical* because the BPR team is trying to isolate the implementation idiosyncrasies that cause inefficiency and reduce value. BPR data flow diagrams/flowcharts may include new symbols and information to illustrate timing, throughput, delays, costs, and value. Given this additional data, the BPR team then attempts to simplify the processes and data flows in an effort to maximize efficiency and return the most value to the organization.

Opportunities for the efficient use of information technology may also be recorded on the physical diagrams. If so, the BPR diagram becomes an input to systems analysis (described next).

Process Modeling during Systems Analysis

In systems analysis and in this chapter, we focus exclusively on *logical* process modeling as a part of business requirements analysis. In your information system framework, logical process models have a process focus and a SYSTEM OWNER and/or SYSTEM USER perspective. They are typically constructed as deliverables of the requirements analysis phase of a project. While logical process models are not concerned with implementation details or technology, they may be constructed (through *reverse engineering*) from existing application software, but this technology is much less mature and reliable than the corresponding reverse *data* engineering technology.

In the heyday of the original structured analysis methodologies, process modeling was also performed in the problem analysis phase of systems analysis. Analysts would build a *physical process model of the current system,* a *logical model of the current system,* and a *logical model of the target system.* Each model would be built top-down—from very general models to very detailed models. While conceptually sound, this approach led to modeling overkill and significant project delays, so much so that even structured techniques guru Ed Yourdon called it "analysis paralysis."

Today, most modern structured analysis strategies focus exclusively on the *logical model of the target system* being developed. Instead of being built either top-down or bottom-up, they are organized according to a commonsense strategy called event partitioning.

> **Event partitioning** factors a system into subsystems based on business events and responses to those events.

This strategy for event-driven process modeling is illustrated in Figure 8.17 and described as follows:

❶ A system **context data flow diagram** is constructed to establish *initial* project scope. This simple, one-page data flow diagram shows only the system's main interfaces with its environment.

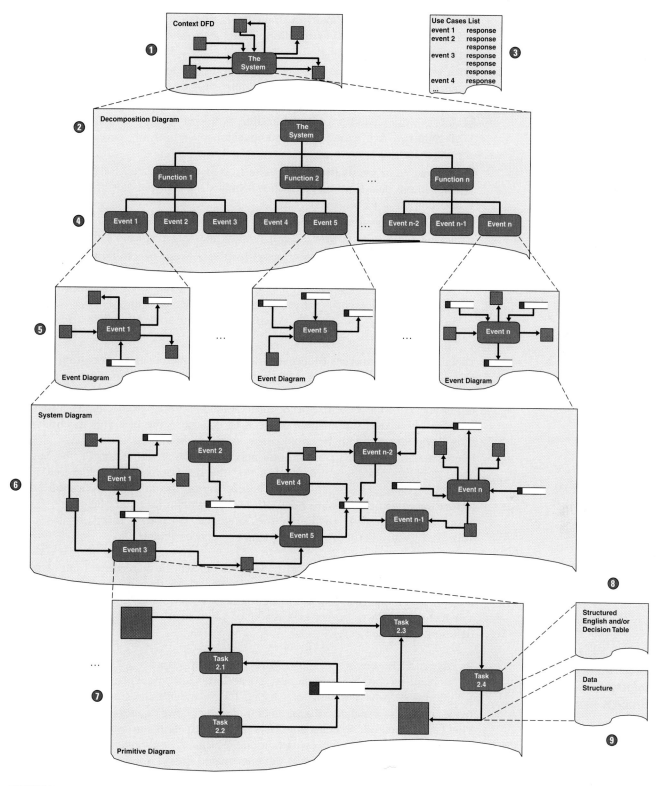

FIGURE 8.17   *Event-Driven Process Modeling Strategy*

❷A **functional decomposition diagram** is drawn to partition the system into logical subsystems and/or functions. (This step is omitted for very small systems.)

❸An **event-response** or **use case list** is compiled to identify and confirm the business events to which the system must provide a response. The list will also describe the required or possible responses to each event.

❹One process, called an **event handler,** is added to the decomposition diagram for each event. The decomposition diagram now serves as the outline for the system.

❺Optionally, an **event diagram** is constructed and validated for each event. This simple data flow diagram shows only the event handler and the inputs and outputs for each event.

❻One or more **system diagrams** are constructed by merging the event diagrams. These data flow diagrams show the "big picture" of the system.

❼**Primitive diagrams** are constructed for those event processes that require additional processing details. These data flow diagrams show all the elementary processes, data stores, and data flows for single events. ❽ The logic of each elementary process, and ❾ the data structure of each elementary data flow, is described using the tools described earlier in the chapter.

The above process models collectively document all the business processing requirements for a system. We'll demonstrate the technique in our SoundStage case study.

The logical process model from systems analysis describes business processing requirements of the system, not technical solutions. Recall from Chapter 5 that the purpose of the decision analysis phase is to determine the best way to implement those requirements with technology. In practice, this decision may have already been standardized as part of an application architecture. For example, the SoundStage application architecture requires that the development team first determine if an acceptable system can be purchased. If not, the current application architecture specifies that software built in-house be written in either Microsoft's *Visual Basic* or *Visual C++* (although the language, Microsoft *InterDev,* has been approved as an alternative for the Member Services Project).

Looking Ahead to Systems Design

During system design, the logical process model will be transformed into a physical process model (called an application schema) for the chosen technical architecture. This model will reflect the technical capabilities and limitations of the chosen technology. Any further discussion of physical process/application design is deferred until Chapter 11.

Fact-Finding and Information Gathering for Process Modeling

Process models cannot be constructed without appropriate facts and information as supplied by the user community. These facts can be collected by a number of techniques such as sampling of existing forms and files, research of similar systems, surveys of users and management, and interviews of users and management. The fastest method of collecting facts and information and simultaneously constructing and verifying the process models is *joint requirements planning (JRP)*. JRP uses a carefully facilitated group meeting to collect the facts, build the models, and verify the models—usually in one or two full-day sessions.

Fact-finding, information gathering, and JRP techniques were explored in Chapter 6.

Computer-Aided Systems Engineering (CASE) for Process Modeling

Like all system models, process models are stored in the repository. Computer-aided systems engineering (CASE) technology, introduced in Chapter 3, provides the repository for storing the process model and its detailed descriptions. Most CASE products support computer-assisted process modeling. Most support decomposition diagrams and data flow diagrams. Some support extensions for business process analysis and redesign.

Using a CASE product, you can easily create professional, readable process models without the use of paper, pencil, eraser, and templates. The models can be easily modified to reflect corrections and changes suggested by end-users. Also, most CASE products provide powerful analytical tools that can check your models for mechanical errors, completeness, and consistency. Some CASE products can even help you analyze the data model for consistency, completeness, and flexibility. The potential time savings and quality are substantial.

CASE tools do have their limitations. Not all process model conventions are supported by all CASE products. Therefore, any given CASE product may force the company to adapt its methodology's process modeling symbols or approach so that it is workable within the limitations of its CASE tool.

All the SoundStage process models in the next section of this chapter were created with Popkin's CASE tool, *System Architect 2001*. For the case study, we provide you the printouts exactly as they came off our printers. We did not add color. The only modifications by the artist were the bullets that call your attention to specific items of interest on the printouts. All the processes, data flows, data stores, and boundaries on the SoundStage process models were automatically cataloged into *System Architect*'s project repository (which it calls an encyclopedia). Figure 8.18 illustrates some of *System Architect*'s screens as used for data modeling.

FIGURE 8.18   *CASE for Process Modeling (using* System Architect 2001 *by Popkin Software & Systems)*

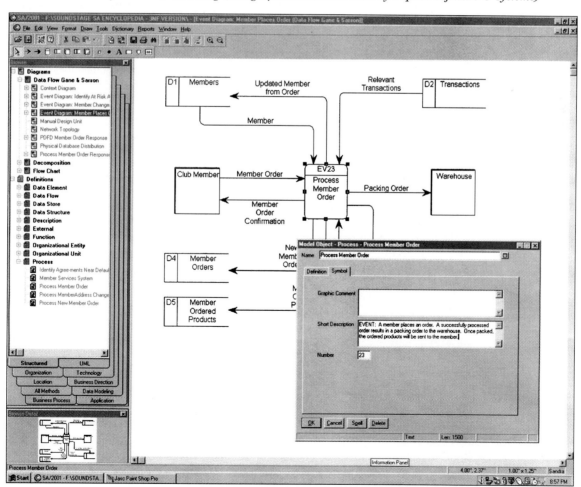

HOW TO CONSTRUCT PROCESS MODELS

As a systems analyst or knowledgeable end-user, you must learn how to draw decomposition and data flow diagrams to model business process requirements. We will use the SoundStage Entertainment Club project to teach you how to draw these process models.

Let's assume the preliminary investigation and problem analysis phases of the project have been completed and the project team understands the current system's strengths, weaknesses, limitations, problems, opportunities, and constraints. The team has also already built the data model (in Chapter 7) to document business data requirements for the new system. Team members will now build the corresponding process models.

The Context Data Flow Diagram

First, we need to document the initial project scope. All projects have scope. A project's scope defines what aspect of the business a system or application is supposed to support. A project's scope also defines how the system being modeled must interact with other systems and the business as a whole. In your information system framework, scope is defined as the INTERFACE focus from the SYSTEM OWNERS' perspective. It is documented with a context data flow diagram.

> A **context data flow diagram** [DeMarco, 1978] defines the scope and boundary for the system and project. Because the scope of any project is always subject to change, the context diagram is also subject to constant change. A synonym is *environmental model* [Yourdon, 1990].

We suggest the following strategy for documenting the system's boundary and scope:

1. Think of the system as a container in order to distinguish the inside from the outside. Ignore the inner workings of the container. This is sometimes called "black box" thinking.
2. Ask your end-users what business transactions a system must respond to. These are the *net inputs* to the system. For each net input, determine its source. Sources will become *external agents* on the context data flow diagram.
3. Ask your end-users what responses must be produced by the system. These are the net outputs to the system. For each net output, determine its destination. Destinations will also become *external agents*. Requirements for reports and queries can quickly clutter the diagram. Consider consolidating them into composite data flows.
4. Identify any *external* data stores. Many systems require access to the files or databases of other systems. They may use the data in those files or databases. Sometimes they may update certain data in those files and databases. But generally, they are not permitted to change the structure of those files and databases—therefore, they are outside of the project scope.
5. Draw your context diagram from all of the preceding information.

If you try to include all the inputs and outputs between a system and the rest of the business and outside world, a typical context data flow diagram might show as many as 50 or more data flows. Such a diagram would have little, if any, communication value. Therefore, we suggest you show only those data flows that represent the main objective or most important inputs and outputs of the system. Defer less common data flows to more detailed DFDs to be drawn later.

The context data flow diagram contains one and only one process (see Figure 8.19). Sometimes, this process is identified by the number "0"; however, our CASE tool did not allow this. External agents are drawn around the perimeter. Data flows define the interactions of your system with the boundaries and with the external data stores.

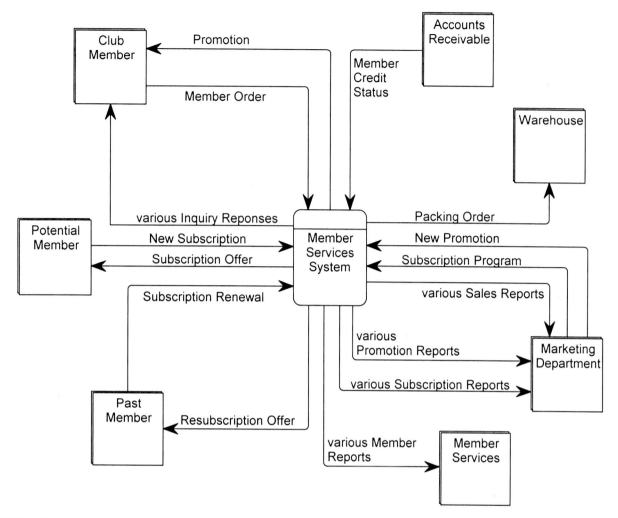

FIGURE 8.19   *The Context Data Flow Diagram (created with* System Architect 2001*)*

As shown in the context data flow diagram, the main purpose of our system is to process NEW SUBSCRIPTIONS in response to SUBSCRIPTION OFFERS, create NEW PROMOTIONS for products, and respond to MEMBER ORDERS by sending PACKING ORDERS to the warehouse to be filled. (Notice that we made all data flow names singular.) Management has also emphasized the need for VARIOUS REPORTS. Finally, the Web extensions to this system require that the system provide members with VARIOUS INQUIRY RESPONSES regarding orders and accounts.

Recall that a decomposition diagram shows the top-down functional decomposition or structure of a system. It also provides us with the beginnings of an outline for drawing our data flow diagrams.

The Functional Decomposition Diagram

Figure 8.20 is the functional decomposition diagram for the SoundStage project. Let's study this diagram. First, notice that the processes are depicted as rectangles, not rounded rectangles. This is merely a limitation of our CASE tool's implementation of decomposition diagrams—you also may have to adapt to your CASE tool.

In many decomposition and data flow diagrams, the processes are not only named, but they also are numbered as part of an identification scheme that uses the following guidelines:

1. The root process MEMBER SERVICES SYSTEM would be numbered 0.

2. The three subsystems would be numbered 1, 2, and 3.

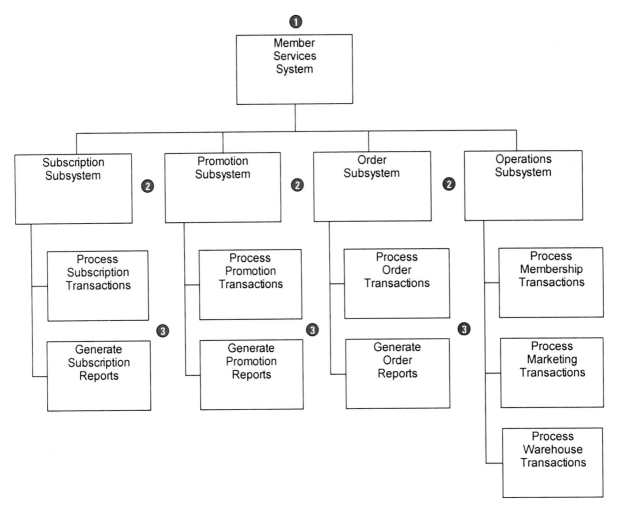

FIGURE 8.20 *A Functional Decomposition Diagram (created with* System Architect 2001*)*

3. The subfunctions of the MEMBER SERVICES SUBSYSTEM would be numbered 1.1 and 1.2.

4. If 1.1 is factored into specific transactions, they would be numbered 1.1.1, 1.1.2, 1.1.3, and so forth. Similarly, specific reports would be numbered 1.2.1, 1.2.2, 1.2.3, and so forth.

Although this scheme is well documented in books, we elected *not* to use the numbers. Readers and users repeatedly misinterpret them as sequential processes and the numbers were never meant to imply sequence. Furthermore, a numbering scheme discouraged us from later reorganizing the system and diagram because no one wants to change all those ID numbers.

The following is an item-by-item discussion of the decomposition diagram. The circled numbers correspond to specific points of interest on the diagram.

1 The root process corresponds to the entire system.

2 The system is initially factored into subsystems and/or functions. These subsystems and functions do not necessarily correspond to organization units on an organization chart. Increasingly, analysts and users are being asked to ignore organizational boundaries and to build cross-functional systems that streamline processing and data sharing.

3 We like to separate the operational and reporting aspects of a system. Thus, we factored each subsystem accordingly. Later, if this structure doesn't make sense, we can change it.

Larger systems might have first been factored into subsystems *and* functions. There is no limit to the number of child processes for a parent process. Many authors used to recommend a maximum of five to nine processes per parent, but any such limit is too artificial. Instead, structure the system such that it makes sense for the business!

Factoring a parent process into a single child process doesn't make sense. It would provide no additional detail. Therefore, if you plan to factor a process, it should be factored into at least two child processes.

The next step is to determine what business events the system must respond to and what responses are appropriate. Events are not hard to find. Some of the inputs on the context diagram are associated with events. But the context diagram rarely shows all the events. Essentially, there are three types of events.

The Event-Response or Use Case List

- **External events** are so named because they are initiated by external agents. When these events happen, an input data flow occurs for the system. For example, the event CUSTOMER PLACES A NEW ORDER is recognized in the form of the input data flow ORDER from the external agent CUSTOMER.
- **Temporal events** trigger processes on the basis of time, or something that merely happens. When these events happen, an input *control flow* occurs. Examples of temporal events might include TIME TO REMIND CUSTOMERS TO PAY PAST INVOICES or END OF MONTH.
- **State events** trigger processes based on a system's change from one state or condition to another. Like temporal events, state events will be illustrated as an input *control flow*.

Information systems usually respond mostly to external and temporal events. State events are usually associated with real-time systems such as elevator or robot control.

One of the more popular and successful approaches for finding and identifying events and responses is a technique called *use cases* developed by Dr. Ivar Jacobson. This technique is rooted in object-oriented analysis but is easily adapted to structured analysis and data flow diagramming.

> **Use case** analysis is the process of identifying and modeling business events, who initiated them, and how the system responds to them.

Use cases provide a method for breaking down the entire scope of system functionality into many smaller statements of system functionality called use cases. Use cases identify and describe necessary system processes from the perspective of users. Each use case is initiated by users or external systems called *actors*.

> An **actor** is anything that needs to interact with the system to exchange information. To the system an actor could be a customer, user, department, organization, or another system.

An actor initiates processes. An actor is not meant to portray a single individual or job title.

Use cases for a system can be discovered in a number of ways. First, we need to identify the actors. The context data flow diagram identifies the key actors as *external agents*. It also identifies *some* of the use cases. The key word is *some*. Recall that the context diagram shows only the main inputs and outputs of a system. There are almost always more inputs and outputs than are depicted—usually many more. Some of the inputs and outputs depicted are really composites of many types and variations on those inputs and outputs (e.g., the "various reports" on our context diagram). Also, the context diagram may not illustrate the many exception inputs and outputs such as errors, inquiries, and follow-ups.

One way to expand the use cases is to interview the external agents (actors) depicted on the diagram. The agents can (1) identify the events (use cases) for which they believe the system may have to provide a response and (2) identify other actors (new external agents) that were not originally shown on the context diagram.

Another way to identify use cases (events) is to study the data model, assuming a data model was developed before drawing data flow diagrams. (Data modeling was covered in Chapter 7.) Essentially, you study the life history of the each entity on that data model. Instances of these entities must be created, updated, and eventually deleted. Events or use cases trigger these actions on the entity. It is not difficult to get users talking about the events that could create, update, and delete entity instances. After all, they live these events daily. We used this approach to build the use case list for the SoundStage project.

A partial table of use cases is illustrated in Figure 8.21 (pp. 342–43). For each use case, we list:

- The actor that initiates the event (which will become an external agent on our DFDs).
- The event (which will be handled by a process on our DFDs).
- The input or trigger (which will become a data or control flow on our DFDs).
- All outputs and responses (which will also become data flows on our DFDs). Notice that we used parentheses to denote temporal events.
- Outputs (but be careful not to imply implementation)—by the way, when we used the term *report* we were not necessarily implying a paper-based document. Notice that our responses include changes to stored data about entities from the data model. These include create new instances of the entity, update existing instances of the entity, and delete instances of the entity.

The number of use cases for a system is usually quite large. This is necessary to ensure that the system designers build a complete system that will respond to all the business events. As a final step, consider assigning each event to one of the subsystems and functions identified in your decomposition diagram (drawn in the previous step).

The complete set of use cases for the SoundStage project may be found in the McGraw-Hill Online Learning Center for this book.

Event Decomposition Diagrams

Now we can further partition our functions in the decomposition diagram. We simply add event handling processes (one per use case) to the decomposition (see Figure 8.22 on page 344.) If the entire decomposition diagram will not fit on a single page, add separate pages for subsystems or functions. The root process on a subsequent page should be duplicated from an earlier page to provide a cross-reference. Figure 8.22 shows only the event processes for the MEMBERSHIPS subsystem. Events for the PROMOTIONS and ORDERS functions would be on separate pages.

There is no need to factor the decomposition diagram beyond the events and reports. That would be like outlining down to the final paragraphs or sentences in a paper. The decomposition diagram, as constructed, will serve as a good outline for the later data flow diagrams.

Event Diagrams

This is an optional, but useful step. Using our decomposition diagram as an outline, we can draw one event diagram for each event process.

An **event diagram** is a context diagram for a single event. It shows the inputs, outputs, and data store interactions for the event.

Event diagrams are easy to draw. More importantly, by drawing an event diagram for each process, users do not become overwhelmed by the overall size of the system. They can examine each use case as its own context diagram.

Before drawing any event diagrams, it can be helpful to have a list of all the data stores available (see margin list). Because SoundStage already completed the data model for this project (Chapter 5), team members simply created a list of the plural for each entity name on that data model. It is useful to review the definition and attributes for each entity/data store on the list.

Most event diagrams contain a single process—the same process that was named to handle the event on the decomposition diagram. For each event, illustrate the following:

— The inputs and their sources. Sources are depicted as external agents. The data structure for each input should be recorded in the repository.

— The outputs and their destinations. Destinations are depicted as external agents. The data structure for each output should be recorded in the repository.

— Any data stores from which records must be "read" should be added to the event diagram. Data flows should be added and named to reflect what data is read by the process.

— Any data stores in which records must be created, deleted, or updated should be included in the event diagram. Data flows to the data stores should be named to reflect the nature of the update.

The sensibility and simplicity of event diagramming makes the technique a powerful communication tool between users and technical professionals!

A complete set of event diagrams for the SoundStage case study would double the length of this chapter without adding substantive educational value. Thus, we will demonstrate the model with three simple examples.

Figure 8.23 on page 345 illustrates a simple event diagram for an external event. Most systems have many such simple event diagrams because all systems must provide for routine maintenance of data stores.

Figure 8.24 on page 345 depicts a somewhat more complex external event, one for the business transaction MEMBER ORDER RESPONSE. Notice that business transactions tend to use and update more data stores and have more interactions with external agents.

Can an event diagram have more than one process on it? The answer is maybe. Some event processes may trigger other event processes. In this case, the combination of events should be shown on a single event diagram. In our experience, most event diagrams have one process. An occasional event diagram may have two or perhaps three processes. If the number of processes exceeds three, you are probably drawing what is called an activity diagram (prematurely), not an event diagram—in other words, you're getting too involved with details. Most event processes do not directly communicate with one another. Instead, they communicate across shared data stores. This allows each event process to do its job without worrying about other processes keeping up.

Figure 8.25 on page 346 shows an event diagram for a temporal event. We added an external entity CALENDAR or TIME to serve as a source for this control flow.

Each event process should be described to the CASE repository with the following properties:

— Event sentence—for business perspective.

— Throughput requirements—the volume of inputs per some time period.

— Response time requirements—how fast the typical event must be handled.

— Security, audit, and control requirements.

— Archival requirements (from a business perspective).

Data Stores (Entities)

AGREEMENTS

MEMBERS

MEMBER ORDERS

MEMBER ORDERED PRODUCTS

PRODUCTS

PROMOTIONS

TITLE PROMOTIONS

Actor	Event (or Use Case)	Trigger	Responses
Marketing	Establishes a new membership subscription plan to entice new members.	NEW MEMBER SUBSCRIPTION PROGRAM	Generate SUBSCRIPTION PLAN CONFIRMATION. Create AGREEMENT in the database.
Marketing	Establishes a new membership resubscription plan to lure back former members.	PAST MEMBER RESUBSCRIPTION PROGRAM	Generate SUBSCRIPTION PLAN CONFIRMATION. Create AGREEMENT in the database.
Marketing	Changes a subscription plan for current members (e.g., extending the fulfillment period)	SUBSCRIPTION PLAN CHANGE.	Generate AGREEMENT CHANGE CONFIRMATION. Update AGREEMENT in the database.
(time)	A subscription plan expires.	(current date)	Generate AGREEMENT CHANGE CONFIRMATION. Logically Delete (void) AGREEMENT in the database.
Marketing	Cancels a subscription plan before its planned expiration date.	SUBSCRIPTION PLAN CANCELATION	Generate AGREEMENT CHANGE CONFIRMATION. Logically Delete (void) AGREEMENT in the database.
Member	Joins the club by subscribing. ("Take any 12 CDs for one penny and agree to buy 4 more at regular prices within two years.")	NEW SUBSCRIPTION	Generate MEMBER DIRECTORY UPDATE CONFIRMATION. Create MEMBER in the database. Create first MEMBER ORDER and MEMBER ORDERED PRODUCTS in the database.
Member	Changes address (including email and privacy code)	CHANGE OF ADDRESS	Generate MEMBER DIRECTORY UPDATE CONFIRMATION. Update MEMBER in the database.
Accounts Receivable	Changes member's credit status	CHANGE OF CREDIT STATUS	Generate CREDIT DIRECTORY UPDATE CONFIRMATION. Update MEMBER in the database.
(time)	90 days after Marketing decides to no longer sell a product.	(current date)	Generate CATALOG CHANGE CONFIRMATION. Logically Delete (deactivate) PRODUCT in the database.

FIGURE 8.21   *A Partial Use Case Table*

Actor	Event (or Use Case)	Trigger	Responses
Member	Wants to pick products for possible purcase. (Logical requirement is driven by vision of Web-based access to information.)	PRODUCT INQUIRY	Generate CATALOG DIESCRIPTION.
Member	Places order.	NEW MEMBER ORDER	Generate MEMBER ORDER CONFIRMATION. Create MEMBER ORDER and MEMBER ORDERED PRODUCTS in the database.
Member	Revises order.	MEMBER ORDER CHANGE REQUEST	Generate MEMBER ORDER CONFIRMATION. Update MEMBER ORDER and/or MEMBER ORDERED PRODUCTS in the database.
Member	Cancels order.	MEMBER ORDER CANCELLATION	Generate MEMBER ORDER CONFIRMATION. Logically Delete MEMBER ORDER and MEMBER ORDERED PRODUCTS in the database.
(time)	90 days after the order.	(current date)	Physically Delete MEMBER ORDER and MEMBER ORDERED PRODUCTS in the database.
Member	Inquires about his/her purchase history (three year time limit)	MEMBER PURCHASE INQUIRY	Generate MEMBER PURCHASE HISTORY
(each) Club	(end of month)	(current date)	Generate MONTHLY SALES ANALYSIS. Generate MONTHLY MEMBER AGREEMENT EXCEPTION ANALYSIS. Generate MEMBERSHIP ANALYSIS REPORT.

FIGURE 8.21   *(Concluded)*

For example, consider the event diagram in Figure 8.23: "A member submits a change of address."

- Occurs 25 times per month.
- Should be processed within 15 days.
- Must protect privacy of addresses unless the member authorizes release.
- Should retain a semipermanent record of some type.

All the above properties can be added to the descriptions associated with the appropriate processes, data flows, and data stores on the model.

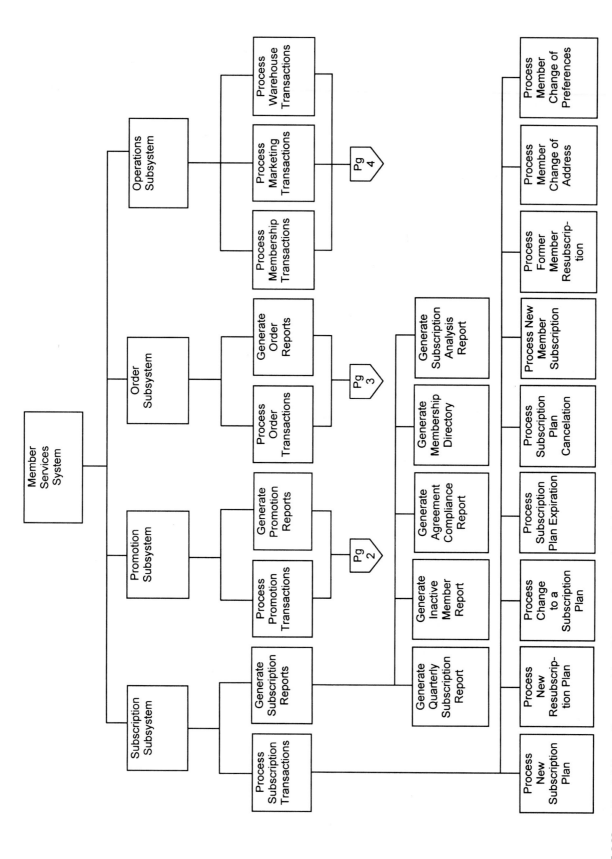

FIGURE 8.22 *A Partial Event Decomposition Diagram (created with System Architect 2001)*

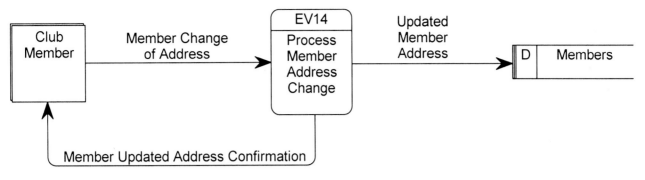

FIGURE 8.23 *A Simple External Event Diagram (created with* System Architect 2001*)*

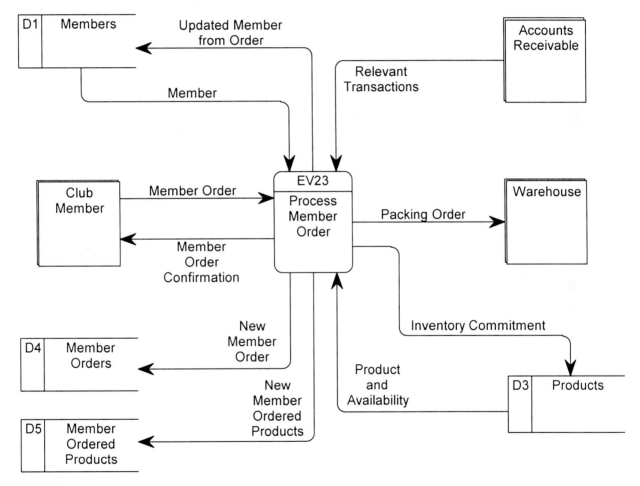

FIGURE 8.24 *A More Complex External Event Diagram (created with* System Architect 2001*)*

The event diagrams serve as a meaningful context for users to validate the accuracy of each event to which the system must provide a response. But these events do not exist in isolation. They collectively define systems and subsystems. It is, therefore, useful to construct one or more system diagrams that show all the events in the system or a subsystem.

The system diagram is said to be "exploded" from the single process that we created on the original context diagram (Figure 8.19). The system diagram shows either (1) all the events for the system on a single diagram or (2) all the events

The System Diagram(s)

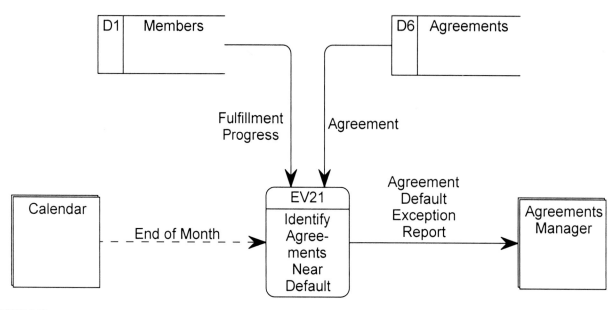

FIGURE 8.25   *A Temporal Event Diagram (created with* System Architect 2001*)*

for a single subsystem on a single diagram. Depending on the size of the system, a single diagram may be too large.

Our SoundStage project is moderate in size, but it still responds to too many events to squeeze all those processes onto a single diagram. Instead, Bob Martinez elected to draw a subsystem diagram for each of the major subsystems. Figure 8.26 (pp. 348–49) shows the subsystem diagram for the ORDERS SUBSYSTEM. It consolidates all the transaction and report-writing events for that subsystem onto a single diagram. (The reporting events may be omitted or consolidated into composites if the diagram is too cluttered.) Notice that the system diagram demonstrates how event processes truly communicate—across and using shared data stores.

If necessary, and after drawing the four subsystem diagrams for this project, Bob could have drawn a system diagram that illustrates only the interactions between those four subsystems. This is a relic of the original top-down data flow diagramming strategy of the original structured analysis methodology. In practice, this higher-level diagram requires so much consolidation of data flows and data stores that its communication value is questionable. To us (and to Bob) it was busywork, and his time was better spent on the next set of data flow diagrams. Before we leave this topic, we should introduce the concept of balancing.

> **Balancing** is the synchronizing of data flow diagrams at different levels of detail to preserve consistency and completeness of the models. Balancing is a quality assurance technique.

Balancing requires that, if you explode a process to another DFD to reveal more detail, you must include the same data flows and data stores on the child diagram that you included in the parent diagram's original process (or their logical equivalents).

We now have a set of event diagrams (one per business event) and one or more system/subsystem diagrams. The event diagram processes are merged into the system diagrams. It is very important that each of the data flows, data stores, and external agents that were illustrated on the event diagrams be represented on the system diagrams. Most CASE tools include facilities to check for balancing errors. Notice that we duplicate data stores and external agents to minimize crossing of lines.

We're almost done! Some event processes on the system diagram may be exploded into a primitive data flow diagram to reveal more detail. This is especially true of the more complex business transaction processes (e.g., order processing). Other events, such as generation of reports, are simple enough that they do not require further explosion.

Event processes with more complex event diagrams should be exploded into a more detailed, primitive data flow diagram such as that illustrated in Figure 8.27 on page 350. This primitive DFD shows detailed processing requirements for the event. This DFD shows several elementary processes *for* the event process. Each elementary process is cohesive—that is, it does only one thing.

When Bob drew this primitive data flow diagram, he had to add new data flows between the processes. In doing so, he tried to practice good data conservation, making sure each process got only the data it truly needed. The data structure for each data flow had to be described in his CASE tool's repository. Also notice that he used data flow junctions to split and merge appropriate data flows on the diagram.

Convince yourself that the incoming and outgoing data flows are logically balanced when compared to Figure 8.26. Note that the primitive DFD contains some new exception data flows that were *not* introduced in Figure 8.26. It should not be hard to imagine a program structure when examining this DFD.

The combination of the context diagram, system diagram, event diagrams, and primitive diagrams completes our process models. Collectively, this *is* the process model. A well-crafted and complete process model can effectively communicate business requirements between end-users and computer programmers, eliminating much of the confusion that often occurs in system design, programming, and implementation!

The data flow diagrams are complete. Where do you go from here? That depends on your choice of methodology. If you are practicing the pure structured analysis methodology (from which data flow diagramming was derived), you must complete the specification. To do so, each data flow, data store, and *elementary* process (meaning one that is not further exploded into a more detailed DFD) must be described to the encyclopedia or data dictionary. CASE tools provide facilities for such descriptions.

Data flows are described by data structures as explained earlier in this chapter. Figure 8.28 (p. 351) demonstrates how *System Architect 2001,* the SoundStage CASE tool, can be used to describe a data flow. Notice that this CASE tool uses an algebraic notation for data structures as described in this chapter. Ultimately, each data element or attribute should also be described in the data dictionary to specify data type, domain, and default value (as was described in Chapter 7). Data stores correspond to all instances of a data entity from our data model. Thus, they are best described in the data dictionary that corresponds to each entity and its attributes as was taught in Chapter 7. Some analysts like to translate each data store's content into a relational data structure similar to that used in Figure 8.28 to describe data flows. We consider this to be busywork—let the entity descriptions from the data model describe the contents of a data store. Besides, defining data structures for data stores could lead to synchronization errors between the data and process models—if you would make any changes to an entity in the data model, you would be forced to remember to make those same changes in the corresponding data store's data structure. This requires too much effort (unless you have a CASE tool capable of doing it automatically for you).

Elementary processes can be described by Structured English and/or decision tables as taught earlier in the chapter. Because they are "elementary," they should be described in one page or less of either tool. Figure 8.29 on page 352 demonstrates how *System Architect 2001* can be used to describe an elementary process.

Primitive Diagrams

A complete set of exploded DFDs and event diagrams for SoundStage can be found at the McGraw-Hill Online Learning Center.

Completing the Specification

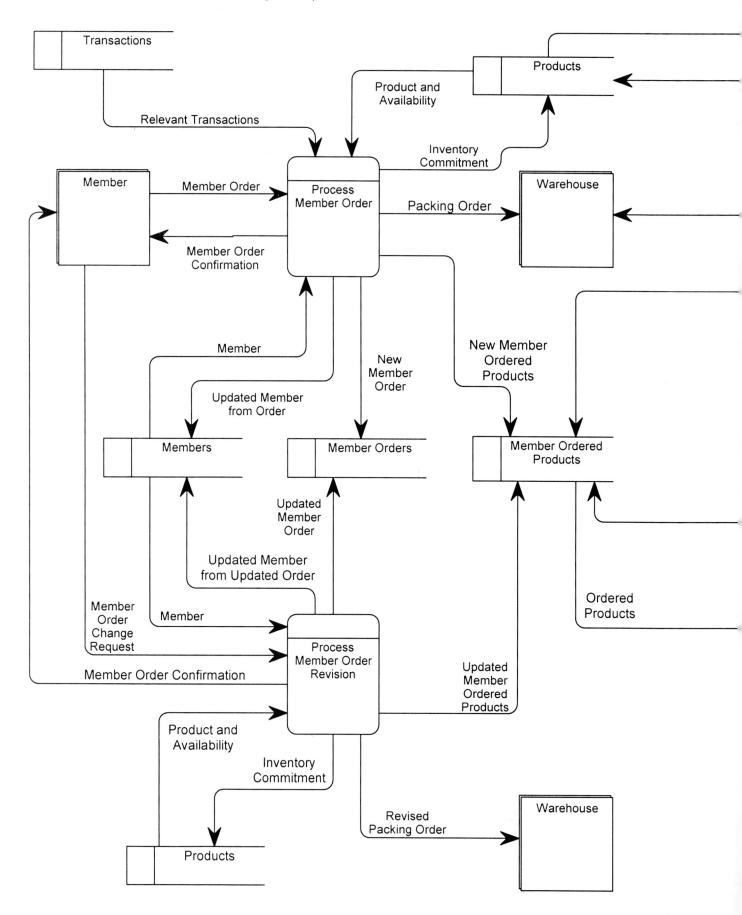

FIGURE 8.26 *A System Diagram (created with* System Architect 2001*)*

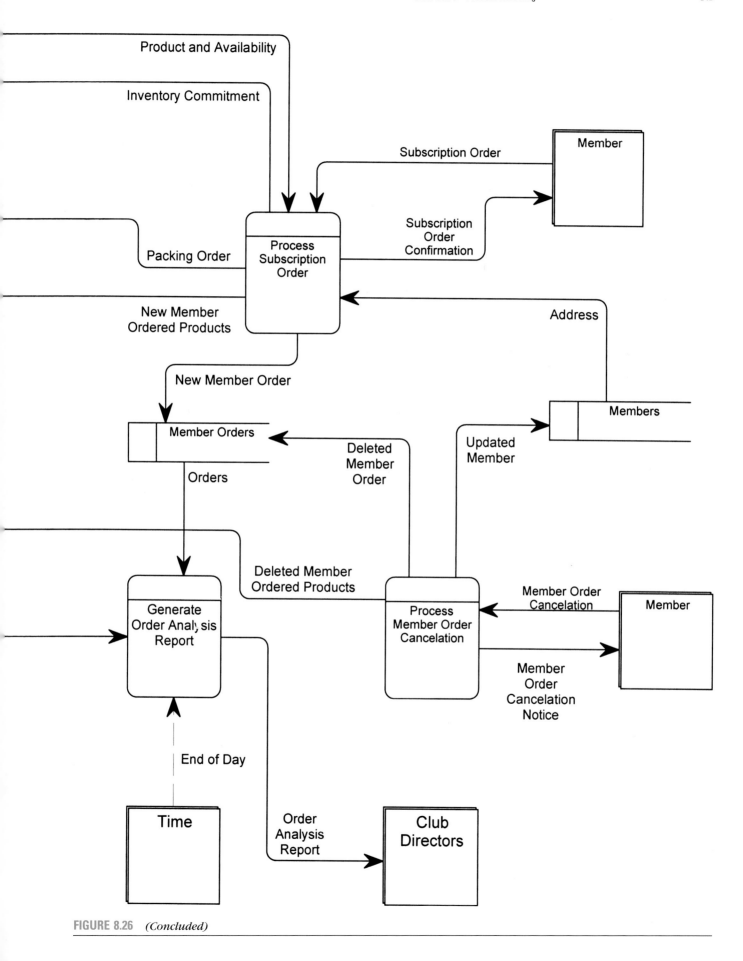

FIGURE 8.26   *(Concluded)*

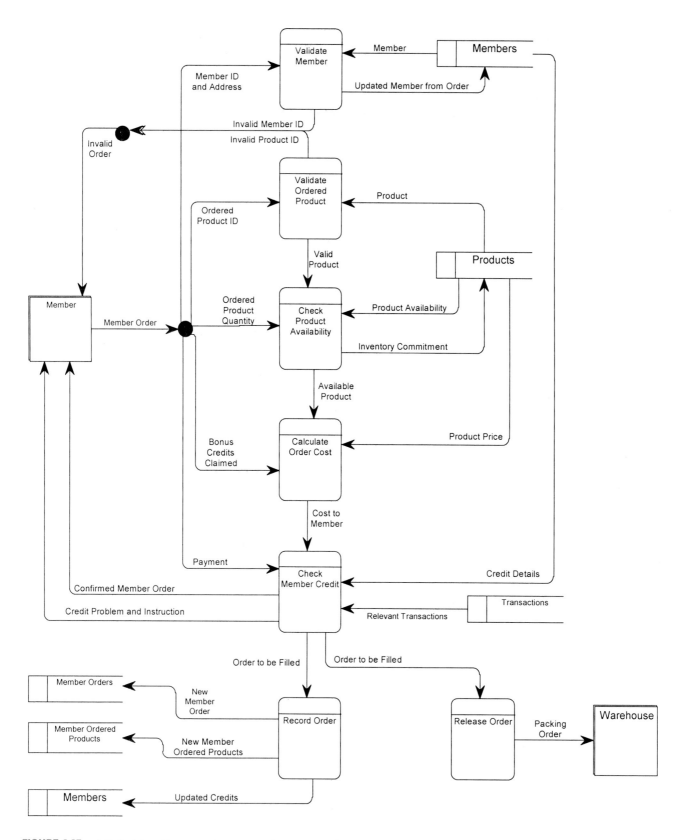

FIGURE 8.27   *A Primitive Diagram (created with System Architect 2001)*

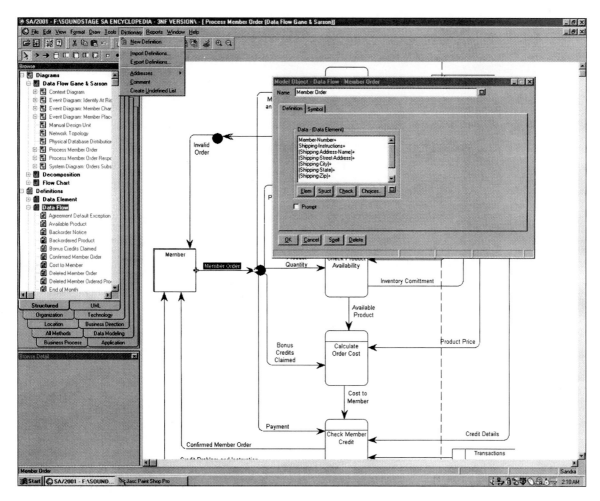

FIGURE 8.28   *A Data Flow (created with* System Architect 2001)

Unfortunately, *System Architect,* like many CASE tools, does not support decision table construction. Fortunately, decision tables are easily constructed using the table features in most word processors and spreadsheets.

Data and process models represent different views of the same system. But these views are interrelated. Modelers need to synchronize the different views to ensure consistency and completeness of the total system specification. In this section, we'll review the basic synchronization concepts for data and process models.

The linkage between data and process models is almost universally accepted by all major methodologies. In short, there should be one data store in the process models for each entity in the data model. Some methodologies exempt associative entities from this requirement, but we believe it is simpler (and more consistent) to apply the rule to all entities on the data model.

Figure 8.30 on page 353 illustrates a typical *data-to-process-CRUD matrix.* The decision to include or not include attributes is based on whether processes need to be restricted as to which attributes they can access.

SYNCHRONIZING OF SYSTEM MODELS

Data and Process Model Synchronization

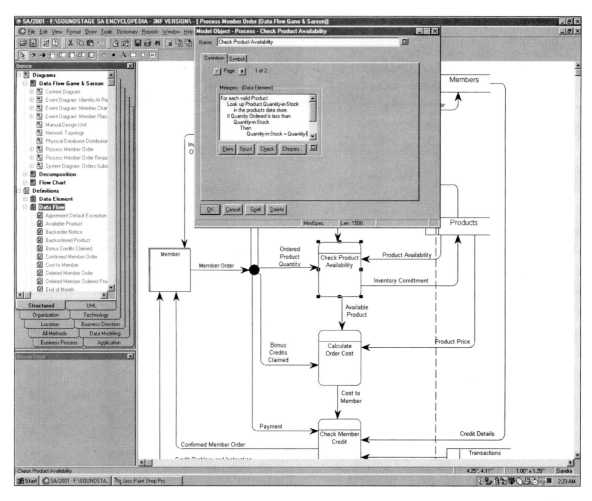

FIGURE 8.29   *An Elementary Process (created with* System Architect 2001*)*

The synchronization quality check is stated as follows:

Every entity should have *at least* one C, one R, one U, and one D entry for system completeness. If not, one or more event processes were probably omitted from the process models. More importantly, users and management should validate that all possible creates, reads, updates, and deletes have been included.

The matrix provides a simple quality check that is simpler to read than either the data or process models. Of course, any errors and omissions should be recorded both on the matrix and in the corresponding data and process models to ensure proper synchronization.

Process Distribution

Process models illustrate the essential work to be performed by the system as a whole. But processes must be distributed to locations where work is to be performed. Some work may be unique to one location. Other work may be performed at multiple locations. Before we design the information system, we should identify and document what processes must be performed at which locations. This can be accomplished through a *process-to-location-association matrix*.

A **process-to-location-association matrix** is a table in which the rows indicate processes (event or elementary processes); the columns indicate locations; and the cells (the intersection of rows and columns) document which processes must be performed at which locations.

Data-to-Process-CRUD Matrix

Entity . Attribute	Process Customer Application	Process Customer Credit Application	Process Customer Change of Address	Process Internal Customer Credit Change	Process New Customer Order	Process Customer Order Cancellation	Process Customer Change to Outstanding Order	Process Internal Change to Customer Order	Process New Product Addition	Process Product Withdrawal from Market	Process Product Price Change	Process Change to Product Specification	Process Product Inventory Adjustment
Customer	C	C			R	R	R	R					
.Customer Number	C	C			R	R	R	R					
.Customer Name	C	C	U		R		R	R					
.Customer Address	C	C	U		RU		RU	RU					
.Customer Credit Rating		C		U	R		R	R					
.Customer Balance Due					RU	U	R	R					
Order					C	D	RU	RU					
.Order Number					C		R	R					
.Order Date					C		U	U					
.Order Amount					C		U	U					
Ordered Product					C	D	CRUD	CRUD		RU			
.Quantity Ordered					C		CRUD	CRUD					
.Ordered Item Unit Price					C		CRUD	CRUD					
Product					R	R	R	R	C	D	RU	RU	RU
.Product Number					R	R	R	R	C			R	
.Product Name					R		R	R	C			RU	
.Product Description					R		R	R	C			RU	
.Product Unit of Measure					R		R	R	C		RU	RU	
.Product Current Unit Price					R		R	R			U		
.Product Quantity on Hand					RU	U	RU	RU					RU

C = create   R = read   U = update   D = delete

FIGURE 8.30   *Sample Data-to-Process-CRUD Matrix*

Figure 8.31 illustrates a typical *process-to-location-association matrix*. Once validated for accuracy, the system designer will use this matrix to determine which processes should be implemented centrally or locally.

Some methodologies and CASE tools may support views of the process model that are appropriate to a location. If so, these views (subsets of the process models) must be kept in sync with the master process models of the system as a whole.

THE NEXT GENERATION

The demand for process modeling skills remains strong. Even though object modeling (Module A) is becoming a new standard in the industry, process modeling skills remain valuable for two reasons:

- The current interest in **business process redesign** requires process models.
- Process models are included in many object modeling strategies such as the *Unified Modeling Language* (UML). Note the object strategy process models are not necessarily DFDs. For example, the UML includes a process-like model called an *activity diagram*.

With respect to the former, business process redesign emphasizes **physical process modeling.** Physical process models include those processes that reflect the current implementation. This may include sequential processes that merely edit, route, copy, or approve a data flow. Physical data flow diagrams also include additional details such as who or what performs each process, the cost of each process, and a critical evaluation of the value returned by each process. Logical data flow diagrams may also be used to optimize redesigned physical business processes.

Leading-edge methodologies are embracing object modeling and *object-oriented analysis* (Part Five, Module A) as the heir apparent for current data and process models. (Actually, and at the risk of oversimplification, object modeling integrates elements of both data and process modeling.) But object modeling and object-oriented analysis techniques are still young and evolving. Also, many structured analysis practitioners are only beginning to practice the more contemporary, event-driven approach you just learned. Thus, process modeling will remain a viable skill for many years.

We look for CASE tools to continue to improve. There is still some interest in CASE tools capable of generating program code directly from process models. Early code generators have experienced mixed results, but if process modeling languages continue to improve in their precision and completeness, we may yet realize the vision of generating "software from pictures." Another aspect of CASE to keep your eyes on is **reverse engineering.** Through reverse engineering, CASE tools read the source code of old programs (such as COBOL) to automatically generate physical process models that can then be *forward engineered* into better systems and programs, possibly generated in other languages.

FIGURE 8.31   *Sample Process-to-Location-Association Matrix*

Process-to-Location-Association Matrix

Process	Customers	Kansas City	Marketing	Advertising	Warehouse	Sales	Accounts Receivable	Boston	Sales	Warehouse	San Francisco	Sales	San Diego	Warehouse
Process Customer Application	X					X			X			X		
Process Customer Credit Application	X						X							
Process Customer Change of Address	X					X			X			X		
Process Internal Customer Credit Change							X							
Process New Customer Order	X					X			X			X		
Process Customer Order Cancellation	X					X			X			X		
Process Customer Change to Outstanding Order	X					X			X			X		
Process Internal Change to Customer Order						X			X			X		
Process New Product Addition			X											
Process Product Withdrawal from Market			X											
Process Product Price Change			X											
Process Change to Product Specification			X	X										
Process Product Inventory Adjustment					X					X				X

WHERE DO YOU GO FROM HERE?

Most of you will proceed directly to Chapter 9, Feasibility Analysis and System Proposal. Given data and process models that describe *logical* system requirements, Chapter 9 will examine the methods and techniques for identifying candidate *physical* solutions that will fulfill the logical requirements, techniques for analyzing the feasibility of each of those solutions, and approaches for presenting the solution that you deem most feasible. This system proposal culminates systems analysis in our FAST methodology.

At your instructor's discretion, some of you may jump to Part Five, Module A, Object-Oriented Modeling and Analysis. This alternative to data and process modeling is rapidly gaining acceptance in industry and may eventually make both data and process modeling obsolete. Of course, object modeling presents many parallels with data and process modeling. You'll see some similar constructs because an object is defined as the encapsulation of related data and the processes that will be allowed to create, read, update, and use that data.

In previous editions of the systems analysis unit we included a chapter on network modeling. It was removed from this edition as part of a downsizing effort. While network modeling is not as broadly applied as data, process, and object modeling, it is no less relevant. Today's information systems are increasingly distributed across networks, including intranets and the Internet. As with data and process modeling, there is a logical network (geography) that should be considered before distributing data and processes (or objects) across the network. We have updated and provided the former network modeling chapter as a supplemental reading module available at the McGraw-Hill Online Learning Center.

If you are interested in how we will use the logical DFDs in this chapter during systems design, you might want to preview Chapter 11, Application Architecture and Process Design. In that chapter, we will teach you how to transform the logical data flow diagrams into physical data flow diagrams that model the technology architecture of a system to be designed and implemented.

SUMMARY

1. A model is a representation of reality. We construct logical models to better understand business problem domains and business requirements. Eventually, logical models will be transformed into physical models to reflect design and implementation decisions.

2. Process modeling is a technique for organizing and documenting the process requirements and design for a system. This chapter focused on a popular process model called a data flow diagram, which depicts the flow of data through a system's processes.

3. Data flow diagramming is a technique for systems thinking.

4. In reality, a system *is* a process. A process is work performed on, or in response to, inputs and conditions.

5. Just as systems can be recursively decomposed into subsystems, processes can be recursively decomposed into subprocesses. A decomposition diagram shows the functional decomposition of a system into processes and subprocesses. It is a planning tool for subsequent data flow diagrams.

6. Logical processes show essential work to be performed by a system without showing how the processes will be implemented. There are three types of logical processes: functions (very high level), events (middle level of detail), and elementary processes (very detailed).

7. Elementary processes are further described by procedural logic. Structured English is a tool for expressing this procedural logic. Structured English is a derivative of structured programming logic constructs married to natural English.

8. More complex elementary processes may be described by policies that are expressed in decision tables. Decision tables show complex combinations of conditions that result in specific actions.

9. Data flows are the inputs to and the outputs from processes. They also illustrate data store accesses and updates.

10. All data flows are comprised of either other data flows or discrete data structures that include descriptive attributes. A data flow should contain only the amount of data needed by a process; this is called data conservation.

11. External agents are entities outside the scope of a system and project but that provide net inputs to or net outputs from a system. As such, they form the boundary of the system.

12. Data stores present files of data to be used and maintained by the system. A data store on a process model corresponds to all instances of an entity on a data model.

13. Process modeling may be used in different types of projects including business process redesign and application

development. For application development projects, this chapter taught an event-driven data flow diagramming strategy as follows:

a. Draw a context data flow diagram that shows how the system interfaces to other systems, the business, and external organizations.

b. Draw a functional decomposition diagram that shows the key subsystems and/or functions that comprise the system.

c. Create an event list that identifies the external and temporal events to which the system must provide a response. External events are triggered by the external agents of a system. Temporal events are triggered by the passing of time.

d. Update the decomposition diagram to include processes to handle the events (one process per event).

e. For each event, draw an event diagram that shows its interactions with external entities, data stores, and, on occasion, other triggers to other events.

f. Combine the event diagrams into one or more system diagrams.

g. For each event on the system diagram, either describe it as an elementary process using Structured English or *explode* it into a *primitive data flow diagram* that includes elementary process that must be subsequently described by either Structured English, decision tables, or both. When exploding processes on data flow diagrams to reveal greater detail, it is important to maintain consistency between the different types of diagrams; this is called *synchronization*.

14. Most computer-aided software engineering tools support both decomposition diagramming and data flow diagramming.

KEY TERMS

actor, p. 339
balancing, p. 346
black hole, p. 315
business process redesign (BPR),
  pp. 310, 353
composite data flow, p. 321
context data flow diagram, pp. 332, 336
control flow, p. 322
converging data flow, p. 327
data attribute, p. 325
data conservation, p. 325
data flow, p. 321
data flow diagram (DFD), p. 308
data store, p. 330
data structure, p. 326
data type, p. 326
decision table, p. 320
decomposition, p. 312
decomposition diagram, p. 313

diverging data flow, p. 327
domain, p. 326
elementary process, p. 315
enterprise process model, p. 331
event, p. 314
event diagram, pp. 334, 340
event handler, p. 334
event partitioning, p. 332
event-response list, p. 334
external agent, p. 329
external entity, p. 329
external event, p. 339
function, p. 314
functional decomposition diagram,
  p. 334
gray hole, p. 315
junction, p. 322
logical model, p. 307
miracle, p. 315

model, p. 307
physical model, p. 307
physical process modeling, p. 354
policy, p. 320
primitive diagram, p. 334
primitive process, p. 315
process, p. 311
process model, p. 308
process modeling, p. 308
process-to-location-association matrix,
  p. 352
reverse engineering, p. 354
state event, p. 339
Structured English, p. 316
system diagram, p. 334
systems thinking, p. 310
temporal event, p. 339
use case list, p. 334
use cases, p. 339

REVIEW QUESTIONS

1. What is the difference between logical and physical modeling? Why is logical modeling more important in systems analysis?

2. What is systems thinking, and how do process models represent systems thinking?

3. What are the four symbols on a data flow diagram?

4. What is a process? Name three types of logical processes.

5. What is process decomposition and what role does it play in process modeling?

6. What purpose does Structured English serve in process modeling?

7. What are the basic constructs of Structured English?

8. What is the relationship between a policy and a decision table?

9. What are the components of a decision table?

10. What is the difference between data flows and data stores? What is the difference between data stores and data entities? What is the difference between data entities and external entities?

11. Differentiate between a data flow and a control flow.

12. What is data conservation?

13. What are the basic constructs of a data structure?

14. Differentiate between an enterprise and application process model.

15. During the survey and study phases, an analyst collected numerous samples, including documents, forms, and reports. Explain how these samples will prove useful for process modeling.

16. What is event partitioning?

17. Name and describe three types of business events.

18. Explain the tasks used to construct an application process model.

PROBLEMS AND EXERCISES

1. Compare and contrast process models with data models. What does each model show? Should you choose between the two modeling strategies? Why or why not?

2. A manager who has noted your use of logical data flow diagrams to document a proposed system's requirements has expressed some concern because of the lack of details that demonstrate the computer's role in the system. Defend your use of logical DFDs. Concisely explain the symbols and how to read a DFD. (Note: The answer to this exercise should be a standard component in any report that will include DFDs. You cannot be certain that the person who reads this report will be familiar with the tool.)

3. Explain why you should exclude implementation details when drawing a logical DFD. Can you think of any circumstances in which implementation details might be useful?

4. Explain why a systems analyst might want to draw logical models of an automated portion of an existing information system rather than simply accepting the existing technical information systems documentation, such as systems flowcharts and program flowcharts.

5. Draw a logical DFD to document the flow of data in your school's course registration and scheduling system.

6. Draw a logical DFD for some day-to-day "system" that you use or observe in use—for instance, your morning routine; making your favorite meal, including appetizer, entrée, side dishes, and dessert; constructing something from scratch.

7. Why is a project repository a valuable systems analysis tool? What are the possible consequences of not creating a project repository during systems analysis?

8. Can you think of any specific times that a project repository might have been helpful when you were writing a computer program? Can you think of a situation in which you misinterpreted a computer program requirement because you didn't know something that could have been recorded in a project repository?

9. Dig out your last computer program. Document the data structures for the following:
   a. Inputs (data flows).
   b. Outputs (data flows).
   c. Files or database (data store).

10. You have compiled a complete project repository for your new inventory control system. It is time to verify the contents of three summary reports specified for your end-users. Each data flow (report), record (data structure), and data attribute should be reviewed. Unfortunately, the entire repository is 373 pages. You can't mark the relevant pages and thumb back and forth between pages during your review. How should the report specifications be presented?

11. Why shouldn't a project repository be organized alphabetically independent of type—for example, data flow, data store, data attribute, and so on—like a traditional repository such as a dictionary?

12. Using the algebraic data structure notation in this chapter, create a project repository entry for the following:
    a. Your driver's license.
    b. Your course registration form.
    c. Your class schedule.
    d. IRS Form 1040 (any version).
    e. An account statement and invoice for a credit card.
    f. Your telephone, electric, or gas bill.
    g. An order form in a catalog.
    h. An application for anything (e.g., insurance, housing).
    i. A retail store catalog.
    j. A typical real estate listing.
    k. A computer printout from a business office or computer course.
    l. A catalog that describes the classes to be offered next semester.
    m. Your checkbook.
    n. Your bank statement.

13. During the study phase of systems analysis, the analyst must gather facts concerning both the manual and automated portions of the system. Why would it be desirable for a systems analyst to obtain samples of the existing computer files and computer-generated outputs? What value would project repository entries for computer files and computer-generated outputs have during *logical* modeling?

14. Visit a local business or school office. Ask for samples of five business forms or regular reports. Specify algebraic data structures for each sample.

15. Obtain a formal statement of a policy and procedure, such as a policy for a credit card. Evaluate the policy and procedure statement in terms of the common natural English specification problems identified earlier in this chapter.

16. Reconstruct the policy and procedure used in exercise 15 with the tools you learned in this chapter. Try to specify the decision table with a minimum number of rules.

17. Write a Structured English procedure for balancing your checkbook.

18. Produce a mini spec to describe how to prepare your favorite recipe, tune a car, or perform some other familiar task. Ask a novice to perform the task, working from your specification.

PROJECTS AND RESEARCH

1. Through information systems trade journals, research a commercial CASE product. Evaluate that package's project repository. Can you define both information and process models? Can you describe data structures to the repository? Can you describe individual data attributes to the repository? What types of analytical reports can be generated from the repository?

2. Enter into a contract with a programming instructor at your school. Convert one or more of his or her programming assignments to include data flow diagrams, data structures, and Structured English as described in this chapter. Have students currently enrolled in the course compare and contrast this form of specification with the original assignment's content and style.

3. Research a process modeling tool for business process redesign. How does it differ in symbology and constructs from logical data flow diagrams?

MINICASES

1. Given the following narrative description, draw a context DFD for the portion of the activities described.

   The purpose of the TEXTBOOK INVENTORY SYSTEM at a campus bookstore is to supply textbooks to students for classes at a local university. The university's academic departments submit initial data about courses, instructors, textbooks, and projected enrollments to the bookstore on a TEXTBOOK MASTER LIST. The bookstore generates a PURCHASE ORDER, which is sent to publishing companies supplying textbooks. Book orders arrive at the bookstore accompanied by a PACKING SLIP, which is checked and verified by the receiving department. Students fill out a BOOK REQUEST that includes course information. When they pay for their books, the students are given a SALES RECEIPT.

2. Given the following narrative description, draw a context DFD for the portion of the activities described.

   The purpose of the PLANT SCIENCE INFORMATION SYSTEM is to document the study results from a wide variety of experiments performed on selected plants. A study is initiated by a researcher who submits a RESEARCH PROPOSAL. After a panel review by a group of scientists, the researcher is required to submit a RESEARCH PLAN AND SCHEDULE. An FDA RESEARCH PERMIT REQUEST is sent to the Food and Drug Administration, which sends back a RESEARCH PERMIT. As the experiment progresses, the researcher fills out and submits EXPERIMENT NOTES. At the conclusion of the project, the researcher's results are reported on an EXPERIMENT HISTOGRAM.

3. Given the following narrative description of a system, draw a context diagram, functional decomposition diagram, and an event. Try to brainstorm events that might not be explicitly described. State any assumptions.

   The purpose of the production scheduling system is to respond to a PRODUCTION ORDER (submitted by the SALES DEPARTMENT) by generating a daily PRODUCTION SCHEDULE, generating RAW MATERIAL REQUISITIONS (sent to the MATERIALS MANAGEMENT DEPARTMENT) for all production orders scheduled for the next day, and generating JOB TICKETS for the work to be completed at each workstation during the next day (sent to the SHOP FLOOR SHIFT SUPERVISOR). The work is described in the following paragraphs.

   The production scheduling problem can be conveniently broken down into three functions: routing, loading, and releasing. For each product on a PRODUCTION ORDER, we must determine which workstations are needed, in what sequence the work must be done, and how much time should be necessary at each workstation to complete the work. This data is available from the PRODUCTION ROUTE SHEETS. This process, which is referred to as ROUTING THE ORDER, results in a ROUTE TICKET.

   Given a ROUTE TICKET (for a single product on the original PRODUCTION ORDER), we then LOAD THE REQUEST. Loading is nothing more than reserving dates and times at specific workstations. The reservations that have already been made are recorded in the WORKSTATION LOAD SHEETS. Loading requires us to look for the earliest available time slot for each task, being careful to preserve the required sequence of tasks (determined from the ROUTE TICKET).

   At the end of each day, the WORKSTATION LOAD SHEETS for each workstation are used to produce a PRODUCTION SCHEDULE. JOB TICKETS are prepared for each task at each workstation. The materials needed are determined from the BILL OF MATERIALS data store, and MATERIAL REQUESTS are generated for appropriate quantities.

4. Health Care Plus is a supplemental health insurance company that pays claims after its policyholders' primary insurance benefits through their employer or another policy have been exhausted. The following narrative partially describes its claims processing system. Draw a logical data flow diagram for the following physical narrative. State any assumptions.

   Policyholders must submit an EXPLANATION OF HEALTH CARE BENEFITS (EOHCB) along with proof that their primary health policy claim has been paid. All CLAIMS are mailed to the claims processing department.

   CLAIMS are initially sorted by the claims screening clerk. This clerk returns all requests that do not include the EOHCB. For those requests returned, a PENDING CLAIM is created, dated, and stored by date. Once each week, the clerk deletes all tickets that are more than 45 days old and sends a letter to the policyholders notifying them that their case has been closed. Requests that include the EOHCB are then sorted according to type of claim. Requests that include an EOHCB REFERENCE NUMBER are matched with an EOHCB form, which is pulled from the OPEN CLAIMS file. At the end of each day, all these claims are forwarded to the preprocessing department.

In the preprocessing department, clerks screen the EOHCB for missing data. They complete the form if possible. Otherwise, a copy of the claim is returned to the policyholder with a letter requesting the missing data. The original EOHCB is placed in the OPEN CLAIMS file, and a PENDING CLAIM is sent to the claims screening clerk. Completed claims are assigned a claim number, and the claim is microfilmed and filed for archival purposes.

A different clerk checks to see if the PROOF OF PRIMARY HEALTH CARE POLICY PAYMENT was included or is on file in the PRIMARY PAYMENT file. If it is not available, the policyholder is sent a letter requesting the proof. The EOHCB is placed in a PENDING PROOF FILE. Claims are automatically purged if they remain in this file for more than 14 days (a letter is sent to policyholders whose claims have been purged).

If proof is available, another clerk pulls the policyholder's policy record from the POLICY file, records policy and action codes on the EOHCB, and refiles the policy. At the end of the day, all preprocessed claims are forwarded to Information Systems.

5. Given the following narrative description, compile an event-response list and draw a context diagram. State any assumptions.

The purpose of the GREEN ACRES REAL ESTATE SYSTEM is to assist agents as they sell houses. Sellers contact the agency, and an agent is assigned to help the seller complete a LISTING REQUEST. Information about the house and lot taken from that request is stored in a file. Personal information about the sellers is copied by the agent into a sellers file.

When a buyer contacts the agency, he or she fills out a BUYER REQUEST. Every two weeks, the agency sends prospective buyers AREA REAL ESTATE LISTINGS and an ADDRESS CROSS REFERENCE LISTING containing actual street addresses. Periodically, the agent will find a particular house that satisfies most or all of a specific buyer's requirements, as indicated in the BUYER'S REQUIREMENTS STATEMENT distributed weekly to all agents. The agent will occasionally photocopy a picture of the house along with vital data and send the MULTIPLE LISTING STATEMENT (MLS) to the potential buyer.

When the buyer selects a house, he or she fills out an OFFER that is forwarded through the real estate agency to the seller, who responds with either an OFFER ACCEPTANCE or a COUNTEROFFER. After an offer is accepted, a PURCHASE AGREEMENT is signed by all parties. After a PURCHASE AGREEMENT is notarized, the agency sends an APPRAISAL REQUEST to an appraiser, who appraises the value of the house and lot. The agency also notifies its finance company with a FINANCING APPLICATION.

6. Given the following narrative description, draw a context diagram and system-level DFD for the portion of the activities described. State any assumptions.

The purpose of the OPEN ROAD INSURANCE SYSTEM is to provide automotive insurance to car owners. Initially, customers are required to fill out an INSURANCE APPLICATION. A DRIVER'S TRAFFIC RECORD REQUEST is requested from the local police department. Also, a VEHICLE TITLE AND REGISTRATION is requested from the Bureau of Motor Vehicles. PROPOSED POLICIES are sent in by various insurance companies that will underwrite those policies based on a quoted fee. The agent determines the best policy for the type and level of coverage desired and gives the customer a copy of the INSURANCE POLICY PROPOSAL AND QUOTE. If the customer accepts, he or she pays the INITIAL PREMIUM and is issued both the policy and a state-required INSURANCE COVERAGE STATEMENT (a card to be carried at all times when driving a vehicle). The customer information is now stored. Periodically, a PREMIUM NOTICE is generated, which—along with POLICY COVERAGE CHANGES—is sent to the customer, who responds by sending in a PREMIUM after which new INSURANCE COVERAGE STATEMENTS are issued.

Both a vehicle owner and the insurance company are required to provide annual PROOF OF LIABILITY INSURANCE to the Bureau of Motor Vehicles.

7. The following case describes how the typical IRS regional center processes your tax return.*

Initially, postal trucks bring tax returns to the regional center. The envelopes are then sorted by type of return—for example, long form versus short form and whether or not the envelope contains a payment. The sorted envelopes are sent to Receipt and Control, where they are further separated into 27 types falling into three general categories: short forms requesting refunds, long forms requesting refunds, and returns containing tax payments.

The documents are sorted twice because of the sheer volume of the returns. It's not unusual for the IRS to receive more than 200,000 returns in one day. The first sort divides that total to make the job more manageable.

Why so many types? Some returns are requests for extensions for filing. Others are quarterly estimated tax payments. There are over 500 official government forms for filing tax returns!

For example, to process short forms requesting refunds, operators submit forms to a machine that scans the returns and stores the data for later processing. The data is read by the main computer. It determines the correct tax, decides whether a refund should be sent, updates taxpayers' files, and prints letters, notices, liens, etc.

The refund information is sent to the National Computing Center, which subsequently triggers the Treasury Department to issue the actual refund checks. Letters, notices, and other communications are sent to local IRS sites around the country, from which appropriate information is sent to taxpayers.

The processing of long forms requesting refunds is similar, but not identical, to the processing of the short forms because the long forms usually include multiple schedules of information, such as itemized deductions. First, returns are sorted into blocks of batches to be processed as single units. Batches are numbered to en-

*"The IRS: How Your Return Is Processed," *USA Today,* January 8, 1986, p. 7A. Copyright 1986, USA TODAY. Reprinted with permission.

sure that no returns are lost or excessively delayed. The batches are then forwarded to examiners. The examiners check for and correct errors and code the returns for processing.

The examiners send back to the taxpayers any returns with incomplete or uncorrectable data. Also, clerks stamp a document locator number on each return for additional tracking capability as the return moves through the system. From this point, the processing is similar to the short form. Returns are input to the computer system. Data is stored for subsequent processing. The data is read by the main computer. It determines the correct tax, decides whether a refund should be sent, updates taxpayers' files, selects returns for possible tax audits, and prints letters, notices, liens, and so on. Refund information is sent to the National Computing Center, which subsequently triggers the Treasury Department to issue the actual refund checks. Notices and information regarding audits are sent to local IRS sites around the country, from which appropriate information is sent to taxpayers.

For returns containing tax payments, examiners check for and correct errors, code the returns for processing, and send back to taxpayers any returns with incomplete or uncorrectable data. Returns are entered into the computer. The computer checks taxpayer calculations and amounts, assigns document locator numbers, and stores the data. Then, the preceding steps are repeated using different operators.

The data from the second operators is checked against the first set for accuracy. Error reports are sent to examiners. Accurate data is stored for subsequent processing. Checks are collected for daily deposit into the Federal Reserve Bank.

Examiners check for errors, correcting any errors they can, and write the taxpayers for any missing information. At this point, the returns follow identical processing as described for the long forms requesting refunds.

Draw the logical data flow diagram for the physical description.

8. Prepare a decision table that accurately reflects the following course grading policy:

A student may receive a final course grade of A, B, C, D, or F. In deriving the student's final course grade, the instructor first determines an initial or tentative grade for the student, which is determined in the following manner:

A student who has scored a total of no lower than 90 percent on the first three assignments and exams and received a score no lower than 70 percent on the fourth assignment will receive an initial grade of A for the course. A student who has scored a total lower than 90 percent but no lower than 80 percent on the first three assignments and exams and received a score no lower than 70 percent on the fourth assignment will receive an initial grade of B for the course. A student who has scored a total lower than 80 percent but no lower than 70 percent on the first three assignments and exams and received a score no lower than 70 percent on the fourth assignment will receive an initial grade of C for the course. A student who has scored a total lower than 70 percent but no lower than 60 percent on the first three assignments and

exams and received a score no lower than 70 percent on the fourth assignment will receive an initial grade of D for the course. A student who has scored a total lower than 60 percent on the first three assignments and exams or received a score lower than 70 percent on the fourth assignment will receive an initial and final grade of F for the course. Once the instructor has determined the initial course grade for the student, the final course grade will be determined. The student's final course grade will be the same as his or her initial course grade if no more than three class periods during the semester were missed. Otherwise, the student's final course grade will be one letter grade lower than his or her initial course grade (for example, an A will become a B).

Are there any conditions for which there was no action specified for the instructor to take? If so, what would you do to correct the problem? Can your decision table be simplified by eliminating impossible rules or consolidating rules?

9. **The Poker Chip Challenge.** Joe, Gordon, and Susan own Granger's Restaurant Supply. They are in dire financial straits. They need $250,000 to meet their debts and cannot get a bank loan because of their poor credit rating. Among them, they can collect only $50,000.

They have decided on a drastic and risky solution to their problem. They will go to Atlantic City and try to gamble their $50,000 into enough money to cover their debts. There is one problem, however. They are lousy gamblers! Within one short hour, they lose the entire $50,000. As they leave the casino, they run into the president of Premier Restaurant & Supply, Inc., their fiercest competitor. He has been trying, unsuccessfully, to buy Granger's for some time. The unlucky trio offers Granger's to the greedy competitor for a bargain basement price. However, their rival, sensing an opportunity to get the business for absolutely nothing, offers the following proposition:

"I have five poker chips in my pocket—three black and two white. I propose to blindfold each of you and then give you each a chip. One by one, I will remove your blindfolds. You will be permitted to see the chip in your colleagues' hands; however, you must keep your own chip concealed in your closed palm. If any one of you can tell me the color of your own chip, then I will give you $1 million, more than enough to ensure the financial future of your business. Each of you has the option of guessing or not guessing. However, if any one of you guesses wrong, you must give me your company, free and clear: Is it a deal?"

The partners have little choice and no other reasonable hope, so they accept the challenge. The competitor then shows them the five chips—three black and two white—and chuckles as he places the blindfolds and gives each person one chip. He returns the two unused chips to his pocket.

The blindfold is removed from Joe, the eldest businessman and a world-class chess master. He looks at his partners' chips but, despite his logical mind, cannot determine the color of his own chip. He responds, "I just cannot give an answer: It's too risky. I'm better off giving my partners the opportunity for a better guess."

The blindfold is removed from Gordon, a graduate of a prestigious business school. After looking at the chips of his two partners, he too is unable to guess the color of his own chip. He passes the opportunity to Susan.

The competitor grins as he starts to remove the blindfold from Susan. He doesn't give her any more of a chance than he gave Joe or Gordon.

Susan interrupts confidently, "You can leave my blindfold on. How about double or nothing!" The competitor laughs aloud, "It's your funeral!"

Susan replies, "I'll take that $2 million in cash! I know from the answers of my colleagues that my chip is_____." She is correct, and the winnings save Granger's from financial ruin.

Construct a decision table that shows how Susan knew the color of her chip.

SUGGESTED READINGS

Copi, I. R. *Introduction to Logic*. New York: Macmillan, 1972. Copi provides a number of problem-solving illustrations and exercises that aid in the study of logic. The poker chip problem in our exercises was adapted from one of Copi's reasoning exercises.

DeMarco, Tom. *Structured Analysis and System Specification*. Englewood Cliffs, NJ: Prentice Hall, 1978. This is the classic book on the structured systems analysis methodology, which is built heavily around the use of data flow diagrams. The progression through (1) *current physical system DFDs*, (2) *current logical system DFDs*, (3) *target logical system DFDs*, and (4) *target physical system DFDs* is rarely practiced anymore, but the essence of DeMarco's pioneering work lives on in event-driven structured analysis. DeMarco created the data structure and logic notations used in this book.

Gildersleeve, T. R. *Successful Data Processing Systems Analysis*. Englewood Cliffs, NJ: Prentice Hall, 1978. The first edition of this book includes an entire chapter on the construction of decision tables. Gildersleeve does an excellent job of demonstrating how narrative process descriptions can be translated into condition and action entries in decision tables. Unfortunately, the chapter was deleted from the second edition.

Harmon, Paul, and Mark Watson. *Understanding UML: The Developers Guide*. San Francisco: Morgan Kaufman Publishers, Inc., 1998. This book does an excellent job of introducing use cases.

Martin, James, and Carma McClure. *Action Diagrams: Towards Clearly Specified Programs*. Englewood Cliffs, NJ: Prentice Hall, 1986. This book describes a formal grammar of Structured English that encourages the natural progression of a process (program) from Structured English to code. Action diagrams are supported directly in some CASE tools.

Matthies, Leslie H. *The New Playscript Procedure*. Stamford, CT: Office Publications, 1977. This book provides a thorough explanation and examples of the weaknesses of the English language as a tool for specifying business procedures.

McMenamin, Stephen M., and John F. Palmer. *Essential Systems Analysis*. New York: Yourdon Press, 1984. This was the first book to suggest event partitioning as a formal strategy to improve structured analysis. The book also strengthened the distinction between logical and physical process models and the increased importance of the logical models (which they called *essential* models).

Robertson, James, and Suzanne Robertson. *Complete Systems Analysis* (Volumes 1 and 2). New York: Dorset House Publishing, 1994. This is the most up-to-date and comprehensive book on the event-driven approach to structured analysis, even though we feel it still overemphasizes the current system and physical models more than the Yourdon book described below.

Seminar notes for *Process Modeling Techniques*. Atlanta: Structured Solutions, Inc., 1991. You probably can't get a copy of these notes, but we wanted to acknowledge the instructors of the *AD/Method* methodology course that stimulated our thinking and motivated our departure from classical structured analysis techniques to the event-driven structured analysis techniques taught in this chapter. Structured Solutions was acquired by Protelicess, Inc.

Wetherbe, James and Nicholas P. Vatarli. *Systems Analysis and Design: Best Practices*. 4th ed. St. Paul, MN: West Publishing, 1994. Jim Wetherbe has always been one of the strongest advocates of system concepts and system thinking as part of the discipline of systems analysis and design. Jim has shaped many minds, including our own. The authors provide a nice chapter on system concepts in this book—and the rest of the book is must reading for those of you who truly want to learn to "systems think."

Yourdon, Edward. *Modern Structured Analysis*. Englewood Cliffs, NJ: Yourdon Press, 1989. This was the first mainstream book to abandon classic structured analysis' overemphasis on the current physical system models and to formalize McMenamin and Palmer's event-driven approach.

Focus on PEOPLE

Focus on DATA

Focus on PROCESSES

Focus on INTERFACES

Focus on DEVELOPMENT

Stakeholders

Activities

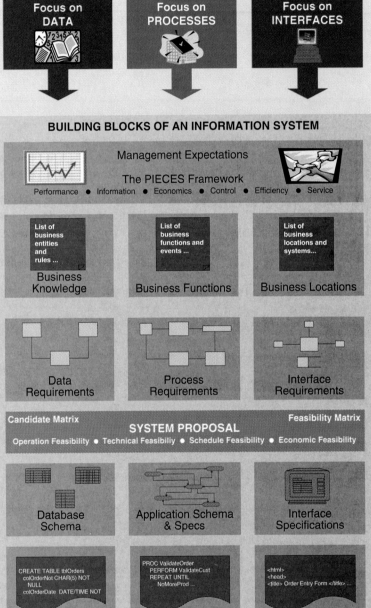

BUILDING BLOCKS OF AN INFORMATION SYSTEM

Management Expectations

The PIECES Framework

Performance ● Information ● Economics ● Control ● Efficiency ● Service

List of business entities and rules ...

Business Knowledge

List of business functions and events ...

Business Functions

List of business locations and systems...

Business Locations

Data Requirements

Process Requirements

Interface Requirements

Candidate Matrix   Feasibility Matrix

SYSTEM PROPOSAL

Operation Feasibility ● Technical Feasibiliy ● Schedule Feasibility ● Economic Feasibility

Database Schema

Application Schema & Specs

Interface Specifications

```
CREATE TABLE tblOrders
colOrderNot CHAR(5) NOT
NULL
colOrderDate DATE/TIME NOT
```

Database Programs

```
PROC ValidateOrder
 PERFORM ValidateCust
 REPEAT UNTIL
   NoMoreProd ...
```

Application Programs

```
<html>
<head>
<title> Order Entry Form </title> ...
```

Interface Programs

INFORMATION TECHNOLOGY & ARCHITECTURE

Database Technology ● Process Technology ● Interface Technology ● Network Technology

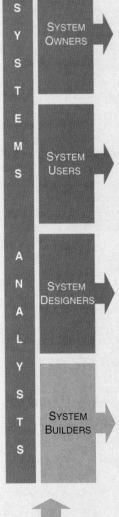

SYSTEMS ANALYSTS

SYSTEM OWNERS

SYSTEM USERS

SYSTEM DESIGNERS

SYSTEM BUILDERS

VENDORS AND CONSULTANTS

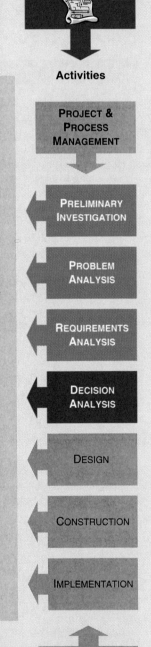

PROJECT & PROCESS MANAGEMENT

PRELIMINARY INVESTIGATION

PROBLEM ANALYSIS

REQUIREMENTS ANALYSIS

DECISION ANALYSIS

DESIGN

CONSTRUCTION

IMPLEMENTATION

OPERATIONS AND SUPPORT

9

FEASIBILITY ANALYSIS AND THE SYSTEM PROPOSAL

CHAPTER PREVIEW AND OBJECTIVES

Systems analysts sell change. Good systems analysts thoroughly evaluate alternative solutions before proposing change. In this chapter you will learn how to analyze and document those alternatives on the basis of four feasibility criteria: operational, technical, schedule, and economics. You will also learn how to make a system proposal in the form of a written report and a formal presentation. You will know that you understand the feasibility analysis and recommendation skills needed by the systems analyst when you can:

— Identify feasibility checkpoints in the systems life cycle.

— Identify alternative system solutions.

— Define and describe four types of feasibility and their respective criteria.

— Perform various cost-benefit analyses using time-adjusted costs and benefits.

— Write suitable system proposal reports for different audiences.

— Plan for a formal presentation to system owners and users.

SOUNDSTAGE

SOUNDSTAGE ENTERTAINMENT CLUB

SCENE

This episode begins soon after the completion of the requirements analysis phase. We join Sandra, Bob, and Terri Hitchcock as they are formulating a plan for completing the subsequent phase for the project—decision analysis.

SANDRA

Good morning! I don't know about the two of you, but I feel this project has been coming along nicely. I really think we've done a great job of determining the business requirements for the new system.

TERRI

I think so too.

BOB

So do I.

SANDRA

Well, now it's time for us to really start earning our money. Our methodology calls for us to complete the decision analysis phase. That phase involves looking at technology to identify alternative computer-based solutions, evaluate them, and choose one to recommend to management.

BOB

I thought we were never going to get to the point where we could talk about technology. I have a great idea for a new system.

SANDRA

That's great. But we need to consider several alternatives. It's quite possible that your idea is an excellent one, but it may or may not be the most feasible. Besides, don't forget that some of the users did express some ideas and opinions of their own. We should show them some courtesy by considering their ideas and suggestions.

TERRI

I don't know why I'm here. This is out of my league. I don't know enough about technology and computers . . .

SANDRA

Oh, I'm sorry, Terri. I didn't ask you to meet with Bob and me with the intent of having you help us specify some technical solutions. I was hoping to get your input on some business matters.

TERRI

That's a relief. What kind of business matters?

SANDRA

Once Bob and I come up with some alternative solutions, we will be conducting a feasibility analysis of each alternative. We will be assessing candidates to determine how feasible they are from an operational, technical, economic, and schedule standpoint. I wanted to get your input on how much weight or importance we should place on each of these criteria. I also thought that you could help us determine what kind of cost-effectiveness analysis techniques we should conduct.

TERRI

No problem. I have some pretty good ideas there and I would be happy to help you with the financial numbers.

SANDRA

There's more . . . I was hoping that you could help us put together a report and presentation of our recommendation.

I've written various technical and business reports, but I know it would be beneficial to have input from someone such as you. Often people like myself can get caught up in the details or technical jargon when we are trying to communicate to a business audience.

BOB

Oh man! We have to write another report?

SANDRA

That's right, and we have to give a formal presentation, defending our recommendation.

BOB

Now I am starting to see why you were so interested in my communication skills during my job interview.

SANDRA

They are indeed very important. Let's face it! We are salespersons. We have to sell change!

DISCUSSION QUESTIONS

1. How would you characterize the focus of the decision analysis phase of this project?

2. Why did Terri initially feel uncomfortable with her participation in the decision analysis phase?

3. Do you think Sandra was correct in wanting to consider the solution ideas and opinions of the users?

4. Do you think it is necessary or beneficial for them to write a report and give an oral presentation of their recommendations for a new system? Why or why not?

FEASIBILTY ANALYSIS AND THE SYSTEM PROPOSAL

In today's business world, it is becoming more and more apparent that analysts must learn to think like business managers. Computer applications are expanding at a record pace. Now more than ever, management expects information systems to pay for themselves. Information is a major capital investment that must be justified, just as marketing must justify a new product and manufacturing must justify a new plant or equipment. Systems analysts are called on more than ever to help answer the following questions: Will the investment pay for itself? Are there other investments that will return even more on their expenditure?

This chapter deals with feasibility analysis issues of interest to the systems analyst and users of information systems. It also emphasizes the importance of making recommendations to management in the form of a system proposal that is a formal written report and/or oral presentation. As is illustrated in the chapter map, feasibility analysis is appropriate to the systems analysis phases but particularly important to the decision analysis phase. The system proposal represents the deliverable and presents the technical DATA, PROCESS, and INTERFACE solution.

Let's begin with a formal definition of feasibility and feasibility analysis.

> **Feasibility** is the measure of how beneficial or practical the development of an information system will be to an organization.

> **Feasibility analysis** is the process by which feasibility is measured.

Feasibility should be measured throughout the life cycle. In earlier chapters we called this a **creeping commitment** approach to feasibility. The scope and complexity of an apparently feasible project can change after the initial problems and opportunities are fully analyzed or after the system has been designed. Thus, a project that is feasible at one point may become infeasible later.

Figure 9.1 shows feasibility checkpoints during the systems analysis phases of our life cycle. The checkpoints are represented by red diamonds. The diamonds indicate that a feasibility reassessment and management review should be conducted at the end of the prior phase (before the next phase). A project may be canceled or revised at any checkpoint, despite whatever resources have been spent.

This idea may bother you at first. Your natural inclination may be to justify continuing a project based on the time and money you've already spent. Those costs are sunk! A fundamental principle of management is never to throw good money after bad—cut your losses and move on to a more feasible project. That doesn't mean the costs already spent are not important. Costs must eventually be recovered if the investment is ever to be considered a success. Let's briefly examine the checkpoints in Figure 9.1.

The first feasibility analysis is conducted during the preliminary investigation phase. At this early stage of the project, feasibility is rarely more than a measure of the urgency of the problem and the first-cut estimate of development costs. It answers the question: Do the problems (or opportunities) warrant the cost of a detailed study and analysis of the current system? Realistically, feasibility can't be accurately measured until the problems (and opportunities) and requirements are better understood.

After estimating benefits of solving the problems and opportunities, analysts will estimate costs of developing the expected system. Experienced analysts routinely increase these costs by 50 percent to 100 percent (or more) because experience tells them the problems are rarely well-defined and user requirements are typically understated.

The next checkpoint occurs after a more detailed study and problem analysis of the current system. Because the problems are better understood, the analysts can make better estimates of development costs and of the benefits to be obtained from a new system. The minimum value of solving a problem is equal to the cost of that problem. For example, if inventory carrying costs are $35,000 over acceptable limits, then the minimum value of an acceptable information system would be $35,000. It is hoped an improved system will be able to do better than that; however, it must return this minimum value!

Development costs, at this point, are still just guesstimates. We have yet to fully define user requirements or to specify a design solution to those requirements.

If the cost estimates significantly increase from the preliminary investigation phase to the problem analysis phase, the likely culprit is scope. Scope has a tendency to increase in many projects. If increased scope threatens feasibility, then scope might be reduced.

Feasibility Analysis—A Creeping Commitment Approach

Systems Analysis—Preliminary Investigation Checkpoint

Systems Analysis—Problem Analysis Checkpoint

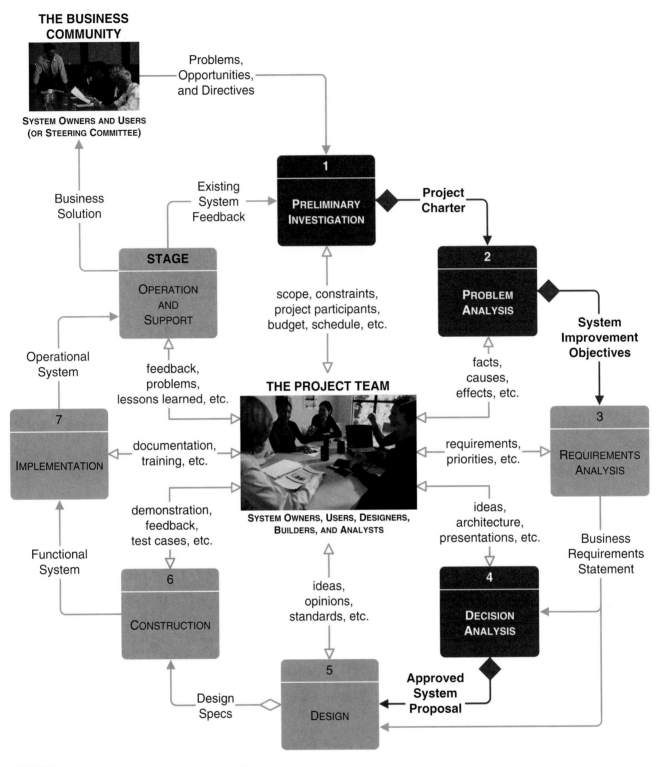

THE BUSINESS COMMUNITY

SYSTEM OWNERS AND USERS
(OR STEERING COMMITTEE)

Problems, Opportunities, and Directives

Business Solution

Existing System Feedback

1
PRELIMINARY INVESTIGATION

Project Charter

STAGE
OPERATION AND SUPPORT

scope, constraints, project participants, budget, schedule, etc.

2
PROBLEM ANALYSIS

System Improvement Objectives

Operational System

feedback, problems, lessons learned, etc.

THE PROJECT TEAM

facts, causes, effects, etc.

7
IMPLEMENTATION

documentation, training, etc.

SYSTEM OWNERS, USERS, DESIGNERS, BUILDERS, AND ANALYSTS

requirements, priorities, etc.

3
REQUIREMENTS ANALYSIS

Functional System

demonstration, feedback, test cases, etc.

ideas, architecture, presentations, etc.

Business Requirements Statement

6
CONSTRUCTION

ideas, opinions, standards, etc.

4
DECISION ANALYSIS

5
DESIGN

Design Specs

Approved System Proposal

FIGURE 9.1 *Feasibility Checkpoints during Systems Analysis*

The decision analysis phase represents a major feasibility analysis activity since it charts one of many possible implementations as the target for systems design.

Problems and requirements should be known by now. During the decision analysis phase, alternative solutions are defined in terms of their input/output methods, data storage methods, computer hardware and software requirements, processing methods, and people implications. The following list presents the typical range of options that can be evaluated by the analyst.

- Do nothing! Leave the current system alone. Regardless of management's opinion or your own opinion of this option, it should be considered and analyzed as a baseline option against which all others can and should be evaluated.
- Reengineer the (manual) business processes, not the computer-based processes. This may involve streamlining activities, reducing duplication and unnecessary tasks, reorganizing office layouts, and eliminating redundant and unnecessary forms and processes, among others.
- Enhance existing computer processes.
- Purchase a packaged application.
- Design and construct a new computer-based system.

After defining these options, each option is analyzed for operational, technical, schedule, and economic feasibility. This chapter will closely examine these four classes of feasibility criteria. One alternative is recommended to system owners for approval. The approved solution becomes the basis for general and detailed design.

So far, we've defined feasibility and feasibility analysis, and we've identified feasibility checkpoints during systems analysis. Most analysts agree that there are four categories of feasibility tests:

- **Operational feasibility** is a measure of how well the solution will work in the organization. It is also a measure of how people feel about the system/project.
- **Technical feasibility** is a measure of the practicality of a specific technical solution and the availability of technical resources and expertise.
- **Schedule feasibility** is a measure of how reasonable the project timetable is.
- **Economic feasibility** is a measure of the cost-effectiveness of a project or solution. This is often called a *cost-benefit analysis*.

Operational and technical feasibility criteria measure the worthiness of a problem or solution. Operational feasibility is people oriented. Technical feasibility is computer oriented.

Economic feasibility deals with the costs and benefits of the information system. Actually, few systems are infeasible. Instead, different options tend to be more or less feasible than others. Let's take a closer look at the four feasibility criteria.

Operational feasibility criteria measure the urgency of the problem (survey and study phases) or the acceptability of a solution (definition, selection, acquisition, and design phases). How do you measure operational feasibility? There are two aspects of operational feasibility to be considered:

1. Is the problem worth solving, or will the solution to the problem work?
2. How do the end-users and management feel about the problem (solution)?

Is the Problem Worth Solving, or Will the Solution to the Problem Work? Do you recall the PIECES framework for identifying problems (Chapters 3, 5, and 6)? PIECES can

Sidebar headings (right margin)

Systems Design— Decision Analysis Checkpoint

FOUR TESTS FOR FEASIBILITY

Operational Feasibility

be used as the basis for analyzing the urgency of a problem or the effectiveness of a solution. The following is a list of the questions that address these issues:

P   *Performance.* Does the system provide adequate throughput and response time?

I   *Information.* Does the system provide end-users and managers with timely, pertinent, accurate, and usefully formatted information?

E   *Economy.* No, we are not prematurely jumping into economic feasibility! The question here is, Does the system offer adequate service level and capacity to reduce the costs of the business or increase the profits of the business?

C   *Control.* Does the system offer adequate controls to protect against fraud and embezzlement and to guarantee the accuracy and security of data and information?

E   *Efficiency.* Does the system make maximum use of available resources including people, time, flow of forms, minimum processing delays, and the like?

S   *Services.* Does the system provide desirable and reliable service to those who need it? Is the system flexible and expandable?

NOTE   The term *system,* used throughout this discussion, may refer either to the existing system or a proposed system solution, depending on which phase you're currently working in.

How Do the End-Users and Managers Feel about the Problem (Solution)?   It's important not only to evaluate whether a system *can* work, but we must also evaluate whether a system *will* work. A workable solution might fail because of end-user or management resistance. The following questions address this concern:

- Does management support the system?
- How do the end-users feel about their role in the new system?
- What end-users or managers may resist or not use the system? People tend to resist change. Can this problem be overcome? If so, how?
- How will the working environment of the end-users change? Can or will end-users and management adapt to the change?

Essentially, these questions address the *political* acceptability of solving the problem or the solution.

Usability Analysis   When determining operational feasibility in the later stages of the system life cycle, **usability analysis** is often performed with a working prototype of the proposed system. This is a test of the system's user interfaces and is measured in how easy they are to learn and to use and how they support the desired productivity levels of the users. Many large corporations, software consultant agencies, and software development companies employ user interface specialists for designing and testing system user interfaces. They have special rooms equipped with video cameras, tape recorders, microphones, and two-way mirrors to observe and record a user working with the system. Their goal is to identify the areas of the system where the users are prone to make mistakes and processes that may be confusing or too complicated. They also observe the reactions of the users and assess their productivity.

How do you determine if a system's user interface is usable? There are certain goals or criteria that experts agree help measure the usability of an interface and they are as follows:

- *Ease of learning*—How long it takes to train someone to perform at a desired level.

- *Ease of use*—You are able to perform your activity quickly and accurately. If you are a first-time user or infrequent user, the interface is easy and understandable. If you are a frequent user, your level of productivity and efficiency is increased.
- *Satisfaction*—You, the user, are favorably pleased with the interface and prefer it over types you are familiar with.

Today, very little is technically impossible. Consequently, technical feasibility looks at what is practical and reasonable. Technical feasibility addresses three major issues:

Technical Feasibility

1. Is the proposed technology or solution practical?
2. Do we currently possess the necessary technology?
3. Do we possess the necessary technical expertise, and is the schedule reasonable?

Is the Proposed Technology or Solution Practical?   The technology for any defined solution is normally available. The question is whether that technology is mature enough to be easily applied to our problems. Some firms like to use state-of-the-art technology, but most firms prefer to use mature and proven technology. A mature technology has a larger customer base for obtaining advice concerning problems and improvements.

Do We Currently Possess the Necessary Technology?   Assuming the solution's required technology is practical, we must next ask ourselves, Is the technology available in our information systems shop? If the technology is available, we must ask if we have the capacity. For instance, Will our current printer be able to handle the new reports and forms required of a new system?

If the answer to either of these questions is no, then we must ask ourselves, Can we get this technology? The technology may be practical and available, and, yes, we need it. But we simply may not be able to afford it at this time. Although this argument borders on economic feasibility, it is truly technical feasibility. If we can't afford the technology, then the alternative that requires the technology is not practical and is technically infeasible!

Do We Possess the Necessary Technical Expertise, and Is the Schedule Reasonable?   This consideration of technical feasibility is often forgotten during feasibility analysis. We may have the technology, but that doesn't mean we have the skills required to properly apply that technology. For instance, we may have a database management system (DBMS). However, the analysts and programmers available for the project may not know that DBMS well enough to properly apply it. True, all information systems professionals can learn new technologies. However, that learning curve will impact the technical feasibility of the project; specifically, it will impact the schedule.

Given our technical expertise, are the project deadlines reasonable? Some projects are initiated with specific deadlines. You need to determine whether the deadlines are mandatory or desirable. For instance, a project to develop a system to meet new government reporting regulations may have a deadline that coincides with when the new reports must be initiated. Penalties associated with missing such a deadline may make meeting it mandatory. If the deadlines are desirable rather than mandatory, the analyst can propose alternative schedules.

Schedule Feasibility

It is preferable (unless the deadline is absolutely mandatory) to deliver a properly functioning information system two months late than to deliver an error-prone, useless information system on time! Missed schedules are bad. Inadequate systems are worse! It's a choice between the lesser of two evils.

Economic Feasibility

The bottom line in many projects is economic feasibility. During the early phases of the project, economic feasibility analysis amounts to little more than judging whether the possible benefits of solving the problem are worthwhile. Costs are practically impossible to estimate at that stage because the end-user's requirements and alternative technical solutions have not been identified. However, as soon as specific requirements and solutions have been identified, the analyst can weigh the costs and benefits of each alternative. This is called a cost-benefit analysis. Cost-benefit analysis is discussed later in this chapter.

The Bottom Line

You have learned that any alternative solution can be evaluated according to four criteria: operational, technical, schedule, and economic feasibility. How do you pick the best solution? It's not always easy. Operational and economic issues often conflict. For example, the solution that provides the best operational impact for the end-users may also be the most expensive and, therefore, the least economically feasible. The final decision can be made only by sitting down with end-users, reviewing the data, and choosing the best overall alternative.

COST-BENEFIT ANALYSIS TECHNIQUES

Economic feasibility has been defined as a cost-benefit analysis. How do you estimate costs and benefits? And how do you compare those costs and benefits to determine economic feasibility? Most schools offer complete courses on these subjects—courses on financial management, financial decision analysis, and engineering economics and analysis. Such a course should be included in your plan of study. This section presents an overview of the techniques.

How Much Will the System Cost?

Costs fall into two categories. There are costs associated with developing the system, and there are costs associated with operating a system. The former can be estimated from the outset of a project and should be refined at the end of each phase of the project. The latter can be estimated only after specific computer-based solutions have been defined. Let's take a closer look at the costs of information systems.

The costs of developing an information system can be classified according to the phase in which they occur. Systems development costs are usually onetime costs that will not recur after the project has been completed. Many organizations have standard cost categories that must be evaluated. In the absence of such categories, the following lists should help:

- *Personnel costs*—The salaries of systems analysts, programmers, consultants, data entry personnel, computer operators, secretaries, and the like who work on the project make up the personnel costs. Because many of these individuals spend time on many projects, their salaries should be prorated to reflect the time spent on the projects being estimated.
- *Computer usage*—Computer time will be used for one or more of the following activities: programming, testing, conversion, word processing, maintaining a project dictionary, prototyping, loading new data files, and the like. If a computing center charges for usage of computer resources such as disk storage or report printing, the cost should be estimated.
- *Training*—If computer personnel or end-users have to be trained, the training courses may incur expenses. Packaged training courses may be charged out on a flat fee per site, a student fee (such as $395 per student), or an hourly fee (such as $75 per class hour).
- *Supply, duplication, and equipment costs.*
- *Cost of any new computer equipment and software.*

Sample development costs for a typical solution are displayed in Figure 9.2.

Almost nobody forgets system development budgets when itemizing costs. On the other hand, it is easy to forget that a system will incur costs after it is operating. The lifetime benefits must recover both the developmental and operating costs. Unlike system development costs, operating costs tend to recur throughout the lifetime of the system. The costs of operating a system over its useful lifetime can be classified as fixed and variable.

Fixed costs occur at regular intervals but at relatively fixed rates. Examples of fixed operating costs include:

— Lease payments and software license payments.
— Prorated salaries of information systems operators and support personnel (although salaries tend to rise, the rise is gradual and tends not to change dramatically from month to month).

Variable costs occur in proportion to some usage factor. Examples include:

— Costs of computer usage (e.g., CPU time used, terminal connect time used, storage used), which vary with the workload.
— Supplies (e.g., preprinted forms, printer paper used, punched cards, floppy disks, magnetic tapes, and other expendables), which vary with the workload.
— Prorated overhead costs (e.g., utilities, maintenance, and telephone service), which can be allocated throughout the lifetime of the system using standard techniques of cost accounting.

Sample operating cost estimates for a solution are also displayed in Figure 9.2. After determining the costs and benefits for a possible solution, you can perform the cost-benefit analysis.

Because benefits or potential benefits become known before costs, we'll discuss benefits first. Benefits normally increase profits or decrease costs, both highly desirable characteristics of a new information system.

As much as possible, benefits should be quantified in dollars and cents. Benefits are classified as tangible or intangible.

Tangible benefits are those that can be easily quantified.

Tangible benefits are usually measured in terms of monthly or annual savings or of profit to the firm. For example, consider the following scenario:

> While processing student housing applications, we discover that considerable data is being redundantly typed and filed. An analysis reveals that the same data is typed seven times, requiring an average of 44 additional minutes of clerical work per application. The office processes 1,500 applications per year. That means a total of 66,000 minutes or 1,100 hours of redundant work per year. If the average salary of a secretary is $6 per hour, the cost of this problem and the benefit of solving the problem is $6,600 per year.

Alternatively, tangible benefits might be measured in terms of unit cost savings or profit. For instance, an alternative inventory valuation scheme may reduce inventory carrying cost by $0.32 per unit of inventory. Some examples of tangible benefits are listed in the margin.

Other benefits are intangible.

Intangible benefits are those benefits believed to be difficult or impossible to quantify.

Unless these benefits are at least identified, it is entirely possible that many projects would not be feasible. Examples of intangible benefits are listed in the margin.

Unfortunately, if a benefit cannot be quantified, it is difficult to accept the validity of an associated cost-benefit analysis that is based on incomplete data. Some analysts dispute the existence of intangible benefits. They argue that all benefits are quantifiable; some are just more difficult than others. Suppose, for example,

---

What Benefits Will the System Provide?

———— ✓ ————

Tangible Benefits

Fewer processing errors
Increased throughput
Decreased response time
Elimination of job steps
Increased sales
Reduced credit losses
Reduced expenses

———— ✓ ————

Intangible Benefits

Improved customer goodwill
Improved employee morale
Better service to community
Better decision making

Estimated Costs for Client-Server System Alternative

DEVELOPMENT COSTS:

Personnel:

2	Systems Analysts (400 hours/ea $50.00/hr)	$40,000
4	Programmer/Analysts (250 hours/ea $35.00/hr)	$35,000
1	GUI Designer (200 hours/ea $40.00/hr)	$8,000
1	Telecommunications Specialist (50 hours/ea $50.00/hr)	$2,500
1	System Architect (100 hours/ea $50.00/hr)	$5,000
1	Database Specialist (15 hours/ea $45.00/hr)	$675
1	System Librarian (250 hours/ea $15.00/hr)	$3,750

Expenses:

4	Smalltalk training registration ($3,500.00/student)	$14,000

New Hardware & Software:

1	Development Server	$18,700
1	Server Software (operating system, misc.)	$1,500
1	DBMS server software	$7,500
7	DBMS Client software ($950.00 per client)	$6,650

Total Development Costs: | $143,275

PROJECTED ANNUAL OPERATING COSTS

Personnel:

2	Programmer/Analysts (125 hours/ea $35.00/hr)	$8,750
1	System Librarian (20 hours/ea $15.00/hr)	$300

Expenses:

1	Maintenance Agreement for Server	$995
1	Maintenance Agreement for Server DBMS software	$525
	Preprinted forms (15,000/year @ .22/form)	$3,300

Total Projected Annual Costs: | $13,870

FIGURE 9.2   *Costs for a Proposed Systems Solution*

that improved customer goodwill is listed as a possible intangible benefit. Can we quantify goodwill? You might try the following analysis:

1. What is the result of customer ill will? The customer will submit fewer (or no) orders.
2. To what degree will a customer reduce orders? Your user may find it difficult to specifically quantify this impact. But you could try to have the end-user estimate the possibilities (or invent an estimate to which the end-user can react). For instance,
   a. There is a 50 percent (.50) chance that the regular customer would send a few orders—fewer than 10 percent of all its orders—to competitors to test their performance.
   b. There is a 20 percent (.20) chance that the regular customer would send as many as half its orders (.50) to competitors, particularly those orders we are historically slow to fulfill.

c. There is a 10 percent (.10) chance that a regular customer would send us an order only as a last resort. That would reduce that customer's normal business with us to 10 percent of its current volume (90 percent or .90 loss).

d. There is a 5 percent (.05) chance that a regular customer would choose not to do business with us at all (100 percent or 1.00 loss).

3. We can calculate an estimated business loss as follows:

$$
\begin{aligned}
\text{Loss} ={}& .50 \times (.10 \text{ loss of business}) \\
&+ .20 \times (.50 \text{ loss of business}) \\
&+ .10 \times (.90 \text{ loss of business}) \\
&+ .50 \times (1.00 \text{ loss of business}) \\
={}& .29 \\
={}& 29\% \text{ statistically estimated loss of business}
\end{aligned}
$$

4. If the average customer does $40,000 per year of business, then we can expect to lose 29 percent or $11,600 of that business. If we have 500 customers, this can be expected to amount to a total of $5,800,000.

5. Present this analysis to management, and use it as a starting point for quantifying the benefit.

Is the Proposed System Cost-Effective?

There are three popular techniques to assess economic feasibility, also called *cost-effectiveness:* payback analysis, return on investment, and net present value.

The choice of techniques should consider the audiences that will use them. Virtually all managers who have come through business schools are familiar with all three techniques. One concept that should be applied to each technique is the adjustment of cost and benefits to reflect the time value of money.

The Time Value of Money   A concept shared by all three techniques is the **time value of money**—a dollar today is worth more than a dollar one year from now. You could invest that dollar today and, through accrued interest, have more than one dollar a year from now. Thus, you'd rather have that dollar today than in one year. That's why your creditors want you to pay your bills promptly—they can't invest what they don't have. The same principle can be applied to costs and benefits *before* a cost-benefit analysis is performed.

Some of the costs of a system will be accrued after implementation. Additionally, all benefits of the new system will be accrued in the future. Before cost-benefit analysis, these costs should be brought back to current dollars. An example should clarify the concept.

Suppose we are going to realize a benefit of $20,000 two years from now. What is the current dollar value of that $20,000 benefit? The current value of the benefit is the amount of money we would need to invest today to have $20,000 two years from now. If the current return on investments is running about 10 percent, an investment of $16,528 today would give us our $20,000 in two years (we'll show you how to calculate this later). Therefore, the current value of the estimated benefit is $16,528—that is, we'd rather have $16,528 today than the promise of $20,000 two years from now.

The same adjustment could be made on costs that are projected into the future. For example, suppose we are projecting a cost of $20,000 two years from now. What is the current dollar value of that $20,000 cost? The current value of the cost is the amount of money we would need to invest today to have $20,000 to pay the cost two years from now. Again, if we assume a 10 percent return on current investments, an investment of $16,528 today would give us the needed $20,000 in two years. Therefore, the current value of the estimated cost is $16,528—that is, we can fulfill our cost obligation of $20,000 in two years by investing $16,528 today.

Why go to all this trouble? Because projects are often compared against other projects that have different lifetimes. Time value analysis techniques have become the preferred cost-benefit methods for most managers. By time-adjusting costs and benefits, you can improve the following cost-benefit techniques.

Payback Analysis   The **payback analysis** technique is a simple and popular method for determining if and when an investment will pay for itself. Because system development costs are incurred long before benefits begin to accrue, it will take some time for the benefits to overtake the costs. After implementation, you will incur additional operating expenses that must be recovered. Payback analysis determines how much time will lapse before accrued benefits overtake accrued and continuing costs. This period of time is called the **payback period.**

In Figure 9.3 we see an information system that will be developed at a cost of $418,040. The estimated net operating costs for each of the next six years are also recorded in the table. The estimated net benefits over the same six operating years are also shown. What is the payback period?

First, we need to adjust the costs and benefits for the time value of money (that is, adjust them to current dollar values). Here's how! The present value of a dollar in year n depends on something typically called a **discount rate.** The discount rate is a percentage similar to interest rates that you earn on your savings account. In most cases the discount rate for a business is the **opportunity cost** of being able to invest money in other projects, including the possibility of investing in the stock market, money market funds, bonds, and the like. Alternatively, a discount rate could represent what the company considers an acceptable return on its investments. This number can be learned by asking any financial manager, officer, or comptroller.

Let's say the discount rate for our sample company is 12 percent. The current value, actually called the **present value,** of a dollar at any time in the future can be calculated using the following formula:

$$PV_n = 1/(1 + i)^n$$

where PV_n is the present value of $1.00 n years from now and i is the discount rate. Therefore, the present value of a dollar two years from now is

$$PV_2 = 1/(1 + .12)^2 = 0.797$$

FIGURE 9.3
Payback Analysis for a Project

	A	B	C	D	E	F	G	H	I
4	Cash flow description	Year 0	Year 1	Year 2	Year 3	Year 4	Year 5	Year 6	
5	Development cost:	($418,040)							
6	Operation & maintenance cost:		($15,045)	($16,000)	($17,000)	($18,000)	($19,000)	($20,000)	
7	Discount factors for 12%:	1.000	0.893	0.797	0.712	0.636	0.567	0.507	
8	Time-adjusted costs (adjusted to present value):	($418,040)	($13,435)	($12,752)	($12,104)	($11,448)	($10,773)	($10,140)	
9	Cumulative time-adjusted costs over lifetime:	($418,040)	($431,475)	($444,227)	($456,331)	($467,779)	($478,552)	($488,692)	
10									
11	Benefits derived from operation of new system:	$0	$150,000	$170,000	$190,000	$210,000	$230,000	$250,000	
12	Discount factors for 12%:	1.000	$0.893	$0.797	$0.712	$0.636	$0.567	$0.507	
13	Time-adjusted benefits (current of present value):	$0	$133,950	$135,490	$135,280	$133,560	$130,410	$126,750	
14	Cumulative time-adjusted benefits over lifetime:	$0	$133,950	$269,440	$404,720	$538,280	$668,690	$795,440	
15		0	1	2	3	4	5	6	
16	Cumulative lifetime time-adjusted costs + benefits:	($418,040)	($297,525)	($174,787)	($51,611)	$70,501	$190,138	$306,748	
17									

Payback Analysis chart showing Dollars versus Year.

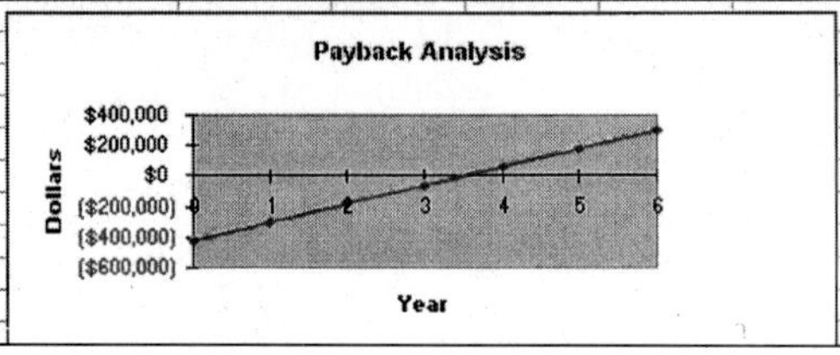

Does that bother you? Earlier we stated that a dollar today is worth more than a dollar a year from now. But it looks as if it is worth less. This is an illusion. The present value is interpreted as follows. If you have 79.7 cents today, it is better than having 79.7 cents two years from now. How much better? Exactly 20.3 cents better since that 79.7 cents would grow into one dollar in two years (assuming our 12 percent discount rate).

To determine the present value of any cost or benefit in year 2, you simply multiply 0.797 times the estimated cost or benefit. For example, the estimated operating expense in year 2 is $16,000. The present value of this expense is $16,000 × 0.797, or $12,752 (rounded up). Fortunately, you don't have to calculate discount factors. There are tables similar to the partial one shown in Figure 9.4 that show the present value of a dollar for different time periods and discount rates. Simply multiply this number times the estimated cost or benefit to get the present value of that cost or benefit. More detailed versions of this table can be found in many accounting and finance books as well as in spreadsheet functions.

Better still, most spreadsheets include built-in functions for calculating the present value of any cash flow, be it cost or benefit. All the examples in this module were done with Microsoft *Excel*. The same tables can be prepared with *Lotus 1-2-3*. The beauty of a spreadsheet is that once the rows, columns, and functions have been set up, you simply enter the costs and benefits and let the spreadsheet discount the numbers to present value. (In fact, you can also program the spreadsheet to perform the cost-benefit analysis.)

Returning to Figure 9.3, we have brought all costs and benefits for our example back to present value. Notice that the discount rate for year 0 is 1.000. Why? The present value of a dollar in year 0 is exactly $1. It makes sense. If you hold a dollar today, it is worth exactly $1!

Now that we've discounted the costs and benefits, we can complete our payback analysis. Look at the cumulative lifetime costs and benefits. The lifetime costs are gradually increasing over the six-year period because operating costs are being incurred. But also notice that the lifetime benefits are accruing at a much faster pace. Lifetime benefits will overtake the lifetime costs between years 3 and 4. By charting the cumulative lifetime time-adjusted Costs + Benefits, we can estimate that the break-even point (when Costs + Benefits = 0) will occur approximately 3.5 years after the system begins operating.

Is this information system a good or bad investment? It depends! Many companies have a payback period guideline for all investments. In the absence of such a guideline, you need to determine a reasonable guideline before you determine the payback period. Suppose that the guideline states that all investments must have a payback period less than or equal to four years. Because

FIGURE 9.4		*Partial Table for Present Value of a Dollar*							
Periods	....	8%	9%	10%	11%	12%	13%	14%	....
1		0.926	0.917	0.909	0.901	0.893	0.885	0.877	
2		0.857	0.842	0.826	0.812	0.797	0.783	0.769	
3		0.794	0.772	0.751	0.731	0.712	0.693	0.675	
4		0.735	0.708	0.683	0.659	0.636	0.613	0.592	
5		0.681	0.650	0.621	0.593	0.567	0.543	0.519	
6		0.630	0.596	0.564	0.535	0.507	0.480	0.456	
7		0.583	0.547	0.513	0.482	0.452	0.425	0.400	
8		0.540	0.502	0.467	0.434	0.404	0.376	0.351	

our example has a payback period of 3.5 years, it is a good investment. If the payback period for the system were greater than four years, the information system would be a bad investment.

It should be noted that you can perform payback analysis without time-adjusting the costs and benefits. The result, however, would show a 2.8-year payback that looks more attractive than the 3.5-year payback that we calculated. Thus, non-time-adjusted paybacks tend to be overly optimistic and misleading.

Return-on-Investment Analysis   The **return-on-investment (ROI) analysis** technique compares the lifetime profitability of alternative solutions or projects. The ROI for a solution or project is a percentage rate that measures the relationship between the amount the business gets back from an investment and the amount invested. The lifetime ROI for a potential solution or project is calculated as follows:

$$\text{Lifetime ROI} = (\text{Estimated lifetime benefits} - \text{Estimated lifetime costs}) \: / \: \text{Estimated lifetime costs}$$

Let's calculate the lifetime ROI for the same systems solution we used in our discussion of payback analysis. Once again, all costs and benefits should be time-adjusted. The time-adjusted costs and benefits were presented in rows 9 and 16 of Figure 9.3. The estimated lifetime benefits minus estimated lifetime costs equal

$$\$795,440 - \$488,692 = \$306,748$$

Therefore, the lifetime ROI is

$$\text{Lifetime ROI} = \$306,748/\$488,692 = .628 = 63\%$$

This is a lifetime ROI, *not* an annual ROI. Simple division by the lifetime of the system ($63 \div 6$) yields an average ROI of 10.5 percent per year. This solution can be compared with alternative solutions. The solution offering the highest ROI is the best alternative. However, as was the case with payback analysis, the business may set a minimum acceptable ROI for all investments. If none of the alternative solutions meets or exceeds that minimum standard, then none of the alternatives is economically feasible. Once again, spreadsheets can greatly simplify ROI analysis through their built-in financial analysis functions.

We could have calculated the ROI without time-adjusting the costs and benefits. This would, however, result in a misleading 129.4 percent lifetime or a 21.6 percent annual ROI. Consequently, we recommend time-adjusting all costs and benefits to current dollars.

Net Present Value   The **net present value** of an investment alternative is considered the preferred cost-benefit technique by many managers, especially those who have substantial business schooling. Once again, you initially determine the costs and benefits for each year of the system's lifetime. And once again, we need to adjust all the costs and benefits back to present dollar values.

Figure 9.5 illustrates the net present value technique. Costs are represented by negative cash flows while benefits are represented by positive cash flows. We have brought all costs and benefits for our example back to present value. Notice again that the discount rate for year 0 (used to accumulate all development costs) is 1.000 because the present value of a dollar in year 0 is exactly \$1.

After discounting all costs and benefits, subtract the sum of the discounted costs from the sum of the discounted benefits to determine the net present value. If it is positive, the investment is good. If negative, the investment is bad. When comparing multiple solutions or projects, the one with the highest positive net present value is the best investment. (This works even if the alternatives have different lifetimes!) In our example the solution being evaluated yields a net present value of \$306,748. This means that if we invest \$306,748 at 12 percent for six years, we will make the same profit that we'd make by implementing this

Cash flow description	Year 0	Year 1	Year 2	Year 3	Year 4	Year 5	Year 6	Total
Net Present Value Analysis for Client-Server System Alternative								
(Numbers rounded to nearest $1)								
Development cost:	($418,040)							
Operation & maintenance cost:		($15,045)	($16,000)	($17,000)	($18,000)	($19,000)	($20,000)	
Discount factors for 12%:	1.000	0.893	0.797	0.712	0.636	0.567	0.507	
Present value of annual costs:	($418,040)	($13,435)	($12,752)	($12,104)	($11,448)	($10,773)	($10,140)	
Total present value of lifetime costs:								($488,692)
Benefits derived from operation of new	$0	$150,000	$170,000	$190,000	$210,000	$230,000	$250,000	
Discount factors for 12%:	1.000	$0.893	$0.797	$0.712	$0.636	$0.567	$0.507	
Present value of annual benefits:	$0	$133,950	$135,490	$135,280	$133,560	$130,410	$126,750	
Total present value of lifetime benefits:								$795,440
NET PRESENT VALUE OF THIS ALTERNATIVE:								$306,748

FIGURE 9.5
Net Present Value Analysis for a Project

information systems solution. This is a good investment provided no other alternative has a net present value greater than $306,748.

Once again, spreadsheets can greatly simplify net present value analysis through their built-in financial analysis functions.

FEASIBILITY ANALYSIS OF CANDIDATE SYSTEMS

During the decision analysis phase of system analysis, the systems analyst identifies candidate system solutions and then analyzes those solutions for feasibility. We discussed the criteria and techniques for analysis in this chapter. In this section, we evaluate a pair of documentation techniques that can greatly enhance the comparison and contrast of candidate system solutions. Both use a matrix format. We have found these matrices useful for presenting candidates and recommendations to management.

Candidate Systems Matrix

The first matrix allows us to compare candidate systems on the basis of several characteristics. The **candidate systems matrix** documents similarities and differences between candidate systems; however, it offers no analysis.

The columns of the matrix represent candidate solutions. Better analysts always consider multiple implementation options. At least one of those options should be the existing system because it serves as our baseline for comparing alternatives.

The rows of the matrix represent characteristics that differentiate the candidates. For purposes of this book, we based some of the characteristics on the information system building blocks. The breakdown is as follows:

- INTERFACES—identify how the system will interact with people and other systems.
- DATA—identify how data stores will be implemented (e.g., conventional files, relational databases, other database structures), how inputs will be captured (e.g., on-line, batch, etc.), how outputs will be generated (e.g., on a schedule, on demand, printed, on screen, etc.).
- PROCESSES—identify how (manual) business processes will be modified, how computer processes will be implemented. For the latter, we have numerous options, including on-line versus batch processes and packaged versus built-in-house software.
- Geography—identify how processes and data will be distributed. Once again, we might consider several alternatives—for example, centralized versus decentralized versus distributed (or duplicated) versus cooperative

(client/server) solutions. Network distribution types and strategies will be discussed in Chapter 11.

The cells of the matrix document whatever characteristics help the reader understand the differences between options. Figure 9.6 demonstrates the basic structure of the matrix.

Before considering any solutions, we must consider any constraints on solutions. Solution constraints take the form of architectural decisions intended to bring order and consistency to applications. For example, a technology architecture may restrict solutions to relational databases or client/server networks.

There are several approaches for identifying candidate solutions, including:

— Recognizing ideas and opinions expressed by users. Throughout a systems project, users may suggest manual or technology-related solutions. However valid these recommendations may appear, they should be given consideration.

— Consulting methodology and architecture standards. Many organizations' development methodology and architecture standards may dictate how technology solutions are to be selected and what technology(ies) may be represented.

— Brainstorming possible solutions. Brainstorming is an effective technique for identifying possible solutions. It is particularly effective when done using an organized approach or framework, such as the IS building blocks or other IS characteristics. Brainstorming should encompass solutions that represent buy, build, and a combination of buy and build options.

— Seeking references. The analyst should solicit ideas and opinions from other persons and organizations that have implemented similar systems.

— Browsing appropriate journals and periodicals. Such literature may feature advertisements and articles concerning automation strategies, successes, failures, and technologies.

A combination of the above approaches could be used independently by the development team members to derive a number of possible alternative system solutions.

A sample, partially completed candidate systems matrix listing three of the five candidates is shown in Figure 9.7. In Figure 9.7, the matrix is used to provide overview characteristics concerning the portion of the system to be computerized, the business benefits, and software tools and/or applications needed. Subsequent pages would provide additional details concerning other characteristics such as those mentioned previously. Two columns can be similar except for their entries in one or two cells. Multiple pages would be used if we were considering more than three candidates. A simple word processing "table" template can be duplicated to create a candidate systems matrix.

Feasibility Analysis Matrix

The second matrix complements the candidate systems matrix with an analysis and ranking of the candidate systems. It is called a **feasibility analysis matrix.**

The columns of the matrix correspond to the same candidate solutions as shown in the candidate systems matrix. Some rows correspond to the feasibility criteria

---

FIGURE 9.6

Candidate Systems Matrix Template

	Candidate 1 Name	Candidate 2 Name	Candidate 3 Name
Interfaces			
Data			
Processes			
Geography			

Characteristics	Candidate 1	Candidate 2	Candidate 3	Candidate ...
Portion of System Computerized Brief description of that portion of the system that would be computerized in this candidate.	COTS package Platinum Plus from Entertainment Software Solutions would be purchased and customized to satisfy Member Services required functionality.	Member Services and warehouse operations in relation to order fulfillment.	Same as candidate 2.	
Benefits Brief description of the business benefits that would be realized for this candidate.	This solution can be implemented quickly because it's a purchased solution.	Fully supports user required business processes for SoundStage Inc. Plus more efficient interaction with member accounts.	Same as candidate 2.	
Servers and Workstations A description of the servers and workstations needed to support this candidate.	Technically architecture dictates Pentium III, MS Windows 2000 class servers and workstations (clients).	Same as candidate 1.	Same as candidate 1.	
Software Tools Needed Software tools needed to design and build the candidate (e.g., database management system, emulators, operating systems, languages, etc.). Not generally applicable if applications software packages are to be purchased.	MS Visual C++ and MS Access for customization of package to provide report writing and integration.	MS Visual Basic 5.0 System Architect 2001 Internet Explorer	MS Visual Basic 5.0 System Architect 2001 Internet Explorer	
Application Software A description of the software to be purchased, built, accessed, or some combination of these techniques.	Package Solution	Custom Solution	Same as candidate 2.	
Method of Data Processing Generally some combination of: on-line, batch, deferred batch, remote batch, and real-time.	Client/Server	Same as candidate 1.	Same as candidate 1.	
Output Devices and Implications A description of output devices that would be used, special output requirements (e.g., network, preprinted forms, etc.), and output considerations (e.g., timing constraints).	(2) HP4MV department laser printers (2) HP5SI LAN laser printers	(2) HP4MV department laser printers (2) HP5SI LAN laser printers (1) PRINTRONIX bar-code printer (includes software & drivers) Web pages must be designed to VGA resolution. All internal screens will be designed for SVGA resolution.	Same as candidate 2.	
Input Devices and Implications A description of input methods to be used, input devices (e.g., keyboard, mouse, etc.), special input requirements (e.g., new or revised forms from which data would be input), and input considerations (e.g., timing of actual inputs).	Keyboard & mouse	Apple "Quick Take" digital camera and software (15) PSC Quickscan laser bar-code scanners (1) HP Scanjet 4C Flatbed Scanner Keyboard & mouse	Same as candidate 2.	
Storage Devices and Implications Brief description of what data would be stored, what data would be accessed from existing stores, what storage media would be used, how much storage capacity would be needed, and how data would be organized.	MS SQL Server DBMS with 100GB arrayed capability.	Same as candidate 1.	Same as candidate 1.	

FIGURE 9.7   *Sample Candidate Systems Matrix*

presented in this chapter. Rows are added to describe the general solution and a ranking of the candidates. The general format is shown in Figure 9.8.

The cells contain the feasibility assessment notes for each candidate. Each row can be assigned a rank or score for each criterion (e.g., for operational feasibility, candidates can be ranked 1, 2, 3, etc.). After ranking or scoring all candidates on each criterion, a final ranking or score is recorded in the last row. Be careful. Not all feasibility criteria are necessarily equal in importance. Before assigning final rankings, you can quickly eliminate any candidates for which any criterion is deemed infeasible. In reality, this doesn't happen very often.

A completed feasibility analysis matrix is presented as Figure 9.9. In Figure 9.9 the feasibility assessment is provided for each candidate solution. In this example, a score is recorded directly in the cell for each candidate's feasibility criteria assessment. Again, this matrix format can be most useful for defending your recommendations to management.

THE SYSTEM PROPOSAL

Recall from Chapter 5 that the decision analysis phase involves identifying candidate solutions, analyzing those solutions, comparing and then selecting the best overall solution, and then recommending a solution. We've just learned how to do the first three tasks. Let's now learn about recommending a solution.

Recommending a solution involves producing a **system proposal.** This deliverable is usually a formal written report or oral presentation intended for system owners and users. Therefore, the systems analysts should be able to write a formal business report and make a business presentation without getting into technical issues or alternatives. Let's survey some important concepts of written reports and presentations.

Written Report

Unfortunately, the written report is the most abused method used by analysts to communicate with system users. We have a tendency to generate large, voluminous reports that look impressive. Sometimes such reports are necessary, but often they are not. If you lay a 300-page technical report on a manager's desk, you can expect that manager will skim it but not read it—and you can be certain it won't be studied carefully.

Length of the Written Report   Report size is an interesting issue. After many bad experiences, we have learned to use the following general guidelines to restrict report size:

- To executive-level managers—one or two pages.
- To middle-level managers—three to five pages.
- To supervisory-level managers—less than 10 pages.
- To clerk-level personnel—less than 50 pages.

FIGURE 9.8
Feasibility Analysis Matrix Template

	Candidate 1 Name	Candidate 2 Name	Candidate 3 Name
Description			
Operational Feasibility			
Technical Feasibility			
Schedule Feasibility			
Economic Feasibility			
Ranking			

Feasibility Criteria	Wt.	Candidate 1	Candidate 2	Candidate 3	Candidate ..
Operational Feasibility **Functionality**. A description of to what degree the candidate would benefit the organization and how well the system would work. **Political**. A description of how well received this solution would be from both user management, user, and organization perspective.	30%	Only supports Member Services requirements and current business processes would have to be modified to take advantage of software functionality Score: 60	Fully supports user required functionality. Score: 100	Same as candidate 2. Score: 100	
Technical Feasibility **Technology**. An assessment of the maturity, availability (or ability to acquire), and desirability of the computer technology needed to support this candidate. **Expertise**. An assessment of the technical expertise needed to develop, operate, and maintain the candidate system.	30%	Current production release of Platinum Plus package is version 1.0 and has only been on the market for 6 weeks. Maturity of product is a risk and company charges an additional monthly fee for technical support. Required to hire or train C++ expertise to perform modifications for integration requirements. Score: 50	Although current technical staff has only Powerbuilder experience, the senior analysts who saw the MS Visual Basic demonstration and presentation have agreed the transition will be simple and finding experienced VB programmers will be easier than finding Powerbuilder programmers and at a much cheaper cost. MS Visual Basic is a mature technology based on version number. Score: 95	Although current technical staff is comfortable with Powerbuilder, management is concerned with recent acquisition of Powerbuilder by Sybase Inc. MS SQL Server is a current company standard and competes with SYBASE in the client/server DBMS market. Because of this we have no guarantee future versions of Powerbuilder will "play well" with our current version of SQL Server. Score: 60	
Economic Feasibility **Cost to develop:** **Payback period (discounted):** **Net present value:** **Detailed calculations:**	30%	Approximately $350,000. Approximately 4.5 years. Approximately $210,000. See Attachment A. Score: 60	Approximately $418,040. Approximately 3.5 years. Approximately $306,748. See Attachment A. Score: 85	Approximately $400,000. Approximately 3.3 years. Approximately $325,500. See Attachment A. Score: 90	
Schedule Feasibility An assessment of how long the solution will take to design and implement.	10%	Less than 3 months. Score: 95	9–12 months Score: 80	9 months Score: 85	
Ranking	100%	60.5	92	83.5	

FIGURE 9.9   *Sample Feasibility Analysis Matrix*

It is possible to organize a larger report to include subreports for managers who are at different levels. These subreports are usually included as early sections in the report and summarize the report, focusing on the bottom line: What's wrong? What do you suggest? What do you want?

Organization of the Written Report   There is a general pattern to organizing any report. Every report consists of both primary and secondary elements.

Primary elements present the actual information that the report is intended to convey. Examples include the introduction and the conclusion.

While the primary elements present the actual information, all reports also contain secondary elements.

Secondary elements package the report so the reader can easily identify the report and its primary elements. Secondary elements also add a professional polish to the report.

As indicated in Figure 9.10, the primary elements can be organized in one of two formats: factual and administrative. The **factual format** is very traditional and best suited to readers who are interested in facts and details as well as conclusions. This is the format we would use to specify detailed requirements and design specifications to system users. But the factual format is not appropriate for most managers and executives.

The **administrative format** is a modern, result-oriented format preferred by many managers and executives. This format is designed for readers who are interested in results, not facts. It presents conclusions or recommendations first. Any reader can read the report straight through, until the point at which the level of detail exceeds their interest.

Both formats include some common elements. The **introduction** should include four components: purpose of the report, statement of the problem, scope of the project, and a narrative explanation of the contents of the report. The **methods and procedures section** should briefly explain how the information contained in the report was developed—for example, how the study was performed or how the new system will be designed. The bulk of the report will be in the **facts section.** This section should be named to describe the type of factual data to be presented (e.g., "Existing Systems Description," "Analysis of Alternative Solutions," or "Design Specifications"). The **conclusion** should briefly summarize the report, verifying the problem statement, findings, and recommendations.

Figure 9.11 shows the secondary, or packaging, elements of the report and their relationship to the primary elements. Many of these elements are self-explanatory. We briefly discuss here those that may not be. No report should be distributed without a **letter of transmittal** to the recipient. This letter should be clearly visible, not inside the cover of the report. A letter of transmittal states what type of action is needed on the report. It can also call attention to any features of the project or report that deserve special attention. In addition, it is an appropriate place to acknowledge the help you've received from various people.

The **abstract or executive summary** is a one- or two-page summary of the entire report. It helps readers decide if the report contains information they need to know. It can also serve as the highest-level summary report. Virtually every manager reads these summaries. Most managers will read on, possibly skipping the detailed facts and appendixes.

Writing the Report   This is not a writing textbook. You should take advantage of every opportunity to improve your writing skills through business and technical writing classes, books, audiovisual courses, and seminars. Writing can greatly influence career paths in any profession. Figure 9.12 illustrates the proper procedure for writing a formal report. Here are some guidelines to follow:

— *Paragraphs should convey a single idea.* They should flow nicely, one to the next. Poor paragraph structure can almost always be traced to outlining deficiencies.

FIGURE 9.10	*Formats for Written Reports*		
	Factual Format		**Administrative Format**
I.	Introduction	I.	Introduction
II.	Methods and procedures	II.	Conclusions and recommendations
III.	Facts and details	III.	Summary and discussion of facts and details
IV.	Discussion and analysis of facts and details	IV.	Methods and procedures
V.	Recommendations	V.	Final conclusion
VI.	Conclusion	VI.	Appendixes with facts and details

Letter of transmittal

Title page

Table of contents

List of figures, illustrations, and tables

Abstract or executive summary
    (The primary elements — the body of the report, in either the factual
    or administrative format — are presented in this portion of the report.)

Appendixes

FIGURE 9.11
*Secondary Elements for
a Written Report*

- *Sentences should not be too complex.* The average sentence length should not exceed 20 words. Studies suggest that sentences longer than 20 words are difficult to read and understand.
- *Write in the active voice.* The passive voice becomes wordy and boring when used consistently.
- *Eliminate jargon, big words, and deadwood.* For example, replace "DBMS" with "database management system," substitute "so" for "accordingly," try "useful" instead of "advantageous," and use "clearly" instead of "it is clear that."

Get yourself a copy of *The Elements of Style* by William S. Strunk, Jr., and E. B. White. This classic paperback may set a record in value-to-cost ratio. Barely bigger than a pocket-sized book, it is a gold mine of information. Anything we might suggest about grammar and style can't be said any more clearly than in *The Elements of Style.*

To communicate information to the many different people involved in a systems development project, a systems analyst is frequently required to make a formal presentation.

Formal Presentation

> **Formal presentations** are special meetings used to sell new ideas and gain approval for new systems. They may also be used for any of these purposes: sell a new system; sell new ideas; sell change; head off criticism; address concerns; verify conclusions; clarify facts; and report progress. In many cases, a formal presentation may set up or supplement a more detailed written report.

Effective and successful presentations require significant preparation. The time allotted to presentations is frequently brief; therefore, organization and format are critical issues. You cannot improvise and expect acceptance.

Presentations offer the advantage of impact through immediate feedback and spontaneous responses. The audience can respond to the presenter, who can use emphasis, timed pauses, and body language to convey messages not possible with the written word. The disadvantage to presentations is that the material presented is easily forgotten because the words are spoken and the visual aids are transient. That's why presentations are often followed by a written report, either summarized or detailed.

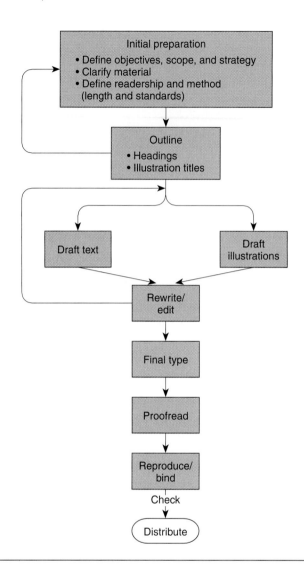

FIGURE 9.12

Steps in Writing a Report

Source: Copyright Keith London.

Preparing for the Formal Presentation   It is particularly important to know your audience. This is especially true when your presentation is trying to sell new ideas and a new system. The systems analyst is frequently thought of as the dreaded agent of change in an organization. As Machiavelli wrote in his classic book *The Prince,*

> There is nothing more difficult to carry out, nor more dangerous to handle, than to initiate a new order of things. For the reformer has enemies in all who profit by the old order, and only lukewarm defenders in all those who would profit from the new order, this lukewarmness arising partly from fear of their adversaries—and partly from the incredulity of mankind, who do not believe in anything new until they have had actual experience of it.[1]

People tend to be opposed to change. There is comfort in the familiar way things are today. Yet a substantial amount of the analyst's job is to bring about change (in methods, procedures, technology, and the like). A successful analyst must be an effective salesperson. It is entirely appropriate (and strongly recommended) for an analyst to formally study salesmanship. To effectively present and sell change, you must be confident in your ideas and have the facts to back them up. Again, preparation is the key!

---

[1] Niccolo Machiavelli, *The Prince and Discourses,* trans. Luigi Ricci (New York: Random House, 1940, 1950). Reprinted by permission of Oxford University Press.

First, define your expectations of the presentation—for instance, that you are seeking approval to continue the project, that you are trying to confirm facts, and so forth. A presentation is a summary of your ideas and proposals that is directed toward your expectations.

Executives are usually put off by excessive detail. To avoid this, your presentation should be carefully organized around the allotted time (usually 30 to 60 minutes). Although each presentation differs, you might try the organization and time allocation suggested in Figure 9.13. This figure illustrates some typical topics of an oral presentation and the amount of time to allow for those topics. Note that this particular outline is for a systems analysis presentation. Other types of presentations might be slightly different.

What else can you do to prepare for the presentation? Because of the limited time, use visual aids—predrawn flipcharts, overhead slides, Microsoft PowerPoint slides, and the like—to support your position. Just like a written paragraph, each visual aid should convey a single idea. When preparing pictures or words, use the guidelines shown in Figure 9.14.

Microsoft PowerPoint contains software guides called wizards to assist the most novice users to create professional-looking presentations. The wizard steps the user through the development process by asking a series of questions and tailoring the presentation based on responses. To hold your audience's attention, consider distributing photocopies of the visual aids at the start of the presentation. This way, the audience doesn't have to take as many notes.

Finally, practice the presentation in front of the most critical audience you can assemble. Play your own devil's advocate or, better yet, get somebody else to raise criticisms and objections. Practice your responses to these issues.

Conducting the Formal Presentation   If you are well prepared, the presentation is 80 percent complete. A few additional guidelines may improve the actual presentation:

— *Dress professionally.* The way you dress influences people. John T. Malloy's books, *Dress for Success* and *The Woman's Dress for Success Book,* are excellent reading for both wardrobe advice and the results of studies regarding the effects of clothing on management.

— *Avoid using the word "I" when making the presentation.* Use "you" and "we" to assign ownership of the proposed system to management.

— *Maintain eye contact with the group and keep an air of confidence.* If you don't show management that you believe in your proposal, why should management believe in it?

FIGURE 9.13	*Typical Outline and Time Allocation for an Oral Presentation*

I. Introduction (one-sixth of total time available)
   A. Problem statement
   B. Work completed to date
II. Part of the presentation (two-thirds of total time available)
   A. Summary of existing problems and limitations
   B. Summary description of the proposed system
   C. Feasibility analysis
   D. Proposed schedule to complete project
III. Questions and concerns from the audience (time here is not to be included in the time allotted for presentation and conclusion; it is determined by those asking the questions and voicing their concerns)
IV. Conclusion (one-sixth of total time available)
   A. Summary of proposal
   B. Call to action (request for whatever authority you require to continue systems development)

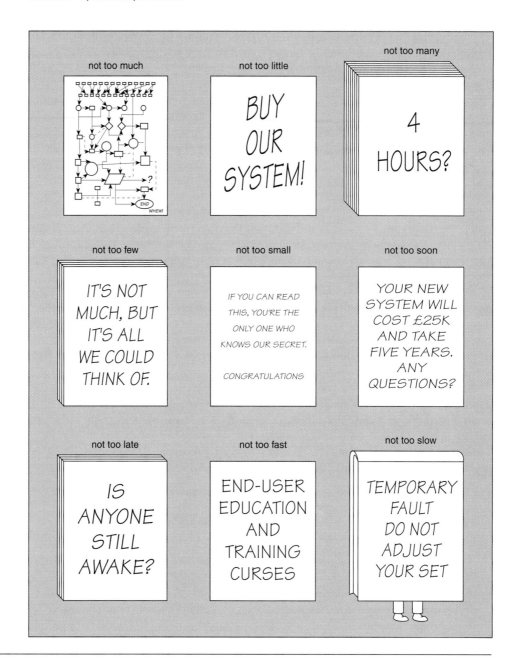

FIGURE 9.14
Guidelines for Visual Aids
Source: Copyright Keith London.

— *Be aware of your own mannerisms.* Some of the most common mannerisms include using too many hand gestures, pacing, and repeatedly saying "you know" or "OK." Although mannerisms alone don't contradict the message, they can distract the audience.

Sometimes while you are making a presentation, some members of the audience may not be listening. This lack of attention may take several forms. Some people may be engaged in competing conversations, some may be daydreaming, some may be busy glancing at their watches, some who are listening may have puzzled expressions, and some may show no expression. The following suggestions may prove useful to keep people listening:

— *Stop talking.* The silence can be deafening. The best public speakers know how to use dramatic pauses for special emphasis.

- *Ask a question, and let someone in the audience answer it.* This involves the audience in the presentation and is a very effective way of stopping a competing conversation.

- *Try a little humor.* You don't have to be a talented comedian. But everybody likes to laugh. Tell a joke on yourself.

- *Use some props.* Use some type of visual aid to make your point clearer. Draw on the chalkboard, illustrate on the back of your notes, create a physical model to make the message easier to understand.

- *Change your voice level.* By making your voice louder or softer, you force the audience to listen more closely or make it easier for them to hear. Either way, you've made a change from what the audience was used to, and that is the best way to get and hold attention.

- *Do something totally unexpected.* Drop a book, toss your notes, jingle your keys. Doing the unexpected is almost always an attention grabber.

Usually a formal presentation will include time for questions from the audience. This time is very important because it allows you to clarify any points that were unclear and draw additional emphasis to important ideas. It also allows the audience to interact with you. However, sometimes answering questions after a presentation may be difficult and frustrating. We suggest the following guidelines when answering questions:

- *Always answer a question seriously, even if you think it is a silly question.* Remember, if you make someone feel stupid for asking a "dumb" question, that person will be offended. Also, other members of the audience won't ask their questions for fear of the same treatment.

- *Answer both the individual who asked the question and the entire audience.* If you direct all your attention to the person who asked the question, the rest of the audience will be bored. If you don't direct enough attention to the person who asked the question, that person won't be satisfied. Try to achieve a balance. If the question is not of general interest to the audience, answer it later with that specific person.

- *Summarize your answers.* Be specific enough to answer the question, but don't get bogged down in details.

- *Limit the amount of time you spend answering any one question.* If additional time is needed, wait until after the presentation is over.

- *Be honest.* If you don't know the answer to a question, admit it. Never try to bluff your way out of a question. The audience will eventually find out, and you will have destroyed your credibility. Instead, promise to find out and report back. Or ask someone in the audience to do some research and present the findings later.

Following Up the Formal Presentation   As mentioned earlier, it is extremely important to follow up a formal presentation because the spoken word and impressive visual aids used in a presentation do not usually leave a lasting impression. For this reason, most presentations are followed by written reports that provide the audience with a more permanent copy of the information that was communicated.

SUMMARY

1. Feasibility is a measure of how beneficial the development of an information system would be to an organization. Feasibility analysis is the process by which we measure feasibility. It is an ongoing evaluation of feasibility at various checkpoints in the life cycle. At any of these checkpoints, the project may be canceled, revised, or continued. This is called a creeping commitment approach to feasibility.
2. There are four feasibility tests: operational, technical, schedule, and economic.
   a. Operational feasibility is a measure of problem urgency or solution acceptability. It includes a measure of how the end-users and managers feel about the problems or solutions.
   b. Technical feasibility is a measure of how practical solutions are and whether the technology is already available within the organization. If the technology is not available to the firm, technical feasibility also looks at whether it can be acquired.
   c. Schedule feasibility is a measure of how reasonable the project schedule or deadline is.
   d. Economic feasibility is a measure of whether a solution will pay for itself or how profitable a solution will be. For management, economic feasibility is the most important of our four measures.
3. To analyze economic feasibility, you itemize benefits and costs. Benefits are either tangible (easy to measure) or intangible (hard to measure). To properly analyze economic feasibility, try to estimate the value of all benefits. Costs fall into two categories: development and operating.
   a. Development costs are onetime costs associated with analysis, design, and implementation of the system.
   b. Operating costs may be fixed over time or variable with respect to system usage.
4. Given the costs and benefits, economic feasibility is evaluated by the techniques of cost-benefit analysis. Cost-benefit analysis determines if a project or solution will be cost-effective—if lifetime benefits will exceed lifetime costs. There are three popular ways to measure cost-effectiveness: payback analysis, return-on-investment analysis, and net present value analysis.
   a. Payback analysis defines how long it will take for a system to pay for itself.
   b. Return-on-investment and net present value analyses determine the profitability of a system.
   c. Net present value analysis is preferred because it can compare alternatives with different lifetimes.
5. A candidate systems matrix is a useful tool for documenting the similarities and differences between candidate systems being considered.
6. A feasibility analysis matrix is used to evaluate and rank candidate systems. Both the candidate systems matrix and the feasibility analysis matrix are useful for presenting the results of a feasibility analysis as part of a system proposal.
7. Written reports are the most common communications vehicle used by analysts. Reports consist of both primary and secondary elements. Primary elements contain factual information. Secondary elements package the report for ease of use. Reports may be organized in either the factual or administrative format. The factual format presents the details before conclusions; the administrative format reverses that order. Managers like the administrative format because it is results-oriented and gets right to the bottom-line question.
8. Formal presentations are a special type of meeting at which a person presents conclusions, ideas, or proposals to an interested audience. Preparation is the key to effective presentations.
9. The system proposal may be a formal written report or oral presentation.

KEY TERMS

abstract or executive summary, p. 382
administrative format, p. 382
candidate systems matrix, p. 377
conclusion, p. 382
creeping commitment, p. 365
discount rate, p. 374
economic feasibility, p. 367
facts section, p. 382
factual format, p. 382
feasibility, p. 365
feasibility analysis, p. 365
feasibility analysis matrix, p. 377
fixed costs, p. 371
formal presentation, p. 383
intangible benefits, p. 371
introduction, p. 382
letter of transmittal, p. 382
methods and procedures section, p. 382
net present value, p. 376
operational feasibility, p. 367
opportunity cost, p. 374
payback analysis, p. 374
payback period, p. 374
present value, p. 374
primary elements, p. 381
return-on-investment (ROI) analysis, p. 376
schedule feasibility, p. 367
secondary elements, p. 381
system proposal, p. 380
tangible benefits, p. 371
technical feasibility, p. 367
time value of money, p. 373
usability analysis, p. 368
variable costs, p. 371

REVIEW QUESTIONS

1. What is the difference between feasibility and feasibility analysis?
2. What are the four tests for project feasibility? How is each test for feasibility measured?
3. How can the PIECES framework be used in operational feasibility analysis? Explain.
4. When performing usability analysis, list the three goals that help measure the usability of an interface.
5. What are the two categories of costs? Give several examples of each.
6. What is the difference between fixed and variable operating costs? Give several examples of each.
7. What is the difference between a tangible and an intangible benefit? Give several examples of each.
8. What are the three popular techniques to assess economic feasibility?
9. Of the three techniques in question 8, which one is most preferred by managers today?
10. List four types of business and technical reports. List the guidelines for the length of the report based on the intended audience.
11. When writing a business or technical report, why should you write in the active voice?
12. When you are asked a question that you don't know the answer to while giving a formal presentation, how should you respond?

PROBLEMS AND EXERCISES

1. Explain what is meant by the creeping commitment approach to feasibility. What feasibility checkpoints can be built into a systems development life cycle?
2. Can you think of any technological trends or products that may be technically infeasible for the small- to medium-sized business at the current time? Defend your reasoning.
3. Whether or not you have information systems experience, you have experience with people who use computers (including friends, relatives, acquaintances, teachers, and fellow employees). Considering their biases for and against computers, identify issues that may make a proposed system operationally infeasible or unacceptable to those individuals.
4. List several intangible benefits. How would you quantify each intangible benefit in terms of dollars and cents (a measure that management can understand)?
5. What are some of the advantages and disadvantages of the payback analysis, return-on-investment analysis, and present value analysis cost-benefit techniques?
6. Identify two formats for a written report. What are the elements common to both formats? Should the length of a written report vary by audience? Why or why not?
7. The secret of effective oral and written communications is to know your audience. What are some things you would want to know about the audience?
8. Why do formal presentations usually accompany written reports?
9. Systems analysts tend to generate large written reports—too large for managers to read. How would you handle a size problem with a technical report?
10. What are some ways you might improve your written communications skills? Identify specific courses that help you improve your skills.

PROJECTS AND RESEARCH

1. Visit a local information systems shop. Try to obtain documentation of the systems development life cycle standards or guidelines. What feasibility checkpoints have been installed? What feasibility checkpoints do you think should be installed?
2. What feasibility criteria does the information systems shop you visited for Project 1 use to evaluate projects? How do the criteria compare to the criteria in this book? Have we omitted any tests that they feel are important? Have they omitted any tests we use?
3. While attending the next lecture in each of your classes, observe the instructor's presentation of class material. If you feel comfortable doing it, discuss your findings with your professor.
4. Try to obtain a systems development report outline or table of contents from an information systems shop. Was the report organized using the factual format or the administrative format? Do you think everybody who should have read that report did read it? Why or why not?
5. Reorganize the report in Question 4 into an alternative outline or table of contents. Be sure to include secondary elements even if they weren't included in the original report.

MINICASES

1. A new production scheduling information system for XYZ Corporation could be developed at a cost of $125,000. The estimated net operating costs and estimated net benefits over five years of operation would be:

Year	Estimated Net Operating Costs	Estimated Net Benefits
0	$125,000	$ 0
1	3,500	26,000
2	4,700	34,000
3	5,500	41,000
4	6,300	55,000
5	7,000	66,000

Assuming a 12 percent discount rate, what would be the payback period for this investment? Would this be a good or bad investment? Why? What is the annual ROI (return on investment) for the project? What is the net present value

of the investment if the current discount rate is 12 percent?

2. The analysis team of Purcref Corporation has completed its feasibility analysis of alternative system solutions to a new investment tracking system. Members of the team disagree about which solution to recommend to management. They all agree that a fair and unbiased feasibility analysis has been done on the three candidates. They simply disagree on which is the best candidate Some believe that candidate A should be recommended because it is the most technically feasible solution. Some feel candidate B should be chosen because it is the most operationally feasible. Others think candidate C should be chosen because it is economically the most feasible of the three. Candidate A can be completed about one month sooner than candidates B and C. Which candidate should be recommended to management? Why? How do you think the team should select the candidate to propose?

SUGGESTED READINGS

Bovee, Courtland L., and John V. Thill. *Business Communications Today*. 2nd ed. New York: Random House, 1989.

Gildersleeve, Thomas R. *Successful Data Processing Systems Analysis*. 2nd ed. Englewood Cliffs, NJ: Prentice Hall, 1985. This book provides an excellent chapter on cost-benefit analysis techniques. Chapter 5 discusses presentations. We are indebted to Gildersleeve for the creeping commitment concept.

Gore, Marvin, and John Stubbe. *Elements of Systems Analysis*. 4th ed. Dubuque, IA: Brown, 1988. The feasibility analysis chapter suggests an interesting matrix approach to identifying, cataloging, and analyzing the feasibility of alternative solutions for a system.

Smith, Randi Sigmund. *Written Communications for Data Processing*. New York: Van Nostrand Publishing, 1976.

Stuart, Ann. *Writing and Analyzing Effective Computer System Documentation*. New York: Holt, Rinehart and Winston, 1984.

Uris, Auren. *The Executive Deskbook*. 3rd ed. New York: Van Nostrand Reinhold, 1988.

Walton, Donald. *Are You Communicating? You Can't Manage Without It*. New York: McGraw-Hill, 1989.

Wetherbe, James. *Systems Analysis and Design: Traditional, Structured, and Advanced Concepts and Techniques*. 2nd ed. St. Paul, MN: West, 1984. Wetherbe pioneered the PIECES framework for problem classification. In this chapter we extended that framework to analyze operational feasibility of solutions.

PART THREE

SYSTEMS DESIGN METHODS

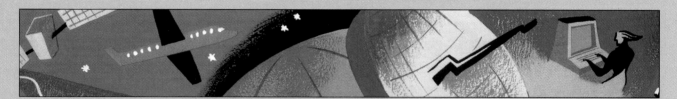

The chapters in Part Three introduce you to systems design methods. Chapter 10, Systems Design, provides the context for all the subsequent chapters by introducing the activities of systems design. Systems design includes the preparation of detailed computer-based specifications that will fulfill the requirements specified during systems analysis and construction of system prototypes. With respect to information systems development, systems design consists of the configuration, procurement, and design and integration phases.

Chapter 11, Application Architecture and Modeling, introduces physical process and data design. It specifically addresses design decisions regarding distribution issues for shared data and processes. This results in an application architecture that consists of design units that can be assigned to different team members for detailed design, construction, and unit testing. The chapter also includes coverage of the new client/server approach.

Chapter 12, Database Design, introduces the design of physical data stores from the data model developed in Chapter 7.

Chapter 13, Output Design and Prototyping, teaches output design and prototyping. Different types, formats, and media for outputs are presented. The use of the most common types of graphs are discussed. The chapter demonstrates how to design and prototype printed and display outputs.

Chapter 14, Input Design and Prototyping, teaches input design and prototyping. Formats, methods, media, human factors, and internal controls for inputs are stressed. The proper usage of screen-based controls for data input on graphical user interface (GUI) screen designs is discussed. The chapter also emphasizes prototyping as a way of finding, documenting, and communicating input design requirements.

Chapter 15, User Interface Design and Prototyping, teaches user interface design and prototyping. You will learn how to develop a friendly and effective interface for an application. The design of the user interface is crucial because user acceptance of the system is frequently dependent on a friendly, easy-to-use interface. A GUI-based interface for obtaining the inputs and outputs designed in Chapters 13 and 14 is demonstrated.

Stakeholders

Activities

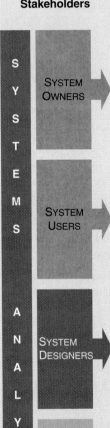

BUILDING BLOCKS OF AN INFORMATION SYSTEM

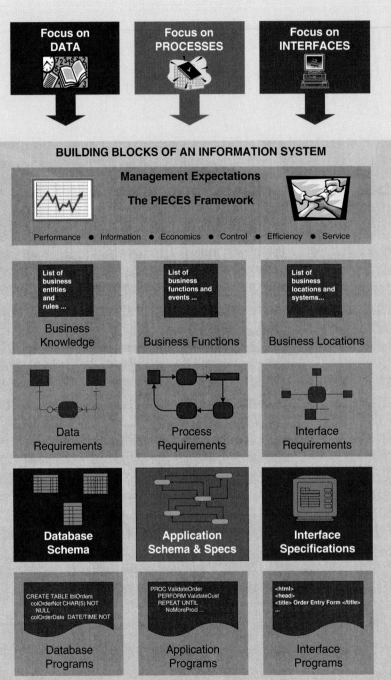

Management Expectations

The PIECES Framework

Performance ● Information ● Economics ● Control ● Efficiency ● Service

List of business entities and rules ...	List of business functions and events ...	List of business locations and systems...
Business Knowledge	**Business Functions**	**Business Locations**
Data Requirements	**Process Requirements**	**Interface Requirements**
Database Schema	**Application Schema & Specs**	**Interface Specifications**
CREATE TABLE tblOrders colOrderNot CHAR(5) NOT NULL colOrderDate DATE/TIME NOT	PROC ValidateOrder PERFORM ValidateCust REPEAT UNTIL NoMoreProd ...	<html> <head> <title> Order Entry Form </title> ...
Database Programs	**Application Programs**	**Interface Programs**

S
Y
S
T
E
M
S

A
N
A
L
Y
S
T
S

SYSTEM OWNERS

SYSTEM USERS

SYSTEM DESIGNERS

SYSTEM BUILDERS

PROJECT & PROCESS MANAGEMENT

PRELIMINARY INVESTIGATION

PROBLEM ANALYSIS

REQUIREMENTS ANALYSIS

DECISION ANALYSIS

DESIGN

CONSTRUCTION

IMPLEMENTATION

VENDORS AND CONSULTANTS

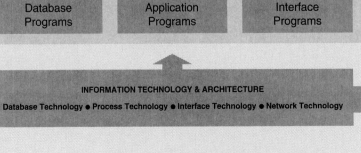

INFORMATION TECHNOLOGY & ARCHITECTURE

Database Technology ● Process Technology ● Interface Technology ● Network Technology

OPERATIONS AND SUPPORT

10

SYSTEMS DESIGN

CHAPTER PREVIEW AND OBJECTIVES

In this chapter you will learn more about the design phase in the FAST systems development methodology. You will know that you understand the process of systems design when you can:

— Describe the design phase in terms of your information building blocks.

— Identify and differentiate between several systems design strategies.

— Describe the design phase tasks in terms of a computer-based solution for an in-house development project.

— Describe the design phase in terms of a computer-based solution involving procurement of a commercial systems software solution.

Although some techniques of systems design are introduced in this chapter, it is not the intent of this chapter to teach the techniques of systems design. This chapter teaches only the process of systems design and introduces you to some techniques that will be taught in later chapters.

SOUNDSTAGE
SOUNDSTAGE ENTERTAINMENT CLUB

SCENE

This episode begins soon after the project's executive sponsor (the person who pays for the system to be built) has approved the business requirements that came out of the requirements analysis phase. Sandra and Bob are planning the design phases for the project. We join Sandra and Bob in a small conference room.

SANDRA

OK, Bob, here's what I think we should do. Our *FAST* methodology calls for the completion of the design phase. Let's agree on what needs to be done and then determine how we're going to get the job done.

BOB

Sounds good to me. Oh yeah ... I wanted to remind you that we better not forget that Dick Krieger (director of Warehouse Operations) told us the warehousing operations are using bar-coding technology. He is really concerned that we are going to come up with some design that does not integrate with their bar-coding system

SANDRA

According to our *FAST* methodology we need to begin by designing the applica-

tion architecture. It was a miserable rainy weekend and I was bored. So I took the liberty to attack this first design task. Here are the physical data flow diagrams I developed to show how the system data and processes are to be distributed across the various business locations. These diagrams are going to be our blueprints for completing the remaining design tasks. We can use the diagrams to split up the design work.

BOB

Great. How do you want to proceed?

SANDRA

First, I hope you agree that we should use a prototyping approach to completing the design phase.

BOB

Do you think the users will be willing to participate in reviewing the screens that we design?

SANDRA

Don't you remember, we already discussed this with the people at our launch meeting? They assured us that they will encourage the users to participate.

BOB

Sure. Actually I would prefer to prototype the design.

SANDRA

Good. I was going to suggest that you build our prototype database. I think that is a one-person job. When you are finished, let me know. We can then meet to decide which inputs and outputs we want to design and how we want to go about designing the overall interface for the system.

BOB

Sounds good to me. I'm hungry. Let's go eat. I want to hurry back and get going on the prototype design.

DISCUSSION QUESTIONS

1. How would you characterize the focus of the design phase of this project?

2. How could Sandra and Bob present their hardware and software to management without overwhelming them with technical jargon?

3. Why would Sandra want to commit to prototyping (building models) as the approach to use in completing the design and integration phase for their project?

WHAT IS SYSTEMS DESIGN?

In Chapter 3 you learned about the systems development process. In that chapter we purposefully limited our discussion to only briefly examine each phase. In this chapter, we take a much closer look at the systems design phase that follows systems analysis.

> Information **systems design** is defined as those tasks that focus on the specification of a detailed computer-based solution. It is also called **physical design.**

Thus, whereas systems analysis emphasized the business problem, systems design focuses on the technical or implementation concerns of the system.

As was illustrated in the chapter map at the start of this chapter, systems design is driven by the technical concerns of SYSTEM DESIGNERS. Hence, it addresses the DATA, PROCESS, and INTERFACE building blocks from the SYSTEM DESIGNERS' perspective. The SYSTEMS ANALYSTS serve as facilitators of systems design.

Most of us define the process of design too restrictively. We envision ourselves drawing blueprints of the computer-based systems to be programmed and developed by ourselves or our own programmers. Thus, we design inputs, outputs, files, databases, and other computer components. Recruiters of computer-educated graduates refer to this restrictive definition as the "not-invented-here syndrome." In reality, many companies purchase more software than they write in-house. That shouldn't surprise you. Why reinvent the wheel? Many systems are sufficiently generic that computer vendors have written adequate—but rarely, if ever, perfect—software packages that can be bought and possibly modified to fulfill end-user requirements.

This chapter examines systems design from the perspectives of both in-house development or "build" projects and software procurement or "buy" projects. Let's begin our study by first examining some overall strategies for systems design.

There are many strategies or techniques for performing systems design. They include *modern structured design, information engineering, prototyping, JAD, RAD, and object-oriented design.* These strategies are often viewed as competing alternative approaches to systems design, but in reality, certain combinations complement one another. Let's briefly examine these strategies and the scope or goals of the projects to which they are suited. The intent is develop a high-level understanding only. The subsequent chapters and Module B will actually teach you the techniques.

> NOTE   Recall from Chapter 3 that methodology "routes" are sometimes defined for these approaches.

Structured design, information engineering, and object-oriented design are examples of model-driven approaches.

> **Model-driven design** emphasizes the drawing of pictorial system models to document the technical or implementation aspects of a new system.

The design models are often derived from logical models that were developed earlier in model-driven analysis (discussed in Chapter 5). Ultimately, the system design models become the blueprints for constructing and implementing the new system.

Today, model-driven approaches are almost always enhanced by the use of automated tools. Some designers draw system models with general-purpose graphics software such as *Visio Professional* or *Corel Flow*. Other designers and organizations require the use of repository-based CASE or modeling tools such as *System Architect 2001, Visio Enterprise, Visible Analyst,* or *Rational ROSE*. CASE tools offer consistency and completeness as well as rule-based error checking.

Let's briefly examine the most commonly encountered model-driven design approaches. Model-driven design approaches are featured in the model-driven methodologies and routes (introduced in Chapter 3).

Modern Structured Design   Structured design techniques help developers deal with the size and complexity of programs.

> **Modern structured design** is a process-oriented technique for breaking up a large program into a hierarchy of modules that result in a computer program that is easier to implement and maintain (change). Synonyms (although technically inaccurate) are top-down program design and structured programming.

The concept is simple. Design a program as a top-down hierarchy of modules. A module is a group of instructions—a paragraph, block, subprogram, or subroutine. The top-down structure of these modules is developed according to various design rules and guidelines. (Thus, merely drawing a hierarchy or structure chart for a program is *not* structured design).

Structured design is considered a process-oriented technique because its emphasis is on the PROCESS building blocks in our information system—specifically, software processes. Structured design seeks to factor a program into the top-down hierarchy of modules that have the following properties:

SYSTEMS DESIGN APPROACHES

Model-Driven Approaches

— Modules should be highly **cohesive;** that is, each module should accomplish one and only one function. This makes the modules reusable in future programs.

— Modules should be loosely **coupled;** in other words, modules should be minimally dependent on one another. This minimizes the effect that future changes in one module will have on other modules.

The software model derived from structured design is called a **structure chart** (Figure 10.1). The structure chart is derived by studying the flow of data through the program. Structured design is performed during systems design. It does not address all aspects of design—for instance, structured design will not help you design inputs, outputs, or databases.

Structured design has lost some of its popularity with many of today's applications that call for newer techniques that focus on *event-driven* and *object-oriented*

FIGURE 10.1 *The End Product of Structured Design*

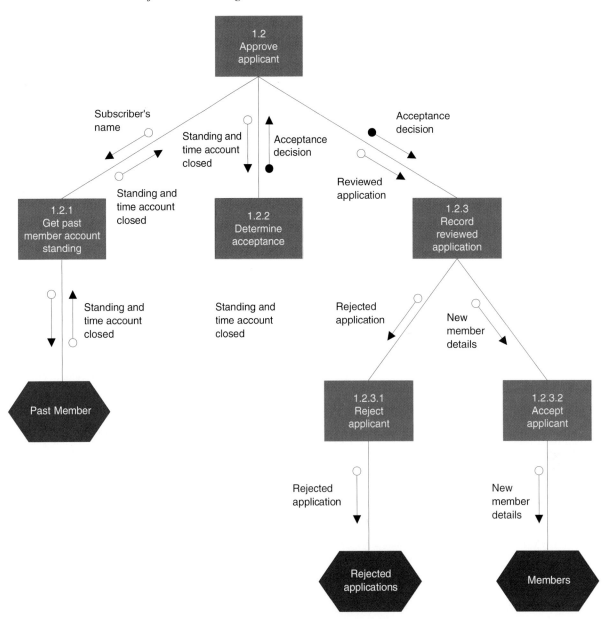

programming techniques. However, it is still a popular technique involving the design of mainframe-based application software and is used to address coupling and cohesion issues at the "system" level.

Information Engineering (IE)   In Chapter 5 you learned that

> **Information engineering (IE)** is a model-driven and DATA-centered, but PROCESS-sensitive technique to plan, analyze, and design information systems.

The primary tool of IE is a data model diagram (see Figure 10.2). IE involves conducting a business area requirements analysis from which information system applications are carved out and prioritized. The applications identified in IE become projects to which other systems analysis *and* design methods are intended to be applied in order to develop the production systems. These methods may include some combination of modern structured analysis (discussed in Chapter 5), modern structured design, prototyping, and object-oriented analysis and design.

Prototyping   Traditionally, physical design has been a paper-and-pencil process. Analysts drew pictures that depicted the layout or structure of outputs, inputs, and databases and the flow of dialogue and procedures. This is a time-consuming

FIGURE 10.2   *Sample Information Engineering Physical Entity Relationship Diagram*

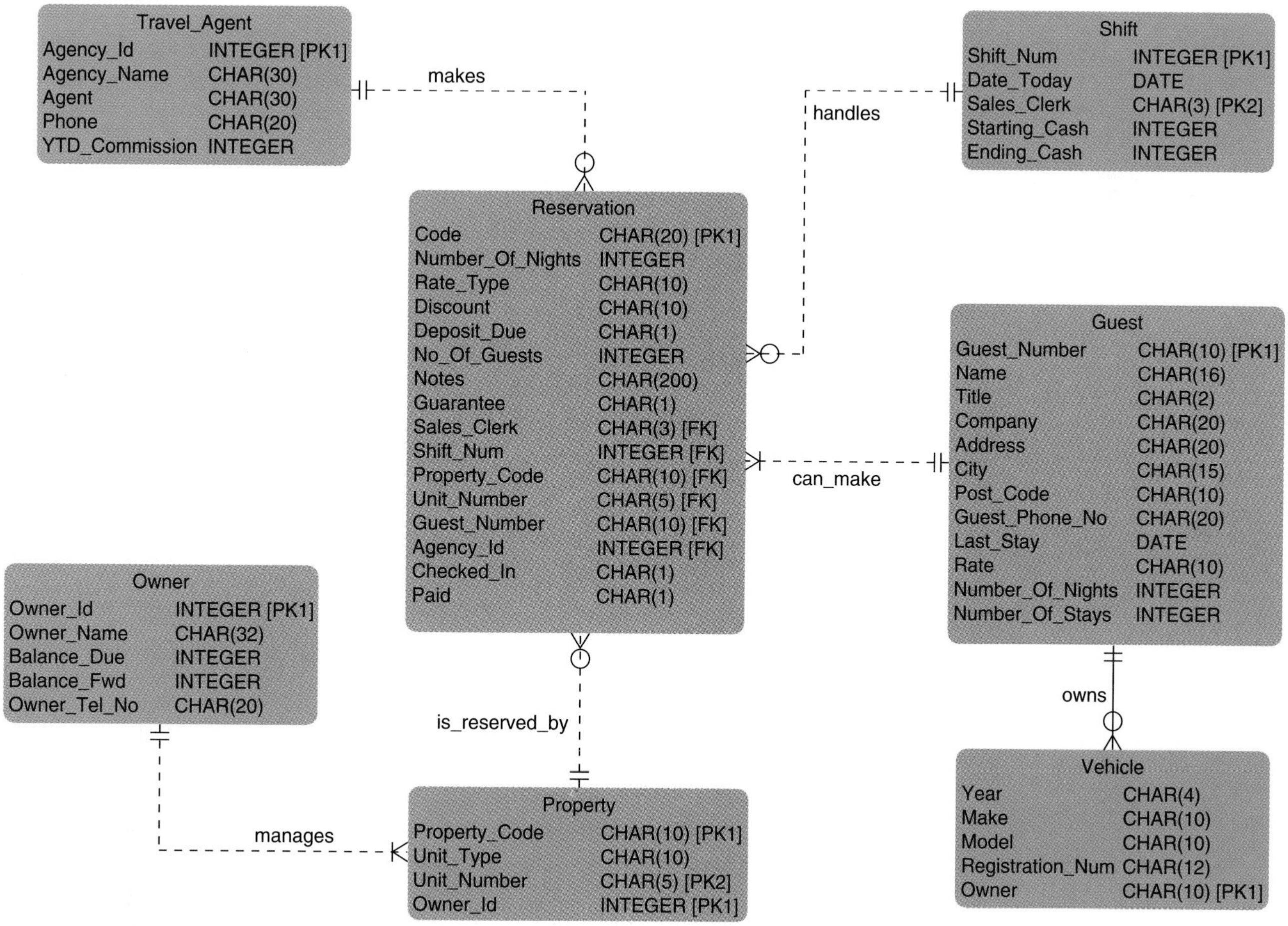

process that is prone to considerable error and omissions. Frequently, the resulting paper specifications were inadequate, incomplete, or inaccurate.

Today, many analysts and designers prefer prototyping, a modern engineering-based approach to design. The prototyping approach is an iterative process involving a close working relationship between the designer and the users. This approach has several advantages.

- Prototyping encourages and requires active end-user participation. This increases end-user morale and support for the project. End-user morale is enhanced because the system appears real to them.
- Iteration and change are a natural consequence of systems development—that is, end-users tend to change their minds. Prototyping better fits this natural situation because it assumes that a prototype evolves, through iteration, into the required system.
- It has often been said that end-users don't fully know their requirements until they see them implemented. If so, prototyping endorses this philosophy.
- Prototypes are an active, not passive, model that end-users can see, touch, feel, and experience.
- An approved prototype is a working equivalent to a paper design specification, with one exception—errors can be detected much earlier.
- Prototyping can increase creativity because it allows for quicker user feedback, which can lead to better solutions.
- Prototyping accelerates several phases of the life cycle, possibly bypassing the programmer. In fact, prototyping consolidates parts of phases that normally occur one after the other.

There are also disadvantages or pitfalls to using the prototyping approach. Most of these can be summed up in one statement: Prototyping encourages ill-advised shortcuts through the life cycle. Fortunately, the following pitfalls can all be avoided through proper discipline.

- Prototyping encourages a return to the "code, implement, and repair" life cycle that used to dominate information systems. As many companies have learned, systems developed in prototyping languages can present the same maintenance problems that have plagued legacy systems developed in languages such as *COBOL*.
- Prototyping does not negate the need for the systems analysis phases. A prototype can just as easily solve the wrong problems and opportunities as a conventionally developed system.
- You cannot completely substitute any prototype for a paper specification. No engineer would prototype an engine without some paper design. Yet many information systems professionals try to prototype without a specification. Prototyping should be used to complement, not replace, other methodologies. The level of detail required of the paper design may be reduced, but it is not eliminated.
- Numerous design issues are not addressed by prototyping. These issues can inadvertently be forgotten if you are not careful.
- Prototyping often leads to premature commitment to a design (usually the first design that is developed).
- When prototyping, the scope and complexity of the system can quickly expand beyond original plans. This can easily get out of control.
- Prototyping can reduce creativity in designs. The very nature of any implementation—for instance, a prototype of a report—can prevent analysts, designers, and end-users from looking for better solutions.

Prototypes often suffer from slower performance than their third-generation language counterparts (albeit this difference is rapidly becoming a non-issue).

Prototypes can be quickly developed using many of the 4GLs and object-oriented programming languages available today. Figure 10.3 depicts a prototype screen for a system. Prototypes can be built for simple outputs, computer dialogues, key functions, entire subsystems, or even the entire system. Each prototype system is reviewed by end-users and management, who make recommendations about requirements, methods, and formats. The prototype is then corrected, enhanced, or refined to reflect the new requirements. Prototyping technology makes such revisions in a relatively straightforward manner. The revision and review process continues until the prototype is accepted. At that point, the end-users are accepting both the requirements and the design that fulfills those requirements.

Design by prototyping doesn't necessarily fulfill all design requirements. For instance, prototypes don't always address important performance issues and storage constraints. Prototypes rarely incorporate internal controls. The analyst or designer must still specify these.

Object-Oriented Design   Object-oriented design (OOD) is the newest design strategy. The concepts behind this strategy (and technology) are covered extensively in Module B, Object-Oriented Design and Modeling, but a simplified introduction is appropriate here. This technique is an extension of the object-oriented analysis strategy presented in Chapter 5 and covered in more detail in Module A, Object-Oriented Analysis and Modeling. Figure 10.4 shows one of the many diagrams used in object-oriented design.

Object technologies and techniques are an attempt to eliminate the separation of concerns about DATA and PROCESS.

FIGURE 10.3   *Sample Prototype Screen*

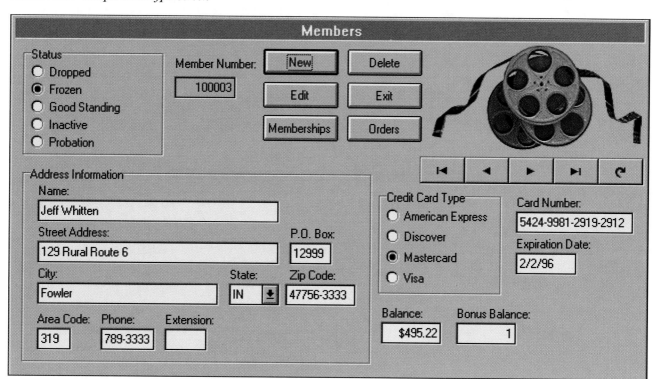

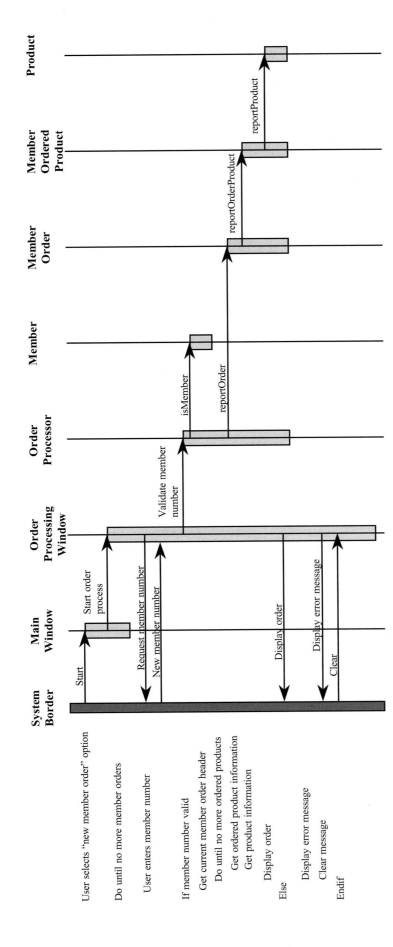

FIGURE 10.4 *Sample Object-Oriented Design Model*

Object-oriented design (OOD) techniques are used to refine the object requirements definitions identified earlier during analysis and to define design specific objects.

For example, based on a design implementation decision, during OOD the designer may need to revise the data or process characteristics for an object that was defined during systems analysis. Likewise, a design implementation decision may necessitate that the designer define a new set of objects that will make up an interface screen that the user(s) may interact with in the new system.

Another popular design strategy used today is rapid application development.

Rapid application development (RAD) is the merger of various structured techniques (especially the DATA-driven information engineering) with *prototyping* techniques and *joint application development* techniques to accelerate systems development.

RAD calls for the interactive use of structured techniques and prototyping to define the users' requirements and design the final system. Using structured techniques, the developer first builds preliminary data and process models of the business requirements. Prototypes then help the analyst and users to verify those requirements and to formally refine the data and process models. The cycle of models, then prototypes, then models, then prototypes, and so forth ultimately results in a combined business requirements and technical design statement to be used for constructing the new system.

The expedition of the design effort is enhanced through the emphasis on user participation in joint application development sessions. Recall that **joint application development (JAD),** introduced in Chapter 5 and discussed in more detail in Chapter 6, is a technique that complements other systems analysis and design techniques by emphasizing *participative development* among SYSTEM OWNERS, USERS, DESIGNERS, and BUILDERS. During the JAD sessions for systems design, the systems designer will take on the role of facilitator for possibly several full-day workshops intended to address different design issues and deliverables. JAD is an essential element contributing greatly to the acceleration emphasis of RAD.

Like most commercial methodologies, our hypothetical *FAST* methodology does not impose a single approach on systems design. Instead, it integrates all the popular approaches introduced in the preceding paragraphs. The SoundStage case study will demonstrate these methods in the context of a typical first assignment for a systems analyst. The systems analysis techniques will be applied within the framework of:

— Your information system building blocks (from Chapter 2).
— The *FAST* phases (from Chapter 3).
— *FAST* tasks that implement a phase (described in this chapter).

Given this context, we can now study systems design. We will begin by studying systems design as it relates to an in-house development or "build" project. Afterward, we will examine how the systems design phases are affected when a decision has been made to acquire or "buy" a commercial software package as solution.

Let's begin by placing systems design for in-house development projects into context relative to the system life cycle. As is illustrated in Figure 10.5, an approved system proposal from the decision analysis phase triggers the design phase. The goal of the design phase is twofold. First, the analyst seeks to design a system that both fulfills requirements and will be friendly to its end-users. Human

Rapid Application Development (RAD)

FAST Systems Design Strategies

SYSTEMS DESIGN FOR IN-HOUSE DEVELOPMENT —THE "BUILD" SOLUTION

engineering will play a pivotal role during design. Second, and still very important, the analyst seeks to present clear and complete specifications to the computer programmers and technicians. As is shown is Figure 10.5 the approved design specifications will trigger the construction phase of our in-house development project.

FIGURE 10.5   *The Context of Systems Design for In-House Development*

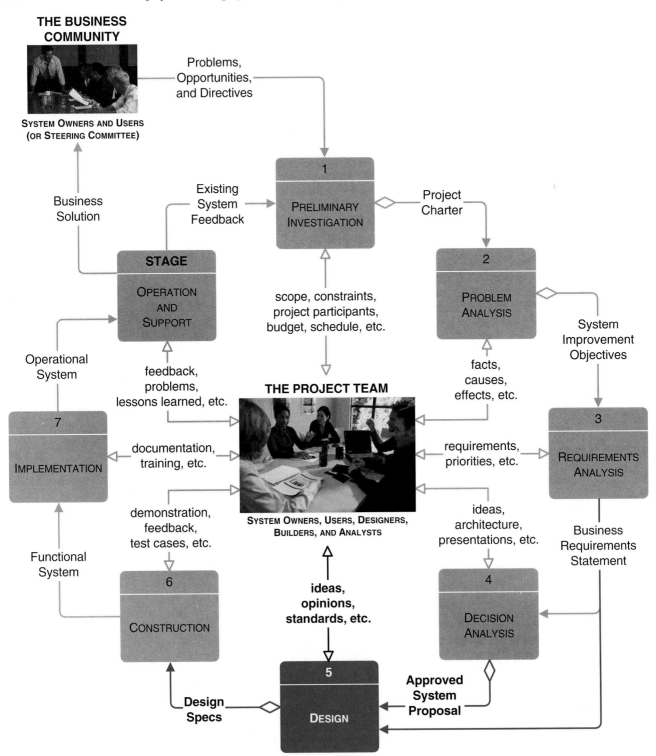

Figure 10.6 is a task diagram depicting the work (= tasks) that should be performed to complete the design phase. This task diagram does not mandate any specific methodology, but we will describe in the accompanying paragraphs those approaches, tools, and techniques you might want to consider for each design task. This task diagram is only a template. The project team and project manager may expand on or alter the template to reflect the unique needs of any given project.

Let's now examine each systems design task in detail.

The purpose of this first design task is to specify an application architecture. An **application architecture** defines the technologies to be used by (and used to build) one, more, or all information systems in terms of its data, processes, interfaces, and network components. Thus, designing the application architecture involves considering network technologies and making decisions on how the systems' DATA, PROCESSES, and INTERFACES are to be distributed among the business locations.

This task is accomplished by analyzing the data models and process models that were initially created during requirements analysis. Given the data models, process models, and target solution, distribution decisions will need to be made. As decisions on how data, processes, and interface are made, they are documented. An example is the **physical data flow diagram (PDFD)** that is used to establish physical processes and data stores (databases) across a network (see Figure 10.7). You will learn about PDFDs to document application architecture in Chapter 11.

To complete this activity, the analyst may involve a number of SYSTEM DESIGNERS and SYSTEM USERS. System users may be involved in this activity to help address

Task 5.1—Design the Application Architecture

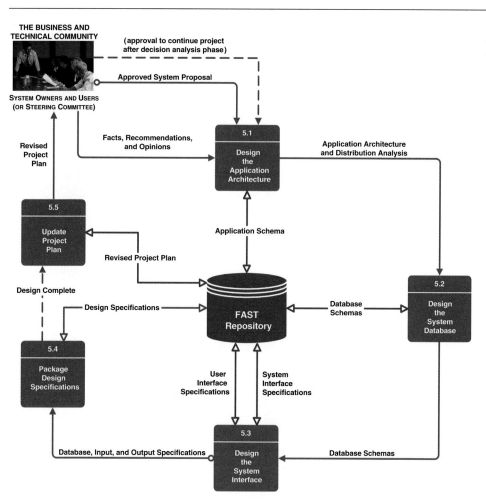

FIGURE 10.6   *The Systems Design Tasks for In-House Development*

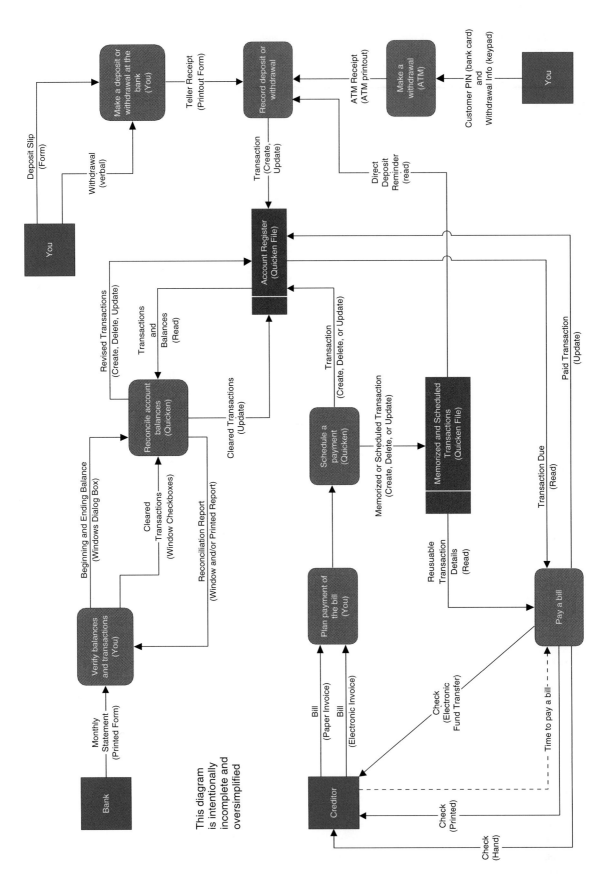

FIGURE 10.7 *A Sample Physical Data Flow Diagram*

This diagram is intentionally incomplete and oversimplified

business data, process, and location issues. Several different SYSTEM DESIGNER specialists may be instrumental in the completion of this activity, including a *data and database administrator, network administrator and engineers, applications administrator,* and various other experts, as needed (e.g., an expert on automatic data capture for addressing bar-coding technology and issues).

The key inputs to this task are the facts, recommendations, and opinions that are solicited from various sources and the approved system proposal from the decision analysis phase. The principal deliverable of the task is the application architecture and distribution analysis that serves as a blueprint for subsequent detailed design phase activities.

Typically the next system design task is to develop the corresponding database design specifications. The design of data goes far beyond the simple layout of records. Databases are a shared resource. Many programs will typically use them. Future programs may use databases in ways not originally envisioned. Consequently, the designer must be especially attentive to designing databases that are adaptable to future requirements and expansion.

The designer must also analyze how programs will access the data in order to improve performance. You may already be somewhat familiar with various programming data structures and their impact on performance and flexibility. These issues affect database organization decisions. Other issues to be addressed during database design include record size and storage volume requirements. Finally, because databases are shared resources, the designer must also design internal controls to ensure proper security and disaster recovery techniques, in case data is lost or destroyed.

The purpose of this task is to prepare technical design specifications for a database that will be adaptable to future requirements and expansion. While the SYSTEMS ANALYSTS who may participate in database modeling facilitate this task, the SYSTEM DESIGNERS are responsible for the completion of this activity. The *data administrator* may participate (or complete) the database design. Recognize that most likely the new system uses some portion of an existing database. This is where the knowledge of the database administrator is crucial. Finally, SYSTEM BUILDERS may also participate when asked to build a prototype database for the project.

As is illustrated in Figure 10.6, a key input to this activity is application architecture and distribution analysis decisions from the prior design task. The deliverable of the task includes the resulting database schemas. An example of a database schema was presented earlier in Figure 10.2. A **database schema** is the structural model for a database. It is a picture or map of the records and relationships to be implemented by the database. You will learn how to develop database schemas in Chapter 12.

Task 5.2—Design the System Database(s)

Once the database has been designed and possibly a prototype built, the systems designer can work closely with system users to develop input, output, and dialogue specifications. Because end-users and managers will have to work with inputs and outputs, the designer must be careful to solicit their ideas and suggestions, especially regarding format. Their ideas and opinions must also be sought regarding an easy-to-learn and easy-to-use dialogue for the new system.

Transaction outputs will frequently be designed as preprinted forms onto which transaction details will be printed. Reports and other outputs are usually printed directly onto paper or displayed on a terminal screen. The precise format and layout of the outputs must be specified. Finally, internal controls must be specified to ensure that the outputs are not lost, misrouted, misused, or incomplete. Figure 10.8 is a sample output design. You will learn how to design outputs in Chapter 13.

For inputs, it is crucial to design the data capture method to be used. For instance, you may design a form on which data to be input will be initially recorded. You

Task 5.3—Design the System Interface

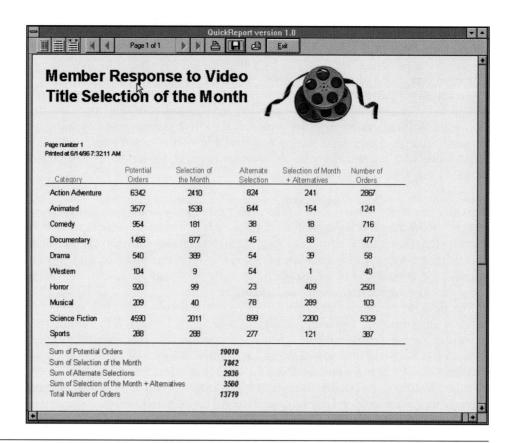

FIGURE 10.8
A Sample Output Prototype Screen

want to make it easy for the data to be recorded on the form, but you also want to simplify the entry of the data from the form into the computer or onto a computer-readable medium. This is particularly true if the data is to be input by people who are not familiar with the business application. Also, any time you input data to the system, you can make mistakes. We need to define editing controls to ensure the accuracy of input data. A sample input prototype screen was depicted earlier in Figure 10.3. You will learn how to design inputs in Chapter 14.

For interface or dialogue design, the design must consider such factors as terminal familiarity, possible errors and misunderstandings that the end-user may have or may encounter, the need for additional instructions or help at certain points, and screen content and layout. You are trying to anticipate every little error or keystroke that an end-user might make—no matter how improbable. Furthermore, you are trying to make it easy for the end-user to understand what the screen is displaying at any given time. Figure 10.9 is a sample interface design. You will learn how to do interface design in Chapter 15.

SYSTEM USERS should be involved in this activity! The inputs, outputs, and interface dialogues are what they will see and work with. The degree to which they are involved is emphasized in design efforts that involve prototyping. They will be asked to provide feedback regarding each input/output prototype. SYSTEM DESIGNERS are responsible for the completion of this activity. They may draw on the expertise of systems designers that specialize in *graphical user interface* design. In addition, SYSTEM BUILDERS may construct the various screen designs for the users to review during design by prototyping.

As was illustrated in Figure 10.6, the key input to this activity is the database schema(s) from the previous task and the user and system interface specifications that are available from the project's repository. The deliverable of the design task is the completed database, input, and output specifications.

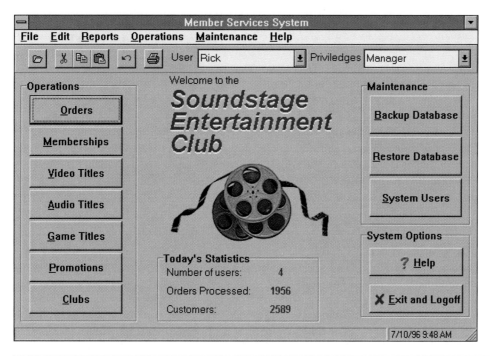

FIGURE 10.9
*A Sample Dialogue Interface
Prototype Screen*

This final design task involves packaging all the specifications from the previous design tasks into a set of specifications that will guide the computer programmer's activities during the construction phase of the systems development methodology.

There is more to this task than packaging, however. How much more depends on two things: (1) where you draw the line between the system designer's and computer programmer's responsibilities, and (2) whether the methodology and solution calls for the design of the overall program structure. Most organizations have adopted accelerated systems development approaches that do not require the latter. Program structure dealt with quality issues that were of concern to developers of systems that used older programming languages and tended to be mainframe-based applications. To learn more about designing program structure we refer you to McGraw-Hill's Online Learning Center.

The SYSTEMS ANALYST, who may be aided by the SYSTEM DESIGNERS, usually completes this task. Before proceeding with the packaging of the design specifications and the construction phase, the systems design should be reviewed with all appropriate audiences. While SYSTEM USERS have already seen and approved the outputs, inputs, and dialogue for the new system, the overall work and data flow for the new system should get a final walkthrough and approval. SYSTEM OWNERS should get a final chance to question the project's feasibility and determine whether the project should be adjusted, terminated, or approved to proceed to construction. At this stage of a project the company's *audit staff* may become heavily involved. The staff will pass judgment on the internal controls in a new system.

As was illustrated in Figure 10.6, the inputs to this task are the various database, input, and output specifications that were created earlier. Once these specifications have been reviewed, approved, and organized as design specifications that are suitable for constructing the new system, they are made available to the team of system builders via the project repository. It is more common for a project manager to make design specifications available via a shared repository than to provide each individual developer with a copy of a printed set of organized specifications.

**Task 5.4—Package Design
Specifications**

Task 5.5—Update the Project Plan

Now that we're approaching the completion of the design phase, we should reevaluate project feasibility and *update the project plan* accordingly. The *project manager* in conjunction with SYSTEM OWNERS and the entire project team facilitate this task. The SYSTEMS ANALYSTS and SYSTEM OWNERS are the key individuals in this task. The analysts and owners should consider the possibility that, based on the completed design work, the overall project schedule, cost estimates, and other estimates may need to be adjusted.

As shown in Figure 10.6, this task is triggered when the project manager determines that the design is complete. The key deliverable of the task is the updated project plan. The updated plan should now include a detailed plan for the construction phase that should follow. Recall that the techniques and steps for updating the project plan were taught in Chapter 4, Project Management.

SYSTEMS DESIGN FOR INTEGRATING COMMERCIAL SOFTWARE—THE "BUY" SOLUTION

Let's now examine systems design for solutions that involve acquiring a commercial-off-the-shelf (COTS) software product. The life cycle for projects that involve purchase or "buy" solutions is illustrated in Figure 10.10. Notice that the business requirements statement (for software) and its integration as a business solution trigger a series of phases absent from the in-house development process we just learned about. The most notable differences between the "buy" versus in-house development projects is the inclusion of a new **procurement phase** and a special decision analysis phase (process labeled 5A) to address software and services.

When new software is needed, the selection of appropriate products is often difficult. Decisions are complicated by technical, economic, and political considerations. A poor decision can ruin an otherwise successful analysis and design. The systems analyst is becoming increasingly involved in the procurement of software packages (as well as peripherals and computers to support specific applications being developed by that analyst). The purpose of the procurement and decision analysis phases is to:

1. Identify and research specific products that could support our recommended solution for the target information system.
2. Solicit, evaluate, and rank vendor proposals.
3. Select and recommend the best vendor proposal.
4. Contract with the awarded vendor to obtain the product.

In this section we will examine the tasks involved in completing the procurement and decision analysis phases for a "buy" solution. As is depicted in Figure 10.10, a "buy" solution affects how other phases in the life cycle are also completed (phases that are impacted are shaded in light blue). After examining the procurement and decision analysis phases, we will explore the impacts that a "buy" solution would have on how those phases would be completed.

Figure 10.11 is a task diagram depicting the work (= tasks) that should be performed to complete the procurement and decision analysis phases for a "buy" project solution. This task diagram does not mandate any specific methodology, but we will describe in the accompanying paragraphs those approaches, tools, and techniques you might want to consider for each design task. This task diagram is only a template. The project team and project manager may expand on or alter the template to reflect the unique needs of any given project.

The first two tasks (4.1 and 4.2) are procurement phase tasks, and the remaining tasks (5A.1, 5A.2, and 5A.3) are decision analysis related tasks. Let's now examine each task in detail.

Task 4.1—Research Technical Criteria and Options

The first task is to research technical alternatives. This task identifies specifications that are important to the software and/or hardware that is to be selected. The task involves focusing on the software and/or hardware requirements established in the requirements analysis phase. These requirements specify the functionality, features,

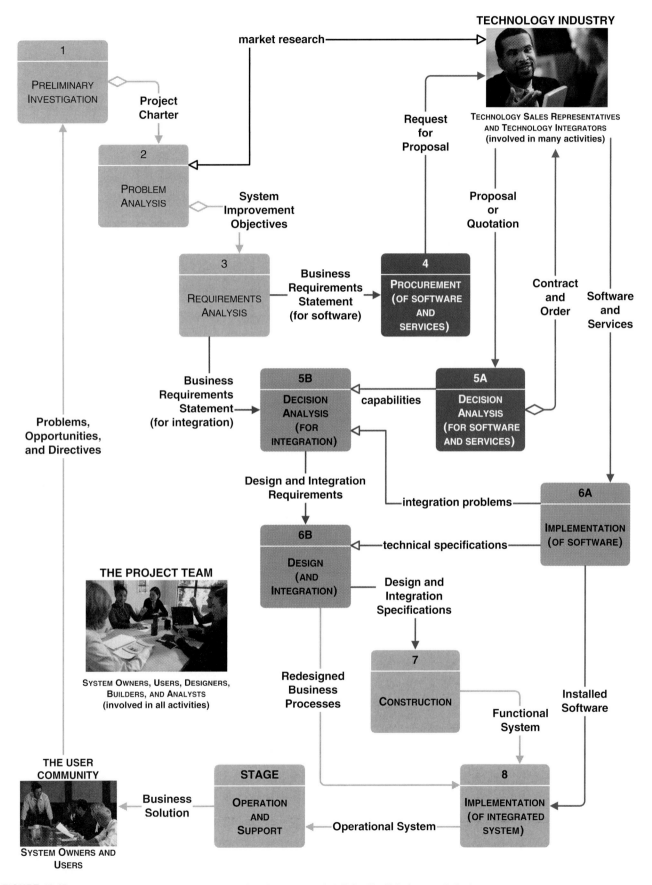

FIGURE 10.10   *The Context of Systems Design for Commercial Off-the-Shelf Software Solution*

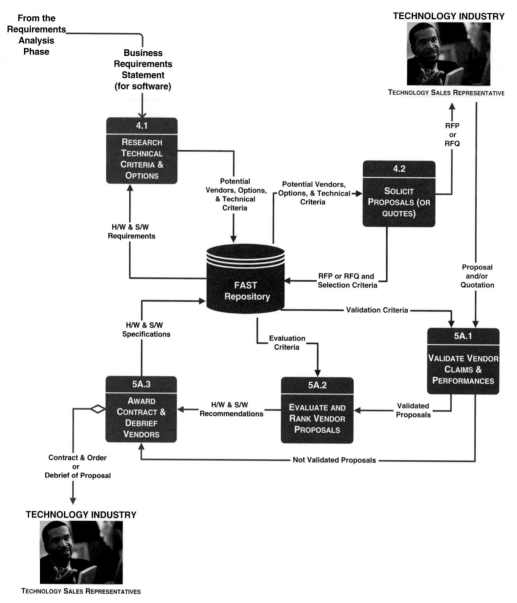

FIGURE 10.11 *Tasks for the Procurement Phase*

and critical performance parameters for our new software/hardware.

Most analysts read appropriate magazines and journals and search the Internet to help them identify those technical and business issues and specifications that will become important to the selection decision. Other sources of information for conducting research include the following:

— *Internal standards* may exist for hardware and software selection. Some companies insist that certain technology will be bought from specific vendors if those vendors offer it. For instance, some companies have standardized on specific brands of microcomputers, terminals, printers, database management systems, network managers, data communications software, spreadsheets, and programming languages. A little homework here can save you a lot of unnecessary research.

— *Information services* are primarily intended to constantly survey the marketplace for new products and advise prospective buyers on what specifications to consider. They also provide information such as the number of installations and general customer satisfaction with the products. Some information services are listed in the margin.

— *Trade newspapers and periodicals* offer articles and experiences on various types of hardware and software that you may be considering. Many can be found in school and company libraries. Subscriptions (sometimes free) are also available.

The research should also identify potential vendors that supply the products to be considered. After the analysts have completed their homework, they will initiate contact with these vendors. Thus, the analysts will be better equipped to deal with vendor sales pitches after doing their research!

The purpose of this task is to research technical alternatives to specify important criteria and options that will be important for the new hardware and/or software that is to be selected. This task is facilitated by the *project manager*. SYSTEM DESIGNERS are responsible for the completion of this task. The designer may seek input from various technical experts including data and database administrators, network administrators, and applications administrators.

As is illustrated in Figure 10.11, a key input to this task is the business requirements statement (for software) established in the requirements analysis phase. The designer will also obtain additional product and vendor facts from various sources. They are careful not to get their information solely from a salesperson—not that sales representatives are dishonest, but the number one rule of salesmanship is to emphasize the product's strengths and deemphasize its weaknesses. The principal deliverable of this task includes a list of potential vendors, product options, and technical criteria.

To complete this task, designers must conduct extensive research to gain important facts concerning the hardware/software product and vendor. They must be careful to screen their various sources. The sources are used to identify potential vendors from which the products might be obtained. This step may be optional if your company has a commitment or contract to acquire certain products from a particular source. Finally, the designer must review the product, vendor, and supplier findings.

The next task is to solicit proposals for quotes from vendors. If your company is committed to buying from a single source (IBM, for example), the task is quite informal. You simply contact the supplier and request price quotations and terms. But most decisions offer numerous alternatives. In this situation, good business sense dictates that you use the competitive marketplace to your advantage.

The solicitation task requires the preparation of one of two documents: a **request for quotations (RFQ)** or a **request for proposals (RFP).** The request for quotations is used when you have already decided on the specific product but that product can be acquired from several distributors. Its primary intent is to solicit specific configurations, prices, maintenance agreements, conditions regarding changes made by buyers, and servicing. The request for proposals is used when several different vendors and/or products are candidates and you want to solicit competitive proposals and quotes. RFPs can be thought of as a superset of RFQs. Both define selection criteria that will be used in a later validation.

The primary purpose of the RFP is to communicate requirements and desired features to prospective vendors. Requirements and desired features must be categorized as mandatory (must be provided by the vendor), extremely important (desired from the vendor but can be obtained in-house or from a third-party vendor), or desirable (can be done without). Requirements might also be classified by two alternate criteria: those that satisfy the needs of the systems and those that satisfy our needs from the vendor (for example, service).

This task is facilitated by the *project manager*. The SYSTEM DESIGNER is also responsible for completing this activity and may seek the input from *data* and

Task 4.2—Solicit Proposals (or Quotes) from Vendors

database administrators, network administrators, and *applications administrators* when writing the RFP or RFQ.

The key input to this task is the potential vendors, options, and technical criteria that resulted from previous research. The principal deliverable of this task is the RFP or RFQ that is to be received by candidate vendors. The quality of an RFP has a significant impact on the quality and completeness of the resulting proposals. A suggested outline for an RFP is presented in Figure 10.12, since an actual RFP is too lengthy to include in this book.

Many of the skills you developed in Part Two, such as process and data modeling, can be very useful for communicating requirements in the RFP. Vendors are very receptive to these tools because they find it easier to match products and options and package a proposal that is directed toward your needs. Other important skills include report writing (discussed in Chapter 9) and questionnaires (covered in Chapter 6).

FIGURE 10.12
Request for Proposals

Request for Proposals (RFP)

I. **Introduction**
   A. **Background**
   B. **Brief summary of needs**
   C. **Explanation of RFP document**
   D. **Call for action on part of vendor**
II. **Standards and instructions**
   A. **Schedule of events leading to contract**
   B. **Ground rules that will govern selection decision**
       1. **Who may talk with whom and when**
       2. **Who pays for what**
       3. **Required format for a proposal**
       4. **Demonstration expectations**
       5. **Contractual expectations**
       6. **References expected**
       7. **Documentation expectations**
III. **Requirements and features**
   A. **Hardware**
       1. **Mandatory requirements, features, and criteria**
       2. **Essential requirements, features, and criteria**
       3. **Desirable requirements, features, and criteria**
   B. **Software**
       1. **Mandatory requirements, features, and criteria**
       2. **Essential requirements, features, and criteria**
       3. **Desirable requirements, features, and criteria**
   C. **Service**
       1. **Mandatory requirements**
       2. **Essential requirements**
       3. **Desirable requirements**
IV. **Technical questionnaires**
V. **Conclusion**

Soon after the RFPs or RFQs are sent to prospective vendors, you will begin receiving proposal(s) and/or quotation(s). Because proposals cannot and should not be taken at face value, claims and performance must be validated. This task is performed independently for each proposal; proposals are not compared with one another.

The purpose of this task is to validate requests for proposals and/or quotations received from vendors. SYSTEM DESIGNERS are responsible for the completion of this activity. Once again, the designer may involve the following individuals in validating the proposals: *data* and *database administrators, network administrators,* and *applications administrators.*

This task is triggered by the receipt of proposal(s) and/or quotation(s) received from prospective vendors. The key outputs of this task are those vendor proposals that proved to be validated proposals or claims and others whose claims were not validated.

To complete this task, the designer must collect and review all facts pertaining to the product requirements and features. They must review the vendor proposals and should eliminate any proposal that does not meet all the mandatory requirements. If the requirements were clearly specified, no vendor should have submitted such a proposal. For proposals that cannot meet one or more extremely important requirements, verify that the requirements or features can be fulfilled by some other means. For each vendor proposal not eliminated, the designer must validate the vendor claims and promises against validation criteria. Claims about mandatory, extremely important, and desirable requirements and features can be validated by completed questionnaires and checklists (included in the RFP) with appropriate vendor-supplied references to user and technical manuals. Promises can be validated only by ensuring that they are written into the contract. Finally, performance is best validated by a demonstration, which is particularly important when you are evaluating software packages. Demonstrations allow you to obtain test results and findings that confirm capabilities, features, and ease of use.

The validated proposals can now be evaluated and ranked. The evaluation and ranking is, in reality, another cost-benefit analysis performed during systems development. The evaluation criteria and scoring system should be established before the actual evaluation occurs so as not to bias the criteria and scoring to subconsciously favor any one proposal.

The executive sponsor ideally should facilitate this task. SYSTEM DESIGNERS are responsible for the completion of this activity. The designer may involve several experts in evaluating and ranking the proposals, including *data* and *database administrators, network administrators,* and *applications administrators.*

The inputs to this task include validated proposals and the evaluation criteria to be used to rank the proposals. The key deliverable of this task is the hardware and/or software recommendations.

The ability to perform a feasibility assessment is an extremely important skill requirement for completing this task. Feasibility assessment techniques and skills were covered in Chapter 9. To complete this task, designers must first collect and review all details concerning the validated proposals. They must then establish an evaluation criteria and scoring system. There are many ways to go about this. Some methods suggest that requirements be weighted on a point scale. Better approaches use dollars and cents! Monetary systems are easier to defend to management than points. One such technique is to evaluate the proposals on the basis of hard and soft dollars. Hard-dollar costs are the costs you will have to pay to the selected vendor for the equipment or software. Soft-dollar costs are additional costs you will incur if you select a particular vendor (for instance, if you select vendor A, you may incur an additional expense to vendor B to

Task 5A.1—Validate Vendor Claims and Performances

Task 5A.2—Evaluate and Rank Vendor Proposals

overcome a shortcoming of vendor A's proposed system). This approach awards the contract to the vendor who fulfills all essential requirements while offering the lowest total hard-dollar cost plus soft-dollar penalties for desired features not provided (for a detailed explanation of this method see Isshiki, 1982 or Joslin, 1977). Once the evaluation criteria and scoring system have been established, the last step toward completing our task is to do the actual evaluation and ranking of the vendor proposals.

Task 5A.3—Award (or Let) Contract and Debrief Vendors

Having ranked the vendor proposals, the next activity usually includes presenting a recommendation to management for final approval. Once again, communication skills, especially salesmanship, are important if the analyst is to persuade management to follow the recommendations. Given management's approval of the recommendation, a contract must then be drawn up and awarded to the winning vendor. This activity often also includes debriefing losing vendors, being careful not to burn bridges.

The purpose of this activity is to negotiate a contract with the vendor who supplied the winning proposal, and to debrief those vendors that submitted losing proposals. Ideally, the executive sponsor who must approve recommendations and project continuation should facilitate the activity. But it is the SYSTEM DESIGNER who must make and defend the recommendation and award the contract. In doing so, the systems designer may involve a company lawyer in drafting the contract. Report writing and presentation skills are important for completing this task.

The key inputs include the hardware and software recommendation and the nonvalidated proposals from the previous evaluation tasks. Pending the approval of the executive sponsor, a contract order would subsequently be produced for the "winning" vendor. A debriefing of proposals would be provided for the losing vendors.

To complete this task, the designer must first present a hardware and software recommendation for final approval. Once the final hardware and software approval decision is made, a contract must then be negotiated with the winning vendor. Certain special conditions and terms may have to be written into the standard contract and order. Ideally, no computer contract should be signed without the advice of a lawyer. The analyst must be careful to read and clarify all licensing agreements. No final decision should be approved without the consent of a qualified accountant or management. Purchasing, leasing, and leasing with a purchase option involve complex tax considerations. Finally, out of common courtesy and to maintain good relationships, provide a debriefing of proposals for losing vendors. The purpose of this meeting is not to allow the vendors a second chance to be awarded the contract; rather, the briefing is intended to inform the losing vendors of precise weaknesses in their proposals and/or products.

Impact of Buy Decision on Remaining Life Cycle Phases

It is not merely enough to purchase or build systems that fulfill the target system requirements. The analyst must integrate or interface the new system to the myriad of other existing systems that are essential to the business. Many of these systems may use dramatically different technology, techniques, and file structures.

The analyst must consider how the target system fits into the federation of systems of which it is a part. The integration requirements that are specified are vital to ensuring that the target system will work in harmony with those systems.

As was depicted in Figure 10.10, the decision to buy a commercial software packaged solution can impact additional phases (denoted in light blue) of the life cycle. Upon completion of the decision analysis (for software and services) phase and its intensive evaluation of the commercial product, we have become knowledgeable of its capabilities (or shortcomings). During decision analysis for integration we will need to make revisions to reflect this new knowledge in our data and process models that comprised the business requirements statement. When software and services are received from the vendor(s), the software must be implemented. During implementation we may encounter integration problems that must also be reflected in our business requirements statement. These capabilities and integration problems are reflected in the design and integration requirements.

Finally, given the design and integration requirements we must now complete the design phase. Completion of the design phase involves many of the same tasks that were discussed earlier in the chapter. The primary difference is that you simply are not "developing" an entire system. Rather, we may be designing technical specifications for developing a small subset of programs, software utilities, and other components necessary for the business processes and the commercial software product to be integrated and work together properly. Let's consider an example. Our existing business system may use bar-coding technology to capture data. Yet, our software product may require that data be entered via the keyboard. We may need to customize the software product to allow data to be entered via the keyboard or from a batch file containing scanned data.

WHERE DO YOU GO FROM HERE?

This chapter provided a detailed overview of systems design for a project. You are now ready to learn some of the systems design skills introduced in this chapter. Because systems design is dependent on requirements specified during systems analysis, we recommend that you first complete Chapters 5–9. Chapter 5 gives you an overview of systems analysis. Chapters 6–9 teach different system analysis tools and techniques that provide for basic inputs to the systems design activities presented in Part Three.

The order of the system design chapters that follow is flexible; however, the authors did present the subsequent techniques in the sequence that they are commonly completed for systems design projects.

SUMMARY

1. Formally, information systems design is defined as those tasks that focus on the specifications of a detailed computer-based solution. Whereas systems analysis emphasizes the business problem, systems design focuses on the technical or implementation concerns of the system.

2. Systems design is driven by the technical concerns of system designers. Therefore, with respect to the information systems building blocks, systems design addresses the DATA, PROCESS, and INTERFACE building blocks from the system designer's perspective.

3. Systems design differs for in-house development or "build" projects versus "buy" projects where a systems software package is bought.

4. There are many popular strategies or techniques for performing systems design. These techniques can be used in combination with one another.
   a. Modern structured design, a technique that focuses on processes.
   b. Information engineering (IE), a technique that focuses on data and strategic planning to produce application projects.
   c. Prototyping, a technique that is an iterative process involving a close working relationship between designers and users to produce a model of the new system.
   d. Joint application development (JAD), a technique that emphasizes participative development among system owners, users, designers, and builders. During JAD sessions for systems design, the system designer takes on the role of the facilitator.
   e. Rapid application development (RAD), a technique that represents a merger of various structured techniques with prototyping and JAD to accelerate systems development.
   f. Object-oriented design (OOD), a new design strategy that follows up object-oriented analysis to refine object requirement definitions and to define new design specific objects.

5. For in-house development (build) projects, the systems design involves developing technical design specifications that will guide the construction and implementation of the new system. To complete the design phase, the system designer must complete the following tasks:

a. Design the application architecture.
b. Design the system database(s).
c. Design the system interface.
d. Package the design specifications.
e. Update the project plan.

6 Systems design for solutions that involve acquiring a commercial-off-the-shelf (COTS) software product include a procurement and decision analysis phase that addresses software and services. Completion of these phases involves the following tasks:
a. Research technical criteria and options.
b. Solicit proposals (or quotes) from vendors.
c. Validate vendor claims and performances.
d. Evaluate and rank vendor proposals.
e. Award (or let) contract and debrief vendors.

7 It is not merely enough to purchase or build systems that fulfill the target system requirements. The analyst must integrate or interface the new system to the myriad of other existing systems that are essential to the business. Many of these systems may use dramatically different technology, techniques, and file structures.

KEY TERMS

application architecture, p. 403
cohesive, p. 396
coupled, p. 396
database schema, p. 405
information engineering (IE), p. 397
joint application development (JAD), p. 401

model-driven design, p. 395
modern structured design, p. 395
object-oriented design (OOD), p. 401
physical data flow diagram (PDFD), p. 403
physical design, p. 394
procurement phase, p. 408

rapid application development (RAD), p. 401
request for proposal (RFP), p. 411
request for quotation (RFQ), p. 411
structure chart, p. 396
systems design, p. 394

REVIEW QUESTIONS

1. What is the difference in *focus* between systems analysis and systems design?
2. Identify several systems design strategies.
3. List several advantages to using prototyping as a systems design approach.
4. List several disadvantages or pitfalls of prototyping as a systems design approach.
5. Explain the relationship between prototyping, JAD, and RAD.
6. What are the tasks for completing systems design for an in-house development project?

7. What are the tasks for completing the procurement and decision analysis of software and services needed for a project involving a "buy" solution?
8. List several possible sources of information when researching software and/or hardware.
9. Is the procurement phase required for all systems projects? Why or why not?
10. Explain the difference between a request for proposal (RFP) and a request for quotation (RFQ).

PROBLEMS AND EXERCISES

1. How can a successful and thorough systems analysis be ruined by a poor systems design? Answer the question relative to these factors:
a. The impact on the subsequent implementation (in other words, the systems implementation phases, which you studied in Chapter 3).
b. The lifetime of the system after it is placed into operation.
c. The impact on future projects.
2. What skills are important during systems design? Create an itemized list of these skills. Identify other computer,

business, and general education courses that would help you develop or improve your skills. Prepare a plan and schedule for taking the courses. (If you are not in school, prepare a plan for using available corporate training resources, reading appropriate books, enrolling in seminars or continuing education courses, etc.) Review your plan with your counselor, adviser, or instructor.
3. What by-products of the systems analysis phases are used in the systems design phases? Why are they important? How are they used? What would happen if they were incomplete or inaccurate?

4. Distinguish between the terms *validation* and *evaluation* as they apply to the configuration phase of computer equipment and software.
5. Explain what you would do if a vendor said the following in response to an RFP.

This thing is not useful to you or me. It rarely tells me what you really want or need. I can do a better job by visiting your business and configuring a system to meet your needs. Also, it takes too long for me to answer all the questions in the RFP. And even if I do, you may not fully understand or appreciate the answers and their implications.

6. A programming assignment in the classroom is a subset of a systems design. Obtain a copy of a programming assignment from a current course. Evaluate the design from the perspective of the system design phase tasks and the completeness of the design specifications.

PROJECTS AND RESEARCH

1. Make an appointment with or write to a hardware and software vendor. Tell the vendor you would like to see and discuss a typical RFP. Ask the vendor how he feels about RFPs. If he doesn't like them, find out why. How could RFPs be improved from the vendor's point of view? Do the vendor's attitudes about RFPs help the vendor, the end-user, or both?
2. Make an appointment to discuss the physical design standards of a local information systems operation. Does it have standards? Does it follow them? Why or why not? Does the company use prototyping during systems design? Why or why not? If it does prototype systems, what products does it use? Has the approach been successful?
3. The city of Granada's art museum recently purchased a Hewlett Packard Vectra XU 5/90 microcomputer. Museum staff members read an article about an art collection inventory system software package that they want to put on that computer. You, having experienced end-users who too hastily purchased software that didn't fulfill promises and expectations, are concerned that they are jumping the gun and should approach the software selection decision with great care. Write a letter to the museum's board of trustees that expresses your concerns and proposes a better approach.
4. Write a letter to your last (or favorite) programming instructor. Suggest a disciplined approach to developing a systems specification to guide the programming assignments for the next term. Your goal should be a system (of programming) specification that will eliminate or drastically reduce the need for students to request clarification from the systems designer, played by the instructor. Defend your approach.

MINICASES

1. Keith Stallard is a relatively new programmer/analyst at Schuster and Petrie, Inc. Having spent two years as a programmer, he was promoted one year ago. The Information Services division of S & P requires job performance reviews twice a year. Tim Hayes, associate director of Financial Systems, has scheduled a job performance review with Keith.

"Well, Keith, do you still want this programmer/analyst job? You've had about six months to get used to your new responsibilities," asks Tim.

"More than ever!" responded Keith. "Now that I've had a taste of systems work, I know it's right for me. I assume this meeting will determine if I'm making progress. How am I doing?"

Tim responds, "You're right. I've discussed your performance on the job cost accounting system project with both your supervisor and your key user contact. Your technical design statement was quite impressive. But I have to ask you, where did you learn to complete such thorough design specifications? The implementation appears to be moving along more smoothly than expected, largely because of your specifications."

Keith answers, "I did a lot of reading in systems analysis and design textbooks at the college library. I also queried both users and programmers about problems with typical specifications. My own experience as a programmer has influenced my specifications. But to be honest, I was really embarrassed by my performance on the account aging project. That's why I did all those things!"

"I don't understand," states a puzzled Tim. "We gave you acceptable ratings. The account aging project was a little off schedule, but that's the only problem I recall."

Keith explains, "It was a little more complicated than that. Bill had done the systems analysis and was supervising me since it was my first experience with systems design. But he had to be called off the project to fix a major flaw in another system. I kept working on the design and passed the design document to Rita [a programmer]. Then lightning struck for a second time. Rita had to go into the hospital, and I had to assume her programming responsibilities. It was the first time I ever had to cut code from my own specifications. There were so many details, and I hadn't documented all of them. Surprisingly, I couldn't even remember all of the thought processes that went into my own specifications. Now I know how the maintenance programmers feel!"

"Eventually I got the system up and running—only to find out that some of the reports were not acceptable to my users. The content was there, but the format was wrong. I had to take certain liberties with the format. In my school days, that was what we did with all programming assignments. I just didn't appreciate the importance of user involvement in the design process. I assumed that systems analysis took care of all the user issues. And then, the Internal Audit department got hold of my design specifications. They didn't like them at all! There weren't enough internal controls to satisfy their standards. By that time, I had half the programs written and tested. I had to redesign many system components and rewrite several affected programs. To make a long story short, I never want to go through that kind of design experience again. So I learned about systems design."

"So . . . and you still want to be an analyst? After all that?" asks Tim with a big smile.

"Yes," answered Keith with a smile and a nod. "Despite the problems, I found the work to be so much more satisfying than programming. I knew it wasn't going to be easy. But it was enjoyable."

Tim takes over the conversation, "Well, we've discussed your strengths. But we do need to work on a few things. First, as you know, most of our older systems are being converted to online systems using databases. I'm sending you to a one-week intensive course for this new *Oracle* database management system. I also want you to go through our user interface course the next time it's offered. I don't know if you've heard that the on-line interface on your accounts aging system hasn't lived up to expectations. You also need to work on your writing and speaking skills. The report you did to sell the new job costing system wasn't well organized, was too wordy, and contained numerous grammatical errors and typos. You

were lucky that Bill got it before your users. You might have lost the sale. And your presentation of that system to management could have gone a little smoother. Public speaking is tough, I realize that. But you did not seem confident. You made a good recommendation! But if you don't seem confident and comfortable with your own recommendation, how will management feel about it? We did get it through, though. But your communications skills need improvement . . . especially if you want my job in the future. You have that potential, Keith. Don't waste it!"

"I understand," replies an accepting Keith. "And I appreciate your honesty. I've suspected the problem. I guess I never took those English and communications courses seriously. I'll get enrolled in some evening continuing education courses for the next term. I'm not going to let poor communications skills get in the way of my future."

"Let's get to the bottom line, Keith," says Tim. "You've shown better than average progress in your new assignment. That's why, effective next month, you'll see a little increase in your paycheck. If you keep up the good work and improve in the areas we've outlined, I'm certain you'll be promoted to systems analyst within two years. Now, let's talk about design specification some more. Do you think we could teach our other analysts to do that?"

a. Think back to your programming courses (or experiences). What are some problems you've had responding to programming assignments?

b. What did Keith learn about working from his own specifications?

c. As a systems analyst working on the design phase of a project, what types of people did Keith have to communicate with? Why does communication become tougher during systems design than during systems analysis?

SUGGESTED READINGS

Application Development Strategies (monthly periodical). Arlington, MA: Cutter Information Corporation. This is our favorite theme-oriented periodical that follows system development strategies, methodologies, CASE, and other relevant trends. Each issue focuses on a single theme.

Boar, Benard. *Application Prototyping: A Requirements Definition Strategy for the 80s.* New York: Wiley, 1984. This is one of the first books to appear on the subject of systems prototyping. It provides a good discussion of when and how to do prototyping, as well as thorough coverage of the benefits that may be realized through this approach.

Coad, Peter, and Yourdon, Edward. *Object-Oriented Design.* 2nd ed. Englewood Cliffs, NJ: Yourdon Press, 1991. Chapter 1 is a great way to expose yourself to objects and the relationship of object methods to everything that preceded them.

Connor, Denis. *Information System Specification and Design Road Map.* Englewood Cliffs, NJ: Prentice Hall, 1985. This book compares prototyping with other popular analysis

and design methodologies. It makes a good case for not prototyping without a specification.

Gane, Chris. *Rapid Systems Development.* Englewood Cliffs: NJ: Prentice Hall, 1989. This book presents a nice overview of RAD that combines model-driven development and prototyping in the correct balance.

Isshiki, Koichiro R. *Small Business Computers: A Guide to Evaluation and Selection.* Englewood Cliffs, NJ: Prentice Hall, 1982. Although it is oriented toward small computers, this book surveys most of the better-known strategies for evaluating vendor proposals. It also surveys most of the steps of the selection process, although they are not put in the perspective of the entire systems development life cycle.

Joslin, Edward O. *Computer Selection.* Rev. ed. Fairfax Station, VA: Technology Press, 1977. Although somewhat dated, the concepts and selection methodology originally suggested in this classic book are still applicable. The book provides keen insights into vendor, customer, and end-user relations.

Lantz, Kenneth E. *The Prototyping Methodology.* Englewood Cliffs, NJ: Prentice Hall, 1986. This book provides excellent coverage of the prototyping methodology.

Wood, Jane, and Denise Silver. *Joint Application Design: How to Design Quality Systems in 40% Less Time.* New York: John Wiley & Sons, 1989. This book provides an excellent in-depth presentation of joint application development (JAD).

Yourdon, Edward. *Modern Structured Analysis.* Englewood Cliffs, NJ: Yourdon Press, 1989. Chapter 4, "Moving into Design," shows how modern structured design picks up from modern structured analysis.

Zachman, John A. "A Framework for Information System Architecture," *IBM Systems Journal* 26, no. 3 (1987). This article presents a popular conceptual framework for information systems design.

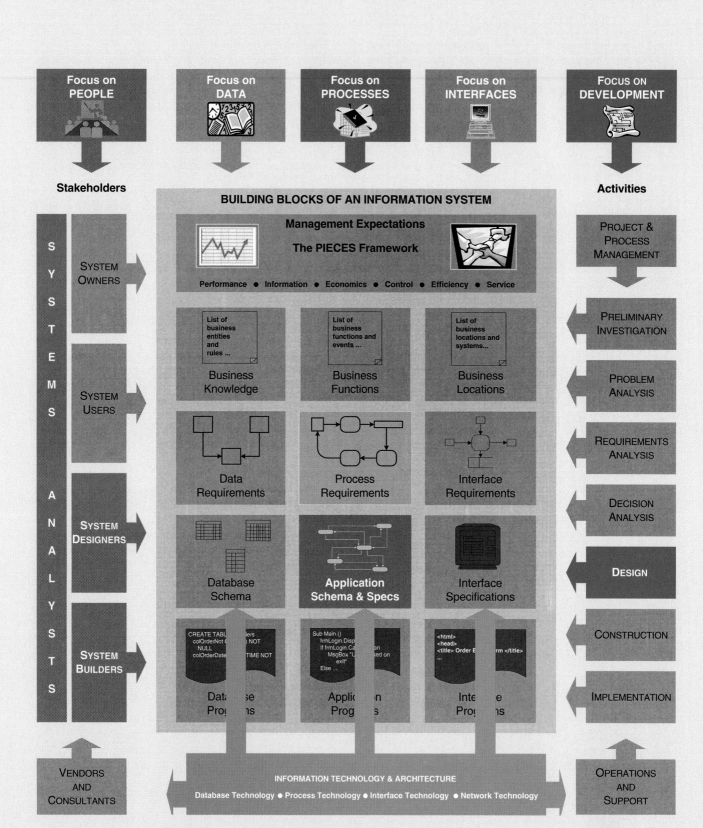

11

APPLICATION ARCHITECTURE AND MODELING

CHAPTER PREVIEW AND OBJECTIVES

This chapter teaches you techniques for designing the overall information system application architecture with a focus on physical process models. Information application architecture and physical process modeling include techniques for distributing data, processes, and interfaces to network locations in a distributed computing environment. Physical data flow diagrams are used to document the architecture and design in terms of design units—cohesive collections of data and processes at specific locations—that can be designed, prototyped, or constructed in greater detail and subsequently implemented as stand-alone subsystems. You will know that you understand application architecture and process design when you can:

- Define an information system's architecture in terms of DATA, PROCESSES, and INTERFACES— the building blocks of all information systems. Consistent with modern trends, these building blocks will be distributed across a NETWORK.

- Differentiate between logical and physical data flow diagrams, and explain how physical data flow diagrams are used to model an information system's architecture.

- Describe both centralized and distributed computing alternatives for information system design, including various client/server and Internet-based computing options.

- Describe database and data distribution alternatives for information system design.

- Describe user and system interface alternatives for information system design.

- Describe various software development environments for information system design.

- Describe strategies for developing or determining the architecture of an information system.

- Draw physical data flow diagrams for an information system's architecture and processes.

SCENE

Bob is putting the finishing touches on the business requirements statement, the final deliverable of systems analysis. Meanwhile, Sandra is attending a meeting of the application architecture variance committee. She has requested a variance from that committee. Attendees include Patricia Wilcox, chair; Gary Bond, representing the technology architecture committee; Helen Grissom, representing the data architecture team; and Brock Peterson, representing the development center.

PATRICIA

Thank you for all coming! I hope that you heard that Gary has been appointed to this committee to replace Jonathan. I think it appropriate to review our charter.

As part of the strategic information planning project that was completed last year, we established an application architecture that standardized on the use of various technologies for all future information systems. But that project team also realized that there may be a need, from time to time, to consider variances from that technology. That is the charge of this committee.

Today, we consider our first such variance. I see that all of you have the written request. I'll ask Sandra Shepherd, the project manager for the Member Services Information System, to present the request now.

SANDRA

Thanks, Patricia. As you know, the current application architecture standards were largely based on our internal expertise at the time of the plan. I am managing the Member Services project as the first strategic project to result from the information systems plan. We have been researching technologies to implement the e-commerce dimension of our Member Services project. In so far as this is the first e-commerce project at SoundStage, we have no technology standards. I am here to request a variance to use a set of Web-based technologies to build our new system.

PATRICIA

So, Sandra, what variance are you requesting?

SANDRA

As you are well aware, our software development environments of choice have been *Visual Basic* and *C++*, with some use of *PowerBuilder* for rapid application development projects. These languages have served us well for our client/server development projects, and we intend to use them again for the client/server dimensions of this project. But they are not well suited to the Web dimension of our project.

GARY

So what technology do you want to use?

SANDRA

We had originally planned to learn and use *Java*. *Java* is certainly in our long-term plans. But *Java* has a significant learning curve, especially when you add in the database connectivity issues. We also considered Microsoft *InterDev*. It offers good database connectivity over the Internet, but we are concerned that it is not suited to building a commerce application.

PATRICIA

What language are you proposing?

SANDRA

Actually, it is more of a full-scale application development environment. It's called *Cold Fusion* and it is well respected as a rapid application development tool for Web-based applications like the one we need to build. It provides database connectivity to our SQL Server databases. It leverages our expertise in HTML, and its procedural scripting language seems somewhat similar to that of *PowerBuilder*.

GARY

As I understand it, *Cold Fusion* is not a pure object-oriented technology.

SANDRA

It guess it depends on how you define object-oriented. Our research, and that of our academic friends, tells us that there are degrees of object-orientation. Everything we've read says that VB is somewhere in the continuum, but not pure object-oriented. It would probably be classified as a component-based language, similar to *Visual Basic* and *PowerBuilder*.

GARY

Component? I thought that was a synonym for object?

SANDRA

Technically, no. Components are pre-built objects with limited or no customizability.

GARY

Will that limit us? The power of objects is their extensibility.

SANDRA

No. *Cold Fusion*'s components are pretty much exactly what we need for the Web requirements for our project. Besides, to fully exploit a pure object language would require us to have specified requirements in an object analysis modeling language. It's too late for that.

GARY

So, if you use *Visual Basic* for the client/server portion of the project, will we experience coordination problems, or worse, interoperability problems between the Windows portion of the system and the Web portion?

SANDRA

I don't think so . . . at least I hope not. The only thing the two physical subsystems have in common is the same database. But they both will access the database via remote procedure calls to the SQL engine. The SQL engine does all the database work. It really shouldn't matter whether we have one or five different technologies to implement the actual programs.

BROCK

So what're your alternatives?

SANDRA

We really don't have any. The Web-based e-commerce solution is a primary requirement. We simply cannot implement that requirement in *Visual Basic* or *PowerBuilder*. I suppose we could write the user interface in pure HTML and then try to connect the Web pages to *Visual Basic* programs to do the application logic. We would need to find some middleware to connect the VB programs to the HTML programs.

PATRICIA

That sounds awfully complex.

HELEN

Can I assume that you are not proposing a variance for the database architecture? You still plan to use our *SQL Server* standard?

SANDRA

Absolutely! Because we are designing a multi-tiered architecture, we may even decide to move some of the business logic to the *SQL Server* as stored procedures.

HELEN

I'd like to see us take advantage of that technology! Sandra, what will we have to do to make VB and this *Cold Fusion* work against an *SQL Server* database server? Will you use middleware such as ODBC? From what I understand, that is never as efficient as having native SQL calls in the programming language.

SANDRA

Actually, obviously VB could use native SQL since VB and *SQL Server* are both Microsoft products! But *Cold Fusion* is not. So we either use native *SQL Server* for the VB programs and ODBC for the *Cold Fusion,* or we use ODBC for both even if we sacrifice some performance on the VB side.

HELEN

Not necessarily. We could get a more efficient set of ODBC drivers. I hear the Intersolv product is excellent.

GARY

No. I think we can grant your variance. But why don't we wait and see if we have a performance problem before we buy the third-party ODBC. On the FLAGS project we thought our Microsoft ODBC wouldn't be able to cut it, but the performance turned out to be good enough.

BROCK

Sandra, I'm willing to do the variance too. But I do have another question about performance. I don't know about *Cold Fusion,* but Web applications are always a little slower than non-Web applications, especially knowing that our members mostly use slow modems for access. So that doesn't worry me. But what about the VB part of your project? If my memory serves, this is one of the

larger projects that we've used VB on. Is VB going to give us the performance we need for a system this large?

SANDRA

I don't think we have a problem. But if we do, I have a fallback solution. We will simply shift some of the processing workload to the database server by writing some of the application logic in stored procedures.

PATRICIA

Sounds like a plan. Do I hear a motion to grant this variance?

DISCUSSION QUESTIONS

1. Why would an organization establish a standard set of technologies for all projects?

2. Why might the committee want to grant variances to its technology standards? Why might it not want to grant such variances?

3. Visit the Internet site of the *Cold Fusion* software. How well does it fit to developing Web applications? Do you think it is well suited to building an e-commerce, transaction processing application.

4. What decision should the committee make? Obviously, there is no right or wrong answer here.

Chapter 10 presented a high-level overview of the entire systems design process. You learned that early during system design you develop an architectural blueprint that will serve as an outline for subsequent internal and external design. This chapter focuses exclusively on that blueprint and current alternatives for application architecture. (Subsequent chapters then focus on the detailed internal and external design of each architectural component.) The architectural blueprint will communicate the following design decisions:

APPLICATION ARCHITECTURE

— The degree to which the information system will be centralized or distributed—Most contemporary systems are distributed across networks, including both intranets and the Internet.

— The distribution of stored data across a network—Most modern databases are either distributed or duplicated across networks, either in a client/server or network computing pattern.

— The implementation technology for all software to be developed in-house— Which programming language and tools will be used?

- The integration of any commercial off-the-shelf software—And the need for customization of that software.
- The technology to be used to implement the user interface—Including inputs and outputs.
- The technology to be used to interface with other systems.

These considerations define the application architecture for the information system.

> An **application architecture** specifies the technologies to be used to implement one or more (possibly all) information systems in terms of DATA, PROCESS, INTERFACE, and how these components interact and communicate across a network. It serves as an outline for detailed design, construction, and implementation.

Let's use a slightly different approach for this chapter. In most chapters, we have initially taught concepts and principles before we introduced tools and techniques. But for this chapter, let's first introduce the primary tool, physical data flow diagrams. This will work for two reasons. First, you already know the system concepts and basic constructs of data flow diagrams from Chapter 8. Second, the tool is an elegant and relatively simple way to introduce the different types of application architecture that we want to teach you.

Although you will learn a new technique in this chapter, physical data flow diagrams, the technique is not as important as the application architecture concepts used to partition an information system across a computer network.

PHYSICAL DATA FLOW DIAGRAMS

Data flow diagrams (DFDs) were introduced in Chapter 8 as a systems analysis tool for modeling the *logical* (meaning nontechnical) business requirements of an information system. With just a few extensions of the graphical language, DFDs can also be used as a systems design tool for modeling the *physical* (meaning technical) architecture and design of an information system.

> **Physical data flow diagrams** model the technical and human design decisions to be implemented as part of an information system. They communicate technical choices and other design decisions to those who will actually construct and implement the system.

In other words, physical DFDs serve as a technical blueprint for system construction and implementation.

Physical data flow diagrams were conceived by Gane, Sarson, and DeMarco as part of a formal software engineering methodology called *structured analysis and design*. This methodology was especially well suited to mainframe *COBOL* transaction-based information systems and software. The methodology required rigorous and detailed specification of both logical and physical representations of an information system. In sequence, systems analysts or software engineers would develop the following system models and associated detailed specifications:

1. *Physical DFDs of the current system.* These physical DFDs were intended to help analysts identify and analyze physical problems in the existing system during the problem analysis phase of systems analysis.
2. *Logical DFDs of the current system.* These logical DFDs were merely a transformation of the above physical DFDs that remove all physical detail. They were used as a point of departure for the requirements analysis phase of systems analysis.
3. *Logical DFDs of the target system.* These logical DFDs and their accompanying specifications (data structures and structured English) were intended to represent the detailed nontechnical requirements for the new system.

4. *Physical DFDs of the target system.* These physical DFDs were intended to propose and model the technology choices and design decisions for all logical processes, data flows, and data stores. These diagrams (the focus of this chapter) are developed during the systems design stage of the project.

5. *Structure Charts of the software elements of the target system.* The above physical DFDs would be transformed into structure charts that illustrate a top-down hierarchy of software modules that would conform to accepted principles of good software design.

As you may have guessed, the above methodology was labor intensive and required significant precision and rigor to accomplish its intended result. Today, the complete structured analysis and design methodology as described above is rarely practiced—it is not as well suited to today's object-oriented and component-based software technologies—but data flow diagramming (both logical and physical) remains a useful and much practiced legacy of the structured analysis and design era of systems development.

Let's examine the graphical conventions for *physical* DFDs. Physical DFDs use the same basic shapes and connections as logical DFDs (Chapter 8), namely: (a) *processes,* (b) *external agents,* (c) *data stores,* and (d) *data flows.* A sample physical DFD is shown in Figure 11.1. For now, just notice that the physical DFD primarily shows more technical and implementation detail than its logical DFD equivalent.

Recall that processes are the key shapes on any DFD. That's why they are called **Physical Processes**
process models. Physical DFDs depict the planned, physical implementation of each process.

> A **physical process** is either a *processor,* such as a computer or person, or the technical implementation of specific work to be performed, such as a computer program or manual process.

Earlier in the project, during requirements analysis, we specified *logical* processes needed to fulfill essential business requirements. These logical processes were modeled in our logical data flow diagrams (Chapter 8). Now, during system design, we must specify how these logical processes will be physically implemented. As implied in the above definition for physical processes, there are two elements to physical data flow diagrams:

— Logical processes are frequently <u>assigned to</u> specific physical processors such as PCs, servers, mainframes, people, or other devices in a computer network. To this end, we might draw a physical DFD to model the network's structure.

— Each logical process must be <u>implemented as</u> one *or more* physical processes. Note that some logical processes must be split into multiple physical processes for one or more of the following reasons:
  – To split the process into that portion to be performed by people and that portion to be performed by the computer.
  – To split the process into that portion to be implemented with one technology and that portion to be implemented with a different technology.
  – To show multiple but different implementations of the same logical process (such as one process for paper orders and a different process for Internet orders).
  – To add processes that are necessary to handle exceptions or to implement security requirements and audit trails.

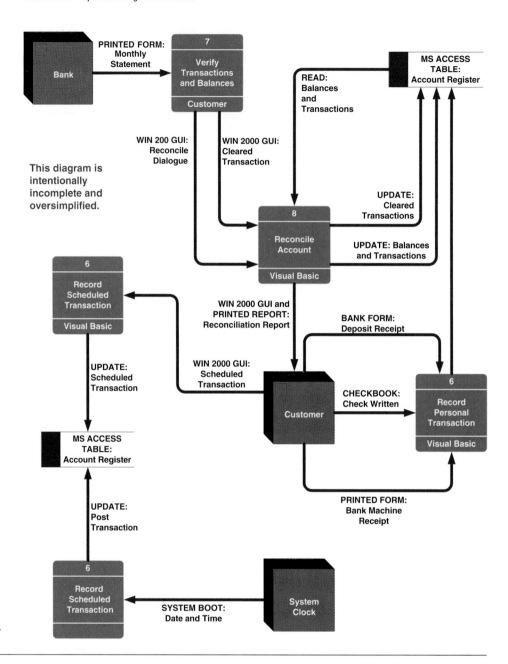

FIGURE 11.1
A Sample Physical Data Flow Diagram

In all cases, if you split a logical process into multiple physical processes, or add additional physical processes, you have to add all necessary data flows to preserve the essence of the original logical process. In other words, the physical processes must still meet the logical process requirements.

Process IDs are optional but can be useful for matching physical processes with their logical counterparts (especially if the logical process is to be implemented with multiple physical processes). Process names use the same action verb + noun/object clause convention as we introduced in Chapter 8. This name is recorded in the center of the shape (see margin). In the bottom of the shape, the implementation is recorded. This convention may have to be adjusted depending on the capabilities of your CASE or automated diagramming tool. The following names demonstrate various possible implementations of the same logical process:

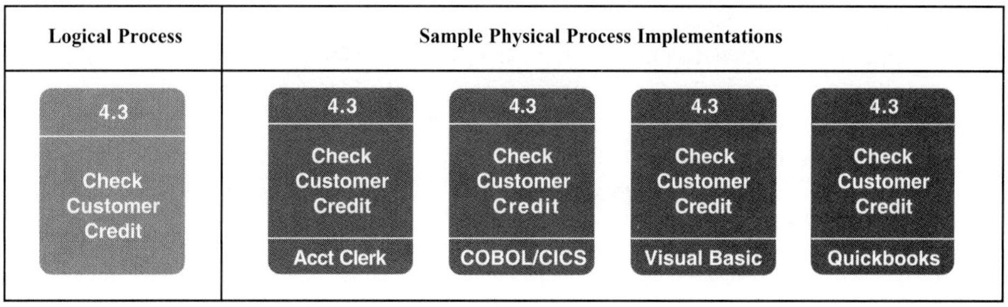

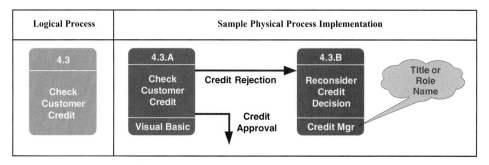

If your CASE tool limits the size of names, you may have to develop and use a set of abbreviations for the technology (and possibly abbreviate your action verbs and object clauses).

If a logical process is to be implemented partially by people and partially by software, it must be split into separate physical processes and appropriate data flows must be added between the physical processes. The name of a physical process to be performed by people, not software, should indicate who would perform that process. We recommend you use titles or roles, not proper names. The following is an example:

Logical Process	Sample Physical Process Implementation

(figure: Logical Process 4.3 "Check Customer Credit"; Sample Physical Process Implementation showing process 4.3.A "Check Customer Credit" (Visual Basic) with "Credit Rejection" flow to 4.3.B "Reconsider Credit Decision" (Credit Mgr) and "Credit Approval" flow, with callout "Title or Role Name")

We didn't just change the manual process, RECONSIDER CREDIT decision, to an external agent, CREDIT MANAGER, because the entire logical process, CHECK CUS-TOMER CREDIT is in the project scope. For that reason, both aspects of the physical implementation are also in the scope. The design is not complete until we specify the process for both the automated and manual aspects of the business requirement.

For computerized processes, the implementation method is, in part, chosen from one of the following possibilities:

— A purchased <u>application</u> software package (e.g., SAP, an enterprise software application, or ARIBA, an Internet-based procurement/purchasing software application).

— A system or utility program (e.g., EXCHANGE SERVER, an e-mail/messaging system, or NETSCAPE COMMERCE SERVER, an electronic commerce framework).

— An existing application program from a program library, indicated simply as LIBRARY or NAME of library.

— A program to be written. Typically, the implementation method specifies the language or tool to be used to construct the program. Example implementation methods include VISUAL BASIC, C++, JAVA, POWERBUILDER, COBOL, VBA, MS ACCESS, PERL, ORACLE DEVELOPER/2000, or FORTE.

One final physical process construct should be introduced, the *multi-process* (see margin). The multi-process indicates multiple implementations of the same physical processor or process. For example, we can use this symbol to indicate multiple PCs, the implementation of a named program on multiple PCs, or the

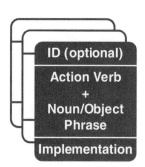

implementation of work to be performed by multiple people. Some CASE tools do not support this construct. If not, you may need to resort to plural names to imply multiplicity of a process or processor.

Many designers prefer a more physical naming convention for computer processes. Instead of a noun + verb phrase, they would substitute the file name of the computer program's physical source code. Consider the following examples:

Logical Process	Sample Physical Process Implementations	
4.3 Check Customer Credit	**4.3.A** CHK_CREDIT.COB COBOL + CICS	**4.3.B** appCheckCredit.vbx Visual Basic

Many organizations have naming conventions and standards for program names.

Again, the number of physical processes on a physical DFD will almost always be greater than the number of logical processes on its equivalent logical DFD. For one thing, processes may be added to reflect data collection, filtering, forwarding, preparation, or quality checks—all in response to the implementation vision that has been selected. Also, some logical processes may be split into multiple physical processes to reflect portions of a process to be done manually versus by a computer, to be implemented with different technology, or to be distributed to clients, servers, or different host computers. It is important that the final physical DFDs reflect all manual and computer processes required for the chosen implementation strategy.

Physical Data Flows

Recall that all processes on any DFD must have at least one input and one output data flow.

> A **physical data flow** represents any of the following: (1) the planned implementation of an input to or output from a physical process; (2) a database command or actions such as create, read, update, or delete; (3) the import of data from or the export of data to another information system across a network; or (4) the flow of data between two modules or subroutines within the same program.

Implementation method: data flow name →

OR

Data flow name (implementation method) →

Physical data flows are named as indicated by the templates in the margin. Figure 11.2 demonstrates the application of one of these naming conventions as applied to several types of physical data flows.

Physical DFDs must also indicate any data flows to be implemented as business forms. For instance, FORM 23: COURSE REQUEST might be a one-part business form used by students to register for classes. Business forms frequently use a multiple (carbon or carbonless) copy implementation. At some point in processing, the different copies are split and travel to different manual processes. This is shown on a physical DFD as a diverging data flow (introduced in Chapter 8). Each copy should be uniquely named. For example, at a restaurant, the customer receives FORM: CREDIT CARD VOUCHER (CUSTOMER COPY) and the merchant retains FORM: CREDIT CARD VOUCHER (MERCHANT COPY).

Most logical data flows are carried forward to the physical DFDs. Some may be consolidated into single physical data flows that represent business forms. Others may be split into multiple flows as a result of having split logical processes into

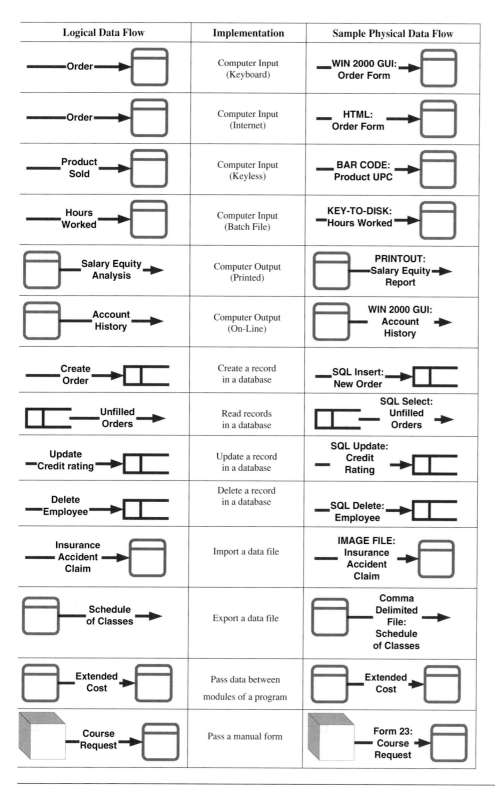

Logical Data Flow	Implementation	Sample Physical Data Flow
Order	Computer Input (Keyboard)	WIN 2000 GUI: Order Form
Order	Computer Input (Internet)	HTML: Order Form
Product Sold	Computer Input (Keyless)	BAR CODE: Product UPC
Hours Worked	Computer Input (Batch File)	KEY-TO-DISK: Hours Worked
Salary Equity Analysis	Computer Output (Printed)	PRINTOUT: Salary Equity Report
Account History	Computer Output (On-Line)	WIN 2000 GUI: Account History
Create Order	Create a record in a database	SQL Insert: New Order
Unfilled Orders	Read records in a database	SQL Select: Unfilled Orders
Update Credit rating	Update a record in a database	SQL Update: Credit Rating
Delete Employee	Delete a record in a database	SQL Delete: Employee
Insurance Accident Claim	Import a data file	IMAGE FILE: Insurance Accident Claim
Schedule of Classes	Export a data file	Comma Delimited File: Schedule of Classes
Extended Cost	Pass data between modules of a program	Extended Cost
Course Request	Pass a manual form	Form 23: Course Request

FIGURE 11.2
Physical Data Flows

multiple physical processes. Still others may be duplicated as multiple flows with different technical implementations. For example, the logical data flow ORDER might be implemented as all of the following: FORM: ORDER, PHONE: ORDER, (verbal order taken over the phone), HTML: ORDER (order submitted over the Internet), FAX: ORDER (order received by fax), and MESSAGE: ORDER (an order submitted via e-mail).

Physical External Agents

External agents are carried over from the logical DFD to the physical DFD unchanged. Why? By definition, external agents were classified during systems analysis as outside the scope of the systems and, therefore, not subject to change. Only a change in requirements can initiate a change in external agents.

Physical Data Stores

From Chapter 8 you know that each data store on the logical DFD now represents all instances of a named entity on an entity relationship diagram (from Chapter 7). Physical data stores implement the logical data stores.

> A **physical data store** represents the implementation of one of the following: (1) a database, (2) a table in a database, (3) a computer file (e.g., *VSAM* or *B-Tree*), (4) a tape or media backup of anything important, (5) any temporary file or batch as needed by a program (e.g., TAX TABLES), or (6) any type of noncomputerized file.

When most people think of data stores, they think of computer files and databases. But many data stores are not computerized. File cabinets of paper records immediately come to mind; however, most businesses are replete with more subtle forms of manual data stores such as address cards, paper catalogs, cheat sheets of various important and reusable information, standards manuals, standard operating procedures manuals, directories, and the like. Despite predictions about the demise of paper files, they will remain a part of many systems well into the foreseeable future—if for no other reasons than (1) there is psychological comfort in paper and (2) the government frequently requires it!

The name of a physical data store uses the format indicated in the margin. Some examples of physical data stores are shown in Figure 11.3.

Some designs require that temporary files be created to act as a queue or buffer between physical processes that have different timing. Such files are documented in the same manner except their name should include some indication of their temporary status.

ID (opt)	Implementation Method: Data Store Name

ID (opt)	Data Store Name (Implementation Method)

Physical processes, data flows, external agents, and data stores make up the physical data flow diagrams. And these physical DFDs model the proposed or planned architecture of an information system application. We can subsequently use that physical model to design the internal and external details for each data store (Chapter 12) and data flow (Chapters 13–15). Now that you understand the basic components of physical DFDs, let's use them to introduce some of today's architectural choices for information system design.

INFORMATION TECHNOLOGY ARCHITECTURE

Information technology (IT) architecture can be a complex subject worthy of its own course and textbook. (See the Suggested Readings at the end of this chapter.) In this section, we will attempt to summarize contemporary IT alternatives and trends that are influencing design decisions as we go to press. It should be noted that new alternatives are continuously evolving. The best systems analysts will not only learn more about these technologies, but will also understand how they work and their limitations. Such a level of detail is beyond the scope of this book, but you should find various technology, database, and networking courses that will provide greater insight. Systems analysts must continuously read popular trade journals to stay abreast of the latest technologies and techniques that will keep their customers and their information systems competitive.

Your information system framework (see the chapter map at the beginning of

Logical Data Store	Implementation	Physical Data Store
Human Resources	A database (multiple tables)	Oracle : Human Resources DB
Marketing	A database view (subset of a database)	SQL Server: Northeast Marketing DB
Purchase Orders	A table in a database	MS Access: Purchase Orders
Accounts Receivable	A legacy file	VSAM File: Accounts Receivable
Tax Rates	Static data	ARRAY: Tax Table
Orders	An off-line archive	TAPE Backup: Closed Orders
Employees	A file of paper records	File Cabinet: Personnel Records
Faculty/Staff Contact Data	A directory	Handbook: Faculty/Staff Directory
Course Enrollments By Date	Archived reports (for reuse and recall)	REPORT MGR: Course Enrollment Reports

FIGURE 11.3
Physical Data Stores

this chapter) provides one suitable framework for understanding IT architecture. A few explanations are in order:

- The grayed blocks represent building blocks and phases completed during systems analysis.

- Our goal in general design is to specify the technical solution (at a high level) for various system components—DATA, PROCESSES, and INTERFACES (in color).

- Architectural standards and/or technology constraints are represented in the bottom row of the framework. Notice that these standards or decisions are determined either as part of a separate architecture project (preferred and increasingly common) or as part of each system development project.

- The upward-pointing arrows indicate the technology standards that will influence or constrain the design models.

In Chapter 2 we introduced DATA, PROCESSES, and INTERFACES as the key building blocks of all information systems. In most organizations, previous stand-alone mainframe and personal computers are being linked, using server computers, to form complex networks. Accordingly, our data, process, and interface building blocks are being distributed or duplicated across networks. We call the approach distributed systems architecture.

Distributed Systems

Today's information systems are no longer monolithic, mainframe computer-based systems. Instead, they are built on some combination of networks to form distributed systems.

> A **distributed system** is one in which the DATA, PROCESS, and INTERFACE components of an information system are distributed to multiple locations in a computer network. Accordingly, the processing workload required to support these components is also distributed across multiple computers on the network.

The opposite of distributed systems are centralized systems.

> In **centralized systems,** a central, multi-user computer (usually a mainframe) hosts all the DATA, PROCESS, and INTERFACE components of an information system. The users interact with this host computer via terminals (or, today, a PC emulating a terminal), but virtually <u>all</u> of the actual processing and work is done on the host computer.

Distributed systems are inherently more complicated and more difficult to implement than distributed solutions. So why is the trend toward distributed systems?

— Modern businesses are already distributed and, thus, they need distributed system solutions.

— Distributed computing moves information and services closer to the customers that need them.

— Distributed computing consolidates the incredible power made possible by the proliferation of personal computers across an enterprise (and society in general). Many of these personal computers are only used to a fraction of their processing potential when utilized as stand-alone PCs.

— In general, distributed system solutions are more user-friendly because they utilize the PC as the user interface processor.

— Personal computers and network servers are much cheaper than mainframes. (But admittedly, the total cost of ownership is at least as expensive once the networking complexities are added in.)

There is a price to be paid for distributed systems. Network data traffic can cause congestion that actually slows performance. Data security and integrity can also be more easily compromised in a distributed solution. Still, there is no arguing the trend toward distributed systems architecture. While many centralized, legacy applications still exist, they are gradually being transformed into distributed information systems.

Figure 11.4 compares the various flavors of distributed systems architecture. Conceptually, any information system application can be mapped to five layers:

— The **presentation layer** is the actual user interface—the presentation of inputs and outputs to the user.

— The **presentation logic layer** is any processing that must be done to generate the presentation. Examples include editing input data and formatting output data.

— The **application logic layer** includes all the logic and processing required to support the actual business application and rules. Examples include credit checking, calculations, data analysis, and the like.

— The **data manipulation layer** includes all the commands and logic required to store and retrieve data to and from the database.

— The **data layer** is the actual stored data in a database.

Figure 11.4 shows these conceptual layers as rows. The columns in Figure 11.4 illustrate how and where these flavors can be implemented in different distributed

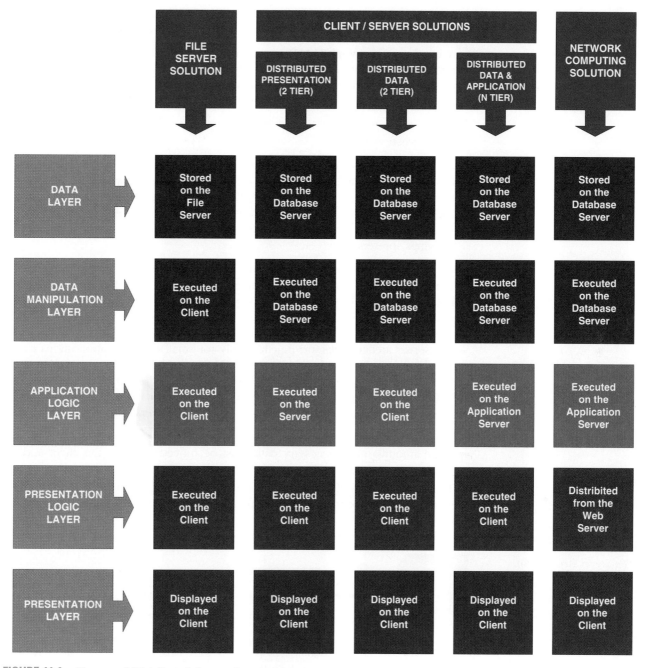

FIGURE 11.4   *Flavors of Distributed Computing and Systems*

information system architectures. There are three flavors of distributed systems architecture:

— *File server* computing.
— *Client/server* computing (various flavors).
— Internet-based computing (called *network computing*).

Let's discuss each in greater detail.

File Server Architecture   Today very few personal computers and workstations are used to support stand-alone information systems. Organizations need to share data and services. Local area networks allow many PCs and workstations to be connected to share resources and communicate with one another.

A **local area network (LAN)** is a set of client computers (usually PCs) connected to one or more servers (usually a more powerful PC or larger computer) through either cable or wireless connections over relatively short distances—for instance, in a single department or in a single building.

In the simplest LAN environments, a file server architecture is used to implement information systems.

A **file server system** is a LAN-based solution in which a server computer hosts only the data layer. All other layers of the information system application are implemented on the client PC. (Note: File servers are also typically used to share other nondatabase files across networks—examples include word processing documents, spreadsheets, images and graphics, engineering drawings, presentations, etc.)

A file server architecture is illustrated in Figure 11.5. This architecture is typical of that used for many PC database engines such as Microsoft *Access* and *FoxPro*. (Think about it! While your *Access* database may be stored on a network server, the actual *Access* program must be installed or executed from each PC that uses the database.)

File server architectures are practical only for small database applications to be shared by relatively few users because if the application wants to examine only one record in the database (e.g., one customer), the entire file or table of records must be first downloaded to the client PC where the data manipulation logic will be executed to read the desired record. There are several disadvantages to this approach.

— Large amounts of unnecessary data must be moved between the client and server. This data traffic can significantly reduce application performance.

— The client PC must be robust (a so-called *fat client*). It is doing virtually all of the actual work including data manipulation, application logic, presentation

FIGURE 11.5
A File Server Architecture

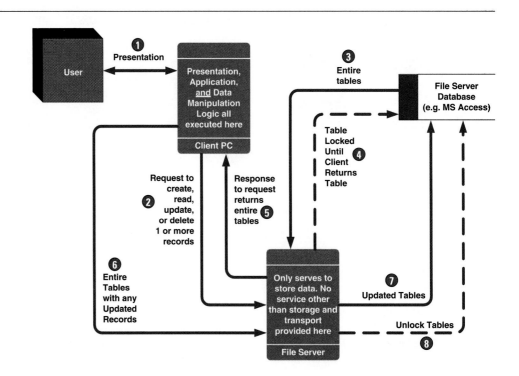

logic, and presentation. It must also have enough disk capacity to store the downloaded tables.

— Database integrity can be easily compromised. Think about it. If any record has been downloaded to be updated, the entire file has been downloaded. Other users (clients) must be prevented (or locked) from making changes to any other record in that file. The greater the number of simultaneous users, the more this locking requirement slows response time.

Very few mission critical information systems can be implemented with file server technology. So why are file server database management systems such as MS *Access* so popular? First, file server tools such as *Access* can be used to develop fairly robust applications for individuals and <u>small</u> work groups. Second, and perhaps more significantly, file server databases such as *Access* can be used to rapidly construct prototypes for more robust client/server architectures.

Client/Server Architectures   The prevailing distributed computing model of the current era is called client/server computing (although it is rapidly giving way to Internet-based models). There are nearly as many definitions of client/server as there are books and products that use the term. At least one technology corporation CEO has sarcastically defined the term as "anything that sells new IT products." For our purposes, let's define the term as follows:

> A **client/server system** is a solution in which the presentation, presentation logic, application logic, data manipulation, and data layers are distributed between client PCs and one or more servers.

The client computers may be any combination of personal computers or workstations, "sometimes connected" notebook computers, handheld computers (e.g., Palm or Windows CE Platforms), Web TVs, or any devices with embedded processors that could connect to the network (e.g., robots or controllers on a manufacturing shop floor). Clients may be thin or fat.

> A **thin client** is personal computer that does not have to be very powerful (or expensive) in terms of processor speed and memory because it only presents the interface (screens) to the user—in other words, it acts only as a terminal. Examples include *Windows* Terminal and *X/Windows*.

In thin client computing, the actual application logic executes on a remote application server (such as Cytrix Server or Microsoft Windows Terminal Server).

> A **fat client** is a personal computer, notebook computer, or workstation that is typically more powerful (and expensive) in terms of processor speed, memory, and storage capacity. Almost all PCs are considered fat clients.

A server in the client/server model must be more powerful and capable than a server in the file server model. In fact, a mainframe computer can play the role of server in a client/server solution. More typical, however, are network servers running client/server capable operating systems such as UNIX, Windows 2000 Server or Enterprise Edition, or Linux. Several types of servers may be used in a client/server solution. These may reside on separate physical servers or be consolidated into fewer servers.

> A **database server** hosts one or more shared databases (like a file server) but also executes all database commands and services for information systems (unlike a file server). Most database servers host an SQL database engine such as *Oracle*, Microsoft *SQL Server*, or IBM *Universal Database* (formerly known as *DB2*).

A **transaction server** hosts services that ultimately ensure that all database updates for a single business transaction succeed or fail as a whole. Examples include IBM *CICS,* BEA *Tuxedo,* and *Microsoft Transaction Server.*

An **application server** hosts application logic and services for an information system. It must communicate on the front end with the clients (for presentation) and on the back end with database servers for data access and update. An application server is often integrated with the transaction server. Most application servers are based on either the *CORBA* object-sharing standard or the Microsoft *COM+* or *DNA* standard.

A **messaging** or **groupware server** hosts services for e-mail, calendaring, and other work group functionality. This type of functionality can actually be integrated into information system applications. Examples include Lotus *Notes* and Microsoft *Exchange Server.*

A **Web server** hosts Internet or intranet Web sites. It communicates with fat and thin clients by returning to them documents (in formats such as *HTML*) and data (in formats such as *XML*). Some Web servers are specifically designed to host e-commerce applications (e.g., Netscape *Commerce Server*).

Client/server architecture itself comes in several flavors, each of which deserves its own explanation. Each of these C/S flavors is also compared to the others in Figure 11.4.

Client/Server—Distributed Presentation   Most centralized (or mainframe) computing applications use an older character user interface (CUI) that is cumbersome and awkward when compared to today's graphical user interfaces (GUI) such as Microsoft *Windows* and UNIX *X/Windows* (not to mention Web browsers such as Netscape *Navigator* and Microsoft *Internet Explorer*). As personal computers rapidly replaced dumb terminals, users became increasingly comfortable with this newer technology. And as they developed familiarity and experience with PC productivity tools such as word processors and spreadsheets, they wanted their centralized, legacy computing applications to have a similar look and feel using the GUI model.
Enter distributed presentation!

A **distributed presentation** client/server system is a solution in which the presentation and presentation logic layers are shifted from the server of a legacy system to reside on the client. The application logic, data manipulation, and data layers remain on the server (usually a mainframe).

Sometimes called the poor person's client/server, this alternative builds on and enhances centralized computing applications. Essentially, the old CUIs are stripped from the legacy applications and regenerated as GUIs that will run on the PC. In other words, only the user interface (or presentation layer) is distributed to the client.
Distributed presentation offers several advantages. First, it can be implemented relatively quickly because most aspects of the legacy application remain unchanged. Second, users get a fast, friendly, and familiar user interface to legacy systems—one that looks at least somewhat familiar to their PC productivity tools. Finally, the useful lifetime of legacy applications can be extended until resources warrant a wholesale redevelopment of the application. The disadvantage is that the application's functionality cannot be significantly improved, and the solution does not maximize the potential of the client's desktop computer by dealing only with the user interface.
A class of CASE tools, sometimes called screen scrapers, automatically read the CUI and generate a first-cut GUI that can be modified by a GUI editor. Figure 11.6 demonstrates this technology. Figure 11.7 shows a physical DFD for a distributed presentation solution.

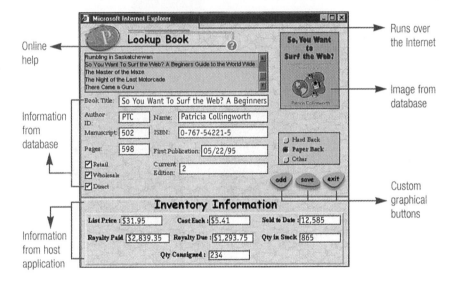

FIGURE 11.6
Building a GUI from a CUI

Client/Server—Distributed Data This is the simplest form of true client/server computing. A local area network usually connects the clients to the server.

A **distributed data** client/server system is a solution in which the data <u>and</u> data manipulation layers are placed on the server(s), and the application logic, presentation logic, and presentation are placed on the clients. This is also called **two-tiered client/server computing.**

A two-tiered, distributed data client/server system is illustrated as a physical DFD in Figure 11.8.

It is important to understand the difference between file server systems and distributed data client/server systems. Both store their actual database on a server. But only client/server systems execute all data manipulation commands (e.g., SQL instructions to create, read, update, and delete records) on a server. Recall that in file server systems, those data manipulation commands must be implemented on the client. Distributed data client/server solutions offer several advantages over file server solutions:

— There is <u>much</u> less network traffic because only the database requests and the database records that are needed are actually transported to and from the client workstations.

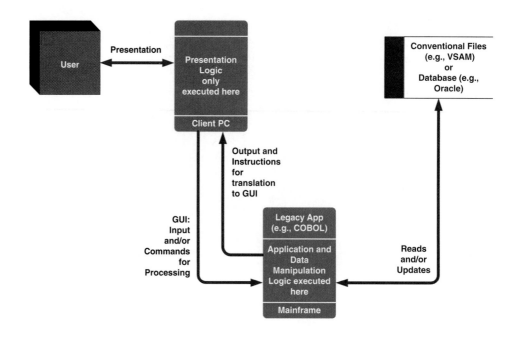

FIGURE 11.7
Client/Server System:
Distributed Presentation

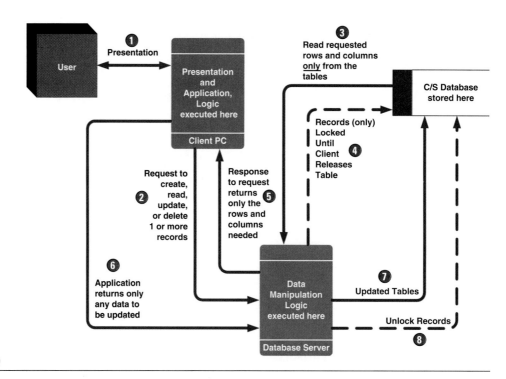

FIGURE 11.8
Client/Server System:
Distributed Data (Two Tiers)

— Database integrity is easier to maintain. Only the records in use by a client must typically be locked. Other clients can simultaneously work on other records in the same table or database.

Unfortunately, the client workstation must still be fairly robust ("fat") to provide the processing for the application logic layer. This logic is usually written in a client/server programming language such as Sybase Corporation's *PowerBuilder,* Microsoft's *Visual Basic,* or *Visual C++*. Those programs must be compiled for and execute on the client. To improve application efficiency and reduce network traffic, some business logic may be distributed to the database server in the form of stored procedures (discussed in the next chapter).

The *database server* is fundamental to this architecture. Database servers store the database, but they also execute the database instructions <u>directly on those servers.</u> The clients merely send their database instructions to the server. The server returns only the result of the database command processing—not entire databases or tables. All high-end database engines such as *Oracle* and Microsoft *SQL Server* use this approach. A distributed data architecture may involve more than one database server. Data may be distributed across several database servers or duplicated on several database servers.

The key potential disadvantage to the two-tiered client/server is that the application logic must be duplicated and thus maintained on all the clients, possibly hundreds or thousands. The designer must plan for version upgrades and provide controls to ensure that each client is running the most current release of the business logic, as well as ensure that other software on the PC (purchased or developed in-house) does not interfere with the business logic.

Client/Server—Distributed Data and Application   When the number of clients grows, two-tiered systems frequently suffer performance problems associated with the inefficiency of executing all the application logic on the clients. Also, in multiple-user transaction processing systems (also called on-line application processing or OLAP), transactions must be managed by software to ensure that all the data associated with the transaction is processed as a single unit. This generally requires a distribution of a multi-tiered client/server approach.

> A **distributed data and application** client/server system is a solution in which: (1) the data <u>and</u> data manipulation layers are placed on their own server(s), (2) the application logic is placed on its own server, and (3) only the presentation logic and presentation are placed on the clients. This is also called **three-tiered** or **n-tiered client/server computing.**

The three-tiered client/server solution uses the same database servers as in the two-tiered approach. Additionally, the three-tiered system introduces an application and/or transaction server. By moving the application logic to its own server, that logic need only be maintained on the server instead of all the clients. The three-tiered solution is depicted as a physical data flow diagram in Figure 11.9.

Three-tiered client/server logic can be written and partitioned across multiple servers using languages such as Microsoft *Visual Basic* and *C++* (in combination with a transaction monitor such as Microsoft *Transaction Server*). High-end tools such as *Forté* provide an even greater opportunity to distribute application logic and data across a complex network. As with the database server solution, some business logic could be distributed to the database server in the form of stored procedures.

In a three-tiered system, the clients execute a minimum of the overall system's components. Only the user interface and some relatively stable or personal application logic need be executed on the clients. This simplifies client configuration and management.

The biggest drawback of the three-tiered client/server is its complexity in design and development. The most difficult aspect of a three-tiered client/server application design is partitioning.

> **Partitioning** is the act of determining how to best distribute or duplicate application components (DATA, PROCESS, and INTERFACES) across the network.

Fortunately, CASE tools are constantly improving to provide greater assistance with partitioning.

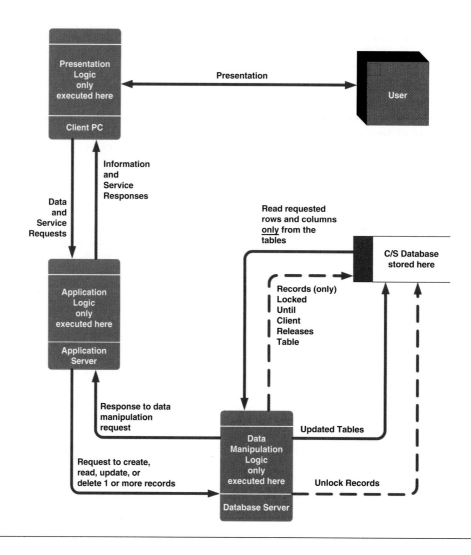

FIGURE 11.9 *Client/Server System: Distributed Data and Application (Three Tiers)*

Internet-Based Computing Architectures Some consider Internet-based system architectures to simply be the latest evolution of client/server. And we would be hard pressed to argue against their case. But we present Internet-based computing alternatives in this section as a fundamentally different form of distributed architecture that is rapidly reshaping the design thought processes of systems analysts and information technologists.

A **network computing system** is a multi-tiered solution in which the presentation and presentation logic layers are implemented in client-side Web browsers using content downloaded from a Web server. The presentation logic layer then connects to the application logic layer that runs on an application server, which subsequently connects to the database server(s) on the backside.

Think about it! All information systems running in browser—financials, human resources, operations—all of them! E-commerce is part of this formula, and as we go to press, e-commerce applications are getting most of the attention. But the same Internet technologies being used to build e-commerce solutions are being used to reshape the internal information systems of most businesses—we call it e-business (although that term is also subject to multiple interpretations). Network computing is, in our view, a fundamental shift <u>away from</u> what we just described as client/server.

Very few new technologies have witnessed as explosive a growth in business and society as the Internet or World Wide Web. The Internet extends the reach of our information and transaction processing systems to include potential customers, customers, partners, remotely located employees, suppliers, the government, and even competitors. Until recently the Internet was largely being used to establish a company's presence in a virtual marketplace and to disseminate public information about products and services and provide a new foundation for customer-focused service. But today most businesses are extremely focused on developing e-commerce solutions that will allow customers to directly interact with and conduct business on the Web (such as direct-to-consumer shopping). We've even seen the invention of the virtual business, a business that "does business" entirely on the Web [e.g., Amazon.com (books and media), ETrade (stocks and bonds), eBay (auctions), and Buy.com (electronics and appliances)]. One of the most intriguing debates is whether these "click-and-mortar" virtual companies can turn a profit and actually compete with more traditional "brick-and-mortar" companies—the latter of which are diversifying rapidly to enter cybermarkets.

But the greatest potential of this Internet technology may actually be its application to traditional information systems applications and development on intranets.

> An **intranet** is a secure network, usually corporate, that uses Internet technology to integrate desktop, work group, and enterprise computing into a single cohesive framework.

Everything runs in (or at least from) a browser—your productivity applications such word processing and spreadsheets; any and all traditional information systems applications you need for your job (e.g., financials, procurement, human resources, etc.); all e-mail, calendaring, and work group services (allowing, for example, virtual meetings and group editing of documents); and of course, all of the external Internet links that are relevant to your job.

The appeal of this concept should not be hard to grasp. Each employee's "start page" is a **portal** into all computer information systems and services he or she needs to do his or her entire job. Because everything runs in a Web browser, there is no longer a need to worry about, or develop for, multiple different computer architecture (Intel versus Motorola versus RISC) or worry about different desktop operating systems (Windows 95 versus NT versus 98 versus 2000). A physical data flow diagram for network computing is shown in Figure 11.10. Notice that a Web server is added to our prior three-tiered model. The DFD also shows both e-commerce (business-to-consumer) and e-business (business-to-business) dimensions of network computing.

Does this all sound too good to true? Something of a cyber-Camelot? By the time you read this, our own institution will have likely implemented its first mission critical e-business information system. Purdue, like all enterprises, has a procurement (or purchasing) function. We buy everything from pencils to furniture to computers to radioactive isotopes—literally tens of thousands of different supplies, materials, and other products. As we go to press, Purdue has redesigned its procurement system to be a combination intranet/Internet/extranet application. Here's how it works:

> The entire application runs in either the Microsoft or Netscape browser. Any employee of the university, once authenticated, can initiate a purchase requisition via his Web browser (the intranet dimension). Employees can even "shop" for items from a Web-mall of approved suppliers with which the university has standing contracts(the Internet dimension). When a requisition is submitted, it will be smart enough to know who must approve it (at every level) based on cost and type of items ordered. The system will be able to automatically check for available funds to pay for the purchase. Employees will be able to audit the electronic flow of the requisition through the approval process and into purchase order status. Managers will be able to revise the approval flow to get additional input as needed. Most final orders will be transmitted electronically over a secure business-to-businesses extranet between the university and its

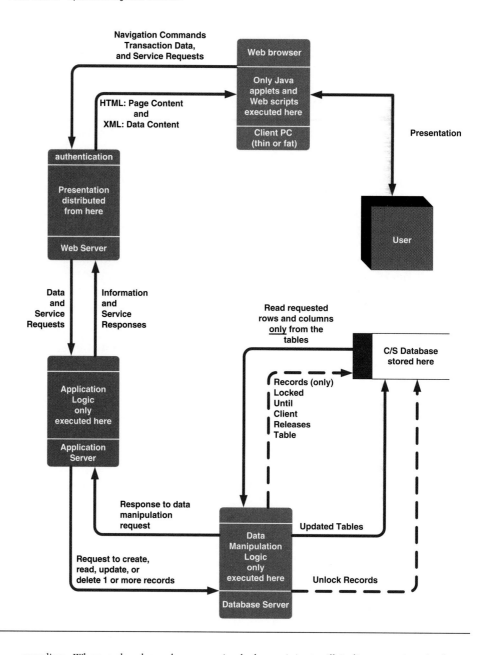

FIGURE 11.10
Network Computing System:
Internet/Intranet

suppliers. When ordered goods are received, the recipient will indicate receipt via the same Web-based system, and payment, via electronic funds in most cases, will be made. At any time, managers will be able to generate useful procurement information for employees, departments, suppliers, whatever! A paperless office? Very close!

Such intriguing system development possibilities are being fueled by some fundamental emerging technologies that you should make a part of your curriculum plan of study if possible:

— The programming language of choice for the application logic in network computing architectures will be *Java*. While still maturing, *Java* has enjoyed the fastest adoption rate of any language in history. Essentially, *Java* is a reasonably platform-independent programming language designed specifically to exploit both Internet and object-oriented programming standards. *Java* is designed to execute in your Internet browser making it less susceptible to differences in computing platforms and operating systems (but it's not yet perfect).

— The interface language of choice for the presentation and presentation logic layers in network computing architectures is currently *HTML,* or hypertext markup language. *HTML* is used to create the pages that run in your browser. Soon another player in this layer will dominate—*XML,* the extensible markup language. This widely embraced standard allows developers to also define the structure of the data to be passed to Web pages, a critical requirement for Web-based e-commerce and intranet-based information systems. *XML* may eventually replace *HTML;* or the two standards may merge into one very powerful language.

- As with traditional information systems, the data and data manipulation layers will likely continue to be implemented with SQL database engines.

- Web browsers will continue to be important. In fact, Web browsers may ultimately be more important than your choice of desktop operating system. Is it any wonder that *Windows,* with each new version, looks and feels more like a browser—check out the Active Desktop option in *Windows.* It looks almost exactly like a Web page into your PC's files and applications. Also check out the user interface for *Quicken,* the personal financial management program. Again, it looks more like a Web application than a *Windows* application, even when you are not connected to the Web.

All of these Internet and intranet solutions involve leading-edge technologies and standards that have no doubt changed even since these words were written. Technology vendors will undoubtedly play a significant role in the evolution of the technology. However the specific technologies play out, we expect the Internet and intranets to become the most common architectural models for tomorrow's information systems.

So where are we in our study of information technology architecture? You've learned that several distributed systems and network options exist for modern information systems. Most can be broadly classified as either client/server or network computing architectures. To be sure, there is much more for you to learn about the underlying communications technology, but that is the subject of at least one additional book. The AIS/ACM/AITP Model Curriculum for Undergraduate Information Systems recommends that all information systems graduates complete at least one course in information technology architecture and one course in fundamental data communications.

Once we've decided on an overall architecture for an information system, there are still fundamental information technology decisions to be made regarding our DATA, PROCESS, and INTERFACE building blocks.

Data Architectures— Distributed Relational Databases

The network provides the map for distributing data to optimal locations. Historically, the desire to control data centrally has been considered essential. As a "shared" resource, centralized data is easier to manage. In the not-too-distant past, the only practical way to accomplish this goal was to store all data on a central computer, with absolute control by a central data administration group.

The underlying technology of client/server and network computing has made it possible to distribute data without loss of control. This control is accomplished through advances in distributed relational database technology.

A **relational database** stores data in a tabular form. Each file is implemented as a table. Each field is a column in the table. Each record in the file is a row in the table. Related records between two tables (e.g., CUSTOMERS and ORDERS) are implemented by intentionally duplicating columns in the two tables (in this example, CUSTOMER NUMBER is stored in both the CUSTOMERS and ORDERS tables).

A **distributed relational database** distributes or duplicates tables to multiple database servers located in geographically important locations (such as different sales regions).

The software required to implement distributed relational databases is called a distributed relational database management system.

> A **distributed relational database management system** (or distributed RDBMS) is a software program that controls access to and maintenance of the stored data in the relational format. It also provides for backup, recovery, and security. It is sometimes called a client/server database management system.

In a distributed RDBMS, the underlying database engine that processes all database commands executes on the database server. The advantage of the latter is to reduce the data traffic on the network. This is a significant advantage for all but the smallest systems (as measured in number of users). A distributed relational DBMS also provides more sophisticated backup, recovery, security, integrity, and processing (although the differences seem to erode with each new PC RDBMS release).

Examples of distributed RDBMSs include Oracle Corporation's *Oracle*, IBM's *Universal Database* family, Microsoft's *SQL Server*, and Sybase Corporation's *Sybase*. Most RDBMSs support two types of distributed data.

> **Data partitioning** truly distributes rows and columns to specific database servers with little or no duplication between servers. Different columns can be assigned to different database servers (called *vertical partitioning*) or different rows in a table can be allocated to different database servers (called *horizontal partitioning*).

> **Data replication** duplicates some or all tables (rows and columns) on more than one database server. Entire tables can be duplicated on some database servers; while subsets of rows in a table can be duplicated to other database servers. The RDBMS with replication technology not only controls access to and management of each database server database, but it also prorogates updates on one database server to any other database server where the data is duplicated.

For a given information system application, the DATA architecture must specify the RDBMS technology and the degree to which data will be partitioned or replicated. One way to document these decisions is to record them in the physical data stores as follows. Notice how we used the ID area to indicate codes for partitioning (P) and replication (M for the master copy and R for the replicated copy). In the case of the former, we should specify which rows and/or columns are to be partitioned to the physical database.

Logical Data Store	Physical Data Stores Using Partitioning	Physical Data Stores Using Replication
1 CUSTOMERS	1P.# Oracle 7: REGION 1 CUSTOMERS 1P.# Oracle 7: REGION 2 CUSTOMERS	Not applicable. Branch offices do not need access to data about customers outside of their own sales region.
2 PRODUCTS	Not applicable. All branch offices need access to data for all products, regardless of sales region.	2M Oracle 8i: PRODUCTS (Master) 2R Oracle 8i: PRODUCTS (Replicated Copy)

An application's DATA architecture is selected based on the desired client/server or network computing model and the database technology needed to support that model. Many organizations have standardized on both their PC RDBMS of choice, as well as their preferred distributed, enterprise RDBMS of choice. For example, SoundStage has standardized on Microsoft *Access* and *SQL Server*. Generally, a qualified database administrator should be included in any discussions about the database technology to be used and the design implications for any databases that will use that technology.

Another fundamental information technology decision must be made regarding inputs, outputs, and intersystem connectivity. The decision used to be simple—batch inputs versus on-line inputs. Today we must consider modern alternatives such as automatic identification, pen data entry, various graphical user interfaces, electronic data interchange, imaging, and voice recognition, among others. Let's briefly examine these alternatives and their physical DFD constructs.

INTERFACE Architectures— Inputs, Outputs, and Middleware

Batch Inputs or Outputs   In batch processing, transactions are accumulated into batches for periodic processing. The batch inputs are processed to update databases and produce appropriate outputs. Most outputs tend to be generated to paper or microfiche on a scheduled basis. Others might be produced on demand or within a specified time period (e.g., 24 hours).

Contrary to popular belief, batch input technologies are not quite obsolete. You rarely see punched cards and tape batches today, but some application requirements lend themselves to batch processing. Perhaps the inputs arrive in natural batches (e.g., mail), or perhaps outputs are generated in natural batches (e.g., invoices). Many organizations still collect and process time cards in batches. There is, however, a definite trend away from batch input to on-line approaches. In the meantime, key-to-disk file is the most common and its physical data flow construct would look as follows. First, notice that the logical name is singular, but the batch name is plural. Also notice that the batch goes into a temporary data store, which is read by a payroll process triggered by date.

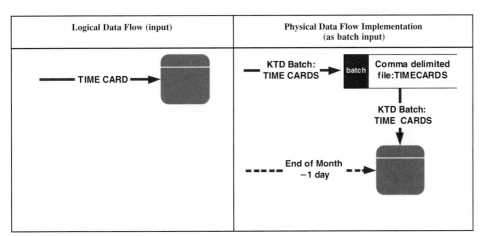

Logical Data Flow (input)	Physical Data Flow Implementation (as batch input)
──▶ TIME CARD ──▶ ▢	KTD Batch: TIME CARDS ──▶ batch ──▶ Comma delimited file:TIMECARDS ↓ KTD Batch: TIME CARDS ↓ End of Month −1 day ──▶ ▢

Batch output is quite another story. Many applications lend themselves to batch output. Examples include generation of invoices, account statements, grade reports, paychecks, W-2 tax forms, and many others. Batch outputs often share one common physical characteristic, the use of a preprinted form. It should not be difficult for you to envision a preprinted form to be loaded in the printer to produce any of the aforementioned output examples. A physical data flow construct would look something like the following. Again, note the plural name reflective of batch processing.

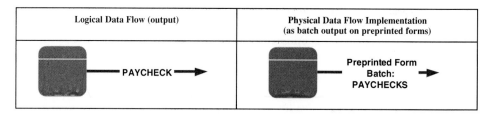

Logical Data Flow (output)	Physical Data Flow Implementation (as batch output on preprinted forms)
PAYCHECK →	Preprinted Form Batch: PAYCHECKS →

As older batch-based systems become candidates for replacement, other physical implementation alternatives should be explored.

On-Line Inputs and Outputs   The majority of systems have slowly evolved from batch processing to on-line or real-time processing. On-line inputs and outputs provide for a more conversational dialogue between the user and computer applications. They also provide near immediate feedback in response to transactions, problems, and inquiries. In today's fast-paced economy, most business transactions and inquiries are best processed as soon as possible. Errors are identified and corrected more quickly because there is no time lapse between data entry and input (as was the case in batch processing). Furthermore, on-line methods permit greater human interaction in decision making.

Today most systems are being designed for on-line processing, even if the data arrives in natural batches. Technically, all GUI and Web applications are on-line or real-time, and since we've already learned that those architectures are preferred in client/server and network computing, then we can expect that most physical data flows will be implemented with some type of GUI technology. The physical data flow constructs would look something like the following. For the physical output, notice that two formats of the physical output are possible. We could have added the junction symbols (Chapter 8) to make the flows mutually contingent (both required) or mutually exclusive (either/or, but not both).

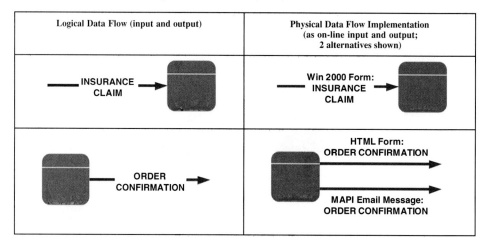

Logical Data Flow (input and output)	Physical Data Flow Implementation (as on-line input and output; 2 alternatives shown)
INSURANCE CLAIM →	Win 2000 Form: INSURANCE CLAIM →
ORDER CONFIRMATION →	HTML Form: ORDER CONFIRMATION → MAPI Email Message: ORDER CONFIRMATION →

Remote Batch   Remote batch combines the best aspects of batch and on-line inputs and outputs. Distributed on-line computers handle data input and editing. Edited transactions are collected into a batch file for later transmission to host computers that process the file as a batch. Results are usually transmitted as a batch back to the original computers.

Remote batch is hardly a new alternative, but personal computers have given the option new life. For example, one of the authors' colleagues uses a Microsoft *Access* program to input and test the feasibility of a schedule of classes for his academic department each semester. When finished, he generates a comma, delimited file to transmit to the academic scheduling unit for batch processing. The entire physical input model looks something like this:

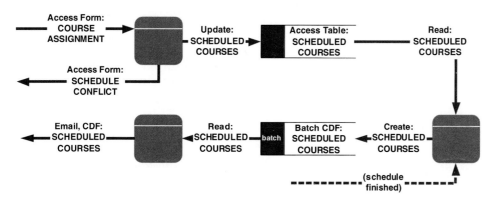

Remote batch using PCs should get another boost with the advances in handheld and subnotebook computer technology. These four-ounce to four-pound computers can be used to collect batches of everything from inventory counts to mortgage applications. The inputs are remotely batched on the device for later transmission as a batch.

Keyless Data Entry (and Automatic Identification)   Keying in data has always been a major source of errors in computer inputs (and inquiries). Any technology that reduces or eliminates the possibility of keying errors should be considered for system design. In batch systems, keying errors can be eliminated through optical character reading (OCR) and optical mark reading (OMR) technology. Both are still viable options for input design. The physical data flow construct is shown below.

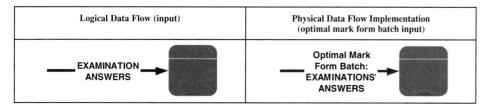

The real advances in keyless data entry are coming for on-line systems in the form of auto-identification systems. For example, bar coding schemes (such as the Universal Product Codes that are common in the retail industry) are widely available for many modern applications. For example, Federal Express creates a bar code-based label for all packages when you take the package to a center for delivery. The bar codes can be read and traced as the package moves across the country to its final destination. Bar code technology is being constantly improved to compress greater amounts of data into smaller labels. The physical data flow construct follows. (The receiving physical process would be named for the function it performs.)

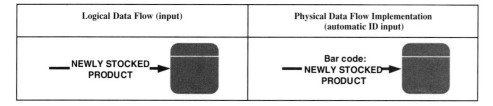

Pen Input   As pen-based operating systems (e.g., the *Palm OS* and Microsoft's *Windows CE*) become more widely available and used, and the tools for building pen-based applications become available and standardized, we expect to see more system designs that exploit this technology.

Some businesses already use this technology for remote data collection. For example, UPS uses pen-based notebook systems to help track packages through the delivery system. The driver calls up the package tracking number on the special tablet computer. The customer signs the pad in the designated area. When the driver returns to the truck and places the tablet computer back in its docking cradle, the updated delivery data is transmitted by cellular modem to the distribution center where the package tracking system updates the database (ultimately enabling the shipper to know that you have received the package via a simple Web inquiry).

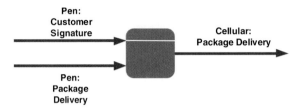

Electronic Messaging and Work Group Technology   Electronic mail has grown up! No longer merely a way to communicate more effectively, information systems are being designed to directly incorporate the technology. For example, Microsoft *Exchange Server* and IBM *Lotus Notes* allow for the construction of intelligent electronic forms that can be integrated into any application. Basic messaging services can also be integrated into applications.

For example, any employee via an e-mail-based form could initiate travel requests. Based on the data submitted on the form, predefined rules could automatically route the request to the appropriate decision makers. For example, less expensive travel requests might be routed directly to a business officer. More expensive requests might be routed first to a department head for approval; then to a business officer. Eventually approved forms can be automatically input to the appropriate reimbursement processing information system for normal processing. And at each step, the messaging system automatically informs the initiator of progress via e-mail. A physical DFD that included an e-mail message implementation was presented earlier.

Electronic Data Interchange   Businesses that operate in many locations and businesses that seek more efficient exchange of transactions and data with other businesses often utilize electronic data interchange.

> **Electronic data interchange (EDI)** is the standardized electronic flow of business transactions or data between businesses. Typically, many businesses must commit to a data format to make EDI feasible.

With EDI, a business can eliminate its dependence on paper documents and mail. For example, most colleges now accept SAT or ACT test scores via EDI from national testing centers. This has been made possible because college registrars have agreed to a standard format for these test scores.

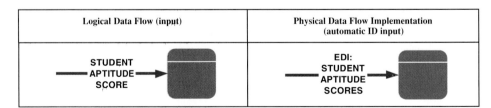

Imaging and Document Interchange   Another emerging I/O technology is based on image and document interchange. This is similar to EDI except that the actual images of forms and data are transmitted and received. It is particularly useful in applications in which the form images or graphics are required. For example, the insurance industry has made great strides in electronically transmitting, storing, and using claims images. Other imaging applications combine data with pictures or graphs. For example, a law enforcement application can store, transmit, and receive photographic images and fingerprints.

Middleware   Most of the above subsections focused on input and output—the user interface. But many system designs require process-to-process physical data flows. Earlier in this chapter, we described various client/server and network computing scenarios that automatically include process-to-process data flows because clients and servers must talk to one another. They do this through middleware.

> **Middleware** is utility software that enables communication between different processors in a system. It may be built into the respective operating systems or added through purchased middleware products. Middleware products allow the programmers to ignore underlying communication protocols.

Middleware is said to be the "slash" in client / server. There are three classes of middleware that happen to correspond to the middle three layers of our distributed systems framework: presentation logic, application logic, and data manipulation logic.

— *Presentation middleware* allows a programmer to build user interface components that can talk to Web browsers or a desktop GUI. For example, HTTP allows the programmer to communicate with a Web browser through a standard application programmer interface (API).

— *Application middleware* enables two programmer-written processes on different processors to communicate with one another in whatever way is best suited to the overall application. Application middleware is essential to multi-tier application development. Examples of application middleware are numerous: remote procedure calls (RPCs), message queues, and object request brokers.

— *Database middleware* allows a programmer to pass SQL commands to a database engine for processing through a standard API.
    Another common type of middleware is ODBC (Object Database Connectivity) and JDBC (Javabean Database Connectivity) that automatically translates SQL commands for one database server for use on a different database server (for example, *Oracle* to *SQL Server,* or vice versa).

On a physical data flow diagram, middleware can be depicted by specifying the middleware class name on the physical data flow (e.g., ODBC).

The PROCESS architecture of an application is defined in terms of the software languages and tools that will be used to develop the business logic and application programs for that process. Typically, this is expressed as a menu of choices because different software development environments are suited to different applications.

PROCESS Architectures— The Software Development Environment

> A **software development environment** (SDE) is a language and tool kit for constructing information system applications. They are usually built around one or more programming languages such as *COBOL, Basic, C++, Pascal, Smalltalk,* or *Java.*

One way to classify SDEs is according to the type of client/server or network computing architecture they support.

SDEs for Centralized Computing and Distributed Presentation   Not that long ago, the software development environment for centralized computing was very simple. It consisted of the following:

— An editor and compiler, usually *COBOL,* to write programs.
— A transaction monitor, usually *CICS,* to manage any on-line transactions and terminal screens.

- A file management system, such as *VSAM,* or a database management system, such as *DB2* to manage stored data.

That was it! Because all these tools executed on the mainframe, only that computer's operating system (more often than not, *MVS*) was critical.

The personal computer brought many new *COBOL* development tools down to the mainframe. A PC-based *COBOL* SDE such as the Micro Focus *COBOL Workbench* usually provided the programmer with more powerful editors and testing and debugging tools at the workstation level. A programmer could do much of the development work at that level and then upload the code to the central computer for system testing, performance tuning, and production. Frequently, the SDE could be interfaced with a CASE tool and code generator to take advantage of process models developed during systems analysis.

Eventually, SDEs provided tools to develop distributed presentation client/server. For example, the Micro Focus *Dialog Manager* provided *COBOL Workbench* users with tools to build *Windows*-based user interfaces that could cooperate with the *CICS* transaction monitors and the mainframe *COBOL* programs.

SDEs for Two-Tier Client/Server   Today the typical SDE for two-tiered client/server applications (also called distributed data) consists of a client-based programming language with built-in SQL connectivity to one or more server database engines. Examples of two-tiered client/server SDEs include Powersoft's *PowerBuilder,* Microsoft's *Visual Basic* (Client/Server Edition), and Borland's *Delphi* (Client/Server Edition). Typically, these SDEs provide the following:

- Rapid application development (RAD) for quickly building the graphical user interface that will be replicated and executed on all the client PCs.
- Automatic generation of the template code for the above GUI and associated system events (such as mouse-clicks, keystrokes, etc.) that use the GUI. The programmer only has to add the code for the business logic.
- A programming language that is compiled for replication and execution on the client PCs.
- Connectivity (in the above language) for various relational database engines and interoperability with those engines. Interoperability is achieved by including SQL database commands (for example, to create, read, update, delete, and sort records) that will be sent to the database engine for execution on the server.
- A sophisticated code testing and debugging environment for the client.
- A system testing environment that helps the programmer develop, maintain, and run a reusable test script of user data, actions, and events against the compiled programs to ensure that code changes do not introduce new or unforeseen problems.
- A report writing environment to simplify the creation of new end-user reports off a remote database.
- A help authoring system for the client PCs.

Today most of these tools come in the bundled SDE, but independent software tool vendors have emerged to produce replacement tools that often provide still greater functionality and/or productivity than those provided in the basic SDE. To learn more about such add-on tools, search the Internet for Programmers Paradise, a software development tool Web storefront.

Some of the process logic of any two-tiered client/server application can be offloaded to the database server in the form of stored procedures. In this case, stored procedures are written in a superset of the SQL language. These procedures are then "called" from the client for execution on the server. Different experts seem to love and hate stored procedures. On the plus side, stored procedures can be made to better enforce data integrity in database tables. They are reusable and verifiable. On the negative side, they blur the distinction between the application and data manipulation layers of our framework—they are application logic that execute on the database servers. Many designers prefer a more cohesive design strategy called clean layering.

Clean layering requires that the presentation, application, and data layers of an application be physically separated.

Clean layering is said to allow components of each layer to be revised and enhanced without affecting other layers in the system.

SDEs for Multi-tier Client/Server   The current state of the art in enterprise application development is occurring in SDEs for three-tiered (and beyond) client/server architectures. Unlike two-tiered applications, n-tiered applications must support more than 100 users with mainframe-like transaction response time and throughput, with 100 gigabyte or larger databases. While the two-tiered SDEs described earlier are trying to expand in this market, a different class of SDEs currently dominates the market. Typically, the SDEs in this class must provide all the capabilities typically associated with two-tiered SDEs plus the following:

— Support for heterogeneous computing platforms, both client and server, including Windows, OS/2, UNIX, Macintosh, and legacy mainframes and minicomputers.
— Code generation and programming for both clients and servers. Most tools in this genre support pure object-oriented languages such as *C++* and *Smalltalk*.
— A strong emphasis on reusability using software application frameworks, templates, components, and objects.
— Bundled minicase tools for analysis and design that interoperate with code generators and editors.
— Tools to help analysts and programmers partition application components between the clients and servers.
— Tools to help developers deploy and manage the finished application to clients and servers. This generally includes security management tools.
— The ability to automatically scale the application to larger and different platforms, client and server. This issue of scalability was always assumed in the mainframe computing era but is relatively new to the client/server computing era.
— Sophisticated software version control and application management.

Examples of n-tiered client/server SDEs include Intersolv's *Allegris*, Dynasty's *Dynasty*, Forté Software's *Forté*, Texas Instruments' *Performer* and *Composer*, Compuware's *Uniface*, and IBM's *VisualAge* (a family of products). Again, a large number of independent software tool vendors are building add-on and replacement tools for these SDEs.

SDEs for Internet and Intranet Client/Server   Rapid application development tools are emerging to enable client/server Internet and intranet applications. Most of these languages are built around four core standard technologies:

HTML (hypertext markup language)—the language used to construct most Internet and intranet page content and hyperlinks.
XML (extensible markup language)—an extensible language for transporting data and properties across the Web.
CGI (Computer Gateway Interface)—a standard for publishing graphical World Wide Web components, constructs, and links.

Java—a general-purpose programming language for creating platform-independent programs, servlets, and applets that can execute from within a browser's *Java Virtual Machine.*

Examples of *Java*-specific SDEs include Symantec's *Visual Café,* Inprise's *Jbuilder,* and Sun's *Java Workbench.* These SDEs can create both Internet, intranet, and non-Internet/intranet applications. Virtually all existing two-tiered and n-tiered SDEs are also evolving to support *HTML, XML, CGI,* and *Java.*

APPLICATION ARCHITECTURE STRATEGIES FOR SYSTEMS DESIGN

Regardless of what it is called, all information systems have an application architecture. Different organizations apply different strategies to determining application architecture. Let's briefly classify the two most common approaches.

The Enterprise Application Architecture Strategy

In the enterprise application architecture strategy, the organization develops an enterprisewide information technology architecture to be followed in all subsequent information systems development projects. This IT architecture defines the following:

- The approved network, data, interface, and processing technologies and development tools (inclusive of hardware and software; and clients and servers).
- A strategy for integrating legacy systems and technologies into the application architecture.
- An ongoing process for continuously reviewing the application architecture for currency and appropriateness.
- An ongoing process for researching emerging technologies and making recommendations for their inclusion in the application architecture.
- A process for analyzing requests for variances from the approved application architecture. (You may recall that SoundStage received such a variance to prototype object technology in the member services system project.)

An initial application architecture is usually developed as a separate project or as part of a strategic information systems planning project. The ongoing maintenance of the application architecture is usually assigned to a permanent information technology research group or to an enterprise application architecture committee.

Subsequent to the approval of the application architecture, every information system development project is expected to use or choose technologies based on that architecture. In most cases, this greatly simplifies the architecture phase of a system development methodology (such as *FAST*). You simply select from the approved technologies according to the architecture's rules or guidelines.

Of course, even if a technology is approved in the application architecture, it is subject to a feasibility analysis as described in the next subsection.

The Tactical Application Architecture Strategy

In the absence of an enterprisewide application architecture, each project must define its own architecture for the information system being developed. There still may exist some sort of information technology research and deployment group.

While the proposed application architecture for any new information system may be influenced by existing technologies, the developers usually have somewhat greater latitude in requesting new technologies. Of course, the final decision must be defended and approved as feasible. IT feasibility usually includes the following aspects:

- Technical feasibility—This can either be a measure of a technology's maturity, a measure of the technology's suitability to the application being designed, or a measure of the technology's ability to work with other technologies.

- Operational feasibility—This is a measure of how comfortable the business management and users are with the technology and how comfortable the technology managers and support personnel are with the technology.
- Economic feasibility—This a measure of both whether or not the technology can be afforded and whether it is cost-effective, meaning the benefits outweigh the costs.

Feasibility criteria and techniques for measuring them were covered in Chapter 9.

The use of logical DFDs to model process requirements is a fairly accepted practice. However, the transition from analysis-oriented logical DFDs to design-oriented physical DFDs has historically been somewhat mysterious and elusive. We desire a high-level general design that can serve as an application architecture for the system and a general design for the processes that make up the system. At the same time, we don't want to get caught up in a counterproductive modeling exercise that slows our progress in systems design and rapid application development. Simply stated, we want a blueprint to guide us through detailed design and construction. The blueprint will identify design units for detailed specification or rapid development, whichever is most productive in our project.

MODELING THE APPLICATION ARCHITECTURE OF AN INFORMATION SYSTEM

The mechanics for drawing physical DFDs are virtually identical to those of logical DFDs. The rules of correctness are also identical. An acceptable design results in

Drawing Physical Data Flow Diagrams

- A system that works.
- A system that fulfills user requirements (specified in the logical DFDs).
- A system that provides adequate performance (throughput and response time).
- A system that includes sufficient internal controls (to eliminate human and computer errors, ensure data integrity and security, and satisfy auditing constraints).
- A system that is adaptable to ever-changing requirements and enhancements.

We could develop a single physical DFD for the entire system, or a set of physical DFDs for the target system. Our methodology suggests the following:

- A physical data flow diagram should be developed for the network architecture. Each process on this diagram is a physical processor (client or server) in the system. Each server is its own processor; however, it is usually impractical to show each client. Instead, each class of clients (e.g., an order entry clerk) is represented by a single processor.
- For each processor on the above model, a physical data flow diagram should be developed to show those event processes (see Chapter 8) that will be assigned to that processor. It is possible that you would choose to duplicate some event processes on multiple processors. For instance, orders may be processed on each region's servers and clients.
- For all but the simplest event processes, they should be factored into design units and modeled as a single physical data flow diagram.

  A **design unit** is a self-contained collection of processes, data stores, and data flows that share similar design attributes. A design unit serves as a subset of the total system whose inputs, outputs, files and databases, and programs can be designed, constructed, and unit tested as a single subsystem. (The concept of design units was first proposed by McDonnell Douglas in its STRADIS methodology.)

An example would be a set of processes (one or more) to be designed as a single program. The design unit could then be assigned to a single programmer (or team) who (which) can work independently of other programmers and teams without adversely affecting the work of the other programmers. The implemented units would then be assembled into the final application system. Design units can also be prioritized for implementing versions of a system.

Prerequisites

Let's set the table by describing the prerequisites to creating physical DFDs. They include:

- A logical data model (entity relationship diagram created in Chapter 7).
- Logical process models (data flow diagrams created in Chapter 8).
- Repository details for all of the above.

Given these models and details, we can distribute data and processes to create a general design. Your general design will normally be constrained by one or more of the following:

- Architectural standards that predetermined the choice of database management systems, network topology and technology, user interface(s), and/or processing methods.
- Project objectives that were defined at the beginning of systems analysis and refined throughout systems analysis.
- The feasibility of chosen or desired technology and methods. Feasibility analysis techniques were covered in Chapter 9.

Within any restrictions of those constraints, the ensuing techniques can be applied.

The Network Architecture

The first physical DFD to be drawn is the network topology DFD.

> A **network architecture** DFD is a physical data flow diagram that allocates processors (clients and servers) and devices (e.g., machines and robots) to a network and establishes (1) the connectivity between the clients and servers and (2) where users will interact with the processors (usually only the clients).

To identify the processors and their locations, the developer utilizes two resources:

- If an enterprise information technology architecture exists, that architecture likely specifies the client/server vision that should be targeted.
- The advice of competent network managers and/or specialists should be solicited to determine what's in place, what's possible, and what impact the system may have on the computer network.

Network architecture DFDs (see Figure 11.11) need to be labeled to show somewhat different information than normal DFDs. They don't show specific data flows per se. Instead, they show highways over which data flows may travel in either direction. Also, network topology DFDs indicate the following:

- Servers and their physical locations. Servers are not always located at the sites indicated on a location connectivity diagram. Network staff access to servers is usually an issue. Some network management tasks can be accomplished remotely, and some tasks also require hands-on access.
- Clients and their physical locations. In this case, the location connectivity diagram is useful in identifying "classes" of like users (e.g., ORDER CLERKS, SALES REPRESENTATIVES, etc.) who will be serviced by similar clients. A single processor should represent the entire class at a single location. The same class may be replicated in multiple locations. For example, you would expect each SALES REGION to have similar types of employees.
- Processor specifications. The repository descriptions of processors can be used to define processor specifications such as RAM, hard disk capacity, and display.
- Transport protocols. Connections are labeled with transport protocols (e.g., TCP/IP) and other relevant physical parameters.

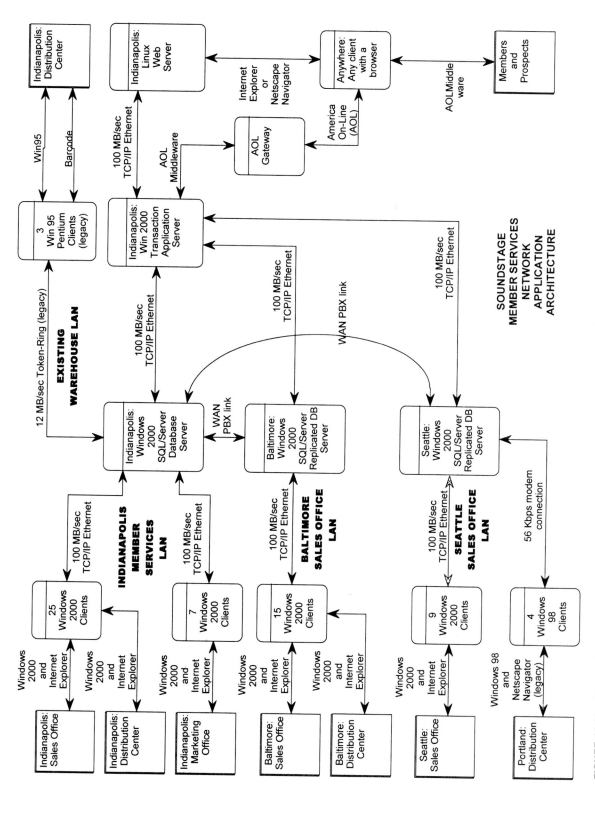

FIGURE 11.11   Network Architecture DFD

The network topology DFD can be used to either design a computer network, or to document the design of an existing computer network. In either case, the network is being modeled so that we can subsequently assign information system processes, data stores, and data flows to servers on the network.

Data Distribution and Technology Assignments

The next step is to distribute data stores to the network processors. The required logical data stores are already known from systems analysis—as data stores on the logical DFDs (Chapter 8) or as entities on the logical ERDs (Chapter 7). We need only determine where each will be physically stored and how they will be implemented.

To distribute the data and assign their implementation methods, the developers utilize three resources:

— If available, the data distribution matrices from systems analysis (Chapters 7 and 8) model the data needs at business locations from a technology-independent perspective.
— If an enterprise information technology architecture exists, that architecture likely specifies the database vision and technologies that should be targeted.
— The advice of data and database administrators should be solicited to determine what's in place, what's possible, and what impact the database may have on the overall system.

The distribution options were described earlier in the chapter and are summarized as follows:

— Store all data on a single server. In this case, the database (consisting of multiple tables) should be named, and that named database and its implementation method (e.g., Oracle: dbmemberServices) should be added to the physical DFD and connected to the appropriate processor.
— Store specific tables on different servers. In this case, and for clarity's sake, we should record each table as a data store on the physical DFD and connect each to the appropriate server.
— Store subsets of specific tables on different servers. In this case we record the tables exactly as above except that we indicate which tables are subsets of the total set of records. For example, the label DB2: ORDERS TABLE (REG SUBSET) would indicate a subset of all orders for a region are stored in a DB2 database table.
— Replicate (duplicate) specific tables or subsets on different servers. In this case, replicated data stores are shown on the physical DFD. One copy of any replicated table is designated as the MASTER, and all other copies are designated as COPY or REPLICANT.

Why distribute data storage? There are many possible reasons. First, some data instances are of local interest only. Second, performance can often be improved by subsetting data to multiple locations. Finally, some data needs to be localized to assign custodianship of that data. The data distribution and technology assignments for the SoundStage case study are shown in Figure 11.12.

Data distribution decisions can be very complex—normally the decisions are guided by data and database professionals and taught in data management courses and textbooks. In this book we want to consider only how to document the partition and duplication decisions.

Process Distribution and Technology Assignments

Information system processes can now be assigned to processors as follows:

— For two-tiered client/server systems, all the logical event diagrams (Chapter 6) are assigned to the client.

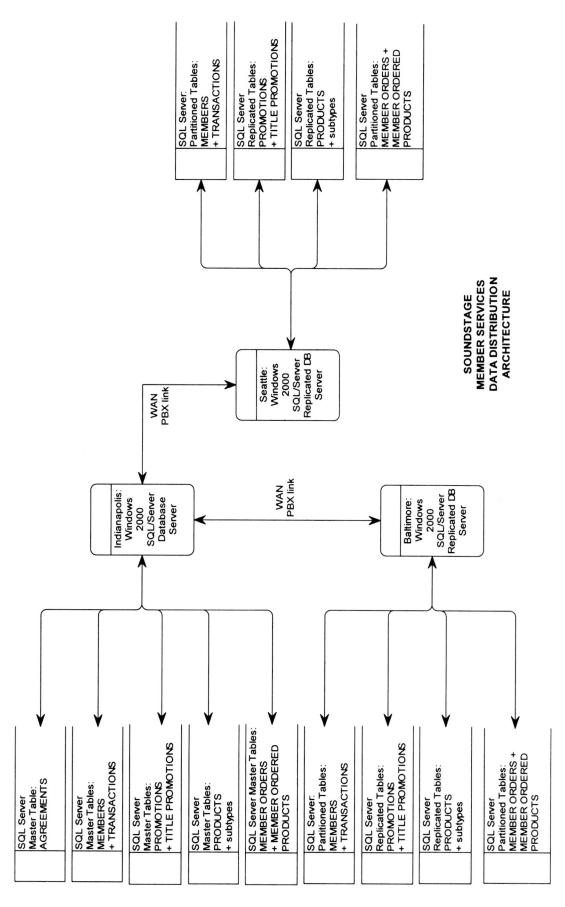

FIGURE 11.12 *Data Distribution and Technology Assignments for SoundStage*

— For three-tiered client/server and network computing systems, you must closely examine each event's primitive (detailed) data flow diagram. You need to determine which primitive processes should be assigned to the client and which should be assigned to an application server. In general, data capture and editing are assigned to clients and other business logic is assigned to servers. If you partition different aspects of a logical DFD to different clients and servers, you should draw separate physical DFDs for the portions on each client and server.

After partitioning, each physical DFD corresponds to a design unit for a given business event. (Business events, or use cases, were discussed in Chapter 8.) For each of these design units, you must assign an implementation method, the SDE that will be used to implement that process. You must also assign implementation methods to the data flows.

SoundStage's Member Services system will be implemented with a multi-tiered client/server and network computing architecture. A sample DFD for one event to be assigned to a client is shown in Figure 11.13. Notice that the data stores are shown even though we know they have been partitioned to a database server. This is for the benefit of the programmers who must implement the DFD.

The Person/Machine Boundaries

The last step of process design is to factor out any portion of the physical DFDs that represent manual, not computerized processes. This is sometimes called "establishing a person/machine boundary." Establishing a person/machine boundary is not difficult, but it is not as simple as you might first think. The difficulty arises when the person/machine boundary cuts through a logical process—in other words, part of the process is to be manual and part is to be computerized. This situation is common on logical DFDs because they are drawn without regard to implementation alternatives.

Figure 11.14 adds the person/machine boundary to a physical DFD. Notice that our boundary cuts through several processes, including the CHECK MEMBER CREDIT process. The solution to this process requires two steps.

1. The manual process portions are pulled out as a separate design unit (see Figure 11.15). All these processes are completely manual. The interfaces of the manual design units to the computerized processes (on Figure 11.14) are depicted as external agents. Ultimately, the manual processes in the design unit must be clearly described to those people who will have to perform them.

2. If necessary, the processes on the original diagram should be renamed to reflect only the computerized portion. (In practice, the processes were already named that way.)

WHERE DO YOU GO FROM HERE?

In this chapter, you have learned how to outline the design of a new information system to fulfill the requirements identified and modeled during systems analysis. This general design for the new system will guide the detailed design and construction of that system.

Most readers will now progress to the detailed design chapters that build on the general design for the new system. For most of you, we recommend you start with Chapter 12, Database Design. Most design-by-prototyping and rapid application development techniques are absolutely dependent on the existence of the planned information system's database. Databases must be carefully designed to ensure adaptability and flexibility during the system's lifetime. Thus, Chapter 12 is the best place to begin your study of detailed design. Subsequently, you can move on to chapters that cover other aspects of detailed design including inputs, outputs, and programs.

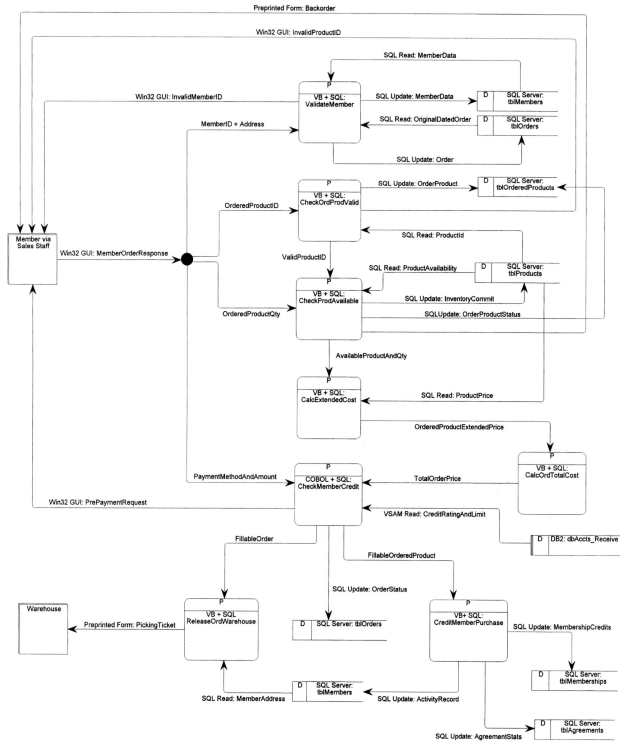

FIGURE 11.13   *A Physical DFD for an Event*

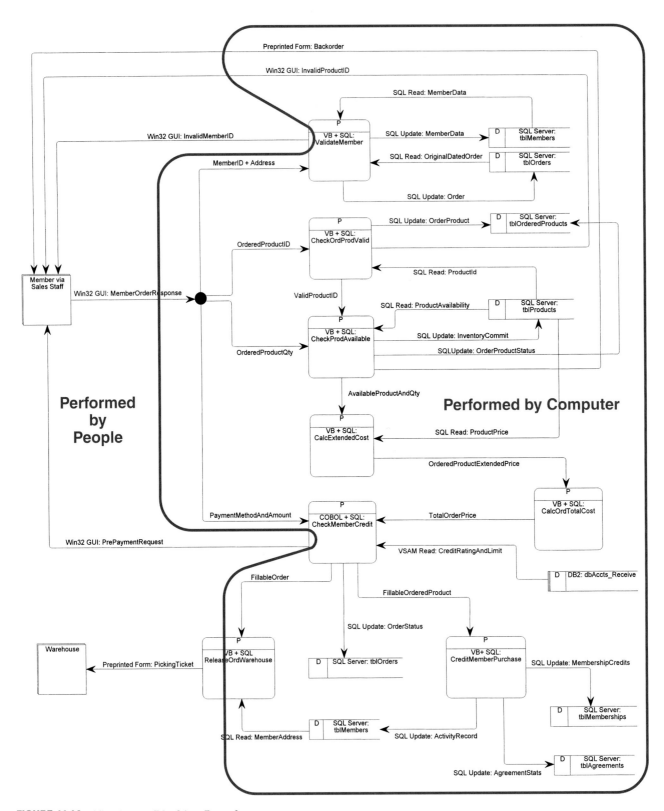

FIGURE 11.14   *The Person/Machine Boundary*

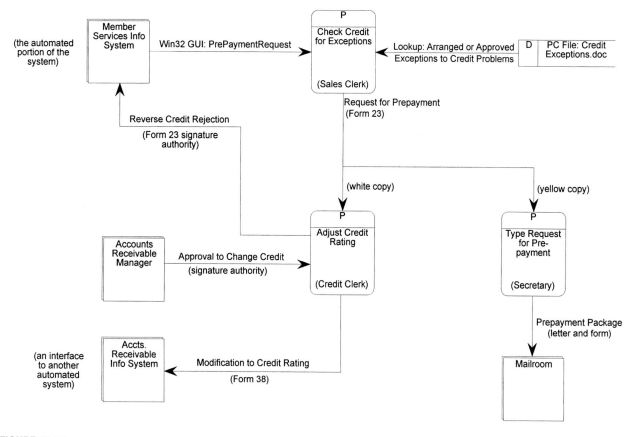

FIGURE 11.15 *A Manual Design Unit*

SUMMARY

1. Physical data flow diagrams model the technical and human design decisions to be implemented as part of an information system. They communicate technical choices and other design decisions to those who will actually construct and implement the system.

2. An information technology architecture defines the technologies to be used by one, more, or all information systems. There are four categories of technology architectures: network, data, interface, and process.

3. A distributed system is one in which the data, process, and interface components of an information system are distributed to multiple locations in a computer network. The five layers of distributed systems architecture are *(a)* presentation, *(b)* presentation logic, *(c)* application logic, *(d)* data manipulation, and *(e)* data.

4. A local area network is a set of client computers connected to one or more servers through either cable or wireless connections over relatively short distances.

5. A file server system is a LAN-based solution in which a server computer hosts only the data layer. All other layers are implemented on the client.

6. The prevailing computing model is currently client/server in which the presentation, presentation logic, application logic, data manipulation, and data layers are distributed between client PCs and one or more servers. Clients are classified by their power as thin or fat. Servers are dedicated to functions such as database, transactions, applications, messaging or work group, or Web.

7. Distributed presentation, distributed data, and distributed data and logic are flavors of client/server.

8. Network computing uses Internet technology to build Internet or intranet applications.

9. Data storage is typically implemented using distributed relational database technology that either partitions data to different servers or replicates data on multiple servers.

10. User interface options include batch, on-line, remote batch, keyless data entry (including optical character/mark and bar coding methods), pen input, electronic messaging, electronic data interchange, and imaging.

11. System interfacing is typically implemented using middleware, software that enables processes to communicate with one another.

12. Processes are implemented using highly integrated tool kits called software development environments.

13. Application architectures may be developed and enforced strategically, or they may tactically evolve on a project-by-project basis.

KEY TERMS

application architecture, p. 424
application logic layer, p. 432
application server, p. 436
centralized system, p. 432
clean layering, p. 451
client/server system, p. 435
database server, p. 435
data layer, p. 432
data manipulation layer, p. 432
data partitioning, p. 444
data replication, p. 444
design unit, p. 453
distributed data, p. 437
distributed data and application, p. 439
distributed presentation, p. 436
distributed relational database, p. 443

distributed relational database management system, p. 444
distributed system, p. 432
electronic data interchange (EDI), p. 448
fat client, p. 435
file server system, p. 434
intranet, p. 441
local area network (LAN), p. 434
messaging or groupware server, p. 436
middleware, p. 449
network architecture, p. 454
network computing system, p. 440
partitioning, p. 439
physical data flow, p. 428
physical data flow diagram, p. 424
physical data store, p. 430

physical process, p. 425
portal, p. 441
presentation layer, p. 432
presentation logic layer, p. 432
relational database, p. 443
software development environment, p. 449
thin client, p. 435
three-tiered or n-tiered client/server computing, p. 439
transaction server, p. 436
two-tiered client/server computing, p. 437
Web server, p. 436

REVIEW QUESTIONS

1. What is an application architecture? What role does it play in information systems development?
2. Differentiate between logical and physical data flow diagrams. When is each relevant?
3. Differentiate between a processor and a process as it relates to physical data flow diagrams.
4. Briefly describe four physical implementations of a data flow.
5. Briefly describe six physical implementations of a data store.
6. Differentiate between centralized and distributed systems.
7. Name five layers suitable for distribution in a system.
8. Differentiate between file servers and database servers? How about file server systems and client/server systems?
9. What are clients and servers? What role does each play in client/server architecture? What is the difference between fat and thin clients?
10. Briefly describe five types of servers that can be used in a client/server system.
11. Differentiate between two-tiered and three-tiered client/server computing.
12. What is partitioning and why is it applicable to client/server computing?
13. Differentiate between client/server and network computing?

14. How do intranets fit into Internet-based computing approaches?
15. Differentiate between e-commerce and e-business? Which one use the Internet and which one exploits an intranet?
16. What is a distributed relational database? How does it differ from a file server database?
17. Differentiate between partitioned and replicated data. Describe two styles of partitioning.
18. Explain the difference between batch, on-line, and remote batch input methods. Define an input and conceive a situation that might call for each of the three methods to be used.
19. Why are keyless data entry alternatives attractive to business? Describe three keyless data entry technologies.
20. Why have graphical user interfaces become an issue for systems analysts designing custom applications (such as order entry and inventory control) for a business?
21. What is electronic data interchange?
22. What is middleware? What role does it play in client/server architecture?
23. What is a software development environment?
24. Differentiate between strategic and tactical architecture.
25. What is a network architecture DFD?
26. What is a design unit?

PROBLEMS AND EXERCISES

1. Your boss proclaims, "We haven't invested in client/server architecture yet. We are a mainframe shop." Tactfully explain to your boss why she or he is wrong. Explain how your boss might evolve the client/server architecture while continuing to use the mainframe.

2. Respond to the following editorial: "Client/server computing is too big a step for us at this time. I realize that PCs and networks may—and I emphasize may—be cheaper to acquire, operate, and maintain, but we can't just get rid of our mainframe and all its legacy applications just like that. We can't rush into this mainframe versus server issue just yet."

3. "We just made the transition to client/server. Now we are talking Internet-based applications. Isn't network computing just another form of client/server?" Defend this statement. Then challenge this statement.

4. You have been hired as a consultant to a small copy shop. The owner proudly proclaims, "We are ahead of our time. We are already doing two-tiered client/server. We use Microsoft *Access* as our database server." Explain to the owner why simply placing the *Access* data on a file server does not implement two-tiered client/server. Explain the advantage of true two-tiered client/server computing. What would it take?

5. What are the options for partitioning business logic across a three-tiered client/server network?

6. Explain how a corporate intranet might serve as a framework for a company's client/server information systems.

7. Why are networking and data communications essential subjects for future systems designers who will build client/server applications? How do you plan to acquire this knowledge?

8. What is the network architecture of your school's class registration system?

9. Develop a set of business guidelines to determine whether data should be centralized, distributed, or replicated (including combination approaches).

10. Describe the software development environment in your last programming course or project. What elements of an SDE as described in this chapter were missing? How might they have helped?

11. Prepare a physical data flow diagram to describe the backup and recovery procedures for a lost or damaged master file or database.

12. Your organization uses a tactical, project-by-project approach to determining application architecture. Write a two-page proposal to management that defends a more strategic approach, and adapt the *FAST* methodology to propose a project approach to develop an initial enterprisewide application architecture. How do you propose to deal with legacy applications?

13. What criteria can be used to measure information system design quality?

PROJECTS AND RESEARCH

1. Recent business reports and editorials have concluded that client/server computing is not the economic panacea that most initially believed. Why might cost not be a factor in the decision to go client/server for all new applications?

2. Research the evolving subject of client/server applications development. Study the costs and benefits and the issues and problems to be addressed. Be sure to find pro and con examples—both exist! Present your findings in a recommendations report to information systems management.

3. Client/server technology continues to evolve. Research and prepare a technology update report on one of the following subjects:

   a. Database management systems for servers in a client/server architecture.

   b. Network architecture and operating systems for a client/server environment.

   c. Distributed computing environment (DCE), an evolving standard for open systems interoperability for client/server computing. (Note: Open systems is a goal whereby dissimilar computers and software can communicate and cooperate in a manner that is transparent to users.)

   d. Client programming languages versus downsized *COBOL/CICS*. Examples of the former include *PowerBuilder, Visual Basic,* and *Delphi*.

   e. Middleware for easy, transparent access to existing data stored using dissimilar structure and different servers and hosts.

4. Make an appointment to visit a local business computing facility. How has the growth in numbers of microcomputers affected the networking strategy of the business? What is the networking strategy? What architecture was chosen and why?

5. Visit a systems analyst or programmer in a local business. Describe the software development environments used in that shop. Were any elements missing?

6. Research the SDEs available for a single programming language such as *Visual Basic, C++, COBOL, Delphi,* or *Java*. Report on third-party extensions available for that environment.

7. At the time of this writing, *Java* was beginning to emerge as a candidate programming language (with SDEs) for

mainstream information systems (as opposed to just Internet and intranet applets). Research the current viability and penetration of this new object-oriented programming language.

8. At the time of this writing, *XML* was beginning to emerge as a candidate standard for data transportation over the Web. Research the current viability of this emerging standard. How is it being used to implement Internet and e-commerce solutions.

9. System flowcharts are a dated predecessor to physical data flow diagrams. But they exist as legacy documentation in most IT shops. Acquire a flowchart from a local business application, an older systems analysis and design textbook, or your instructor (who probably has access to many in his or her course archives). Convert that systems flowchart to a physical data flow diagram.

MINICASES

1. Chapter 7 provided a data model. Using your school or business's technical environment as a constraint (or one provided by your instructor), use the process analysis and design technique in this chapter to derive a design. The design should be documented using physical data flow diagrams. (Note: Your technical environment should at least specify processors, operating systems, languages, network topology, network operating systems and protocols, and database management systems.)

2. Pacific Imports, a wholesale distributor of a wide variety of imported products, is located in Los Angeles, California. Pacific has just completed the automation of its mail-order system. This minicase reviews the design and implementation.

   a. At approximately 8:30 each morning, the Order Entry Department receives all new sales order forms. These forms include orders received both by mail and over the phone. Order Entry enters and edits all orders on personal computers, creating a batch of orders on a network file server.

   b. The batch of orders is transmitted from the server's disk to an AS/400 midrange computer via a network gateway over a TCP/IP point-to-point network connection. There the order-processing program checks the inventory master file (actually, a database table) to validate products ordered and to determine the availability and price of the products ordered. For any products that are out of stock, a backorder is created in the database and printed as a notice for the customer. Fillable orders (and partial orders) are then processed as follows.

   c. Another program checks the customer accounts master file to check credit. For customers who have a poor credit rating, a prepayment request is printed. The or-

der is placed on hold in the order master file. Also, a report of these notices is prepared for management. The program prints an order confirmation letter for those orders that will be filled.

   d. Another program produces a four-part warehouse order form that includes a picking copy, a packing copy, a shipping copy, and an invoicing copy. The processed order is moved to a sales master file, a copy of which is archived on magnetic tape for backup and later use.

   The system also includes an on-line program that allows the sales manager both to query the inventory master file to obtain prices and to query the customer accounts master file to obtain credit information needed to manually override a credit rejection. This program is available to the sales manager from 8 A.M. until noon each working day.

   e. Customer order cancellations are processed immediately upon receipt (by mail or phone). When the request is received by a clerk, he or she enters the order cancellation using the PC as a terminal into the AS/400. An on-line program reads the sales master file to determine the order's status. If the order has not been filled, the clerk phones the warehouse to have the order terminated. At the end of the day, the program generates cancellation notices for customers and a cancellation report for the sales manager.

   i.   What are the limitations of the English language for describing a design to programmers who must implement that design?

   ii. Did you spot any errors of omission in the narrative description—that is, important questions or processes that were not covered?

   iii. Draw a physical data flow diagram for this case.

SUGGESTED READINGS

Berstein, Phillip, and Eric Newcomer. *Principles of Transaction Processing: For the Systems Professional.* San Francisco: Morgan Kaufman Publishers, Inc., 1997. This book covers virtually every transaction processing model, transaction monitor, and transaction server currently implemented.

Gane, Chris, and Trish Sarson. *Structured Systems Analysis: Tools and Techniques.* Englewood Cliffs, NJ: Prentice Hall, 1979. This classic on process modeling became the basis of physical data flow diagrams.

Goldman, James. *Applied Data Communications: A Business-Oriented Approach*. 2nd ed. New York: John Wiley & Sons, Inc., 1998. Our colleague at Purdue has written an excellent textbook for those seeking to learn about data communications and networking from a business perspective.

Goldman, James; Phillip Rawles; and Julie Mariga. *Client/Server Information Systems: A Business-Oriented Approach*. New York: John Wiley & Sons, Inc., 1999. Our colleagues have written an outstanding textbook that introduces students to information technology architecture for information systems.

Kara, Dan. "Why Partition? Multitiered Application Architecture." In *Application Development Trends*. Natick, MA: Software Productivity Group, May 1997, pp. 38–46. This article stimulated our interest and research in the need to develop and teach partitioning techniques as part of this book.

Kara, Daniel A., et al. "Enterprise Application Development: Seminar Notes." Chicago: Software Productivity Group, November 12, 1996. This seminar and the writings of the SPG have strengthened our understanding of two-tiered and n-tiered software application development techniques and technologies.

Orfali, Robert; Dan Harkey; and Jeri Edwards. *Client/Server Survival Guide*. 3rd ed. New York: John Wiley & Sons, Inc., 1999. This professional reference manual has served us well for three editions of client/server technology and terminology evolution.

Renaud, Paul. *Introduction to Client/Server Systems*. 2nd ed. New York: John Wiley & Sons, Inc., 1996. This is another reference book on the primary distributed computing architecture of our time.

Smith, Patrick, and Steve Guengerich. *Client/Server Computing*. 2nd ed. Indianapolis, IN: SAMS Publishing, 1994. This professional book has been used to teach the basics of client/server technology and architecture to our students at Purdue. Given the rapid evolution of this technology, there may now exist a third edition. Check out the technology case studies in the appendixes.

Theby, Stephen E. "Derived Design: Bridging Analysis and Design." McDonnell Douglas Professional Services: Improved System Technologies, 1987. The techniques described in this paper are the basis for a phase in STRADIS (Structured Analysis, Design, and Implementation of Information Systems), a systems development methodology. The technique was altered and simplified to make it suitable to the level of this textbook. As authors, we were quite impressed with the full derived design technique as advocated in the STRADIS methodology.

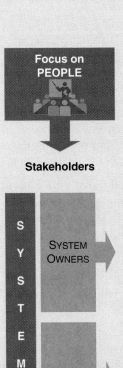

Focus on
PEOPLE

Focus on
DATA

Focus on
PROCESSES

Focus on
INTERFACES

Focus on
DEVELOPMENT

Stakeholders

Activities

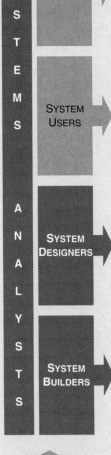

S Y S T E M S   A N A L Y S T S

SYSTEM
OWNERS

SYSTEM
USERS

SYSTEM
DESIGNERS

SYSTEM
BUILDERS

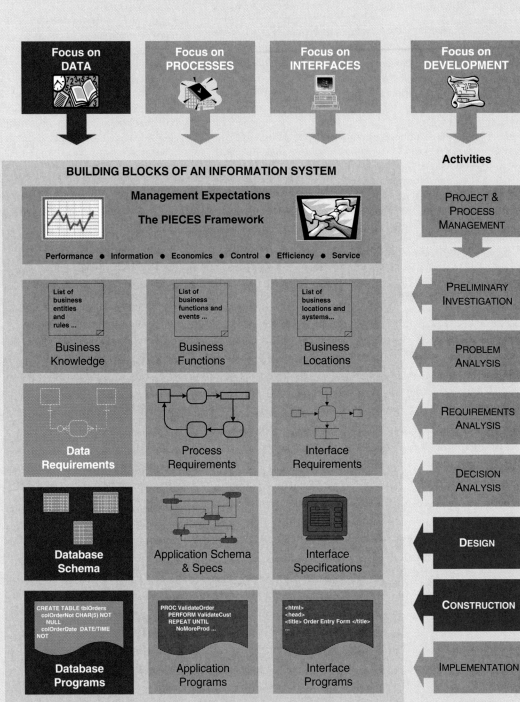

BUILDING BLOCKS OF AN INFORMATION SYSTEM

Management Expectations

The PIECES Framework

Performance ● Information ● Economics ● Control ● Efficiency ● Service

List of business entities and rules ...	List of business functions and events ...	List of business locations and systems...
Business Knowledge	**Business Functions**	**Business Locations**
Data Requirements	**Process Requirements**	**Interface Requirements**
Database Schema	**Application Schema & Specs**	**Interface Specifications**

```
CREATE TABLE tblOrders
  colOrderNot CHAR(5) NOT
  NULL
  colOrderDate  DATE/TIME
NOT
```
Database Programs

```
PROC ValidateOrder
  PERFORM ValidateCust
  REPEAT UNTIL
    NoMoreProd ...
```
Application Programs

```
<html>
<head>
<title> Order Entry Form </title>
...
```
Interface Programs

PROJECT &
PROCESS
MANAGEMENT

PRELIMINARY
INVESTIGATION

PROBLEM
ANALYSIS

REQUIREMENTS
ANALYSIS

DECISION
ANALYSIS

DESIGN

CONSTRUCTION

IMPLEMENTATION

VENDORS
AND
CONSULTANTS

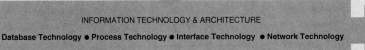

INFORMATION TECHNOLOGY & ARCHITECTURE

Database Technology ● Process Technology ● Interface Technology ● Network Technology

OPERATIONS
AND
SUPPORT

12

DATABASE DESIGN

CHAPTER PREVIEW AND OBJECTIVES

Data storage is a critical component of most information systems. Some people consider it to be the critical component. This chapter teaches the design and construction of physical databases. You will know that you have mastered the tools and techniques of database design when you can:

— Compare and contrast conventional files and modern, relational databases.

— Define and give examples of fields, records, files, and databases.

— Describe a modern data architecture that includes files, operational databases, data warehouses, personal databases, and work group databases.

— Compare the roles of systems analyst, data administrator, and database administrators as they relate to databases.

— Describe the architecture of a database management system.

— Describe how a relational database implements entities, attributes, and relationships from a logical data model.

— Transform a logical data model into a physical, relational database schema.

— Generate SQL code to create the database structures in a schema.

SCENE

The application architecture and general system design for the new Member Services System have been completed and approved. Sandra and Bob are now working with Steve Watson, the database administrator, to design the database for the new system.

BOB

I have finished the first draft of the database schema for the new system. As specified in the application architecture, the entire database will be replicated at each distribution center. Each database will be stored on a Compaq quad-processor Pentium Pro with RAID disks. The network operating system will be *Windows* NT, and the database engine will be *SQL Server*.

STEVE

How much disk, Bob?

BOB

We need to calculate that, Steve. I want to get the database schema completed before I calculate the storage space requirements.

SANDRA

Besides, it should not be difficult to calculate the storage. In the logical data model we recorded the number of instances of each entity and expected growth. We also recorded the type and size of each data attribute. As soon as we complete the schema, we can calculate each table's storage requirement from those numbers.

STEVE

Thanks, you're doing me a big favor. I will add an appropriate buffer for data management overhead and growth. We need to get those servers ordered.

SANDRA

I just put that on my action item list. Bob, why don't you share the preliminary database design with us.

BOB

Right! This was really quite straightforward. After confirming that each entity on the data model was normalized, I used our CASE tool, *System Architect*, to generate the first cut database design. I had to tweak the design based on

SA's generation algorithm, but it turned out to be pretty complete.

[Bob hands out copies of a diagram—Figure 12.A.]

This is the current schema.

SANDRA

This will sure come in handy in my screen design meeting Monday. Do you mind if I generate a prototype of these database tables?

BOB

Don't bother; I already generated an *Access* prototype in anticipation that you would need it. It's in the project directory in a subdirectory called "Test Data"—but I haven't entered any test data yet.

SANDRA

Thanks, I'll take care of that.

STEVE

You know, if the prototype is solid, we might test that new Microsoft wizard that converts an *Access* database into an *SQL Server* database.

BOB

Wouldn't you rather have the standard SQL code instead?

STEVE

Yes, but that takes a while to write.

BOB

Unnecessary! *System Architect* can automatically generate the SQL code to construct the database as soon as you sign off on this database schema.

STEVE

Righteous! This is getting easier all the time.

BOB

Essentially, there is one table for each entity in the approved data model. All primary and secondary keys will be indexes into the tables. Notice that all attributes have been set to the correct data types and null options . . .

STEVE

I don't mean to interrupt, but these foreign keys won't work.

BOB

Excuse me?

STEVE

This is not a complaint. Many books and methodologies allow identical names for the primary keys and their corresponding foreign keys, especially in the logical model. But that becomes a problem in the physical model. Ideally, every attribute is stored in one and only one table. That way, whenever anyone refers to the attribute, you know what they are talking about. Looking at your model, if I refer to MEMBER_NUMBER, what table am I talking about? The MEMBERS table? The MEMBER ORDERS table? The MEMBERSHIPS table? You see? You and I know about foreign keys, but users do not.

SANDRA

I see your point, but we can't delete foreign keys from the tables that are related to members—that's how the database establishes links between the tables.

STEVE

Yes, and I'm sure that using the same names for the primary and foreign keys was useful during logical systems analysis. It helped the users see how you intended to use the foreign keys. But *SQL Server* does not require identical names for the primary and foreign keys. We've found it useful to change the physical name of the foreign key to describe the *role* that the foreign key plays in that table.

BOB

You know, I saw that property, "role," in the System Architect screens but did not bother to look it up.

STEVE

Actually, that's great news. It means that your CASE tool supports roles. Sandra, you look puzzled. Maybe an example would help. Consider the table members. What is the role of MEMBER_NUMBER in that table?

SANDRA

It's the primary key; therefore, the role is to uniquely identify a member.

STEVE

Correct. Now what is the role of MEMBER_NUMBER in the MEMBER ORDERS table?

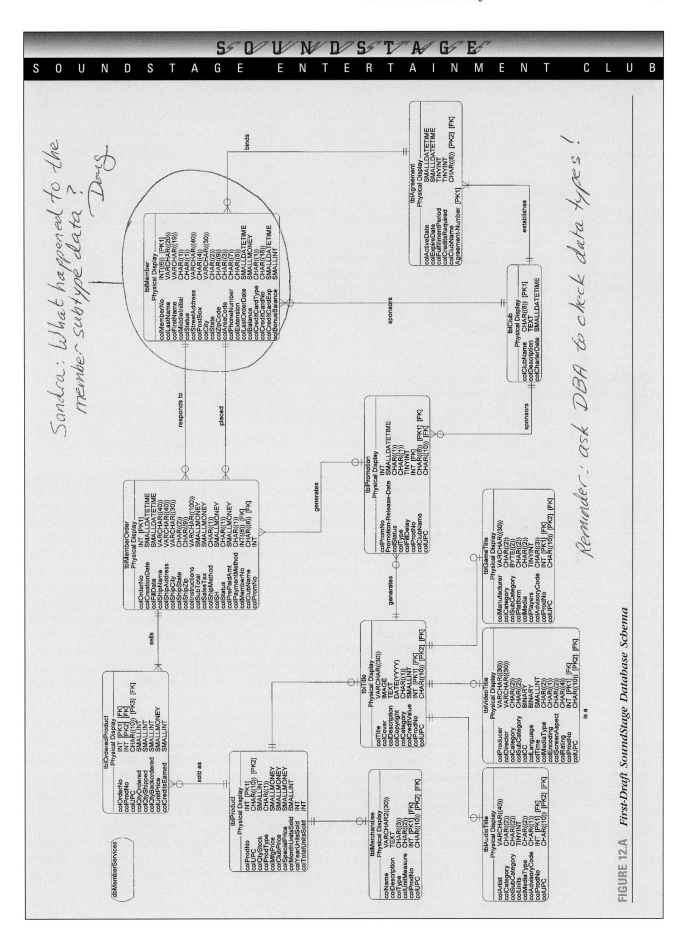

FIGURE 12.A   *First-Draft SoundStage Database Schema*

SANDRA

Obviously, you are looking for a different answer ... Let's see ... OK, I would say its role is to identify the member *that placed an order.*

STEVE

Absolutely correct! So why not change the foreign key name to ordered_by_ member_number?

BOB

I get it! And I'm on it. I'll have all these fixed as soon as I figure out how *System Architect* handles it.

STEVE

Great! And I'll bet you credits to navy beans that *System Architect* can automatically generate the correct SQL code for these roles ...

DISCUSSION QUESTIONS

1. What role does the logical data model (Chapter 7) play in database design?
2. How do the entities, attributes, and relationships from the data model

impact disk storage capacity in database design?
3. What value would a PC database prototype (e.g., Microsoft *Access*) return to a project that has targeted to a high-end database system (e.g., *Oracle* or *SQL Server*)?
4. What is a database schema and how is it different from a logical data model?

CONVENTIONAL FILES VERSUS THE DATABASE

All information systems create, read, update, and delete (sometimes abbreviated *CRUD*) data. This data is stored in files and databases.

A **file** is a collection of similar records.

Examples include a customer file, order file, and product file.

A **database** is a collection of *interrelated* files.

The key word is *interrelated*. A database is *not* merely a collection of files. The records in each file must allow for relationships (think of them as "pointers") to the records in other files. For example, a sales database might contain order records that are "linked" to their corresponding customer and product records.

Let's compare the file and database alternatives. Figure 12.1 illustrates the fundamental difference between the file and database environments. In the file environment, data storage is built around the applications that will use the files. In the database environment, applications will be built around the integrated database. Accordingly, the database is not necessarily dependent on the applications that will use it. In other words, given a database, new applications can be built to share that database. Each environment has its advantages and disadvantages.

As shown in the chapter map, this chapter is concerned with the design and (initial) construction of the database for an information system. The prerequisite is a data (requirements) model from Chapter 7. The deliverables are a database (design) schema and database (definition) program.

The Pros and Cons of Conventional Files

In most organizations, many existing information systems and applications are built around conventional files. You may already be familiar with various conventional file organizations (e.g., indexed, hashed, relative, and sequential) and their access methods (e.g., sequential and direct) from a programming course. These conventional files will likely be in service for quite some time.

Conventional files are relatively easy to design and implement because they are normally designed for use with a single application or information system, such as accounts receivable or payroll. If you understand the end-user's output requirements for that system, you can easily determine the data that will have to be captured and stored to produce those outputs and define the best file organization for those requirements.

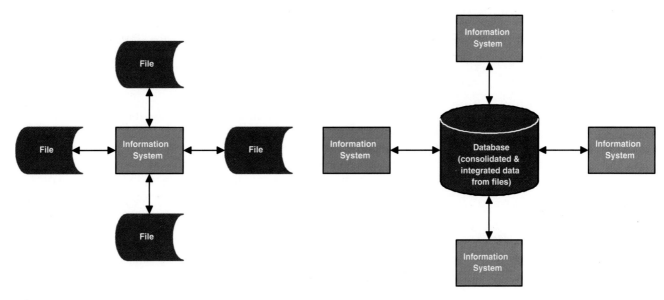

FIGURE 12.1   *Conventional Files versus the Database*

Historically, another advantage of conventional files has been processing speed. They can be optimized for the access of the application. At the same time, they can rarely be optimized for shared use by different applications or systems. Still, files have generally outperformed their database counterparts; however, this limitation of database technology is rapidly disappearing, thanks to cheaper and more powerful computers and more efficient database technology.

Conventional files also have numerous disadvantages. Duplication of data items in multiple files is normally cited as the principal disadvantage of file-based systems. Files tend to be built around single applications without regard to other (future) applications. Over time, because many applications have common data needs, the common data elements get stored redundantly in many different files. This duplicate data results in duplicate inputs, duplicate maintenance, duplicate storage, and possibly data integrity problems (different files showing different values for the same data item).

And what happens if the data format needs to change? Consider the problem faced by many firms if all systems must support a nine-digit ZIP code or four-digit years (to accommodate the year 2000). Because these fields may be stored in many files (with different names), each file would have to be studied and identified. Subsequently, all of the programs that use these ZIP code and date fields would have to be changed.

A significant disadvantage of files is their inflexibility and nonscalability. Files are typically designed to support a single application's *current* requirements and programs. Future needs—such as new reports and queries—often require these files to be restructured because the original file structure cannot effectively or efficiently support the new requirements. But if we elect to restructure those files, all programs using those files would also have to be rewritten. In other words, the current programs have become dependent on the current files, and vice versa. This usually makes reorganization impractical; therefore, we elect to create new, redundant files (same data; structured differently) to meet the new requirements. But that exacerbates the aforementioned redundancy problem. Thus, the inflexibility and redundancy problems tend to complicate one another!

As legacy file-based systems and applications become candidates for reengineering, the trend is overwhelmingly in favor of replacing file-based systems and applications with database systems and applications. For that reason, we have elected to focus this chapter on database design. (Those who encounter the need to learn and apply older file design techniques will find an abundance of reference material in the form of both *COBOL* file processing and older systems analysis and design textbooks in their local college library. Early editions of this textbook also covered file design.)

The Pros and Cons of Database

We've already stated the principal advantage of database—the ability to share the same data across multiple applications and systems. A common misconception about the database approach is that you can build a single superdatabase that contains all data items of interest to an organization. This notion, however desirable, is not currently practical.[1] The reality of such a solution is that it would take forever to build such a complex database. Realistically, most organizations build several databases, each one sharing data with several information systems. Thus, there will be *some* redundancy between databases. However, this redundancy is both greatly reduced and, ultimately, controlled.

Database technology offers the advantage of storing data in flexible formats. This is made possible because databases are defined separately from the information systems and application programs that will use them. Theoretically, this allows us to use the data in ways not originally specified by the end-users. Care must be taken to truly achieve this *data independence*. If the database is well designed, different combinations of the same data can be easily accessed to fulfill future report and query needs. The database scope can even be extended without changing existing programs that use it. In other words, new fields and record types can be added to the database without affecting current programs.

Database technology provides superior *scalability,* meaning that the database and the systems that use it can be grown or expanded to meet the changing needs of an organization. Database technology provides better technology for client/server and network computing architectures. Typically, such architectures require the database to run on its own server. Client/server and network computing was covered in Chapter 11.

On the other hand, database technology is more complex than file technology. Special software, called a *database management system* (*DBMS*), is required. While a DBMS is still somewhat slower than file technology, these performance limitations are rapidly disappearing. Considering the long-term benefits described earlier, most new information systems development is using database technology.

But the advantages of data independence, greatly reduced data redundancy, and increased flexibility come at a cost. Database technology requires a significant investment. The cost of developing databases is higher because analysts and programmers must learn how to use the DBMS. Finally, to achieve the benefits of database technology, analysts and database administrators and experts must adhere to rigorous design principles.

Another potential problem with the database approach is the increased vulnerability inherent in the use of shared data. You are placing all your eggs in one basket. Therefore, backup and recovery and security and privacy become important issues in the world of databases.

Despite the problems discussed, database usage is growing by leaps and bounds. The technology will continue to improve, and performance limitations will all but disappear. Design methods and tools will also improve. For these reasons, this chapter will focus on database design as an important skill for systems analysts.

---

[1] Enterprise resource planning (ERP) applications such as *SAP R/3, PeopleSoft,* and *Oracle* provide a common, customizable database that truly supports almost all the core, operational, and managerial data required in many organizations. For example, *SAP R/3* provides several thousand tables.

We should begin with a disclaimer. Many of the concepts and issues that are important to database design are also taught in database and data management courses. Most information systems curricula include at least one such course. It is not our intent in this chapter to replace that course. Students of information systems should actively seek out courses that *focus* on data management and database techniques; those courses will cover many more relevant technologies and techniques than we can cover in this single chapter.

That said, we will first introduce (or, for some of you, review) those database concepts and issues that are pertinent to the systems analyst's responsibilities in information system design. Although the chapter focus is on database design, experienced readers will immediately notice that many of the concepts transcend the choice between files and databases.

Fields are common to both files and databases.

Fields

A **field** is the physical implementation of a data attribute (introduced in Chapter 7). Fields are the smallest unit of *meaningful* data to be stored in a file or database.

There are four types of fields that can be stored: *primary keys, secondary keys, foreign keys,* and *descriptive* or *nonkey fields.*

A **primary key** is a field whose values identify one and only one record in a file. (This concept was introduced previously in Chapter 7.)

For example, CUSTOMER NUMBER uniquely identifies a single CUSTOMER record in a database, and ORDER NUMBER uniquely identifies a single ORDER record in a database. Also recall from Chapter 7 that a primary key might be created by combining two or more fields (called a **concatonated key**).

A **secondary key** is an *alternate* identifier for a database. A secondary key's value may identify either a single record (as with a primary key) or a subset of all records (such as all ORDERS that have the ORDER STATUS of "backordered").

A single file in a database may only have one primary key, but it may have several secondary keys. To facilitate searching and sorting, an **index** is frequently created for keys.

Foreign keys (also introduced in Chapter 7) are pointers to the records of a different file in a database. Foreign keys are how the database links the records of one type to those of another type.

For example, an ORDER RECORD contains the foreign key CUSTOMER NUMBER to "identify" or "point to" the CUSTOMER record that is associated with the ORDER. Notice that a foreign key in one file requires the existence of the corresponding primary key in another table—otherwise, it does not "point" to anything! Thus, the CUSTOMER NUMBER in an ORDERS file requires the existence of a CUSTOMER NUMBER in the CUSTOMERS file in order to link those files.

A **descriptive field** is any other (nonkey) field that stores business data.

For example, given an EMPLOYEES file, some descriptive fields include EMPLOYEE NAME, DATE HIRED, PAY RATE, and YEAR-TO-DATE WAGES.

The business requirements for both keys and descriptors were defined when you performed data modeling in systems analysis (Chapter 7).

Fields are organized into records. Records are common to both files and databases.

Records

A **record** is a collection of fields arranged in a predefined format.

For example, a CUSTOMER RECORD may be described by the following fields (notice the common notation):

CUSTOMER (NUMBER, LAST-NAME, FIRST-NAME, MIDDLE-INITIAL, POST-OFFICE-BOX-NUMBER, STREET-ADDRESS, CITY, STATE, COUNTRY, POSTAL-CODE, DATE-CREATED, DATE-OF-LAST-ORDER, CREDIT-RATING, CREDIT-LIMIT, BALANCE, BALANCE-PAST-DUE . . .)

During systems design, records will be classified as either fixed-length or variable-length records. Most database technologies impose a **fixed-length record structure,** meaning that each record instance has the same fields, same number of fields, and same logical size. Some database systems will, however, compress unused fields and values to conserve disk storage space. The database designer must generally understand and specify this compression in the database design.

In your prior programming courses (especially COBOL), you may have encountered **variable-length record structures** that allow different records in the same file to have different lengths. For example, a variable-length order record might contain certain common fields that occur once for every order (e.g., ORDER NUMBER, ORDER DATE, and CUSTOMER NUMBER) and other fields that repeat some number of times based on the number of products sold on the order (e.g., PRODUCT NUMBER and QUANTITY ORDERED). Database technologies typically disallow (or at least discourage) variable-length records. This is not a problem, as we'll show later in the chapter.

When a computer program "reads" a record from a database, it actually retrieves a group or *block* (or *page*) of records at a time. This approach minimizes the number of actual disk accesses.

> A **blocking factor** is the number of *logical records* included in a single read or write operation (from the computer's perspective). A block is sometimes called a *physical record.*

Today, the blocking factor is usually determined and optimized by the chosen database technology, but a qualified database administrator may be allowed to fine-tune that blocking factor for performance. Database tuning considerations are best deferred to a database course or textbook.

Files and Tables

Similar records are organized into groups called files. In database systems, a file is frequently called a *table.*

> A **file** is the set of all occurrences of a given record structure.

> A **table** is the *relational* database equivalent of a file. Relational database technology will be introduced shortly.

Some types of conventional files and tables include:

- **Master files** or tables contain records that are relatively permanent. Thus, once a record has been added to a master file, it remains in the system indefinitely. The values of fields for the record will change over its lifetime, but the individual records are retained indefinitely. Examples of master files and tables include CUSTOMERS, PRODUCTS, and SUPPLIERS.
- **Transaction files** or tables contain records that describe business events. The data describing these events normally has a limited useful lifetime. For instance, an INVOICE record is ordinarily useful until the invoice has been paid or written off as uncollectible. In information systems, transaction records are frequently retained *on-line* for some period of time. Subsequent to their useful lifetime, they are *archived* off-line. Examples of transaction files include ORDERS, INVOICES, REQUISITIONS, and REGISTRATIONS.
- **Document files** and tables contain stored copies of historical data for easy retrieval and review without the overhead of regenerating the document.
- **Archival files** and tables contain master and transaction file records that have been deleted from on-line storage. Thus, records are rarely deleted; they are merely moved from on-line storage to off-line storage. Archival

requirements are dictated by government regulation and the need for subsequent audit or analysis.

- **Table look-up files** contain relatively static data that can be shared by applications to maintain consistency and improve performance. Examples include SALES TAX TABLES, ZIP CODE TABLES, and INCOME TAX TABLES.

- **Audit files** are special records of updates to other files, especially master and transaction files. They are used in conjunction with archival files to recover "lost" data. Audit trails are typically built into better database technologies.

In the not too distant past, file design methods required the analyst to specify precisely how the records in a database should be sequenced (called **file organization**) and accessed (called **file access**). In today's database environment, the database technology itself usually predetermines and/or limits the file organization for all tables contained in the database. Once again, a trained database administrator may be given some control over organization, storage location, and access methods for the purpose of performance tuning.

Databases

As described earlier, stand-alone, application-specific files were once the lifeblood of most information systems; however, they are being slowly but surely replaced with databases. Recall that a database may loosely be thought of as a set of interrelated files. By interrelated, we mean that records in one file may be associated or linked with the records in a different file.

For example, a STUDENT record may be linked to all of that student's COURSE records. In turn, a COURSE record may be linked to the STUDENT records that indicate completion of that course. This two-way linking and flexibility allow us to eliminate *most* of the need to redundantly store the same fields in the different record types. Thus, in a very real sense, multiple files are consolidated into a single file—the database.

The idea of relationships between different collections of data was introduced in Chapter 7. In that chapter, you learned to discover a system's data requirements and model those requirements as *entities* and *relationships*. The database now provides for the technical implementation of those entities and relationships.

So many applications are now being built around database technology that database design has become an important skill for the analyst. The history of information systems has led to one inescapable conclusion:

Data is a resource that must be controlled and managed!

Few, if any, information systems staffs have avoided the frustration of uncontrolled growth and duplication of data stored in their systems. As systems were developed, implemented, and maintained, the common data needed by the different systems was duplicated in multiple conventional files. This duplication carried with it a number of costs: extra storage space required, duplicated input to maintain redundantly stored data and files, and data integrity problems (e.g., the ADDRESS for a specific customer not matching in the various files that contained that customer's address).

Out of necessity, database technology was created so an organization could maintain and use its data as an integrated whole instead of as separate data files. We can now develop a shared data resource that can be used by several information systems.

Data Architecture   Data becomes a business resource in a database environment. Information systems are built around this resource to give both computer programmers and end-users flexible access to data.

A business's **data architecture** defines how that business will develop and use both files and databases to store all of the organization's data; the file and database technology to be used; and the administrative structure set up to manage the data resource.

Figure 12.2 illustrates the data architecture into which many companies have evolved. As shown in the figure, most companies still have numerous conventional file-based information system applications, most of which were developed before the emergence of high-performance database technology. In many cases, the processing efficiency of these files or the projected cost to redesign these files has slowed conversion of the systems to database.

As shown in Figure 12.2, **operational (or *transactional*) databases** are developed to support day-to-day operations and business transaction processing for major information systems. These systems are developed (or purchased) over time to replace the conventional files that formerly supported applications. Access to these databases is limited to computer programs that use the DBMS to process transactions, maintain the data, and generate regularly scheduled management reports. Some query access may also be provided.

Many information systems shops hesitate to give end-users access to operational databases for queries and reports. The volume of unscheduled reports and queries could overload the computers and hamper business operations that the databases were intended to support. Instead, data warehouses are developed, possibly on separate computers.

FIGURE 12.2   *A Typical, Modern Data Architecture*

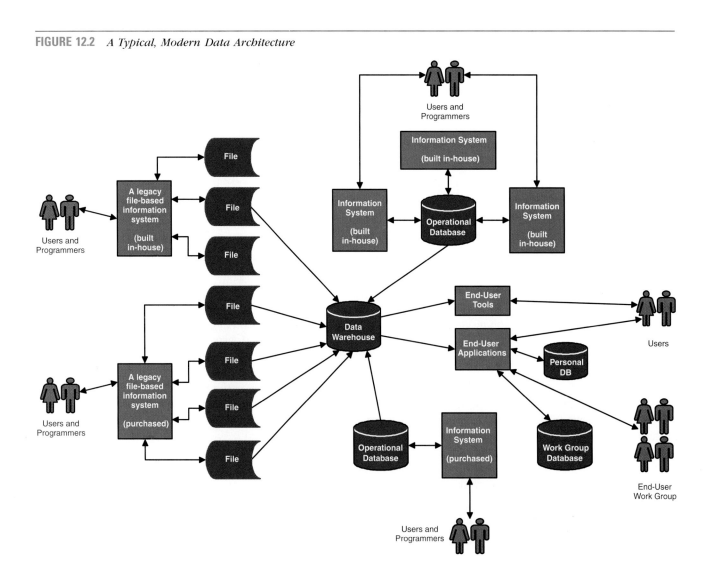

Data warehouses store data extracted from the operational databases and conventional files. Fourth-generation programming languages, query tools, and decision support tools are then used to generate reports and analyses off these data warehouses. These tools often allow users to extract data from both conventional files and operational databases. This is sometimes called **data mining.**

Figure 12.2 also shows **personal** and **work group** (or departmental) **databases.** Personal computer and local network database technology has rapidly matured to allow end-users to develop personal and departmental databases. These databases may contain unique data, or they may import data from conventional files, operational databases, and/or data warehouses. Personal databases are built using PC database technology such as *Access, dBASE, Paradox,* and *FoxPro.*

Contemporary data architecture also allows for Internet-enabled database technology. For example, *Oracle 8i* provides special tools and facilities for Web-enabling a database.

Admittedly, this overall scenario is advanced, but many firms are currently using variations of it. To manage the enterprisewide data resource, a staff of database specialists may be organized around the following administrators:

A **data administrator** is responsible for the data planning, definition, architecture, and management.

One or more **database administrators (DBAs)** are responsible for the database technology, database design and construction consultation, security, backup and recovery, and performance tuning.

In smaller businesses, these roles may be combined or assigned to one or more systems analysts.

Database Architecture   So far, we have made several references to the *database technology* that makes the above data architecture possible.

Database architecture refers to the database technology including the database engine, database utilities, database CASE tools for analysis and design, and database application development tools.

The control center of a database architecture is its database management system.

A **database management system (DBMS)** is specialized computer software available from computer vendors that is used to create, access, control, and manage the database. The core of the DBMS is often called its **database engine.** The engine responds to specific commands to create database structures and then to create, read, update, and delete records in the database.

The database management system is purchased from a database technology vendor such as Oracle, IBM, Microsoft, or Sybase.

Figure 12.3 depicts a typical database management system architecture. A systems analyst, or database analyst, designs the structure of the data in terms of record types, fields contained in those record types, and relationships that exist between record types. These structures are defined to the database management system using its data definition language.

Data definition language (DDL) is used by the DBMS to physically establish those record types, fields, and structural relationships. Additionally, the DDL defines views of the database. Views restrict the portion of a database that may be used or accessed by different users and programs.

Most database management systems store both **user data** and **metadata**—the data (or specifications) about the data—such as record and field definitions, synonyms, data relationships, validation rules, help messages, and so forth. Some metadata is stored in the actual database, while other metadata is stored in CASE tool repositories.

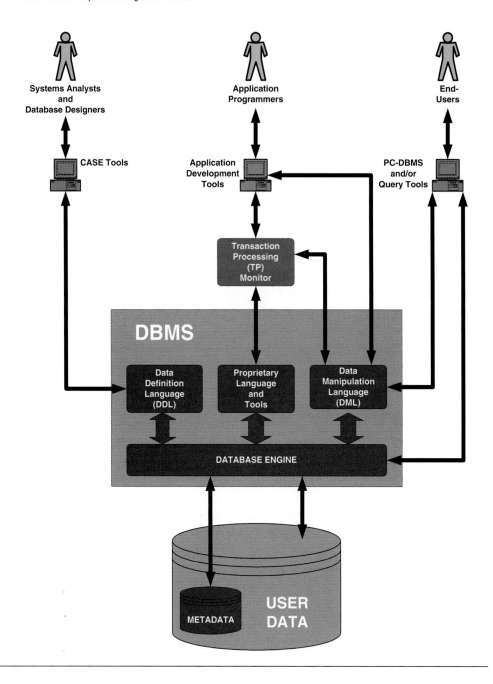

FIGURE 12.3
A Typical Database Management System Architecture

To help design databases, CASE tools may be provided either by the database technology vendor (e.g., Oracle's *Designer 2000*) or from a third-party CASE tool vendor (e.g., Popkin's *System Architect 2001,* Visio's *Visio Enterprise,* or Computer Associates' *ERwin,* etc.).

The database management system also provides a data manipulation language to access and use the stored data in applications.

> A **data manipulation language (DML)** is used to create, read, update, and delete records in the database and to navigate between different records and types of records—for example, from a CUSTOMER record to the ORDER records for that customer. The DBMS and DML hide the details concerning how records are organized and allocated to the disk.

In general, the DML is very flexible in that it may be used by itself to create, read, update, and delete records; or its commands may be "called" from a separate host programming language such as *COBOL, Visual Basic,* or *Java.*

Many DBMSs don't require the use of a DDL to construct the database or a DML to access the database. Instead (or in addition), they provide their own proprietary tools and commands to perform those tasks. This is especially true of PC-based DBMSs such as Microsoft *Access. Access* provides a simple graphical user interface to create the tables and both a form-based environment and scripting language (*Visual Basic for Applications*) to access, browse, and maintain the tables.

Many DBMSs also include proprietary report writing and inquiry tools to allow users to access and format data without directly using the DML. Many high-end DBMSs are designed to interact with popular third-party transaction processing monitors such as IBM's *CICS* and Microsoft's *Transaction Server.*

All of the above technology is illustrated in Figure 12.3. Today, almost all new database development is using relational database technology.

Relational Database Management Systems   There are several types of database management systems. They can be classified according to the way they structure records. Early database management systems organized records in hierarchies or networks implemented with indexes and linked lists. You may study these further in a database course. But today, most successful database management systems are based on relational technology.

> **Relational databases** implement data in a series of two-dimensional tables that are "related" to one another via foreign keys. Each table (sometimes called a *relation*) consists of named columns (which are fields or attributes) and any number of unnamed rows (which correspond to records).

Figure 12.4 illustrates a logical data model. Figure 12.5 is the physical, relational database implementation of that data model (called a **schema**). In a relational database, files are seen as simple two-dimensional tables, also known as relations. The rows are records. The columns correspond to fields.

The following shorthand notation for tables is commonly encountered in systems design and database books.

CUSTOMERS (<u>CUSTOMER-NUMBER</u>, CUSTOMER-NAME, CUSTOMER-BALANCE, . . .)

ORDERS (<u>ORDER-NUMBER</u>, CUSTOMER-NUMBER (FK), . . .)

ORDERED-PRODUCTS (<u>ORDER-NUMBER</u> (FK), <u>PRODUCT-NUMBER</u> (FK), QUANTITY-ORDERED, . . .)

PRODUCTS (<u>PRODUCT-NUMBER</u>, PRODUCT-DESCRIPTION, QUANTITY-IN-STOCK, . . .)

Both the DDL and DML of most relational databases is called **SQL** (pronounced *S-Q-L* by some and *sequel* by others). SQL supports complete database creation, maintenance, and usage. To access data in tables and records, SQL provides the following basic commands:

— SELECT specific records from a table based on specific criteria (e.g., SELECT CUSTOMER WHERE BALANCE > 500.00).

— PROJECT out specific fields from a table (e.g., PROJECT CUSTOMER TO INCLUDE ONLY CUSTOMER-NUMBER, CUSTOMER-NAME, BALANCE).

— JOIN two or more tables across a common field—a primary and foreign key (JOIN CUSTOMER AND ORDER USING CUSTOMER-NUMBER).

FIGURE 12.4
A Simple, Logical Data Model

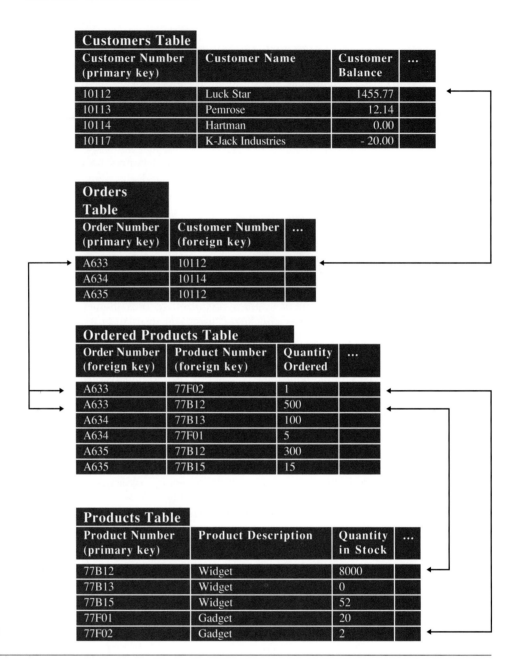

FIGURE 12.5

A Simple, Physical Database Schema

When used in combination, these basic commands can address most database requirements. A fundamental characteristic of SQL is that commands return a set of records, not necessarily just a single record (as in nonrelational database and file technology). SQL databases also provide commands for creating, updating, and deleting records, as well as sorting records.

High-end relational databases also extend the SQL language to support triggers and stored procedures.

Triggers are programs embedded within a table that are automatically invoked by updates to another table. For example, if a record is deleted from a PASSENGER AIRCRAFT table, a trigger can force the automatic deletion of all corresponding records in a SEATS table for that aircraft.

Stored procedures are programs embedded within a table that can be called from an application program. For example, a data validation algorithm might be embedded in a table to ensure that new and updated records contain valid data *before* they are stored. Stored procedures are written in a proprietary extension of SQL such as Microsoft's *Transact SQL* or Oracle's *PL/SQL*.

Both triggers and stored procedures are reusable because they are stored with the tables themselves (as metadata). This eliminates the need for application programmers to create the equivalent logic within each application that uses the tables.

The SQL language for defining, using, and manipulating a relational database is covered extensively in most database courses and textbooks. All high-end relational database management systems (e.g., *Oracle, UDB/DB2,* and *SQL Server*) and many personal computer relational database management systems (such as Microsoft *Access*) support the SQL language standards.

Examples of high-performance relational DBMSs include Oracle Corporation's *Oracle,* IBM's *Universal Database,* Microsoft's *SQL Server* (being used in the Sound-Stage project), and Sybase Corporation's *Sybase*. Many of these databases run on mainframes, minicomputers, and network database servers. Additionally, most personal computer DBMSs are relational (or at least partially so). Examples include Microsoft's *Access* and *Foxpro*. These database engines can run on both stand-alone personal computers and local area network file servers. Figure 12.6 illustrates a relational database management system's user interface.

Database textbooks and courses offer entire chapters and units on relational databases. We encourage you to learn more!

In Chapter 7 you learned how to model data requirements for an information system. That model took the form of a fully attributed entity relationship diagram and a repository of metadata. Chapter 7 also taught a technique called data analysis or normalization. This technique was used to produce a data model that meets the following quality criteria:

— *A good data model is simple.* As a general rule, the data attributes that describe an entity should describe only that entity.

PREREQUISITE FOR DATABASE DESIGN—NORMALIZATION

What Is a Good Data Model?

FIGURE 12.6
User/Designer Interface for a Relational PC DBMS (Microsoft Access*)*

— *A good data model is essentially nonredundant.* This means that each data attribute, other than foreign keys, describes at most one entity.

— *A good data model should be flexible and adaptable to future needs.* In the absence of this criteria, we would tend to design databases to fulfill only *today's* business requirements.

So how do we achieve the above goals? How can you design a database that can adapt to future requirements that you cannot predict? The answer lies in data analysis.

Recall that normalization is a three-step technique that places the data model into first normal form, second normal form, and third normal form.

— An entity is in **first normal form (1NF)** if there are no attributes that can have more than one value for a single instance of the entity (frequently called repeating groups). Any attributes that can have multiple values actually describe a separate entity.

— An entity is in **second normal form (2NF)** if it is already in 1NF and if the values of all nonprimary key attributes are dependent on the full primary key—not just part of it. Any nonkey attributes that are dependent on only part of the primary key should be moved to any entity where that partial key becomes the full key.

— An entity is in **third normal form (3NF)** if it is already in 2NF and if the values of its nonprimary key attributes are not dependent on any other nonprimary key attributes. Any nonkey attributes that are dependent on other nonkey attributes must be moved or deleted.

There are other higher degrees of normalization that are best left to database textbooks and courses. Database design should proceed only if the underlying logical data model is in at least 3NF. For a more detailed explanation, we encourage you to review Chapter 7.

CONVENTIONAL FILE DESIGN

The focus of this chapter is on database design; however, we would be remiss to not say a few words about conventional file design. First, file design is simplified because of its orientation to a single application. Typically, the output and input designs (Chapters 13 and 14) would be completed first since the file design is dependent on supporting those application requirements.

Most fundamental entities from the data model would be designed as master or transaction records. The master files are typically fixed-length records. Associative entities from the data model are typically joined into the transaction records to form variable-length records (based on the one-to-many relationships). Other types of files (not represented in the data model) are added as necessary.

Two important considerations of conventional file design are *file access* and *organization.* The systems analyst usually studies how each program (from Chapter 10) will access the records in the file (sequentially or randomly) and then select an appropriate file organization (e.g., sequential, indexed, hashed, etc.). In practice, many systems analysts select an indexed sequential (or ISAM/VSAM) organization to support the likelihood that different programs will require different access methods into the records.

MODERN DATABASE DESIGN

The design of any database will usually involve the DBA and database staff. They will handle the technical details and cross-application issues. Still, it is useful for the systems analyst to understand the basic design principles for relational databases.

The design rules presented here are, in fact, guidelines. We cannot cover every idiosyncrasy. Also, because SoundStage has elected to use Microsoft's *SQL Server* as its database management system, our design will be constrained by that technology. Each relational DBMS presents its own capabilities and constraints.

Fortunately, the guidelines presented here are fairly generic and applicable to most DBMS environments. Database courses and textbooks tend to cover a wider variety of technology and issues.

Computer-assisted systems engineering (CASE) has been a continuing theme throughout this book. There are specific CASE products that address database analysis and design (e.g., Computer Associates' *ERwin*). Also, most general-purpose CASE tools now include database design tools. In this example, we continued to use Popkin's *System Architect 2001* CASE product for the SoundStage case study. Finally, most CASE tools (including *System Architect*) can automatically generate SQL code to construct the database structures for the most popular database management systems. This code generation capability is an enormous time-saver.

The goals of database design are as follows:

Goals and Prerequisites to Database Design

— A database should provide for the efficient storage, update, and retrieval of data.

— A database should be reliable—the stored data should have high integrity to promote user trust in that data.

— A database should be adaptable and scalable to new and unforeseen requirements and applications.

The system's logical data model—in our case, a fully attributed and normalized entity relationship diagram (ERD)—serves as the prerequisite. This model from Chapter 7 is reproduced in Figure 12.7. Every attribute in that model must be defined as to its data type, domain, and default. These properties were also covered in Chapter 7.

The design of a database is depicted as a special model called a database schema.

The Database Schema

A **database schema** is the *physical* model or blueprint for a database. It represents the technical implementation of the logical data model. (*System Architect 2001* calls it a *physical data model.*)

NOTE   We should acknowledge some potentially confusing terminology here. We are using the terms *logical* and *physical* in a manner consistent with earlier chapters in this book. Unfortunately, most database books use the terms *conceptual* (our *logical*) and logical (our *physical*). We apologize for this unavoidable industry confusion.

A relational database schema defines the database structure in terms of tables, keys, indexes, and integrity rules. A database schema specifies details based on the capabilities, terminology, and constraints of the chosen database management system. Each DBMS supports different data types, integrity rules, and so forth.

The transformation of the logical data model into a physical relational database schema is governed by some fairly generic rules and options. These rules and guidelines are summarized as follows:

1. Each fundamental, associative, and weak entity is implemented as a separate table. Table names may have to be formatted according to the naming rules and size limitations of the DBMS. For example, a logical entity named MEMBER ORDERED PRODUCT might be changed to a physical table named tblMemberOrdProd. The prefix and compression of spaces is consistent with contemporary naming standards and guidelines in modern programming languages.
   a. The primary key is identified as such and implemented as an index into the table.
   b. Each secondary key is implemented as its own index into the table.
   c. An index should be created for any nonkey attributes that were identified as subsetting criteria requirements (Chapter 7).
   d. Each foreign key will be implemented as such. The inclusion of these

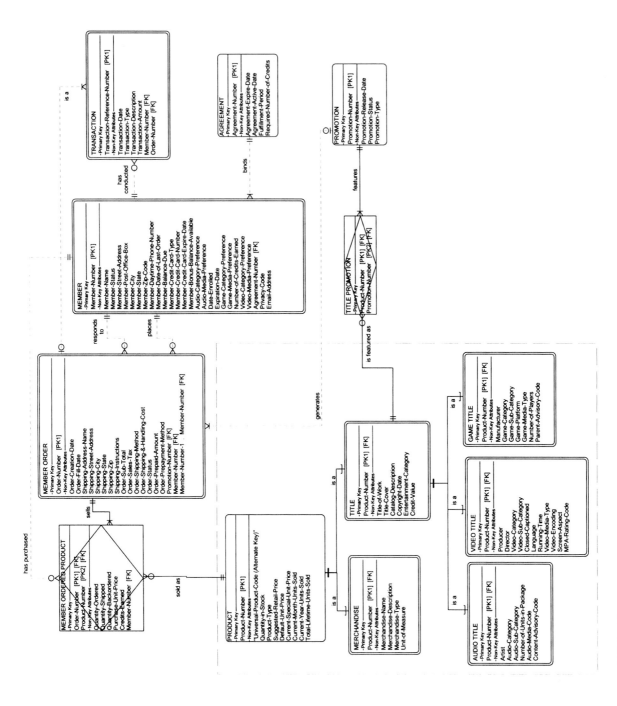

FIGURE 12.7 *SoundStage Logical Data Model in Third Normal Form*

foreign keys implements the relationships on the data model and allows tables to be JOINED in SQL and application programs.

e. Attributes will be implemented with fields. These fields correspond to columns in the table. The following technical details must usually be specified for each attribute. (These details may be automatically inferred by the CASE tool from the logical descriptions in the data model.)

Field names may have to be shortened and reformatted according to DBMS constraints and internal rules. For example, in the logical data model, most attributes might be prefaced with the entity name (e.g., MEMBER NAME). In the physical database, we might simply use NAME.

i. *Data type.* Each DBMS supports different data types and terms for those data types. Figure 12.8 shows different physical data types for a few different database management systems.

ii. *Size of the field.* Different DBMSs express precision of real numbers differently. For example, in *SQL Server,* a size specification of NUMBER (3,2) supports a range from −9.99 to 9.99.

iii. *NULL or NOT NULL.* Must the field have a value before the record can be committed to storage? Again, different DBMSs may require different

---

FIGURE 12.8   *Partial List of Physical Data Types for Different Database Technologies*

Logical Data Type (to be stored in field)	Physical Data Type Microsoft *Access*	Physical Data Type Microsoft *SQL Server*	Physical Data Type Oracle
Fixed length character data *(use for fields with relatively fixed length character data)*	TEXT	CHAR (size) *or* character (size)	CHAR (SIZE)
Variable length character data *(use for fields that require character data but for which size varies greatly —such as ADDRESS)*	TEXT	VARCHAR (max size) *or* character varying (max size)	VARCHAR (max size)
Very long character data *(use for long descriptions and notes—usually no more than one such field per record)*	MEMO	TEXT	LONG VARCHAR *or* LONG VARCHAR2
Integer number	NUMBER	INT (size) *or* integer *or* smallinteger *or* tinyinteger	INTEGER (size) *or* NUMBER (size)
Decimal number	NUMBER	DECIMAL (size, decimal places) *or* NUMERIC (size, decimal places)	DECIMAL (size, decimal places) *or* NUMBER (size, decimal places) *or* NUMBER
Financial number	CURRENCY	MONEY	*see decimal number*
Date (with time)	DATE/TIME	DATETIME *or* SMALLDATETIME *Depending on precision needed*	DATE
Current time *(use to store the date and time from the computer's system clock)*	*not supported*	TIMESTAMP	*not supported*
Yes or No; *or* True or False	YES/NO	BIT	*use CHAR(1) and set a yes or no domain*
Image	OLE OBJECT	IMAGE	LONGRAW
Hyperlink	HYPERLINK	VARBINARY	RAW
Can designer define new data types?	*NO*	*YES*	*YES*

reserved words to express this property. By definition, primary keys can never be allowed to have NULL values.

iv. *Domains.* Many database management systems can automatically edit data to ensure that fields contain legal data. This can be a great benefit to ensuring data integrity independent from the application programs. If the programmer makes a mistake, the DBMS catches the mistake. But for DBMSs that support data integrity, the rules must be precisely specified in a language that is understood by the DBMS.

v. *Default.* Many database management systems allow a default value to be automatically set in the event that a user or programmer creates a record containing fields with no values. In some cases, NULL serves as the default.

vi. Again, many of the above specifications were documented as part of a complete logical data model. If that data model was developed with a CASE tool, the CASE tool may be capable of automatically translating the data model into the physical language of the chosen database technology.

2. Supertype/subtype entities present additional options as follows:
   a. Each supertype and subtype can be implemented with a separate table (all having the same primary key).
   b. Alternatively, if the subtypes are of *similar* size and data content, a database administrator may elect to collapse the subtypes into the supertype to create a single table. This presents certain problems for setting defaults and checking domains. In a high-end DBMS, these problems can be overcome by embedding the default and domain logic into *stored procedures* for the table.
   c. Alternatively, the supertype's attributes could be duplicated in a table for each subtype.
   d. Some combination of the above options could be used.

3. Evaluate and specify referential integrity constraints (described in the next section).

The SoundStage database schema was automatically generated from the logical data model by our CASE tool *System Architect 2001.* It is illustrated in Figure 12.9. We call your attention to the following numbered bullets on the figure.

❶ Each rounded rectangle defines a table. The named rows in the rectangle actually correspond to the named columns that will be created for the table.

❷ SoundStage has defined a standard naming convention for tables and columns. The conventions are based on the programming guidelines called *Hungarian Notation.* Each object is named without spaces, dashes, or underscores. And each object is given a prefix that defines all similar objects. For database objects, the following standards were used:

tbl   Indicates a database table.

col   Indicates a column in the table.

Although not depicted on the schema, other common database prefixes may be included in the schemas underlying data dictionary (repository) such that those prefixes may be used to generate correct code. Possibilities include:

db   Indicates the database itself.

idx   Indicates an index built for a table.

dom   Indicates a domain that can be applied to one or more fields.

❸ Logical relationships, both identifying and nonidentifying, are transformed in *constraints* that are implemented using the foreign keys.

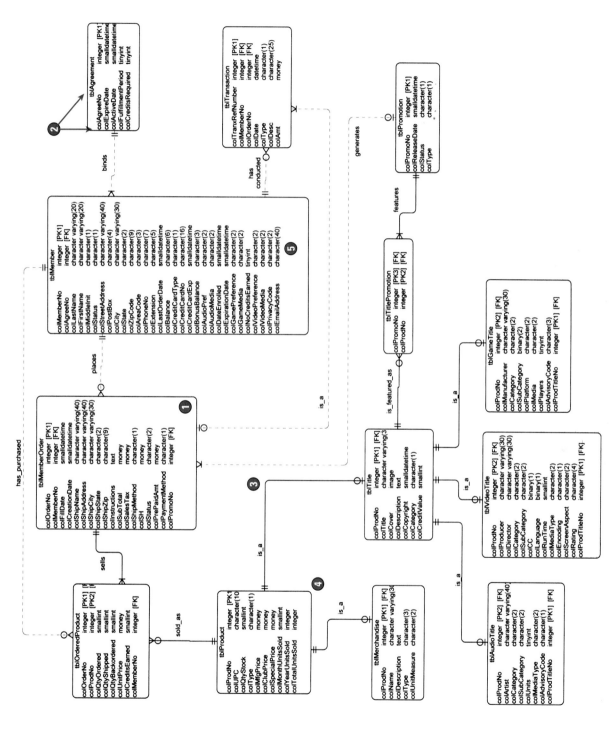

FIGURE 12.9 *Initial SoundStage Physical Database Schema*

❹ We elected to make each supertype and subtype entity in the logical, generalization hierarchy into its own physical table. (This was the default option for *System Architect 2001*'s physical data model generator.)

❺ Notice that *System Architect 2001* automatically inferred physical data types for each field based on (1) the selection of Microsoft *SQL Server 7* as the target database management system and (2) the logical data types we had defined for each entity's attributes during systems analysis. The generated physical data types can be changed to reduce storage space required, improve data integrity, or better represent all the possible values included in the domain.

Although not depicted on the database schema, the schema generator also creates an index for each primary key indicated in the schema. You can add additional indexes for unique secondary keys (such as the Universal Product Code or UPC field in tblProduct) or for any nonkey attribute that can be used to subset all records in a table (such as tblTransaction.colType). These indexes can improve the performance of the final database.

Some CASE tools generate database schemas with considerably more detail than our example. For example, some database schemas indicate for each field whether or not the field must take on a value:

- NULL means the field does not have to have a value.
- NOT NULL means the field must have a value. Because primary keys are used to uniquely access records, no PK field may take on NULL values.

Would you ever want to compromise the third normal form entities when designing the database? For example, would you ever want to combine two third normal form entities into a single table (that would, by default, no longer be in third normal form)? Usually not! Although a database administrator may create such a compromise to improve database performance, he or she should carefully weigh the advantages and disadvantages. Although such compromises may mean greater convenience through fewer tables or better overall performance, such combinations may also lead to the possible loss of data independence—should future new fields necessitate resplitting the table into two tables, programs will have to be rewritten. As a general rule, combining entities into tables is not recommended.

Data and Referential Integrity

Database integrity is about trust. Can the business and its users trust the data stored in the database? Data integrity provides necessary internal controls for the database. There are at least three types of data integrity that must be designed into any database.

Key Integrity   Every table should have a primary key (which may be concatenated). The primary key must be controlled such that no two records in the table have the same primary key value. (Note that for a concatenated key, the concatenated value must be unique—not the individual values that make up the concatenation.)

Also, the primary key for a record must never be allowed to have a NULL value. That would defeat the purpose of the primary key, to uniquely identify the record.

If the database management system does not enforce these rules, other steps must be taken to ensure them. Most DBMSs do enforce key integrity.

Domain Integrity   Appropriate controls must be designed to ensure that no field takes on a value that is outside of the range of legal values. For example, if GRADE POINT AVERAGE is defined to be a number between 0.00 and 4.00, then controls must be implemented to prevent negative numbers and numbers greater than 4.00.

Not long ago, application programs were expected to perform all data editing. Today, most database management systems are capable of enforcing domain rules. For the foreseeable future, the responsibility for data editing will continue to be shared between the application programs and the DBMS.

Referential Integrity The architecture of relational databases implements relationships between the records in tables via *foreign keys*. The use of foreign keys increases the flexibility and scalability of any database, but it also increases the risk of referential integrity errors.

> A **referential integrity** error exists when a foreign key value in one table has no matching primary key value in the related table.

For example, an INVOICES table usually includes a foreign key, CUSTOMER NUMBER, to "reference back to" the matching CUSTOMER NUMBER primary key in the CUSTOMERS table. What happens if we delete a CUSTOMER record? There is the *potential* that we may have INVOICE records whose CUSTOMER NUMBER has no matching record in the CUSTOMERS table. Essentially, we have compromised the referential integrity between the two tables.

How do we prevent referential integrity errors? One of two things should happen. When considering the deletion of CUSTOMER records, either we should automatically delete all INVOICE records that have a matching CUSTOMER NUMBER (which doesn't make much business sense), or we should disallow the deletion of the CUSTOMER record until we have deleted all INVOICE records.

Referential integrity is specified in the form of deletion rules as follows:[2]

- *No restriction.* Any record in the table may be deleted without regard to any records in any other tables.

  In looking at the final SoundStage data model, we could not apply this rule to any table.

- *Delete:Cascade.* A deletion of a record in the table must be automatically followed by the deletion of matching records in a related table. Many relational DBMSs can automatically enforce delete:cascade rules using triggers.

  In the SoundStage data model, an example of a valid delete:cascade rule would be from MEMBER ORDER to MEMBER ORDERED PRODUCT. In other words, if we delete a specific MEMBER ORDER, we should automatically delete all matching MEMBER ORDERED PRODUCTS for that order.

- *Delete:Restrict.* A deletion of a record in the table must be disallowed until any matching records are deleted from a related table. Again, many relational DBMSs can automatically enforce delete:restrict rules.

  For example, in the SoundStage data model, we might specify that we should disallow the deletion of any PRODUCT so long as there exists MEMBER ORDERED PRODUCTS for that product.

- *Delete:Set null.* A deletion of a record in the table must be automatically followed by setting any matching keys in a related table to the value NULL. Again, many relational DBMSs can enforce such a rule through triggers.

  The Delete:Set null option was not used in the SoundStage data model. It is used only when you are willing to delete a master table record, but you don't want to delete corresponding transaction table records for historical reasons. By setting the foreign key to NULL, you are acknowledging that the record does not point back to a corresponding master record, but at least you don't have it pointing to a nonexisting master record.

The final database schema, complete with referential integrity rules, is illustrated in Figure 12.10. This is the blueprint for writing the SQL code (or equivalent) to create the tables and data structures.

Roles

Some database standards insist that no two fields have *exactly* the same name. This constraint simplifies documentation, help systems, and metadata definitions. This presents an obvious problem with foreign keys. By definition, a foreign key

---

[2] Knowledgeable database students know that there are also insertion and update rules for referential integrity. A full discussion of these rules is deferred to database courses and textbooks.

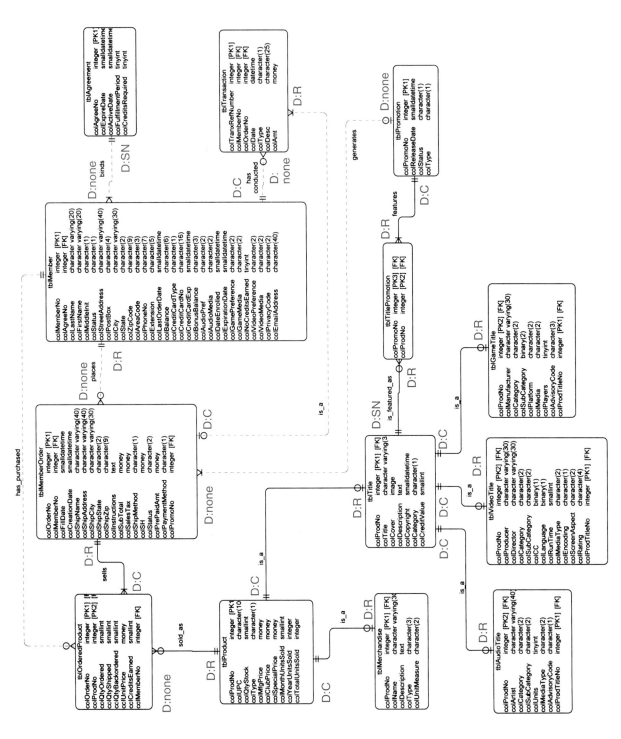

FIGURE 12.10 *Final SoundStage Physical Database Schema*

490

must have a corresponding primary key. During *logical* data modeling, using the same name suited our purpose of helping the users understand that these foreign keys allow us to match related records in different entities. But in a *physical* database, it is not always necessary or even desirable to have these redundant field names in the database.

To fix this problem, foreign keys can be given role names.

> A **role name** is an alternate name for a foreign key that clearly distinguishes the purpose that the foreign key serves in the table.

For example, in the SoundStage database schema, PRODUCT_NUMBER is a primary key for the PRODUCTS table and a foreign key in the MEMBER ORDERED PRODUCTS table. The name should not be changed in the PRODUCTS table. But it may make sense to rename the foreign key to ORDERED_PRODUCT_NUMBER to more accurately reflect its role in the MEMBER ORDERED PRODUCTS table.

The decision to require role names or not is usually established by the data or database administrator.

In Chapter 7, Data Modeling and Analysis, we briefly introduced the concept of logical data distribution analysis.

> **Data distribution analysis** establishes which business locations need access to which logical data entities and attributes.

Database Distribution and Replication

We used a simple matrix in Chapter 7 to map entities and attributes to locations. Many CASE tools, including *System Architect 2001*, include facilities for building such a matrix. We should give some consideration now to the impact of data distribution analysis on database design.

In today's multi-tier, client/server, network-centric world, information systems and databases are rarely centralized. Instead, they are distributed across a network that may span many buildings, cities, states, or countries. Accordingly, we may need to partition, distribute, or replicate all or part of a database design to different physical database servers in different physical locations. Basically, we need to perform a physical database distribution analysis that takes into consideration what we learned during our logical data distribution analysis.

Essentially, we have a number of distribution options available to us:

- **Centralization** of the database. In other words, we would implement the database on a single server regardless of the number of physical locations that may require access to it. This solution is simple and the easiest to maintain; however, it violates a data management rule that has become important to many data administrators and users—data should be located as closely as possible to its users.

- **Horizontal distribution** of the data. In this option, each table (or entire rows in a table) would be assigned different database servers and locations. This option results in efficient access and security because each location has only those tables and rows required for that location. Unfortunately, data cannot always be easily recombined for management analysis across sites.

- **Vertical distribution** of the data. In this option, specific columns of tables are assigned to specific databases and servers. The advantages and disadvantages are very similar to that of horizontal distribution.

- **Replication** of the data. Replication refers to the physical duplication of entire tables to multiple locations. Most high-end, enterprise database management systems include replication technology that coordinates updates to the duplicated tables and records to maintain data integrity. This solution offers performance and accessibility advantages and reduces network traffic, but it also increases the complexity of data integrity and requires more physical storage capacity.

These alternatives are not mutually exclusive. The designer must carefully plan degrees of data distribution and replication.

Given our physical database schema, we can define *views* that correspond to specific geographic locations (and subviews for different users and applications). A database view may be very selective. It may include a specific subset of tables, a specific subset of columns in tables, or even a specific subset of records in tables. Each view must be carefully synchronized with the master database schema such that changes to the master schema can, if appropriate, be propagated to the views. CASE tools can be very helpful in defining views and keeping all views in sync.

For the SoundStage project, we plan to replicate the entire database in each of three cities. The data integrity for common tables will be implemented using *SQL Server*'s replication technology. The systems analyst will not typically program the replication rules. A qualified database analyst or administrator will do that. Since we will implement the entire physical database schema on each city's server, there is no need to define views for our project.

Database Prototypes

Prototyping is not an alternative to carefully thought-out database schemas. On the other hand, once the schema is completed, a prototype database can usually be generated very quickly. Most modern DBMSs include powerful, menu-driven database generators that automatically create a DDL and generate a prototype database from that DDL. A database can then be loaded with test data that will prove useful for prototyping and testing outputs, inputs, screens, and other systems components.

Database Capacity Planning

A database is stored on disk. Ultimately, the database administrator will want an estimate of disk capacity for the new database to ensure that sufficient disk space is available. Database capacity planning can be calculated with simple arithmetic as follows. This simple formula ignores factors such as packing, coding, and compression, but by leaving out those possibilities, you are adding slack capacity.

1. For each table, sum the *field* sizes. This is the *record size* for the table. Avoid the implications of compression, coding, and packing—in other words, assume that each stored character and digit will consume one byte of storage. Note that formatting characters (e.g., commas, hyphens, slashes, etc.) are almost never stored in a database. Those formatting characters are added by the application programs that will access the database and present the output to the users.
2. For each table, multiply the *record size* times the number of entity instances to be included in the table. It is recommended that growth be considered over a reasonable time period (e.g., three years). This is the *table size*.
3. Sum the *table sizes*. This is the *database size*.
4. Optionally, add a slack capacity buffer (e.g., 10 percent) to account for unanticipated factors or inaccurate estimates above. This is the *anticipated database capacity*.

Database Structure Generation

CASE tools are frequently capable of generating SQL code for the database directly from a CASE-based database schema. This code can be exported to the DBMS for compilation. Even a small database such as the SoundStage model can require 50 pages or more of SQL data definition language code to create the tables, indexes, keys, fields, and triggers. Clearly, a CASE tool's ability to automatically generate syntactically correct code is an enormous productivity advantage. Furthermore, it almost always proves easier to modify the database schema and regenerate the code than to maintain the code directly. Figure 12.11 is a two-page sample of code generated by *System Architect 2001* from the SoundStage database schema. The SoundStage example actually generated more than 30 pages of SQL code to

```
/* SQL Product = SQL Server V7 */

CREATE DATABASE dbMemberServices

CREATE TABLE tblAgreement(
        colAgreementNo                  integer NOT NULL,
        colExpireDate                   smalldatetime NULL,
        colActiveDate                   smalldatetime NOT NULL,
        colFulfillmentPeriod            tinyint NOT NULL,
        colCreditsRequired              tinyint NOT NULL)
go


ALTER TABLE tblAgreement ADD CONSTRAINT AGREEMENT_PK
    PRIMARY KEY  (colAgreementNo)
go

CREATE TABLE tblAudioTitle(
        colArtist                       character varying(40) NULL,
        colCategory                     character(2) NOT NULL,
        colSubCategory                  character(2) NOT NULL,
        colUnits                        tinyint NOT NULL                DEFAULT 1,
        colMediaType                    character(2) NOT NULL,
        colAdvisoryCode                 character(1) NULL,
        colProductNo                    integer NOT NULL)
go

sp_bindrule domAudioCategory, 'tblAudioTitle.colCategory'
go


sp_bindrule domAudioCategory, 'tblAudioTitle.colSubCategory'
go


sp_bindefault defOne, 'tblAudioTitle.colUnits'
go


sp_bindrule domPositive, 'tblAudioTitle.colUnits'
go


sp_bindrule domAudioMedia, 'tblAudioTitle.colMediaType'
go


sp_bindrule domContentAdvisory, 'tblAudioTitle.colAdvisoryCode'
go


ALTER TABLE tblAudioTitle ADD CONSTRAINT AUDIO_TITLE_PK
    PRIMARY KEY  (colProductNo)
go

CREATE TABLE tblGameTitle(
        colManufacturer                 character varying(30) NOT NULL,
        colCategory                     character(2) NOT NULL,
        colSubCategory                  binary(2) NULL,
        colPlatform                     character(2) NOT NULL,
        colMedia                        character(2) NOT NULL,
        colPlayers                      tinyint NOT NULL,
        colAdvisoryCode                 character(3) NULL,
```

FIGURE 12.11   *Partial SQL Code to Construct the SoundStage Database*

```
        colProductNo                        integer NOT NULL)
go

sp_bindrule domGameCategories, 'tblGameTitle.colCategory'
go

sp_bindrule domGameCategories, 'tblGameTitle.colSubCategory'
go

sp_bindrule domGamePlatforms, 'tblGameTitle.colPlatform'
go

sp_bindrule domGameMediaTypes, 'tblGameTitle.colMedia'
go

sp_bindefault defOne, 'tblGameTitle.colPlayers'
go

ALTER TABLE tblGameTitle ADD CONSTRAINT GAME_TITLE_PK
   PRIMARY KEY  (colProductNo)
go

CREATE TABLE tblMember(
        colMemberNo                         integer NOT NULL,
        Last_Name                           character varying((20)) NOT NULL,
        First_Name                          character varying((19)) NOT NULL,
        Middle_Initial                      character((1)) NOT NULL,
        colStatus                           character(1) NOT NULL,
        colStreetAddress                    character varying(40) NOT NULL,
        colPostBox                          character(4) NULL,
        colCity                             character varying(30) NOT NULL,
        colState                            character(2) NOT NULL,
        colZipCode                          character(9) NOT NULL,
        Area_Code                           character((3)) NOT NULL,
        Phone_Number                        character((7)) NOT NULL,
        Extension                           character((5)) NOT NULL,
        colLastOrderDate                    smalldatetime NULL,
        colBalance                          character(6) NOT NULL        DEFAULT 0.00,
        colCreditCardType                   character(1) NULL,
        colCreditCardNo                     character(16) NULL           UNIQUE,
        colCreditCardExp                    smalldatetime NULL,
        colBonusBalance                     character(3) NULL            DEFAULT 0,
        colAudioPref                        character(2) NOT NULL,
        colAudioMedia                       character(2) NOT NULL,
        colDateEnrolled                     smalldatetime NOT NULL,
        colExpirationDate                   smalldatetime NULL,
        colGamePreference                   character(2) NOT NULL,
        colGameMedia                        character(2) NOT NULL,
        colNoCreditsEarned                  tinyint NOT NULL,
        colVideoPreference                  character(2) NOT NULL,
        colVideoMedia                       character(2) NOT NULL,
        colAgreementNo                      integer NOT NULL,
        colPrivacyCode                      character(2) NULL,
        colEmailAddress                     character(40) NULL)
go
```

FIGURE 12.11 *Concluded*

create the database in Figure 12.9. Can you imagine the effort required to hand-code that much SQL (with reasonable accuracy)? Clearly, CASE tool generation of SQL code can be very productive—however, the code generated is only as good and complete as the data model.

THE NEXT GENERATION OF DATABASE DESIGN

Relational database technology is widely deployed and used in contemporary information systems shops. The skills taught in this chapter will remain viable well into the foreseeable future. But one new technology is slowly emerging that could ultimately change the landscape dramatically—*object* database management systems.

Object database management systems store true objects—that is, encapsulated data and all of the processes that can act on that data. Because relational database management systems are so widely used, we don't expect this change to happen quickly. Furthermore, the relational DBMS vendors are not likely to give up their market share without a fight. It is expected that these vendors will either build object technology into their existing relational DBMSs or they will create new, object DBMSs and provide for the transition between relational and object models. Regardless, this is one technology to keep an eye on.

WHERE DO YOU GO FROM HERE?

Let's begin with the obvious! If you have information systems career aspirations, you had better plan to take one or more true database courses. The topics presented in this chapter represent only the tip of the iceberg as it relates to database technology, development, and management. Most IS curricula include at least one good database or data management course to add value to your education. Take it!

You have only begun your journey through system design. The database is the *brain* of a new system or application. The subsequent chapters focus on the design of other crucial body parts. Chapters 13 through 15 teach input, output, and user interface design, respectively. Think of inputs, outputs, and interfaces as the *soul* of the system.

SUMMARY

1. The data captured by an information system is stored in files and databases. A file is a collection of similar records. A database is a collection of interrelated files.

2. Many legacy systems were built with file technology. Because files were built for specific applications, their design was optimized for those applications. This close relationship between the files and their applications made it difficult to restructure the files to meet future requirements. And because many applications use the same data, it is not uncommon to find redundant files with data values that do not always match.

3. As the above legacy systems are slowly reengineered, they are usually converted to database technology. Well-designed databases share nonredundant data and overcome all the limitations of conventional files.

4. Database design is the process of translating logical data models (Chapter 7) into physical database schemas.

5. The smallest unit of meaningful data that can be stored is called a field. There are four types of fields.

   a. A primary key is a field that uniquely identifies one and only one record in a file or table.

   b. A secondary key is a field that may either uniquely identify one and only one record in a file or table or identify a set of records with some common, meaningful characteristic.

   c. A foreign key is a field that points to a related record in a different table.

   d. All other fields are called descriptive fields.

6. Fields are organized into records, and similar records are organized into files or tables.

7. A database is a collection of tables (files) with logical pointers that relate records in one table to records in a different table.

8. The data architecture that has evolved in most organizations includes conventional files, operational databases, data warehouses, and personal and work group databases. To coordinate this complex infrastructure, many organizations assign a data administrator to plan and

manage the overall data resource and database administrators to implement and manage specific databases and database technologies.

9. A database architecture is built around a database management system (DBMS) that provides the technology to define the database structure and then to create, read, update, and delete records in the tables that make up that structure. A DBMS provides a data language to accomplish this. That language provides at least two components:

   a. A data definition language to create and maintain the database structure and rules.

   b. A data manipulation language to create, read, use, update, and delete records in the database.

10. Today, relational database management systems are used to support the development and reengineering of the overwhelming number of information systems. Relational databases store data in a collection of tables that are related via foreign keys.

   a. The data definition and manipulation languages of most relational DBMSs are consolidated into a standard language known as SQL.

   b. High-end relational database management systems support triggers and stored procedures, programs that are stored with the tables and callable from other SQL-based programs.

11. Data analysis and normalization are techniques for removing impurities from a data model as a preface to designing the database. These impurities can make a database unreliable, inflexible, and nonscalable.

12. Normalization involves checking each entity (table) for first, second, and third normal form impurities.

   a. An entity is in first normal form if it contains no repeating attributes (that is, attributes that can have multiple values for a single instance of the entity).

   b. An entity is in second normal form if it contains no partial dependencies (that is, a nonkey attribute whose value is dependent only on part of the entity's primary key).

   c. An entity is in third normal form if it contains no derived attributes (that is, calculated or logic-based attributes) or no transitive dependencies (that is, a nonkey attribute whose value is dependent on another nonkey attribute).

13. Distribution and replication decisions should be made before database design. Each unique database should be represented by its own logical data submodel.

14. A database schema is the physical model for a database based on the chosen database technology. The rules for transforming a logical data model into a physical database schema are generalized as follows:

   a. Each entity becomes a table.

   b. Each attribute becomes a field (column in the table).

   c. Each primary and secondary key becomes an index into the table.

   d. Each foreign key implements a possible relationship between instances of the table.

15. Database integrity should be checked and, if necessary, improved to ensure that the business and its users can trust the stored data.

   a. Key integrity ensures that every record will have a unique, non-NULL primary key value.

   b. Domain integrity ensures that appropriate fields will store only legitimate values from the set of all possible values.

   c. Referential integrity ensures that no foreign key value points to a nonexistent primary key value. A deletion rule should be specified for every relationship with another table. The deletion rules either cascade the deletion to related records in other tables, disallow the deletion until related records in other tables are first deleted, or allow the deletion but set any foreign keys in related tables to NULL.

16. SQL-DDL is written or generated to create the database.

KEY TERMS

REVIEW QUESTIONS

1. Differentiate between conventional files and databases.
2. What is a database? What is the difference between an operational database and a personal database?
3. Explain the advantages and disadvantages of conventional files versus databases.
4. Define the terms *field, record,* and *file.*
5. Differentiate between primary, secondary, and foreign keys. What important role do foreign keys serve?
6. Identify six types of files, and give several examples of each.
7. Differentiate between fixed- and variable-length records. What impact does a record storage format have on a file design?
8. Differentiate between file access and file organization.
9. Differentiate between an operational database and a data warehouse. What types of applications does each serve?
10. Differentiate between a data and a database administrator. What is the relationship between these positions and the systems analyst?
11. Briefly explain the differences between a data definition language, a host programming language, and a data manipulation language.
12. List and briefly describe the three table operations used to manipulate relational tables.
13. Differentiate between triggers and stored procedures.
14. Give three characteristics of a good data model.
15. List and briefly describe the three steps of normalization.
16. Describe four alternatives for distributed database design.

PROBLEMS AND EXERCISES

1. Eudrup University's current student registration system consists of the following ISAM files: STUDENTS, COURSES, INSTRUCTORS, REGISTRATIONS, SPACE, and SCHEDULES. The first three files are fixed-length master files. The last three are variable-length transaction files. The administration would like to integrate all this data into a single database. Write a step-by-step project plan to accomplish this. (Hint: You may need to review Chapter 7.)
2. Kevin, an inventory manager, is considering a DBMS for his microcomputer. He's not certain that he really understands what a database is. In college, he took an introductory computer course and learned about files. He assumed database is the current buzzword for a collection of files. Write him a memo explaining the difference between a file and database environment. What are the advantages and disadvantages of each environment?
3. Draw a data model for your school's course registration system and then convert the data model into a relational database schema.
4. If databases were created with the ability to solve many of the problems characteristic of conventional file-based systems, why aren't all information systems shops operating in a database environment?
5. Transform the following entities into 3NF entities. Draw the original and final data models, and state any reasonable assumptions. (Note: Primary keys are underlined.)
   AIRCRAFT (AIRCRAFT ID NUMBER, AIRCRAFT CODE, AIRCRAFT DESCRIPTION, NUMBER OF SEATS).
   FLIGHT (FLIGHT NUMBER, DEPARTURE CITY, 1 { ARRIVAL CITY } n, MEAL CODE, 1 { FLIGHT DATE, CURRENT SEAT PRICE } m).
   PASSENGER (PASSENGER NUMBER, PASSENGER NAME, FREQUENT FLYER NUMBER, 1 { FLIGHT NUMBER, SEAT NUMBER, QUOTED SEAT PRICE, AMOUNT PAID, BALANCE DUE }).
6. How are relationships appearing on an entity relationship diagram implemented by a relational database management system?
7. How would you implement a recursive relationship in a relational database system?
8. Explain the role of data analysis in database design. Why not just go straight to database design?
9. What is the difference between a table in first normal form (1NF) and second normal form (2NF)? Give an example of an entity in 1NF and show its conversion to 2NF.
10. What is the difference between a data entity in second normal form (2NF) and third normal form (3NF)? Give an example of an entity in 2NF and show its conversion to 3NF.

PROJECTS AND RESEARCH

1. Visit a local information systems shop that uses a DBMS. Describe the existing database environment. Does it have production-oriented databases or end-user databases? What host programming language(s) is utilized to load, maintain, and use the data? Ask the systems analyst or database administrator to give you a brief orientation on the physical and logical structures supported by the DBMS. To what extent are conventional files used?
2. Visit a local information systems shop that operates in a strictly conventional file environment. Ask the systems analyst for information describing several of the master and transaction files. Do some of the files contain duplicate data? Is this data input several times? What impact has the duplicated data had on maintenance? Do they experience problems with data integrity? Have the analyst explain the impact of changing the format of one of the files.

MINICASES

1. Sunset Valley Distributors recently completed a major conversion project. Several months ago, Sunset decided to move into the database era. Many of its computer-based files had become unreliable, difficult to maintain, and too inflexible to be used to fulfill many end-user reporting and inquiry requests. A DBMS seemed to be the obvious solution. Two systems analysts were primarily responsible for the conversion project, which took several months to complete. The systems analysts had decided to simply implement each of the computer-based files as a separate table in their relational database. Once the conversion was completed, the same problems that existed with the file-based system reappeared in the database system. Reports contained inaccurate data, report and inquiry requests could not easily be obtained, and data maintenance was still difficult. A consultant was hired to investigate the problems. The consultant acknowledged that many of the problems resulted because the analysts failed to do data modeling. Explain the importance of doing data modeling ahead of time when designing databases.

2. Design the logical schema for a relational database using the entity relationship diagram that follows. The primary keys of the entities are as follows:

Data Structure			Field Size
EMPLOYEE	=	SOCIAL SECURITY NUMBER (PK)	9
	+	EMPLOYEE NAME	32
	+	EMPLOYEE STREET ADDRESS	32
	+	EMPLOYEE CITY	12
	+	EMPLOYEE STATE	2
	+	EMPLOYEE ZIP CODE	9
	+	EMPLOYEE HOME PHONE NUMBER	10
	+	EMPLOYEE EMAIL ADDRESS	15
	+	1 { DEPARTMENT CODE +	2
		OFFICE LOCATION +	3
		OFFICE PHONE NUMBER } 4	5
	+	DATE EMPLOYED	8
	+	DATE OF BIRTH	8
	+	(SPOUSE NAME +	
		SPOUSE DATE OF BIRTH)	20 + 8
	+	0 { DEPENDENT NAME +	20
		DEPENDENT RELATIONSHIP +	1
		DEPENDENT DATE OF BIRTH }n	8
	+	[MONTHLY SALARY |	5 or 3.2
		HOURLY PAY RATE]	
	+	VACATION DAYS DUE	2
	+	SICK DAYS DUE	2
	+	GROSS PAY YEAR-TO-DATE	6.2
	+	FEDERAL TAX WITHHELD YEAR-TO-DATE	5.2
	+	STATE TAX WITHHELD YEAR-TO-DATE	4.2
	+	FICA TAX WITHHELD YEAR-TO-DATE	5.2

x.y indicates number of digits to the left and right of the decimal point, respectively

3. Design the 3NF logical schema for a relational database using the entity relationship diagram that follows:

Data Structure			Field Size
SUPPLIER	=	SUPPLIER IDENTIFICATION NUMBER	12
	+	SUPPLIER NAME	30
	+	SUPPLIER STREET ADDRESS	30
	+	SUPPLIER CITY	15
	+	SUPPLIER STATE	2
	+	SUPPLIER PHONE NUMBER	10
	+	1 { PAYMENT METHOD +	1
		EARLY DISCOUNT PERIOD +	2
		EARLY DISCOUNT RATE +	0.2
		PAYMENT DEADLINE } 3	2
	+	1 { MATERIAL NUMBER +	9
		MATERIAL DESCRIPTION +	30
		UNIT PRICE +	5.2
		QUANTITY DISCOUNT THRESHOLD +	3
		QUANTITY DISCOUNT PERCENTAGE } n	0.2

x.y indicates number of digits to the left and right of the decimal point, respectively

4. Given the sample form that appears at the top of page 499, prepare a list of entities and their associated data attributes as determined from the document. Then, completely normalize the entities to 3NF and draw a hypothetical ERD. Your instructor should be the final interpreter for the form.

5. Design the relational database schema for the following entity relationship diagram at the bottom of page 499.

6. Precious Jewels Diamond Centers is a franchised jewelry store that specializes in diamonds and other gems, custom selected by and for customers. Gems are custom set into rings, pendants, and other pieces. Precious Jewels also serves as a diamond broker, providing gems to other franchises and jewelry stores. These gems are sent out on approval. The stores have the option of purchasing the gems or returning them.

Jeff Kassels, vice president, is looking for a consultant to improve information systems for its IBM and Compaq microcomputers. About two years ago, executives decided to purchase two microcomputers. On the recommendation of the computer superstore, they also bought Microsoft *Excel* (a spreadsheet), Borland *Paradox* (a PC-DBMS), and Lotus *Word Pro* (a word processor). Initially and unfortunately, they didn't invest in the training to exploit these packages, especially the *Paradox* database package.

Eventually, they hired some young students who were into PCs. They wrote some *Paradox* programs and *Excel* macros for inventory control and sales. The programs seemed to work. Precious Jewels entered lots of data into the system, and it generated several reports.

Later, Precious Jewels staff realized the need for new reports and inquiries. They tried to generate them themselves, but they just didn't understand the report writer in *Paradox*. The original students were unavailable, so they hired a woman who does *Paradox* programming on the side. She couldn't seem to generate the reports from the stored data—even though the data is in there.

For Minicase 4

| PURCHASING REQUISITION Form 12 Rev.1988 | INSTRUCTIONS — INCLUDE IN EACH REQUISITION ONLY SUCH ARTICLES AS MAY BE PURCHASED FROM ONE FIRM. IF SPECIAL HANDLING IS DESIRED, NOTE. SEE REVERSE SIDE FOR SPECIAL COMMENTS BY REQUESTOR. |

| DEPARTMENT COMPLETES UNSHADED AREA | | PURCHASING COMPLETES SHADED AREA | ORDER NO. |

DEPT. OR FUNCTION: Computer Information Systems

COMMITMENT NO.

COMMODITY CODE | ORDER TYPE

M F C	RES CODE.	ACCOUNT NUMBER				DEPT. REFERENCE	AMOUNT	FUND EXPIRATION DATE	SHIP TO STAFF MEMBER	ORDER DATE
		FUND	CENTER DEPT.— PROJ.	OBJECT					Jonathan Doe	
1				5-6207			8,736.00		DEPT. 242	FOLLOW UP
2				5-6106			399.00		BUILDING & ROOM	PRICING METHOD
3				5-6107			84.00		Administration	□ RQ #

REQUISITIONER'S PHONE NO.: 555-4545

MATERIAL WILL BE USED FOR

VENDOR SUGGESTED: IBM, Main Street, Somewhere, IN 47906

VENDOR NAME

VENDOR NUMBER

PRICING METHOD:
□ RQ #
□ 1 Phone/Verbal Quote
□ 2 Agreement/Contract
□ 3 Price List on File
□ 4 Repair Negotiation
□ 5 None of the above
BUYER

FOB: □ 1 DESTINATION □ 2 DESTINATION PREPAY & ADD □ 3 SHIPPING POINT □ 4 SHIPPING POINT FREIGHT ALLOWED □ 5 SEE BELOW   VIA   TERMS

ITEM #	ITEM DESCRIPTION	MFC	QUANTITY	UNIT	UNIT PRICE	EXTENDED PRICE	DELIVER ON	EST.	COMM.
	IBM PS/2 Model 70 86 8570-121	1	1		7,995.00	4,797.00			
	IBM PS/2 2-8 MB Memory Module Expansion Option #5211	1	1		1,695.00	1,017.00			
	IBM PS/2 2MB Memory Module Kit #5213	1	3		1,395.00	2,511.00			
	IBM 8513 PS/2 Color Display	1	1		685.00	411.00			
	IBM 8770 PS/2 Mouse	2	1		95.00	57.00			
	IBM PC Network Adapter II/A #150122	2	1		570.00	342.00			
	IBM DOS 3.3	3	1		120.00	84.00			

REQUESTED — HEAD OF DEPT.: *Thomas J. Mathien*   DATE: 6-3-99

APPROVED — FOR THE COMPTROLLER   DATE

BYPASS APPROVAL REQUESTED □
APPROVAL SIGNATURE/DATE

PURCHASING APPROVALS: PA | AD | DIR

RECOMMENDED — DEAN OR ADMINISTRATOR   DATE

APPROVED — FOR THE EXECUTIVE VICE PRESIDENT AND TREASURER   DATE

OCGBA PREAUDIT
BY: DATE:

For Minicase 5

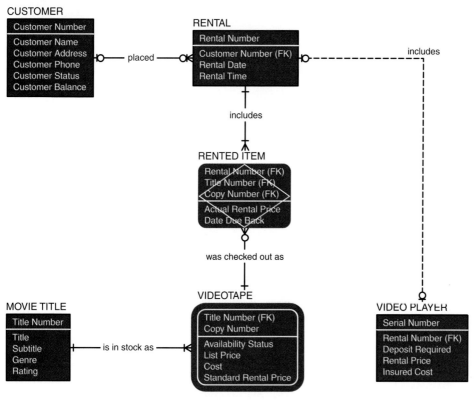

To compound matters, there are growing problems with the data already in the system. As Precious Jewels staff regenerated the original reports, they noticed that records existed that should have been deleted a long time ago. And to make matters worse, they found records of gems for which there was no associated purchase order. This caused insurance problems.

When they asked their new consultant to add some new fields to existing programs, she informed Precious Jewels management that many of the existing programs would have to be rewritten because fields would need to be moved to different or new files for efficiency.

a. What went wrong and why?

b. What benefits do you think can be derived from studying data before you study output needs and processing requirements?

c. Why do you think that consultants—and experienced analysts—so frequently ignore or do not adequately consider the future implications of the databases they design?

SUGGESTED READINGS

Bruce, Thomas. *Designing Quality Databases with IDEF1X Information Models.* New York: Dorset House Publishing, 1992. This has rapidly become our favorite practical database design book. Incidentally, the foreword was written by John Zachman whose *Framework for Information Systems Architecture* inspired our own information system building blocks framework.

McFadden, Fred; Jeffrey Hoffer; and Mary Prescott. *Modern Database Management.* 5th ed. Reading, MA: Addison-Wesley, 1994. For those seeking to expand their overall data management and database education, this is one of the most popular introductory textbooks on the market and our own favorite. These authors do a particularly thorough job of explaining distributed database design (in much greater detail than is possible in our book).

Teorey, Toby. *Database Modeling & Design: The Fundamental Principles.* 2nd ed. San Francisco: Morgan Kaufman Publishers, Inc., 1990. This is our favorite database design conceptual book. Appendix A provides a concise review of the SQL language.

**Focus on
PEOPLE**

**Focus on
DATA**

**Focus on
PROCESSES**

**Focus on
INTERFACES**

**Focus on
DEVELOPMENT**

Stakeholders

Activities

S
Y
S
T
E
M
S

A
N
A
L
Y
S
T
S

SYSTEM
OWNERS

SYSTEM
USERS

SYSTEM
DESIGNERS

SYSTEM
BUILDERS

PROJECT &
PROCESS
MANAGEMENT

PRELIMINARY
INVESTIGATION

PROBLEM
ANALYSIS

REQUIREMENTS
ANALYSIS

DECISION
ANALYSIS

DESIGN

CONSTRUCTION

IMPLEMENTATION

BUILDING BLOCKS OF AN INFORMATION SYSTEM

Management Expectations

The PIECES Framework

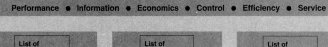

Performance • Information • Economics • Control • Efficiency • Service

List of business entities and rules ...	List of business functions and events ...	List of business locations and systems...
Business Knowledge	Business Functions	Business Locations

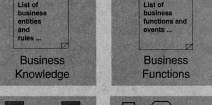

Data Requirements	Process Requirements	Interface Requirements

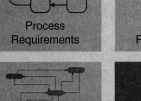  

Database Schema	Application Schema & Specs	Interface Specifications

```
CREATE TABLE tblOrders
colOrderNot CHAR(5) NOT
NULL
colOrderDate DATE/TIME NOT
```

Database
Programs

```
PROC ValidateOrder
   PERFORM ValidateCust
   REPEAT UNTIL
      NoMoreProd ...
```

Application
Programs

```
<html>
<head>
<title> Order Entry Form </title>
...
```

Interface
Programs

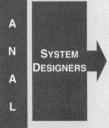

VENDORS
AND
CONSULTANTS

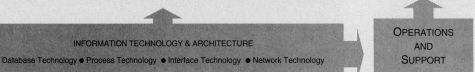

INFORMATION TECHNOLOGY & ARCHITECTURE

Database Technology ● Process Technology ● Interface Technology ● Network Technology

OPERATIONS
AND
SUPPORT

13

OUTPUT DESIGN AND PROTOTYPING

CHAPTER PREVIEW AND OBJECTIVES

In this chapter you will learn how to design and prototype computer outputs. You will know how to design and prototype outputs when you can:

— Distinguish between internal, external, and turnaround outputs.

— Differentiate between detailed, summary, and exception reports.

— Identify several output implementation methods.

— Differentiate among tabular, zoned, and graphic formats for presenting information.

— Distinguish among area, bar, column, pie, line, radar, donut, and scatter charts and their uses.

— Describe several general principles that are important to output design.

— Design and prototype computer outputs.

Output and input design represent something of a "chicken or egg" sequencing problem. Which do you do first? In this edition, we present output design first. Classic system design prefers this approach as something of a system validation test—design outputs and then make sure the inputs are sufficient to produce the outputs. In practice, this sequencing of tasks becomes less important because modern systems analysis techniques sufficiently predefine logical input and output requirements. You and your instructor may safely swap Chapters 13 and 14 if you prefer.

Note: The authors gratefully acknowledge the contributions and recommendations of Carlin "Rick" Smith at Purdue University. Rick developed most of the prototypes that demonstrate the concepts and techniques taught in this chapter. Rick also contributed ideas for further improving chapter currency and topics.

SCENE

This episode begins in the conference room where Sandra and Bob scheduled a Tuesday morning meeting to review output designs with the Order Processing staff. Sally Hoover required her entire staff to be present to become familiar with the new outputs that would be generated by the new system. Dick Krieger, warehouse manager, has also been invited to review an output that Order Processing will send to his area.

SANDRA

OK, let's call this meeting to order. I think we should go ahead and get started so each of you can get back to normal jobs as quickly as possible. First, Bob and I would like to thank you for coming this morning. Today, we want to review some of the system outputs. Let's begin with a picking ticket form that is sent to the Warehouse. We'll discuss that output first so that Dick can then be excused while we discuss other outputs that are specific to the Order Processing staff.

[Sandra adjusts the computer project pad so that the screen image is in clear focus on the pull-down projection screen.]

Based on the requirements statement we created earlier, I've developed a prototype of what the form might look like. If it's OK, I'd like to confirm a few facts before we review this output.

[Hearing no objections, Sandra continues.]

These forms will be printed each day, but I need to know how many could be printed in a single day.

SALLY

Currently, we never process more than 4,000 a day.

ANN

[A member of the Order Processing staff.]

But what about growth? We expect to continue adding 100 new members a month. Maybe we had better plan on 5,000 orders a day.

[Sandra makes some notations on the form.]

SALLY

Sandra, what's the form you're writing on?

SANDRA

This is a list of output specifications I'm making for the information systems people. It helps them to anticipate the impact the system outputs will have on their facilities. It will also help us to choose the best printer to meet your needs.

SALLY

OK, I can't wait to comment. This looks nothing like our old picking ticket form. It looks like a sheet of mailing labels with bar codes.

SANDRA

That's not too far from the truth. This form will go to Dick's people in the warehouse. Notice that there will be a peel-off bar code for each product on the picking ticket (customer order) and a bar code with the customer and order information. Is this what you had in mind, Dick?

DICK

Absolutely! Our warehouse personnel can run the scanner over the product bar code and it will instantaneously tell them if the item is in stock and precisely where they should go in the warehouse to pick it up. As I understand it from our previous conversation, when they actually retrieve the product they will scan the label again to reflect the fact that the product has been picked, and then remove and attach the bar code sticker. I assume the name and order label would also be scanned and attached to the packaged order.

SANDRA

That's correct. Also, if the product is not available, an automatic backorder would be generated. And the partially completed picking ticket would be processed again at a later time. That's why we needed two separate labels, each containing the customer and order information. We may need one to go

on the partial order shipment, and the other for subsequent backorder processing.

ANN

What do you mean by automatic?

SANDRA

Well, as the information is scanned, necessary data will be fed into the computer regarding the status of the order, what products were picked, what products are backordered, and other information that the Warehouse has to write in on the picking ticket forms that they currently receive and send back to you folks. This way, you can simply track the status of customer orders by calling them up on the computer. No paper exchange is necessary. Thus, the form serves both to communicate information as well as to subsequently collect information.

SALLY

That sounds great! Between us and the Warehouse, we've lost a fair number of those completed picking tickets in the past.

SANDRA

Dick, what do you think?

DICK

I think this will work. I'm eager to try it.

SANDRA

I guess if this is OK by you, you're free to leave. I know you have to get back to other duties. Thanks for coming.

DICK

Thanks for the invitation. I must say that this should not only streamline my operations, but I think it will do a lot to help the two different groups communicate much more effectively.

[Dick leaves]

SANDRA

Now that Dick is gone, let's take a look at some outputs that the system will generate for you folks. Bob came up with prototypes of a number of printed reports that we understand you will find useful in running your operations.

SOUNDSTAGE

S O U N D S T A G E   E N T E R T A I N M E N T   C L U B

BOB

Don't forget the screens. I also prototyped several screens that are intended to provide information that you will likely request on an ad hoc basis.

SALLY

Before we get started, did you ever receive the memo or complaint list that I sent you regarding the reports that we are currently receiving? I don't want to experience the same problems with the new system. I'm tired of receiving reports when they are too outdated to be of use. I want to receive them on time. I want them to be accurate. And I want them to be relevant. I can't have myself and my people making decisions with poor, or absent, information.

SANDRA

Yes, I did receive it. I sent you a copy, did I not, Bob?

BOB

I got it. That's part of what this meeting is about today. As I show you the various reports and screens, give me feedback. I want to ensure the content and style are acceptable. I also need you to indicate who will receive the reports and when. Realize that some of the screens and reports are new. Some you requested, some Sandra and I thought of. So be sure to tell us whether or not they would be of value. We have no desire to overwhelm you with unnecessary information.

SALLY

Great. Let's see those screens and reports.

DISCUSSION QUESTIONS

1. Sandra and Bob designed the picking ticket output such that it could also be used to aid in data collection. What would be the advantages of this solution?

2. Why was Sandra concerned with knowing the volume of picking tickets that would be generated?

3. Why would Sandra and Bob want Warehouse and Order Processing representatives present to talk about the picking ticket output?

4. Do you think the kinds of concerns expressed by Sally regarding the current outputs they receive are common?

Outputs present information to system users. Outputs are the most visible component of a working information system. As such, they are often the basis for the users' and management's final assessment of the system's value. During requirements analysis, you defined *logical* output requirements. During decision analysis, you *may* have considered different alternative physical implementation alternatives. In this chapter, you will learn how to physically design the outputs.

Today, most outputs are designed by rapidly constructing prototypes. These prototypes may be simple computer-generated mockups with dummy data, or they may be generated from prototype databases such Microsoft *Access,* which can be rapidly constructed and populated with test data. These prototypes are rarely fully functional. They won't contain security features or optimized data access that will be necessary in the final version of a system. Furthermore, in the interest of productivity, we may not include every button or control feature that would have to be included in a production system.

During requirements analysis, outputs were modeled as data flows that consist of data attributes. Even in the most thorough of requirements analysis, we will miss requirements. Output design may introduce new attributes or fields to the system.

We begin with a discussion of types of outputs. Outputs can be classified according to two characteristics: (1) their distribution and audience and (2) their implementation method. Figure 13.1 illustrates this taxonomy. The characteristics are discussed briefly in the following sections.

One way to classify outputs is according to their distribution inside or outside the organization and the people who read and use them.

Internal outputs are intended for the internal system owners and system users within an organization. They only rarely find their way outside the organization.

OUTPUT DESIGN CONCEPTS AND GUIDELINES

Distribution and Audience of Outputs

Distribution / Delivery	Internal Output (reporting)	Turnaround Output (external; then internal)	External Output (transactions)
Printer	Detailed, summary, or exception information printed on hard-copy reports for internal business use. Common examples: management reports	Business transactions printed on business forms that will eventually be returned as input business transactions. Common examples: phone bills and credit card bills	Business transactions printed on business forms that conclude the business transactions. Common examples: paychecks and bank statements
Screen	Detailed, summary, or exception information displayed on monitors for internal business use. Reports may be tabular or graphical. Examples: on-line management reports and responses to inquiries	Business transactions displayed on monitors in forms or windows that will also be used to input other data to initiate a related transaction. Examples: Web-based display of stock prices with the point-and-click purchase option	Business transactions displayed on business forms that conclude the business transactions Examples: Web-based report detailing banking transactions
Point-of-Sale Terminals	Information printed or displayed on special-purpose terminals dedicated to specific internal business functions. Includes wireless communication information transmission. Examples: end-of-shift cash register balancing report	Information printed or displayed on a special-purpose terminal for the purpose of initiating a follow-up business transaction. Examples: Grocery store monitor that allows customer to monitor scanned prices to be followed by input of debit or credit card payment authorization	Information printed or displayed on special-purpose terminals dedicated to customers. Examples: Account balances display at an ATM machine or printout of lottery tickets; also, account information displayed via television over cable or satellite
Multimedia (audio or video)	Information transformed into speech for internal users. Not commonly implemented for internal users	Information transformed into speech for external users who respond with speech or tone input data. Examples: telephone touch-tone class schedule as part of course registration system	Information transformed into speech for external users. Examples: movie trailer for prospective on-line buyers of DVDs or telephone response to mortgage payoff query
E-mail	Displayed messages related to internal business information. Examples: e-mail messages announcing availability of new on-line business report	Displayed messages intended to initiate business transaction. Examples: e-mail messages whose responses are required to continue processing a business transaction	Messages related to business transactions. Examples: e-mail message confirmations of business transactions conducted via e-commerce on the Web
Hyperlinks	Web-based links to internal information that is enabled via HTML or XML formats. Examples: Integration of all information system reports into a Web-based archival system for on-line archival and access	Web-based links incorporated into Web-based input pages to provide users with access to additional information. Examples: On a Web auction page, hyperlinks into a seller's performance history with an invitation to add a new comment	Web-based links incorporated into Web-based transactions Examples: hyperlinks to privacy policy or an explanation as to how to interpret or respond to information in a report or transaction
Microfiche	Archival of internal management reports to microfilm that requires minimal physical storage space. Examples: Computer output on microfilm (COM)	Not applicable unless there is an internal need to archive turnaround documents. Examples: Computer output on microfilm (COM)	Not applicable unless there is an internal need for copies of external reports. Examples: Computer output on microfilm (COM)

FIGURE 13.1   *A Taxonomy for Computer-Generated Outputs*

Internal outputs either support day-to-day business operations or management monitoring and decision making. Figure 13.2 illustrates three basic sub-classes of internal outputs:

- **Detailed reports** present information with little or no filtering or restrictions. The example in Figure 13.2(a) is a listing of all purchase orders that were generated on a particular date. Other examples of detail reports would be a detailed listing of all customer accounts, orders, or products in inventory. Some detailed reports are historical. Other detailed reports are regulatory, that is, required by government.

- **Summary reports** categorize information for managers who do not want to wade through details. The sample report in Figure 13.2(b) summarizes the month's and year's total sales by product type and category. The data for summary reports is typically categorized and summarized to indicate trends and potential problems. The use of graphics (charts and graphs) on summary reports is also rapidly gaining acceptance because it more clearly summarizes trends at a glance.

— **Exception reports** filter data before it is presented to the manager as information. Exception reports only include exceptions to some condition or standard. The example in Figure 13.2(c) depicts the identification of delinquent member accounts. Another classic example of an exception report is a report that identifies items that are low in stock.

The opposite of internal outputs is external outputs.

External outputs leave the organization. They are intended for customers, suppliers, partners, and regulatory agencies. They usually conclude or report on business transactions.

Examples of external outputs are invoices, account statements, paychecks, course schedules, airline tickets, boarding passes, travel itineraries, telephone bills, purchase orders, and mailing labels.

Figure 13.3 illustrates a sample external output for SoundStage Entertainment Club. This sample, like many external outputs, is initially created as a blank, preprinted form that is designed and duplicated by forms manufacturers for use with computer printers.

Some outputs are both external and internal. They begin as external outputs that exit the organization but ultimately return (in part or in whole) as an internal input.

(a) Detailed reports

FIGURE 13.2a
Levels of Report Detail

SoundStage Entertainment Club Detailed

PRODUCTS ORDERED ON 1/25/2000

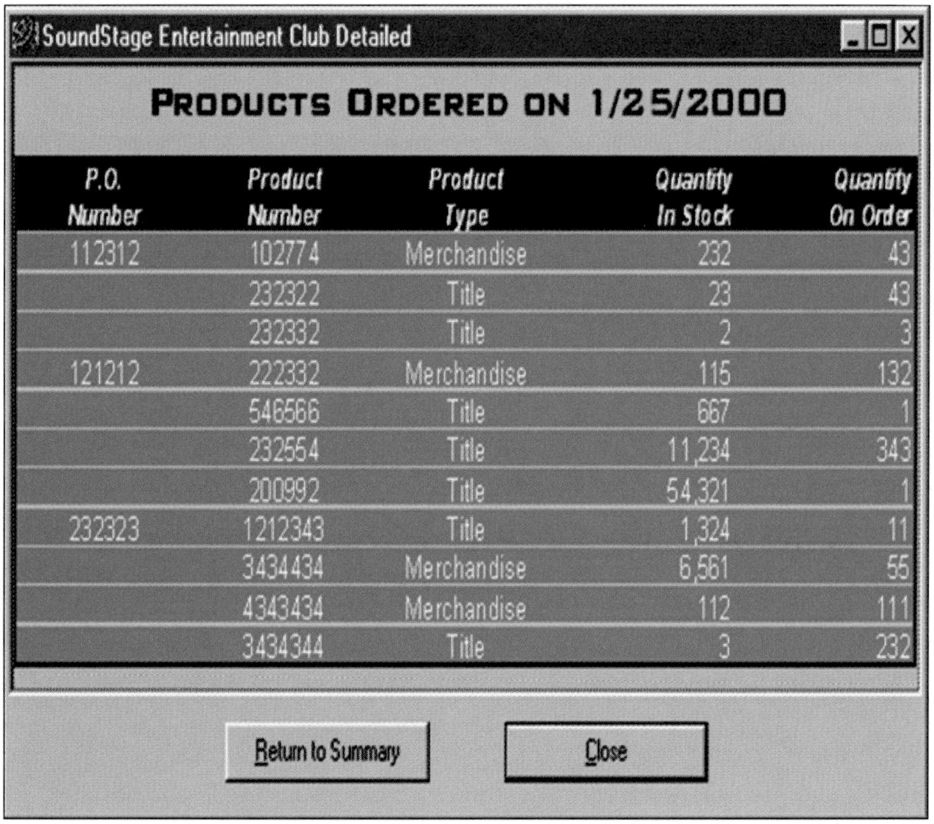

P.O. Number	Product Number	Product Type	Quantity In Stock	Quantity On Order
112312	102774	Merchandise	232	43
	232322	Title	23	43
	232332	Title	2	3
121212	222332	Merchandise	115	132
	546566	Title	667	1
	232554	Title	11,234	343
	200992	Title	54,321	1
232323	1212343	Title	1,324	11
	3434434	Merchandise	6,561	55
	4343434	Merchandise	112	111
	3434344	Title	3	232

Return to Summary   Close

(b) Summary reports

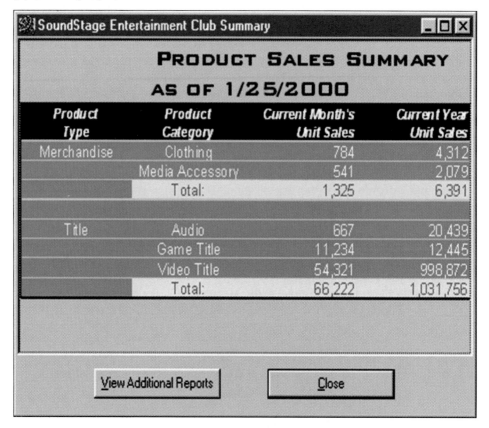

(c) Exception reports

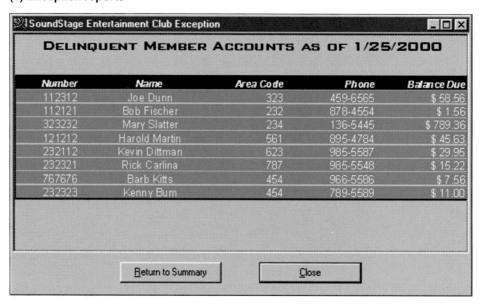

FIGURE 13.2b and 13.2c
Levels of Report Detail

Turnaround outputs are those external outputs that eventually reenter the system as an inputs.

Figure 13.4 demonstrates a turnaround document. Notice that the invoice has upper and lower portions. The top portion is to be detached and returned with the customer payment as an input.

SoundStage Entertainment Club
Fax 317-494-5222

PURCHASE ORDER

The following number must appear on all related correspondence, shipping papers, and invoices:
P.O. NUMBER: 712812

To:
CBS Fox Video Distribution
26253 Rodeo DR
Hollywood, CA

Ship To:
SoundStage Entertainment Club
Shipping/Receiving Station
Building A
2630 Darwin Drive
Indianapolis, IN 45213

P.O. DATE	REQUISITIONER	SHIP VIA	F.O.B. POINT	TERMS
5-3-01	LDB	UPS		Net 30

QTY	DESCRIPTION	UNIT PRICE	TOTAL
20000	Star Wars: The Phantom Menace (VHS)	15.99	319,800.00
3000	Star Wars: The Phantom Menace (DVD Dolby Digital)	19.99	59,970.00
500	Star Wars: The Phantom Menace (DVD DTS)	24.99	12,495.00
8000	Star Wars: The Phantom Menace (PlayStation II)	16.99	135,920.00
400	Star Wars: The Phantom Menace Soundtrack (CD)	16.99	6,796.00
600	Star Wars: The Phanton Menace Theater Poster	4.99	2,994.00

	Subtotal	537,975.00
	Tax	37,658.25
	Total	575,633.25

1. Please send two copies of your invoice.

2. Enter this order in accordance with the prices, terms, delivery method, and specifications listed above.

3. Please notify us immediately if you are unable to ship as specified.

Madge Worthy   5-4-01
Authorized by   Date

FIGURE 13.3
Typical External Document

Implementation Methods for Outputs

We assume you are familiar with different output devices, such as printers, plotters, computer output on microfilm (COM), and PC display monitors. These are standard topics in most introductory information systems courses. In this chapter, we are more concerned with the actual output than with the device. A good systems analyst will consider all available options for implementing an output. Let's briefly examine implementation methods and formats. You should continue to reference Figure 13.1 as we complete this introduction to the output taxonomy.

Printed Output   The most common medium for computer outputs is paper—printed output. Currently, paper is the cheapest medium we will survey. Although the paperless office has been predicted for many years, it has not yet become a reality. Perhaps there is a psychological dependence on paper as a medium. In any case, paper output will be with us for a long time.

Printed output may be produced on impact printers, but increasingly it is printed on laser printers, which have become increasingly cost effective. Internal outputs are typically printed on blank paper (called *stock paper*). External outputs and

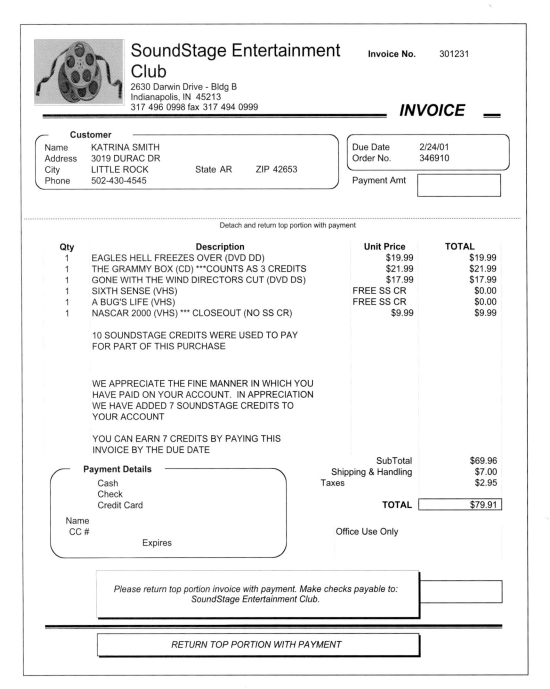

FIGURE 13.4 *Typical Turnaround Document*

turnaround documents are printed on *preprinted forms*. The layout of preprinted forms (such as blank checks and W-2 tax forms) is predetermined and the blank documents are mass-produced. The preprinted forms are run through the printer to add the variable business data (such as *your* paycheck and W-2 tax form).

Perhaps the most common format for printed output is tabular.

Tabular output presents information as columns of text and numbers.

Most of the computer programs you've written probably generated tabular reports. The sample detailed, summary, and exception reports illustrated earlier in the chapter (Figure 13.2) were all tabular.

An alternative to tabular output is zoned output.

Zoned output places text and numbers into designated areas or boxes of a form or screen.

Zoned output is often used in conjunction with tabular output. For example, an order output contains zones for customer and order data in addition to tables (or rows of columns) for ordered products.

Screen Output   The fastest-growing medium for computer outputs is the on-line display of information on a visual display device, such as a CRT terminal or PC monitor. The pace of today's economy requires information on demand. Screen output is most suited to this requirement.

While screen output provides the system user with convenient access to information, the information is only temporary. When the information leaves the screen, that information is lost unless it is redisplayed. For this reason, printed output options are usually added to screen output designs.

Thanks to screen output technology, tabular reports—especially summary reports—can be presented in graphical formats.

Graphic output is the use of a pictorial chart to convey information in ways that demonstrate trends and relationships not easily seen in tabular output.

To the system user, a picture can be more valuable than words. There are numerous types and styles of charts for presenting information. Figure 13.5 summarizes various types of charts that can be output with today's technology. Report writing technology and spreadsheet software can quickly transform tabular data into charts that enable the reader to more quickly draw conclusions.

The popularity of graphic output has also been stimulated by the availability of low-cost, easy-to-use graphics printers and software, especially in the PC industry. Later in this chapter we will show you an alternate graphic design for a Sound-Stage output.

Point-of-Sale Terminals   Many of today's retail and consumer transactions are enabled or enhanced by point-of-sale (POS) terminals. The classic example is the automated teller machine (ATM). POS terminals are both input and output devices. In this chapter, we are interested only in the output dimension. ATMs display account balances and print transaction receipts. POS cash registers display prices and running totals as bar codes are scanned, and they also produce receipts. Lottery POS terminals generate random numbers and print tickets. All are examples of outputs that must be designed.

Multimedia   *Multimedia* is a term coined to collectively describe any information presented in a format other than traditional numbers, codes, and words. This includes graphics, sound, pictures, and animation. It is usually presented as a contemporary extension to screen output. Increasingly, multimedia output is being driven by the transition of information systems applications to the Internet and intranets.

We've already discussed graphical output. But other multimedia formats can be integrated into traditional screen designs. Many information systems offer film and animation as part of the output mix. Product descriptions as well as installation and maintenance instructions can be integrated into on-line catalogs using multimedia tools. Sound bites can also be integrated.

But multimedia output is not dependent on screen display technology. Sound, in the form of telephone Touch-Tone–based systems, can be used to implement an interesting output alternative. Many banks offer their customers Touch-Tone access to a wide variety of account, loan, and transaction data.

E-mail   E-mail has transformed communications in the modern business, if not society as a whole. New information systems are expected to be message-enabled.

	Sample	Selection Criteria
Line Chart		**Line charts** show one or more series of data over a period of time. They are useful for summarizing and showing data at regular intervals. Each line represents one series or category of data.
Area Chart	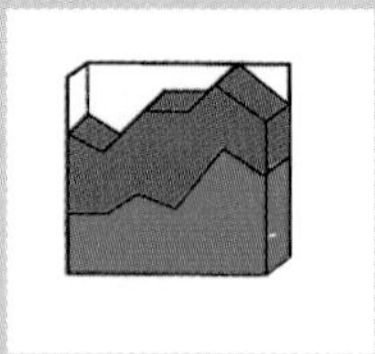	**Area charts** are similar to line charts except that the focus is on the area under the line. That area is useful for summarizing and showing the change in data over time. Each line represents one series or category of data.
Bar Chart	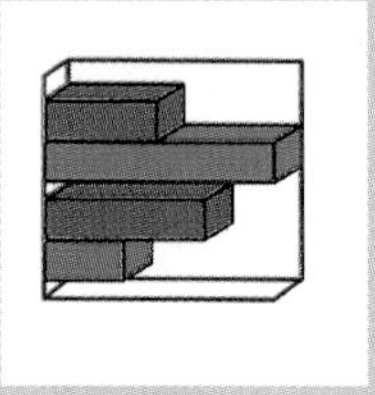	**Bar charts** are useful for comparing series or categories of data. Each bar represents one series or category of data.
Column Chart	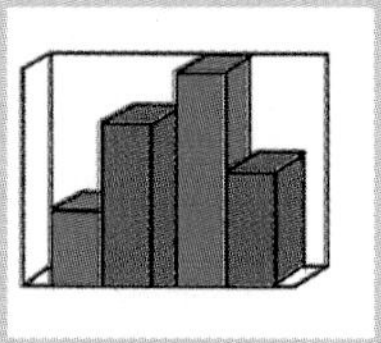	**Column charts** are similar to bar charts except that the bars are vertical. Also, a series of column charts may be used to compare the same categories at different times or time intervals. Each bar represents one series or category of data.
Pie Chart	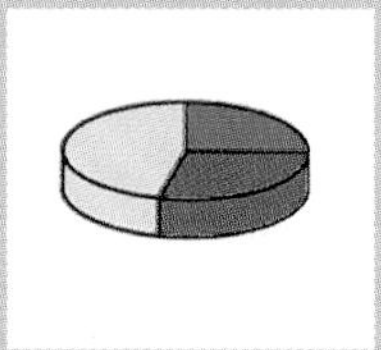	**Pie charts** show the relationship of parts to a whole. They are useful for summarizing percentages of a whole within a single series of data. Each slice represents one item in that series of data.
Donut Chart	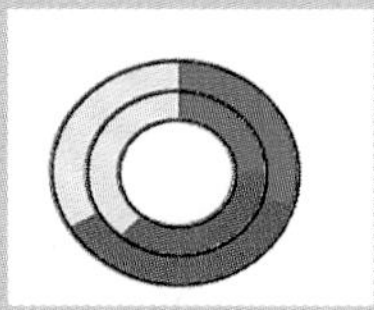	**Donut charts** are similar to pie charts except that they can show multiple series or categories of data, each as its own concentric ring. Within each ring, a slice of that ring represents one item in that series of data.
Radar Chart	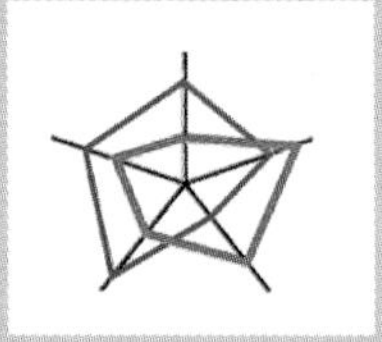	**Radar charts** are useful for comparing different aspects of more than one series or category of data. Each data series is represented as a geometric shape around a central point. Multiple series are overlaid so that can be compared.
Scatter Chart	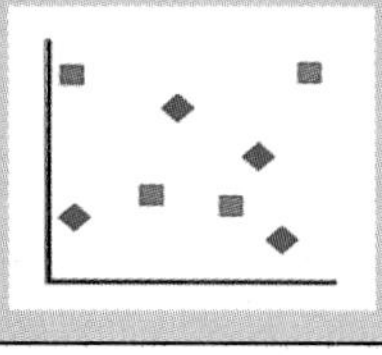	**Scatter charts** are useful for showing the relationship between two or more series or categories of data measured at uneven intervals of time. Each series is represented by data points using either different colors or bullets.

FIGURE 13.5 *Chart Types and Selection Criteria*

How does this impact output design? Transactional systems are increasingly Web enabled. When you purchase products over the Web, you almost always receive automated e-mail output to confirm your order. Follow-up e-mail may inform you of order fulfillment progress and initiate customer follow-up (a form of turnaround output).

Internal outputs may also be e-mail enhanced. For example, a system can push notification of the availability of new reports to interested users. Only those users who truly need the report will access the report and print it. This can generate a significant cost savings over mass distribution.

Hyperlinks   Many outputs are now Web enabled. Many databases and consumer ordering systems are now Web enabled. Web hyperlinks allow users to browse lists of records or search for specific records and retrieve various levels of detailed information on demand. Obviously, this medium can and is extended to computer inputs.

Technology exists to easily transform internal reports into HTML or XML formats for distribution via intranets. This reduces dependence on printed reports and screen reports that require a specific operating system or version (such as Windows). Essentially, all the recipient requires is a current browser that can run on any computer platform (*Windows, Mac, Linux,* or *UNIX*).

But Web-enabled output goes beyond presenting traditional outputs via the Internet and intranets. Many businesses have invested in Web-based internal report systems that consolidate weeks, months, and years of traditional internal reports into an organized database from which the reports can be recalled and displayed or printed. These systems don't create new outputs. They merely reformat previous reports for access via a browser. Think of it as an on-demand, Web-enabled report archival system. Examples of such reporting systems include DataWatch *Monarch/ES* and NSA *Report.Web.*

Microfilm   Paper is bulky and requires considerable storage space. To overcome the storage problem, many businesses use microfilm as an output medium. The first film medium is microfilm. More commonly, they turn to **microfiche,** small sheets of **microfilm** capable of storing dozens or hundreds of pages of computer output. The use of film presents its own problems; microfiche and microfilm can be produced and read only by special equipment.

This completes our introduction to output concepts. If you study Figure 13.1 carefully, you can see that implementation and distribution options can be combined to develop very creative, user-friendly, and exciting outputs.

HOW TO DESIGN AND PROTOTYPE OUTPUTS

In this section, we'll discuss and demonstrate the process of output design and prototyping. We'll introduce some tools for documenting and prototyping output design, and we'll also apply the concepts you learned in the last section. We will demonstrate how automated tools can be used to design and prototype outputs and layouts to system users and programmers. As usual, each step of the output design technique will be demonstrated using examples drawn from our Sound-Stage Entertainment Club case study.

Automated Tools for Output Design and Prototyping

In the not-too-distant past, the primary tools for output design were *printer spacing charts* (see Figure 13.6) and *display layout charts.* Today, this approach is not practiced much. It is a tedious process that is not conducive to today's preferred prototyping and rapid application development strategies, which use automated tools to accelerate the design process.

Before the availability of automated tools, analysts could sketch only rough drafts of outputs to get a feel for how system users wanted outputs to look. With

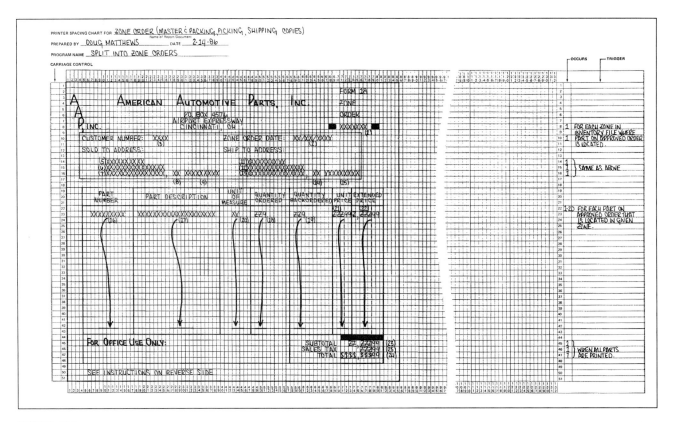

FIGURE 13.6 *Printer Spacing Chart*

automated tools, we can develop more realistic prototypes of these outputs. Perhaps the least expensive and most overlooked prototyping tool is the common spreadsheet. Examples include Lotus *1-2-3* and Microsoft *Excel*. A spreadsheet's tabular format is ideally suited to the creation of tabular output prototypes. And most spreadsheets include facilities to quickly convert tabular data into a variety of popular chart formats. Consequently, spreadsheets provide an unprecedented way to quickly prototype graphical output for system users.

Arguably, the most commonly used automated tool for output design is the PC-database application development environment. Many of you have no doubt learned Microsoft *Access* in either a PC literacy or database development course. While *Access* is not powerful enough to develop most enterprise-level applications, you may be surprised at how many designers use *Access* to prototype such applications. First, it provides rapid development tools to quickly construct a single-user (or few-user) database and test data. That data can subsequently feed the output design prototypes to increase realism. Designers can use *Access*'s report facility to lay out proposed output designs and test them with users.

Many CASE tools include facilities for report and screen layout and prototyping using the project repository created during requirements analysis. *System Architect 2001*'s screen design facility is demonstrated in Figure 13.7.

The above automated tools have significantly accelerated and enhanced the output design process. But the ultimate output design process would not only prototype the output's design but also serve as the final implementation of that output. This more sophisticated solution is found in report writing output tools such as Seagate's *Crystal Reports* and Actuate's *e.Reporting Suite*. These products create the actual "code" to be integrated in the operational information system. Figure 13.8 illustrates two screens from the *Crystal Reports* tool being used to create a SoundStage report from a prototype database.

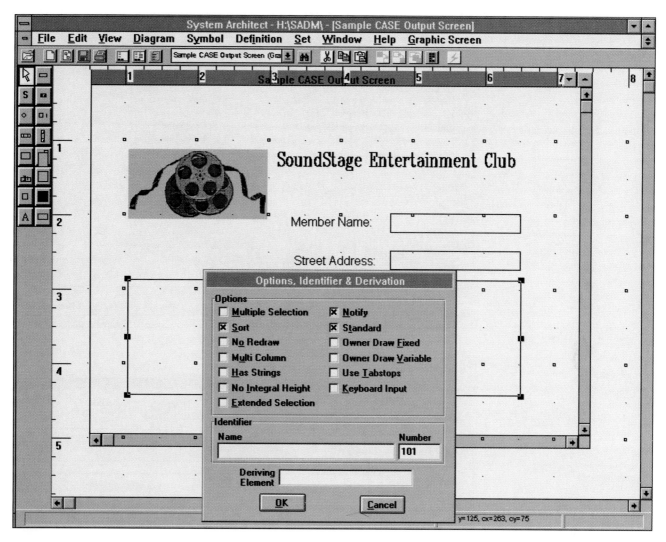

FIGURE 13.7   *CASE Tool for Output Design*

Many issues apply to output design. Most are driven by human engineering concerns—the desire to design outputs that will support the ways in which system users work. The following general principles are important for output design:

Output Design Guidelines

1. *Computer outputs should be simple to read and interpret.* These guidelines may enhance readability:

   a. Every output should have a title.

   b. Every output should be dated and time stamped. This helps the reader appreciate the currency of information (or lack thereof).

   c. Reports and screens should include sections and headings to segment information.

   d. In form-based outputs, all fields should be clearly labeled.

   e. In tabular-based outputs, columns should be clearly labeled.

   f. Because section headings, field names, and column headings are sometimes abbreviated to conserve space, reports should include or provide access to legends to interpret those headings.

   g. Only required information should be printed or displayed. In on-line outputs, use information hiding and provide methods to expand and contract levels of detail.

   h. Information should never have to be manually edited to become usable.

(a)

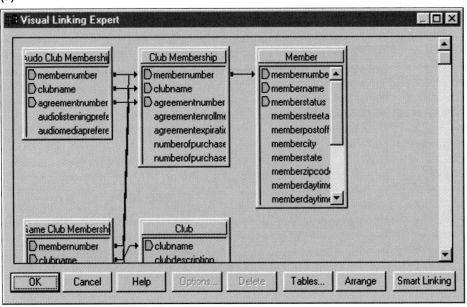

(b)

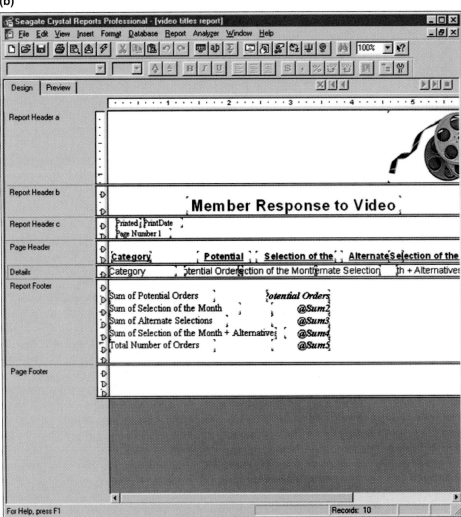

FIGURE 13.8

Report Writer Tool for Report Design

i. Information should be balanced on the report or display—not too crowded; not too spread out. Also, provide sufficient margins and spacing throughout the output to enhance readability.

j. Users must be able to easily find the output, move forward and backward, and exit the report.

k. Computer jargon and error messages should be omitted from all outputs.

2. *The timing of computer outputs is important.* Output information must reach recipients while the information is pertinent to transactions or decisions. This can affect how the output is designed and implemented.

3. *The distribution of (or access to) computer outputs must be sufficient to assist all relevant system users.* The choice of implementation method affects distribution.

4. *The computer outputs must be acceptable to the system users who will receive them.* An output design may contain the required information and still not be acceptable to the system user. To avoid this problem, the systems analyst must understand how the recipient plans to use the output.

The Output Design Process

Output design is not a complicated process. Some steps are essential, and others are dictated by circumstances. The steps are:

1. Identify system outputs and review logical requirements.
2. Specify physical output requirements.
3. As necessary, design any preprinted external forms.
4. Design, validate, and test outputs using <u>some combination of</u>:
   a. Layout tools (e.g., hand sketches, printer/display layout charts, or CASE).
   b. Prototyping tools (e.g., spreadsheet, PC DBMS, 4GL).
   c. Code generating tools (e.g., report writer).

In the following subsections, we examine these steps and illustrate a few examples from the SoundStage project.

Step 1: Identify System Outputs and Review Logical Requirements   Output requirements should have been defined during requirements analysis. Physical data flow diagrams (or design units; both described in Chapter 11) are a good starting point for output design. Those DFDs both identify the net outputs of the system (process-to-external agent) and the implementation method.

Depending on your system development methodology and standards, each of these net output data flows may also be described as a logical data flow in a data dictionary or repository (see data structures, Chapter 8). The data structure for a data flow specifies the attributes or fields to be included in the output. If those requirements are specified in the relational algebraic notation, you can quickly determine which fields repeat, which fields have optional values, and so on. Consider the following data structure:

Data Structure Defining Logical Requirements	Comment
INVOICE = INVOICE NUMBER	← Unique identifier of the output.
+ INVOICE DATE	← One of many fields that must take on a value. Lack of parentheses indicates a value is required.
+ CUSTOMER NUMBER	
+ CUSTOMER NAME	
+ CUSTOMER BILLING ADDRESS = ADDRESS >	← Pointer to a related definition.
+ 1 { SERVICE DATE +	← Begins group of fields that repeats 1 − n times.
SERVICE PROVIDED +	
SERVICE CHARGE } N	
+ PREVIOUS BALANCE DUE	← More required fields with single values.
+ PAYMENTS RECEIVED	
+ TOTAL NEW SERVICE CHARGES	
+ INTEREST CHARGES	

Data Structure Defining Logical Requirements   **Comment**

    + NEW BALANCE DUE
    + MINIMUM PAYMENT DUE
    + PAYMENT DUE DATE
    + (DEFAULT CREDIT CARD NUMBER)   ← Field does not have to have value.
    + ([CREDIT MESSAGE | PAYMENT MESSAGE]) ← Field does not have to have value, but
                                     if it does, it will only provide one of
                                     two possible field options.

Without such precise requirements, discovery prototypes may exist that were created during requirements analysis. In either case, a good requirements statement should be available in some format.

Step 2: Specify Physical Output Requirements   Recall that the decision analysis phase should have established some expectation of how most output data flows will eventually be implemented. Relative to outputs, the decisions were made by determining the best medium and format for the design and implementation based on:

— Type and purpose of the output.
— Operational, technical, and economic feasibility.

Because feasibility is important to more than just outputs, the techniques for evaluating feasibility were covered separately (in Chapter 10). The first set of criteria, however, is described in the following list.

— Is the output for internal or external use?
— If it's an internal output, is it a detailed, summary, or exception report?
— If it's an external report, is it a turnaround document?

After assuring yourself that you understand what type of report the output is and how it will be used, you need to address several design issues.

1. What implementation method would best serve the output? Various methods were discussed earlier in the chapter. You will have to understand the purpose or use of the output to determine the proper method. You can select more than one method for a single output; for instance, screen output with optional printout. Clearly, these decisions are best addressed with the system users.
   a. What would be the best format for the report? Tabular? Zoned? Graphic? Some combination?
   b. If a printout is desired, you must determine what type of form or paper will be used. Stock paper comes in three standard sizes (all specified in inches): 8½ × 11, 11 × 14, and 8½ × 14 inches. Many printers can now easily compress 132 columns of print into an 8-inch width. You need to determine the capabilities and limitations of the intended printer.
   c. For screen output, you need to understand the limitations of the users' display devices. Despite the increase in larger 19- and 21-inch high-resolution monitors, most users still have 15- and 17-inch displays (especially as you reach out directly to consumers in e-commerce applications). It is still recommended that screen outputs (including forms or pages within your application) be designed for the lowest common denominator.
   d. Form images can be stored and printed with modern laser printers, thereby eliminating the need for dealing with forms manufacturers in some businesses.
2. How frequently is the output generated? On demand? Hourly? Daily? Monthly? For scheduled outputs, when do system users need the report?
   a. Users generate many reports on demand. It can be helpful to use automated e-mail to notify users that new versions are available.

b. If reports are to be printed by the information services department, they must be worked into the information systems operations schedule. For instance, a report the system user needs by 9:00 A.M. on Thursday may have to be scheduled for 5:30 A.M. Thursday. No other time may be available.

3. How many pages or sheets of output will be generated for a single copy of a printed output? This information may be necessary to accurately plan paper and forms consumption.

4. Does the output require multiple copies? If so, how many?
   a. Impact printers are usually required to print all copies of a multicopy form at the same time.
   b. Laser printers can only print multiple copies of a form one after the other. This means that if the copies are different in color or fields, the preprinted forms must be collated before final printing.

5. For printed outputs, have distribution controls been finalized? For on-line outputs, access controls should be determined.

These design decisions should be recorded in the data dictionary/project repository. Let's consider an example from our SoundStage Entertainment Club case.

Let's look at one example. One output for SoundStage is the MEMBER RESPONSE SUMMARY REPORT. This report was requested to provide internal management with information regarding customer responses to the monthly promotional offers. The following design requirements were established.

1. The manager will request the report from his or her own workstation. It was determined that the information should be presented as a screen output in both tabular and graphical formats (to be determined via prototypes).
   a. All managers have 17-inch or larger display monitors.
   b. Managers should have the option of obtaining a laser printer output via their LAN configuration. Printouts should be on 8½ × 11-inch stock paper.

2. Managers must be able to display the report on demand. Managers have requested automatic e-mail notification of the availability of any newly generated version of the report. A hyperlink to the latest version of the report should also be made available in the standard home page of every Member Services manager, level 3 and above.

3. Graphical output should be displayable in a single screen and printable on a single page. Tabular data may be printed on one to two pages. The volume of pages is not considered significant for this report.

4. The report must be restricted in access to managers whose network accounts carry level 3 or higher account privileges. The report should include a "Confidential" watermark and a message that prohibits external distribution or information sharing without the written permission of Internal Audit.

Step 3: Design Any Preprinted Forms   External and turnaround documents are separated here for special consideration because they contain considerable constant and preprinted information that must be designed before designing the final output. In most cases, the design of a preprinted form is subcontracted to a forms manufacturer. The business, however, must specify the design requirements and carefully review design prototypes. The design requirements address issues such as the following.

— What preprinted information must appear on the form? This includes contact information, headings, labels, and other common information to appear on all copies of the form.

— Should the form be designed for mailing? If so, address locations become important based on whether or not windowed envelopes will be used.

— How many forms will be required for printing each day? Week? Month? Year?

— What will be the form's size? Form size, along with volume (above) can impact mailing costs.

— Will the form be perforated to serve as a turnaround document? Also, for turnaround documents the location of the address becomes more critical because the return address for the external output becomes the mailing address for the returned document.

— What legends, policies, and instructions need to be printed on the form (both front and back)?

— What colors will be used, and for which copies?

For external documents, there are also several alternatives. Carbon and chemical carbon are the most common duplicating techniques. Selective carbons are a variation whereby certain fields on the master copy will not be printed on one or more of the remaining copies. The fields to be omitted must be communicated to the forms manufacturer. Two-up printing is a technique whereby two sets of forms, possibly including carbons, are printed side by side on the printer.

A SoundStage preprinted output form was previously displayed as Figure 13.3.

Step 4: Design, Validate, and Test Outputs After design decisions and details have been recorded in the project repository, we must design the actual format of the report. The format or layout of an output directly affects the system user's ability to read and interpret it. The best way to lay out the format is to sketch or, better still, generate a sample of the report or document. We need to show that sketch or prototype to the system user, get feedback, and modify the sample. It's important to use realistic or reasonable data and demonstrate all control breaks.

The most important issue during the design step is format. Figure 13.9 summarizes a number of design issues and considerations for printed and tabular reports. Many of these considerations apply equally to screen outputs. Also, screen output offers a number of special considerations that are summarized in Figure 13.10.

FIGURE 13.9 *Tabular Report Design Principles*

Design Issue	Design Guideline	Examples
Page Size	At one time, most reports were printed on oversized paper. This required special binding and storage. Today, the page sizes of choice are standard (8½" × 11") and legal 8½" × 14"). These sizes are compatible with the predominance of laser printers in the modern business.	Not applicable.
Page Orientation	Page orientation is the width and length of a page as it is rotated. The *portrait* orientation (e.g. 8½ W × 11 L) is often preferred because it is oriented the way we orient most books and reports; however, *landscape* (e.g., 11 W × 8½ L) is often necessitated for tabular reports because more columns can be printed.	portrait landscape
Page Headings	Page headers should appear on every page. At a minimum, they should include a recognizable report title, date and time, and page numbers. Headers may be consolidated into one line or use multiple lines	JAN 4, 2001 PAGE 4 of 6 OVERSUBSCRIPTIONS BY COURSE
Report Legends	A legend is an explanation of abbreviations, colors, or codes used in a report. In a printed report, a legend can be printed on only the first page or on every page. On a display screen, a legend can be made available as a pop-up dialogue box.	REPORT LEGEND SEATS NUMBER OF SEATS IN THE CLASSROOM LIM COURSE ENROLLMENT LIMIT REQ NUMBER OF SEATS REQUESTED BY DEPARTMENT RES NUMBER OF SEATS RESERVED FOR DEPARTMENT USED NUMBER OF SEATS USED BY DEPARTMENT AVL NUMBER OF SEATS AVAILABLE FOR DEPARTMENT OVR NUMBER OF OVERSUBSCRIPTIONS FOR DEPARTMENT

Design Issue	Design Guideline	Examples
Column Headings	Column headings should be short and descriptive. If possible, avoid abbreviations. Unfortunately, this is not always possible. If abbreviations are used, include a legend (see Report Legends).	Self-explanatory
Heading Alignments	The relationship of column headings to the actual column data under those headings can greatly affect readability. Alignment should be tested with users for preferences with a special emphasis on the risk of misinterpretation of the information. See examples for possibilities (that can be combined).	Left justification (good for longer and variable length fields) NAME ================================= XXXXXXX X XXXXXXXX XXXXX Right justification (good for some numeric fields, especially monetary fields). Be sure to align decimal points. AMOUNT ========= $$$,$$$.cc Center (good for fixed length fields and some moderate length fields). STATUS ====== XXXX XXXX
Column Spacing	The spacing between columns impacts readability. If the columns are too close, users may not properly differentiate between the columns. If they are spaced too far apart, the user may have difficulty following a single row all the way across a page. As a general rule of thumb, place 3–5 spaces between each column.	Self-explanatory
Row Headings	The first one or two columns should serve as the identification data that differentiates each row. Rows should be sequenced in a fashion that supports their use. Frequently rows are sorted on a numerical key or alphabetically.	By number. STUDENT ID STUDENT NAME ========== ======================== 999-38-8476 MARY ELLEN KUKOW 999-39-5857 By alpha: SERVICE CANCEL SUBSCR TOTAL ====== ====== ====== ====== HBO 45 345 7665
Formatting	Data is often stored without formatting characters to save storage space. Outputs should reformat that data to match the users' norms.	As stored: As output: 307877262 307-87-7262 8004445454 (800) 444-5454 02272000 Feb 27, 2000

Control Breaks	Frequently, rows represent groups of meaningful data. Those groups should be logically grouped in the report. The transition from one group to the next is called a *control break* and is frequently followed by subtotals for the group.	RANK NAME SALARY
		==== ============== ======
		CPT JANEWAY, K 175,000
		CPT KIRK, J 225,000
		CPT PICARD, J 200,000
		CPT SISKO, B 165,000
		                            -------- 	
		         CAPTAINS TOTAL 765,000 ← a control break
		LTC CHAKOTAY 110,000
		LTC DATA 125,000
		LTC RIKER, W 140,000
		LTC SPOCK, S 155,000
		                            --------- 	
		         EXEC OFFCR TOTAL 530,000

End of Report	The end of a report should be clearly indicated to ensure that users have the entire report.	*** END OF REPORT ***

FIGURE 13.9   *(concluded)*

The SoundStage management expressed concern that the MEMBER RESPONSE SUMMARY output could potentially become too lengthy. Often the manager is interested in seeing only information pertaining to member responses for one or a few different product promotions. Thus, it was decided that the manager needed the ability to "customize" the output. The screens used to allow the manager to specify the customization desired should be prototyped as well as the report and graph containing the actual information. Figure 13.11(a) represents the prototype of the screen the user can use to choose a particular report (or graph) and customize its content. The following points should be noted:

❶ A tab dialogue box is used to allow the user to select between obtaining a report or graph. A tab control is used to present a series of related

---

FIGURE 13.10   *Screen Output Design Principles*

Screen Design Consideration	Design Guidelines
Size	Different displays support different resolutions. The designer should consider the "lowest common denominator." The default window size should be less than or equal to the worst resolution display in the user community. For instance, if some users will have only a 640 × 480 pixel resolution display, don't design windows to open at an 800 × 600 pixel resolution
Scrolling	On-line outputs have the advantage of not being limited by the physical page. This can also be a disadvantage if important information such as column headings scrolls off the screen. If possible, freeze important headings at the top of a screen.
Navigation	Users should always have a sense of where they are in a network of on-line screens. Given that, users also require the ability to navigate between screens. WINDOWS: Outputs appear in windows called *forms*. A form may display one record or many. The scroll bar should indicate where you are in the report. Buttons are frequently provided to move forward and backward through records in the report, and to exit the report. INTERNET: Outputs appear in windows called *pages*. A page may display one record or many. Buttons or hyperlinks may be used to navigate through records. Custom search engines can also be used to navigate to specific locations within a report.
Partitioning	WINDOWS: *Zones* are forms within forms. Each form is independent on the other but can be related. The zones can be independently scrollable. The Microsoft *Outlook* bar is one example. Zones can be used for legends or control breaks that take the user to different sections within a report. INTERNET: *Frames* are pages within pages. Users can scroll independently within pages. Frames can enhance reports in many ways. They can be used for a legend, table of contents, or summary information.

Information Hiding	On-line applications such as those that run under *Windows* or within an Internet browser offer capabilities to hide information until it is either needed or becomes important. Examples of such information hiding include:
	 − Drill-down controls that show minimal information and provide readers with simple ways to expand or contract the level of detail displayed.
	   − In *Windows* outputs the use of a small plus- or minus-sign in a small box to the left of a data record offers the option of expanding or contracting the record into more or less detail. All of this expansion and contraction occurs within the output's window.
	   − In *intranet* applications, any given piece of summary information can be highlighted as a hyperlink to expand that information into greater detail. Typically, the expanded information is opened in a separate window so the reader can use the browser's forward and backward buttons to switch between levels of detail.
	 − Pop-up dialogue boxes may be triggered by information.

Highlighting	Highlighting can be used in reports to call users' attention to erroneous data, exception data, or specific problems. Highlighting can also be a distraction if misused. Ongoing human factors research will continue to guide our future use of highlighting. Examples of highlighting include: − Color (avoid colors that color-blind persons cannot distinguish) − Font and case (changing case can draw attention) − Justification (left, right, or centered) − Hyphenation (not recommended in reports) − Blinking (can draw attention or become annoying) − Reverse video
Printing	For many users, there is still comfort in printed reports. Always provide users the option to print a permanent copy of the report. For Internet use, reports may need to be made available in industry standard formats such as Adobe *Acrobat*, which allows users to open and read those reports using free and widely available software.

(a) Report customization prototype

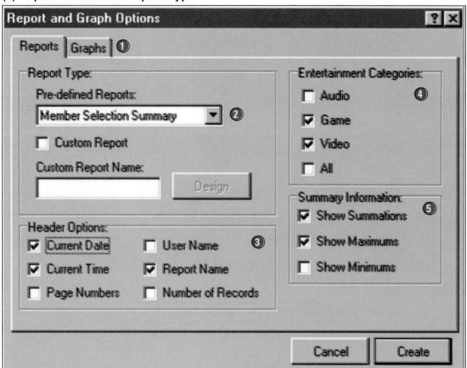

(b) Tabular report prototype

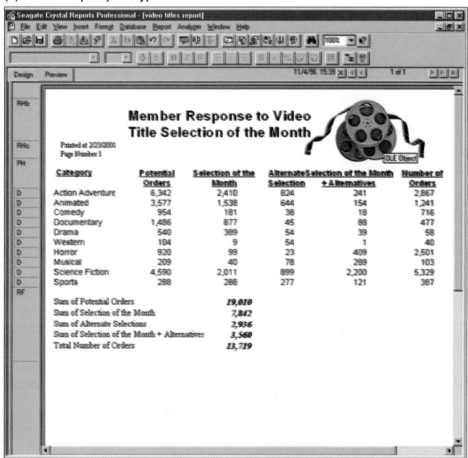

FIGURE 13.11

Report Customization and
Tabular Report Prototypes

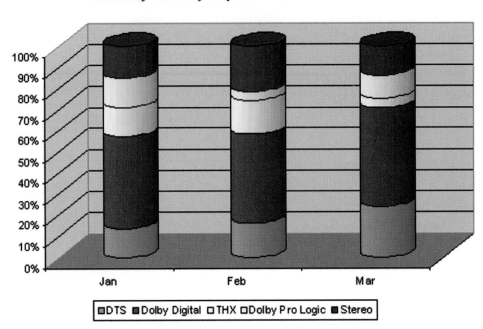

FIGURE 13.12
Graphical Report Prototype

information. If the user clicks on the tab labeled "Graphs," information would be displayed for customizing the output as a graph.

❷ A drop-down list is used to select the desired report. The user can click on the downward arrow to obtain a list of possible reports to choose from.

❸ The user is provided with a series of check boxes that correspond to general options for customizing the selected report. The user simply "checks" those options he wishes to appear on the report.

❹ A group of check boxes is also used to allow the user to select one or more product categories she wishes to include on the report.

❺ Once again, a group of check boxes is used to allow the user to further customize the report. Here the user is allowed to indicate the type of summary information or totals desired for each product category.

Let's now look at a prototype of the report that will result from the previous report customization dialogue. Figure 13.11(b) is a prototype of a screen output version of the actual report. Examine the content and appearance of the tabular design. Notice that the user is allowed to scroll vertically and horizontally to view the entire report. In addition, buttons are provided to allow the user to toggle forward and backward to view different report pages.

Finally, let's look at a prototype of a graphic version of the MEMBER RESPONSE SUMMARY output (see Figure 13.12). Note the following:

— The graph is clearly labeled along the vertical and horizontal axes.

— A legend has been provided to aid in interpreting the graph bars.

When prototyping outputs, it is important to involve the user to obtain feedback. The user should be allowed to actually "exercise" or test the screens. Part of that experience should involve demonstrating how the user may obtain appropriate help or instructions, drill-down to obtain additional information, navigate through pages, request different formats that are available, size the outputs, and perform test customization capabilities. All features should be demonstrated or tested.

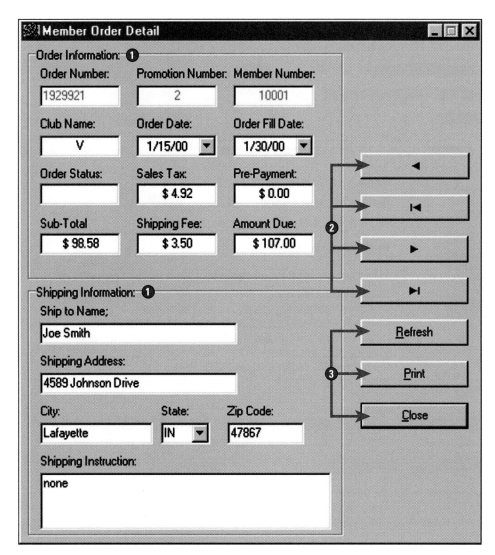

FIGURE 13.13
*Single Record Output
Prototype*

Thus far, we have presented only samples of a tabular and a graphical report. Another type of output is a record-at-a-time report. Users can browse forward and backward through individual records in a file. A sample screen for a record-at-a-time output is shown in Figure 13.13. We call your attention to the following:

❶ Each field is clearly labeled.

❷ Buttons have been added for navigation between records. The almost universally accepted buttons are for FIRST RECORD, NEXT RECORD, PREVIOUS RECORD, and LAST RECORD.

❸ We added buttons for the user to get a printed copy of the output, as well as to exit the report when finished. (Consistent with prototyping, the programmer will write the code for exiting later.)

The last output design considerations we want to address concern Web-based outputs. The SoundStage project will add various e-commerce and e-business capabilities to the Member Services information system. Some of these capabilities will affect output design.

One logical output requirement for the project is catalog browsing. Members should be able to browse and search catalogs, presumably as a preface to placing

Web-Based Outputs and E-Business

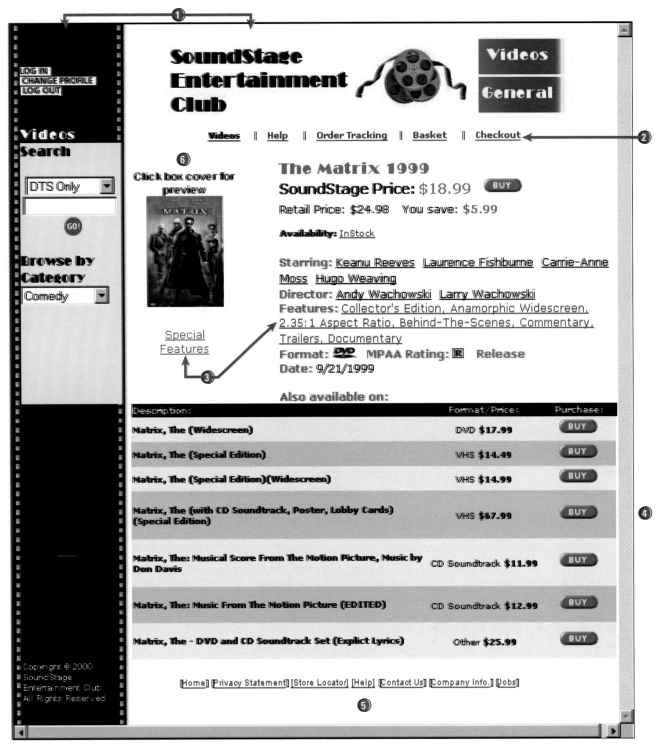

FIGURE 13.14   *Web Database Output*

orders. The catalog itself is the output. Figure 13.14 is a prototype screen for the physical catalog output. Note the following:

❶ This output uses frames to allow the user to focus separately on navigation and output.

❷ The screen uses hyperlinks to provide navigation through complex menu structures that are related to the output.

FIGURE 13.15   *Windows* Media Player *Output*

❸Hyperlinks also allow the user to get additional information. This functionality is referred to as "drill-down."

❹Shading is used to separate each detail line. This practice reflects the more artistic approach used to design Web-based outputs. Also, the "BUY" buttons have effectively transformed this output into a trigger for subsequent inputs. This is the e-commerce virtual equivalent of a turnaround document!

❺ Most Web-based output screen designs require standard footers on screen to provide additional navigation.

❻ A picture can be a selectable object. In this case it represents another type of drill-down where the user is able to obtain additional information.

Another output requirement is to allow members to play video trailers and audio sound bites for products to preview candidate purchases. The preview will be triggered by a hyperlink in the previous screen, and it will activate a multimedia player as shown in Figure 13.15 on the previous page. Such output extensions are expected to become the norm as Internet- and intranet-based applications grow in popularity.

❶ Web-based outputs frequently use plug-ins. This output screen has the standard buttons associated with a typical audio or video player.

❷ Web-based outputs also commonly provide appropriate plug-in or plug-in versions needed for the session.

WHERE DO YOU GO FROM HERE?

This chapter provided a detailed overview of the design and prototyping of computer outputs for a systems development project. It is recommended that you now complete Chapter 14 and not skip to Chapter 15. Chapter 14 deals with designing and prototyping an application's inputs. Chapter 15 deals with designing and prototyping an application's overall interface. As such, Chapter 15 involves tying together and presenting the applications functions addressed in Chapters 13 and 14.

SUMMARY

1. Outputs can be classified according to two characteristics:
   a. Their distribution inside or outside the organization and the people who read and use them.
   b. Their implementation method.
2. Internal outputs are intended for the system owners and users within an organization. They only rarely find their way outside the organization. There are three sub-classes of internal outputs:
   a. Detailed reports—present information with little or no filtering or restrictions.
   b. Summary reports—categorize information for managers who do not want to wade through details.
   c. Exception reports—filter data before it is presented to the manager as information.
3. External outputs leave the organization. They are intended for customers, suppliers, partners, and regulatory agencies. They usually conclude or report on business transactions.
4. Some outputs are both external and internal. They begin as external outputs that exit the organization but return in part or in whole.
5. Turnaround outputs are those external outputs that eventually reenter the system as inputs.
6. A good systems analyst will consider all available options for implementing an output. Several methods and formats exist.

   a. The most common medium for computer outputs is paper—printed output. Internal outputs are typically printed on blank paper (called stock paper). External outputs and turnaround documents are printed on preprinted forms.
      i. Perhaps the most common format for printed output is tabular. Tabular output presents information as columns of text and numbers.
      ii. An alternative to tabular output is zoned output. Zoned output places text and numbers into designated areas or boxes of a form or screen.
   b. Screen output is most suited to the pace of today's economy that requires information on demand. Screen output technology allows reports to be presented in graphical formats. Graphic output is the use of a pictorial chart to convey information in ways that demonstrate trends and relationships not easily seen in tabular output.
   c. Many of today's retail and consumer transactions are enabled or enhanced by point-of-sale (POS) terminals.
   d. *Multimedia* is a term coined to collectively describe any information presented in a format other than traditional numbers, codes, and words. This includes graphics, sound, pictures, and animation.

e. E-mail is becoming a very popular output medium as a means to reach large audiences and generate significant cost savings.

f. Web hyperlinks allow users to browse lists of records or search for specific records and retrieve various levels of detailed information on demand.

g. Paper requires considerable storage space. To overcome the storage problem, many businesses use microfilm as an output medium.

7. The most commonly used automated tool for output design is the PC-database application development environment. Many CASE tools also include facilities for report and screen layout and prototyping using the project repository created during requirements analysis.

8. The following general principles are important for output design:

a. Computer outputs should be simple to read and interpret.

b. The timing of computer outputs is important—their recipients must receive output information while the information is pertinent to transactions or decisions.

c. The distribution of (or access to) computer outputs must be sufficient to assist all relevant system users.

d. The computer outputs must be acceptable to the system users who will receive them.

9. Output design is not a complicated process. Some steps are essential, and others are dictated by circumstances. The steps are:

a. Identify system outputs and review logical requirements.

b. Specify physical output requirements.

c. As necessary, design any preprinted external forms.

d. Design, validate, and test outputs using some combination of:

i. Layout tools (e.g., hand sketches, printer/display layout charts, or CASE).

ii. Prototyping tools (e.g., spreadsheet, PC DBMS, 4GL).

iii. Code generating tools (e.g., report writer).

KEY TERMS

area chart, p. 512
bar chart, p. 512
column chart, p. 512
detailed report, p. 506
donut chart, p. 512
exception report, p. 507
external output, p. 507

graphic output, p. 511
internal output, p. 505
line chart, p. 512
microfiche, p. 513
microfilm, p. 513
pie chart, p. 512
radar chart, p. 512

scatter chart, p. 512
summary report, p. 506
tabular output, p. 510
turnaround output, p. 508
zoned output, p. 511

REVIEW QUESTIONS

1. What two characteristics can be used to classify outputs?
2. What is the difference between external and internal outputs? Give examples of each.
3. What are the three subclasses of internal outputs? Give examples of each.
4. What are turnaround documents? Give several examples.
5. List and briefly describe several output implementation methods.
6. Identify three types of tools that can be used to prototype computer outputs.
7. What is the most common format for printed output?
8. What output format places text and numbers into designated areas or boxes of a form or screen?
9. Which output medium is most suited to providing information on demand?
10. List and briefly describe several types of charts that can be used for graphical outputs.
11. Explain how e-mail and hyperlinks can be used for designing outputs.
12. Explain how automated tools have accelerated and enhanced the output design process.
13. Identify four general principles for output design.
14. Identify the steps involved in output design.
15. Briefly explain how physical data flow diagrams aid in output design.

PROBLEMS AND EXERCISES

1. To what extent should system users be involved in output design? How would you get system users involved? What would you ask them to do for themselves?
2. What are several commonly used media for outputs? What are the advantages and disadvantages of each?
3. Obtain sample outputs of each of the following format types: zoned, tabular, and graphic. Was the format type of each output the most effective of the four alternatives? If not, why? What format type, or combination of format types, would you have chosen to implement the output? Sketch the layout you would have chosen.
4. Explain how a data flow diagram is useful in identifying outputs that need to be designed.
5. Prepare an expanded data dictionary to describe the following outputs:
   a. Your driver's license.
   b. Your course schedule.
   c. Your bank statement.
   d. Your phone bill.
   e. Your W-2 statement.
   f. A bank or credit card account statement and invoice.
   g. An external document printed on a computer
6. Using a spreadsheet, CASE tool, or 4GL, prototype one of the outputs from exercise 5.
7. What effects may be caused by the lack of well-defined internal controls for outputs?
8. Develop sample outputs that utilize an area chart, bar chart, column chart, pie chart, line chart, radar chart, donut chart, and scatter chart. Explain why each chart was chosen over the other chart styles to communicate the data relationship.
9. Talk to a user regarding a computer-generated output that he or she receives. Compare the comments to the general principles for output design presented in this chapter.
10. Explain the importance of good input and database design to output design.

PROJECTS AND RESEARCH

1. The sales manager for The One And Only Music Club has requested a daily report. This report should describe the nearly 1,000 customer order responses received for a given day. A response is a member decision on whether to accept the record-of-the-month selection, request an alternate selection, request both, or request that no selection be sent that month. The report is to be sequenced by MEMBERSHIP NUMBER and CATALOG NUMBER. The data repository for the report follows:

```
ORDER RESPONSE REPORT =
        DATE * of the report
     +  PAGE NUMBER
     +  1 { CUSTOMER NUMBER * 5 digits +
            CUSTOMER LAST NAME * 15 characters +
            CUSTOMER FIRST NAME * 15 characters +
            (CUSTOMER MIDDLE INITIAL) * 1 character +
            MUSICAL PREFERENCE * possible values are:
                easy listening, pop/rock, classical,
                country, jazz, rhythm and blues +
            SELECTION OF MONTH DECISION * yes or no } 1000
     +  1 { CATALOG NUMBER * 5 digits +
            MEDIA * values are: CD, cassette, D +
            NUMBER OF CREDITS * values = 1-5 } 15
     +  AGREEMENT EXPIRATION DATE
```

What type of output was being requested by the sales manager? Prepare an expanded data repository for the output. Prototype the requested output. Verify the output with your instructor (serving as the sales manager or system user). Be sure to include appropriate report headings, edit masks, and timing entries.

2. The sales manager has also requested that the sales staff be able to obtain information concerning a particular customer's order response at any time during normal working hours. Prepare an expanded data repository for the output. Prototype the requested output. Verify the output with your instructor (serving as the sales manager or system user).
3. The order filling operation for a local pharmacy is to be automated. Customer prescriptions are to be entered on-line by pharmacists. The pharmacists insist that they be able to immediately obtain information pertaining to all previously filled prescriptions for a given customer. Using the following data that was captured each time a customer's prescription was filled, prototype the output screen(s).

```
PRESCRIPTION = CUSTOMER NAME * 20 characters
            + DOCTOR NAME * 20 characters
            + 1 { DRUG NAME * 30 characters +
                  QUANTITY PRESCRIBED * 4 digits +
                  MEDICAL INSTRUCTIONS * 120 characters } 10
            + RX NUMBER
            + 1 { DRUG NUMBER * 6 digits +
                  LOT NUMBER * 6 characters +
                  DOSAGE FORM * 1 character +
                  UNIT OF MEASURE * 1 character +
                  QUANTITY DISPENSED * 3 digits +
                  NUMBER OF REFILLS * 2 digits +
                  (EXPIRATION DATE) } 10
```

where DOSAGE FORMS are: P = PILL, C = CAPSULE, L = LIQUID, R = LOTION.
where UNIT OF MEASURES are: G = GRAMS, O = OUNCES, M = MILLILITERS.

4. A moving company maintains data concerning fuel tax liability for its fleet of trucks. When truck drivers return from a trip, they submit a journal describing mileage, fuel purchases, and fuel consumption for each state traveled through. This data is to be input daily to maintain records on trucks and fuel stations. Using the following data, brainstorm and then prototype detailed, exception, and summary reports. Explain the usefulness (value) of each report; do so by describing the types of decisions that are supported by each report.

TRIP JOURNAL = TRUCK NUMBER * **4 characters**
      + DRIVER NUMBER * **9 characters**
      + CO-DRIVER NUMBER * **9 characters**
      + TRIP NUMBER * **3 characters**

      + DATE DEPARTED
      + DATE RETURNED
      + 1 { STATE CODE * **2 characters** +
           MILES DRIVEN * **5 digits** +
           FUEL RECEIPT NO * **9 digits** +
           GALLONS PURCHASED +
           TAXES PAID * **4 digits, 2 dec.** +
           STATION NAME * **10 characters** +
           STATION LOCATION * **15 characters** } **20**

5. Conduct research on a report writer. Can the report writer be used to design both printed and screen display outputs? Can it be used to develop windows and Web-based outputs? Does the product interface with a CASE tool repository or a database?

MINICASES

1. Sharon Miller, a programmer/analyst for SaveBig, was meeting with Betty Carlton, the Accounts Payable assistant manager, to discuss problems with a new system that Sharon was largely responsible for developing. It was no secret that Betty was displeased with some aspects of the new system.

BETTY   Sharon, I guess you heard that I'm having a little trouble with this new payables system. You told me that these computer reports would make my job much easier, and that just hasn't happened.

SHARON   I'm really sorry it hasn't worked for you. But I'm willing to work overtime to make this system work the way you need it to work. What isn't working? I thought I included all the fields you requested. Is it that you're not getting the reports on time?

BETTY   That's not the problem. The information is there. It's just that I find the report difficult to use. For example, this report doesn't tell me what I need to know. I spend most of my time trying to interpret it.

SHARON   Why don't you tell me how you use the report?

BETTY   Well, first I read down the report, line by line, and attempt to count and classify the number of invoices that are less than 10 days old, between 10 and 30 days old, and between 31 and 60 days old. You see, I have this graph here that shows the totals for each category over the past 12 months. I compare the new totals from this report with the totals on the graph to identify any significant trends in payment activities.

SHARON   Why?

BETTY   Because we pay some invoices off early to gain a 2 percent discount. Others we defer to the final due date of the supplier, which is usually 30 or 60 days after receipt of the invoice.

SHARON   I see. Go on.

BETTY   I also use the report to identify the costs of discounts that we did not take, choosing instead to defer payment until close to the final due date. To do that, I locate the invoice on the report and look up that same invoice on the previous copy of the report. Then I log the amount as a lost discount to a particular supplier. At the end of the week, I sum the lost discounts by supplier and send the report to my superiors. It really helps. I can think of several occasions when my superiors authorized a change in payment policy for a particular supplier in order to take better advantage of discounts.

SHARON   When I wrote down specifications for this report, I assumed I understood what information you were requesting. It sounds as if there is some additional data that should have been included on the report. You're wasting a lot of time looking for information that could be automatically generated by the system.

BETTY   I'm glad that you are sympathetic to my problem. I hope you can do something about it. I'm pretty frustrated with that report.

SHARON   I'm sure I can have it fixed in no time. I'm really sorry. I guess I just didn't understand what information you wanted and how you would be using it. Let's see, how should we proceed? I don't want to spend a lot of time designing a new report that isn't precisely what you need. I've got an idea. I just got a new package on my office microcomputer. It's called *Excel*. I'm pretty sure that I can quickly mock up and, if necessary, change a sample of the report you want. I can even simulate this pie chart graph you want. Once we get the reports and graphs looking the way you want, we'll use them as models for the programmers.

BETTY   That's a good idea.

SHARON   OK, what about this other report?

BETTY   I'm getting some flak about this report I asked you to produce for my clerks. It was supposed to list those supplier accounts that we owe payment on. Anyhow, my clerks claim the report is too cluttered and is difficult to use. All they really need to know is the supplier's account number, current balance due, discount date, and the final due date. As you can see, there are a number of unnecessary fields, and the report lists two supplier accounts on the same line. Could you clean this report up too?

SHARON   No problem. Again, I'm sorry I was so off target on those reports. I can probably use *Excel* to mock up these reports as well.

a. What type of reports was Betty receiving?

b. What type of reports does Betty really need?

c. What did you think of Sharon's new strategy of using *Excel* for mocking up new reports? Does the approach sound similar to a strategy that has been frequently described in this book?

d. Why do you suppose the report for Betty's subordinates was inappropriate? Who's to blame, Betty or Sharon? Why?

2. You have been assigned to a project team that is working on the in-house development of a new purchasing system. Team members disagree about which task should be completed next. Some members think inputs should be defined first, then outputs, and the overall user interface. Other members disagree with the sequence in which those tasks should be completed. Take a stance and defend it!

SUGGESTED READINGS

Andres, C. *Great Web Architecture*. Foster City, CA: IDG Books Worldwide, Inc., 1999. Books on effective Web interface design are beginning to surface. The science of human engineering for Web interfaces has not yet progressed as far as client/server interfaces (e.g., *Windows*). Here is an early title that explores many dimensions of Web architecture and interfaces using real-world examples.

Application Development Strategies (monthly periodical). Arlington, VA: Cutter Information Corporation. This is our favorite theme-oriented periodical that follows system development strategies, methodologies, CASE, and other relevant trends. Each issue focuses on a single theme. This periodical will provide a good foundation for how to develop prototypes.

Galitz, W. O. *User-Interface Screen Design*. New York: John Wiley & Sons, Inc., 1993. This is our favorite book on overall user interface design.

Shelly, G., T. Cashman, and H. Rosenblatt. *Systems Analysis and Design*. 3rd ed. Cambridge, MA: Course Technology, 1998. We mention our competitors for their excellent coverage of tabular, printed output design. They afford many more pages of coverage and examples than we could afford in our latest edition.

**Focus on
PEOPLE**

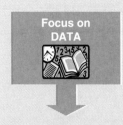

**Focus on
DATA**

**Focus on
PROCESSES**

**Focus on
INTERFACES**

**Focus on
DEVELOPMENT**

Stakeholders

Activities

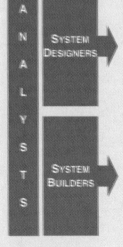

SYSTEMS ANALYSTS

SYSTEM
OWNERS

SYSTEM
USERS

SYSTEM
DESIGNERS

SYSTEM
BUILDERS

BUILDING BLOCKS OF AN INFORMATION SYSTEM

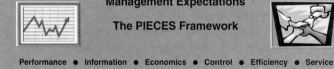

Management Expectations

The PIECES Framework

Performance ● Information ● Economics ● Control ● Efficiency ● Service

List of business entities and rules ... **Business Knowledge**	List of business functions and events ... **Business Functions**	List of business locations and systems... **Business Locations**

**Data
Requirements**

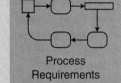

**Process
Requirements**

**Interface
Requirements**

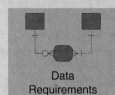

**Database
Schema**

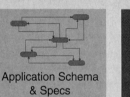

**Application Schema
& Specs**

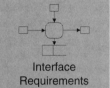

**Interface
Specifications**

```
CREATE TABLE tblOrders
  colOrderNot CHAR(5) NOT
  NULL
  colOrderDate  DATE/TIME NOT
```
**Database
Programs**

```
PROC ValidateOrder
  PERFORM ValidateCust
  REPEAT UNTIL
    NoMoreProd ...
```
**Application
Programs**

```
<html>
<head>
<title> Order Entry Form </title>
...
```
**Interface
Programs**

PROJECT &
PROCESS
MANAGEMENT

PRELIMINARY
INVESTIGATION

PROBLEM
ANALYSIS

REQUIREMENTS
ANALYSIS

DECISION
ANALYSIS

DESIGN

CONSTRUCTION

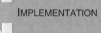

IMPLEMENTATION

VENDORS
AND
CONSULTANTS

INFORMATION TECHNOLOGY & ARCHITECTURE

Database Technology ● Process Technology ● Interface Technology ● Network Technology

OPERATIONS
AND
SUPPORT

14

INPUT DESIGN AND PROTOTYPING

CHAPTER PREVIEW AND OBJECTIVES

In this chapter you will learn how to design computer inputs. It is the second of three chapters that address the design of on-line systems using a graphical user interface for either client/server or Web-based systems. You will know how to design inputs when you can:

— Define the appropriate format and media for a computer input.

— Explain the difference between data capture, data entry, and data input.

— Identify and describe several automatic data collection technologies.

— Apply human factors to the design of computer inputs.

— Design internal controls for computer inputs.

— Select proper screen-based controls for input attributes that are to appear on a GUI input screen.

— Design a Web-based input interface.

Output and input design represent something of a "chicken or egg" sequencing problem. Which do you do first? In this edition, we present output design first. Classic system design prefers this approach as something of a system validation test—design outputs and then make sure the inputs are sufficient to produce the outputs. In practice, this sequencing of tasks becomes less important because modern systems analysis techniques sufficiently predefine logical input and output requirements. You and your instructor may safely swap Chapters 13 and 14 if you prefer.

Note: The authors gratefully acknowledge the contributions and recommendations of Carlin "Rick" Smith at Purdue University. Rick developed most of the prototypes that demonstrate the concepts and techniques taught in this chapter. Rick also contributed ideas for further improving chapter currency and topics.

SCENE

This episode begins in the conference room where Sandra and Bob have scheduled a Saturday morning meeting to review input design screens with the Order Processing staff. Sally Hoover suggested a Saturday morning overtime meeting because she wanted her entire staff to become familiar with the new input and on-line methods that would be used in the new system.

SANDRA

OK, let's call this meeting to order. I think we should get started so we don't take away too much of everyone's weekend. I know all of you are eager to enjoy your weekend, so we'll make this as brief as possible. First, Bob and I would like to thank you for your cooperation so far. We realize that we have spent a great deal of time defining what you need in order to do your jobs. We have been deliberately avoiding specific details of how to do things because we wanted to build a system that would make your jobs easier. It is your system, and it is shaping up very well. Today, we want to review system inputs. Let's begin with a member order.

[Sandra adjusts the computer project pad so the screen image is in clear focus on the pull-down projection screen.]

This is the proposed member order response form that you will be receiving from our customers. Bob has begun to build a screen that you folks will use to enter data from these forms. We want your feedback. It is not the intent of this meeting to focus on the overall appearance of this screen. Rather, we simply want to make sure that we are capturing all the data that we need to capture about our customer orders. We are also concerned with the manner in which that data is captured. So, Bob, why don't you take it from here.

BOB

Thanks, Sandra. First, notice that you will not have to enter the order date anymore; the system will do that for you by using the current date. In fact, you'll notice that the background color of that

field is grayed to suggest that you can't change the date.

SALLY

Can't change the date? What if the order date is different from the date when the order is entered?

SANDRA

Why would they be different?

SALLY

Sometimes orders are sent to the wrong department, or they don't get entered on the day they are received due to a high volume of orders. Since we generate automatic orders to members based on the date, it's very important that the date on a member's order be correct.

BOB

That's my fault. Sally and I discussed this, but I forgot to make the change to the specifications. If we have the system fill the date field with the current date but allow the Order Processing staff to override that value we can solve that problem and save them a lot of typing at the same time.

SALLY

I have another question, Sandra. Do we have to enter all those fields in order to begin processing an order?

SANDRA

The order number is automatically generated by the system. Also, when you enter the member number, the member's name will appear for you to verify. In addition to the order date we just talked about, the selection of the month accepted field is represented by a check box that will have a check mark to indicate a default answer of yes. And to save you some more typing, the product number will be automatically inserted.

CLERK A

But 70 percent of our members reject the selection of the month. That means we will have to change those fields most of the time.

SANDRA

That's interesting. We will change the default value to no and leave the

product number blank. Are there any more questions or concerns?

BOB

If not, what we need to do now is to verify the size of each field and get an idea of the range of values each field can assume. This will help us write programs to check the accuracy of the data before it is processed.

CLERK B

Do I have to use a computer? In the past, I've primarily screened paper orders and passed them off to somebody else for data entry. But they tell me that I have to enter orders. I have to tell you, I'm not that comfortable with a computer. You "kids" find them easy. I find them intimidating.

SANDRA

Don't worry. I remember my first experience with a computer. I was very intimidated. I thought that if I pressed the wrong key, the entire business would fold like a bad poker hand. First, we are going to develop an overall interface that will be very intuitive. As much as possible, we will try to make you forget that you are talking with a machine. Also, you will be asked to give us feedback as we attempt to develop such screens. What would you think if there wasn't any reference guide and everything you needed to know was right on the screen?

CLERK B

What happens if I enter something several times and the computer won't take it, but I don't understand what I'm doing wrong?

BOB

That's a good question. As you can see with this screen example, we will try to provide you with the best mechanism for ensuring that correct data is selected and entered. For example, notice the field with the label "Closed Caption"? That field contains two possible values, yes or no. I provided a check box where all you do is click to toggle its value from yes to no. If a box contains a check mark, that means its value is yes. Over here you see what we call radio buttons. Each button corresponds with a possible value of this field. And here we used a box with a list of possible values. So in either case, you could not possibly enter any other value than one that is provided. I should also point out that anytime you need more information about a particular field, you will be able to get immediate help.

CLERK B

Well, if all the screens work like that I think there's hope for me.

CLERK A

What about the Web interface?

SANDRA

Well, we made some assumptions that those users are probably comfortable with their computers. We have built a first draft of the Web interface. Let me show you . . . This will take a second to come up on the screen.

CLERK A

While we're waiting, I agree that most users are comfortable with their computers. But will they be comfortable buying our products on the Web? A lot of people are concerned about privacy and security. And some of the Web sites my daughter visits are incredibly difficult to browse. It's like a maze. That would not be good for business.

BOB

Maybe we should just do a brief demo and give you folks the secret URL so you can play with it. It would be interesting to get both your perspective and that of some of your family members.

DISCUSSION QUESTIONS

1. Why did Sandra and Bob want to focus on the content of the input screen and its individual fields? What would be the advantages of this emphasis?

2. Do users of applications that have a graphical interface such as those provided by *Windows*-based products really need help features? Explain.

3. Why would Sandra and Bob be concerned with having the computer provide values for fields where possible, rather than the users?

4. What benefits are derived by Sandra and Bob involving the users during input design?

5. What factors and concerns do you have about ordering over the Web? How can good input and interface design alleviate your concerns?

6. What do you think of Bob's closing idea?

"Garbage in! Garbage out!" This overworked expression is no less true today than it was when we first studied computer programming. Management and users make important decisions based on system outputs (Chapter 13). These outputs are produced from data that is either input or retrieved from databases. And any data in the databases must have been first input. In this chapter, you are going to learn how to design computer inputs. Input design serves an important goal—capture and get the data into a format suitable for the computer.

Today most inputs are designed by rapidly constructing prototypes. These prototypes may be simple computer-generated mockups, or they may be generated from prototype database structures such as those developed for Microsoft *Access*. These prototypes are rarely fully functional. They won't contain security features, data editing, or data updates that will be necessary in the final version of a system. Furthermore, in the interest of productivity, they may not include every button or control feature that would have to be included in a production system.

During requirements analysis, inputs were modeled as data flows that consist of data attributes. Even in the most thorough of requirements analysis, we will miss requirements. Input design may introduce new attributes or fields to the system. This is especially true if output design (Chapter 13) introduced new attributes to the outputs—the inputs must always be sufficient to produce the outputs!

We begin with a discussion of types of inputs. Inputs can be classified according to two characteristics: (1) how the data is initially captured, entered, and processed and (2) the method and technology used to capture and enter the data. Figure 14.1 illustrates this taxonomy. The characteristics are discussed briefly in the following sections.

When you think of "input," you usually think of input devices, such as keyboards and mice. But input begins long before the data arrives at the device. To actually input business data into a computer, the systems analyst may have to design source

INPUT DESIGN CONCEPTS AND GUIDELINES

Data Capture, Data Entry, and Data Processing

Process / Method	Data Capture	Data Entry	Data Processing
Keyboard	Data usually captured on a business form that becomes the source document for input. Data can be collected real-time (over the phone).	Data entered via keyboard. This is the most common input method, but also the most prone to errors.	OLD: Data can be collected into batch files (disk) for processing as a batch. NEW: Data processed as soon as it has been keyed.
Mouse	Same as above.	Used in conjunction with keyboard to simplify data entry. Mouse serves as a pointing device for a screen. Can be with geographical user interfaces to reduce errors through point-and-click choices.	Same as above, but the use of a mouse is most commonly associated with on-line and real-time processing.
Touch Screen	Same as above.	Data entered on a touch screen display or hand held device. Data-entry users either touch commands and data choices or enter data using handwriting recognition.	On PCs, touch screen choices are processed same as above. On handheld computers, data is stored on the handheld for later processing as a remote batch.
Point-of-Sale	Data is captured as close to the point of sale (or transaction) as humanly possible. No source documents.	Data is often entered directly by the customer (e.g., ATM) or by an employee directly interacting with the customer (e.g., retail cash register). Input requires specialized, dedicated terminals that utilize some combination of the other techniques in this table.	Data is almost always processed immediately as a transaction or inquiry.
Sound	Data is captured as close to the source as possible, even when the customer is remotely located (e.g., at home or their place of employment).	Data is entered using Touch-Tones (typically from a telephone). Usually requires fairly rigid command menu structure and limited input options.	Data is almost always processed immediately as a transaction or inquiry.
Speech	Same as sound.	Data (and commands) are spoken. This technology is not as mature and much less reliable and common than other techniques.	Data is almost always processed immediately as a transaction or inquiry.
Optical Mark	Data is recorded on optical scan sheets as marks or precisely formed letters, numbers, and punctuation. This is the oldest form of automatic data capture.	Eliminates the need for data entry. (Very commonly used in education for test scoring, course evaluations, and surveys.)	Data is almost always processed as a batch.
Magnetic Ink	Data is usually prerecorded on forms that are subsequently completed by the customer. The customer records additional data on the form.	A magnetic ink reader reads the magnetized data. The customer-added data must be entered using another input method. This technique is used in applications requiring high accuracy and security, the most common of which is bank checks (for check number, account number, bank ID)	Data is almost always processed as a batch.
Electromagnetic	Data is recorded directly on the object to be described by data.	Data is transmitted by radio frequency.	Data is almost always processed immediately.
Smart Card	Data is recorded directly on a device to be carried by the customer, employee, or other individual that is described by that data.	Data is read by smart card readers.	Data is almost always processed immediately.
Biometric	Unique human characteristics become data.	Data is read by biometric sensors. Primary applications are security and medical monitoring.	Data is processed immediately.

Figure 14.1 *An Input Taxonomy*

documents, input screens, and methods and procedures for getting the data into the computer (from customer to form to data-entry clerk to computer).

This brings us to our first fundamental question. What is the difference between data capture and data entry? Data happens! It accompanies business events called transactions. Examples include ORDERS, TIME CARDS, RESERVATIONS, and the like. We must determine when and how to capture the data when "it happens."

> **Data capture** is the identification and acquisition of new data.

When is easy! It's always best to capture the data as soon as possible after it originates. *How* is another story! Historically, special paper forms called source documents were used.

> **Source documents** are forms used to record business transactions in terms of data that describes those transactions.

Display screens that can duplicate the appearance of almost any paper-based form are gradually replacing the paper forms. This trend is being accelerated by Web-based e-commerce and e-business. Still, business forms are commonly used as source documents for data entry. Design of source documents requires care. The layout and readability will affect speed of data entry.

> **Data entry** is the process of translating the source data or document into a computer-readable format.

Because data entry used to be 100 percent keyboard-based, businesses employed armies of data entry clerks. As on-line computing became more common, the responsibility for data entry shifted directly to system users. Today another transformation is occurring. Thanks to personal computers and the Internet, some data entry has shifted directly to the consumer. In all cases, data entry produces input for data processing.

Entered data must subsequently be processed—data processing. In this chapter, we are not concerned with how the data is transformed into outputs. But we are interested in the timing of input processing. When does the input data get processed?

Batch Processing   Batch processing used to be the dominant form of data processing.

> In **batch processing,** the entered data is collected into files called batches. Each file is processed as a batch of many transactions.

Contrary to popular belief, some data is still processed in batches. Time cards are the classic example. Most batches are recorded as disk files (hence the term *key-to-disk*). Some older systems may still record batches on magnetic tape (*key-to-tape*).

On-Line Processing   Today most (but not all) information systems have been converted to on-line processing.

> In **on-line processing,** the captured data is processed immediately.

Initially, data was entered at terminals. Today, that same data is captured on PCs and workstations to take advantage of their ability to perform some of the data validation and editing before it gets sent to the server computers. Because of PCs, we rarely hear the term *on-line processing* any more. We usually hear the term *client/server,* where the PC is the client.

Most of today's applications present the user with a PC-based **graphical user interface (GUI).** Microsoft *Windows* is the dominant GUI in today's businesses.

But the emergence of the Web as a platform for Internet and intranet applications may make a Web browser the most important user interface in the future. Microsoft Internet Explorer and Netscape Navigator are the dominant browser interfaces in today's market. This chapter will address input design techniques for both the *Windows* client/server interface and the browser interface.

Remote Batch   Batch and on-line represent extremes on the processing spectrum. A combination solution also exists—the remote batch.

> In **remote batch processing,** data is entered using on-line editing techniques; however, the data is collected into a batch instead of being immediately processed. Later, the batch is processed.

Modern remote batch can take several forms. A simple example uses a PC-based front-end application to capture and store the data. The data can later be transmitted across a network for batch processing. A more contemporary example of remote batch processing uses disconnected laptop or handheld computers (or devices) to collect data for later processing. If you've recently received a package from UPS or Federal Express, you've seen such devices used by the drivers to record pickups and deliveries.

Now that we've covered the basic data capture, data entry, and data processing techniques, we can more closely examine the input methods shown as rows in Figure 14.1.

Input Methods and Implementation

Different input devices, such as keyboards and mice, are covered in most introductory information systems courses. In this section, we are more interested in the method and its implementation than in the technology. In particular, we are interested in how the choice of a method affects data capture, entry, and processing as described in the previous section. You should continue to study Figure 14.1 as we introduce these methods.

Keyboard   Keyboard data entry remains the most common form of input. Unfortunately, it requires the most data editing because people make mistakes keying data from source documents. Fortunately, graphical user interfaces such as Microsoft *Windows* and Web browsers now make it possible to design on-line screens that reduce errors by forcing correct choices on the user. We will explore several useful GUI controls for such interfaces in the next section.

Mouse   A mouse is a pointing device used in conjunction with graphical user interfaces. The mouse has made it easy to navigate on-line forms and click on commands and input options. For example, the legitimate values for an attribute can be recorded on a screen as "clickable" boxes or buttons that eliminate the need to key in that data. This results in fewer data-entry errors. We will explore mouse-based controls in our input designs for this chapter.

Touch Screen   An emerging technology that will greatly impact input design in the near future is the touch screen display. Such displays are common in handheld and palm-top computers that are finding their way into countless information system applications. A Symbol Technologies handheld computer based on the *Palm Operating System* is shown in the margin. Such devices simplify many data collection activities in a warehouse and on a manufacturing shop floor. Touch screen buttons can be programmed to collect the data. Most such devices support handwriting recognition as well. The Symbol Technologies unit depicted also can scan and read bar codes (discussed shortly).

A Handheld Computer

Point-of-Sale   Point-of-sale (POS) terminals have been with us for some time. They have all but replaced old-fashioned cash registers. These terminals capture data at the point of sale and provide time-saving ways to enter data, perform transactional calculations, and produce some output. Like the handhelds just described, most can scan and read bar codes to eliminate keying errors. Automatic teller machines (ATMs), another form of POS terminal, are operated directly by the consumer.

Sound and Speech   Sound represents another form of input. You might have used a Touch-Tone telephone-based system to register for this course. Such tone-based systems require special input/output technology that drives the design. Those systems are beyond the scope of this book.

A more sophisticated form of this input method uses voice recognition technology to make it possible to input data. Currently this technology is relatively immature and unreliable. It is best utilized to input commands, not data. But the time may come when voice recognition technology replaces the keyboard as the principle means by which we enter data.

The remaining input methods are broadly classified as **automatic data capture (ADC).** With advancements in today's input technology, we can eliminate much (and sometimes all) human intervention associated with the input methods discussed in the previous section. By eliminating human intervention we can decrease the time delay and errors associated with human interaction.

Optical Mark   **Optical mark recognition** (OMR) technology for input has existed for several decades. It is primarily batch processing-oriented. The classic example is the optical mark forms used for objective-based questions (e.g., multiple choice) on examinations. The technology is also useful in surveys and questionnaires or any other application where the number of possible data values is relatively limited and highly structured. Most applications that could benefit from this input method have probably already exploited it.

Optical character recognition (OCR) is less prevalent despite its maturity. It requires the user or customer to carefully handwrite input data on a business form. If the letters and numbers are properly scribed, an OCR reader can process the forms without human intervention. Obviously, this depends on the handwriting of the user or customer. But it does work. Columbia House Record Club used to use an OCR form for customer responses to orders. Like most OCR applications, the number of fields to be input was very small (reducing the possibility of errors). Processing methods must be implemented for any inputs rejected due to illegibility.

Today the most prevalent form of optical technology involves bar coding. **Bar codes** are on almost every product we buy, but bar-coding technology is not limited to retail sales. You can create bar codes for almost any business application. You can even integrate bar codes into *Windows*-based applications as shown in Figure 14.2.

Magnetic Ink   Magnetic ink ADC technology is one you will likely recognize. It usually involves using magnetic stripe cards, but it also may include the use of magnetic ink character recognition (MICR). Over 1 billion magnetic stripe cards are in use today! They have found their way into a number of business applications, such as credit card transactions, building security access control, and employee attendance tracking. MICR is most widely used in the banking industry.

Electromagnetic Transmission   Electromagnetic ADC technology is based on the use of radio frequency to identify physical objects. This technology involves attaching a tag and antenna to the physical object that is to be tracked. The tag contains memory that is used to identify the object being tracked. The tag can be read by a reader whenever the object resides within the electromagnetic field generated by the reader. This identification technology is becoming very popular in applications that involve tracking physical objects that are out of sight and on the move.

Figure 14.2
Bar Codes in a Windows
Application

For example, electromagnetic ADC is being used for public transportation tracking and control, tracking manufactured products, and tracking animals, to name a few.

A Smart Card

Smart Cards   Smart card technology has the ability to store a massive amount of information. Smart cards are similar to, but slightly thicker than, credit cards. They also differ in that they contain a microprocessor, memory circuits, and a battery. Think of it as a credit card with a computer on board. They represent a portable storage medium from which input data can be obtained. While this technology is only beginning to make inroads in the United States, smart cards are used on a daily basis by over 60 percent of the French population. Smart card applications are particularly promising in the area of health records where a person's blood type, vaccinations, and other past medical history can be made readily available. Other uses may include such applications as passports, financial information for point-of-sale transactions, and pay television, to name a few. Another future application could be a combination debit card that automatically maintains and displays your account balance. A smart card used in a security application is shown in the margin.

Biometric   Biometric ADC technology is based on unique human characteristics or traits. For example, individuals can be identified by their own unique fingerprint, voice pattern, or pattern of certain veins (retina or wrist). Biometric ADC systems consist of sensors that capture an individual's characteristic or trait, digitize the image pattern, and then compare the image to stored patterns for identification. Biometric

ADC is popular because it offers the most accurate and reliable means for identification. This technology is particularly popular for systems that require security access.

Because inputs originate with system users, human factors play a significant role in input design. Inputs should be as simple as possible and designed to reduce the possibility of incorrect data being entered. The needs of system users *must* be considered. With this in mind, several human factors should be evaluated.

The volume of data to be input should be minimized. The more data that is input, the greater the potential number of input errors and the longer it takes to input that data. Thus, numerous considerations should be given to the data that is captured for input. These general principles should be followed for input design:

— *Capture only variable data.* Do not enter constant data. For instance, when deciding what elements to include in a SALES ORDER input, we need PART NUMBERS for all parts ordered. However, do we need to input PART DESCRIPTIONS for those parts? PART DESCRIPTION is probably stored in a database table. If we input PART NUMBER, we can look up PART DESCRIPTION. Permanent (or semipermanent) data should be stored in the database. Of course, inputs must be designed for maintaining those database tables.

— *Do not capture data that can be calculated or stored in computer programs.* For example, if you input QUANTITY ORDERED and PRICE, you don't need to input EXTENDED PRICE, which is equal to QUANTITY ORDERED × PRICE. Another example is incorporating FEDERAL TAX WITHHOLDING data in tables (arrays) instead of keying in that data every time.

— *Use codes for appropriate attributes.* Codes were introduced earlier. Codes can be translated in computer programs by using tables.

Second, if source documents are used to capture data they should be easy for system users to complete and subsequently enter into the system. The following suggestions may help:

— *Include instructions for completing the form.* Remember that people don't like to have to read instructions printed on the back side of a form.

— *Minimize the amount of handwriting.* Many people suffer from poor penmanship. The data-entry clerk or CRT operator may misread the data and input incorrect data. Use check boxes wherever possible so the system user only needs to check the appropriate values.

— *Data to be entered (keyed) should be sequenced so it can be read like this book, top to bottom and left to right.* Figure 14.3(a) demonstrates a good flow. The system user should not have to move from right-to-left on a line or jump around on the form, as shown in Figure 14.3(b), to enter data.

— *When possible, use designs based on known metaphors.* The classic example of this is the personal finance application *Quicken.* The program's ease of use is greatly enhanced by its on-screen re-creation of the checkbook metaphor. The user writes checks by filling in a graphical representation of the check. And the check register looks exactly like its paper equivalent. Not all inputs lend themselves to metaphors, but some are greatly enhanced by the imitation (see Figure 14.4).

— There are several other guidelines and issues specific to data input for GUI screen designs. We'll introduce these guidelines as appropriate when we discuss GUI controls for input design later in this chapter, as well as in the chapters on output design and user interface design.

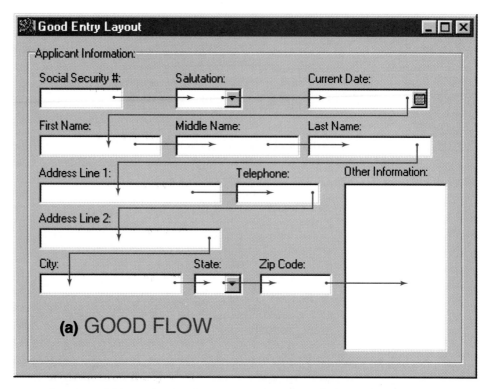

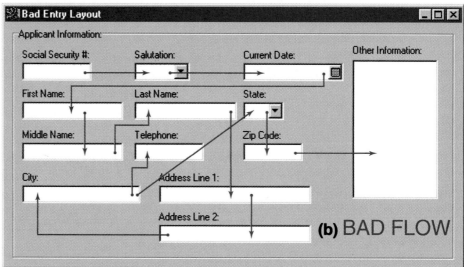

Figure 14.3
Good and Bad Flow in a Form

Internal Controls—Data Editing for Inputs

Internal controls are a requirement in all computer-based systems. Internal input controls ensure that the data input to the computer is accurate and that the system is protected against accidental and intentional errors and abuse, including fraud. The following internal control guidelines are offered:

1. The number of inputs should be monitored. This is especially true with the batch method, because source documents may be misplaced, lost, or skipped.

   – In batch systems, data about each batch should be recorded on a batch control slip. Data includes BATCH NUMBER, NUMBER OF DOCUMENTS, and CONTROL TOTALS (e.g., total number of line items on the documents). These

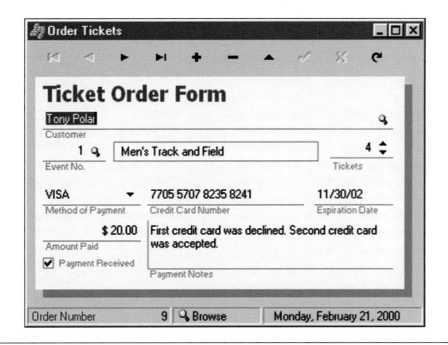

Figure 14.4
Metaphoric Screen Design

totals can be compared with the output totals on a report after processing has been completed. If the totals are not equal, the cause of the discrepancy must be determined.

– In batch systems, an alternative control would be one-for-one checks. Each source document would be matched against the corresponding historical report detail line that confirms the document has been processed. This control check may be necessary only when the batch control totals don't match.

– In on-line systems, each input transaction should be logged to a separate audit file so it can be recovered and reprocessed if there is a processing error or if data is lost.

2. Care must also be taken to ensure that the data is valid. Two types of errors can infiltrate the data: data-entry errors and invalid data recorded by system users. Data-entry errors include copying errors, transpositions (typing 132 as 123), and slides (keying 345.36 as 3453.6). The following techniques are widely used to validate data:

– **Existence checks** determine whether all <u>required</u> fields on the input have actually been entered. Required fields should be clearly identified as such on the input screen.

– **Data type checks** ensure that the correct type of data is input. For example, alphabetic data should not be allowed in a numeric field.

– **Domain checks** determine whether the input data for each field falls within the legitimate set or range of values defined for that field. For instance, an upper-limit range may be put on PAY RATE to ensure that no employee is paid at a higher rate. (These checks should have been specified during requirements analysis—Chapters 7 and 8 in this book.)

– **Combination checks** determine whether a known relationship between two fields is valid. For instance, if the VEHICLE MAKE is Pontiac, then the VEHICLE MODEL must be one of a limited set of values that comprises cars manufactured by Pontiac (Firebird, Grand Prix, and Bonneville, to name a few).

- **Self-checking digits** determine data-entry errors on primary keys. A check digit is a number or character that is appended to a primary key field. The check digit is calculated by applying a formula, such as Modulus 11, to the actual key. The check digit verifies correct data entry in one of two ways. Some data-entry devices can automatically validate data by applying the same formula to the data as it is entered by the system user. If the check digit entered doesn't match the check digit calculated, an error is displayed. Alternatively, computer programs can also validate check digits by using readily available subroutines.

- **Format checks** compare data entered against the known formatting requirements for that data. For instance, some fields may require leading zeros while others don't. Some fields use standard punctuation (e.g., Social Security numbers or phone numbers). A value "A4898 DH" might pass a format check, while a similar value "A489 ID8" would not.

In Chapter 12, you learned that most database management systems perform data validation checks similar to those described in the above list. So why do we need input controls? Simple! Most applications today are networked. Erroneous data is both a network traffic bottleneck and a detractor for transaction throughput and response time. It is always best to capture and correct input errors as close as possible to the source—hence the emphasis on input controls and validation.

GUI CONTROLS FOR INPUT DESIGN

As mentioned earlier, most new applications being developed today include a graphical user interface (GUI). Most are based on Microsoft *Windows*, but the pervasive adoption of the Internet combined with Web-based e-commerce is quickly driving some interfaces to the Web browser. While GUI designs provide a more user-friendly interface, they also present more complex design issues than their predecessors. This chapter will not attempt to address all the GUI design issues; entire books have been written on the subject. Several of our favorites are listed in the suggested readings.

Rather, this chapter will focus on selecting the proper screen-based controls for entering data on a GUI screen. Think of controls as "widgets" for building a user interface. They are included in most contemporary application development environments such as Microsoft *Access* and *Visual Basic*, Sybase's *PowerBuilder*, InPrise's *JBuilder*, Symantec's *Visual Café*, IBM's *Visual Age*, and many others. Many of these tools share controls (and code) via the repository. This approach is called **repository-based programming.**

Figure 14.5 illustrates access to a repository that contains input controls and code. The approach is based on the object-oriented and component-based programming techniques that have become pervasive in application development. This figure depicts controls that could be used by various systems analysts or programmers to prototype an interface. The developers can, in a single location, define most of the properties and constraints for a reusable field and the data validation code for that field. Once defined, the object or control can be used by any number of other systems analysts and programmers in the organization. This repository-based approach guarantees that every instance of the field will be used in a consistent manner. Furthermore, the repository entries can be changed if business rules dictate and no additional changes to the applications will be required.

Common GUI Controls for Inputs

This section examines some of the most common controls used in GUI-based input forms. We address the purpose, advantages, disadvantages, and guidelines for each control. Given this understanding, we are then in a good position to make decisions concerning which controls should be considered for each data attribute that will be input on our screens. We will defer the transitions between our screen designs until Chapter 15, User Interface Design.

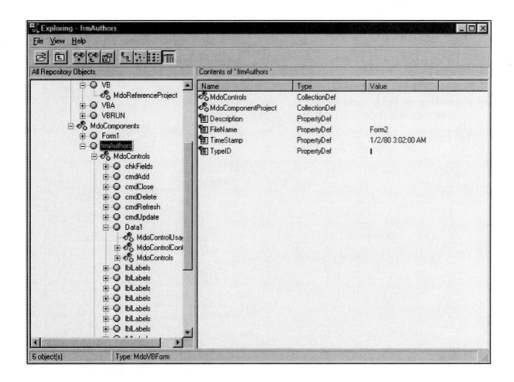

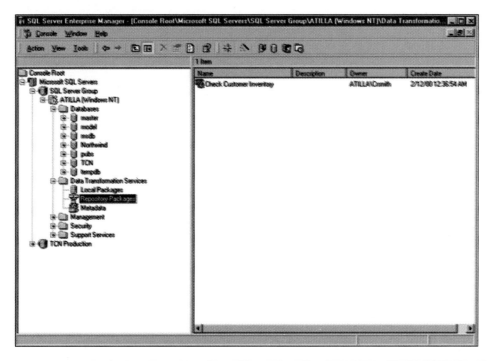

Figure 14.5
Repository-Based Prototyping and Development

Refer to Figure 14.6 as a library of the most common screen-based controls for input data. Each of the controls will be discussed. They are equally applicable to both *Windows-* and Web-based interfaces.

Text Box ❶   Perhaps the most common control used for input of data is the text box. A **text box** consists of a rectangular shaped box that is usually accompanied by a caption. This control requires the user to type the data inside the box. A text box can allow for single or multiple lines of data characters to be entered. When

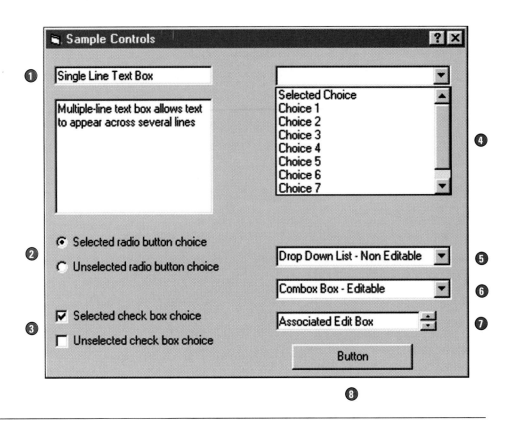

Figure 14.6
Common GUI Input Controls

a text box contains multiple lines of data, scrolling features are also normally included.

A text box is most appropriately used when the input data values are unlimited in scope and the analyst is unable to provide the users with a meaningful list of values from which they can select. For example, a single-line text box would be an appropriate control for capturing a new customer's last name because the possibilities for the customer's last name are virtually impossible to predetermine. A text box would also be appropriate for capturing data about shipping instructions that describe a particular order placed by a customer. Once again, the possible values for shipping instructions are virtually unlimited. In addition, the multiple-line text box would be appropriate due to the unpredictable length of the shipping instructions. In those cases where the text box is not large enough to view the entire input data values, the text box may use scrolling and word-wrap features.

Numerous guidelines should be followed when using a text box on an input screen. A text box should be accompanied by a descriptive, meaningful caption. Avoid using abbreviations for captions. Only the first character of the caption's text should be capitalized.

The location of the caption is also significant. The user should be able to clearly associate the caption with the text box. Therefore, the caption should be located to the left of the actual text box or left-aligned immediately above the text box. Finally, it is also generally accepted that the caption be followed by a colon to help the user visually distinguish the caption from the box.

Generally, the size of the text box should be large enough for all characters of fixed-length input data to be entered and viewed by the user. When the length of the data to be input is variable and could become quite long, the text box's scrolling and word-wrapping features should be applied.

Radio Button ❷   Radio buttons provide the user with an easy way to quickly identify and select a particular value from a value set. A **radio button** consists of a small circle and an associated textual description that corresponds to the value choice. The circle is located to the left of the textual description of the value choice. Radio buttons normally appear in groups—one radio button per value choice. When a user selects the appropriate choice from the value set, the circle corresponding to that choice is partially filled to indicate it has been selected. When a choice is selected, any default or previously selected choice's circle is deselected. Radio buttons also allow the user the flexibility of selecting via the keyboard or mouse.

Radio buttons are most appropriate when a user may be expected to input data that has a limited predefined set of mutually exclusive values. For example, a user may be asked to input an ORDER TYPE and GENDER. Each of these has a limited, predefined, mutually exclusive set of valid values. For example, when the users are to input an ORDER TYPE, they might be expected to indicate one and only one value from the value set "regular order," "rush order," or "standing order." For GENDER, the user would be expected to indicate one and only one value from the set "female," "male," or "unknown."

There are several guidelines to consider when using radio buttons as a means for data input. First, radio buttons should present the alternatives vertically aligned and left-justified to aid the user in browsing. If necessary the choices can be presented where they are aligned horizontally, but adequate spacing should be used to help visually distinguish the choices. Also, the group of choices should be visually grouped to set them off from other input controls appearing on the screen. The grouping should also contain an appropriate meaningful caption. For example, radio buttons for male, female, and unknown might be vertically aligned and left-justified with the heading/caption "Gender" left-justified above the set.

The sequencing of the choices should also be given consideration. The larger the number of choices the more thought should be given to the ease of scanning and identifying the choices. For example, in some cases it may be more natural for the user to locate choices that are presented in alphabetical order. In other cases, the frequency in which a value is selected may be important in regards to where it is located in the set of choices.

Finally, it is not recommended that radio buttons be used to select the value for an input data whose value is simply a Yes/No (or On/Off state). Instead, a check box control should be considered.

Check Box ❸   As with text boxes and radio buttons, a **check box** also consists of two parts. It consists of a square box followed by a textual description of the input field for which the user is to provide the Yes/No value. Check boxes provide the user the flexibility of selecting the value via the keyboard or mouse. An input data field whose value is yes is represented by a square that is filled with a "3." The absence of a "3" means the input field's value is no. The user simply toggles the input field's value from one value/state to the other as desired.

Often a user needs to input a data field whose value set consists of a simple yes or no value. For example, a user may be asked for a Yes/No value for such items as the following input data: CREDIT APPROVED? SENIOR CITIZEN? HAVE YOU EVER BEEN CONVICTED OF FRAUD? and MAY WE CONTACT YOUR PREVIOUS EMPLOYER? In each situation a check box control could be used. A check box control offers a visual and intuitive means for the user to input such data.

The previous example represented a simplified scenario for the use of a stand-alone check box. On a single input screen it may be desirable to ask a user to enter values for a number of related input fields having a Yes/No value. For example, a receptionist at a health clinic may be entering data from a completed patient form. On a section of that form, the patient may have been asked about a number of illnesses. They may have been asked their past medical history and instructed to "check all that apply" from a list of types of various illnesses. If properly designed, the receptionist's input screen would represent each illness as a separate input field using a check box control. The controls would be physically associated into a group on the screen. The group would also be given an appropriate heading/caption. Recognize that even though the check boxes may be visually grouped on the screen, each check box operates as a separate independent input field.

Following these recommended guidelines will improve the use of check box controls. Make sure the textual description is meaningful to the user. Look for opportunities to group check boxes for related Yes/No input fields and provide a descriptive group heading.

To aid in the user's browsing and selecting from a group of check boxes, arrange the group of check box controls where they are aligned vertically and left-justified. If necessary, align horizontally and be sure to leave adequate space to visually separate the controls from one another. Finally, provide further assistance to the user by appropriately sequencing the input fields according to their textual description. In most cases, where the number of check box controls is large, the sequencing should be alphabetical. In those cases where the text description describes dollar ranges or some other measurement, the sequencing may be according to the numerical order. In other cases such as those where a very limited number of controls are grouped, the basis for sequencing may be according to the frequency that a given input data field's Yes/No value is selected. (All input data fields represented using a check box have a default value.)

List Box ❹   A list box is a control that requires the user to select a data item's value from a list of possible choices. The **list box** is rectangular and contains one or more rows of possible data values. The values may appear as either a textual description or graphical representation. List boxes having a large number of possible values may consist of scroll bars to navigate through the row of choices.

It is also common for a list box's row to contain more than one column. For example, a list box could simply contain rows having a single column of permissible values for an input data item called JOB CODE. However, it may be asking too much to expect the user to recognize what each job code actually represented. In this case, to place the values of JOB CODE into a meaningful perspective, the list box could include a second column containing the corresponding JOB TITLE for each job code.

How does one choose between a radio button and a list box control? Both controls are useful in ensuring that the user enters the correct value for a data item. Both are also appropriate when it is desirable to have the value choices constantly visible to the user.

The decision is normally driven by the number of possible values for the data item and the amount of screen space available for the control. Scrolling capabilities make list boxes appropriate for use in those cases where there is limited screen space available and the input data item has a large number of predefined, mutually exclusive values from which to choose.

There are several guidelines to consider when using a list box as a means for data input. A list box should be accompanied by a descriptive caption. Avoid using abbreviations for captions and capitalize only the first character of the caption's text. It is also generally accepted that the caption be followed by a colon to help the user visually distinguish the caption from the box.

The location of the caption is also significant. The user should be able to clearly associate the caption with the list box. Therefore, the caption should appear left-justified immediately above the actual list box.

There are also several guidelines relating to the list box. First, it is recommended that a list box contain a highlighted default value. Second, consider the size of the list box. Generally, the width of the list box should be large enough for most characters of fixed-length input data to be entered and viewed by the user. The length of the box should allow for at least three choices and be limited in size to containing about seven choices. In both cases scrolling features should be used to suggest additional choices are available to the user.

If graphical representations are used for value choices, make sure the graphics are meaningful and truly representative of the choice. If textual descriptions are used, use mixed-case letters and ensure that the descriptions are meaningful. It is important that these decisions or judgments be based on the perspective and opinions of the user!

You should also give careful thought to the ease with which a user can scan and identify the choices appearing in the list box. The list of choices should be left-justified to aid in browsing. Be sure to involve the user when addressing the order in which choices will appear in the list. In some cases it may be natural to the user if the list of choices appeared in alphabetical order. In other cases, the frequency in which a value is selected may be important in regard to where it is located in the list.

Drop-Down List ❺   A drop-down list is another control that requires the user to select a data item's value from a list of possible choices. A **drop-down list** consists of a rectangular selection field with a small button connected to its side. The small button contains the image of a downward pointing arrow and bar. This button is intended to suggest to the user the existence of a hidden list of possible values for a data item.

When requested, the hidden list appears to "drop or pull down" beneath the selection field to reveal itself to the user. The revealed list has characteristics similar to the list box control mentioned in the previous section. When the user selects a value from the list of choices, the selected value is displayed in the selection field and the list of choices once again becomes hidden from the user.

A drop-down list should be used in those cases where the data item has a large number of predefined values and screen space availability prohibits the use of a list box. One disadvantage of a drop-down list is that it requires extra steps by the user, in comparison to the previously mentioned controls.

Many of the guidelines for using list boxes directly apply to drop-down lists. One exception is the placement of the caption. The caption for a drop-down list is generally either left-aligned immediately above the selection field portion of the control or located to the left of the control.

Combination Box ❻   A **combination box,** often simply called a combo box, is a control whose name reflects the fact that it combines the capabilities of a text box and list box. A combo box gives the user the flexibility of entering a data item's value (as with a text box) or selecting its value from a list (as with a list box).

At first glance, a combo box closely resembles a drop-down list control. Unlike the drop-down list control, however, the rectangular box can serve as an entry field for the user to directly enter a data item's value. Once the small button is selected, a hidden list is revealed. The revealed list appears slightly indented beneath the rectangular entry field.

When the user selects a value from the list of choices, the selected value is displayed in the entry field and the list of choices once again becomes hidden from the user.

A combo box is most appropriately used where limited screen space is available and it is desirable to provide the user with the option of selecting a value from a list or typing a value that may or may not appear as an option in the list.

The same guidelines for using drop-down lists directly apply to combo boxes.

Spin Box ❼   A **spin box** is a screen-based control that consists of a single-line text box followed immediately by two small buttons. The two buttons are vertically aligned. The top button has an arrow pointing upward and the bottom button has an arrow pointing down. This control allows the user to enter data directly into the associated text box or to select a value by using the mouse to scroll (or "spin") through a list of values using the buttons. The buttons have a unit of measure associated with them. When the user clicks on one of the arrow buttons, a value will appear in the text box. The value in the text box is manipulated by clicking on the arrow buttons. The upward pointing button will increase the value in the text box by a unit of measure, whereas the downward pointing button will decrease the value in the text box by the same unit of measure.

A spin box is most appropriately used to allow the user to make an input selection by using the buttons to navigate through a small set of meaningful choices or by directly keying the data value into the textbox. The data values for a spin box should be capable of being sequenced in a predictable manner.

Spin boxes should contain a label or caption that clearly identifies the input data item. This label should be located to the left of the text box or left-aligned immediately above the text box portion of the control. Finally, spin boxes should always contain a default value in the text box portion of the control.

Buttons ⑧   Strictly speaking, buttons are not input controls. They do not contribute to the selection of or input of actual data. Nonetheless, input form design is incomplete without them. Buttons serve several purposes. They allow a user to commit all of the data to be processed, or cancel a transaction, or get help. They can be used to navigate between instances of the same form.

Many more screen-based controls are available for designing graphical user interfaces. The above are the most common controls for capturing input data. There are others, and you should become familiar with them and their proper usage for inputting data. In later chapters you will be exposed to several other controls used for other purposes. Keep on top of developments in the area of GUI as new controls are sure to be made available.

Advanced Input Controls

Figure 14.7 (a) and (b) illustrates additional controls for data input. These advanced controls can be used in *Windows* interfaces to create a more sophisticated look and feel. Equivalent controls are likely available for Web-based applications, but most Web-based e-commerce applications aspire to simpler formats. We will not discuss these controls in detail, but they are summarized as follows:

— *Drop-down calendar.* A field is illustrated in Figure 14.7(a). Clicking the down arrow next to the date creates the pop-up calendar shown in Figure 14.7(b). The familiar calendar is another example of metaphoric design.
— *Slider edit calendar.* This is a nonnumeric way to select a value.
— *Masked edit control.* This control builds the format checks described earlier right into the field.
— *Ellipsis control.* Clicking on the three dots causes a pop-up dialogue to appear for data entry. It might be used for a field that consists of several parts (such as an address—street, city, state, and ZIP code.
— *Alternate numeric spinner.* This is a different type of input spinner.
— *Internet hyperlink.* Similar in function to a button, a hyperlink can be linked to Web pages, but it can also be linked to other *Windows* forms. This is an effective way to hide related input forms that do not apply to all or most users.
— *Check list box.* This control is useful for combining several check boxes in situations where several boxes may be applicable.
— *Check tree list box.* This control is useful for presenting data options that need to be hierarchically organized into a tree-like structure.

HOW TO DESIGN AND PROTOTYPE INPUTS

How do you design on-line inputs? Traditionally, the designer was concerned with the overall content, appearance, and functionality of the input screen—in relative isolation of other screens that needed to be designed. The designers knew they

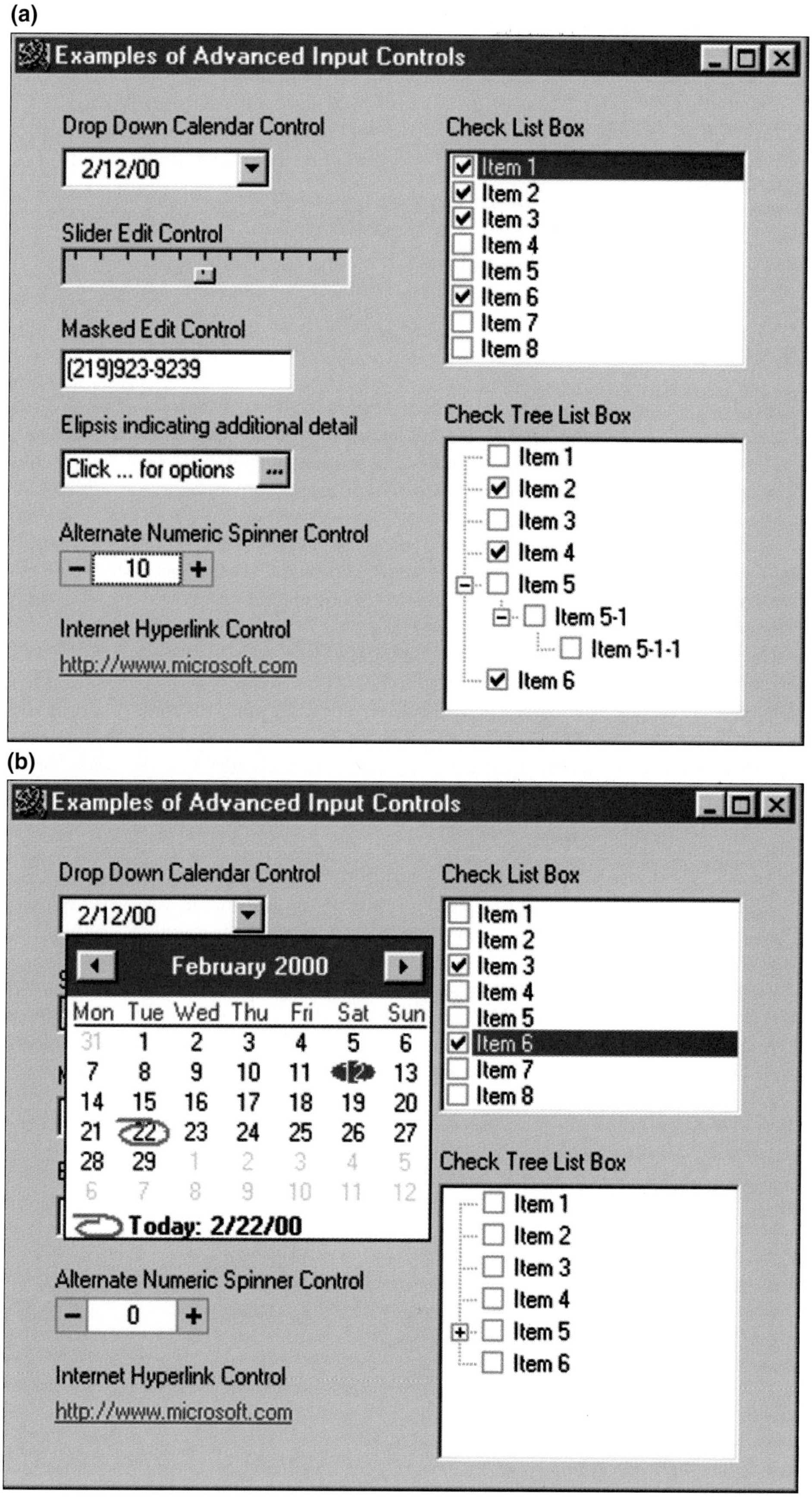

Figure 14.7
Advanced GUI Input Controls

would simply design a subsequent set of menu screens from which the users would select an option that would lead them to the appropriate input screen. Simple enough. However, given today's graphical environments, there is an emphasis on developing an overall system that blends well into the user's workplace environment. This emphasis rarely results in a hierarchical, menu-driven application interface that characterized the more traditional text- or command-based applications of old.

The following sections will demonstrate how the first stage of input design is completed. We will draw on examples from our SoundStage case study. We will examine both client/server, *Windows*-based inputs and Web-based, e-commerce inputs that run in a browser. Later, in Chapter 15 we will integrate the outputs (from Chapter 13) and inputs from this chapter into an overall user interface and dialogue.

Automated Tools for Input Design and Prototyping

In the recent past, the primary tools for input design were *record layout charts* and *display layout charts*. Today, this "sketching" approach is not often practiced. It is a tedious process that is not conducive to today's preferred prototyping and rapid application development strategies, which use automated tools to accelerate the design process.

Before the availability of automated tools, analysts could sketch only rough drafts of inputs to get a feel for how system users wanted outputs to look or how the batch records would be structured. With automated tools, we can develop more realistic prototypes of these inputs.

Arguably, the most commonly used automated tool for input design is the PC-database application development environment. While Microsoft *Access* is not powerful enough to develop most enterprise-level applications, you may be surprised at how many designers use *Access* to prototype such applications. Given a database structure (easily specified in *Access*), you can quickly generate or create forms to input data. You can include most of the GUI controls we described in this chapter. The users can subsequently exercise those forms and tell you what works and what doesn't.

Many CASE tools include facilities for report and screen layout and prototyping using the project repository created during requirements analysis. *System Architect 2001*'s screen design facility was previously demonstrated in Chapter 13, Figure 13.7.

Most GUI-based programming languages such as *Visual Basic* can easily be used to construct nonfunctional prototypes of inputs. The key term here is *nonfunctional*. The forms will look real, but there will be no code to implement any of the buttons or fields. That is the essence of rapid prototyping.

The Input Design Process

Input design is not a complicated process. Some steps are essential, and others are dictated by circumstances. The steps are:

1. Identify system inputs and review logical requirements.
2. Select appropriate GUI controls.
3. Design, validate, and test inputs using some combination of:
   a. Layout tools (e.g., hand sketches, printer/display layout charts, or CASE).
   b. Prototyping tools (e.g., spreadsheet, PC DBMS, 4GL).
4. If necessary, design the source document.

In the following subsections, we examine these steps and illustrate a few examples from the SoundStage project.

Step 1: Identify System Inputs and Review Logical Requirements   Input requirements should have been defined during requirements analysis. Physical data flow diagrams (or design units; both described in Chapter 11) are a good starting point for input design. Those DFDs identify both the net outputs of the system (external agent-to-process) and the implementation method.

Depending on your system development methodology and standards, each of these net input data flows may also be described as a logical data flow in a data dictionary or repository (see Chapter 8). The data structure for a data flow specifies the attributes or fields to be included in the output. If those requirements are specified in the relational algebraic notation, you can quickly determine which fields repeat, which fields have optional values, and so on. Consider the following data structure:

Data Structure Defining Logical Requirements

ORDER = ORDER NUMBER	← Unique identifier of the output.
+ ORDER DATE	← One of many fields that must take on a value. Lack of parentheses indicates a value is required.
+ CUSTOMER NUMBER	
+ CUSTOMER NAME	
+ CUSTOMER SHIP ADDRESS = ADDRESS >	← Pointer to a related definition.
+ (CUSTOMER BILLING ADDRESS = ADDRESS >)	
+ 1 { PRODUCT NUMBER +	← A group of fields that repeats 1 − n times. Parentheses indicates optional value.
QUANTITY ORDERED} N	
+ (DEFAULT CREDIT CARD NUMBER)	← An optional field, meaning one that does not have to have a value.

In the absence of such precise requirements, there may exist discovery prototypes that were created during requirements analysis. In either case, a good requirements statement should be available in some format.

After reviewing input requirements specified during requirements analysis for the SoundStage case study, it was determined that there were three inputs that pertained to the subject VIDEOTAPE. It was determined that a single input screen could be used to support the three inputs NEW VIDEO TITLE, DISCONTINUED VIDEO TITLE, and VIDEO TITLE UPDATE. The data content for the three inputs should capture or display the following data:

PRODUCT NUMBER +

UNIVERSAL PRODUCT CODE +

QUANTITY IN STOCK +

PRODUCT TYPE +

MANUFACTURER'S SUGGESTED RETAIL UNIT PRICE +

CLUB DEFAULT UNIT PRICE +

CURRENT SPECIAL UNIT PRICE +

CURRENT MONTH UNITS SOLD +

CURRENT YEAR UNITS SOLD +

TOTAL LIFETIME UNITS SOLD +

TITLE OF WORK +

CATALOG DESCRIPTION +

COPYRIGHT DATE +

CREDIT VALUE +

PRODUCER +

DIRECTOR +

VIDEO CATEGORY

The attributes PRODUCT NUMBER, MONTHLY UNIT SALES, YEAR UNIT SALES, and TOTAL UNIT SALES are not to be entered by the user. Rather, these attributes are to be automatically generated by the system. Also, for the TITLE COVER, the user will be expected to simply specify a bitmap file that will contain an actual image of the new video title.

Step 2: Select Appropriate GUI Controls   Now that we have an idea of the content for our input, we can address the proper screen-based control to use for each attribute to appear on our screen. Using the repository-based programming approach, we would first check to see if such decisions and other attribute characteristics have already been made and recorded as repository entries. If so, we would simply reuse those repository entries that correspond to the attributes we will use on our input screens. In those cases where there is no repository entry, we will have to simply create them.

To choose the correct control for our attributes, we must begin by examining the possible values for each attribute. Here are some preliminary decisions regarding our input attributes identified in the previous step:

- PRODUCT NUMBER, CURRENT MONTH UNITS SOLD, CURRENT YEAR UNITS SOLD, TOTAL LIFETIME UNITS SOLD, UNIVERSAL PRODUCT CODE, MANUFACTURER'S SUGGESTED RETAIL UNIT PRICE, CLUB DEFAULT UNIT PRICE, CURRENT SPECIAL UNIT PRICE, PRODUCER, and DIRECTOR attributes all have input data values that are unlimited in scope or noneditable. Since the designer is unable to provide the user with a meaningful list of values from which to choose, a single-line text box was chosen. Since the attribute CATALOG DESCRIPTION also fits this criteria, a multiple-line text box (referred to as a memo box by some products) was selected.

- PRODUCT TYPE, LANGUAGE, VIDEO ENCODING, SCREEN ASPECT, and VIDEO MEDIA TYPE all contain a limited predefined set of values. Therefore, it was determined that radio buttons would be the preferred screen-based control for these input items.

- It was determined CLOSED CAPTION? is an input attribute that contains a yes/no value. Therefore, a check box was selected as the control for this attribute.

- QUANTITY IN STOCK, RUNNING TIME, COPYRIGHT DATE, and CREDIT VALUE contain data values that can be sequenced in a predictable manner. Thus, a spin box with an associated text box would be a good choice for these attributes.

- The attributes VIDEO CATEGORY and VIDEO SUBCATEGORY contain a large number of predefined values. With so many attributes to display on our screen, it was determined that a drop-down list would be the best control choice.

- TITLE COVER presented an interesting challenge. Its value is actually a drive, directory, and name of a file that contains a bitmap image of the cover of the video title. This attribute will make use of an advanced control called an image box to store a picture of the video title cover. When this object is selected by the user, a set of controls and special dialogue (user interaction) will be used to capture the input for this item. We'll illustrate this input later in step 3.

Once again, there are many other screen-based controls that could be used to input data. Our examples focus on the most commonly used controls. How well you complete this activity will be a function of how knowledgeable you are with these common controls and other more advanced controls.

Step 3: Design, Validate, and Test Inputs   This step involves developing prototype screens for users to review and test. Their feedback may result in the need to return to steps 1 and 2 to add new attributes and address their characteristics.

Let's take a look at a couple of SoundStage screen prototypes. Figure 14.8 represents a possible prototype screen for handling NEW VIDEO TITLE, DISCONTINUED VIDEO TITLE, and VIDEO TITLE UPDATE. The logo appearing in the upper-right portion of the screen was included to adhere to a company standard—all screens must display the company logo. The buttons also appearing in the upper center and right portion of the screen were added because of the decision to combine the three inputs into a single screen. They were needed to allow the

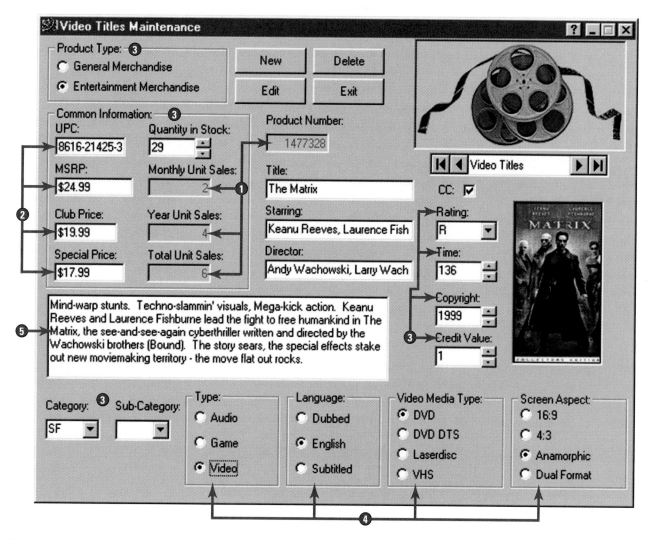

Figure 14.8   *Input Prototype for Video Title Maintenance*

user the option of selecting the desired type of input and record action. We will discuss these buttons and other command and navigation controls and their use in Chapter 15.

Note the following issues in Figure 14.8:

❶ The PRODUCT NUMBER, MONTHLY UNIT SALES, YEAR UNIT SALES, and TOTAL UNIT SALES are screened in a special color as a visual clue to the user that these fields are locked and they cannot enter data into them. These fields are automatically generated by the system. Other fields appearing on the screen have a white background as a visual clue that they can be edited.

❷ Edit masks were specified for these input fields. The UNIVERSAL PRODUCT CODE field contains dashes in specified locations. The user does not actually enter these dashes. Rather, the user simply types in the numbers and afterward the entire content is redisplayed according to the specified edit mask. The same is true for the MANUFACTURER'S SUGGESTED RETAIL PRICE, CLUB DEFAULT UNIT PRICE, and CURRENT SPECIAL UNIT PRICE fields. For example, in either of these three fields the user could type the number 9, press enter, and the content would be redisplayed (according to the edit mask) with a dollar sign and decimal point.

❸ Each field on a screen has been given a label that is meaningful to the users. Feedback from users indicated "CC" was a commonly recognized abbreviation for "closed caption." Also, the users indicated that a label was not necessary for CATALOG DESCRIPTION.

❹ Related radio buttons have been arranged in a group box that contains a descriptive label. Group boxes are frequently used to visually associate a variety of controls that are related. For example, the fields inside the group box labeled "Common Information" were grouped because the user associates these attributes to any type of SoundStage product. Also, each label that corresponds to a radio button option is not what is actually input and stored in the database. Rather, what you see is the meaning of the value. The actual value that is stored is a code. For example, the code value E would actually be stored instead of "English" if the user selects the radio button labeled "English" for the attribute LANGUAGE.

❺ The multiple-line text box has a vertical scroll bar feature. This is a visual clue that there is additional text not appearing inside the CATALOG DESCRIPTION field.

When prototyping input screens, it is important to let the user exercise or test the screens. Part of that experience should involve demonstrating how the user may obtain appropriate help or instructions. New versions of Microsoft products use what is called "tooltips" to provide a brief description of buttons and boxes that appear on a screen. The tooltip description displays when the user positions the mouse over the top of the object. Also, the F1 key is universally accepted as initiating context sensitive help. A help button is another option. Whichever approach(es) you use, it is not necessary to actually implement the help in a prototype.

Finally, prototypes need not display all details to a user unless they are requested (or triggered by a user action). For example, the drop-down list for Motion Picture Association of America RATING code displays only a default value. However, the downward pointing arrow is a visual clue that a list box containing possible values exists. The list box may be viewed by simply clicking on the downward pointing arrow. The result of that action is illustrated in the margin.

A Drop-Down Menu

The previous example was fairly simple because it contained only data that might be updated in one database table. But what if an input includes data to be updated in more than one table? And suppose there is a one-to-many relationship between the tables. Consider MEMBER ORDER that has a one-to-many relationship to MEMBER ORDERED PRODUCTS. How do we design a single input to capture the data for both tables?

Figure 14.9 represents a prototype screen for entering MEMBER and MEMBER ORDERED PRODUCTS on a single form. The form is segmented into two window-panes. MEMBER data is in the top pane, and MEMBER ORDERED PRODUCT data is in the bottom pane. You may be wondering what happens if the number of MEMBER ORDERED PRODUCTS exceeds the space allotted for that pane. In other words, where is the scroll bar for the bottom pane? Many *Windows* GI controls are "intelligent." If the number of rows in the bottom pane exceeds the space, a vertical scroll bar will automatically appear.

As one last *Windows* example, Figure 14.10 represents a single screen design that consolidates three different or similar inputs from our data flow diagrams: NEW MEMBER, MEMBER CANCELLATION, and MEMBER UPDATE. This form also uses the standard input controls that we've discussed in this chapter. The consolidation of logical and physical data flows into single screen designs is very common.

Step 4: If Necessary, Design the Source Document   If a source document will be used to capture data, we must also design that document. The source document is for the system user. In its simplest form, the prototype may be a simple sketch or an industrial artist's rendition.

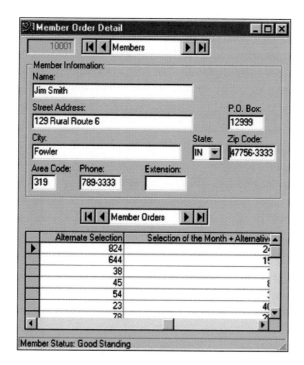

Figure 14.9
Input Prototype for Member Order

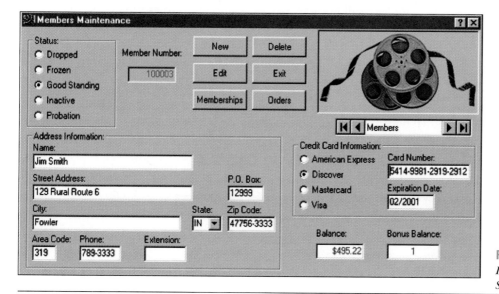

Figure 14.10
Input Prototype for Member Shopping

A well-designed source document will be divided into zones. Some zones are used for identification; these include company name, form name, official form number, date of last revision (an important attribute that is often omitted), and logos. Other zones contain data that identifies a specific occurrence of the form, such as form sequence number (possibly preprinted) and date. The largest portion of the document is used to record transaction data. Data that occurs once and data that repeats should be logically separated. Totals should be relegated to the lower portion of the form because they are usually calculated and, therefore, not input. Many forms include an authorization zone for signatures. Instructions should be placed in a convenient location, preferably not on the back of the form.

Prototyping tools have become more advanced in recent years. Spreadsheet programs such as Microsoft's *Excel* can make very realistic models of forms. These tools give you outstanding control over font styles and sizes, graphics for logos, and the like. Laser printers can produce excellent printouts of the prototypes.

Another way to prototype source documents is to develop a rough model using a word processor. Pass the model to one of the growing number of desktop publishing systems that can transform the rough model into impressive-looking forms (so impressive, in fact, that some companies now develop forms this way instead of subcontracting their design to a forms manufacturer).

Web-Based Inputs and E-Business

The last input design considerations we want to address concern Web-based outputs. The SoundStage project will add various e-commerce and e-business capabilities to the Member Services information system. Some of these capabilities will require Web-based inputs that must be designed.

One logical output requirement for the project is Web-based MEMBER ORDER. We just showed you the client/server version. Now let's look at the Web-based version. It is common to present a Web storefront (Figure 14.11). In addition to providing the member with information about SoundStage products (an output), the member can click the "buy" button to initiate a purchase. That takes the member to what has become a common metaphor screen in e-commerce applications, the *shopping cart* screen (Figure 14.12). Web interfaces tend to be somewhat more artistic than *Windows* interfaces. Perhaps that is part of the appeal. The interface needs to be visually appealing to entice the customer to purchase products in the absence of a verbal sales pitch. In Figure 14.12:

❶ The shopping cart "frame" is independent of the general navigation frame (on the left). The latter allows the user to search and browse the entire website, it is hoped to find additional products to add to the shopping cart.

❷ Buttons, text boxes, hyperlinks, drop-down boxes, and other common controls are here applied to a Web interface instead of a *Windows* interface.

❸ A checkout hyperlink sends the member to the next "page" to complete the transaction.

The Web interface offers several advantages such as the automatic ability for members to use their forward and backward buttons to navigate different inventory and order pages at the website.

WHERE DO YOU GO FROM HERE?

This chapter provided a detailed overview of the input design tasks of a project. If you haven't covered output design, you should go back and read Chapter 13 next. Otherwise, the next logical unit is Chapter 15, User Interface Design. User interface design ties the input and output screens together into an overall user experience. As we did in this chapter, we will address both client/server, *Windows*-based user interfaces and Web-based, e-business solutions.

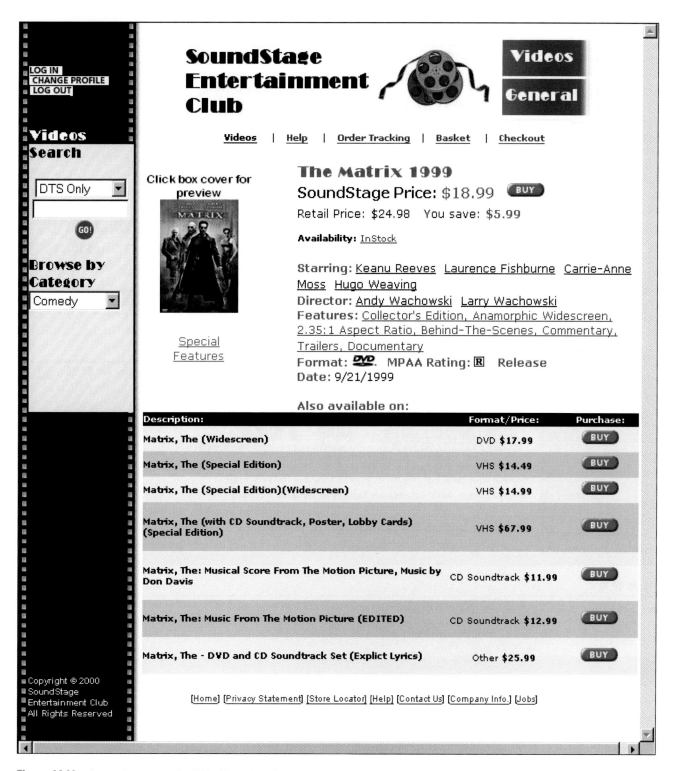

Figure 14.11   *Input Prototype for Web Shopping Cart*

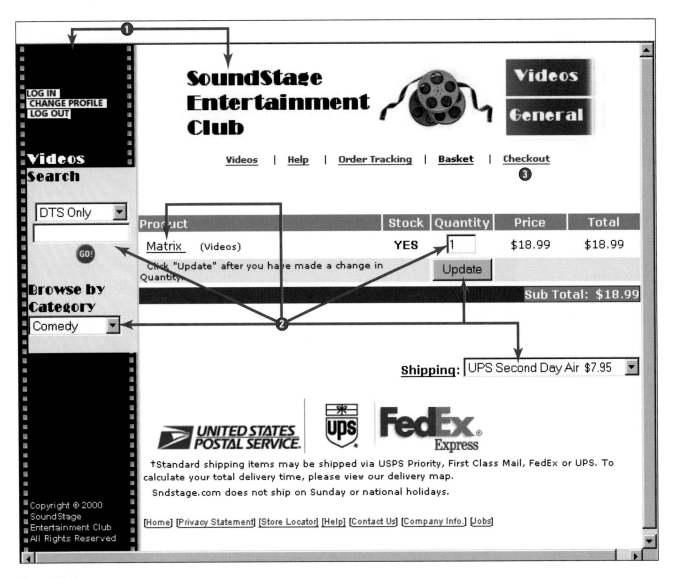

Figure 14.12

SUMMARY

1. Several concepts are important to input design. One of the first things you must learn is the difference between data capture, data entry, and data processing. Alternative input media and methods must also be understood before designing the inputs. And because accurate data input is so critical to successful processing, file maintenance, and output, you should also learn about human factors and internal controls for input design.

2. Data happens! It accompanies business events called transactions. Examples include orders, time cards, reservations, and the like. This is an important concept because system designers must determine when and how to

capture the data. The designer must understand the difference between the following:

– Data capture is the identification and acquisition of new data.

– A source document is a paper form used to record business transactions in terms of data that describes those transactions.

– Data entry is the process of translating the source data into a computer-readable format. That format may be a magnetic disk, an optical-mark form, a magnetic tape, or a floppy diskette, to name a few.

3. Data must be processed using one of the following techniques.
   – In batch processing, the entered data is collected into files called batches that are processed later.
   – In on-line processing, the captured data is processed immediately.
   – In remote batch processing, data is entered using on-line editing techniques; however, the data is collected into batches for later processing.
4. The systems analyst usually selects the method and medium for all inputs. Input methods include:
   – Keyboard   – Optical mark
   – Mouse   – Magnetic ink
   – Touch screen   – Electromagnetic signature
   – Point-of-sale   – Smart cards
   – Sound and speech   – Biometrics
5. Most new applications being developed today consist of screens having a "graphical" looking appearance. This type of appearance is referred to as a graphical user interface (GUI).

6. Inputs should be as simple as possible and designed to reduce the possibility of incorrect data being entered. Furthermore, the needs of data-entry clerks must also be considered. With this in mind, system designers should understand human factors that should be evaluated during input design.
7. Input controls ensure that the data input to the computer is accurate and that the system is protected against accidental and intentional errors and abuse, including fraud.
8. When designing input screens for an application that will contain a GUI appearance, the designer must be careful to select the proper control object for each input attribute. Each control serves a specific purpose, has certain advantages and disadvantages, and should be used according to guidelines. Some of the most commonly used screen-based controls for inputting data include: text box, radio button, check box, list box, drop-down list, combination box, and spin box.

KEY TERMS

REVIEW QUESTIONS

1. What is the difference between data capture, data entry, and data processing?
2. What is a source document and what is its role in input design?
3. Explain the difference between batch, on-line, and remote batch methods.
4. Should all new applications be built using on-line input methods? Explain your answer.
5. Why are automatic data collection (ADC) technologies finding their way into batch and on-line applications? Identify five types of ADC technology and the types of business applications where they may be utilized.
6. List and describe six input data validation techniques.
7. What is the purpose of a text box control? When should a text box be used to input a data attribute? Identify some guidelines that should be followed when using a text box control on an input screen.
8. What is the purpose of a radio button box control? When should a radio button be used to input a data attribute? Identify some guidelines that should be followed when using a radio button control on an input screen.
9. What is the purpose of a check box control? When should a check box be used to input a data attribute? Identify some guidelines that should be followed when using a check box control on an input screen.
10. What is the purpose of a list box control? When should a list box be used to input a data attribute? Identify some guidelines that should be followed when using a list box control on an input screen.
11. What is the purpose of a drop-down list control? When should a drop-down list be used to input a data attribute? Identify some guidelines that should be followed when using a drop-down list control on an input screen.

12. What is the purpose of a combo box control? When should a combo box be used to input a data attribute? Identify some guidelines that should be followed when using a combo box control on an input screen.

13. What is the purpose of a spin box control? When should a spin box be used to input a data attribute? Identify some guidelines that should be followed when using a spin box control on an input screen.

PROBLEMS AND EXERCISES

1. Define an appropriate input method and medium for each of the following inputs:
   a. Customer magazine subscriptions.
   b. Hotel reservations.
   c. Bank account transactions.
   d. Taking inventory in a warehouse
   e. Course registration.
   f. Package deliveries.
   g. Mail orders.
   h. Customer order cancellations.
   i. Employee weekly time cards.

2. For each type of internal control described in this chapter, describe an example in which data integrity errors could occur in the absence of that control?

3. Determine the data attributes that would be required for the following inputs and design client/server or Web screens for them.
   a. New student at the college or university.
   b. Course request for registration.
   c. Ticket request for sporting events at a music hall, stadium, or arena.
   d. Pay-per-view television programs. Do both sides, the television (the consumer) and the cable company (the provider).

4. What implications would input design likely have on output design? How about the other way around?

5. What are some advantages to graphical screen interfaces over traditional, text-based screens?

6. Why are source documents appropriate for on-line inputs?

7. Describe how Web-based inputs are similar to and different from client/server input forms.

8. Select an input screen from a *Windows*-based product and redesign it with a Web interface. Remember, Web interfaces tend to be more artistic.

9. Select an input screen from a Web interface and redesign it with a *Windows* interface. Remember, *Windows* interfaces tend to be less artistic. In fact, the design goal of most *Windows* interfaces is to emulate the customer's office productivity suite—Microsoft *Office*, IBM/Lotus *SmartSuite*, or Corel *PerfectOffice*.

10. Using a spreadsheet package, prototype a course enrollment report that presents the following information. For each course, we want to show all departments that requested space in and/or used space in the course. For each department, we need to know the number of spaces requested (even if zero), the number of reservations given, the number of spaces actually used so far, and the number of spaces under or over the reservations given. For the course as a whole, we need to know the course limit (sum of reservations given), number of seats used (sum), a number of spaces under or over the total reservations, and the number of physical seats in the classroom.

11. What is the worst report you have ever seen or used? What made it hard to read or use? How would you personally improve the design?

PROJECTS AND RESEARCH

1. Review some sample GUI input screens of an application. Evaluate that application's use of GUI controls. Suggest at least three improvements to the input screens.

2. Obtain a copy of an application form—such as a loan, housing, or school form—or any other document used to capture data (such as a course scheduling form, credit card purchase slip, or time card). Do not be concerned whether the application is currently input to a computer system. How do the people who initiate or process the form feel about it? Comment on the human engineering. How well is it divided into zones? Comment on the suitability of the application for data entry. Are data fields that wouldn't be keyed properly located? What changes would you make to the form?

3. Research the following current issue. With the emergence of e-commerce, considerable research is being conducted on Web interfaces that enhance and detract from the shopping experience. Researchers are exploring everything from the number of clicks required to find and order a product, to the impact of different shopping metaphors. Write a five-page report that consolidates and summarizes the latest thinking. Be sure to include appropriate footnotes and references.

4. The order-filling operation for a local pharmacy is to be automated. Customer prescriptions are to be entered on-line by pharmacists. The pharmacy expects the new system to consist of user-friendly, graphical screens. The content of the input PRESCRIPTION contains the following

attributes. For each input attribute, indicate a proper screen-based control to use on the screens that will be used to enter a PRESCRIPTION.

A PRESCRIPTION contains the following attributes:
  CUSTOMER NAME—20 characters
  DOCTOR NAME—20 characters
  1 to 10 occurrences of the following:
    DRUG NAME—30 characters
    QUANTITY PRESCRIBED—4 digits
    MEDICAL INSTRUCTIONS—120 characters
    RX NUMBER—a federal licensing number of 6 digits
  1 to 10 occurrences of the following added by pharmacist:
    DRUG NUMBER—a number that uniquely identifies a prescription drug; 6 characters
    LOT NUMBER—a number that uniquely identifies the lot from which a chemical was produced; 6 characters
    DOSAGE FORM—the form of the medication issued, such as "pill." P5 = PILL C5 = CAPSULE L5 = LIQUID I5 = INJECTION R5 = LOTION
    UNIT OF MEASURE—G5 = grams O5 = ounces M5 = milliliters
    QUANTITY DISPENSED—4 digits
    NUMBER OF REFILLS—2 digits

and optionally:
  EXPIRATION DATE—date prescription expires

5. A moving company maintains data concerning fuel-tax liability for its fleet of trucks. When truck drivers return from a trip, they submit a journal describing mileage, fuel purchases, and fuel consumption for each state traveled through. This data is to be input daily to maintain records on trucks and fuel stations. For each TRIP JOURNAL data attribute, indicate the GUI control to be used on an input screen.

A TRIP JOURNAL consists of the following attributes:
  TRUCK NUMBER—4 characters
  DRIVER NUMBER—9 characters
  CODRIVER NUMBER—9 characters
  TRIP NUMBER—3 characters
  DATE DEPARTED
  DATE RETURNED
  1 to 20 of the following:
    STATE CODE—2 characters
    MILES DRIVEN—5 digits
    FUEL RECEIPT NUMBER—9 characters
    GALLONS PURCHASED—3 digits (1 decimal)
    TAXES PAID—4 digits (2 decimals)
    STATION NAME—10 characters
    STATION LOCATION—15 characters

MINICASES

1. For years, managers at Eudrup, Inc., have expressed frustration with the budget management information system. When it was implemented, every manager would receive a two-inch binder of financial reports every year. Most managers found the reports difficult to navigate, especially when they needed to trace detailed expenditures from summary reports. They begged IT for an on-line system. When they finally got one, they basically received the same collection of summary, detailed, and exception reports as they had received when those reports came in the two-inch binder. They could not easily navigate from summary to detailed.

   At a recent management retreat, managers attempted to articulate their vision of a much improved system. The managers envisioned an on-line home screen that would list each of their accounts and the current balance. They would like to be able to highlight any one account and then drill down to all of the individual transactions that had occurred against that account in either of two intervals: a month or the fiscal year to date. And if they then had any questions about any of those transactions, they would like to be able to drill down to the transaction detail record for that account. A transaction detail record could be any of the following: a purchase order, stores order, work order, transportation order, printing services order, or journal voucher (the latter for interdepartment funds transfers).

   How would you design this output(s)? How would you ensure that the managers would finally be satisfied

by the solution? Design prototypes for a *Windows*-based solution. Design prototypes for a Web-based solution.

2. Kathleen Smathers, a programmer/analyst for the Wholesale Cost-Plus Club, had an early afternoon appointment with Linda Pratney, the Accounts Payable assistant manager. Wholesale Cost-Plus Club is a large, citywide warehouse outlet that sells virtually any type of merchandise to club members for a cost very close to wholesale price (significantly below the retail prices charged by grocery stores, drugstores, department stores, and other retail stores). Kathleen had been largely responsible for the Accounts Payable system implemented last fall. Accounts Payable pays off invoices from the suppliers of the club's merchandise. The meeting's purpose was a mandatory postimplementation review of the new system. (Post-implementation reviews occur one month after a new system replaces an old system.) It was no company secret that Linda was displeased with some aspects of the new system.

   LINDA   Kathleen, I guess you heard that I'm having a little trouble with this new payables system. You told me that the computer screens would make my people's job much easier, and that just hasn't happened. I'm also receiving a lot of complaints from our suppliers.

   KATHLEEN I'm really sorry it hasn't worked for you. But I'm willing to work overtime to make this system work the way you need it to work. What isn't working? I thought I

LINDA included all the fields you requested for entering the invoices. Is it that I didn't provide adequate training for your people?

LINDA That's not the problem. Everyone is saying that the system is simply not easy to use, that it takes them longer than it should to enter invoices. They're asking why the interface doesn't have the same type of simple looking and easy-to-use screens as their other PC programs.

KATHLEEN They must be talking about their *Windows*-based products. They are all running the standard Microsoft Office products. That would have been difficult to do because I'm not up to speed on these new programming languages that allow you to develop those kinds of graphical screens.

LINDA I don't get it, Kathleen. Why couldn't you learn? Because of that, my people have to put up with this unforgiving system. I think we all would rather have waited for you to learn.

KATHLEEN You are right. But I thought the data-entry screens were easy to use, even if they aren't graphical. In fact, I was pretty proud of them. I used whatever text-based functions I could to make the input screens ex-

citing. I used a lot of fancy blinking screens, reverse video, and other things like that to make their job of entering invoices more interesting. And besides, what do supplier complaints have to do with my system?

LINDA According to my people, the new system isn't recording the supplier invoices accurately. They . . .

KATHLEEN That's impossible! That can't be. The program asks the data-entry clerks to enter all the information that's on the supplier's invoice forms. If the suppliers are sending us a correct invoice, then there shouldn't be any problems.

LINDA Kathleen, you're denying that there are problems. I'm telling you that there are. And I expect you to resolve them. If you can't or are unwilling, then I'll get someone else.

a. Do you think Linda is overreacting at the end of their meeting?

b. Was Kathleen right in choosing to develop an interface that was not graphical?

c. Why might the supplier invoices be captured incorrectly?

d. Does the use of blinking screens, reverse video, and other fancy functions ensure a good input screen design?

SUGGESTED READINGS

Andres, C. *Great Web Architecture.* Foster City, CA: IDG Books Worldwide, Inc., 1999. Books on effective Web interface design are beginning to surface. The science of human engineering for Web interfaces has not yet progressed as far as client/server interfaces (e.g., *Windows*). Here is an early title that explores many dimensions of Web architecture and interfaces using real-world examples.

Application Development Strategies (monthly periodical). Arlington, MA: Cutter Information Corporation. This is our favorite theme-oriented periodical that follows system development strategies, methodologies, CASE, and other relevant trends. Each issue focuses on a single theme. This periodical will provide a good foundation for how to develop input prototypes.

Dunlap, Duane. *Understanding and Using ADC Technologies. A White Paper for the ADC Industry.* A SCAN TECH 1995 Presentation. October 23, 1995, Chicago. We are indebted to our friend and colleague. Professor Dunlap is a leader in

the field of ADC. This paper was the basis for much of our discussion on the trends in ADC technology.

Fitzgerald, Jerry. *Internal Controls for Computerized Information Systems.* Redwood City, CA: Jerry Fitzgerald & Associates, 1978. This is our reference standard on the subject of designing internal controls into systems. Fitzgerald advocates a unique and powerful matrix tool for designing controls. This book goes far beyond any introductory systems textbook; it is must reading.

Galitz, W. O. *User-Interface Screen Design.* New York: John Wiley & Sons, Inc., 1993. This is our favorite book on overall user interface design. The author offers several flowcharts of the decision process in applying GUI controls to inputs.

Kozar, Kenneth. *Humanized Information Systems Analysis and Design.* New York: McGraw-Hill, 1989. A good user-oriented treatment of input design.

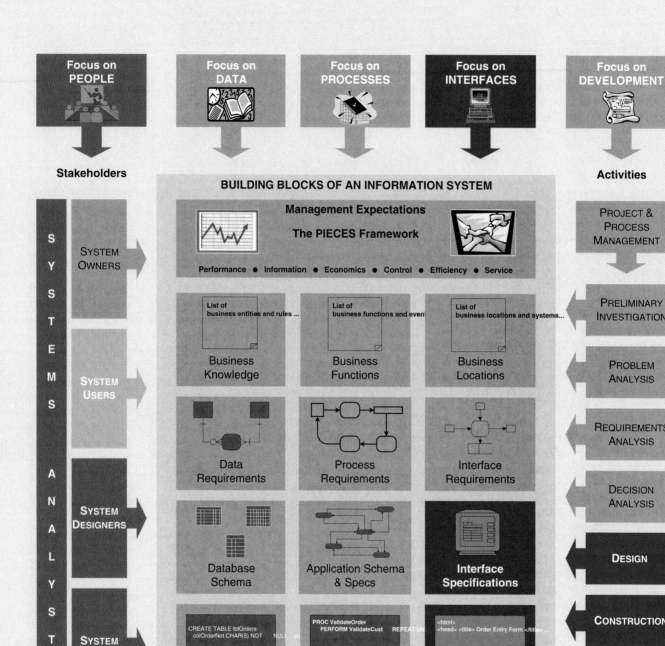

Focus on PEOPLE

Focus on DATA

Focus on PROCESSES

Focus on INTERFACES

Focus on DEVELOPMENT

Stakeholders

Activities

BUILDING BLOCKS OF AN INFORMATION SYSTEM

Management Expectations

The PIECES Framework

Performance ● Information ● Economics ● Control ● Efficiency ● Service

SYSTEM OWNERS

SYSTEM USERS

SYSTEM DESIGNERS

SYSTEM BUILDERS

SYSTEMS ANALYSTS

List of business entities and rules ...

Business Knowledge

List of business functions and event

Business Functions

List of business locations and systems...

Business Locations

Data Requirements

Process Requirements

Interface Requirements

Database Schema

Application Schema & Specs

Interface Specifications

CREATE TABLE tblOrders
colOrderNot CHAR(5) NOT NULL co

Database Programs

PROC ValidateOrder
PERFORM ValidateCust REPEAT UN

Application Programs

<html>
<head> <title> Order Entry Form </title> ...

Interface Programs

PROJECT & PROCESS MANAGEMENT

PRELIMINARY INVESTIGATION

PROBLEM ANALYSIS

REQUIREMENTS ANALYSIS

DECISION ANALYSIS

DESIGN

CONSTRUCTION

IMPLEMENTATION

VENDORS AND CONSULTANTS

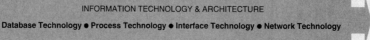

INFORMATION TECHNOLOGY & ARCHITECTURE

Database Technology ● Process Technology ● Interface Technology ● Network Technology

OPERATIONS AND SUPPORT

15

USER INTERFACE DESIGN

CHAPTER PREVIEW AND OBJECTIVES

In this chapter you will learn how to design and prototype the user interface for a system. The user interface should provide a friendly means by which the user can interact with the application to process inputs and obtain outputs. In Chapters 13 and 14, you learned how to design and prototype outputs and inputs. User interface design and prototyping address the overall presentation of the application and may require revisions to those preliminary output and input prototypes. Today there are two commonly encountered interfaces: terminals (or microcomputers behaving as terminals) used in conjunction with mainframes and the more common display monitors connected to microcomputers. There are also several strategy styles for designing the user interface for systems. You will know that you've mastered user interface design when you can:

— Distinguish between different types of computer users and design considerations for each.

— Identify several important human engineering factors and guidelines and incorporate them into a design of a user interface.

— Integrate output and input design into an overall user interface that establishes the dialogue between users and computer.

— Understand the role of operating systems, Web browsers, and other technologies for user interface design.

— Apply appropriate user interface strategies to an information system. Use a state transition diagram to plan and coordinate a user interface for an information system.

— Describe how prototyping can be used to design a user interface.

Note: The authors gratefully acknowledge the contributions and recommendations of Carlin "Rick" Smith at Purdue University. Rick developed most of the prototypes that demonstrate the concepts and techniques taught in this chapter. Rick also contributed ideas for further improving chapter currency and topics.

SCENE

This episode occurs in the conference room where Sandra and Bob are reviewing a prototype of the new system interface with members of the Order Processing staff. The meeting is one week after they met with the users to review input and output design screens.

SANDRA

OK, let's call this meeting to order. I should remind you that when Bob and I previously met with you, we reviewed several rough prototypes of screens that you would be using to enter business transaction data and to view various report information. Since that time, Bob and I made some final revisions to the content of those screens. We will show those revised screens to you. However, we then developed additional screens used to present the overall application and its functions to you. It is the purpose of this meeting to demonstrate the screens that each of you will see when you use the new system. We'd like very much to have your input.

SALLY

I'm not sure we understand what these new screens are?

BOB

Let me address that, Sandra. Sandra and I realize that it would be unreasonable for us to expect you folks to know how to access each of the many individual screens that we reviewed during our last system. Therefore, we developed sort of a menu screen where each of you will simply select the appropriate business function you wish to process. Given that selection, we could have the system then display the appropriate screens to you. It's sort of like your ATM machine. Most ATM machines ask you to enter an access ID and if successful then ask you what you want to do: deposit, withdraw, transfer, or make a payment. Based on your answer, the ATM then knows how to interact with you.

SANDRA

Good analogy, Bob. But if that's unclear, let's get right to the very first screen that each one of you would see. Bob, if you will please activate the new system so that the first screen appears on our projector . . . thanks. As you can see, the first screen you will see is this screen prompting you to enter your ID and password. As with the ATM machine, no one gets past this point without proper authorization!

SALLY

I assume that I will have some say regarding whom those individuals will be.

BOB

Absolutely. I've already drafted a memo asking for those people's names.

SANDRA

Next screen, Bob. OK, on successfully logging onto the new system you will then see this screen. Notice that the screen provides a menu of choices that correspond to the many business functions that are supported by the new Member Services system, again, much like an ATM that would respond with a menu of transaction choices.

SALLY

OK. Now I understand. That's a cool screen!

CLERK A

That is a nice screen. It looks so simple and intuitive. But what is that button with the picture of the man holding a sheet of paper?

CLERK B

Yeah! And what's the button with the balloons?

SANDRA

Bob? You designed this screen. Do you want to address those questions?

BOB

Well, the man with the paper was intended to represent "reports." You'd press that button to obtain a listing of possible reports. The balloons were used to represent "promotions."

CLERK A

That's funny, I thought the man with the reports might have something to do with promotions. He looks like he is selling or promoting something.

CLERK B

I would never have guessed that the balloons implied promotions. There are a few others that don't make sense to me either.

SANDRA

Bob, it sounds like we may have trouble using pictures to clearly suggest most of the available functions. I think we should use clear meaningful text labels for our buttons rather than pictures. That's the problem with pictures. Unless the picture is universally recognized among the people who will use the new system, now and in the future, it's only going to cause confusion.

SALLY

What about the color? Some earlier screens were one color and now this screen is a different color.

BOB

I did that on purpose. Which did you prefer? We will need to make that decision today. I'll make all screens the same color.

SALLY

Definitely the color used on those earlier screens we saw. This bright purple is nice, but it is already annoying me.

SANDRA

I just noticed something, Bob. You have a button labeled "Quit." On a previous screen you called a similar button "Exit." Which is it? We need to be consistent.

BOB

My mistake. According to our standards, it should be called "Exit" throughout.

SANDRA

Let's have one of you volunteer to come up front and take over the keyboard and mouse. Let's test the system. We all can learn better by having each of you give the system a test run.

SALLY

I hate to pull rank, but I'll go first.

BOB

Great! Here's a test ID and password. Let's start the system as if you just showed up for work . . .

―――――――――――――――――

DISCUSSION QUESTIONS

1. What benefits would Sandra and Bob (as well as the users) receive by having the users test the application?

2. Why is it important that the application have consistency regarding colors, labels of buttons, and other features?

3. Why might pictures (graphic representations) not always be preferred on a screen?

4. Why would it be appropriate for Sandra and Bob to have the users view the entire application—since they had already reviewed input and output screens?

USER INTERFACE DESIGN CONCEPTS AND GUIDELINES

In the two previous chapters, we addressed output and input design. In this chapter, we integrate output and input design into an overall user interface that establishes the dialogue between users and computer. The dialogue determines everything from starting the system or logging into the system, to setting options and preferences, to getting help. And the presentation of the outputs and inputs is also part of the interface. We need to examine the screen-to-screen transitions that can occur. In client/server applications (e.g. network-based *Windows*) and Web applications (e.g., Internet- or intranet-based browsers), the user has many alternative paths through menus, hyperlinks, dialogues, and the like. This makes for very accommodating and friendly user interfaces, but greatly complicates design and programming.

Today most user interfaces are designed by rapidly constructing prototypes. These prototypes are generated using rapid application development environments such as Microsoft's *Visual Basic,* Inprise's *JBuilder* (for *Java*), or IBM's *Visual Age* (for various languages). These prototypes are rarely <u>fully</u> functional, but they do contain enough functionality to demonstrate the interface. For example, a help system prototype may be functional to the extent that it calls up a few sample screens to demonstrate levels of assistance. Or a security system might have just enough functionality to demonstrate representative log-in errors even though it is not actually authenticating users. When we get to the construction phase of the life cycle, programmers and analysts will complete the functionality.

We begin our study by examining types of users, human factors, and human engineering guidelines that affect user interface design.

Types of Computer Users

Nowhere are human factors as important as they are in user interface design. Just ask the typical systems analyst who spends half the day answering phone calls from system users who are having difficulty using information systems and computer applications. The overriding consideration of user interface design is the same as for business and technical writing—understand your audience. In interface design, the audience is the SYSTEM USER.

In the chapter map at the beginning of this chapter, we highlighted SYSTEMS ANALYSTS, SYSTEM DESIGNERS, and SYSTEM BUILDERS as the stakeholders that actually <u>perform</u> user interface design, but we also highlighted SYSTEM USERS as the stakeholders for which the user interface design is intended. Users must <u>test</u> and <u>evaluate</u> that design. The chapter map also illustrates that user interface design is performed during the DESIGN and CONSTRUCTION phases and that the activities result in the INTERFACE SPECIFICATIONS (prototypes) and PROGRAM building blocks.

SYSTEM USERS can be broadly classified as either expert or novice—and either nondiscretionary or discretionary.

An **expert user** is an experienced computer user who has spent considerable time using specific application programs. The use of a computer is usually considered nondiscretionary. In the mainframe computing era, this was called a *dedicated user.*

Expert users generally are comfortable with (but not necessarily an expert in) the application's operating environment (e.g., *Windows* or a Web browser). They have invested time in learning to use the computer. They will invest the time in overcoming less-than-friendly user interfaces. In general, they have memorized routine operations to an extent that they neither seek nor want excessive computer feedback and instructions. They want to be able to accomplish their task in as few actions and keystrokes as possible.

> The **novice user** (sometimes called a *casual user*) is a less experienced computer user who will generally use a computer on a less frequent, or even occasional, basis. The use of a computer <u>may</u> be viewed as discretionary (although this is becoming less and less true).

Stated simply, the novice users need more help than the expert users. Help takes many forms including menus, dialogues, instructions, and help screens. Most managers, despite their increasing computer literacy, fall into the novice category. They are paid to recognize and solve problems, exploit opportunities, and create plans and manage the vision—not to learn and use computers. Computers are tools to the modern manager. When the need arises, they want to realize their benefit as quickly as possible and move on.

Expert and novice users are actually extremes in the continuum of all users. The totally novice user who hasn't used a computer is becoming less common. Few college curricula don't require computer literacy for all majors, and students in all majors have discovered the value of increased interdisciplinary computer expertise (sometimes called *informatics*). Novice users also usually graduate to expert users through practice and experience. The net societal impact of the Internet is that more people are becoming increasingly comfortable with computers—creating a class of users that is less novice and more expert with each passing year. Is it any wonder that user interface design is racing toward Web browser-like interfaces, even within *Windows* applications?

It is difficult to imagine today's young students and professionals being uncomfortable with computers. Regardless, most of today's systems are designed for the novice system user, but adapting to the expert user. The focus is on user friendliness or human engineering.

Human Factors

Before designing user interfaces, it may be useful to understand those elements that frequently cause people to have difficulty with computer systems. Our favorite user interface design expert, Wilbert Galitz (see the Suggested Readings) offers the following interface problems:

- Excessive use of computer jargon and acronyms.
- Nonobvious or less-than-intuitive design.
- Inability to distinguish between alternative actions ("What do I do next?").
- Inconsistent problem-solving approaches.
- Design inconsistency.

According to Galitz, these problems result in confusion, panic, frustration, boredom, misuse, abandonment, and other undesirable consequences.

To solve these problems, Galitz offers the following overriding "commandments" of user interface design:

- *Understand your users and their tasks.* This becomes increasingly difficult as we extend our information systems to implement business-to-consumer (B2C) and business-to-business (B2B) functionality using the Internet.

- *Involve the user in interface design.* Find out what the users like and dislike in their current applications. Involve them in screen design and dialogue from the beginning. This commandment is easily enabled with today's PC database and rapid application development technology.

- *Test the system on actual users.* Observation and listening are the key skills here. After initial training, try to avoid excessive coaching and forcing the user to learn the system. Instead, observe their actions and mistakes, and listen to their comments and questions to better understand their interaction with the user interface.

- *Practice iterative design.* The first user interface will probably be unsatisfactory. Expect any user interface design to go through multiple design iterations and testing. When is the interface finished? Probably never! But Galitz suggests that a good goal is that 95 percent of the typical users (be they novice or expert) can perform intended tasks (be they routine or less common) without difficulty or help.

Given the type of user, a number of important human engineering factors should be incorporated into the design:

Human Engineering Guidelines

- *The system user should always be aware of what to do next.* The system should always provide instructions on how to proceed, back up, exit, and the like. Several situations require some type of feedback:
  - *Tell the user what the system expects right now.* This can take the form of a simple message such as READY, TYPE COMMAND, SELECT ONE OR MORE OPTIONS, or TYPE DATA.
  - *Tell the user that data has been entered correctly.* This can be as simple as moving the cursor to the next field in a form or displaying a message such as DATA OK.
  - *Tell the user that data has not been entered correctly.* Short, simple messages about the correct format are preferred. Help functions can supplement these messages with more extensive instructions and examples.
  - *Explain to the user the reason for a delay in processing.* Some actions require several seconds or minutes to complete. Examples include sorting, indexing, printing, and updating. Simple messages such as SORTING—PLEASE STAND BY, or INDEXING—THIS MAY TAKE A FEW MINUTES or PLEASE WAIT tell the user that the system has not failed. The *Windows* hourglass or the *Internet Explorer* revolving globe are iconic clues that processing is occurring.
  - *Tell the user that a task was completed or was not completed.* This is especially important in the case of delayed processing, but it is also important in other situations. A message such as PRINTING COMPLETE or PRINTER NOT READY—TRY AGAIN OR CONTACT YOUR NETWORK ADMINISTRATOR will suffice.

- *The screen should be formatted so that the various types of information, instructions, and messages always appear in the same general display area.* This way, the system user knows approximately where to look for specific information. In most windowing environments, standards often dictate the location of status messages or pop-up dialogue windows.

— *Messages, instructions, or information should be displayed long enough to allow the system user to read them.* Most experts recommend important messages be displayed until the user acknowledges them.

— *Use display attributes sparingly.* Display attributes, such as blinking, highlighting, and reverse video, can be distracting if overused. Judicious use allows you to call attention to something important—for example, the next field to be entered, a message, or an instruction.

— *Default values for fields and answers to be entered by the user should be specified.* In windowing environments, valid values are frequently presented in a separate window or dialogue box as a scrollable region. The default value, if applicable, should usually be first and clearly highlighted.

— *Anticipate the errors users might make.* System users will make errors, even when given the most obvious instructions. If it is possible for the user to execute a dangerous action, let it be known (for example, a message or dialogue box could read ARE YOU SURE YOU WANT TO DELETE THIS FILE?). An ounce of prevention goes a long way!

— *With respect to errors, a user should not be allowed to proceed without correcting an error.* Instructions (and examples) on how to correct the error should be displayed. The error can be highlighted with sound or color and then explained in a pop-up window or dialogue box. A HELP option can be defined to trigger display of additional instructions.

— *If the user does something that could be catastrophic, the keyboard should be locked to prevent any further input, and an instruction to call the analyst or technical support should be displayed.*

Dialogue Tone and Terminology

The overall flow of screens and messages is called a **dialogue.** The tone and terminology of a dialogue are very important human factors in user interface design. With respect to the tone of the dialogue, the following guidelines are offered:

— *Use simple, grammatically correct sentences.* It is best to use conversational English rather than formal, written English.

— *Don't be funny or cute!* When someone has to use the system 50 times a day, the intended humor quickly wears off.

— *Don't be condescending.* Don't insult the intelligence of the system user. For instance, don't offer repeated praise or rewards.

With respect to the terminology used in a computer dialogue, the following suggestions may prove helpful:

— *Don't use computer jargon.*

— *Avoid most abbreviations.* Abbreviations assume that the user understands how to translate them. Check first!

— *Use simple terms.* Use NOT CORRECT instead of INCORRECT. There is less chance of misreading or misinterpretation.

— *Be consistent in your use of terminology.* For instance, don't use both EDIT and MODIFY to mean the same action.

— *Carefully phrase instructions—use appropriate action verbs.* The following recommendations should prove helpful:
  – Use SELECT or CHOOSE instead of PICK when referring to a list of options. Be sure to indicate whether the user can select only one or more than one option from the list of available options.
  – Use TYPE, not ENTER, to request the user to input specific data or instructions. The term ENTER may be confused with the enter key.

– Use PRESS, not HIT or DEPRESS, to refer to keyboard actions. Whenever possible, refer to keys by the symbols or identifiers that are actually printed on the keys. For instance, the ⏎ key is used on some keyboards to designate the RETURN or ENTER key.
– When referring to the on-screen mouse cursor, use the term POSITION THE CURSOR, not POINT THE CURSOR.

USER INTERFACE TECHNOLOGY

Most of today's user interfaces are graphical. The basic structure of the **graphical user interface** (or **GUI**) is provided within either the computer operating system or the Internet browser of choice. In *client/server information systems,* the user interface client is implemented to execute within the PC operating system. In *Internet and intranet information systems,* the user interface is implemented to execute within the PC's Web browser (that, in turn, executes within the PC operating system).

Operating Systems and Web Browsers

The dominant GUI-based operating system for today's client computers (as in a client/server network) is Microsoft *Windows* (various versions). Apple's *Macintosh* and the various flavors of *UNIX* (including *Linux*) also hold market share. For the growing numbers of handheld and palm-top client computers, the current dominant operating system is Palm's *Palm OS*. Microsoft *Windows CE* also holds a percentage of that market.

Increasingly, the operating system is not the key technology factor in user interface design. Internet and intranet applications run within a Web browser. Most browsers run in many operating systems making it possible to design a user interface that is less dependent on the computer itself. The advantages of this computer **platform independence** should be obvious. Instead of writing a user interface for each anticipated computing platform and operating system, you write it for one or two browsers. As we go to press, the dominant Web browsers are Microsoft *Internet Explorer* and Netscape *Navigator,* but version problems within browsers can exist in the user community.

In addition to the operating systems and browsers, the overall design of a user interface is enhanced or restricted by the available features of the users' display monitor, keyboard, and pointing devices. Let's briefly examine some of the other considerations.

Display Monitor

The size of the display area is critical to user interface design. Not all displays are PC monitors! A number of non-PC *terminals* still exist. Terminals are non-PC displays that merely display data and information transmitted by a remote computer, usually a mainframe. And while many terminals have been replaced by PCs, users are still frequently forced to interface with the legacy mainframe applications using *terminal emulators* that open a window on the screen that still displays information and instructions in the original, pre-*Windows* terminal format. For these terminals and terminal emulators, the two most common display areas were 25 lines by 80 columns and 25 lines by 132 columns.

Fortunately, the personal computer monitor has replaced most terminals, and most newer and reengineered applications are being written to a graphical interface. For PC monitors, we don't measure the display in terms of lines and columns. And while diagonal measures such as inches are often quoted, the more relevant measure is graphical resolution. Graphical resolution is measured in pixels, the number of distinct points of light displayed on the screen. Today's most common resolution is 800,000 horizontal pixels by 600,000 vertical pixels in a 17-inch diagonal display. Larger display sizes support even more pixels; however, the designer should generally design the user interface with the assumption of the lowest common or reasonable denominator.

Obviously, handheld and palm-top computers and specialized terminal displays (such as those in cash registers and ATMs) support much smaller displays that must be considered in user interface design.

The manner in which the display area is shown to the user is controlled by both the technical capabilities of the display and the operating system capabilities. Paging and scrolling are the two most common approaches to showing the display area to the user.

> **Paging** displays a complete screen of characters at a time. The complete display area is known as a page (or screen). The page is replaced on demand by the next or previous page; much like turning the pages of a book.

The other common alternative to paging is called scrolling.

> **Scrolling** moves the displayed information up or down on the screen, one line at a time. This is similar to the way movie and television credits scroll up the screen at the end of a movie.

Once again, PC displays offer a wider range of paging and scrolling options.

Keyboards and Pointers

Most (but not all) terminals and monitors are integrated with keyboards. The obvious exception is palm computers such as the PalmPilot. The critical features of the keyboard include character set and function keys.

The character set of most PC keyboards is fairly standard. These character sets can be extended with software to support additional characters and symbols. For specialized terminals or workstations, the manufacturer can design custom keyboards. Most keyboards contain special keys called function keys. PC keyboards usually have 12 such function keys. Terminals have been known to include as many as 32 function keys.

> **Function keys** (usually labeled F1, F2, and so on) can be used to program certain common, repetitive operations in a user interface (for example, HELP, EXIT, and UPDATE).

In an operating system, function keys are often predefined, but application developers can customize them for specific systems. Function keys should be used consistently. That is, any information system's programs should consistently use the same function keys for the same purposes. For example, F1 is commonly used as the help key in both operating systems and applications.

Most GUIs (including operating systems and browsers) use pointing devices including mice, pens, and touch-sensitive screens. Obviously, the most common pointer is the mouse.

> A **mouse** is a small hand-sized device that sits on a flat surface near the terminal. It has a small roller on the underside. As you move the mouse on the flat surface, it causes the pointer to move across the screen. Buttons on the mouse allow you to select objects or commands to which the cursor has been moved.

Driven by the need to scroll through Web pages and other documents, many mice now include a wheel that allows a user to more easily scroll through pages and documents without using the scroll bars.

> **Pens** are becoming important in applications that use handheld devices (such as PalmPilots). Because such devices frequently don't include keyboards, the user interface may need to be designed to allow "typing" on a keyboard displayed on the screen or using a handwriting standard such as *Graffiti* or *Jot*. Prebuilt components exist to implement these common features.

As previously noted, the most common user interface is graphical—either

Windows-based or Web browser-based. The remainder of this chapter will focus on graphical user interface design.

User interface design is the specification of a dialogue or conversation between the system user and the computer. This dialogue generally results in data input and information output. There are several styles of graphical user interfaces. Traditionally these styles were viewed as alternatives, but they are increasingly blended. This section presents an overview of several different styles or strategies used for designing graphical user interfaces and how they are being incorporated into today's applications. We will demonstrate these styles with popular software applications.

The basic construct of a GUI (both operating system- and browser-based) is the **window.** A window is a rectangular, bordered area. A title (and optionally a file name) is displayed at the top of each window.

A window can be smaller or larger than the actual display monitor's viewable area. It usually includes standardized controls in the upper right-hand corner to *maximize* itself to the display screen's size, *minimize* itself to an icon (at the bottom of the screen), toggle to a previous size, and *exit* (or *close*).

The file, form, or document displayed within a window may or may not fit in that window. When the file, form, or document exceeds the window size, **scroll bars** on the right-hand side and bottom of the window are used to navigate that file, form, or document and indicate the current position of the cursor relative to the entire file, form, or document.

A window may be divided into zones called **frames.** Each frame can act independently of the other frames in the same window, using features such as paging, scrolling, display attributes, and color. Each frame can be defined to serve a different purpose. Frames are common in both *Windows* and Web browsers.

Within a window or frame, you can use all of the **user interface controls** that were used in the previous two chapters (such as *text boxes, radio buttons, check boxes, drop-down lists, buttons,* etc.). Additionally, many other user interface type controls will be introduced later in the chapter.

Finally, a window frequently has a **task bar** or tray across the bottom of the window. This task bar can be used to display messages, progress, or special tools (to be discussed later).

The oldest and most commonly employed dialogue strategy is menu selection. Different types of menus cater to novice and expert users.

> **Menu-driven** strategies require the user to select an action from a menu of alternatives.

Menu-driven dialogues actually predate GUIs. A typical pre-GUI hierarchical menu is illustrated in Figure 15.1. Menu options can be logically grouped into high-level options to simplify presentation. As shown in the figure, if the main menu option DISPLAY WARRANTY REPORTS is selected, the submenu WARRANTY SYSTEM REPORT MENU will appear. Then, if the PART WARRANTY SUMMARY option is selected, the report customization and report screens are displayed in sequence. There is no technical limit to how deeply hierarchical menus can be nested. However, the deeper the nesting, the greater the need for direct paths to deeply rooted menu options for the expert user who may find navigating through multiple levels annoying (called *screen thrashing*). And most users also require ways to escape back to the main or higher-level submenus without backtracking through each of the original screens.

GRAPHICAL USER INTERFACE STYLES AND CONSIDERATIONS

Windows and Frames

Menu-Driven Interfaces

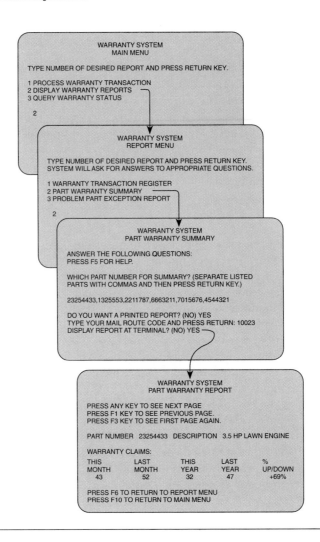

FIGURE 15.1
A Classical Hierarchical Menu Dialogue

Pre-GUI hierarchical menus were relatively easy to design. A dialogue chart such as the one shown in Figure 15.2 (taken from an earlier edition of our book) was used to map the screen-to-screen transitions and ensure consistency and completeness. But the arrival of graphical user interfaces greatly complicated menu design.

In operating system GUIs such as Microsoft *Windows,* user dialogues are not hierarchical. Think about it! Between the time you start and exit a *Windows* application, such as your word processor, the number of different actions and paths you take through your application is seemingly endless. The dialogue is hardly hierarchical! Such dialogue design cannot be modeled as easily by hierarchically based dialogue charts. Let's examine how GUI menus work.

Pull-Down and Cascading Menus In a GUI, menus are usually implemented with pull-down and cascading menus from a **menu bar** as shown in Figure 15.3(a). Each menu option is actually a group of related commands and actions. A menu template is shown in the margin. Many of these menu groups are common to many or all applications. For example, *Windows*-based applications typically include the following menu groups:

Menu template

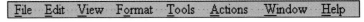

Users can select a menu group either using the mouse or a keyboard shortcut (e.g., simultaneously pressing the Alt-key plus the underlined letter, called a *mnemonic, shortcut,* or *hot-key*).

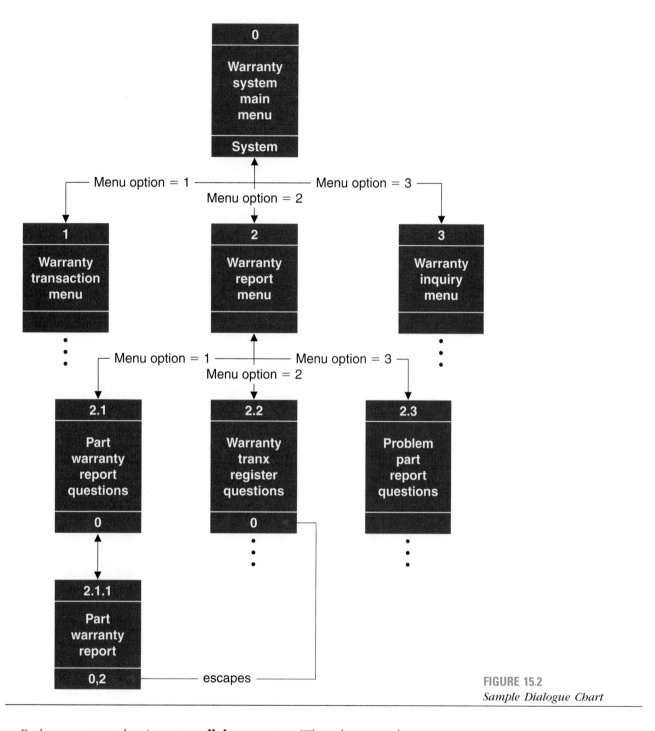

FIGURE 15.2
Sample Dialogue Chart

Each menu group has its own **pull-down menu.** When the user selects a group from the menu bar, a submenu is pulled down.

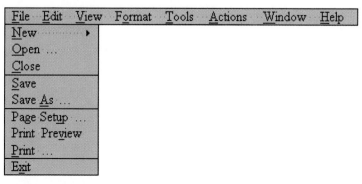

Notice that the submenu choices may be subgrouped by horizontal lines (e.g., grouping all SAVE or print submenu commands). In some cases, a named submenu action is followed by ellipses (three dots) indicating that a **dialogue box** (window) (see Figure 15.3b) will subsequently appear (pop up) to present additional

---

FIGURE 15.3

(a) Pull-down and cascading menus

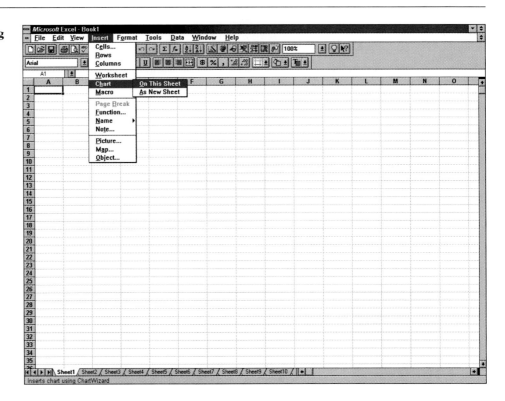

(b) Dialogue box

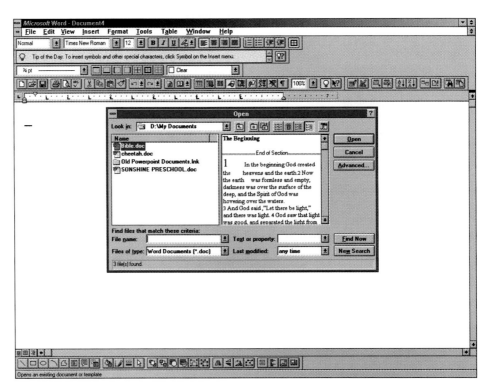

options or collect additional instructions. In other cases, a named submenu action will have a small arrow indicating yet another submenu. This is called a **cascading menu.**

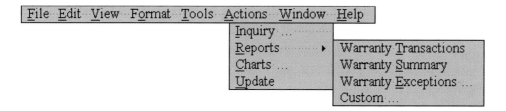

Tear-Off and Pop-Up Menus   Not all menus are relegated to the menu bar. Some GUIs (such as IBM's *OS/2 Presentation Manager* and UNIX's *OSF/Motif*) allow **tear-off menus.** With a tear-off menu, the user can select a drop-down menu or cascaded menu, "drag it off" the menu bar, and relocate it elsewhere on the screen. This is especially useful if the menu must be continually used. Only a copy of the original menu is actually torn off.

A **pop-up menu** is context sensitive and dependent on a pointing device. Activated by the user's clicking of the right mouse button, a menu pops up from nowhere (see Figure 15.4). That menu that pops up depends on the location of the cursor on the screen. The cursor may be pointing to a blank area, a field, a cell, a word, or an object. The right button click will bring up a menu displaying only those actions that apply to whatever is at that cursor location—hence the term *context sensitive.* Pop-up menus may also cascade. Pop-up menus are primarily for expert users because there is no visual clue to their presence.

FIGURE 15.4
Pop-Up Menus

Toolbar and Iconic Menus   **Toolbars** consist of **icons** (pictures) that represent menu shortcuts for actions and commands that are normally embedded in the drop-down and cascading menus (see Figure 15.5). In *Windows* applications, a tool-bar of commonly used actions is found immediately beneath the menu bar. The user can click on any of these tools or icons to immediately invoke that action without going through the menus. Toolbars can be created for any application. Application developers can provide users with some flexibility to customize those toolbars.

While the default location for most toolbars is immediately under the menu bar, many applications allow toolbars to be relocated to the left, right, or bottom of the window at the convenience of the user. This is called *docking* the toolbar. Also, some toolbars can be made to *float* (or move) within any convenient loca-tion <u>inside</u> the window.

> NOTE   In Web-based applications, the toolbar is provided by the browser and cannot be cus-tomized to specific applications. The most important icons on the browser toolbar are the PAGE FORWARD, PAGE BACKWARD, and HOME PAGE icons that are standard to all Web-based Internet and intranet navigation.

Iconic menus use pictures to represent menu options in the main <u>body</u> of the window. In *Windows* applications, these iconic menus are frequently used to provide a control center (of main functions and activities) for a computer application or to document the business steps in using a computer application. Figure 15.6 demonstrates an iconic menu. Each button represents an intuitive menu choice.

Iconic menus are very popular in Web-based applications because those appli-cations run in the browser—browsers do not allow the developer to alter the menu commands in the browser's menu bar. Instead, Web applications frequently use clickable pictures, icons, and buttons to represent the menu options.

FIGURE 15.5
Toolbars

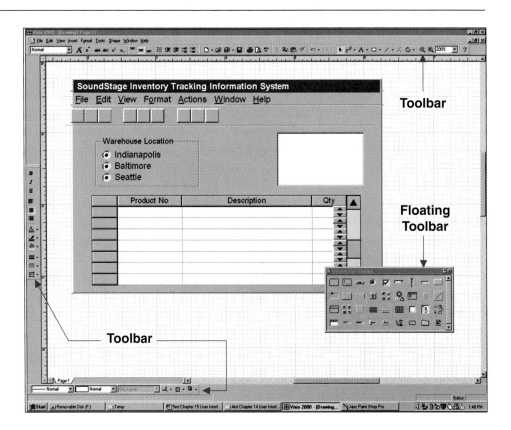

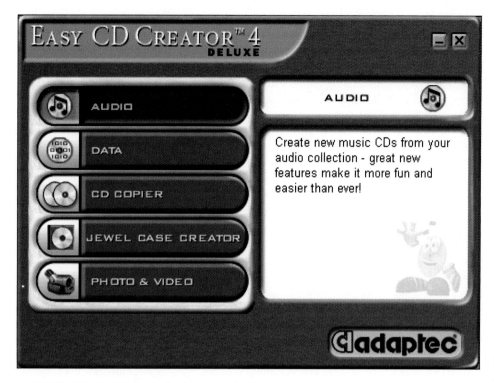

FIGURE 15.6
Iconic Menu

The popularity of Web-like interfaces is significantly influencing *Windows* user interface design. Most client/server information systems have implemented the client user interface to emulate the user's most commonly used PC tools such as the word processor and spreadsheet. The user is familiar with those tools; therefore, it makes sense to design other applications to mimic those menus, toolbars, dialogue boxes, and the like. But the popularity of Web-based applications has given rise to a new consumer-style interface.

Like Web pages, **consumer-style interfaces** for *Windows* applications are somewhat more artistic. While menu bars may still be used, the primary look and feel of the window is more Web-like; thus, it is more consumer friendly. The interface consists of clickable icons and buttons that replace more traditional *Windows* menu approaches. When not overly complicated, this can be a friendlier "face" for *Windows* applications than is traditionally seen in applications such as Microsoft *Office* and Lotus *SmartSuite*. A consumer-style *Windows* interface is illustrated in Figure 15.7.

Notice the absence of the traditional menu bar. Also notice that the buttons (in the left frame) do not conform to traditional *Windows* size and style. The background image is more artistic, as is the use of fonts and color. (Many organizations include a graphic designer as part of the team to develop consumer-style interfaces.) We expect such consumer-friendly styles to be embraced by future *Windows*-based information systems.

Hypertext and Hyperlink Menus   **Hypertext** and **hyperlinks** are products of contemporary Web-based user interfaces. Hypertext and hyperlinks were originally created to navigate within and between Web pages and sites. A word, term, or phrase is marked as a hyperlink (usually formatted as underlined text, usually with color). Clicking on the hyperlink navigates the user to the associated page (or bookmark in a page).

This technology can be easily extended and adapted to implement menus in Web-based Internet and intranet applications. Because these applications run in the browser, and because the browser's menu bar and commands are fixed, we cannot easily implement custom menus as we do in *Windows* applications. Instead, we use hypertext and hyperlinks to implement those menus in the body of the Web page. Each menu option is a hypertext phrase (or a hyperlinked icon or button) that invokes actions or forms on other Web pages. Essentially, this approach creates hierarchical menu structures similar to those that were introduced earlier in Figure 15.1. It is something of an irony that menu design for Web-based applications is being driven by an approach that returns to a style that was extensively used in legacy mainframe applications!

Hypertext and hyperlinks are no longer exclusive to Internet and intranet applications. Many contemporary *Windows* applications have embraced the popularity of the Web by presenting a **hybrid *Windows*/Web** user interface. For example, Figure 15.8 demonstrates the user interface for Intuit's *Quicken,* the popular personal finance program. With its many hyperlinks, it looks like a Web page. While it does include many optional Web-enabled features, it is actually a *Windows* application! The first clue is that it runs in its own window, not the browser's window. It also has its own *Windows* menu bar, complete with all the custom pull-down and cascading menus that are common to *Windows* applications. We expect this hybrid interface to become increasingly pervasive as businesses embrace the Internet and intranets as the fundamental foundation for all information systems.

Instruction-Driven Interfaces

Instead of menus, or in addition to menus, some applications are designed using a dialogue based on an **instruction set** (also called a *command language interface*). Because the user must learn the syntax of the instruction set, this approach is most suitable for expert users. Three types of syntax can be defined. Determining which type should be used depends on the available technology.

— A **language-based syntax** is built around a widely accepted command language that can be used by the user to invoke actions. Examples include *Query by Example (QBE)* and *Structured Query Language (SQL),* both of

FIGURE 15.7
Consumer-Style Interface

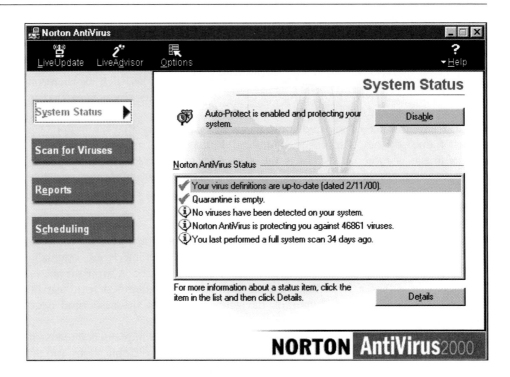

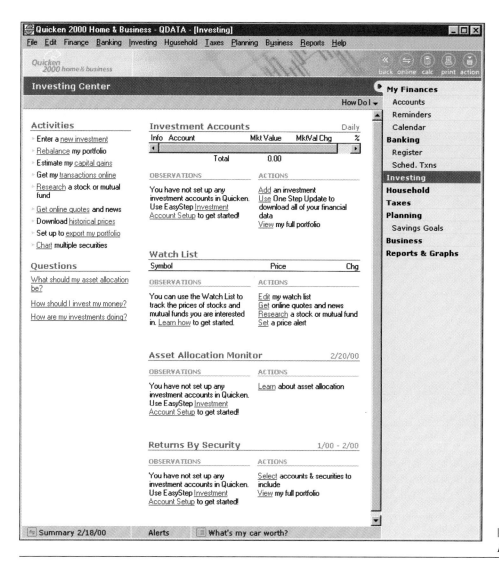

FIGURE 15.8
Hybrid Windows/Web Interface

which are database languages that can be used by the end-user to access data and create custom reports.

— A **mnemonic syntax** is built around commands defined for custom information system applications. Users are provided a screen console in which they can enter commands that will invoke actions and responses from the computer user. Ideally, the commands should be meaningful to the user (including any abbreviations allowed).

— **Natural language syntax** allows users to enter questions and commands in their native language. The system interprets these commands against a known syntax and requests clarification if it doesn't understand what the user wants.

Instruction-driven styles were common to legacy mainframe applications and early DOS-based PC applications. But this style of interaction can still be found in today's graphical applications. For example, Microsoft's *Access* database product contains a query facility that allows the developer to visually (point-and-click) develop a query (see Figure 15.9). The developer simply selects from database tables, columns, and rows to include in a query as shown in Figure 15.9(a). Then, if desired, the developer can view and edit the command-level SQL code that implements the query, as shown in Figure 15.9(b). Once again, the instruction set approach requires a degree of user expertise, experience, and know-how.

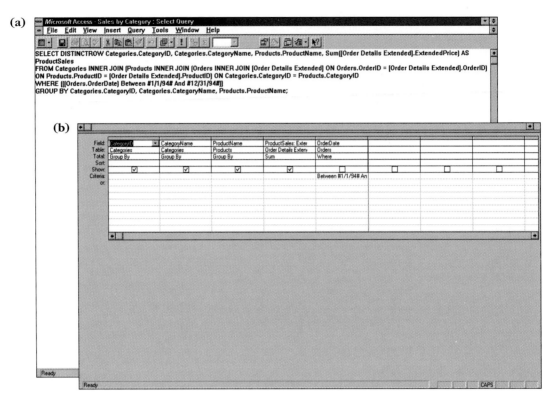

FIGURE 15.9 *Instruction-Driven Interface*

Question-Answer Dialogues

A **question-answer dialogue** style is primarily used to supplement either menu-driven or instruction-driven dialogues. Users are prompted with questions to which they supply answers. The simplest questions involve YES or NO answers—for instance:

DO YOU WANT TO SEE ALL PARTS? [NO].

Notice how the user was offered a default answer! Questions can be more elaborate. For example, the system could ask:

WHICH PART NUMBER ARE YOU INTERESTED IN?

This strategy requires that you consider all possible correct answers and deal with the actions to be taken if incorrect answers are entered. Question-answer dialogue is difficult because you must try to consider everything that the system user might do wrong!

Question-answer dialogues are very popular in Web-based applications. For example, a car reservation system may ask a series of questions to define what type of car and rental agreement you require.

WHERE DO YOU WANT TO PICK UP YOUR RENTAL VEHICLE?

WHERE DO YOU PLAN TO RETURN YOUR RENTAL VEHICLE?

WHAT IS THE PICKUP DATE AND TIME?

WHAT IS THE RETURN DATE AND TIME?

WHAT TYPE AND SIZE OF VEHICLE DO YOU NEED?

DO YOU HAVE ANY PROMOTIONAL COUPONS? . . .

A drop-down list of alternative answers may accompany each question. Together, these questions and answers define a business transaction.

In addition to establishing a user interface style, there are certain special considerations for user interface design. How will users be recognized and authenticated to use the system? Are there any security or privacy considerations to be accommodated in the user interface? Finally, how will users get help via the user interface?

Special Considerations for User Interface Design

Internal Controls—Authentication and Authorization   In most environments, system users must be authenticated and authorized by the system before they are permitted to perform certain actions. In other words, system users must "log into" the system. Most log-ins require both a USER ID and a PASSWORD. System users should not be required to learn and memorize multiple USER IDS and PASSWORDS. Ideally, they should be required to use the same log-in as is used for their local area network account. (*Windows NT* and *2000* allow for this authentication to occur without the need to retype either field.)

Figure 15.10(a) demonstrates the user interface for the SoundStage log-in. The USER ID and PASSWORD will be authenticated against the network accounts file. Notice that the password is printed as asterisks as the user types it in, a common security and privacy measure. Should the user ID or password fail to be authenticated, the security authorization dialogue in Figure 15.10(b) will be displayed.

Authentication is only half of the solution. Once authenticated, the user's access and service privileges for <u>this</u> information system must be established. There are many models for establishing and managing privileges. An important guideline is to assign privileges to *roles,* not to individuals. In most businesses, people change jobs routinely—they are reassigned and promoted to new job responsibilities and roles. Also, job descriptions and roles change from time to time. Finally, people leave the business and some are terminated. For all of these reasons, privileges

FIGURE 15.10

(a) Authentication log-in screen

(b) Authentication error screen

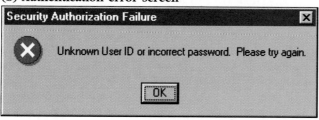

should be assigned to roles. Then it is a simple matter of identifying those roles that any USER ID can assume.

For each role, the specific privileges that should be assigned to the role need to be defined. Privileges may include permission to read specific tables or views; permission to create, update, or delete records (rows) in specific tables or views; permission to generate and view specific reports; permission to execute specific transactions; and the like. Although not technically part of the interface, defining these roles and permissions is needed to both design an appropriate log-in interface and to functionally specify the complete authentication and authorization security model for the system.

Different user views could actually be applied to customize the user interface for different categories of users. For example, it is fairly easy to "ghost" (change the font from black to gray) and disable those menu options and dialogue boxes that are to be restricted from certain classes of system users.

With the emergence of e-commerce, consumers and other businesses must have confidence that we are who we claim to be. Consumers may be providing credit card numbers and other private information for transmission over the Internet. For this reason, SoundStage purchased a Web certification to authenticate itself to its club members and prospective members. At anytime, using the browser interface, SoundStage members can view the authentication certificate in Figure 15.11. With this certification, the SoundStage Web site will display a "Secure Server Certification" icon (see margin—the *padlock*) that will tell consumers their data will be encrypted (securely scrambled) to ensure that their credit card and personal data is not being intercepted or accessed by others when passed along the network.

Internet Browser Security Indicator

FIGURE 15.11
Server Security Certificate

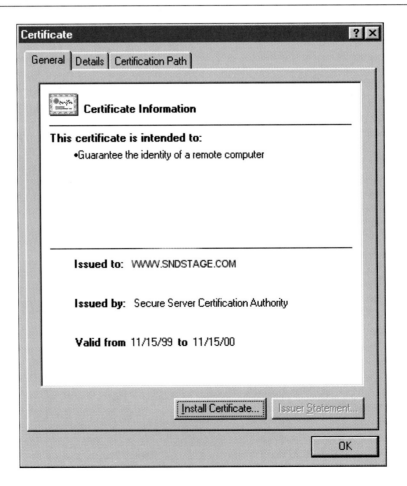

On-Line Help   Nobody wants to read the manuals anymore! At least that's the way it seems. And to some degree, it's justified. Manuals are essentially sequential files of information. People want immediate, direct access to context-sensitive help, that is, help that is smart enough to figure out what they might be trying to do. There is definitely a trend to build help systems and tutorials directly into the application. On-line help becomes part of the user interface.

The general-purpose help for an application is built into the <u>H</u>elp menu for *Windows* applications. For Web applications, help is usually built as separate pages, usually "pop-up" pages in separate windows, so the user can also remain focused on the page that initiated the need for help.

Today, HTML (<u>Hypertext Markup Language</u>) is gradually becoming the universal language for constructing help systems for graphical user interfaces—both Web and *Windows* applications. For example, the entire help system for Microsoft *Office 2000* is written in HTML.

The design, construction, and testing of a help system is simplified by today's automated tools. A complete **help system** includes a table of contents, numerous instructions, examples, and a thorough index. Many **help authoring packages,** such as Blue Sky's *RoboHelp,* leverage the help author's word processor to help with the planning, outlining, writing, indexing, and hypertext linking aspects of authoring a complete help system.

A well-designed help system will implement a wide range of help elements. Perhaps the most commonly encountered types of help are those that users must initiate. As mentioned earlier, the F1 function key is almost universally accepted as a help request command. Likewise, a standard <u>H</u>elp menu bar option is commonly used to organize and present different types and levels of help in most *Windows*-based applications (commercial or custom-built). Finally, as is illustrated in Figure 15.12, *Windows* and Web-based interfaces frequently use **tool tip** controls to provide pop-up help associated with specific tool and object icons. Tool tips appear when the user momentarily positions the cursor over the icon (or object) on the screen. Tool tips are appropriate for all icons because the user

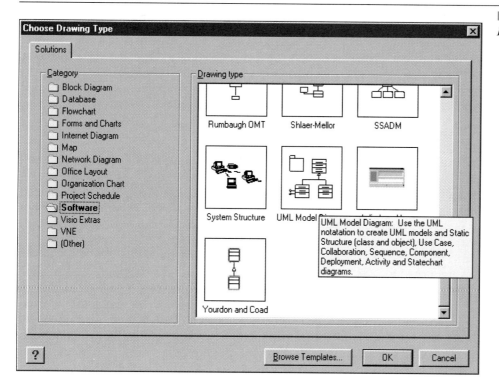

FIGURE 15.12
Help Tool Tip

interface designer can never be assured that the image or label appearing on an icon is going to be meaningful to the system user.

Two additional and common help features particularly effective for the more novice user are help wizards and help agents (or assistants). As is illustrated in Figure 15.13(a), a **help wizard** guides the users through complex processes by presenting a sequence of dialogue boxes that require user input and system feedback. We call your attention to the following:

❶ As is typical of help wizards, the dialogue usually includes a series of instructions or questions for the user to respond to.

❷ The wizard contains explanations to aid in the user's understanding and decision making.

❸ The wizard also provides a button to request more detailed help in completing the task.

❹ The "Next" button suggests additional or subsequent steps to be supported by the help wizard. (The "Next" button is usually changed to "Finish" once a sequence of dialogue boxes is complete.) Figure 15.13(b) represents the resulting screen and subsequent step supported by the help wizard.

Microsoft and third-party software control vendors actually sell wizards to help developers construct wizards!

Agents are another technology with applications to help systems.

Agents are reusable software objects that can operate across different software applications and even across networks.

Microsoft's **help agent** (referred to as an assistant) provides a common help assistant across all *Office 97* and *Office 2000* applications. In its default form, it presents itself as an animated paper clip (see margin). (Microsoft's help agent can be programmed into custom applications, both for *Windows* and for the *Internet Explorer* Web browser.) A single user click on this help agent initiates help.

The Microsoft help agent is complemented by natural language processing technology (see margin) that allows the user to write an inquiry in natural language phrases that are interpreted by the agent to present the most likely help responses. The user can then select one of those responses or enter into the more detailed help index.

The overriding theme for designing a good help system is that the designer should anticipate system user errors. When designing the user interface to report such errors, the designer should always provide the system user with help to resolve the error. After leaving any help session, users should always be returned to where they were in the application before requesting or receiving the help.

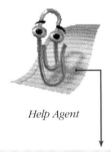

Help Agent

What would you like to do?

💡 Change the default chart type

💡 Troubleshoot charts

💡 Change the display of chart labels, data tables, legends, gridlines, or axes

💡 Change the view of a 3-D chart

💡 Change chart labels, titles, and other text

💡 Change the way data is plotted

How do I customize the X-Axis on my chart

[Options]   [Search]

Natural Language Processing

HOW TO DESIGN AND PROTOTYPE A USER INTERFACE

Today's graphical environments create an emphasis on developing an overall system that blends well into the user's workplace environment. The following sections will demonstrate how to design a user interface for a graphical environment. We will draw on examples from the SoundStage case study. We will examine both client/server, *Windows*-based inputs and Web-based, e-commerce inputs that run in a browser.

Automated Tools for User Interface Design and Prototyping

The automated tools to support user interface design and prototyping are the same as those tools we identified in Chapters 13 and 14 for output and input design. The most commonly used automated tool for user interface design is the PC-database application development environment. Most PC-database products such as Microsoft's *Access* are not powerful enough to develop most enterprise-level applications, but they are more than adequate to use in prototyping an application's user interface screens. Given a database structure (easily specified in *Access*), you can

(a)

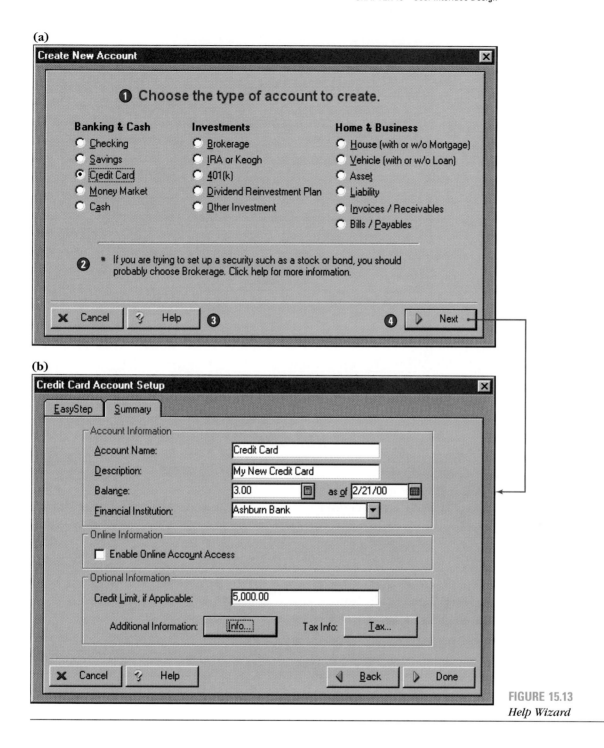

(b)

FIGURE 15.13
Help Wizard

quickly generate or create forms to input data. You can include <u>most</u> of the GUI controls we described in this chapter. The users can subsequently exercise those forms and tell you what works and what doesn't.

Many CASE tools also include facilities for screen layout and prototyping using the project repository created during requirements analysis. *System Architect 2001*'s screen design facility was previously demonstrated in Chapter 13, Figure 13.7.

Most GUI-based programming languages such as *Visual Basic* can easily be used to construct nonfunctional prototypes of user interface screens. The key term here is *nonfunctional*. The forms will look real, but there will be no code to implement any of the buttons or fields. That is the essence of <u>rapid</u> prototyping. For example, Figure 15.14(a) demonstrates a *Visual Basic* dialogue to build a simple menu. Figure 15.14(b) is the resultant menu.

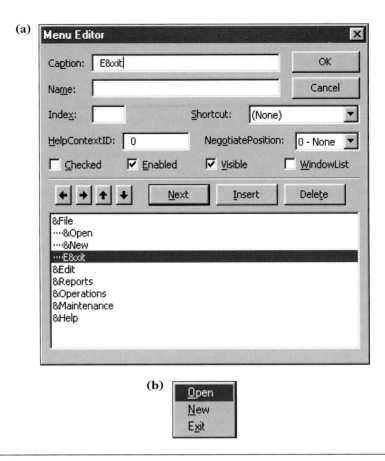

FIGURE 15.14
Visual Basic *Menu Construction*

In Chapter 14 we introduced a number of input controls that could be included in any window. The number of controls available to the interface designer is limited only by the applications development environment that will be used to construct the interface. Figure 15.15 illustrates a few additional controls that are available in the *Visual Basic* environment, including outlook bars, sortable columns with headings, gauge controls, directory list boxes, and noninput drop-down lists.

The User Interface Design Process

User interface design is not a complicated process. The basic steps involved are:

1. Chart the user interface dialogue.
2. Prototype the dialogue and user interface.
3. Obtain user feedback.
4. If necessary, return to Step 1 or 2.

In reality, the above steps are not strictly sequential in practice. Instead, the steps are iterative—for example, as prototypes are developed they are reviewed by the system users who provide feedback that may require revisions or a new prototype. In the following subsections, we examine these steps in a single iteration and illustrate a few examples from the SoundStage project.

Step 1: Chart the Dialogue   A typical user interface may involve many possible screens (which may consist of several windows), perhaps hundreds! Each screen can be designed and prototyped. But what about the coordination of these screens?

Screens typically occur in a specific order. You may also be able to toggle among the screens. Additionally, some screens may appear only under certain conditions. To make matters even more difficult, some screens may occur repet-

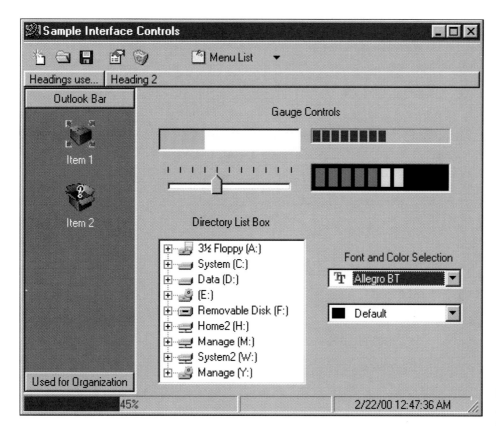

FIGURE 15.15
Additional User Interface Controls

itively until some condition is fulfilled. This sounds almost like a programming problem, doesn't it? We need a tool to coordinate the screens that can occur in a user interface. A **state transition diagram** is used to depict the sequence and variations of screens that can occur when the system user sits at the terminal. (The authors are using the term *screen* in a general sense. When designing graphical interfaces, the term may refer to an entire display screen, a window, or a dialogue box.) You can think of it as a road map. Each screen is analogous to a city. Not all roads go through all cities. Rectangles are used to represent display screens. Arrows represent the flow of control and the triggering event causing the screen to become active or receive focus. The rectangles only describe what can appear during the dialogue. The direction of the arrows indicates the order in which these screens occur. A separate arrow, each with its own label, is drawn for each direction because different actions trigger flow of control from and flow of control to a given screen.

Let's examine a dialogue that is under construction for the SoundStage project (see Figure 15.16). The partially completed SoundStage state transition diagram is being developed using a CASE product, Popkin's *System Architect*. Note the following:

❶ The partial state transition diagram includes references to some of the SoundStage input screens developed in Chapter 14.

❷ The diagram also includes references to some of the output screens designed in Chapter 13.

❸ The MEMBER SERVICES SYSTEM screen will be a new screen that will need to be designed and prototyped. This screen will serve as the application's main window. It will play a major role in providing the user with the ability to get access to the system's input and output screens that were designed earlier. It will also provide the user with the ability to complete a number of

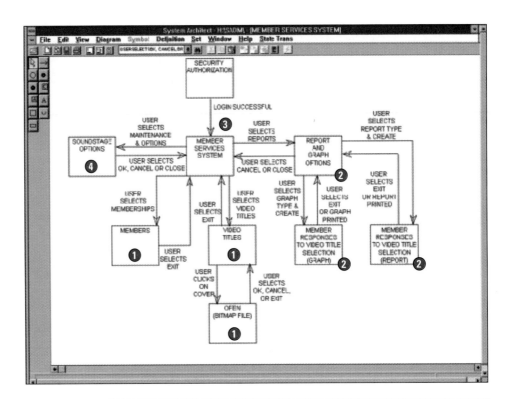

FIGURE 15.16
SoundStage Partial State Transition Diagram

additional functions (beyond input and output processing) that are commonly established during user interface design. It will be accessible only when the users have first been provided with the SECURITY AUTHORIZATION screen and have successfully logged into the system.

❹ The SOUNDSTAGE OPTIONS screen is another new screen to be created. This screen will allow users to set various user options and defaults to be used during their session. For example, selecting a printer, zooming, and many other options.

State transition diagrams such as the one presented in Figure 15.16 can become quite large, especially when all input, output, help, and other screens are added to the diagram. Therefore, it is common to partition the diagram into a set of separate simpler and easier-to-read diagrams.

Step 2: Prototype the Dialogue and User Interface   Recall that we have some new screens to design and prototype. Some of these new screens were identified to bring together the application and its input and output screens that were designed earlier. Some screens were identified to provide the users some flexibility in customizing the application's interaction to suit their own preferences. Still others may have been identified to deal with system controls, such as backup and recovery.

Let's look at some new screens that were to be created for the SoundStage Member Services System. System users would first be presented with the authentication log-in screen that was discussed earlier in the chapter (see Figure 15.10). According to the state transition diagram, the successful log-in of a user results in the SoundStage Member Services System main menu screen depicted in Figure 15.17. Notice the following:

❶ The users and their access privileges are confirmed. Based on their access privilege, certain functions will be enabled and disabled.

FIGURE 15.17   *SoundStage Main Menu*

❷ Through a menu bar selection *or* through a vertical menu of buttons, the user can complete common Member Services business operations. These buttons will lead to screens that allow the user to process appropriate transactions via input screens designed and prototyped earlier. Text labels were used for buttons because the analyst was unable to establish icons (pictures) that all users could readily identify with as a representation of the operations. The menu bar and buttons contain hot keys to provide the user with the flexibility of selecting via the keyboard or mouse. A group box was used to visually associate the buttons that represent related operations.

❸ The user has the ability to complete various routine maintenance operations.

Via the menu bar of the MEMBER SERVICES SYSTEM screen, users can choose to set options for their work session. This new screen is depicted in Figure 15.18.

❶ This screen utilizes **tabs** as a means of allowing the user to alter four related sets of options.

❷ A **slider** control is used to allow the user to adjust the priority for background queries. This control is often used for items whose values are best presented as a spatial representation and when an approximate rather than precise value is sufficient.

In reality, the analyst would need to prototype the content and appearance of the "General," "Print," and "View" tabs as well as the "Database" tab. According to the state transition diagram, this screen will return control to the parent window, MEMBER SERVICES SYSTEM.

According to the state transition diagram, system users are also to be provided with the opportunity to specify report customization preferences. Figure 15.19 depicts a prototype screen that allows SoundStage users to choose a particular report (or graph) and customize its content.

Study the state transition diagram and the screens that we just examined to see how this portion of the overall system dialogue would work. By studying the entire collection of screens, you may discover the need to revise some screens. Such issues as color, naming consistencies of common buttons and menu options, and other look-and-feel conflicts may need to be resolved. Once again, adherence to any standards governing GUIs should be confirmed.

FIGURE 15.18

SoundStage Options and Preferences Screen

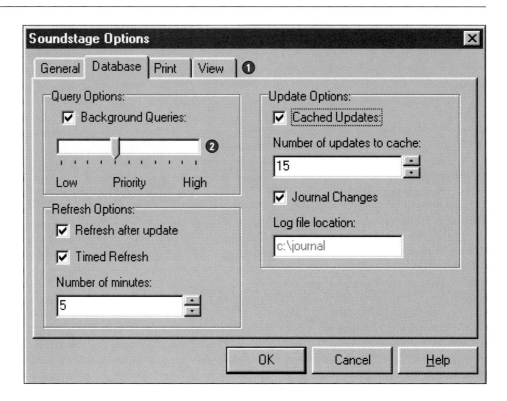

FIGURE 15.19

SoundStage Report Customization Dialogue Screen

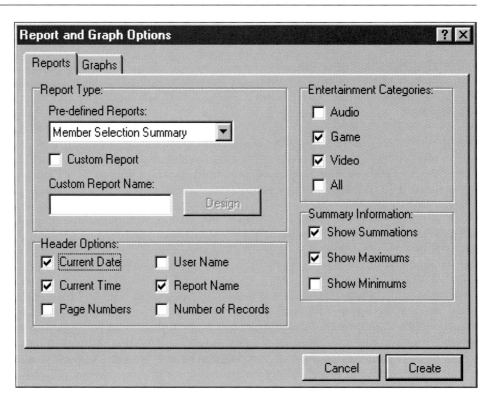

Step 3: Obtain User Feedback   Exercising (or testing) the user interface is a key advantage of all these prototyping environments we have alluded to throughout this chapter.

> **Exercising (or testing) the user interface** means that system users experiment with and test the interface design before extensive programming and actual implementation of the working system. Analysts can observe this testing to improve on the design.

In the absence of prototyping tools, the analyst should at least simulate the dialogue by walking through the screen sketches with system users. User feedback is essential in user interface design. The analyst should encourage the user to participate in testing the application's interface. Finally, the analyst should expect to revisit steps 1 and 2 as needed changes become known.

WHERE DO YOU GO FROM HERE?

This chapter provided a detailed overview of user interface design and prototyping. You are now ready to learn how to construct and implement a system. The construction phase will cover how to build and test the networks, databases, and programs for the new system. The implementation phase will cover how to conduct a system test, develop a conversion plan, install databases, train users, and complete the conversion from the old to the new system. You will learn about the construction and implementation phases in the next chapter. Before proceeding, we recommend that you first review Chapter 3 to see where software design falls into the overall systems development process.

SUMMARY

1. User interface design is concerned with the dialogue between a user and the computer. It is concerned with everything from starting the system or logging into the system to the eventual presentation of desired outputs and inputs.
2. Most user interfaces are designed by rapidly constructing prototypes using rapid application development environments. Such prototypes are rarely fully functional.
3. Relative to user interface design, the system users can be broadly classified as either expert or novice.
   a. An expert user is an experienced computer user who will spend considerable time using specific application programs. The use of a computer is usually considered nondiscretionary.
   b. The novice user is a less experienced computer user who will generally use a computer on a less frequent, or even occasional basis. The use of a computer <u>may</u> be viewed as discretionary.
4. Several human factors frequently cause people to have difficulty with computer systems, including these interface problems:
   a. Excessive use of computer jargon and acronyms.
   b. Nonobvious or less-than-intuitive design.
   c. Inability to distinguish between alternative actions ("What do I do next?").

   d. Inconsistent problem-solving approaches.
   e. Design inconsistency.
5. Galitz offers the following overriding "commandments" of user interface design:
   a. Understand your users and their tasks.
   b. Involve the user in interface design.
   c. Test the system on actual users.
   d. Practice iterative design.
6. Given the type of user for a system, there are a number of important human engineering factors that should be incorporated into the design:
   a. The system user should always be aware of what to do next.
   b. The screen should be formatted so that the various types of information, instructions, and messages always appear in the same general display area.
   c. Messages, instructions, or information should be displayed long enough to allow the system user to read them.
   d. Use display attributes sparingly.
   e. Default values for fields and answers to be entered by the user should be specified.
   f. Anticipate the errors users might make.
   g. A user should not be allowed to proceed without correcting an error.

h. The system user should never get an operating system message or fatal error.

i. If the user does something that could be catastrophic, the keyboard should be locked to prevent any further input, and an instruction to call the analyst or technical support should be displayed.

7. The overall flow of screens and messages is called a dialogue. With respect to the tone of the dialogue, the following guidelines are offered:

a. Use simple, grammatically correct sentences.

b. Don't be funny or cute!

c. Don't be condescending.

With respect to the terminology used in a computer dialogue, the following suggestions may prove helpful:

a. Don't use computer jargon.

b. Avoid most abbreviations.

c. Use simple terms.

d. Be consistent in your use of terminology

e. Carefully phrase instructions—use appropriate action verbs.

8. Most of today's user interfaces are graphical. The basic structure of the graphical user interface (or GUI) is provided within either the computer operating system or in the Internet browser.

9. The overall design of a user interface is enhanced or restricted by the available features of the user's display monitor, keyboard, and pointing devices.

10. There are several styles of graphical user interfaces, including menu-driven, instruction-driven, and question-answer dialogues.

11. Menu-driven strategies require the user to select an action from a menu of alternatives. GUI menu implementation may include:

a. Pull-down and cascading menus.

b. Tear-off and pop-up menus.

c. Toolbar and iconic menus.

d. Hypertext and hyperlink menus.

12. Instruction-driven interfaces are designed using a dialogue based on an instruction set. Three types of syntax may be used for the instruction set:

a. Language-based syntax, which is built around a widely accepted command language that can be used by the user to invoke actions.

b. Mnemonic syntax, which is built around commands defined for the custom information system applications.

c. Natural language syntax, which allows users to enter questions and commands in their own native language.

13. A question-answer dialogue style is primarily used to supplement either menu-driven or instruction-driven dialogues. Users are prompted with questions to which they supply answers.

14. Internal controls and on-line help are some special considerations that should go into user interface design.

15. User interface design consists of three, iterative steps:

a. Chart the user interface dialogue.

b. Prototype the dialogue and user interface.

c. Obtain user feedback.

KEY TERMS

agent, p. 590
cascading menu, p. 581
consumer-style interface, p. 583
dialogue, p. 574
dialogue box, p. 580
exercising the user interface, p. 597
expert user, p. 572
frames, p. 577
function keys, p. 576
graphical user interface (GUI), p. 575
help agent, p. 590
help authoring package, p. 589
help system, p. 589
help wizard, p. 590
hybrid *Windows*/Web, p. 584

hyperlink, p. 583
hypertext, p. 583
icon, p. 582
iconic menu, p. 582
instruction set, p. 584
language-based syntax, p. 584
menu bar, p. 578
menu driven, p. 527
mnemonic syntax, p. 585
mouse, p. 576
natural language syntax, p. 585
novice user, p. 572
paging, p. 576
pens, p. 576
platform independence, p. 575

pop-up menu, p. 581
pull-down menu, p. 579
question-answer dialogue, p. 586
scroll bar, p. 577
scrolling, p. 576
slider, p. 595
state transition diagram, p. 593
tab, p. 595
task bar, p. 577
tear-off menu, p. 581
toolbars, p. 582
tool tip, p. 589
user interface control, p. 577
window, p. 577

REVIEW QUESTIONS

1. What is user interface design?

2. Differentiate between an expert user and a novice user.

3. What are five elements that frequently cause people to have difficulty with the interface of computer systems?

4. List the four "commandments" of user interface design.

5. List several human engineering factors/guidelines that should be incorporated into a user interface design.

6. What is a dialogue? List several do and don't guidelines with respect to the tone of a dialogue. Give examples of each.

7. What are the two providers of the basic structure for any graphical user interface and how do they differ?

8. What is the advantage of platform independence with respect to user interface design.

9. What are the two most common approaches to showing the display area on a display monitor to the user?

10. What are function keys and how are they typically used?

11. List and differentiate between several user interface design styles.

12. What is the basic construct of a GUI?

13. List and briefly describe several types of menus.

14. What are the three types of syntax that may be used for an instruction set?

15. What are two important special considerations for user interface design?

16. What are the steps in user interface design? Are the steps sequential? Explain your answer.

17. What tool is used to chart a user interface design?

18. What is meant by "exercising the user interface?"

PROBLEMS AND EXERCISES

1. To what extent should the system user be involved during user interface design? What would you do for the user? What would you ask the user to do for you? Detail a strategy that consists of specific steps you and the system users would follow.

2. What documentation prepared during input design and output design is needed during user interface design? How does that input and output design documentation relate to user interface design?

3. Explain the difference between an expert and a novice user. How would your strategy for designing user interfaces for an expert user differ from that for designing user interfaces for a novice user?

4. Display properties can be overused and frequently hinder a user's performance during a computer session. Cite some examples in which display properties are appropriate and in which display properties may hinder a user's performance.

5. Describe several strategies (styles) for designing user interfaces. What criteria would you consider when choosing between the strategies?

6. What is the worst user interface design you have ever seen or used? What made it difficult for you to use? How would you personally improve the design?

7. Select a dialogue screen from a *Windows*-based product and redesign it with a Web interface. Remember, Web interfaces tend to be more artistic.

8. Select a screen from a Web interface and redesign it with a *Windows* interface. Remember, *Windows* interfaces tend to be less artistic. In fact, the design goal of most *Windows* interfaces is to emulate the customer's office productivity suite such as Microsoft *Office*.

PROJECTS AND RESEARCH

1. Arrange to study a personal computer or network application. It may be either a business system (such as an inventory, accounts receivable, or personnel system) or a productivity tool (such as a word processor, spreadsheet, or database system). Analyze the human engineering of the user interface. Analyze the human engineering of the display screens. If possible, discuss the dialogue and screens with system users. What do they like and dislike about the design?

2. Redesign the application in Project 1 to improve the user interface and screens. If possible, discuss your improved design with users. Do they like your new design better? Did they raise any concerns?

3. Obtain documentation or magazine reviews on an automated screen design aid. If possible, arrange for a demonstration. How would the product improve your productivity? How would the product decrease your productivity? What features do you dislike or would you prefer to see?

4. *Windows* user interface design is increasingly governed by standards and guidelines developed from years of application development. But Web-based user interface design is still very much an art form. Research the latest trends and recommendations for effective Web user interface design.

5. Electronic commerce is emerging as an exciting new business-to-consumer, business-to-business, and employee-to-employee platform for transaction processing and management. But little is known about what works and what doesn't when attempting to entice users to want to initiate business transactions on the Web. Research this topic and present recommendations for effective e-commerce user interface design.

6. The paperless office has long been a popular, but elusive goal. The combination of Web-based and work flow technology is moving us closer to that goal. Research a product called *Ariba* (developed by Purdue University alumni) to learn how this technology is changing the way businesses support procurement/purchasing. Write a report describing the role of user interface design in such a product. To what other types of applications would the integration of Web and interface technology be useful?

MINICASES

1. Richards & Sons, Inc., is a large investment company located in Tampa, Florida, that buys stocks, bonds, commodities, and various other assets for its clients. The company also manages clients' investment portfolios for a variety of investment objectives. Finding new people with money to invest is crucial to its success, so keeping track of clients and potential clients is very important. Morgan Adamson is the senior analyst in charge of the new sales prospect and contact management system at Richards & Sons, Inc. The new system has just been installed and the project team is working with the system users who are doing acceptance testing. Morgan is talking to Kevin Brock, the junior analyst who was responsible for the design of the system's user interface.

MORGAN   I guess you've heard that some of the system users are not very happy with the new contact tracking system. In particular, they are expressing dissatisfaction with the user interface that you designed.

KEVIN   I don't understand what the problem could be—I put a lot of time and effort into that design. What are the specific problems?

MORGAN   Some of the users are complaining that they don't know what to do next or how to use some of the screens. Are you sure that all of the screens are consistent throughout this system?

KEVIN   I didn't think it was important where the information was as long as I clearly labeled it. Besides, the users should be expected to read the screen. I deliberately put lots of highlighting, blinking, and reverse video fields on the screens to draw attention to important information.

MORGAN   Yes, I know. Some of the users claim that there is so much highlighted information that it distracts from the purpose of the screen. I've also had complaints that proper default values were not specified for some of the fields.

KEVIN   Default values were specified for the most common fields, but the users should expect to have to type some of the information in—that's their job, isn't it?

MORGAN   In some cases, maybe several possible default values should have been provided in a pop-up window to eliminate possible keying errors. Remember, we want to reduce the amount of user entry keystrokes as much as possible. Some of the clerks say that they don't understand some of the terminology and abbreviations on some of the screens.

KEVIN   Oops! I must not have checked the screens carefully enough to catch some of the computer terminology. That's no big deal. I can fix that right away. Are there any other problems?

MORGAN   Other users have indicated that the use of certain menu options is not consistent across all of the screens. Didn't you use the same labels and screen buttons or menus for the same actions throughout the entire interface?

KEVIN   No, I didn't. I thought that I could reuse the function keys to mean different things as long as I clearly labeled them on each screen.

MORGAN   Why?

KEVIN   Some of the keyboards have only 12 function keys, and I was afraid I might run out of keys to assign unless I reused them.

MORGAN   You need to realize that the system users don't always read the instructions that you provide. Whether that is right or wrong is not important; you need to be consistent so that the users don't have to learn a different set of function keys for each screen.

KEVIN   I guess I really didn't think that through very well; it shouldn't be too difficult to make all the function key assignments consistent.

MORGAN   There have also been some complaints about an insufficient amount of help messages for some of the input screens. For instance, one clerk said the contact entry screen consistently refused to accept the date of contact he tried to enter. The system did output an error message indicating that the date was incorrect and should be reentered, but the clerk doesn't understand why the date was invalid. He says the system doesn't provide any information about the correct format of the date or any valid examples. You did design help screens and messages for each of the input screens, didn't you?

KEVIN   Uh, well, I'm not exactly sure what you mean by a "help" screen. I did very thorough input error checking so that invalid data could not be entered. I created error *messages* for each of the edited fields. I thought that would be sufficient for the users to identify the input error and make the necessary correction.

MORGAN   I think I'm beginning to understand the problem. We need to talk about some very important human engineering guidelines that you need to follow whenever you are designing a user interface. First . . .

a. What did Kevin do wrong in designing the user interface? What are some of the other mistakes that an analyst might make when designing a user interface?

b. How could these mistakes have been avoided? (What would you have done differently?) What role should the system user play in interface design?

SUGGESTED READINGS

Andres, Clay. *Great Web Architecture*. Foster City, CA: IDG Books Worldwide. This is an interesting title. It uses a "design-by-example" approach based on input from "top Web architects" to illustrate and discuss Web-based systems, including many with e-commerce and e-business aspects. This is not an academic title, but it is nonetheless interesting.

Galitz, Wilbert. *User-Interface Screen Design*. New York: Wiley QED, 1993. Ignore the date. This book remains our favorite user interface design book because it is so conceptually and fundamentally sound. He teaches workstation, PC, and mainframe interface design here, based on well-thought-out principles and guidelines. We can't wait for the update. Would that we could afford to develop an entire elective course built around this outstanding book!

Galitz, Wilbert. *It's Time to Clean Your Windows: Designing GUIs That Work*. New York: John Wiley & Sons, 1994. This is another excellent book that provides an unbiased reference on designing graphical interfaces.

Hix, Deborah, and H. Rex Hartson. *Developing User Interfaces: Ensuring Usability Through Product & Process*. New York: John Wiley & Sons, 1993. John Wiley & Sons must have the corner on user interface design books. These authors have academic roots. The book is somewhat hard to read, but nonetheless very well organized and written. We especially like the integration with systems analysis and design.

Horton, William K. *Designing & Writing Online Documentation: Help Files to Hypertext*. New York: John Wiley & Sons, 1990. We were only able to provide cursory coverage of this important topic.

Mandel, Theo. *Elements of User Interface Design*. New York: John Wiley & Sons, 1997. Here is a somewhat newer and very comprehensive book that includes some of the early design foundations for the Web.

Martin, Alexander, and David Eastman. *The User Interface Design Book for the Applications Programmer*. New York: John Wiley & Sons, 1996.

Microsoft Corporation. *Microsoft Windows User Experience: Official Guidelines for User Interface Developers and Designers*. Redmond, WA: Microsoft Press, 1999. This is the official standard for designing *Windows* user interfaces. There are many insights to Microsoft's intentions for maximizing the user experience.

Schmeiser, Lisa. *Web Design Templates Sourcebook*. Indianapolis, IN: New Riders Publishing, 1997. This is not an academic title, or even a traditional professional market title. It caught our eye because it uses an HTML-based template approach (over 300) to present designs that can ultimately evolve into finished products. From our perspective, this represents an intriguing twist on the prototyping model.

Weinschenk, Susan, and Sarah C. Yeo. *Guidelines for Enterprise-Wide GUI Design*. New York: John Wiley & Sons, 1995. Clearly, John Wiley & Sons is the market's leading publisher for this subject.

BEYOND SYSTEMS ANALYSIS AND DESIGN

Part Four introduces you to the systems implementation and systems support phases of the systems development life cycle. Two chapters make up this unit. First, Chapter 16, Systems Construction and Implementation, presents systems implementation, the process of putting the design specifications for the new information system in actual operation. Two implementation phases are discussed: construction and delivery. Each of these phases is discussed in terms of purpose, activities, roles, inputs and outputs, techniques, and steps.

Chapter 17, Systems Operations and Support, discusses four types of systems support for an application. This ongoing maintenance of a system after it has been placed into production consists of correcting errors, recovering the system, assisting users, and adapting the system. Systems support is very important because it is likely that young systems analysts will be responsible for maintaining legacy systems. This chapter concludes our exploration of the systems development life cycle.

Focus on PEOPLE

Focus on DATA

Focus on PROCESSES

Focus on INTERFACES

Focus on DEVELOPMENT

Stakeholders

Activities

SYSTEMS ANALYSTS

SYSTEM OWNERS

SYSTEM USERS

SYSTEM DESIGNERS

SYSTEM BUILDERS

BUILDING BLOCKS OF AN INFORMATION SYSTEM

Management Expectations

The PIECES Framework

Performance ● Information ● Economics ● Control ● Efficiency ● Service

PROJECT & PROCESS MANAGEMENT

 List of business entities and rules ...

Business Knowledge

 List of business functions and events ...

Business Functions

 List of business locations and systems...

Business Locations

PRELIMINARY INVESTIGATION

PROBLEM ANALYSIS

Data Requirements

Process Requirements

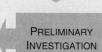

Interface Requirements

REQUIREMENTS ANALYSIS

DECISION ANALYSIS

Database Schema

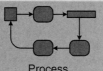

Application Schema & Specs

Interface Specifications

DESIGN

CONSTRUCTION

```
CREATE TABLE tblOrders
colOrderNot CHAR(5) NOT
NULL
colOrderDate  DATE/TIME NOT
```
Database Programs

```
PROC ValidateOrder
  PERFORM ValidateCust
  REPEAT UNTIL
    NoMoreProd
```
Application Programs

```
<html>
<head>
<title> Order Entry Form </title>
...
```
Interface Programs

IMPLEMENTATION

VENDORS AND CONSULTANTS

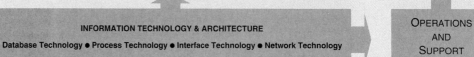

INFORMATION TECHNOLOGY & ARCHITECTURE

Database Technology ● Process Technology ● Interface Technology ● Network Technology

OPERATIONS AND SUPPORT

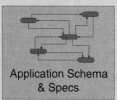

16

SYSTEMS CONSTRUCTION AND IMPLEMENTATION

CHAPTER PREVIEW AND OBJECTIVES

In this chapter you will learn more about the construction and implementation phases of systems development. These two phases construct and deliver the final system into operation. You will know that you understand the processes of constructing and implementing a system when you can:

— Explain the purpose of the construction and implementation phases of the systems life cycle.

— Describe the systems construction and implementation phases in terms of your information building blocks.

— Describe the systems construction and implementation phases in terms of major tasks, roles, inputs, and outputs.

— Explain several application program and system tests.

— Identify several system conversion strategies.

— Identify those chapters in this textbook that can help you actually perform the tasks of systems construction and implementation.

Although some of the techniques of systems construction and implementation are introduced in this chapter, it is *not* the intent of this chapter to teach the *techniques*. This chapter teaches only the process of construction and implementation.

SOUNDSTAGE
SOUNDSTAGE .ENTERTAINMENT CLUB

SCENE

When we last visited SoundStage, Sandra and Bob were busy finishing the technical design specifications for their project. Sandra and Bob have since progressed to the construction and implementation of the new system. They are meeting to plot the strategy for completing these final phases to deliver the new system into production.

SANDRA

Well, Bob, I think we are finally starting to see the light at the end of the tunnel. I'm getting really excited, but at the same time a little nervous. We're going to be cutting it close on our deadline.

BOB

You're telling me. I worked all weekend to try to get caught up. I'm tired, but I feel a lot better about where I'm at with construction of the final pieces for the new system.

SANDRA

I thought that I was the only one working this weekend. I spent most of the weekend planning how we are going to complete the delivery of the new system once we finished the construction of all the final pieces. You've been working on their construction. So where do we stand in that respect?

BOB

I think we're in pretty good shape! Fortunately the new system is to use one of our existing LANs. So we gained some time by not having to oversee the installation and testing of a LAN. Since we used a prototyping approach to development, the database was already built and tested before we built any screens. The new system didn't involve purchasing a software package, so we didn't need to install and test any vendor packages. All that left us with was a few new programs that needed to be written and tested. So . . . now you know what I did this weekend.

SANDRA

Sounds like we're in better shape than I thought.

BOB

Yes. I just have a few more new programs to write and test. Then I'll have to run a system test to ensure that they all work together properly.

SANDRA

Great. I've been generating test data for when we do a system test. I've also been coming up with a conversion plan. We'll do a staged conversion. We will install it at our pilot warehouse location and then transport it to others later. In talking with the managers, we concluded that the conversion should be a parallel conversion. They really wanted to continue using the existing system at the same time as the new system.

BOB

That's probably a very good idea since this system represents a critical component of the business.

SANDRA

Since you were working on the new database, I thought I'd ask you to proceed with installing that database.

BOB

No problem. I can quickly crank out some programs to extract data from the production databases to load into the new one. There may be some data that needs to be keyed in from other sources though.

SANDRA

I was afraid of that, so I asked permission to hire a few people from a local temporary employment company to help us out. I've studied the training needs of the users.

BOB

Since we used a prototyping approach to involve the users throughout the development process, they should be relatively familiar with the new system. Don't forget, I provided content-sensitive help inside the application. So they should be relatively easy to take care of when they use the application.

SANDRA

That's true. That strategy should pay off. However, what about any future employee who will be a user of the system?

BOB

Good point. I guess we will have to schedule a few training sessions as needed.

SANDRA

I agree. We should provide training sessions now and for any new employee. But also, I have been outlining a user manual that highlights the application for the user.

BOB

What else do we have left to do?

SANDRA

That's it . . . except the actual conversion itself. Here, take a look at my draft of the conversion plan and tell me what you think.

DISCUSSION QUESTIONS

1. How would you characterize the focus of the implementation phase of this project?

2. What were some of the benefits that Sandra and Bob realized during systems implementation by following the prototyping approach to systems development?

3. What impact would the need to build a network or install a purchased software package have on Sandra and Bob's implementation of the new system?

4. How does the system test that Bob and Sandra mentioned differ from the testing that Bob did over the weekend?

5. What type of concerns would the managers have with discarding the existing system and immediately placing the final system into production?

Let's begin with a definition of systems construction and implementation.

Systems construction is the development, installation, and testing of system components. Unfortunately, *systems development* is a common synonym. (We dislike that synonym since it is more frequently used to describe the *entire* life cycle.)

Systems implementation is the delivery of that system into production (meaning day-to-day operation).

Relative to the information systems building blocks, systems construction and implementation addresses DATA, PROCESSES, and INTERFACES primarily from the system builders' perspective (see the chapter map).

Figure 16.1 illustrates the construction and implementation phases. Notice that the trigger for systems construction phase is the approval of the technical design specifications resulting from the design phase. Given the design specifications, we can construct and test system components for that design. Eventually, we will have built the functional system. The functional system can then be implemented or delivered as an operational system.

This chapter examines each of these phases in detail.

WHAT IS SYSTEMS CONSTRUCTION AND IMPLEMENTATION?

THE CONSTRUCTION PHASE

The purpose of the construction phase is to develop and test a functional system that fulfills business and design requirements and to implement the interfaces between the new system and existing production systems. Programming is generally recognized as a major aspect of the construction phase. But with the trend toward system solutions that involve acquiring or purchasing software packages, the implementation and integration of software components is becoming an equally, if not more, common and visible aspect of the construction phase.

In this section you will learn about several tasks involved in the construction phase of a typical systems development project. Figure 16.2 is an activity diagram depicting the various tasks for the construction phase. Let's examine each construction phase task in greater detail.

Task 6.1—Build and Test Networks (If Necessary)

Recall that in the requirements analysis phase of systems analysis, we established network requirements. Subsequently, during the design phase we developed distributed data and process models. Using these technical design specifications to implement the network architecture for an information system is a prerequisite for the remaining construction and implementation activities.

In many cases, new or enhanced applications are built around existing networks. If so, skip this task. However, if the new application calls for new or modified networks, they must normally be implemented before building and testing databases and writing or installing computer programs that will use those networks. Thus, the first task of the construction phase may be to build and test networks.

This phase involves analysts, designers, and builders. A network designer and network administrator assume the primary responsibility for completing this task. The network designer is a specialist in the design of local and wide area networks and their connectivity. The network administrator has the expertise for building and testing network technology for the new system. He or she will also be familiar with network architecture standards that must be adhered to for any possible new networking technology. This person is also responsible for security. (The network designer and network administrator may be the same person.) While the systems analyst may be involved in the completion of this task, the analyst's role is more that of a facilitator and ensures that business requirements are not compromised by the network solution.

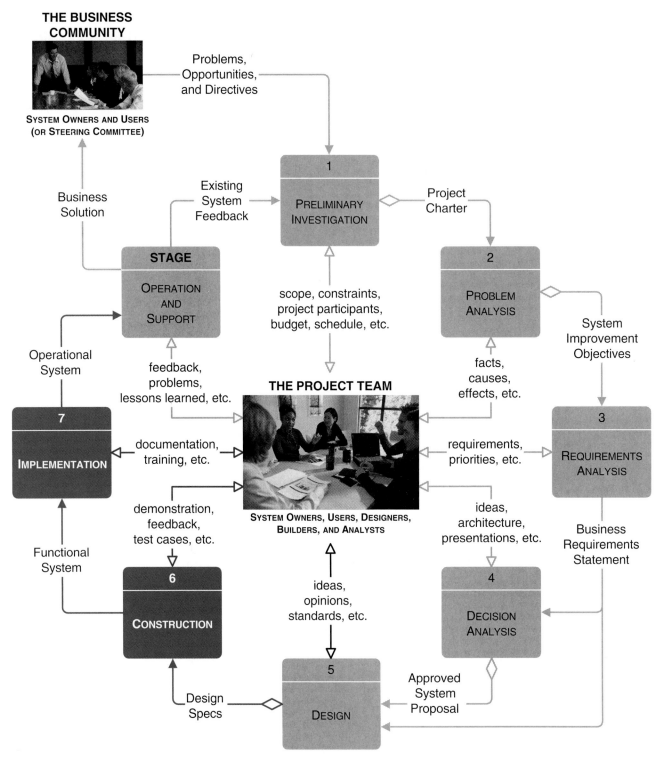

Figure 16.1   *The Context of Systems Construction and Implementation*

Building and testing databases are unfamiliar tasks for many students, who are accustomed to having an instructor provide them with the test databases. This task must immediately precede other programming activities because databases are the resources shared by the computer programs to be written. If new or modified databases are required for the new system, we can now build and test those databases.

This task involves systems users, analysts, designers, and builders. The same system specialist that designed the databases will assume the primary responsibility in completing this task. System users may also be involved in this task by providing or approving the test data to be used in the database. When the database to be built is a noncorporate, applications-oriented database, the systems analyst often completes this task. Otherwise, systems analysts mostly ensure business requirements compliance. The database designer will often become the system builder responsible for the completion of this activity. The task may involve database programmers to build and populate the initial database and a database administrator to tune the database performance, add security controls, and provide for backup and recovery.

The primary inputs to this task are the database schema(s) specified during systems design. Sample data from production databases may be loaded into tables for testing the databases. The final product of this task is an unpopulated *database structure* for the new database. The term *unpopulated* means the database structure is implemented but data has not been loaded into the database structure. As you'll soon see, programmers will eventually write programs to populate and maintain those new databases. Revised database schema and test data details are also produced during this task and placed in the project repository for future reference.

Some systems solutions may have required the purchase or lease of software packages. If so, once networks and databases for the new system have been built, we can install and test the new software. This new software will subsequently be placed in the software library.

This activity typically involves systems analysts, designers, builders, and vendors and consultants. This is the first task in the life cycle that is specific to the applications programmer. The systems analyst typically participates in the testing of the software package by clarifying requirements. Likewise, the system designer may be involved in this task to clarify integration requirements and program documentation that is to be used in testing the software. Network administrators may be involved in actually installing the software package on the network server. Finally, this task typically involves participation from the software vendor and consultants who may assist in the installation and testing process.

The main input to this task is the new software packages and documentation that is received from the system vendors. The applications programmer will complete the installation and testing of the package according to integration requirements and program documentation developed during system design. The principal deliverable of this task is the installed and tested software package that is made available in the software library. Any modified software specifications and new integration requirements that were necessary are documented and made available in the project repository to provide a history and serve as future reference.

We are now ready to develop (or complete) any in-house programs for the new system. Recall that prototype programs are frequently constructed in the design phase. These prototypes are included as part of the technical design specifications for completing systems construction and implementation. However, these prototypes are rarely fully functional or complete. Therefore, this activity may involve developing or refining those programs.

Task 6.2—Build and Test Databases

Task 6.3—Install and Test New Software Packages (If Necessary)

Task 6.4—Write and Test New Programs

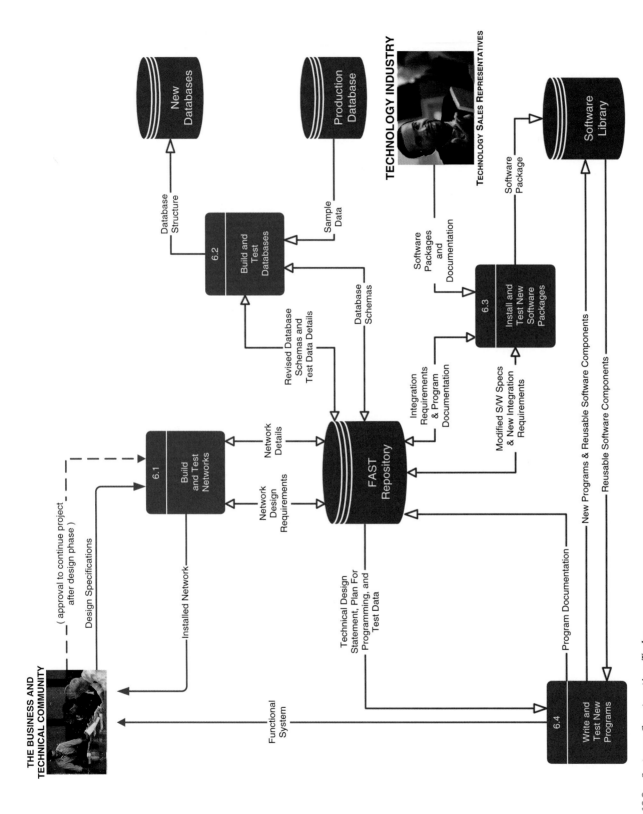

Figure 16.2 *Systems Construction Tasks*

This task involves the systems analysts, designers, and builders. The systems analyst typically clarifies business requirements to be implemented by the programs. The designer may have to clarify the program design, integration requirements, and program documentation (developed during systems design) that is used in writing and testing the programs. The system builders will assume the primary responsibility for this activity. The applications programmer (builder) is responsible for writing and testing in-house software. Most large programming projects require a team effort. One popular organization strategy is the use of *chief programmer teams.* The team is managed by the *chief programmer,* a highly proficient and experienced programmer who assumes overall responsibility for the program design strategy, standards, and construction. The chief programmer oversees all coding and testing activities and helps with the most difficult aspects of the programs. Other team members include a *backup chief programmer, program librarian, programmers,* and *specialists.* The applications programmer is often aided by an application or software tester who specializes in building and running *test scripts* that are consistently applied to programs to test all possible events and responses.

The primary inputs to this activity are the technical design statement, plan for programming, and test data developed during systems design. Since any new programs or program components may have already been written and be in use by other existing systems, the experienced applications programmer will know to first check for possible reusable software components available in the software library. The principal deliverables of this activity are the new programs and reusable software components that are placed in the software library. This activity also results in program documentation that may need to be approved by a quality assurance group. Some information systems shops have a quality assurance group staffed by specialists who review the final program documentation for conformity to standards. This group will provide appropriate feedback regarding quality recommendations and requirements. The final program documentation is then placed in the project repository for future reference.

Testing is an important skill that is often overlooked in academic courses on computer programming. Testing should not be deferred until after the entire program has been written! There are three levels of testing to be performed: stub testing, unit or program testing, and systems testing.

> **Stub testing** is the test performed on individual events or modules of a program. In other words, it is the testing of an isolated subset of a program.

> **Unit or program testing** is a test whereby all the events and modules that have been coded and stub tested for a program are tested as an integrated unit; it is the testing of an entire program.

> **Systems testing** ensures that application programs written and tested in isolation work properly when they are integrated into the total system.

Just because a single program works properly doesn't mean that it works properly with other programs. The integrated set of programs should be run through a systems test to make sure one program properly accepts, as input, the output of other programs. Once the system test is complete and determined to be successful, we can proceed to the implementation of the system.

What's left to do? New systems usually represent a departure from the way business is currently done; therefore, the analyst must provide for a smooth transition from the old system to the new system and help users cope with normal start-up problems. Thus, the implementation phase delivers the production system into operation.

The functional system from the construction phase is the key input to the implementation phase (see Figure 16.1). The deliverable of the implementation phase

THE IMPLEMENTATION PHASE

(and the project) is the operational system that will enter the *operation and support* stage of the life cycle.

In your information system framework, the implementation phase considers the same building blocks as the construction phase (see the chapter map). In this section you will learn about several tasks involved in the implementation phase for a typical systems development project. Figure 16.3 is an activity diagram depicting the various tasks for the implementation phase. Let's examine each implementation phase task in greater detail.

Task 7.1—Conduct System Test

Now that the software packages and in-house programs have been installed and tested, we need to conduct a final system test. All software packages, custom-built programs, and any existing programs that comprise the new system must be tested to ensure that they all work together.

This task involves analysts, owners, users, and builders. The systems analyst facilitates the completion of this task. The systems analyst typically communicates testing problems and issues with the project team members. The system owners and system users hold the ultimate authority on whether or not a system is operating correctly. System builders, of various specialties, are involved in the systems testing. For example, applications programmers, database programmers, and networking specialists may need to resolve problems revealed during systems testing.

The primary inputs to this task include the software packages, custom-built programs, and any existing programs comprising the new system. The system test is done using the system test data that was developed earlier by the systems analyst. As with previous tests that were performed, the system test may result in required modifications to programs, thus, once again prompting the return to a construction phase task. This iteration would continue until a successful system test was experienced.

Task 7.2—Prepare Conversion Plan

Once a successful system test has been completed, we can begin preparations to place the new system into operation. Using the design specifications for the new system, the systems analyst will develop a detailed conversion plan. This plan will identify databases to be installed, end-user training and documentation that need to be developed, and a strategy for converting from the old system to the new system.

The project manager facilitates the activity. Systems analyst, system designer, and system builder roles are not typically involved unless deemed necessary by the project manager. Finally, many organizations require that all project plans be formally presented to a steering body (sometimes called a steering committee) for final approval.

This activity is triggered by the completion of a successful system test. Using the design specifications for the new system, a detailed conversion plan can be assembled. The principal deliverable of this activity is the conversion plan that will identify databases to be installed, end-user training and documentation that need to be developed, and a strategy for converting from the old system to the new system.

The conversion plan may include one of the following commonly used installation strategies:

– **Abrupt cut-over.** On a specific date (usually a date that coincides with an official business period such as month, quarter, or fiscal year), the old system is terminated and the new system is placed into operation. This is a high-risk approach because there may still be major problems that won't be uncovered until the system has been in operation for at least one business period. On the other hand, there are no transition costs. Abrupt cut-over may be necessary if, for instance, a government mandate or business policy becomes

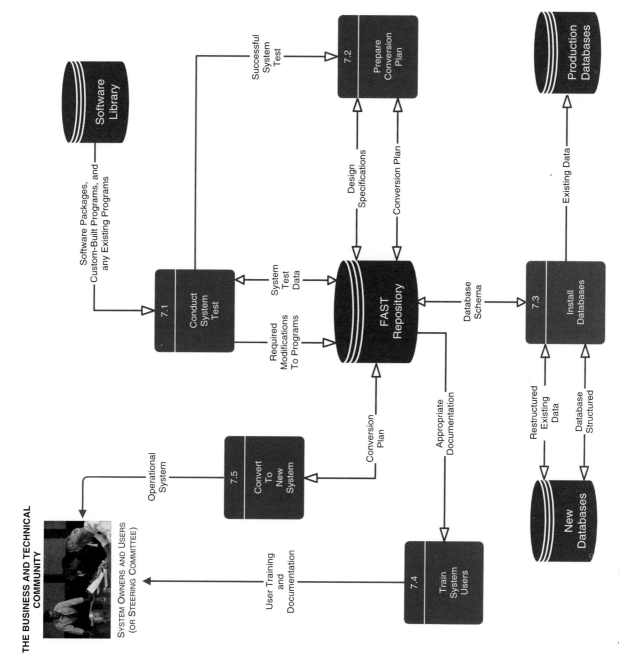

THE BUSINESS AND TECHNICAL COMMUNITY

SYSTEM OWNERS AND USERS (OR STEERING COMMITTEE)

Software Library

Software Packages, Custom-Built Programs, and any Existing Programs

7.1 Conduct System Test

Successful System Test

7.2 Prepare Conversion Plan

Design Specifications

Conversion Plan

System Test Data

Required Modifications To Programs

FAST Repository

Conversion Plan

Appropriate Documentation

Database Schema

7.3 Install Databases

Production Databases

Existing Data

Restructured Existing Data

Database Structured

New Databases

7.5 Convert To New System

Operational System

7.4 Train System Users

User Training and Documentation

Figure 16.3 *Systems Implementation Tasks*

613

effective on a specific date and the system couldn't be implemented before that date.

— **Parallel conversion.** Under this approach, both the old and new systems are operated for some time period. This ensures that all major problems in the new system have been solved before the old system is discarded. The final cut-over may be either abrupt (usually at the end of one business period) or gradual, as portions of the new system are deemed adequate. This strategy minimizes the risk of major flaws in the new system causing irreparable harm to the business; however, it also means the cost of running two systems over some period must be incurred. Because running two editions of the same system on the computer could place an unreasonable demand on computing resources, this may be possible only if the old system is largely manual.

— **Location conversion.** When the same system will be used in numerous geographical locations, it is usually converted at one location first (using either abrupt or parallel conversion). As soon as that site has approved the system, it can be farmed to the other sites. Other sites can be cut over abruptly because major errors have been fixed. Furthermore, other sites benefit from the learning experiences of the first test site. The first production test site is often called a beta test site.

— **Staged conversion.** Like location conversion, staged conversion is a variation on the abrupt and parallel conversions. A staged conversion is based on the version concept introduced earlier. Each successive version of the new system is converted as it is developed. Each version may be converted using the abrupt, parallel, or location strategies.

The conversion plan also typically includes a systems acceptance test plan. The systems acceptance test is the final opportunity for end-users, management, and information systems operations management to accept or reject the system. A **systems acceptance test** is a final system test performed by end-users using real data over an extended time period. It is an extensive test that addresses three levels of acceptance testing: verification testing, validation testing, and audit testing.

— **Verification testing** runs the system in a simulated environment using simulated data. This simulated test is sometimes called alpha testing. The simulated test is primarily looking for errors and omissions regarding end-user and design specifications that were specified in the earlier phases but not fulfilled during construction.

— **Validation testing** runs the system in a live environment using real data. This is sometimes called beta testing. During this validation, a number of items are tested.

  a. *Systems performance.* Is the throughput and response time for processing adequate to meet a normal processing workload? If not, some programs may have to be rewritten to improve efficiency or processing hardware may have to be replaced or upgraded to handle the additional workload.

  b. *Peak workload processing performance.* Can the system handle the workload during peak processing periods? If not, improved hardware and/or software may be needed to increase efficiency or processing may need to be rescheduled—that is, consider doing some of the less critical processing during nonpeak periods.

  c. *Human engineering test.* Is the system as easy to learn and use as anticipated? If not, is it adequate? Can enhancements to human engineering be deferred until after the system has been placed into operation?

  d. *Methods and procedures test.* During conversion, the methods and procedures for the new system will be put to their first real test. Methods and procedures may have to be modified if they prove to be awkward and inefficient from the end-users' standpoint.

  e. *Backup and recovery testing.* All backup and recovery procedures should be tested. This should include simulating a data loss disaster and testing the time required to recover from that disaster. Also, a before-and-after comparison of the data should be performed to ensure that data was properly recovered. It is crucial to test these procedures. Don't wait until the first disaster to find an error in the recovery procedures.

— **Audit testing** certifies that the system is free of errors and is ready to be placed into operation. Not all organizations require an audit. But many firms have an independent audit or quality assurance staff that must certify a system's acceptability and documentation before that system is placed into final operation. There are independent companies that perform systems and software certification for end-users' organizations.

Task 7.3—Install Databases

Recall that in a previous phase you built and tested databases. To place the system into operation, you will need fully loaded (or "populated") databases. Therefore, the next task we'll survey is installation of databases. The purpose of this task is to populate the new systems databases with existing data from the old system.

At first, this activity may seem trivial. But consider the implications of loading a typical table, say, MEMBER. Tens or hundreds of thousands of records may have to be loaded. Each must be input, edited, and confirmed before the database table is ready to be placed into operation.

Systems builders play a primary role in this activity. The task will normally be completed by application programmers who will write the special programs to extract data from existing databases and programs to populate the new databases. Systems analysts and designers may play a small role in completing this activity. Their primary involvement will be the calculation of database sizes and estimating of time required to perform the installation. Finally, often data-entry personnel or hired help may be assigned to do data entry.

Special programs will have to be written to populate the new databases. Existing data from the production databases coupled with the database schema(s) models and database structures for the new databases will be used to write computer programs to populate the new databases with restructured existing data. The principal deliverable of this task is the restructured existing data that has been populated in the databases for the new system.

Task 7.4—Train Users

Change may be good, but it's not always easy. Converting to a new system necessitates that system users be trained and provided with documentation (user manuals) that guides them through using the new system.

Training can be performed one on one; however, group training is generally preferred. It is a better use of your time, and it encourages group-learning possibilities. Think about your education for a moment. You really learn more from your fellow students and colleagues than from your instructors. Instructors facilitate learning and instruction, but you master specific skills through practice with large groups where common problems and issues can be addressed more effectively. Take advantage of the ripple effect of education. The first group of trainees can then train several other groups.

The task is completed by the systems analyst and involves system owners and users. Given appropriate documentation for the new system, the systems analysts will provide end-user documentation (typically in the form of manuals) and training for the system users. The system owners must support this activity. They must be willing to approve the release time necessary for people to obtain the training needed to become successful users of the new system. Remember, the system is for the user! User involvement is also important in this activity because the end-users will inherit the successes and failures from this effort. Fortunately, user involvement during this task is rarely overlooked. The most important aspect of their involvement is training and advising the users. They must be trained to use equipment and to follow the procedures required of the new system. But no matter how good the training is, users will become confused at times. Or perhaps they will find mistakes or limitations. Thus, it is the responsibility of the analyst to help the users through the learning period until they become comfortable with the new system.

Given appropriate documentation for the new system, the systems analyst will provide the system users with documentation and training needed to properly use the new system. The principal deliverable of this task is user training and documentation. Many organizations hire special systems analysts who do nothing but write user documentation and training guides. If you have a skill for writing clearly, the demand for your services is out there! Figure 16.4 is a typical outline for a training manual. The Golden Rule should apply to user manual writing: "Write unto others as you would have them write unto you." You are not a business expert. Don't expect the reader to be a technical expert. Every possible situation and its proper procedure must be documented.

Task 7.5: Convert to New System

Conversion to the new system from the old system is a significant milestone. After conversion, the ownership of the system officially transfers from the analysts and programmers to the end-users. The analyst completes this task by carefully carrying out the conversion plan. Recall that the conversion plan includes detailed instal-

---

Figure 16.4 *An Outline for a Training Manual*

Training Manual End-Users Guide Outline

I. **Introduction.**

II. **Manual.**

   A. The manual system (a detailed explanation of people's jobs and standard operating procedures for the new system).

   B. The computer system (how it fits into the overall workflow).

      1. Terminal/keyboard familiarization.

      2. First-time end-users.

         a. Getting started.

         b. Lessons.

   C. Reference manual (for nonbeginners).

III. **Appendixes.**

   A. Error messages.

lation strategies to follow for converting from the existing to the new production information system. This task also involves completing a systems audit.

The task involves the systems owners, users, analysts, designers, and builders. The project manager who will oversee the conversion process facilitates it. The system owners provide feedback regarding their experiences with the overall project. They may also provide feedback regarding the new system that has been placed into operation. The system users will provide valuable feedback pertaining to the actual use of the new system. They will be the source of the majority of the feedback used to measure the system's acceptance. The systems analysts, designers, and builders will assess the feedback received from the system owners and users once the system is in operation. In many cases, that feedback may stimulate actions to correct identified shortcomings. Regardless, the feedback will be used to help benchmark new systems projects down the road.

The key input to this activity is the conversion plan that was created in an earlier implementation phase task. The principal deliverable is the operational system that is placed into production in the business.

WHERE DO YOU GO FROM HERE?

This chapter provided a detailed overview of the construction and implementation phases of systems development. You are now ready to learn systems operation and support, covered in Chapter 17.

Before proceeding, you may wish to revisit Chapter 3 and its introduction to the systems development process. This review will help you to first understand how systems operations support fits into the overall systems development process.

SUMMARY

1. Systems construction is the development, installation, and testing of system components.
2. Systems implementation is the delivery of that system into production (meaning day-to-day operation).
3. The purpose of the construction phase is to develop and test a functional system that fulfills business and design requirements and to implement the interfaces between the new system and existing production systems.
4. The construction phase consists of four tasks: build and test networks, build and test databases, install and test new software packages, and write and test new programs.
5. Three levels of testing are performed on new programs:
   a. Stub testing is the test performed on individual modules, whether they be main program, subroutine, subprogram, block, or paragraph.
   b. Unit or program testing is a test whereby all the modules that have been coded and stub tested are tested as an integrated unit.
   c. Systems testing ensures that application programs written in isolation work properly when they are integrated into the total system.
6. The purpose of the implementation phase is to smoothly convert from the old system to the new system.
7. The systems implementation consists of the following activities: Conducting a system test, preparing a systems conver-

sion plan, installing databases, training system users, and converting from the old system to the new system.
8. There are several commonly used strategies for converting from an existing to a new production information system, including:
   a. Abrupt cut-over. On a specific date, the old system is terminated and the new system is placed into operation.
   b. Parallel conversion. Both the old and new systems are operated for some time period to ensure that all major problems in the new system are solved before the old system is discarded.
   c. Location conversion. When the same system will be used in numerous geographical locations, it is usually converted at one location and, following approval, farmed to the other sites.
   d. Staged conversion. Each successive version of the new system is converted as it is developed. Each version may be converted using the abrupt, parallel, or location strategies.
9. The systems acceptance test is the final opportunity for end-users, management, and information systems operations management to accept or reject the system. A systems acceptance test is a final system test performed by end-users using real data over an extended period. It is an

extensive test that addresses three levels of acceptance testing: verification testing, validation testing, and audit testing.

a. **Verification testing** runs the system in a simulated environment using simulated data.

b. **Validation testing** runs the system in a live environment using real data. This is sometimes called beta testing.

c. **Audit testing** certifies that the system is free of errors and is ready to be placed into operation.

abrupt cut-over, p. 612
audit testing, p. 615
location conversion, p. 614
parallel conversion, p. 614
staged conversion, p. 614

stub testing, p. 611
systems acceptance test, p. 614
systems construction, p. 607
systems implementation, p. 607
systems testing, p. 611

unit or program testing, p. 611
validation testing, p. 614
verification testing, p. 614

REVIEW QUESTIONS

1. Define systems construction and explain its purpose.
2. What are the key inputs and deliverables of the construction phase?
3. Define systems implementation and explain its purpose.
4. What are the key deliverables of the implementation phase?
5. What are the four major tasks in constructing the new system?
6. Who are the role players in the construction phase?
7. What are the five major tasks in implementing a new system?
8. Who are the role players in the implementation phase?
9. List and briefly describe three levels of testing that may be performed on new programs.
10. List and briefly describe four strategies commonly used to convert from an old system to a new production system.
11. What is a systems acceptance test? When is this test performed?
12. What are three levels of acceptance testing?

PROBLEMS AND EXERCISES

1. How can successful and thorough systems planning, analysis, and design be ruined by poor systems construction and implementation? How can poor systems analysis or design ruin a smooth construction and implementation? For both questions, list some consequences.
2. What skills are important during systems construction and implementation? Create an itemized list. Identify computer, business, and general education courses that would help you develop or improve those skills.
3. How do your information systems building blocks aid in systems construction and implementation? Examine each building block and address issues and relevance to the construction and implementation phases.
4. What products of the systems design phases are used in the systems construction phases? Why are they important? How are they used? What would happen if they were incomplete or inaccurate?
5. Differentiate between the types of testing done on application programs and the types of tests conducted for an overall system?
6. Why should a systems analyst perform a post-implementation review? What types of benefits can be derived from it?

PROJECTS AND RESEARCH

1. You are preparing to meet with your end-users to discuss converting from their old system to a new system. In this meeting you wish to discuss possible strategies. Prepare a brief description of the alternative strategies along with a description of situations for which each approach would be preferred and required.

2. Visit a local information systems shop. Ask to see samples of the conversion plan for a recent implementation of a new system. What items are detailed in that plan? Can you think of any items that should have been included? What type of conversion strategy was used?

3. Visit a local company and obtain a copy of a user manual for a system. Evaluate the manual. Is it written clearly using nontechnical terminology? How do the users feel about the manual? Did the manual help with the transition to the new system? Why or why not?

MINICASES

1. Tim Stallard is a systems analyst at Beck Electronic Supply. He has been a systems analyst for only six months. Unusual personnel turnover thrust him into the position after only 18 months as a programmer. Now it is time for his semiannual job performance review. *(Tim enters the office of Ken Delphi. Ken is the assistant director of MIS at Beck.)*

KEN Another six months! It hardly seems that long since your last job performance review.

TIM I personally feel very good about my progress over the last six months. This new position has been an eye-opener. I didn't realize that analysts do so much writing. I enrolled in some continuing education writing classes at the local junior college. The courses are helping . . . I think.

KEN I wondered what you did. It shows in everything from your memos to your reports. More than any technical skills, your ability to communicate will determine your long-term career growth here at Beck. Now, let's look at your progress in other areas. Yes, you've been supervising the materials requirements planning project construction and implementation for the last few months. This is your first real experience with the entire process, right?

TIM Yes. You know, I was a programmer for 18 months. I thought I knew everything there was to know about systems construction and implementation. But this project has taught me otherwise.

KEN How's that?

TIM The computer programming tasks have gone smoothly. In fact, we finished the entire system of programs six weeks ahead of schedule.

KEN I don't mean to interrupt, but I just want to reaffirm the role your design specifications played in accelerating the computer programming tasks. Bob has told me repeatedly that he had never seen such thorough and complete design specifications. The programmers seem to know exactly what to do.

TIM Thanks! That really makes me feel good. It takes a lot of time to prepare design specifications like that, but I think that it really pays off during construction and implementation. Now, what was I going to say? Oh yes. Even though the programming and testing were completed ahead of schedule, the system still hasn't been placed into operation; it's two weeks late.

KEN That means you lost the six-week buffer plus another two weeks. What happened?

TIM Well, I'm to blame. I just didn't know enough about the nonprogramming tasks. First, I underestimated the difficulties of training. My first-draft training manual made too many assumptions about computer familiarity. My end-users didn't understand the instructions, and I had to rewrite the manual. I also decided to conduct some training classes for the end-users. My instructional delivery was terrible, to put it mildly. I guess I never really considered the possibility that, as a systems analyst, I'd have to be a teacher. I think I owe a few apologies to some of my former instructors. I can't believe how much time needs to go into preparing for a class.

KEN Yes, especially when you're technically oriented and your audience is not.

TIM Anyway, that cost me more time than I had anticipated. But there are still other problems that have to be solved. And I didn't budget time for them!

KEN Like what?

TIM Like getting data into the new databases. We have entered several thousand new records. And to top it off, management is insisting that we operate the new system in parallel with the old system for at least two months. Then, and only then, will they be willing to allow the old system to be discarded.

KEN Well, Tim, I think you're learning a lot. Obviously, we threw you to the wolves on this project. But I needed Bob's experience and attention elsewhere. I knew when I pulled Bob off the project that it could introduce delays—I call it the rookie factor. Under normal circumstances, I would never have let you work on this alone. But you're doing a good job and you're learning. We have to take the circumstances into consideration. You'll obviously feel some heat from your end-users because the implementation is behind schedule, and I want you to deal with that on your own. I think you can handle it. But don't hesitate to call on Bob or me for advice. Now, let's talk about some training and job goals for the next six months.

a. Above and beyond programming, what tasks do you think make up systems construction and implementation? Can you think of any tasks that weren't described in this minicase?

b. Why is training so difficult? How do you feel about the prospects of becoming a "teacher"? How long do you think it takes to prepare for one hour of classroom instruction? What tasks do you think would be involved in preparing for a lesson plan?

c. A 3,000-record (row) database table must be created for a new system. Each record consists of 15 fields/attributes. The record length is 200 bytes. How long do you suppose it would take to create that database table? If necessary, use your own typing speed as a

performance gauge. What factors would affect how long it may take to get the table up and running?

d. What assumption did Tim make about transition from the old system to the new system? Why was it wrong? Can you think of any circumstances under which it would be correct?

SUGGESTED READINGS

Bell, P., and C. Evans. *Mastering Documentation*. New York: John Wiley & Sons, 1989.

Brooks, F. P. *The Mythical Man-Month*. Reading, MA: Addison-Wesley. 1995.

Boehm, Barry. "Software Engineering." *IEEE Transactions on Computers, C-25,* December 1976. This classic paper demonstrated the importance of catching errors and omis-

sions before programming begins.

Metzger, Philip W. *Managing a Programming Project*. 2nd ed. Englewood Cliffs, NJ: Prentice Hall, 1981. This is one of the few books to place emphasis solely on systems implementation.

Mosely, D. J. *The Handbook of MIS Application Software Testing*. Englewood Cliffs, NJ: Yourdon Press, 1993.

Focus on PEOPLE

Focus on DATA

Focus on PROCESSES

Focus on INTERFACES

Focus on DEVELOPMENT

Stakeholders

Activities

BUILDING BLOCKS OF AN INFORMATION SYSTEM

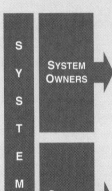

S Y S T E M S   A N A L Y S T S

SYSTEM OWNERS

SYSTEM USERS

SYSTEM DESIGNERS

SYSTEM BUILDERS

Management Expectations

The PIECES Framework

Performance ● Information ● Economics ● Control ● Efficiency ● Service

List of business entities and rules ...

Business Knowledge

List of business functions and events ...

Business Functions

List of business locations and systems...

Business Locations

Data Requirements

Process Requirements

Interface Requirements

Database Schema

Application Schema & Specs

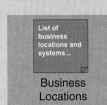

Interface Specifications

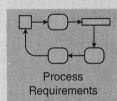

CREATE TABLE tblOrders
colOrderNot CHAR(5) NOT
NULL
colOrderDate DATE/TIME NOT

Database Programs

PROC ValidateOrder
PERFORM ValidateCust
REPEAT UNTIL
NoMoreProd

Application Programs

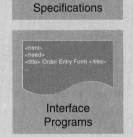

<html>
<head>
<title> Order Entry Form </title>

Interface Programs

PROJECT & PROCESS MANAGEMENT

PRELIMINARY INVESTIGATION

PROBLEM ANALYSIS

REQUIREMENTS ANALYSIS

DECISION ANALYSIS

DESIGN

CONSTRUCTION

IMPLEMENTATION

VENDORS AND CONSULTANTS

PRODUCTION INFORMATION SYSTEMS

Database Technology ● Process Technology ● Interface Technology ● Network Technology

OPERATIONS AND SUPPORT

17

SYSTEMS OPERATIONS AND SUPPORT

CHAPTER PREVIEW AND OBJECTIVES

Once a system has been implemented, it enters operations and support. Systems operation is the ongoing function in which the system operates until it is replaced. Systems support involves servicing, maintaining, and improving a functional information system through its lifetime. Systems operation and support occur in parallel. In this chapter, you will learn more about systems operation and support. In particular, it is useful to understand the different types of systems support provided for a production system. You will know that you understand the process of systems support when you can:

— Define systems operations and support.

— Describe the relative roles of a repository, program library, and database in systems operations and support.

— Differentiate between maintenance, recovery, technical support, and enhancement as system support activities.

— Describe the tasks required to maintain programs in response to bugs.

— Describe the role of benchmarking in system maintenance.

— Describe the systems analyst's role in system recovery.

— Describe forms of technical support provided by a systems analyst for the user community.

— Describe the tasks that should be and may be performed in system enhancement and the relationship between the enhancement and original systems development process.

— Describe the role of reengineering in system enhancement. Describe three types of reengineering.

Although some of the techniques of systems support are introduced in this chapter, this chapter does not teach these techniques. This chapter teaches only the *process* of systems support as it relates to the development processes you have been studying throughout this book.

SOUNDSTAGE
SOUNDSTAGE ENTERTAINMENT CLUB

SCENE

The new Member Services information system is operational. Orders are coming in. The Web interface is generating new business. Sandra and Bob are waiting for their next assignment. In the meantime, both have been providing ongoing support for the Member Services system. We join them for their morning kickoff.

SANDRA

I'll bet you are waiting for your next project assignment, aren't you, Bob. I have some news for you. You got the procurement redesign project. I'm jealous. Obviously, that team sees your Web experience with Member Services as an asset.

BOB

Really! I was hoping for procurement. That is a major e-business project. When do I start?

SANDRA

Jim Miles needs time to establish the statement of work and shore up executive sponsorship. It looks like about one month. In the meantime, I still need you here. The Member Services system has been working really well since we put it into production—with the exception of a few PCs crashing every once in a while and the inevitable problems with browser compatibility once you get outside of SoundStage.

BOB

Browser incompatibility? What is that about?

SANDRA

It seems that a lot of members are using very old browsers. Then they send e-mail to complain that our system does not work.

BOB

But browsers are free. I thought we gave them the hyperlinks to update their browser version.

SANDRA

We did. They're missing it or ignoring it. We don't know yet. But management is asking if there is something we can do

to better steer these members and avoid the public relations backlash.

BOB

Hmmm. I wonder if we could test for browser version when they authenticate themselves as members, and if they aren't running one of the supported browsers, we can walk them through the upgrade process. As I recall, we may not have let them progress too far into the system before they encountered a feature that might be incompatible with their browser.

SANDRA

My thoughts exactly. I used my daughter's Macintosh to get into my back door account last night. And I got all the way into the checkout process before I crashed. I tried to think about how I'd feel about having placed that much of an order before the crash. I think I would have been pretty upset.

BOB

Here's another thought. I wonder if we can't take snapshots along the way and store in-process orders in the database. Then, if they encounter a problem, even if it isn't caused by the system, they can return to an order state that includes their order up to the time they crashed.

SANDRA

The idea is good, but you lost me at the restructure the database part. I always get nervous when any enhancement restructures the database because that could break the current system's functionality. I think that idea may exceed your one month left on this project.

BOB

You may be right . . . What about the system crashes you mentioned earlier?

SANDRA

I got a phone call—a couple actually—this past Friday. You weren't around so I handled it by myself. That's all taken care of, but we have a few other problems that need to be corrected as quickly as possible. There are a few errors that will require some program cor-

rections. Since you did most of the programming, I thought you should take care of them before your replacement comes over. Here, I wrote up the problems for you.

BOB

[Bob reviews the problem descriptions.]

I can't believe these! We conducted all sorts of program and system tests before and after the system was installed.

SANDRA

Don't get too upset. The problems are relatively minor. But I do want to get them corrected as soon as possible. I want the users to know that we intend to provide top-notch support for their new system. Besides, in all my years of working on projects, I've never seen a new system placed into operation that did not have a few errors that slip by. It is nearly impossible to test a system for all possible errors.

BOB

That's comforting to know. I'll get right on these corrections. Anything else?

SANDRA

Yes, but I will take care of it. We do have a user returning from maternity leave who still isn't up to speed on the new system. She has been calling occasionally for assistance. I gave her the user manual and other users have been trying to help her get up to speed. I will be meeting with her today to do some further training.

BOB

What about enhancements to the new system? If you recall, we froze the specifications. But there were a few new requirements that the users were asking us to implement.

SANDRA

I thought we should give the system a little more time to work out the bugs and allow time for the users to become proficient with the new system. But you're right. We will need to begin planning for those enhancements. But you'll be gone to your juicy new assignment by then, and I'll be coaching a new hire.

BOB

I'm going to miss working together, too. I've learned an awful lot from working with you, Sandra. I'm sure I can get this first round of fixes done before I go to the new assignment. If you get any

more feedback concerning errors, let me know.

___

DISCUSSION QUESTIONS

1. How would you characterize the focus of the support phase of this project?

2. What are some activities that Bob will

need to perform to make corrections to the system?

3. What other types of user assistance might be expected of Bob and Sandra?

4. What types of activities are likely to be included in Bob and Sandra's plan for making enhancements to the new system?

Systems support was introduced in Chapter 3.

> **Systems support** is the ongoing technical support for users, as well as the maintenance required to fix any errors, omissions, or new requirements that may arise.

Before an information system can be supported, it must first be in operation.

> **Systems operation** is the day-to-day, week-to-week, month-to-month, and year-to-year execution of an information system's businesses processes and application programs.

An operational system (not to be confused with an operating system) is frequently called a **production system.** Systems operation and support is often ignored in systems analysis and design textbooks. Young analysts are often surprised to learn that half of their duties (or more) are associated with supporting production systems.

In the chapter map, you can see that systems operation and support uses all the information system building blocks since the information system is operational. All the system knowledge and working components of the system are important to its ongoing operation and support. This is reinforced in Figure 17.1, which demonstrates that systems operations and support often require developers to revisit the activities, and hence the building blocks that were developed during systems analysis, design, and implementation.

Figure 17.2 illustrates systems development, systems operations, and systems support as separate processes. Until this chapter, the entire book has focused on systems development. Systems development responds to problems, opportunities, and directives by developing improved business solutions. That process has been the focus of the first 16 chapters of this book. During systems development, we have accumulated considerable system knowledge and constructed programs and databases. The figure illustrates these three data stores:

— The **repository** is a data store(s) of accumulated system knowledge—system models, detailed specifications, and any other documentation that has been accumulated during the system's development. This knowledge is reusable and critical to the production system's ongoing support. The repository is implemented with various automated tools, and it is often centralized as an enterprise business and IT resource.

— The **program library** is a data store of all application programs. The source code for these programs must be maintained for the life of system. Almost

THE CONTEXT OF SYSTEMS OPERATION AND SUPPORT

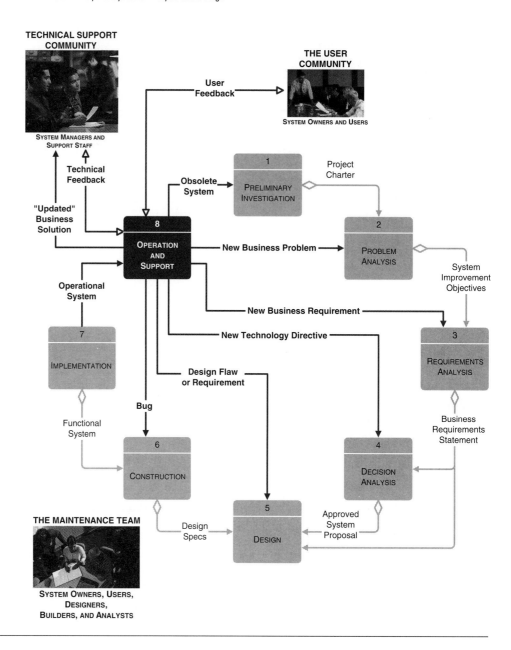

FIGURE 17.1
*The Context of Systems
Operation and Support*

always, the software-based librarian will control access to (via check-in and
checkout) the stored programs as well as track changes and maintain several
previous versions of the programs in case a problem in a new version forces
a temporary use of a prior version. Examples of such *software configuration
tools* are Intersolv's *PVCS*, IBM's *SCM,* and Microsoft's *SourceSafe.*

— The **business data** includes all the actual business data created and maintained
by the production application programs. This includes conventional files,
relational databases, data warehouses, and any object databases. This data is
under the administrative control of the data administrator who is charged with
backup, recovery, security, performance, and the like.

Unlike systems analysis, design, and implementation, systems support cannot
sensibly be decomposed into actual phases that a support project must perform.
Rather, systems support consists of four ongoing activities. Each activity is a type
of support project that is triggered by a particular problem, event, or opportunity
encountered with the implemented system.

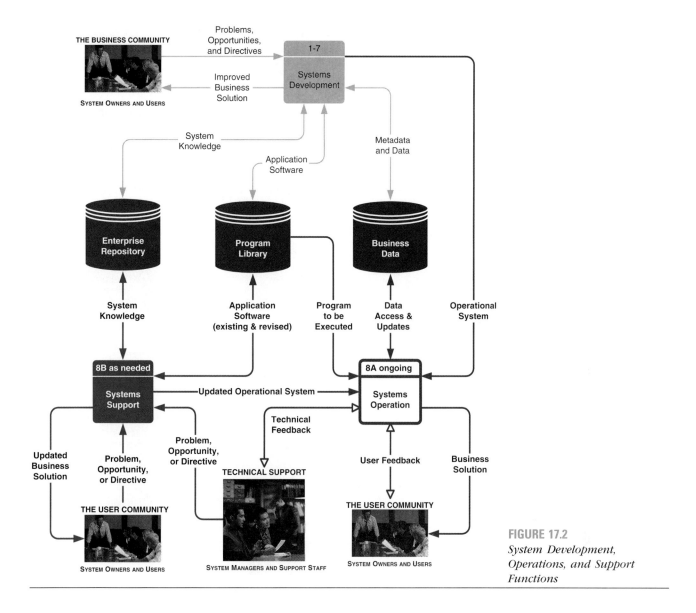

FIGURE 17.2
System Development, Operations, and Support Functions

Figure 17.3 is an activity diagram that illustrates the four types of support activities. The activities are:

- **Program maintenance.** Unfortunately, most systems suffer from software defects or bugs, errors that slipped through the testing of software.
- **System recovery.** From time to time, a system failure may result in a program "crash" and/or loss of data. Human error or a hardware or software failure may have caused this. The systems analyst or technical support specialists may then be called on to recover the system—that is, to restore a system's files and databases and to restart the system.
- **Technical support.** Regardless of how well the users have been trained and how good the end-user documentation is, users will eventually require additional assistance—unanticipated situations arise, new users are added, and so forth.
- **System enhancement.** New requirements may include new business problems, new business requirements, new technical problems, or new technology requirements.

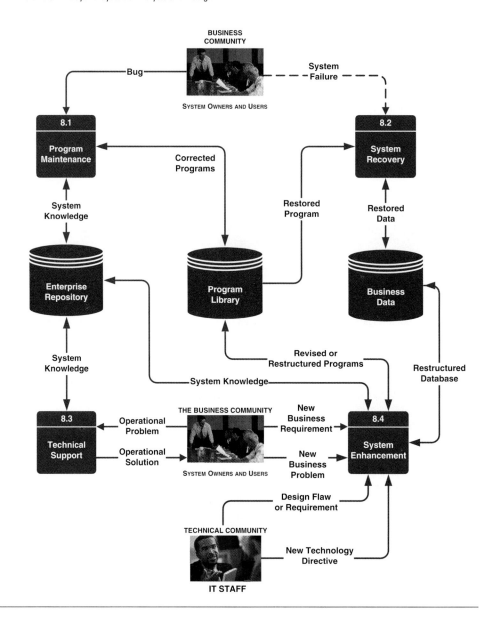

FIGURE 17.3
Systems Support Activities

This chapter examines each of these support activities in various levels of detail, but it abandons the approach used in previous systems analysis, design, and implementation overview chapters. Not all activities will be decomposed into tasks because some of these activities are fairly routine. Also, in some cases, an activity will involve returning to the system development activities that were discussed in earlier chapters.

SYSTEM MAINTENANCE

Regardless of how well designed, constructed, and tested a system or application may be, errors or bugs will inevitably occur. **Bugs** can be caused by any of the following:

— Poorly validated requirements.
— Poorly communicated requirements.
— Misinterpreted requirements.
— Incorrectly implemented requirements or designs.
— Simple misuse of the programs.

The fundamental objectives of system maintenance are:

— To make predictable changes to existing programs to correct errors that were made during systems design or implementation.

— To preserve those aspects of the programs that were correct, and to avoid the possibility that "fixes" to programs cause other aspects of those programs to behave differently.

— To avoid, as much as possible, degradation of system performance. Poor system maintenance can gradually erode system throughput and response time.

— To complete the task as quickly as possible without sacrificing quality and reliability. Few operational information systems can afford to be down for any extended period. Even a few hours can cost millions of dollars.

To achieve these objectives, you need an appropriate understanding of the programs you are fixing and of the applications in which those programs participate. Lack of this understanding is often the downfall of systems maintenance!

How does system maintenance map to your information system building blocks?

— DATA—System maintenance may improve input data editing or correct a structural problem in the database.

— PROCESSES—Most system maintenance is program maintenance.

— INTERFACES—System maintenance may involve correcting problems related to how the application interfaces with the users or another system.

Figure 17.4 illustrates typical system maintenance tasks. Let's examine these tasks.

Task 8.1.1—Validate the Problem

System maintenance miniprojects are triggered by the identification of the problem, usually called a bug. Most such bugs are identified by users when they discover some aspect of the system that does not appear to be working as it should. The first task of the systems analyst or programmer is to *validate the problem*.

Working with the users, the team should attempt to validate the problem by reproducing it. If the problem cannot be reproduced, the project should be suspended until the problem recurs and the user can explain the circumstances under which it occurred. The "as is" program is executed, as closely as possible approximating the circumstances and the data that was present when the problem was first encountered. In most cases, the user who encountered the problem should be the one who re-creates it.

One possible output is an unsubstantiated bug. In this scenario, the user (even with help from the analyst) could not re-create the error. To maintain a productive relationship with the user, the analyst should never make the user feel that it was his or her fault. Instead, the analyst should respect the possibility that the bug is still real and that it will eventually be validated. Should the error recur, the user should be coached to immediately document the exact sequence of events in detail or to call the analyst. Most importantly, the user should be instructed not to perform any subsequent action, if possible, that may prevent the error from being validated.

In some cases the analyst confirms the error but recognizes it as a simple misuse or mistaken use of the program. In such cases, the appropriate output is corrective instructions to the user.

The third possibility is that the bug is real. The analyst should do two things. First, the context of the bug should be examined by studying all relevant documentation (system knowledge) in the repository. In other words, don't try to fix what you don't understand. Second, *all subsequent maintenance will be performed on a copy of the program. The original program remains in the program library and can be used in production systems until it is fixed.*

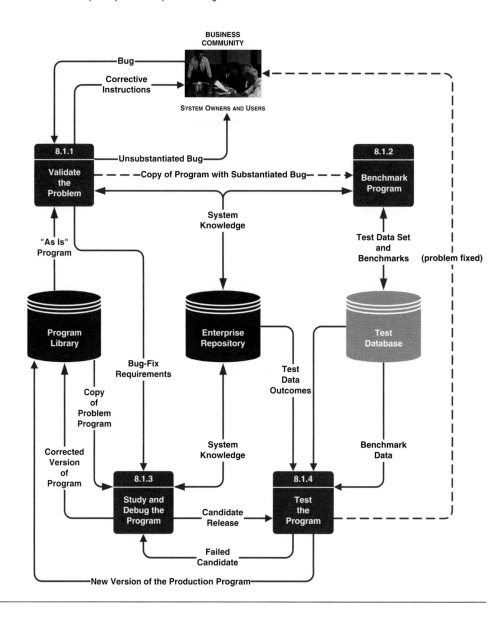

FIGURE 17.4
System Maintenance Tasks

Task 8.1.2—Benchmark Program

Given a copy of the program with a substantiated bug, the analyst should *benchmark the program*. A program isn't all bad, or it would have never been placed into operation. System maintenance can result in unpredictable and undesirable side effects that impact the program's or application's overall functionality and performance. In other words, the solution can cause unexpected side effects. For this reason, before making any changes to programs, the programs should be executed and tested to establish a baseline against which the modified programs and applications can later be measured.

This task can be performed by either the systems analyst and/or programmer. Users should also participate to ensure the test is conducted as closely as possible under circumstances that simulate a normal working environment.

Test cases can be defined in either of two ways. First, past test data may exist in the repository as a form of system knowledge. If so, that data should be reexecuted to establish or verify the benchmark. Usually, the test data should also be analyzed for completeness and, if necessary, revised. New test scenarios may have been identified since the system went into operation. Any revised test data should be recorded in the repository for subsequent maintenance projects.

Alternatively, test data can be automatically captured using a software testing tool. As users enter test data, that data is recorded in a special type of repository as a **test script.** Later the analyst and user can document each test case in the same repository. Ultimately, the test script is executed against the program to test that the program executes properly and also to measure the program's response time and/or throughput. *SQA Suite* by SQA, Inc., is an example of one automated tool for testing. As shown in the figure, the test data and benchmarks are stored in a test database (not the production database) for the next task.

The analyst or programmer needs to have good testing skills (usually taught in programming courses) and may require training in test tools. Neither is explicitly taught in this textbook.

The primary task in system maintenance is to make the required changes to the programs. This task, performed by the application programmer, is not dissimilar from that described in the previous chapter on systems implementation. Essentially, the programmer responds to "bug-fix" requirements that establish the expectation for fixing the problem. The programmer "debugs" (edits) a copy of the problem program. Changes are not made to the production program. The result is a corrected version of the program. This is a candidate release, meaning a candidate to become the next production version of the program.

Usually the original programmer is not making these changes. In fact, several programmers may have written parts of any given program that is now being debugged. Those programmers may no longer be available for clarification. Even if they are available, their memory of the application may not be sufficient or accurate. For this reason, the effective maintenance programmer requires system knowledge. Ideally, this knowledge comes from the repository, but that assumes that the knowledge has been properly maintained throughout the system's lifetime. Too often this is not true, especially for older systems. The programmer may need to seek out this knowledge or, in some cases, reconstruct the knowledge through analysis of the program.

Application and program knowledge usually comes from studying the source code. Program understanding can take considerable time. This activity is slowed by some combination of the following limitations:

- Poor program structure. Examples include *COBOL* programs written with nonstructured techniques or *Visual Basic* or *Java* programs written with nonobject-oriented techniques.
- Unstructured logic (from pre-*structured era* coding styles).
- Prior maintenance (quick fixes and poorly designed extensions).
- Dead code (instructions that cannot be reached or executed—often leftovers from prior testing and debugging).
- Poor or inadequate documentation.

The purpose of application understanding is to see the big picture—that is, how the programs fit into the total application and how they interact with other programs. The purpose of program understanding is to gain insight into how the program works and doesn't work. You need to understand the fields (variables) and where and how they are used, and you need to determine the potential impact of changes throughout the program. Program understanding can also lead to better estimates of the time and resources that will be required to fix the errors.

There is a big difference between editing a new program and editing an existing program. As the designer and creator of a new program, you are probably intimately familiar with the structure and logic of the program. By contrast, as the editor of the existing program, you are not nearly as familiar (or current) with that program. Changes that you make may have an undesirable ripple effect through

Task 8.1.3—Study and Debug the Program

Task 8.1.4—Test the Program

other parts of the program or, worse still, other programs in the application and information system.

A candidate release of the program must be tested before it can be placed into operation as the next new version of the production program. The following tests are essential or recommended:

— **Unit testing** (essential) ensures that the stand-alone program fixes the bug without undesirable side effects to the program. The test data and current performance that you recovered, created, edited, or generated when the programs were benchmarked are used here.
— **System testing** (essential) ensures that the entire application, of which the modified program was a part, still works. Again, the test data and current performance are used here.
— **Regression testing** (recommended) extrapolates the impact of the changes on program and application throughput and response time from the before-and-after results using the test data and current performance.

Failed candidates are returned for additional debugging. Successful candidates are released for production. Older versions of the program are retained in the program library for version control.

> **Version control** is a process whereby a librarian (usually software-based) keeps track of changes made to programs. This allows recovery of prior versions of the programs if new versions cause unexpected problems. In other words, version control allows users to return to a previously accepted version of the system.

One example of version control software is INTERSOLV's *PVCS*.

The high cost of system maintenance is due, in large part, to failure to update system knowledge (in the repository) and program source code documentation (in the program library). Application documentation is usually the responsibility of the systems analyst who supports that application. Program documentation is usually the responsibility of the programmer who made the program changes.

SYSTEM RECOVERY

From time to time a system failure is inevitable. It generally results in an aborted or "hung" program (also called an ABEND or "crash") and may be accompanied by loss of transactions or stored business data. The systems analyst often fixes the system or acts as intermediary between the users and those who can fix the system. This section summarizes the analyst's role in system recovery.

System recovery activities can be summarized as follows:

1. In many cases the analyst can sit at the user's terminal and recover the system. It may be something as simple as pressing a specific key or rebooting the user's personal computer. The systems analyst may need to provide users with corrective instructions to prevent the crash from recurring. In some cases the analyst may arrange to observe the user during the next use of the program or application.

2. In some cases the analyst must contact systems operations personnel to correct the problem. This is commonly required when servers are involved. An appropriate network administrator, database administrator, or webmaster usually oversees such servers.

3. In some cases the analyst may have to call data administration to recover lost or corrupted data files or databases. When recovering business data, it is not only the database that must be restored.

   a. Any transactions that occurred between the last backup and the database's recovery must be reprocessed. This is sometimes called a *roll forward*.

b. If the crash occurred during a transaction, and that transaction was partially completed, then any transactional updates to the database that occurred before the crash must be undone before reprocessing the complete transaction. This is sometimes called a *roll back.*

Database management systems and transaction monitors provide facilities for transaction roll forward and roll back. These data backup and recovery techniques are beyond the scope of this book and are deferred to database courses and textbooks.

4. In some cases the analyst may have to call network administration to fix a local, wide, or internetworking problem. Network professionals can usually log out an account and reinitialize programs.

5. In some cases the analyst may have to call technicians or vendor service representatives to fix hardware problems.

6. In some cases the analyst will discover a possible software bug caused the crash. The analyst attempts to quickly isolate the bug and trap it (automatically or by coaching users to manually avoid it) so that it can't cause another crash. Bugs are then handled as described in the previous section of this chapter.

TECHNICAL SUPPORT

Another relatively routine ongoing activity of systems support is technical support. No matter how well users have been trained or how well documentation has been written, users will require additional assistance. The systems analyst is generally on call to assist users with the day-to-day use of specific applications. In mission critical applications, the analyst must be on call day and night.

The most typical tasks include:

— Routinely observing the use of the system.
— Conducting user-satisfaction surveys and meetings.
— Changing business procedures for clarification (written and in the repository).
— Providing additional training as necessary.
— Logging enhancement ideas and requests in the repository.

SYSTEM ENHANCEMENT

Adapting an existing system to new requirements is the norm for all information systems. Business is change! The pace of change in today's economy is accelerated and rapid response is the expectation. System enhancement requires the systems analyst to evaluate a new requirement to either effect change or direct the change request through an appropriate subset of the original systems development process. In some cases, the analyst may need to recover the system's existing physical structure as a preface to directing the change through systems redevelopment. In this section we will examine two types of adaptive maintenance—system enhancement and systems reengineering.

System enhancement is an adaptive process. Most such enhancement is in response to one of the following events as shown in Figure 17.5.

— *New business problems*—a new or anticipated business problem will make a portion of the current system unusable or ineffective.
— *New business requirements*—a new business requirement (e.g., new report, transaction, policy, or event) is needed to sustain the value of the current system.
— *New technology requirements*—a decision to consider or use a new technology (e.g., new software or version, or different type of hardware) in an existing system needs to be made.

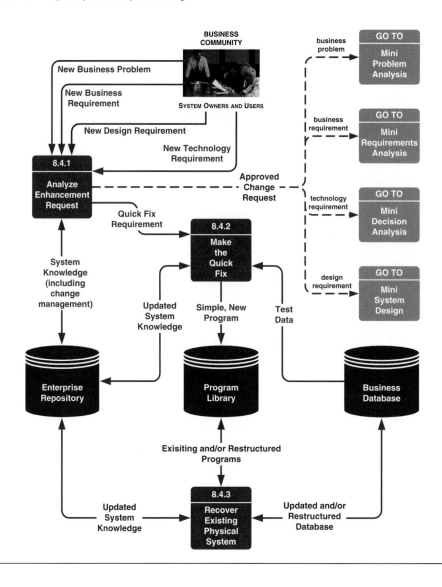

FIGURE 17.5
System Enhancement Tasks

- *New design requirements*—an element of the existing system needs to be redesigned against the same business requirements (e.g., add new database tables or fields, add or change to a new user interface, etc.).

Systems enhancement is reactionary in nature—fix it when it breaks or when users or managers request change. System enhancement extends the useful life of an existing system by adapting it to inevitable change. This objective can be linked to your information system building blocks as follows:

- DATA—Many system enhancements are requests for new information (reports or screens) that can be derived from existing stored data. But some data enhancements call for the restructuring of stored data.
- PROCESSES—Most system enhancements require the modification of existing programs or the creation of new programs to extend the overall application system. But some enhancement requests can be accomplished through careful redesign of existing business processes.
- INTERFACES—Many enhancements require modifications to how the users interface with the system, and how the system interfaces with other systems.

Figure 17.5 expands on the activities of system enhancement. In this section, we briefly describe each activity, participants and roles, inputs and outputs, and techniques.

This activity determines the appropriate course of action to achieve a system enhancement requirement. The initial task is to analyze the request against all other outstanding change requests to determine priority. The system knowledge in the repository is invaluable here, especially if it is current. At a minimum, the importance of the requirement must be assessed against the time and cost of a solution.

Task 8.4.1—Analyze Enhancement Request

At the end of Chapter 6, you were introduced to the ongoing function of change management. Requests for change almost always outnumber resources needed to facilitate change. Change management systems formally capture all change requests in the repository so that change can be prioritized. Changes may be batched so that they can be implemented at optimal times.

If immediate change is needed, the approved change request(s) must be directed to solutions according to the type(s) of change required.

— *New business problems* must be directed to a downsized version of the problem analysis phase. From there, the enhancement will be directed through appropriately downsized versions of requirements analysis, decision analysis, design, construction, and implementation.
— *New business requirements* must be directed to downsized requirements analysis, decision analysis, design, construction, and implementation.
— *New technical requirements* must usually be directed to a decision analysis before design, construction, and implementation. The decision analysis determines if the proposed technology would be feasible in the new system. This is exceedingly important because radical technical change may be costly and complex.
— *New design requirements* must obviously be directed to design, construction, and implementation.

To prioritize and plan enhancement projects, the systems analyst should be skilled in project management (Chapter 4) and feasibility techniques (Chapter 9).

Some system enhancements can be accomplished quickly by writing new simple programs or making very simple changes to existing programs. Simple programs and changes are those that can be made without restructuring stored data (changing the database structure), without updating stored data, and without input of new data (for purposes of storing that data). In other words, these programs generate new (or revise existing) reports and outputs. New program requirements represent many of today's enhancement requests.

Task 8.4.2—Make the Quick Fix

> NOTE   It is our belief that any new program requirements that exceed our definition of simple should be treated as new business requirements and subjected to systems analysis and design to more fully consider implications within the complete application system's structure.

Most such programs can be easily written with 4GLs or report writing tools such as *SAS, Brio,* or *Crystal Reports*. With these tools, programs can be completed within hours. Since they generally do not enter or update data stores, testing requirements are not nearly as stringent.

The quick fix might also be as simple as changing business processes so that they can work with existing information system processes. For example, the analyst may suggest ways in which existing reports can be manually adapted to support different needs.

In Figure 17.5, we see that quick fix requirements are identified and studied. The analysts or programmers use existing business data as test data to write simple new programs that may or may not be added to the program library. Of course, updated system knowledge should be added to the repository.

Task 8.4.3—Recover Existing Physical System

Sometimes the repository does contain up-to-date or accurate system knowledge. Frequently, documentation is out of date. And in some cases, systems were developed without rigorous development processes or enforced documentation standards. In still other cases, systems were purchased; purchased systems are notorious for poor and inadequate documentation. In all of these instances, the analyst may be asked to recover the existing physical structure of a system as a preface to subsequent system enhancement. In some cases, reengineering technology exists to physically restructure and improve system components without altering their functionality. Let's briefly examine some recovery and restructuring possibilities.

Database Recovery and Restructuring   Sometimes systems analysts help in the reengineering of files and databases. Many of today's data stores are still implemented with traditional file structures (such as *VSAM*) or early database structures (such as hierarchical *IMS* structures). Today's database technology of choice is SQL-based relational databases. Tomorrow, object database technology may present yet another paradigm shift. A more common requirement is the changing of versions in an existing database structure (such as *Oracle 7* to *Oracle 8*).

A migrating of data structures from one data storage technology or version to another is a major endeavor, filled with opportunities to corrupt essential, existing business data and programs. Thus, reengineering file and database structures has become an important task.

Database reengineering is usually covered more extensively in data and database management courses and textbooks; however, a brief explanation is in order here. The key player in database restructuring is the database administrator. The systems analyst plays a role because of the potential impact on existing applications. Network analysts may also be involved if databases are (to be) distributed across computer networks.

All databases store metadata about their structure. This metadata can be read and transformed into a physical data model. This data model can be stored in the repository (as system knowledge) to assist analysts in the redesign or use of the database. In some cases, an updated and/or restructured database could be generated, but great care must be taken because many programs use the existing database structure. While database technology theoretically separates data structure from program structure, significant restructuring of databases can still cripple programs.

If the database requires such restructuring, it might be better to identify the change in the repository as another new design requirement that should be directed through an appropriate system redevelopment to determine and react to the impact the new structure may have on existing programs.

Program Analysis, Recovery, and Restructuring   Many businesses are questioning the return on investment in corrective and adaptive maintenance to software. If complex and high-cost software can be identified, it might be reengineered to reduce complexity and maintenance costs. One possibility is to analyze program library and maintenance costs. This task almost always requires software capable of performing the analysis.

Software tools such as Viasoft's *VIA/Recap* measures your software library using a variety of widely accepted software metrics.

Software metrics are mathematically proven measurements of software quality and productivity.

Examples of software metrics applicable to maintenance include:

— **Control flow knots**—the number of times logic paths cross one another. Ideally, a program should have zero control flow knots. (We have seen knot counts in the thousands on some older, poorly structured programs.)

— **Cycle complexity**—the number of unique paths through a program. Ideally, the fewer, the better.

Software metrics, in combination with cost accounting (on maintenance efforts), can help identify those programs that would benefit from restructuring.

The input to program analysis is existing programs in the program library. The software may generate restructured programs or merely add system knowledge about the programs to be used later for either restructuring or enhancement. Program analysis was a critical first step in solving the year 2000 software problem. Businesses had to analyze programs to determine where dates were used and what impact changes might have as the programs were enhanced to accommodate the millennium rollover.

Program recovery is similar to program analysis. Existing program code is read from the library. It is then transformed into some sort of physical model appropriate to the software. For example, many CASE tools can reverse engineer a *COBOL* program into a structure chart for that program or reverse engineer a *Visual Basic* program into an object model (discussed in Modules A and B, following this chapter). These models are added to the repository as new system knowledge to assist with enhancement of the system.

Be very careful not to misuse recovery technology. With purchased software applications, the software license usually prohibits reverse engineering. At best, reverse engineering of purchased software may be unethical. In the worst case, reverse engineering of such software may be illegal and a violation of copyright law and trade secrets.

Finally, some reengineering tools actually support the restructuring of programs. There are three distinct types of program restructuring.

— **Code reorganization** restructures the modular organization and/or logic of the program. Logic may be restructured to eliminate control flow knots and reduce cycle complexity.

— **Code conversion** translates the code from one language to another. Typically, this translation is from one language version to another. There is a debate on the usefulness of translators between different languages. If the languages are sufficiently different, the translation may be very difficult. If the translation is easy, the question is "why change?"

— **Code slicing** is the most intriguing program-reengineering option. Many programs contain components that could be factored out as subprograms. If factored out, they would be easier to maintain. More importantly, if factored out, they would be reusable. Code slicing cuts out a piece of a program to create a separate program or subprogram. This may sound easy, but it is not! Consider your average *COBOL* program. The code you want to slice out may be located in many paragraphs and have dependent logic in many other paragraphs. Furthermore, you would have to simultaneously slice out a subset of the data division for the new program or subprogram.

The candidate program for restructuring is copied from the program library. It is reengineered using one or more of the preceding methods, it is thoroughly tested (as described earlier in the chapter), and the reengineered program is returned to the program library where it is available for production. Any new data, process, and/or network models are updated in the repository.

SYSTEM OBSOLESCENCE

At some point, it will not be cost-effective to support and maintain an information system. All systems degrade over time. And when support and maintenance become cost-ineffective, a new systems development project must be started to replace the system. At this time we come full circle to Chapters 3–16 of this book.

WHERE DO YOU GO FROM HERE?

This chapter provided a detailed overview of the systems support phase of systems development. You learned about the different types of systems support: maintenance, enhancement, reengineering, and design recovery.

If you have been covering the chapters in order, you are now prepared to do systems development. Otherwise, you may wish to return to previous chapters to learn more about the tools and techniques used in systems development. Completion of this book does not guarantee your future success in systems development. Systems development approaches, tools, and techniques continue to evolve. Thus, your learning will be an ongoing process.

SUMMARY

1. Systems support is the ongoing maintenance of a system after it has been placed into operation. This includes program maintenance and system improvements.
2. Systems support involves solving different types of problems. There are several types of systems support: system maintenance, system recovery, end-user assistance, system enhancement, and reengineering.
3. Regardless of how well designed, constructed, and tested a system or application may be, errors or bugs will occur. The corrective action that must be taken is called system maintenance.
4. From time to time a system failure is inevitable. It generally results in an aborted or "hung" program (also called an ABEND or crash) and possible loss of data. The systems analyst often fixes the system or acts as intermediary between the users and those who can fix the system; this is referred to as system recovery.
5. Another relatively routine ongoing activity of systems support is end-user assistance. No matter how well users have been trained or how well documentation has been written, users will require additional assistance.

The systems analyst is generally on call to assist users with the day-to-day use of specific applications. In mission critical applications, the analyst must be on call day and night.
6. Most adaptive maintenance is in response to new business problems, new information requirements, or new ideas for enhancement. It is reactionary in nature—fix it when it breaks or when users make a request. We call this system enhancement. The objective of system enhancement is to modify or expand the application system in response to constantly changing requirements. Another type of reactionary maintenance deals with changing technology. Information system staffs have become increasingly reluctant to wait until systems break. Instead, they choose to analyze their program libraries to determine which applications and programs are costing the most to maintain or which ones are the most difficult to maintain. These systems might be adapted to reduce the costs of maintenance. The preceding examples of adaptive maintenance are classified as reengineering.

KEY TERMS

bug, p. 628
business data, p. 626
code conversion, p. 637
code reorganization, p. 637
code slicing, p. 637
control flow knots, p. 637
cycle complexity, p. 637
production system, p. 625

program library, p. 625
program maintenance, p. 627
regression testing, p. 632
repository, p. 625
software metrics, p. 637
system enhancement, p. 627
systems operation, p. 625
system recovery, p. 627

system testing, p. 632
systems support, p. 625
technical support, p. 627
test script, p. 631
unit testing, p. 632
version control, p. 632

REVIEW QUESTIONS

1. Define systems operations and systems support. What is the relationship between the two phases?
2. What is a production system?
3. What three data stores are common to systems development, systems operation, and systems support? Briefly describe the contents of each.
4. What are the four different systems support activities? Define the purpose of each activity.
5. How do application development projects and application maintenance projects differ in their use of a project repository and program library?
6. Describe five types of software bugs.
7. What are the objectives of system maintenance? How do the information systems building blocks apply to system maintenance?
8. How does a systems analyst or programmer validate a bug? Why do they validate a bug?
9. What is the purpose of program benchmarking as it pertains to system maintenance?
10. What are the inhibitors to program understanding?
11. What are three types of program tests?
12. Define software metrics. Give two examples of software metrics. How are software metrics used in systems reengineering?
13. What are three types of reengineering that can be performed on a program?
14. What is version control? Why is version control a necessity in system maintenance?
15. Describe three variations of software reengineering.

PROBLEMS AND EXERCISES

1. How is a systems support request handled differently from a project request for a new information system?
2. Obtain a copy of a computer program. Analyze the program for control flow knots and cycle complexity.

PROJECTS AND RESEARCH

1. A number of CASE tools are specifically oriented to system maintenance, redevelopment, and reengineering. Through research, do a market survey to identify the characteristics and capabilities of these tools. Briefly compare several products. Select one product and write the vendor. Request product literature, company information, success stories, and the like. Complete your report with an in-depth discussion of the CASE tool.
2. A number of commercial methodologies include system maintenance or redevelopment within their approach. Do a market survey to identify the characteristics and capabilities of these methodologies. Select one methodology and write the vendor. Request product literature, company information, success stories, and the like. Compare and contrast the methodology with the generic methodology presented in this chapter. Identify at least three major features or capabilities that you like better and three that you have some concerns about.
3. Make an appointment with a systems analyst or programmer. Conduct an interview on either the problems and issues encountered in system maintenance or the person's standard approach or methodology for system maintenance. Compare and contrast this with the information and approach presented in this chapter. Write a report of your findings.

MINICASES

1. The Minnesota State University is a large public university located within 20 miles of four cities in Minnesota. [Scene: *Kurt Wilson, director of Administrative Information Management, is meeting with Paula Teague, assistant director of Applications Development.*]

KURT   Good morning, Paula. How's the cold?

PAULA   Much better, thank you. It's good to be back. I assume this is the meeting I had to cancel when I got sick?

KURT   Right. As you know, the administrative information systems master plan will be complete within the next three months. Assuming the executive committee approves the plan, the real work begins—delivering the new business processes and applications outlined in the plan.

PAULA   I've been wondering when you were going to address that issue. We can't keep up with new systems development requests as it is. Am I going to get additional staff?

KURT   I'm afraid not. In this era of staff downsizing, I suspect that we'll be lucky to hold on to what we have.

PAULA   Well, I know we can increase productivity using CASE tools driven off the planning models your staff has recorded in the new repository. But there is a learning curve with CASE technology, as well as the new methodology. Also, these new applications call for a greater degree of adaptability and integration than we have historically expected. I just don't see how we can deliver more systems with the same or fewer people.

KURT   I've got an idea. I've been running some numbers against the time accounting system. According to our own records, we are using 19.3 FTE [full-time equivalent staff] to simply support existing system maintenance.

PAULA   That wouldn't surprise me. Legacy code is the anchor that inhibits new systems development in all shops. Don't tell me you are going to eliminate existing systems support? I think there would be an immediate and fatal backlash from the user community.

KURT   True, but that's not exactly what I had in mind. Don't have a cardiac, but what would you say if I told you that I wanted you to reduce your maintenance effort to 8.5 FTE? *[Paula does not respond.]* Your silence indicates that you are concerned.

PAULA   And rightfully so, don't you think? The user community will scream for my head on a platter! You are talking about cutting support by more than 50 percent.

KURT   Actually, Paula, I'm asking you to cut support by less than 25 percent, but to use 50 percent fewer people!

PAULA   And how am I supposed to do that?

KURT   I have a couple of ideas for you to consider. First, according to my analysis of change request forms, almost two-thirds of all requests fall into the category of enhancements. Half of those enhancements can be characterized as desirable, not essential. It seems to me that we could declare a temporary moratorium on such maintenance projects.

PAULA   I can't confirm your numbers, but I'll agree that we are going to have to be pickier about what we choose to do and not do. I'd like to see your data after this meeting.

KURT   No problem! Second, I'd like you to consider a SWAT team approach to maintenance.

PAULA   SWAT? Like the police?

KURT   SWAT stands for "Specialists With Automated Tools." With only 8.5 FTE for maintenance, it seems to me that it would be a mistake to assign one or two persons per existing development team. Instead, I see a maintenance SWAT team that takes over all maintenance.

PAULA   It might work. I'd want to make sure the SWAT team had at least one member from each current development team to preserve application knowledge. But what's the "automated tools" angle?

KURT   Glad you asked that. Our luncheon meeting will be with a sales representative from a company called Viasoft. They sell what amounts to CASE tools for maintaining, enhancing, and reengineering existing *COBOL* programs. They promise to increase productivity of maintenance programmers who specialize in the tools.

PAULA   Now we're talking. There would be the usual learning curve, but once the team is comfortable with the technology, we might just be able to do 75 percent of our existing maintenance with less than 50 percent of the existing staff.

KURT   I think so! In fact, I'll make a bolder prediction. In time, I think you'll eventually be able to increase support over today's levels with less than 50 percent staff. Anyway, you're a pretty creative person. Give some more thought to ways we can make this work. We need to go to our luncheon date.

   a. Why does system support consume up to 80 percent of some systems development budgets?

   b. Can you think of other ways to provide adequate systems support while reducing systems support staff?

   c. Can you think of any aspects of systems support that have not been addressed by Kurt's SWAT and technology proposals?

SUGGESTED READINGS

Arnold, Robert. *Software Reengineering*. Los Alamitos, CA: IEEE Computer Society Press, 1993. This is a reference book and research compilation on the subject of business process, database, and software reengineering.

Hammer, M., and J. Champy. *Reengineering the Corporation*. New York: Harper Business, 1993. Mike Hammer is widely regarded as the father of business process reengineering.

Martin, E.W.; D.W. DeHayes; J.A. Hoffer; and W.C. Perkins. *Managing Information Technology: What Managers Need to Know*. New York: Macmillan, 1994.

ADVANCED ANALYSIS AND DESIGN METHODS

Parts Two and Three introduced you to classical structured tools and techniques for systems analysis and design. These methods have dominated the practice of systems analysis and design for many years and are still used extensively. Through the years, these techniques have evolved to accommodate changing technologies including object-based technologies such as graphical user interfaces and languages such as *C++*, *Visual Basic*, and *Java*. But methodologists have long realized that to fully exploit object technologies we must eventually create and use object-oriented tools and techniques.

The two modules in Part Five introduce you to advanced systems analysis and design methods, namely, *object-oriented systems analysis and design.* These methods represent the cutting edge of model-driven systems development and they actively support rapid and iterative development technologies based on objects and components.

Module A, Object-Oriented Analysis and Modeling, introduces diagrams and techniques for requirements analysis based upon object concepts. Increasingly, systems analysts are discovering that "objects" are an

intuitive way to conceptualize, validate, and analyze user requirements—regardless of whether or not the intent is to implement the models using object technologies. Indeed, there now exist object-modeling approaches for nontechnology projects such as information systems strategy planning and business process redesign.

During the formative years of object-oriented analysis, numerous and competing modeling notations emerged. This slowed the acceptance of object methods for systems analysis and design. But the key object modeling tools and techniques ultimately converged. This book teaches the standardized object modeling notation called the *Unified Modeling Language* (or *UML*). As we go to press, the UML has gained widespread acceptance, but different methodologies based upon th UML have become common. We expect these methodologies to also converge, or at least reduce in number, through commercial acceptance. But given the acceptance and growing practice of object-oriented systems analysis, your early understanding of the concepts and techniques should prove to be a career advantage.

Increasingly, information systems are constructed with object technologies. Object technologies were introduced in Chapters 1 and 11 as emerging technologies to accelerate productivity and improve quality of systems development, enhancement, and maintenance. Module B, Object-Oriented Design, provides an overview of the transition of requirements-oriented object models from Module A, to design-oriented object models. These models are usually developed iteratively, in conjunction with rapid prototyping that implements the object models using object technologies. Once again, a UML-based approach is taught.

To be sure, the data and process model-driven approaches taught in Parts Two and Three of the book continue to be widely practiced. But eventually, we fully expect that best and preferred practices will be based on object models (using the UML) as opposed to process models (such as DFDs) and data models (such as ERDs). These two modules will introduce you to these advanced systems analysis and design methods.

Focus on PEOPLE

Focus on OBJECTS

Focus on DEVELOPMENT

Stakeholders

Activities

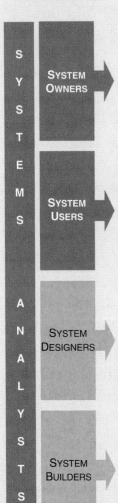

SYSTEM OWNERS

SYSTEM USERS

SYSTEM DESIGNERS

SYSTEM BUILDERS

BUILDING BLOCKS OF AN INFORMATION SYSTEM

Management Expectations

The PIECES Framework

Performance ● Information ● Economics ● Control ● Efficiency ● Service

List of business concepts and rules ...

Business Knowledge

List of business events and responses ...

Business Functions

List of business locations and systems ...

Business Locations

Static Requirements

Dynamic Requirements

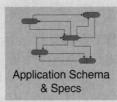

Interface Requirements

Database Schema

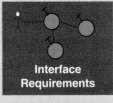

Application Schema & Specs

Interface Specifications

```
CREATE TABLE tblOrders
   colOrderNot CHAR(5) NOT
      NULL
   colOrderDate DATE/TIME
NOT
```

Database Programs

```
PROC ValidateOrder
   PERFORM ValidateCust
   REPEAT UNTIL
      NoMoreProd ...
```

Application Programs

```
<html>
<head>
<title> Order Entry Form </title>
...
```

Interface Programs

PROJECT & PROCESS MANAGEMENT

PRELIMINARY INVESTIGATION

PROBLEM ANALYSIS

REQUIREMENTS ANALYSIS

DECISION ANALYSIS

DESIGN

CONSTRUCTION

IMPLEMENTATION

VENDORS AND CONSULTANTS

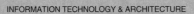

INFORMATION TECHNOLOGY & ARCHITECTURE

Database Technology ● Process Technology ● Interface Technology ● Network Technology

OPERATIONS AND SUPPORT

A

OBJECT-ORIENTED ANALYSIS AND MODELING

MODULE PREVIEW AND OBJECTIVES

This is the first of two modules on object-oriented tools and techniques for systems development. This module focuses on object modeling during systems analysis. You will know object modeling as a systems analysis technique when you can:

— Define object modeling and explain its benefits.

—Recognize and understand the basic concepts and constructs of an object model.

— Read and interpret an object model.

— Describe object modeling in the context of systems analysis.

— Explain the unified modeling language and list its diagrams.

— Construct a use case model.

— Discover objects and classes and their relationships.

— Construct an object class diagram.

— Construct a state diagram to model an object's behavior

AN INTRODUCTION TO OBJECT MODELING

In earlier chapters you were introduced to activities that called for drawing system models. You learned that system models play an important role in systems development by providing a means for dealing with unstructured problems. This module will present object modeling during systems analysis as a technique for defining business requirements for a new system. The approach of using object modeling during systems analysis is called object-oriented analysis.

Object-oriented analysis (OOA) techniques are used to (1) study existing objects to see if they can be reused or adapted for new uses and (2) define new or modified objects that will be combined with existing objects into a useful business computing application.

Object-oriented analysis techniques are best suited to projects that will implement systems using emerging object technologies to construct, manage, and assemble those objects into useful computer applications. The object-oriented approach is centered around a technique referred to as object modeling.

Object modeling is a technique for identifying objects within the systems environment and the relationships between those objects.

The object modeling approach prescribes the use of methodologies and diagramming notations that are completely different from the ones used for data modeling and process modeling. In the late 80s and early 90s many object-oriented methods were being used throughout industry. The most notable of these were Grady Booch's *Booch Method,* James Rumbaugh's *Object Modeling Technique (OMT),* and Ivar Jacobson's *Object-Oriented Software Engineering (OOSE).* The existence of so many methods and associated modeling techniques was a major problem for the object-oriented, system development industry. It was not uncommon for a developer to have to learn several object modeling techniques depending on what was being used at the time. This use of different object modeling techniques limited the sharing of models across projects (reduced reusability) and development teams. It hampered communication between team members and the users, which led to many errors being introduced into the project. These problems and others led to the effort to design a standard modeling language.

In 1994 Grady Booch and James Rumbaugh joined forces to merge their respective object-oriented development methods with the goal of creating a single, standard process for developing object-oriented systems. Ivar Jacobson joined them in 1995, and the three altered their focus to create a standard object modeling language instead of a standard object-oriented approach or method.[1] Referencing their own work as well as countless others in the object-oriented industry, the Unified Modeling Language (UML) version 1.0 was released in 1997.

The **Unified Modeling Language** is a set of modeling conventions that is used to specify or describe a software system in terms of objects.

The UML does not prescribe a method for developing systems—only a notation that is now widely accepted as a standard for object modeling. The Object Management Group (OMG), an industry standards body, adopted the UML in November 1997. In this module and the next we will present only an introduction to the UML and some of its diagrams. Other excellent books are dedicated to the detailed use of the UML and many are listed at the end of this module.

There are many underlying concepts for object modeling. In the next section you will learn about those concepts. Afterward, you will learn how to apply those concepts while developing object models during systems analysis.

---

[1] At the time of this writing, Booch, Rumbaugh, and Jacobson have developed and marketed an object modeling methodology called the Unified Method or Objectory.

Object-oriented analysis (OOA) is based on several concepts. Some of these concepts require a new way of thinking about systems and the development process. As depicted in the module map at the beginning of this module, the DATA, PROCESS, and INTERFACE focuses have been integrated into a single focus on OBJECTS. These OOA concepts have presented a formidable challenge to veteran developers who must relearn how they have traditionally viewed systems. But, as you will soon see, these concepts are not foreign to how you have already come to view your own environment.

The object-oriented approach to system development is based on the concept that objects exist within a system's environment. Objects are everywhere. Let's consider your environment. Look around. What objects are present within your environment? Perhaps you see a door, a window, or the room itself. What about this book? It's an object, and so is the page you are reading. And don't forget you are an object, too. Perhaps you also have a student workbook with you and there are other individuals in the room. You may also see a phone, a chair, and perhaps a table. All these are objects that may be clearly visible within your immediate environment.

But let's stop for a moment and consider the *Webster's Dictionary* definition of an object: "Something that is or is capable of being seen, touched, or otherwise sensed."

The objects mentioned earlier were those that one could see or touch. But what about objects that you might sense? Perhaps you are waiting for a phone call. That phone call is something that you are sensing. You may be waiting for a meeting. Once again, that meeting is something that you can identify, relate to, and anticipate even though you can't actually see that meeting. Thus, according to *Webster's Dictionary,* an anticipated phone call or meeting may be considered an object.

The previous examples pertained to objects that may exist within your immediate environment. Similarly, in the object-oriented approach to system development, it is important to identify those objects that exist within a system's environment. But those objects are considered as being much more than simply "something that is or is capable of being seen, touched, or otherwise sensed." In object-oriented approaches to system development, the definition of an object is as follows:

An **object** is something that is or is capable of being seen, touched, or otherwise sensed, and about which users store data and associate behavior.

Three portions of this definition need to be examined. First, let's consider the term *something.* That something can be characterized as a type of object much like the objects that we identified within your current environment. The types of objects may include a *person, place, thing,* or *event.* An employee, customer, vendor, and student are examples of person objects. A particular warehouse, regional office, building, and room are examples of place objects. Examples of thing objects include a product, vehicle, equipment, videotape, or a window appearing on a user's display monitor. Finally, examples of event objects include an order, payment, invoice, application, registration, and reservation.

Now let's consider the "data" portion of our definition. In object-oriented circles, this part of our definition refers to what are called attributes.

Attributes are the data that represents characteristics of interest about an object.

For example, we might be interested in the following attributes for the person object customer: CUSTOMER NUMBER, FIRST NAME, LAST NAME, HOME ADDRESS, WORK ADDRESS, TYPE OF CUSTOMER, HOME PHONE, WORK PHONE, CREDIT LIMIT, AVAILABLE CREDIT, ACCOUNT BALANCE, and ACCOUNT STATUS. In reality, there may be many customer objects for which we would be interested in these attributes. Each

Objects, Attributes, Methods, and Encapsulation

individual customer is referred to as an object instance. An **instance** (or *object instance*) of an object consists of the values for the attributes that describe a specific person, place, thing, or event. For example, for each customer the attributes would assume values specific to that customer—such as 123456, Lonnie, Bentley, 2626 Darwin Drive, West Lafayette, Indiana, 47906, and so forth. Let's consider your current environment. Perhaps there's another person in the room. Each of you represents an instance of a person object. Each of you can be described according to some common attributes such as LAST NAME, SOCIAL SECURITY NUMBER, PHONE NUMBER, and ADDRESS. But each of you can be described in terms of your own last name, Social Security number, phone number, and address.

Thus, object-oriented approaches to system development are concerned with identifying attributes that are of interest regarding an object. With advances in technology, attributes have evolved to include more than simple data characteristics as represented in the previous example. Today, objects may include newer attribute types, such as a bitmap, a picture, sound, or even video.

Let's now consider the last portion of our definition for an object—the "behavior" of an object.

> **Behavior** refers to those things that the object can do and that correspond to functions that act on the object's data (or attributes). In object-oriented circles, an object's behavior is commonly referred to as a *method, operation,* or *service* (we may use the terms interchangeably throughout our discussion).

This represents a substantially different way of viewing objects! When you look at the "door" object within your environment, you may simply see a motionless object that is incapable of thinking—much less carrying out some action. In object-oriented approaches to system development, that door can be associated with behavior that it is assumed can be performed. For example, the door can *open,* it can *shut,* it can *lock,* or it can *unlock.* All these behaviors are associated with the door and are accomplished by the door and no other object.

Let's consider the phone object. What behaviors could be associated with a phone? With advances in technology we actually have phones that are voice activated and can *answer, dial, hang up,* and carry out other behaviors that can be associated with a phone. Thus, object-oriented approaches to system development simply require an adjustment to how we commonly perceive objects.

Another important object-oriented principle is that an object is solely responsible for carrying out any functions or behaviors that act on its own data (or attributes). For example, only YOU (an object) may CHANGE (behavior) your LAST NAME and HOME ADDRESS (attributes about you). This leads us to an important concept for understanding objects, called encapsulation.

> **Encapsulation** is the packaging of several items together into one unit (also referred to as *information hiding*).

Applied to objects, both attributes and behavior of the object are packaged together. They are considered part of that object. The only way to access or change an object's attributes is through that object's specified behaviors.

In object-oriented development, models depicting objects are often drawn. Let's examine the UML modeling notation used to represent an object in these object models. Figure A.1 (a) depicts the symbol for representing an object instance using the UML modeling notation. An object is represented using a rectangle. The name of the object instance and its classification are underlined and appear at the top of the symbol. The attribute values for the object instance are optionally recorded within the symbol and are separated from the object name with a line.

The name of an object instance is the value of the attribute that uniquely identifies it. In Figure A.1 (a) the attribute CUSTOMER NUMBER, whose value is 412209, uniquely identifies that instance of CUSTOMER. Thus, 412209 is the name of the object instance and CUSTOMER is its classification.

(a)

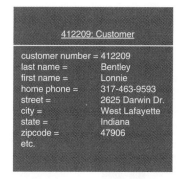

```
    412209: Customer

customer number = 412209
last name =        Bentley
first name =       Lonnie
home phone =       317-463-9593
street =           2625 Darwin Dr.
city =             West Lafayette
state =            Indiana
zipcode =          47906
etc.
```

(b)

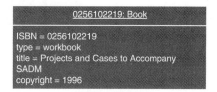

```
    0256101329: Book

ISBN = 0256101329
type = textbook
title = Systems Analysis & Design Methods
copyright = 1996
```

```
    0256102219: Book

ISBN = 0256102219
type = workbook
title = Projects and Cases to Accompany
SADM
copyright = 1996
```

(c)

```
       Customer

customer number
last name
first name
home phone
street
city
state
zipcode
etc.

changeAddress()
```

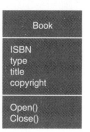

```
        Book

ISBN
type
title
copyright

Open()
Close()
```

FIGURE A.1
Objects and Classes

Classes, Generalization, and Specialization

Another important concept of object modeling is the concept of categorizing objects into classes.

A **class** is a set of objects that share the same attributes and behavior. A class is sometimes referred to as an *object class*.

Let's consider some objects within your current environment. It would be natural for you to classify the textbook and workbook as books [see Figure A.1 (b)]. The textbook and workbook objects represent thing-objects that have some similar attributes and behavior. For example, similar attributes might be ISBN NUMBER, TYPE OF BOOK, TITLE, COPYRIGHT DATE, etc. Likewise, they have similar behavior, such as being able to OPEN and CLOSE. There may be several other objects within your environment that could be classified because of their similarities. For example, you and other individuals in the room might be classified as PERSONS.

How are classes represented in object modeling using the UML approach? As depicted in Figure A.1 (c), a class is represented using a rectangle. The rectangle is divided into three portions. The top portion contains the name of the class. The middle portion contains the name of the common attributes of interest. The lower portion contains the common behavior (or methods). To simplify the appearance of diagrams containing numerous class symbols, sometimes the classes are drawn without including the list of behaviors and attributes.

We can also recognize subclasses of objects [see Figure A.2 (a)]. For example, some of the individuals in the room might be classified as STUDENTS and others

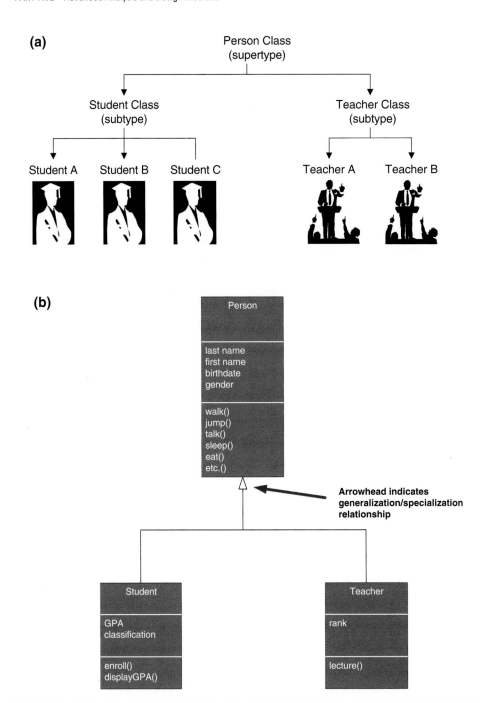

FIGURE A.2

Supertype and Subtype Relationships Between Object Classes

as TEACHERS. Thus, STUDENT and TEACHER object classes are members of the class PERSON. When levels of classes are identified, the concept of inheritance is applied.

Inheritance means that methods and/or attributes defined in an object class can be inherited or reused by another object class.

The approach that seeks to discover and exploit the commonalities between objects/classes is referred to as generalization/specialization.

Generalization/specialization is a technique wherein the attributes and behaviors that are common to several types of object classes are grouped into their own class, called a *supertype*. The attributes and methods of the supertype object class are then inherited by those object classes.

In our example, the object class person is referred to as a *supertype* (or generalization class) whereas student and teacher are referred to as *subtypes* (or specialization class).

> A class **supertype** is an object whose instances store attributes that are common to one or more class subtypes of the object.

The class supertype will have one or more *one-to-one* relationships to object class *subtypes*. These relationships are sometimes called "is a" relationships (or "was a" or "could be a") because each instance of the supertype "is <u>also</u> an" instance of one or more subtypes.

> A class **subtype** is an object class whose instances inherit some common attributes from a class supertype and then add other attributes that are unique to an instance of the subtype.

In object-oriented system development, objects are categorized according to classes and subclasses. Identifying classes realizes numerous benefits. For example, consider the fact that a new attribute of interest, called gender, needs to be added to teacher and student objects. Because the attribute is common to both, the attribute could be added once, with the class person—implying both objects within its class share that attribute. Looking down the road toward program maintenance, the implication is substantial. Program maintenance is enhanced by the need to simply make modifications in one place.

How is generalization/specialization (supertype, subtype classes) depicted using the UML approach? Figure A.2 (b) illustrates how to depict the supertype, subtype relationship between the PERSON, STUDENT, and TEACHER object classes. All the attributes and behaviors of the PERSON object are inherited by the STUDENT and TEACHER objects. Those attributes and behaviors that uniquely apply to a STUDENT or TEACHER are recorded directly in the subtype class symbol.

Object/Class Relationships

Conceptually, objects and classes do not exist in isolation. The things they represent interact with and impact one another to support the business mission. Thus, we introduce the concept of an object/class relationship.

> An **object/class relationship** is a natural business association that exists between one or more objects/classes.

You, for example, interact with this textbook, the phone, the door, and perhaps other individuals in the room. Similarly, objects and classes of objects within a systems environment interact. Consider, for example, the object classes CUSTOMER and ORDER that may exist in a typical information system. We can make the following business assertions about how customers and orders are associated (or interact):

- A CUSTOMER <u>PLACES</u> zero or more ORDERS.
- An ORDER <u>IS PLACED BY</u> one and only one CUSTOMER.

We can graphically illustrate this relationship between CUSTOMER and ORDER as shown in Figure A.3 (a). The connecting line represents a relationship between the classes. The UML refers to this line as an *association* and we will use this term throughout the remaining parts of the module. The verb phrase describes the association. Notice that all associations are implicitly bidirectional, meaning they can be interpreted in both directions (as suggested by the above business assertions).

Figure A.3 (a) also shows the complexity or degree of each association. For example, in the above business assertions, we must also answer the following questions:

(a)

(b)

Multiplicity	UML Multiplicity Notation	Association with Multiplicity		Association Meaning
Exactly 1	1 **or** *leave blank*	Employee —Works for— 1 Department / Employee —Works for— Department		An employee works for one and only one department.
Zero or one	0..1	Employee —Has— 0..1 Spouse		An employee has either one or no spouse.
Zero or more	0..* **or** *	Customer —Makes— 0..* Payment / Customer —Makes— * Payment		A customer can make no payment up to many payments.
One or more	1..*	University —Offers— 1..* Course		A university offers at least 1 course up to many courses.
Specific range	7..9	Team —Has scheduled— 7..9 Game		A team has either 7, 8, or 9 games scheduled.

FIGURE A.3
Object/Class Associations and Multiplicity Notations

- Must there exist an instance of CUSTOMER for each instance of ORDER? Yes!
- Must there exist an instance of ORDER for each instance of CUSTOMER? No!
- How many instances of ORDER can exist for each instance of CUSTOMER? Many!
- How many instances of CUSTOMER can exist for each instance of ORDER? One!

We call this concept multiplicity.

Multiplicity defines how many instances of one object/class can be associated with one instance of another object/class.

Because all associations are bidirectional, multiplicity must be defined in both directions for every relationship. The possible UML graphical notations for multiplicity between classes is shown in Figure A.3 (b).

Let's now consider a special kind of relationship that may exist between objects/classes. Sometimes objects/classes are made up of other objects/classes. This special type of relationship is called aggregation. It is also sometimes referred to as "whole-part" or "part-of" relationships. For example, consider this TEXTBOOK object. This TEXTBOOK contains several objects, including: COVER, TABLE OF CONTENTS, CHAPTER, and INDEX OBJECTS. Furthermore, the CHAPTER object contains PAGE objects, which in turn contain PARAGRAPH objects, which in turn contain WORD objects, and so forth.

By identifying aggregation relationships, we can partition a very complex object and assign behaviors and attributes to the individual objects within it. Figure A.4(a) depicts the UML graphical notation for specifying **composition aggregate relationships** among object classes. Notice that multiplicity is also specified for

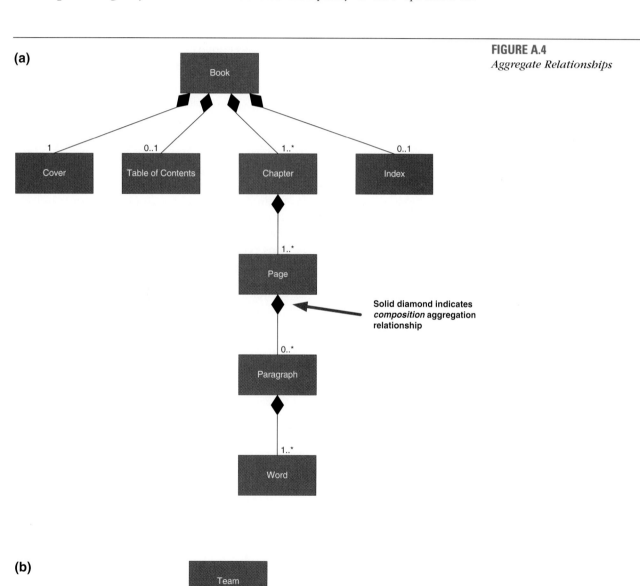

FIGURE A.4
Aggregate Relationships

aggregate relationships. For example, a BOOK is composed of one and only one COVER, zero or one TABLE OF CONTENTS, one or more CHAPTERS, and zero or one INDEX. Figure A.4 (a) also depicts multilevel aggregation wherein CHAPTER consists of one or more PAGES, PAGES consists of zero or more PARAGRAPHS, and so forth. A composition aggregation owns its parts, or in other words, the objects (in this case COVER, TABLE OF CONTENTS, etc.) live in the whole (BOOK). If the whole were to be destroyed, so would its parts.

Another type of aggregation relationship is *shared aggregation* depicted in Figure A.4 (b). This type of aggregation implies that the parts may be shared by many wholes. In Figure A.4 (b) a PLAYER may be a member of more than one TEAM and a TEAM consists of zero to many PLAYERS.

Messages

We just learned that objects/classes interact. But how? Objects/classes interact or "communicate" with one another by passing messages.

> A **message** is passed when one object invokes one or more of another object's methods (behaviors) to request information or some action.

Recall the concept of encapsulation wherein an object is a package of attributes and behavior. Only that object can perform its behavior and act on its data. That is, if you want to secure the room that you are sitting in, the door object must carry out the following behaviors: *close* and *lock*. Thus, if YOU (an object) want the ROOM to become secure, YOU must send a message to the DOOR (an object) requesting it to execute the *close* and *lock* behaviors.

Let's consider our CUSTOMER and ORDER objects mentioned earlier. A CUSTOMER object checking the current status of an ORDER sends a message to an ORDER object by invoking the ORDER object's *display status* behavior (a behavior that accesses and displays the ORDER STATUS attribute).

The object sending a message does not need to know how the receiving object is organized internally or how the behavior is to be accomplished, only that it responds to the request in a well-defined way. This concept of messaging is illustrated in Figure A.5. A message can be sent only between two objects that have an association between them. You will learn how to document messages in Module B.

Polymorphism

An important concept that is closely related to messaging is polymorphism.

> **Polymorphism** means "many forms." Applied to object-oriented techniques, it means that the same named behavior may be completed differently for different objects/classes.

FIGURE A.5
Messaging

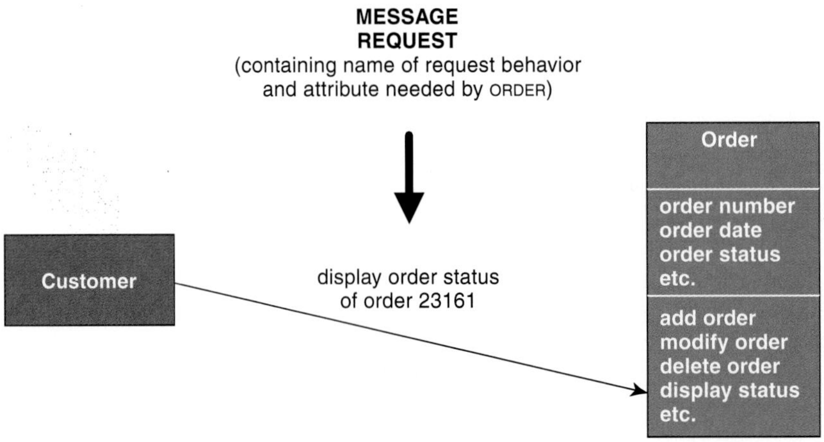

MESSAGE REQUEST
(containing name of request behavior and attribute needed by ORDER)

Customer — display order status of order 23161 — Order (order number, order date, order status, etc. / add order, modify order, delete order, display status, etc.)

Let's consider the WINDOW and DOOR objects within your environment. Both objects have a common behavior that they may perform—close. But how a DOOR object carries out that behavior may differ substantially from the way in which a WINDOW carries out that behavior. A door "swings shut" and windows "slide downward." Thus, the behavior close may take on two different forms. Once again, let's consider the WINDOW object. Not all windows would actually accomplish the close behavior in the same way. Some window objects, like door objects, "swing shut!" Thus, the close behavior may take on different forms even within a given object class.

So how is polymorphism related to message sending? Once again, the requesting object knows what service (or behavior) to request and from which object. However, the requesting object does not need to worry about how a behavior is accomplished.

UML offers five different groups of diagrams to model a system much like a set of blueprints used for constructing a house. Whereas blueprints typically provide the builder perspectives for plumbing, electricity, heating, air-conditioning, and so on, each UML diagram provides the development team a different perspective of the information system. The various UML diagrams and their purposes are as follows:

THE UML DIAGRAMS

- **Use Case Diagrams.** Use case diagrams graphically depict the interactions between the system and external systems and users. In other words, they graphically describe who will use the system and in what ways the user expects to interact with the system. The **use case narrative** is used in addition to textually describe the sequence of steps of each interaction.
- **Class Diagrams.** Class diagrams depict the system's object structure. They show object classes that the system is composed of as well as the relationships between those object classes.
- **Object Diagrams.** Object diagrams are similar to class diagrams, but instead of depicting object classes, they model actual object instances—showing the current values of the instance's attributes. The object diagram provides the developer with a "snapshot" of the system's objects at one point in time. This diagram is not used as often as a class diagram, but when used, it can help a developer better understand the structure of the system.
- **Sequence Diagrams.** Sequence diagrams graphically depict how objects interact with each other via messages in the execution of a use case or operation. They illustrate how messages are sent and received between objects and in what *sequence*.
- **Collaboration Diagrams.** Collaboration diagrams are similar to sequence diagrams but do not focus on the timing or "sequence" of messages. Instead, they present the interaction (or *collaboration*) between objects in a network format.
- **State Diagrams.** State diagrams are used to model the dynamic behavior of a particular object. They illustrate an object's life cycle—the various states that an object can assume and the events that cause the object to transition from one state to another.
- **Activity Diagrams.** Activity diagrams are used to graphically depict the sequential flow of *activities* of either a business process or a use case. They also can be used to model actions that will be performed when an operation is executing as well as the results of those actions.
- **Component Diagrams.** Component diagrams are used to graphically depict the physical architecture of the system. They can be used to show how programming code is divided into modules (or *components*).
- **Deployment Diagrams.** Deployment diagrams describe the physical architecture of the hardware and software in the system. They depict the

software components, processors, and devices that make up the system's architecture.

In this module we will focus only on the use case, class, and state diagrams. Many of the other diagrams will be discussed in Module B.

THE PROCESS OF OBJECT MODELING

In performing object-oriented analysis (OOA), like any other systems analysis method, the purpose is to gain a better understanding of the system and its requirements. In other words, OOA requires that we identify the objects, their data attributes, associated behavior, and associations that support the business system requirements. We perform object modeling to document the identified objects, the data and behavior they encapsulate, plus their relationships with other objects.

There are four general activities when performing object-oriented analysis and they are as follows:

1. Modeling the functions of the system.
2. Finding and identifying the business objects.
3. Organizing the objects and identifying their relationships.
4. Modeling the behavior of the objects.

Modeling the Functional Description of the System

The approach that is commonly used to model the functional aspects of the system is called use case modeling.

> **Use case modeling** is the process of modeling a system's functions in terms of business events, who initiated the events, and how the system responds to the events.

Use case modeling identifies and describes the system functions from the perspective of external users using a tool called use cases.

> A **use case** is a behaviorally related sequence of steps (a scenario), both automated and manual, for the purpose of completing a single business task.

Use cases describe the system functions from the perspective of external users and in the manner and terminology in which they understand. To accurately and thoroughly accomplish this demands a high level of user involvement.

Use cases are the results of decomposing the scope of system functionality into many smaller statements of system functionality. The creation of use cases has proven to be an excellent technique in order to better understand and document system requirements. A use case itself is not considered a functional requirement, but the story (scenario) the use case tells captures the essence of one or more requirements. Use cases are initiated or triggered by external users or systems called actors.

> An **actor** represents anything that needs to interact with the system to exchange information. An actor is a user, a role, which could be an external system as well as a person.

An actor initiates system activity, a use case, for the purpose of completing some business task. An actor represents a role fulfilled by a user interacting with the system and is not meant to portray a single individual or job title. Let's use the example of a college student enrolling for the fall semester's courses. The actor would be the *student* and the business event, or use case, would be *enrolling in course*.

In many information systems business events are triggered by the calendar or the time on a clock. Consider the following examples:

- The billing system for a credit card company automatically generates its bills on the fifth day of the month (billing date).

Actor symbol

Use case symbol

— A bank reconciles its check transactions every day at 5:00 P.M.

— On a nightly basis a report is automatically generated listing which courses have been closed to enrollment (no open seats available) and which courses are still open.

These events are examples of temporal events.

A **temporal event** is a system event that is triggered by time.

Who would be the actor? All of the example events listed previously were performed (or triggered) automatically—when it became a certain date or time. Thus, the actor of a temporal event is time.

Use cases are used during the entire object-oriented development process. During analysis, the use cases are used to model functionality of the proposed system and are the starting point for identifying the objects of the system. During the whole development process, use cases are continually refined in parallel with the process designing the objects. Because use cases contain an enormous amount of system functionality detail, they will be a constant resource for validating and testing development of the system design.

Use cases provide the following benefits:

— As a basis to help identify objects and their high-level relationships and responsibilities.

— A view of system behavior from an external person's viewpoint.

— An effective tool for validating requirements.

— An effective communication tool.

— As a basis for a test plan.

— As a basis for a user's manual.

There are many approaches to begin use case modeling. They include prototyping, user and business analyst interviews, plus countless other fact-finding techniques. Usually it involves a combination of several fact-finding techniques. Let's now examine the steps involved in use case modeling during systems analysis.

Step 1: Identify Any Additional Actors and Use Cases   In Chapter 6, we presented the technique of interviewing users in order to identify the actors and use cases of the system. Another excellent approach to find potential actors and use cases is by analyzing the context model diagram of the system. Recall that a system context model (defined in Chapter 8), illustrates the external parties that interact with the system to provide inputs and receive outputs. In doing so, it identifies the system's scope and boundaries. Figure A.6 (p. 658) is a context model for the SoundStage Member Services System. Find the external parties that provide inputs to the system. Does the external party initiate the input or is the input a response to a request from the system? If the external party initiates the input, it is considered an actor. Some of the inputs and outputs are self-explanatory, but others may be misleading. It is always wise to confirm your findings with the system's users or business analyst. Figure A.7 on page 659 presents a partial list of our findings of analyzing the context diagram with the business analyst.

Step 2: Construct a Use Case Model   Once the use cases and actors have been identified, a **use case model diagram** can be used to graphically depict the system scope and boundaries. The use case model diagram for the use cases listed in Figure A.7 is shown in Figure A.8 on page 660. It was created using Popkin Software's *System Architect* and represents the relationships between the actors and use cases defined for each business subsystem. The subsystems (UML's package symbol) represent logical functional areas of business processes. The partitioning of system behavior into subsystems is very important in understanding the system

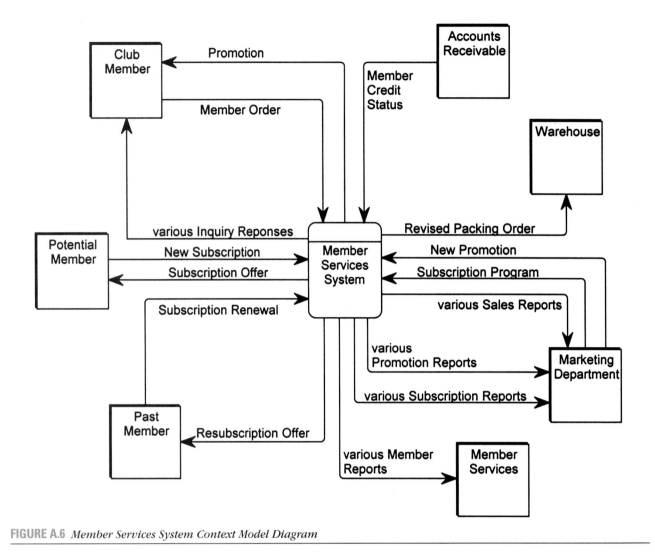

FIGURE A.6 *Member Services System Context Model Diagram*

architecture and is key to defining your development strategy—which use cases will be developed first and by whom.

Step 3: Document the Use Case Course of Events   For each use case identified, we must now document the use case's narrative, which includes the typical course and its alternate courses. During this step the use cases are documented to contain only general information about the business event. The goal is to quickly document all the business events (use cases) in order to define and validate requirements. Each use case will be continually refined in later steps to include more and more detail based on the facts we learn throughout the development process. This initial version of the use case is called a **requirements use case.** The process of documenting a requirements use case was presented in Chapter 6.

Step 4: Define the Analysis Use Cases   Once all requirements use cases have been reviewed and approved by the users, each use case will be refined to include more information in order to specify the system functionality in detail. The resulting use cases are called **analysis use cases** and still should be free of any implementation details. These use cases will be further refined in object-oriented design to specify the <u>how</u> or implementation specifics.

When refining each use case, you may discover the need for additional use

Actor	Use Case Name	Use Case Description
Potential Member	SUBMIT NEW SUBSCRIPTION	Potential member joins the club by subscribing. ("Take any 12 CDs for one penny and agree to buy 4 more at regular prices within two years.")
Club Member	PLACE NEW MEMBER ORDER	Club member places order.
Club Member	MAKE ACCOUNT INQUIRY	Club member wants to examine his or her account history. (90-day time limit)
Club Member	MAKE PURCHASE INQUIRY	Club member inquires about his/her purchase history. (three-year time limit)
Club Member	MAINTAIN MEMBER ORDER	Club member wants to revise an order or cancel an order.
Club Member	SUBMIT CHANGE OF ADDRESS	Club member changes address. (including e-mail and privacy code)
Past Member	SUBMIT RESUBSCRIPTION	Past member rejoins the club by resubscribing.
Marketing	SUBMIT NEW MEMBER SUBSCRIPTION PROGRAM	Marketing establishes a new membership subscription plan to entice new members.
Marketing	SUBMIT PAST MEMBER RESUBSCRIPTION PROGRAM	Marketing establishes a new membership resubscription plan to lure back former members.
Marketing	SUBMIT NEW PROMOTION	Marketing initiates a promotion. (Note: A promotion features specific titles, usually new, that company is trying to sell at a special price. These promotions are integrated into a catalog sent (or communicated) to all members.
Time	GENERATE QUARTERLY PROMOTION ANALYSIS	Print quarterly promotion analysis report.
Time	GENERATE QUARTERLY SALES ANALYSIS	Print annual sales analysis report.
Time	GENERATE QUARTERLY MEMBERSHIP ANALYSIS	Print annual membership analysis report.
Time	GENERATE ANNUAL SALES ANALYSIS	Print annual sales analysis report.
Time	GENERATE ANNUAL MEMBERSHIP ANALYSIS	Print annual membership analysis report.

FIGURE A.7
Partial Listing of Actors and Use Cases for SoundStage Member Services System

cases. A use case may contain complex functionality consisting of several steps that are difficult to understand. To simplify the use case and make it more easily understood, we could extract the more complex steps into their own use cases. This type of use case is called an **extension use case** in that it extends the functionality of the original use case. An extension use case can be invoked only by the use case it is extending. More commonly, you may discover two or more use cases that perform steps of identical functionality. It is best to extract these common steps into their own separate use case called an **abstract use case.** An abstract use case represents a form of "reuse." An abstract use case is available for referencing (or use) by any other use case that requires its functionality. Figure A.9 (p. 661) is an example of graphically depicting a use case with its extensions and abstractions. Figure A.10 on page 662 is an example of an analysis use case, which includes abstract and extension use cases.

After you have defined the analysis use cases, it is now time to start to identify the objects involved in the use cases. These objects represent things or entities in the business domain—things we are interested in and would like to capture information about. At this point we will concentrate on describing these objects with a sentence or two. Later, we will expand our definitions to contain more detailed facts that we learn about each object.

Finding and Identifying the Business Objects

In trying to identify objects, many methodology experts recommend searching the requirements document or other associated documentation and underlining the nouns that may represent potential objects. This could be a monumental task! There are just too many nouns. Use case modeling provides a solution to this

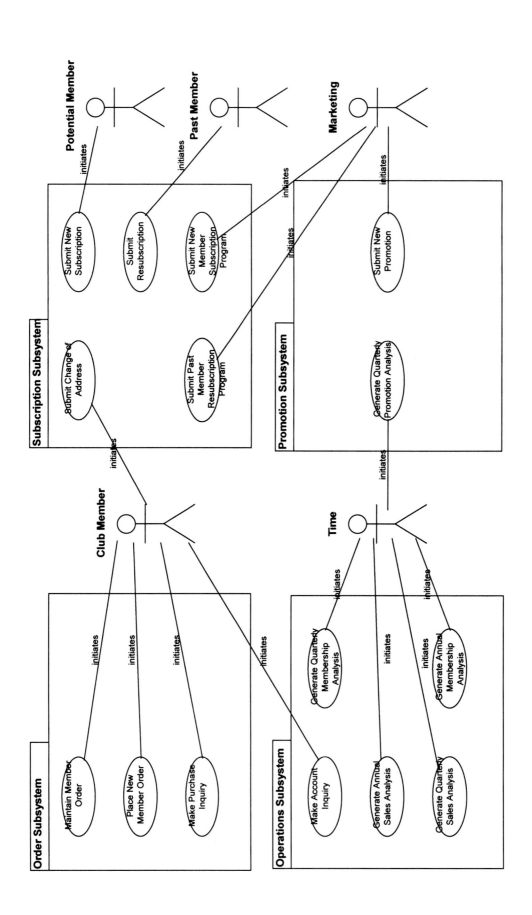

FIGURE A.8 *Member Services System Use Case Model Diagram*

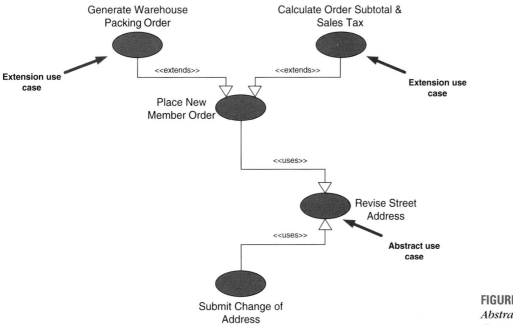

FIGURE A.9
Abstract and Extension Use Cases

problem by breaking down the entire scope of system into use cases. This smaller format simplifies and makes more efficient the technique of underlining the nouns. Let's now examine the steps involved in use case modeling to identify and find business objects for object modeling during systems analysis.

Step 1: Find the Potential Objects   This step is accomplished by reviewing each use case to find nouns that correspond to business entities or events. For example, Figure A.11 on p. 663 depicts the use case PLACE NEW MEMBER ORDER with all the nouns highlighted. Each noun that is found in reviewing the use case is added to a list of potential objects that will be analyzed further (see Figure A.12 on p. 664).

Step 2: Select the Proposed Objects   Not all the candidates on our list represent good business objects. At this time we need to clean up our list by removing the nouns that represent:

— Synonyms.
— Nouns outside the scope of the system.
— Nouns that are roles without unique behavior or are external roles.
— Unclear nouns that need focus.
— Nouns that are really actions or attributes.

Figure A.13 on p. 665 shows the process of "cleaning up" our list of candidate objects. An ✖ marks the candidates we are discarding and a ✓ marks the candidates we are keeping as objects. Also listed is the explanation of why we are keeping or discarding the candidate. Finally, Figure A.14 on p. 666 presents the results of our cleaning up process, plus we have included other objects we have found on the other use cases.

Once we have identified the business objects of the system, it is time to organize those objects and document any major conceptual relationships between the objects. A **class diagram** is used to graphically depict the objects and their

Organizing the Objects and Identifying Their Associations

ANALYSIS USE CASE

Author: S. Shepherd   **Date: 10/25/2000**

USE CASE NAME:	Place New Member Order
ACTOR(S):	Club Member
DESCRIPTION:	This use case describes the process of a club member submitting a new order for SoundStage products. On completion, the club member will be sent a notification that the order was accepted.
REFERENCES	MSS-1.0

TYPICAL COURSE OF EVENTS:	**Actor Action**	**System Response**
	Step 1: This use case is initiated when a member submits an order to be processed.	**Step 2**: The member's personal information such as address and phone number is validated against what is currently on file.
		Step 3: For each product being ordered, validate the product number.
		Step 4: For each product being ordered, check the availability in inventory and record the ordered product information such as the quantity being ordered.
		Step 5: Invoke extension use case *Calculate Order Subtotal & Sales Tax*.
		Step 6: The member's credit card information is verified based on the amount due and Accounts Receivable transaction data is checked to make sure no payments are outstanding.
		Step 7: Invoke extension use case *Generate Warehouse Packing Order*.
		Step 8: Generate an order confirmation notice indicating the status of the order and send it to the member.
	Step 9: This use case concludes when the member receives the order confirmation notice.	

ALTERNATE COURSES:	**Step 2**: If the club member has indicated an address or telephone number change on the order, invoke abstract use case *Revise Street Address*.
	Step 3: If the product number is not valid, send a notification to the member requesting the member to submit a valid product number.
	Step 4: If the product being ordered is not available, record the ordered product information and mark the order as "backordered."
	Step 6: If member's credit card information is invalid or if member is found to be in arrears, a credit problem notice is sent to the member. Modify the order's status to be "on hold pending payment."

PRECONDITION:	Orders can only be submitted by members.
POSTCONDITION:	Member order has been recorded and the Packing Order has been routed to the Warehouse.
ASSUMPTIONS:	None at this time.

FIGURE A.10   *An Example of an Analysis Use Case*

ANALYSIS USE CASE

Author: S. Shepherd   **Date: 10/26/2000**

USE CASE NAME:	Place New Member Order
ACTOR(S):	Club Member
DESCRIPTION:	This use case describes the process of a club member submitting a new order for SoundStage products. On completion, the club member will be sent a notification that the order was accepted.
REFERENCES	MSS-1.0

TYPICAL COURSE OF EVENTS:	**Actor Action**	**System Response**
	Step 1: This use case is initiated when a member submits an order to be processed.	**Step 2**: The member's personal information such as address and phone number is validated against what is currently on file.
		Step 3: For each product being ordered, validate the product number.
		Step 4: For each product being ordered, check the availability in inventory and record the ordered product information such as the quantity being ordered.
		Step 5: Invoke extension use case *Calculate Order Subtotal & Sales Tax.*
		Step 6: The member's credit card information is verified based on the amount due and Accounts Receivable transaction data is checked to make sure no payments are outstanding.
		Step 7: Invoke extension use case *Generate Warehouse Packing Order.*
		Step 8: Generate an order confirmation notice indicating the status of the order and send it to the member.
	Step 9: This use case concludes when the member receives the order confirmation notice.	

ALTERNATE COURSES:	**Step 2**: If the club member has indicated an address or telephone number change on the order, invoke abstract use case *Revise Street Address.*
	Step 3: If the product number is not valid, send a notification to the member requesting the member to submit a valid product number.
	Step 4: If the product being ordered is not available, record the ordered product information and mark the order as "backordered."
	Step 6: If member's credit card information is invalid or if member is found to be in arrears, a credit problem notice is sent to the member. Modify the order's status to be "on hold pending payment."

PRECONDITION:	Orders can only be submitted by members.
POSTCONDITION:	Member order has been recorded and the Packing Order has been routed to the Warehouse.
ASSUMPTIONS:	None at this time.

FIGURE A.11   *Sample Use Case Description with Nouns Highlighted*

POTENTIAL OBJECT LIST
Accounts Receivable Department
Amount Due
Club Member
Credit Card Information
Credit Problem Notice
Credit Status
File
Marketing Department
Member Address
Member Order
Member Phone Number
Member Services Department
Member Services System
Order
Order Confirmation Notice
Order Sales Tax
Order Status
Order Subtotal
Ordered Product
Ordered Product Information
Ordered Product Quantity
Past Member
Payments
Potential Member
Product
Product Inventory
Product Number
Street Address
Transaction
Warehouse
Warehouse Packing Order

FIGURE A.12
*Member Services System
Potential Object List*

associations or relationships. On this diagram we will also include multiplicity, associations, generalization/specialization relationships, and aggregation relationships.

When constructing the diagram, we will use UML notation. Our examples are created with Popkin Software's *System Architect 2001* CASE tool.

Step 1: Identify Associations and Multiplicity   In this step we need to identify associations that exist between objects/classes. Recall that an association between two objects/classes is what one object/class "needs to know" about the other. This allows for one object/class to cross-reference another object. Once the association has been identified, the multiplicity that governs the association must be defined.

It is very important that the analyst not just identify associations that are obvious or recognized by the users. One way to help ensure that possible relationships are identified is to use an object/class matrix. This matrix lists the object/class as column headings as well as row headings. The matrix can then be used as a checklist to ensure that each object/class appearing on a row is checked against *each* object/class appearing in a column for possible associations. The name of the association and the multiplicity can be recorded directly in the intersection cell of the matrix.

POTENTIAL OBJECT LIST		REASON
Accounts Receivable Department	✖	Not relevant for current project
Amount Due	✖	Attribute of "MEMBER ORDER"
Club Member	✓	Type of "MEMBER"
Credit Card Information	✖	Attribute of "MEMBER"
Credit Problem Notice	✖	Potential Interface item to be addressed in object-oriented design
Credit Status	✖	Attribute of "MEMBER"
File	✖	Not relevant for current project
Marketing Department	✖	Not relevant for current project
Member Address	✖	Attribute of "MEMBER"
Member Order	✓	"MEMBER ORDER"
Member Phone Number	✖	Attribute of "MEMBER"
Member Services Department	✖	Not relevant for current project
Member Services System	✖	Not relevant for current project
Order	✖	Another name for "MEMBER ORDER"
Order Confirmation Notice	✖	Potential Interface item to be addressed in object-oriented design
Order Sales Tax	✖	Attribute of "MEMBER ORDER"
Order Status	✖	Attribute of "MEMBER ORDER"
Order Subtotal	✖	Attribute of "MEMBER ORDER"
Ordered Product	✓	"MEMBER ORDERED PRODUCT"
Ordered Product Information	✖	Unclear noun
Ordered Product Quantity	✖	Attribute of "MEMBER ORDERED PRODUCT"
Past Member	✓	Type of "MEMBER"
Payments	✓	Type of "TRANSACTION"
Potential Member	✓	Type of "MEMBER"
Product	✓	"PRODUCT"
Product Inventory	✖	Attribute of "PRODUCT"
Product Number	✖	Attribute of "PRODUCT"
Street Address	✖	Attribute of "MEMBER"
Transaction	✓	"TRANSACTION"
Warehouse	✖	Not relevant for current project
Warehouse Packing Order	✖	Potential Interface item to be addressed in object-oriented design

FIGURE A.13
Analyzing the Potential Object List

Step 2: Identify Generalization/Specialization Relationships   Once we have identified the basic associations and their multiplicity, we must determine if any **generalization/specialization** relationships exist. Recall that generalization/specialization relationships, also known as classification hierarchies or "is a" relationships, consist of superobjects and subobjects. The superobject is general in that it contains the common attributes and behaviors of the hierarchy. The subobjects are specialized in that they contain attributes and behaviors unique to that object, but they inherit the superobject's attributes and behaviors also.

Generalization/specialization relationships may be discovered by looking at the class diagram. Do any associations exist between two objects that have a one-to-one multiplicity? If so, can you say the sentence "object X *is an* object Y" and it is true? If it is true, you may have a generalization/specialization relationship. Also look for objects that have common attributes and behaviors. It may be possible to combine the common attributes and behaviors into a new superobject. Generalization/specialization relationships allow us to take advantage of inheritance, which facilitates the reuse of objects and programming code.

PROPOSED OBJECT LIST

CLUB MEMBER

MEMBER ORDER

MEMBER ORDERED PRODUCT

PAST MEMBER

PAYMENT

POTENTIAL MEMBER

PRODUCT

TRANSACTION

– PLUS –

AGREEMENT

AUDIO TITLE

GAME TITLE

MERCHANDISE

PROMOTION

RETURN

TITLE

VIDEO TITLE

FIGURE A.14
Member Services System
Proposed Object List

Step 3: Identify Aggregation Relationships   In this step we must determine if any aggregation or composition relationships exist. Recall that aggregation is a unique type of association in which one object "is part of" another object. It is often referred to as a **whole/part relationship** and can be read as object A *contains* object B and object B *is part of* object A. Aggregation relationships are asymmetric, in that object B is part of object A, but object A is not part of object B. Aggregation relationships do not imply inheritance, in that object B does not inherit behavior or attributes from object A. Aggregation relationships propagate behavior in that behavior applied to the whole is automatically applied to the parts. For example, if I want to send object A to a customer, object B would be sent also.

Step 4: Prepare the Class Diagram   Figure A.15 is a UML class diagram for the Member Services System. The diagram was constructed using Popkin Software's *System Architect*. Notice that the model depicts business objects/classes within the domain of the SoundStage Member Services System. The object/class notation on the model does not depict behaviors (methods). These will be identified and defined in the next module.

The diagram also reflects the class associations and multiplicity that were identified in step 1, four generalization/specialization relationships that were discovered in step 2, and one aggregate (composite) relationship discovered in step 3. Notice at the bottom of each class symbol the word *persistent* appears.

A class is said to be **persistent** if it outlives the execution of the program.

Typically, this means the objects the class describes will be stored in a database.

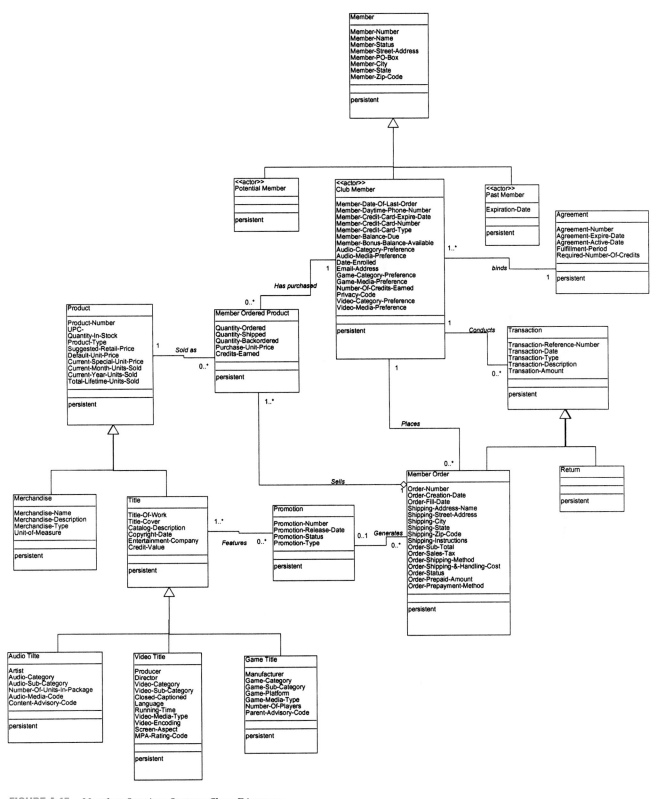

FIGURE A.15 *Member Services System Class Diagram*

Possible "States" of the Space Shuttle

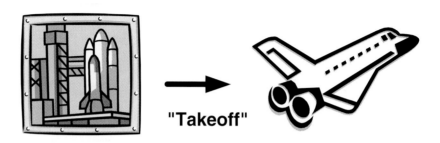

"Takeoff"

"PRE-LAUNCH"
state

"FLIGHT"
state

FIGURE A.16
Object State Example

Modeling the Behavior of the Objects

All objects are said to have **state**—the value of its attributes at one point in time. An object changes state when something happens or when the value of one of its attributes changes. This change in state is triggered by an **event.** Figure A.16 shows the space shuttle resting on the launching pad in a state of *Pre-launch.* After the shuttle takes off (an event) it changes state (state transition), and while it is in the air it is in a state of *Flight.* We could have shown additional states such as *Landed, Checkout,* or *Refurbish* if the requirements specified it. Many of the objects in business systems have complex behaviors or go through many states and types of state.

A **state diagram** models the life cycle of a single object. It depicts the different states an object can have, the events that cause the object to change state over time, and the rules that govern the object's transition between states. In other words it specifies from which state an object is allowed to transition to another state. Figure A.17 is a state diagram constructed using Popkin's *System Architect* for the MEMBER ORDER object of the Member Services System. It begins with an initial state (solid circle) and transitions through a life cycle of different states (rounded corner rectangles) until it reaches its final state (a solid circle inside of a hollow one). Each arrow represents an event that triggers the MEMBER ORDER to change from one state to another.

State diagrams are not required for all objects. Typically, a state diagram is constructed only for those objects that clearly have identifiable states and complex behavior.

W H E R E   D O   Y O U   G O   F R O M   H E R E ?

This module introduced the newer object-oriented approach to system development. Specifically, this module focused on object modeling tools and techniques for systems analysis. You are now ready to learn about the object-oriented approach as it applies to systems design. Object-oriented design is covered in Module B. In Module B you will learn how the object models developed in this chapter are expanded to include design decisions for a new system.

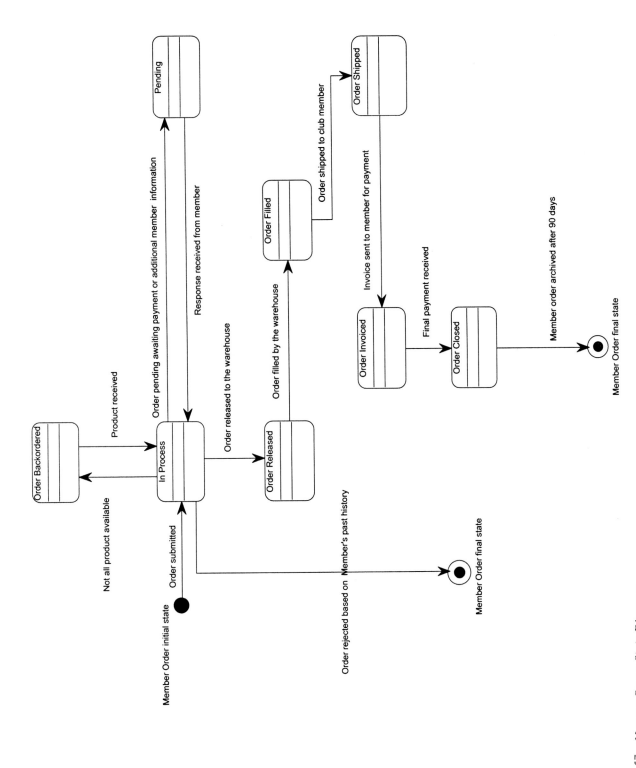

FIGURE A.17 *Member Order State Diagram*

SUMMARY

1. The approach of using object modeling during systems analysis and design is called object-oriented analysis. **Object-oriented analysis (OOA)** techniques are used to (1) study existing objects to see if they can be reused or adapted for new uses, and (2) define new or modified objects that will be combined with existing objects into a useful business computing application.

2. The object-oriented approach is centered around a technique referred to as object modeling. Object modeling is a technique for identifying objects within the systems environment and the relationships between those objects.

3. There are many underlying concepts for object modeling, including:

   a. Systems consist of objects wherein an object is something that is or is capable of being seen, touched, or otherwise sensed, and about which users store data and associate behavior. The data, or attributes, represent characteristics of interest about an object. The behavior of an object refers to those things that the object can do and corresponds to functions that act on the object's data (or attributes). Each object encapsulates the attributes and behavior together as a single unit.

   b. Objects can be categorized into classes. A class is a set of objects that share common attributes and behavior. Objects may be grouped into multiple levels of classes. The most general class in the grouping is the supertype (or generalization of the class). The more refined class is referred to as the subtype class (or specialization class). All subtype classes "inherit" the attributes and behavior of the supertype class.

   c. Objects and classes have relationships. A relationship is a natural business association that exists between one or more objects/classes. The degree, or multiplicity, of an association specifies the business rules governing the association. Some associations are more "structural"—meaning that a class may be related to another class in that one class may represent an as-

   sembly of one or more other class types. This type of association is referred to as an aggregation structure.

   d. Objects communicate by passing messages. A message is passed when one object invokes another object's behavior to request information or some action.

   e. A type of behavior may be completed differently for different objects/classes. This concept is referred to as polymorphism.

4. The UML consists of five groupings of diagrams for modeling the static and dynamic aspects of a system.

5. One of the most critical aspects of performing object-oriented development is correctly identifying the objects and their relationships early in the development process. Use case modeling is a popular approach to assist in object identification.

6. In trying to identify objects, many methodology experts recommend searching the requirements document or other associated documentation and underlining the nouns that may represent potential objects. This could be a monumental task! There are just too many nouns. Use case modeling solves this problem by breaking down the entire scope of system functionality into many smaller statements of system functionality called use cases. This smaller format simplifies and makes more efficient the technique of underlining the nouns.

7. Use case modeling utilizes two constructs: actors and use cases. An actor represents anything that needs to interact with the system to exchange information. An actor is a user, a role that could be an external system as well as a person. A use case is a behaviorally related sequence of steps (a scenario), both automated and manual, for the purpose of completing a single business task.

8. A class diagram is used to organize the objects found as a result of use case modeling and to document the associations between the objects.

9. State diagrams are used to model an object's clearly identifiable states and complex behavior.

KEY TERMS

REVIEW QUESTIONS

1. What is the approach of using object modeling during systems analysis called?
2. Define object modeling.
3. What is an object? Give several examples.
4. What is an attribute? Give several examples of attributes for an object.
5. What is a behavior? Give examples of behaviors for an object.
6. The packaging of an object's attributes and behaviors into a single unit is referred to as what?
7. What is a class? Give an example.
8. Give an example of a supertype and corresponding subtype class. Explain how the concept of inheritance can be applied to the examples.
9. What is the technique called wherein the attributes and behaviors that are common to several types of objects are grouped into their own class, called a supertype?
10. Define object/class association and give an example.
11. What purpose does specifying multiplicity play in defining associations?
12. Give an example of an aggregation structure.
13. Explain the concept of message sending for objects/classes.
14. Define the term *polymorphism*. Give an example.
15. Name the five categories of UML diagrams and describe the individual diagrams within each category.
16. What are the four general activities when performing object-oriented analysis?
17. Describe the process of using use cases to find potential objects.
18. Define a state diagram.

PROBLEMS AND EXERCISES

1. Why is it desirable to categorize objects into classes?
2. What are some objects within your current environment? What classes can be formed from those objects?
3. Draw a class diagram to depict relationships between the objects you identified in problem 2. Be sure to specify the multiplicity. Were you able to identify any aggregate relationships?
4. What are some benefits that can be realized through the object-oriented concepts of inheritance and encapsulation?
5. What are the UML notations for multiplicity? Give examples of object/class relationships that demonstrate each type of multiplicity.
6. Draw a class diagram to communicate the following: ABC Corporation has several employees. For each employee, the company keeps track of that employee's LAST NAME, FIRST NAME, MIDDLE NAME, GENDER, HOME ADDRESS, DATE HIRED, and BIRTHDATE. Some employees are salaried employees. For those employees, the company is interested in their ANNUAL SALARY. Other employees are hourly employees. For hourly employees, the company is interested in knowing their HOURLY PAY RATE.
7. Why is it important to correctly identify the objects and their relationships early in the development process?
8. Describe why it is important to do use case modeling to identify objects.
9. Prepare a state diagram depicting the "marital states" of a person.

PROJECTS AND RESEARCH

1. Research publications to obtain a copy of an object model. Redraw that object model using the UML notation. Be aware that some object models depict information that is not shown on a UML object model.
2. Surf the Internet and make a list of links to sites containing information about object-oriented analysis.
3. Research the Unified Modeling Language (UML) by Rational Software Corporation. Who are the authors of the language? What methodologies is the new language based on? Visit the Rational Software Corporation website. Get a copy of the UML document.
4. Schedule an interview with the business analyst responsible for your school's registration system. Based on the information provided, perform the following:
   a. Draw a context model diagram.
   b. Identify the actors and the use cases they initiate.
   c. Complete a use case description for the event of a student enrolling for a course. For a student dropping a course.
   d. Use the use cases you have previously prepared and construct an object association model diagram.

SUGGESTED READINGS

Booch, G. *Object-Oriented Design with Applications*. Menlo Park, CA: Benjamin Cummings, 1994. Many Booch concepts were integrated into the UML.

Coad, P., and E. Yourdon. *Object-Oriented Analysis*. 2nd ed. Englewood Cliffs, NJ: Prentice Hall, 1991. This book provides a very good overview of object-oriented concepts. However the object model techniques are somewhat limited by comparison to UML and other object-oriented modeling approaches.

Eriksson, Hans-Erik, and Magnus Penker. *UML Toolkit*. New York: John Wiley & Sons, 1998. This book provides detailed coverage of the UML.

Fowler, Martin, and Kendall Scott. *UML Distilled—Applying the Standard Object Modeling Language*. Reading, MA: Addison-Wesley, 1997. A good short guide introducing the concepts and notation of the UML.

Harman, Paul, and Mark Watson. *Understanding UML—The Developer's Guide*. San Francisco: Morgan Kaufmann Publishers, 1997. This book is an excellent reference book. The examples were prepared using Popkin's *System Architect*.

Jacobson, Ivar; Magnus Christerson; Patrik Jonsson; and Gunnar Overgaard. *Object-Oriented Software Engineering—A Use Case Driven Approach*. Wokingham, England: Addison-Wesley, 1992. This book presents detailed coverage of the process of use case modeling and how it's used to identify objects.

Larman, Craig. *Applying UML and Patterns—An Introduction to Object-Oriented Analysis and Design*. Englewood Cliffs, NJ: Prentice Hall, 1997. This is an excellent reference book explaining the concepts of OO development utilizing the UML.

Rumbaugh, James; Michael Blaha; William Premerlani; Frederick Eddy; and William Lorensen. *Object-Oriented Modeling and Design*. Englewood Cliffs, NJ: Prentice Hall, 1991. This book presents detailed coverage of the object modeling technique (OMT) and its application throughout the entire systems development life cycle. Many OMT constructs are now in the UML.

Taylor, David A. *Object-Oriented Information Systems—Planning and Implementation*. New York: John Wiley & Sons, 1992. This book is a very good entry-level resource for learning the concepts of object-oriented technology and techniques.

Focus on PEOPLE

Focus on OBJECTS

Focus on DEVELOPMENT

Stakeholders

Activities

S Y S T E M S   A N A L Y S T S

SYSTEM OWNERS

SYSTEM USERS

SYSTEM DESIGNERS

SYSTEM BUILDERS

VENDORS AND CONSULTANTS

BUILDING BLOCKS OF AN INFORMATION SYSTEM

Management Expectations

The PIECES Framework

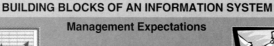

Performance • Information • Economics • Control • Efficiency • Service

List of business concepts and rules ...

Business Knowledge

List of business events and responses ...

Business Functions

List of business locations and systems ...

Business Locations

Static Requirements

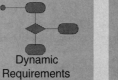

Dynamic Requirements

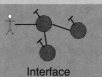

Interface Requirements

Database Schema

Application Schema & Program Specs

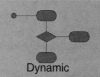

Interface Specifications

```
CREATE TABLE tblOrders
  colOrderNot CHAR(5) NOT
  NULL
  colOrderDate DATE/TIME
NOT
```

Database Programs

```
PROC ValidateOrder
  PERFORM ValidateCust
  REPEAT UNTIL
    NoMoreProd ...
```

Application Programs

```
<html>
<head>
<title> Order Entry Form </title>
...
```

Interface Programs

INFORMATION TECHNOLOGY & ARCHITECTURE
Database Technology • Process Technology • Interface Technology • Network Technology

PROJECT & PROCESS MANAGEMENT

PRELIMINARY INVESTIGATION

PROBLEM ANALYSIS

REQUIREMENTS ANALYSIS

DECISION ANALYSIS

DESIGN

CONSTRUCTION

IMPLEMENTATION

OPERATIONS AND SUPPORT

B

OBJECT-ORIENTED DESIGN AND MODELING

MODULE PREVIEW AND OBJECTIVES

This is the second of two modules on object-oriented tools and techniques for systems development. This module focuses on **object-oriented modeling** tools and techniques that are used during systems design. You will know **object-oriented systems** design when you can

— Differentiate between entity, interface, and control objects.

— Understand the basic concept of *object responsibility* and how it is related to message sending between object types.

— Explain the importance of considering object reuse during systems design.

— Describe three activities involved in completing object design.

— Construct an ideal object model diagram, CRC card, and sequence diagram.

— Construct a design object class diagram.

— Identify activity and implementation diagrams.

AN INTRODUCTION TO OBJECT-ORIENTED DESIGN

This module will present object-oriented techniques for designing a new system. The approach of using object-oriented techniques for designing a system is referred to as **object-oriented design.** Recall that object-oriented development approaches are best suited to projects that will implement systems using emerging object technologies to construct, manage, and assemble those objects into useful computer applications. Object-oriented design is the continuation of object-oriented analysis, continuing to center the development focus around object modeling techniques.

There are many underlying concepts for object modeling during systems design. In this section you will learn about those concepts. Afterward, you will learn how to apply those concepts while learning the process of object-oriented design.

Design Objects

In object-oriented analysis we concentrated on identifying the objects that represented actual data within the business domain. These objects are called **entity objects.** During object-oriented design we continue to refine these entity objects while identifying other types of objects that will be introduced as the result of physical implementation decisions for the new system. Two additional types of objects will be introduced during design. New objects will be introduced to represent a means through which the user will interface with the system. These objects are called **interface objects.** Other types of objects that are introduced are objects that hold application or business rule logic. These objects are called **control objects.**

The structuring of an object-based system into three types of objects was proposed by Dr. Ivar Jacobson and is similar to the mechanism used in *Smalltalk* programming called **model-view-controller (MVC).** The underlying principle for using three types of objects is that responsibilities and behaviors needed to support a system's functionality are distributed across these three types of objects that work together to carry out the service. This practice makes the maintenance, enhancement, and abstraction of those objects simpler and easier to manage than having entity objects encapsulate all required data and behavior, a practice used by some object-oriented design technologies.

The three object types also correlate well with the client/server model. The client is responsible for the application logic (control objects) and presentation method (interface objects), and the server is responsible for the repository (entity objects). Let's further examine each object type.

Interface Objects It is through interface objects that the users communicate with the system. The use case functionality that describes the user directly interacting with the system should be placed in interface objects. The symbol appearing in the margin is used to represent an interface object. The responsibility of the interface object is twofold:

1. It translates the user's input into information that the system can understand and use to process the business event.
2. It takes data pertaining to a business event and translates the data for appropriate presentation to the user.

Each actor or user needs its own interface object to communicate with the system. In some cases the user may need multiple interface objects. Take for example the ATM machine. Not only is there a display for presenting information, but there is also a card reader, money dispenser, and receipt printer. All of these would be considered interface objects.

Entity Objects Entity objects usually correspond to items in real life and contain information, known as attributes, that describes the different instances of the entity. They also encapsulate those behaviors that maintain its information or attributes. The symbol appearing in the margin is used to represent an entity object.

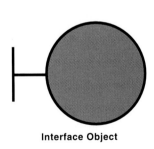

Interface Object

Entity Object

An entity object is said to be **persistent**—meaning the object typically "lives on" after the execution of a method. An entity object exists between method executions because the information about that entity object is typically stored in a database (allowing for later retrieval and manipulation).

Control Objects   When distributing behavior and responsibilities among objects, there is often behavior that does not naturally reside in either the interface or entity objects. In other words, the behavior is not related to how the user interacts with the system, nor is it related to how the data in the system is handled. Rather, such behavior is related to the management of the interactions of objects to support the functionality of the use case. Control objects serve as the "traffic cop" containing the application logic or business rules of the event for managing or directing the interaction between the objects. Control objects allow the scenario to be more robust and simplify the task of maintaining that process once it is implemented. As a general rule, within a use case, a control object should be associated with one and only one actor. Control objects are depicted using the symbol appearing in the margin.

Control Object

In object-oriented systems, objects encapsulate both data and behaviors. In design we focus on identifying the behaviors a system must support and, in turn, design the methods to perform those behaviors. Along with behaviors, we determine the responsibilities an object must have.

Object Responsibilities

In Module A you learned that objects have behaviors, or things that they can do. In object-oriented design it is important to recognize an object has responsibility.

> An **object responsibility** is the obligation that an object has to provide a service when requested, thus collaborating with other objects to satisfy the request if required.

Object responsibility is closely related to the concept of objects being able to send and/or respond to messages. For example, an ORDER object may have the responsibility to display a customer's order, but it may need to collaborate with the CUSTOMER object to get the customer data, the PRODUCT object to get the product data, and the ORDER LINE object to get specific order data about each product being ordered. Thus, CUSTOMER, PRODUCT, and ORDER LINE have an obligation to provide the requested service (provide requested data) to the ORDER object.

The number one driving force for developing systems using object-oriented technology is the potential for **object reusability.** Developers and managers strive to create quality applications cheaper and in less time. Object technology appears to aid in accomplishing that goal. Several studies have documented the success of object reuse. An article that appeared in *ComputerWorld* tells how Electronic Data Systems (EDS) initiated two projects to develop the same system using two different programming languages.[1] One project used a traditional 3GL called *PL/1,* and the other used *Smalltalk,* an object-oriented-based language. The results were impressive as indicated in the following table:

Object Reusability

PL/1	19 calendar months	152 person months	265,000 lines of code
Smalltalk	3.5 calendar months	10.4 person months	22,000 lines of code

Similar studies have produced similar results. To maximize the ability to reuse objects, objects have to be correctly designed initially. This means they have to be defined within a good generalization/specialization hierarchy. The goal is to make objects general enough to be easily used in other applications. For example,

---

[1] "White Paper on Object Technology: A Key Software Technology for the 90s," *ComputerWorld,* May 11, 1992.

when designing a STUDENT object, make sure it is general enough to use in a student registration system as well as in a system that tracks student financial aid or student housing. Any attributes that may be related to a student, but unique to a particular type of student, should be abstracted and placed in a newer, more specialized object.

Many companies achieve their highest level of reuse by exploiting object frameworks or components.

> An **object framework** is a set of related, interacting objects that provide a well-defined set of services for accomplishing a task.

An example of an object framework is a calendar routine, used for calculating or displaying dates. Routines used for charting, printing, or any type of application utility would be good candidates for object frameworks.

> A **component** is a group of objects packaged together into one unit. An example of a component is a dynamic link library (DLL) or executable file.

Through the use of components the developer can easily package and distribute the programming code to others. By using object frameworks and components, developers can concentrate on developing the logic that is new or unique to the application, thus reducing the overall time required to build the entire system.

THE PROCESS OF OBJECT-ORIENTED DESIGN

In performing object-oriented analysis (OOA) we identified objects and use cases based on ideal conditions and independent of any hardware or software solution. During object-oriented design (OOD) we want to refine those objects and use cases to reflect the actual environment of our proposed solution.

Object-oriented design includes the following activities:

1. Refining the use case model to reflect the implementation environment.
2. Modeling object interactions and behavior that support the use case scenario.
3. Updating the object model to reflect the implementation environment.

In the following sections we will review each of these activities to learn what steps, tools, and techniques are used to complete object-oriented design.

Refining the Use Case Model to Reflect the Implementation Environment

In this iteration of use case modeling, the use cases will be refined to include details of how the actor (or user) will actually interface with the system and how the system will respond to that stimulus to process the business event. The manner in which the user accesses the system—via a menu, window, button, bar code reader, printer, and so on—should be described in detail. The contents of windows, reports, and queries should also be specified within the use case. While refining use cases is often time consuming and tedious, it must be completed. These use cases will be the basis on which subsequent user manuals and test scripts are developed during systems implementation. In addition, these use cases will be used by programmers to construct application programs during systems implementation.

In the following steps we will adapt each use case to the implementation environment or "reality" and document the results. It is important that each use case be highly detailed in describing the user interaction with the system. These refined use cases can be used by the user to validate systems design and by the programmer for process and interface specifications.

Step 1: Transforming the "Analysis" Use Cases to "Design" Use Cases   In Module A you learned how to do use case modeling during systems analysis to document user requirements for a given business scenario. In this step, we refine each of those use cases to reflect the physical aspects of the implementation environment for our new system.

Figure B.1 illustrates the refinement of the "Place New Member Order" use case that was originally defined during systems analysis. This copy is labeled as a design use case to distinguish it from the analysis version previously completed. We want to keep the original analysis use cases separate from the refined design use cases to allow maximum flexibility in reusing use cases for variations of different physical implementations. We draw your attention to the following refinements to our use case description in Figure B.1.

❶ We have included an entry that specifies the system user. In analysis, we concentrated on the actor—the party that initiates the business event. In design, we begin to think in terms of "how" the business event is accomplished and by whom. Thus, we are concerned with identifying the party or "system user" that is involved in processing the business event or interacting with the system. In some cases, the actor and the system user may be the same person.

In this implementation of the use case, the actor, "club member," is not the party actually using the system. The club member simply triggers the business event by supplying information to the "order specialist" who then enters it into the system. The system could be designed to allow club members to input orders themselves (via a World Wide Web page or an automated phone system), thus participating as a system user.

❷ Step 4 now describes the windows that will be displayed to the user and the contents (i.e., field names) of the windows.

❸ Descriptions of error messages, special action buttons, possible cursor movements, and other window characteristics should be included in each design use case step.

❹ The design use case step includes references to extension and abstract use cases. Recall that **extension use cases** extend the functionality of the original use case by extracting complex or hard to understand logic into its own use case. **Abstract use cases** are those that contain steps that are used by more than one design use case.

Step 2: Updating the Use Case Diagram and Other Documentation to Reflect Any New Use Cases   After all the analysis use cases have been transformed to design use cases, it is possible that new use cases or even actors have been discovered. It is very important that we keep our documentation accurate and current. Thus, in this step the use case model diagram and the actor and use case glossaries should be updated to reflect any new information introduced in step 1.

In the previous section we refined the use cases to reflect the implementation environment. In this activity we want to identify and categorize the design objects required by the functionality that was specified in each use case and identify the object interactions, their responsibilities, and their behaviors.

Modeling Object Interactions and Behaviors That Support the Use Case Scenario

Step 1: Identify and Classify Use Case Design Objects   Earlier we learned there are three categories of design objects: interface, control, and entity. In this step we examine each design use case to identify and classify the types of objects required by the logic of the use case or business scenario. Figure B.2 depicts the result of analyzing the use case shown in Figure B.1. We draw your attention to the following:

❶ The interface object column contains a list of objects mentioned in the use case with which the users directly interface, such as screens, windows, and printers. The only way an actor or user can interface with a system is via an interface object. Therefore, there should be at least one interface object per actor or user.

DESIGN USE CASE

Author: S. Shepherd   **Date: 11/25/2000**

USE CASE NAME:	Place New Member Order
ACTOR(S):	Club Member – System user: Order Specialist ← ❶
DESCRIPTION:	This use case describes the process of a club member submitting a new order for SoundStage products. On completion, the club member will be sent a notification that the order was accepted.
REFERENCES	MSS-1.1

TYPICAL COURSE OF EVENTS:	Actor Action	System Response
	The main window is currently displayed on the screen waiting for the order specialist to select a menu option. **Step 1**: When a new order is received from the club member, this use case is initiated when the order specialist selects the option "Process New Member Order." **Step 3**: The promotion order specialist enters the MEMBER NUMBER and clicks [OK].	**Step 2**: The system displays a dialogue window requesting the club member's number be entered. **Step 4**: The system verifies that the number is valid and if it is, retrieves and displays the following club member information: MEMBER NAME, MEMBER STATUS , MEMBER STREET ADDRESS, MEMBER P. O. BOX, MEMBER CITY, MEMBER STATE , MEMBER ZIP, MEMBER DAYTIME PHONE NUMBER, and MEMBER EMAIL ADDRESS, along with a menu button labeled "Update Member." The system also displays two "read-only" windows containing order related fields (all currently blank). The first window contains ORDER NUMBER, ORDER CREATION DATE, SUBTOTAL COST, SALES TAX, and ORDER STATUS . The second window contains the individual ordered products. This is a multi-lined, scrollable window containing the fields: PRODUCT NUMBER, DESCRIPTION, QUANTITY AVAILABLE , QUANTITY ORDERED, QUANTITY BACKORDERED, SUGGESTED RETAIL PRICE, PURCHASED UNIT PRICE, EXTENDED COST, and CREDITS EARNED.
	Step 5: The order specialist checks to see if the club member has made any address or phone number changes on the order. If changes have been made the order specialist clicks the [Update Member] menu button - invoke abstract use case *Revise Street Address*. Otherwise, the order specialist clicks [Enter Ordered Product] menu button. **Step 7**: The order specialist enters the PRODUCT NUMBER of the item being ordered and then presses [Enter].	**Step 6**: The system displays a dialogue window prompting the user to enter the PRODUCT NUMBER. **Step 8**: The system validates the product number and then retrieves and displays the DESCRIPTION, CREDIT VALUE , QUANTITY IN STOCK, and SUGGESTED RETAIL PRICE. The cursor is then advanced to the QUANTITY ORDERED field.
	Step 9: The order specialist enters the QUANTITY ORDERED and a PURCHASE UNIT PRICE if it differs from the SUGGESTED RETAIL PRICE and presses [Enter]. If PURCHASE UNIT PRICE is not entered it will default to the SUGGESTED RETAIL PRICE.	**Step 10**: The system validates the inputs then calculates the new EXTENDED COST by multiplying the PURCHASE UNIT PRICE times the QUANTITY ORDERED. If the QUANTITY ORDERED is greater than the QUANTITY IN STOCK, the system calculates QUANTITY BACKORDERED by subtracting QUANTITY IN STOCK from QUANTITY ORDERED. The system displays all fields in the ordered product read-only window.

FIGURE B.1 *Design Use Case*

	Step 12: If the order specialist clicks [Yes] go back to **Step 5**.	**Step 11**: The system displays a window with the message "Is the Ordered Product Information Correct?" If no go to **Step 7**. If yes the system displays another window with the message "Additional Products to be Entered?" **Step 13**: The system assigns a unique ORDER NUMBER, and an ORDER STATUS of "O," then calculates ORDER SUBTOTAL and ORDER SALES TAX - invoke extension use case *Calculate Order Subtotal & Sales Tax*. The system displays all fields in the Order read-only window and then displays the message "Commit Order?" – [Yes] or [Cancel]. ⟵ ③
	Step 14: The order specialist clicks [Yes].	**Step 15**: The system reduces the QUANTITY IN STOCK by the QUANTITY ORDERED for each product being ordered (replace any negative result with zero) and then saves all entries. Then the system verifies the member's status and creditworthiness – invoke abstract use case Verify Club Member Status. **Step 16**: The system generates a packing order - Invoke extension use case *Generate Warehouse Packing Order,* and generates an order confirmation notice indicating the status of the order – invoke abstract use case *Generate Order Confirmation Notice.* If there were any invalid entries submitted by the club member invoke abstract use case *Generate Order Error Report.* ⟵ ④
	Step 17: This use case concludes when the system displays a message "Order 1237645 has been processed successfully, would you like to enter another?" If the order specialist clicks [Yes] the use case is repeated. If the order specialist clicks [No], the main system window is displayed.	
ALTERNATE COURSES:	**Step 4**: If the MEMBER NUMBER is invalid, the system displays a window with the error message "Member Number Not on File." The order specialist can then re-enter the number or cancel the transaction. **Step 8**: If the PRODUCT NUMBER is not valid, the system displays a window with the error message "Product Number not valid." The order specialist either corrects the entry or advances to the next entry. In a later step an error report will be generated for any invalid entries. **Step 10**: If the QUANTITY ORDERED or PURCHASE UNIT PRICE is not valid, the system highlights the field(s) in error and then displays a window with the error message "Highlighted fields invalid." The order specialist either corrects the entry or advances to the next entry. In a later step an error report will be generated for any invalid entries. ⟵ ③ **Step 13**: If any of the products being ordered is not available, record the order status as "P" (partially filled). **Step 16**: If member's credit card information is invalid or if the member is found to be in arrears, a credit problem notice is sent to the member. Modify the order's status to be "H" - on hold pending payment. Exit transaction.	
PRECONDITION:	Order specialist has logged on the system and has authorization for this transaction.	
POSTCONDITION:	Member order has been recorded, the Packing Order has been routed to the Warehouse, and a confirmation notice has been generated and sent to the club member.	
ASSUMPTIONS:	None at this time.	

FIGURE B.1   *(concluded)*

❶ INTERFACE OBJECTS	❷ CONTROLLER OBJECTS	❸ ENTITY OBJECTS
Member Services Main Window	Order Processor	CLUB MEMBER
Club Member Number Dialogue Window	Packing Order Generator	MEMBER ORDER
Order Processing Main Window		MEMBER ORDERED PRODUCT
Club Member Window		PRODUCT
Order Window		
Member Ordered Products Window		
Product Dialogue Window		
Message Window		
Warehouse Printer		

FIGURE B.2   *Interface, Control, and Entity Objects of "Place New Member Order" Use Case*

❷ The control object column contains a list of objects that encapsulate application logic or business rules. As a reminder, a use case should reveal one control object per unique user or actor.

❸ The entity object column contains a list of objects that correspond to the business domain objects whose attributes were referenced in the use case.

Step 2: Identify Object Attributes   During both analysis and design, object attributes may be discovered. In efforts to transform analysis use cases into design use cases, we begin referencing the attributes in the use case text. In this step we examine each use case for additional attributes that haven't been previously identified, and we update our object class diagram to include those attributes.

Step 3: Model High-Level Object Interactions for a Use Case   After identifying and categorizing the design objects involved in a use case, we need to model those objects and their interactions. Such models are called **ideal object model diagrams** and are a type of a use case model in the UML. Figure B.3 depicts a partial ideal object model diagram for the "Place New Member Order" use case. This diagram was developed using Popkin Software's *System Architect* CASE tool. An ideal object model diagram includes symbols to represent actors, interface, control, and entity objects, and lines that represent messages or communication between the objects. We draw your attention to the following:

❶ We have included both the actor that "initiated" the use case (the CLUB MEMBER) and the actor who is the direct "user" of the system (the ORDER SPECIALIST). Notice the ORDER SPECIALIST actor is represented by the actor icon enclosed in an interface object icon. This type of symbol is referred to as a "case worker" in the UML. This communicates to the reader that the primary actor (CLUB MEMBER) does not directly use the system but interfaces with the system through another actor (ORDER SPECIALIST). Another actor represented is the WAREHOUSE. The WAREHOUSE is a "receiver." In this scenario, the warehouse receives the packing order information from the system.

❷ Actors may interact with the system only via interface objects. In our example above, the ORDER SPECIALIST (direct user of the system) interacted with the system by way of the MEMBER SERVICES MAIN WINDOW and ORDER PROCESSING MAIN WINDOW interface objects. Notice that the WAREHOUSE (a receiver of the system) interacts with the WAREHOUSE PRINTER interface object.

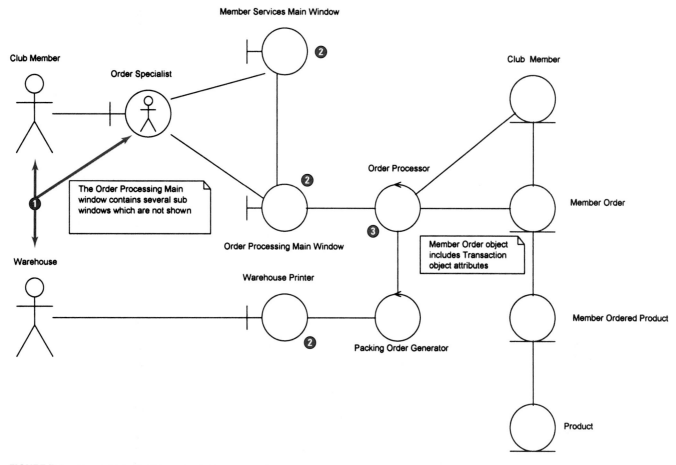

FIGURE B.3   *Partial Ideal Object Model Diagram for "Place New Member Order" Use Case*

❸ By following the logic of the use case we can construct a high-level, first-cut view of how the objects need to interact with each other to support the functionality of the use case. Notice that the ORDER PROCESSOR control object sends messages (depicted using lines) to the CLUB MEMBER entity object. Based on the text of the use case, we know that CLUB MEMBER information needs to be retrieved, displayed, and updated. Therefore, the messages between ORDER PROCESSOR and CLUB MEMBER objects request the CLUB MEMBER object to fulfill its object responsibility by providing those services.

At this point the ideal object diagram is our first attempt at determining how the objects should interact with each other to support the business event. Later we will use this diagram to model more detailed object interactions.

Step 4: Identify Object Behaviors and Responsibilities   Once we have identified all the objects needed to support the functionality of the use case, we shift our attention to defining their specific behaviors and responsibilities. This step involves the following tasks:

1. Analyze the use cases to identify required system behaviors.
2. Associate behaviors and responsibilities with objects.
3. Examine object model for additional behaviors.
4. Verify classifications.

In Module A you learned objects encapsulate data and behavior. Our first task in identifying the object behaviors and responsibilities is accomplished by once again examining our use case. The use case description is examined to identify all *action verb phrases*. Action verb phrases suggest behaviors that are required to complete a use case scenario. These verb phrases correlate to the system behaviors required to respond to a business event of a club member submitting a promotion order. Each use case should be separately examined to identify behaviors associated with the use case.

Once the behaviors have been identified, we must determine if the behaviors are manual or if they will be automated. If they are to be automated, they must be associated with the appropriate object type that will have the responsibility of carrying out that behavior. Figure B.4 lists each verb phrase or behavior in the "Place New Member Order" use case, whether it's manual or to be automated, and the design object type it is to be associated with.

In Figure B.5 we have condensed the behavior list to show only the behaviors that need to be automated. Recall that the object types were defined earlier in step 1. We will use this list as the source of behaviors to be allocated in the next task.

FIGURE B.4

Summary of "Place New Member Order" Use Case Behaviors

BEHAVIORS	AUTOMATED/MANUAL	OBJECT TYPE
Process new member order	Manual/Automated	Control
Submit new order	Manual	
Select process new member order option	Manual	
Display club member number dialogue window	Automated	Interface
Enter member number	Manual	
Verify member number	Automated	Entity
Retrieve club member information	Automated	Entity
Display club member information	Automated	Interface
Display blank order window	Automated	Interface
Display blank member ordered products window	Automated	Interface
Check for address changes	Manual	
Click button	Manual	
Display product number dialogue window	Automated	Interface
Enter product number & press enter	Manual	
Validate product number	Automated	Entity
Retrieve product information	Automated	Interface
Advance cursor	Automated	Interface
Enter order information	Manual	
Validate input fields	Automated	Entity
Calculate extended cost	Automated	Entity
Display ordered product information	Automated	Interface
Display message windows (various)	Automated	Interface
Assign unique order number	Automated	Control
Update order status	Automated	Entity
Calculate order subtotal & sales tax	Automated	Entity
Reduce quantity in stock	Automated	Entity
Calculate quantity backordered	Automated	Entity
Verify member status and creditworthiness	Automated	Entity
Generate warehouse packing order	Automated	Control
Generate order confirmation notice	Automated	Control
Generate order error report	Automated	Control

BEHAVIORS	OBJECT TYPE
Process new member order	Control
Assign unique order number	Control
Generate warehouse packing order	Control
Generate order confirmation notice	Control
Generate order error report	Control
Verify member number	Entity
Retrieve club member information	Entity
Validate product number	Entity
Validate input fields	Entity
Calculate extended cost	Entity
Update order status	Entity
Calculate order subtotal & sales tax	Entity
Reduce quantity in stock	Entity
Calculate quantity backordered	Entity
Verify member status and creditworthiness	Entity
Display club member number dialogue window	Interface
Display club member information	Interface
Display blank order window	Interface
Display blank member ordered products window	Interface
Display product number dialogue window	Interface
Retrieve product information	Interface
Advance cursor	Interface
Display ordered product information	Interface
Display message windows (various)	Interface

FIGURE B.5
Condensed Behavior List for "Place New Member Order" Use Case

Object Name: Member Order	
Sub Object:	
Super Object:	
Behaviors and Responsibilities	**Collaborators**
Report order information	Member Ordered Product
Calculate subtotal cost	
Calculate sales tax	
Update order status	
Create ordered product	
Delete ordered product	

FIGURE B.6   *CRC Card for MEMBER ORDER Object*

Our third task is aimed toward identifying all behaviors that can be associated with an object type and identifying collaborations among those objects. A popular tool for documenting the behaviors and collaborations for an object is the **class responsibility collaboration (CRC) card.**[2] A CRC card for the object type MEMBER ORDER is depicted in Figure B.6. The CRC card contains all use case behaviors and responsibilities that have been associated with the object type MEMBER ORDER. Analysis of the use case scenarios may not reveal all behaviors for any given object type. On the other hand, by examining the object class diagram, you may find additional behaviors (not mentioned in the use case scenarios) that need to be assigned to an object type. For example, analyze the relationships between the objects in Figure B.7. How are those relationships created or deleted? Which

---

[2] CRC cards were pioneered by Kent Beck and Ward Cunningham.

object should be assigned that responsibility? As a rule, the object that controls the relationship should be responsible for creating or deleting the relationship. Draw your attention in Figure B.7 to the relationship between MEMBER ORDER and MEMBER ORDERED PRODUCT. By designing the system to have the MEMBER ORDER object have a behavior to "add ordered product," we have effectively given the MEMBER ORDER object control to create this relationship. Also, recall from Module A that there are four "implicit" behaviors that can be associated with any object class. Those are the ability to create new instances, change its data or attributes,

FIGURE B.7   *Object Class Diagram for "Place New Member Order" Use Case*

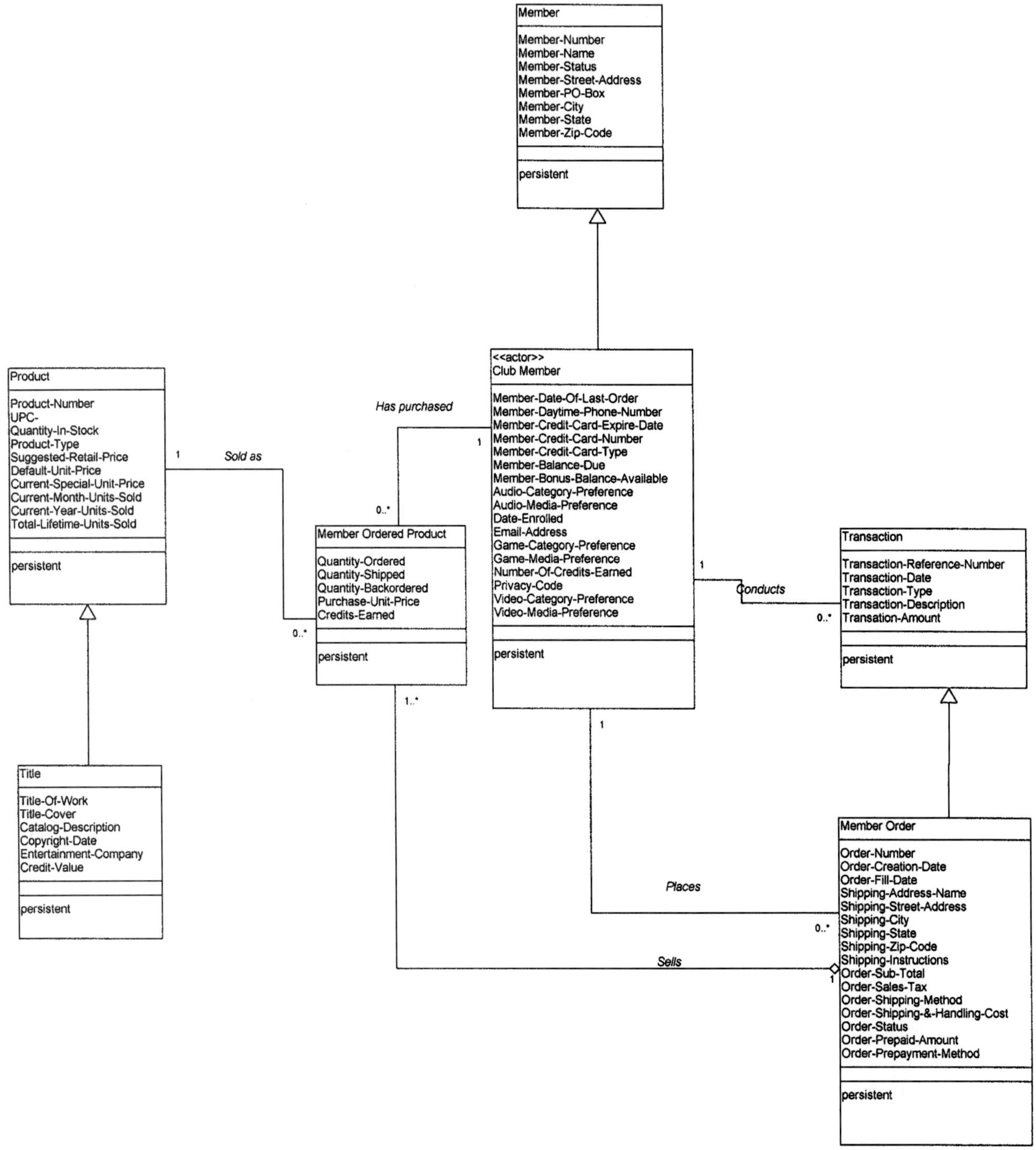

delete instances, and display information about the object class. While examining the use cases to identify and associate behaviors with object types, we also focus on identifying the **collaboration** or cooperation that is necessary between object types. In Figure B.6 the MEMBER ORDER object needs collaboration from the MEMBER ORDERED PRODUCT object to retrieve information about each of the products being ordered. Remember if an object needs another object's attribute to accomplish a behavior, the collaborating object needs to have a behavior or method to provide that attribute.

Identifying the collaboration of object types is necessary to ensure that all use case objects work in harmony to complete the processing required for the business event that triggers the use case scenario.

Finally, our last task is to verify the results from the previous tasks. This consists of conducting walkthroughs with the appropriate users. One verification approach that is commonly used is **role playing.** In role playing, the use case scenarios are acted out by the participants. The participants may assume the role of actors or object types that collaborate to process a hypothetical business event. Message sending is simulated by using an item such as a ball that is passed (or sometimes thrown) between the participants. Role playing is quite effective in discovering missing objects and behaviors, as well as verifying the collaboration among objects.

Step 5: Model Detailed Object Interactions for a Use Case   Once we have determined the objects' behaviors and responsibilities, we can create a detailed model of how the objects will interact with each other to provide the functionality specified in each design use case. The UML provides two types of diagrams to graphically depict these interactions—a sequence diagram and a collaboration diagram. **Sequence diagrams** show us in great detail how the objects interact with each other over time, and **collaboration diagrams** show us how objects collaborate in message sequence to satisfy the functionality of a use case. Collaboration diagrams are similar in appearance to ideal object diagrams except the UML object symbol is used and the interactions between objects are numbered in the sequence they occur. For this reason we will not use one here but instead concentrate on the sequence diagram.

Figure B.8 is a partial sequence diagram for the "Place New Member Order" use case. It is read from top to bottom, following the logic of the use case. Each object referenced in the use case is symbolized by a vertical line. The behaviors or operations that each object needs to fulfill an obligation are represented by a gray box. These boxes represent program code. The arrows between the lines represent interactions or messages being sent to a particular object to invoke one of its operations to satisfy a request.

Once we have designed the objects and their required interactions, we can refine our object model to include the behaviors or implementation methods it needs to possess. Figure B.9 is a partial view of our object class diagram that correlates to the objects used in the "Place New Member Order" use case. We have given each behavior or method a name. Normally these names reflect the programming language used to develop the system.

Updating the Object Model to Reflect the Implementation Environment

The UML offers three additional diagrams to model design and implementation aspects of the system—activity diagrams, component diagrams, and deployment diagrams. **Activity diagrams** are similar to flowcharts in that they graphically depict the sequential flow of *activities* of either a business process or a use case. They are different from flowcharts in that they provide a mechanism to depict activities that occur in parallel. Because of this they are very useful for modeling actions that will be performed when an operation is executing as well as the results

ADDITIONAL UML DESIGN AND IMPLEMENTATION DIAGRAMS

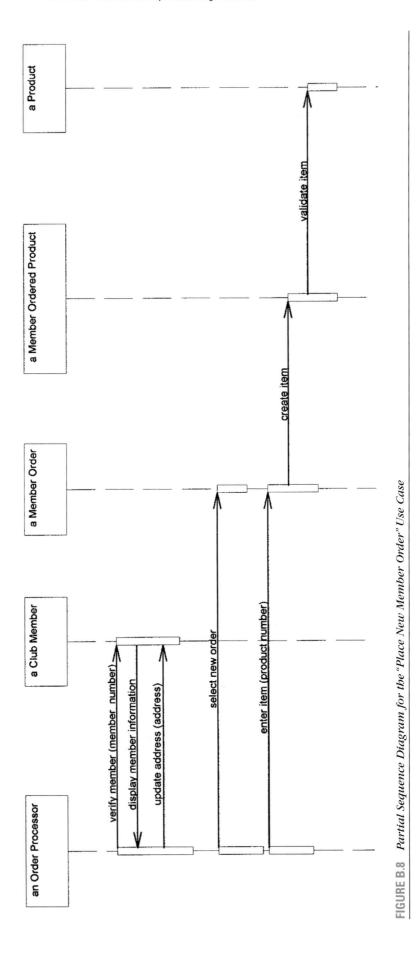

FIGURE B.8   *Partial Sequence Diagram for the "Place New Member Order" Use Case*

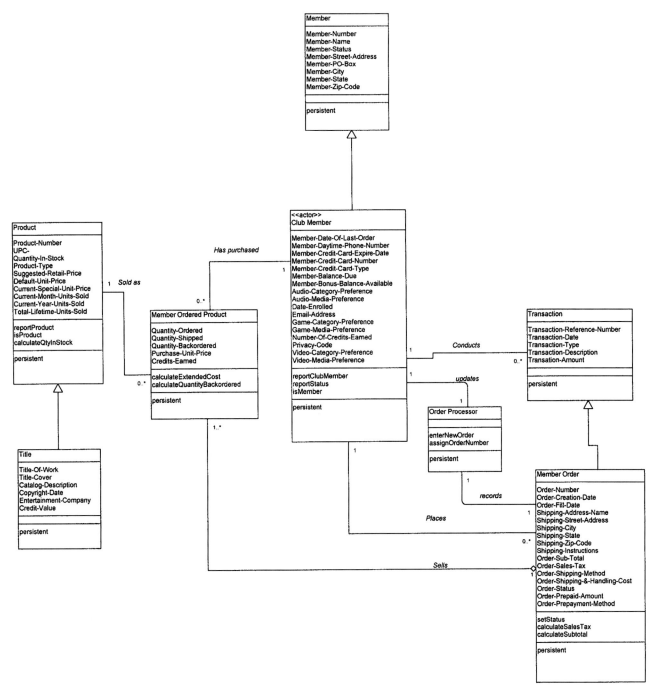

FIGURE B.9   *Partial Object Class Diagram Correlating to "Place New Member Order" Use Case Objects*

of those actions—such as modeling the events that cause windows to be displayed or closed. Activity diagrams are flexible in that they can be used during analysis and design. Figure B.10 is an example of an activity diagram depicting the business process of a manager approving or disapproving a purchase order request. Notice that it includes the following items:

❶ A solid dot represents the start of the process.

❷ A rounded corner rectangle represents an activity or task that needs to be performed.

❸ Arrows depict triggers that initiate activities.

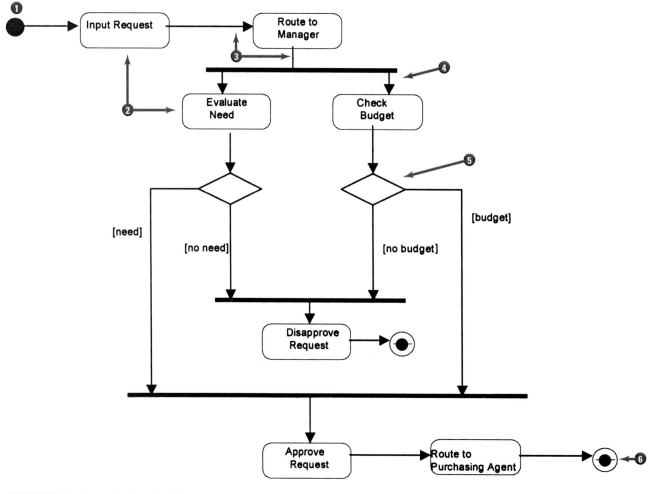

FIGURE B.10   *Example Activity Diagram*

❹ A solid black bar is called a synchronization bar. This symbol allows you to depict activities that can occur in parallel. In Figure B.10 the activities EVALUATE NEED and CHECK BUDGET occur in parallel and can be performed in any order, but both have to be performed before the activity APPROVE REQUEST can be performed.

❺ A diamond represents a decision activity.

❻ A solid dot inside a hollow circle represents the termination of the process.

Component diagrams are implementation type diagrams and are used to graphically depict the physical architecture of the software of the system. They can be used to show how programming code is divided into modules (or *components*) and depict the dependencies between those components. Figure B.11 shows an example of a component diagram constructed in Popkin's *System Architect*. A component is represented in the UML as a rectangle with two smaller rectangles to the left. The dependency between the two components shows how they communicate with each other.

Deployment diagrams are also implementation type diagrams that describe the physical architecture of the hardware and software in the system. They depict the software components, processors, and devices that make up the system's architecture. Figure B.12 shows an example of a deployment diagram constructed in Popkin's *System Architect*. Each box in the diagram is the symbol for a node,

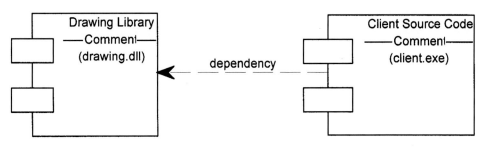

FIGURE B.11
Example Component Diagram

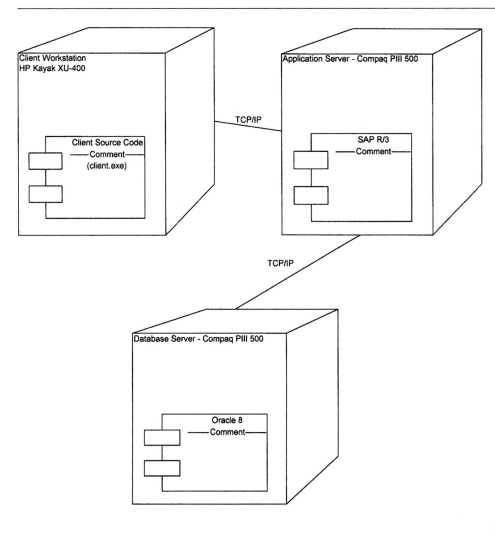

FIGURE B.12
Example Deployment Diagram

which in most cases is a piece of hardware. The hardware may be a PC, main-frame, printer, or even a sensor. Software that resides on the node is represented by the component symbol. The lines connecting the nodes indicate a communication path between the devices. In Figure B.12 the connections are labeled with the type of communication protocols being used.

You have been presented with a short overview of the activity, component, and deployment diagrams. If you desire a more in-depth description of the purpose and the use of these diagrams, excellent books are available documenting the use of the UML. Many are listed at the end of this module.

WHERE DO YOU GO FROM HERE?

This module provided an introduction to the newer object-oriented approach to system design. Since prototyping is an integral part of object-oriented design, it is recommended that you learn about prototyping during system design. Thus, if you haven't already covered these chapters, it is recommended that you now read Chapters 13, 14, and 15 to learn how to prototype inputs, outputs, and user interfaces (respectively).

To provide a better understanding of object-oriented design and its impact on the subsequent construction and implementation of a new system, it is recommended that you consider learning about object-oriented programming (OOP). Consider taking a course dealing with object-oriented programming. Numerous books have recently been written on object-oriented programming. Many of these books demonstrate how the object-oriented concepts explained in Modules A and B are implemented in an object-oriented programming language environment.

SUMMARY

1. Using object-oriented techniques for designing a system is referred to as object-oriented design.

2. Object-oriented design is concerned with identifying and classifying three object types—interface, entity, and control. Interface and control object types are objects that are introduced as a result of implementation decisions that were made during systems design. Entity objects are identified during systems analysis and usually correspond to items in real life and contain information, known as attributes, that describes the different instances of the entity. Interface objects are objects that are introduced to represent a means through which the user will interface with the system. The responsibility of the interface object is twofold: (1) it translates the user's input into information that the system can understand and use to process the business event, and (2) it takes data pertaining to a business event and translates the data for appropriate presentation to the user. Control objects are those that hold application or business rule logic. Controller objects serve as the "traffic cop" containing the application logic or business rules of the event for managing or directing the interaction between the objects.

3. Object responsibility is the obligation that an object has to provide a service when requested, thus collaborating with other objects to satisfy the request if required. Object responsibility is closely related to the concept of objects being able to send and/or respond to messages.

4. Object-oriented design includes the following activities: refining the use case model to reflect the implementation environment, modeling object interactions and behavior that support the use case scenario, and updating the object class diagram to reflect the implementation environment.

5. An ideal object model diagram is used to document objects and their interactions. An ideal object model diagram includes symbols to represent actors, interface, control, and entity objects, and arrows that represent messages or communication between the objects.

6. During systems design, use case descriptions are examined to identify all action verb phrases. Action verb phrases suggest behaviors required to complete a use case scenario. These behaviors must be associated with a system object.

7. A popular tool for documenting the behaviors and collaborations for an object is the class responsibility collaboration (CRC) card.

8. A sequence diagram is used to document the detailed interactions between objects.

KEY TERMS

abstract use case, p. 679
activity diagram, p. 690
class responsibility collaboration (CRC) card, p. 685
collaboration, p. 687
collaboration diagram, p. 690
component, p. 678

control object, p. 676
deployment diagram, p. 690
entity object, p. 676
extension use case, p. 679
ideal object model diagram, p. 682
interface object, p. 676
model-view-controller (MVC), p. 676

object framework, p. 678
object-oriented design, p. 676
object responsibility, p. 677
object reusability, p. 677
persistent, p. 677
role playing, p. 687
sequence diagram, p. 687

REVIEW QUESTIONS

1. What is the approach of using object modeling during systems design called?
2. Differentiate between entity, interface, and control objects.
3. Explain the concept of model-view-controller.
4. What does it mean when an entity object is said to be persistent?
5. Define the term *object responsibility*. Give an example.
6. Explain the importance of object reuse.
7. What is an object framework? Give an example. What is a component? Give an example.
8. Identify the three object-oriented design activities.
9. What is an ideal object model diagram used for in object-oriented design?
10. Explain the difference between an "initiator" actor and a "receiver" actor. Give an example of each actor.
11. Explain how CRC cards are used during object-oriented design.
12. What do verb phrases in a use case suggest?
13. What are four implicit behaviors for any entity object?
14. What is the purpose of a sequence diagram?
15. What is the purpose of an activity diagram?
16. Identify two types of implementation diagrams.

PROBLEMS AND EXERCISES

1. What are the benefits of structuring an object-based system into entity, interface, and control objects?
2. What correlation can be drawn between the three object types in problem 1 and the client/server model?
3. What correlation can be drawn between object responsibility and message sending?
4. What correlation can be drawn between an abstract use case and object reusability?
5. In transforming the analysis use cases to design use cases, what types of refinement might be done? Can such refinements result in new objects?
6. Analyze the following use case to identify different object types.

USE CASE NAME:	Purchasing a soda
ACTOR:	Customer
DESCRIPTION:	This use case describes the scenario of a customer purchasing a can of soda from a soda machine. The soda machine has six selections of soda, is able to accept coins and $1 bills, and has a digital display.
BASIC COURSE:	**Step 1.** This use case is initiated when the customer wants to purchase a can of soda. **Step 2.** The customer inserts a coin into the machine's coin receptacle. **Step 3.** The machine verifies whether the coin is legitimate, records the amount deposited, then updates the display with the total amount deposited. Go to **Step 2** if another coin is deposited. **Step 4.** The customer pushes the selection button of the type of soda desired. **Step 5.** The machine calculates whether change is due, then releases the can of soda from its holding bin, and it drops to the can retrieval receptacle. Machine records transaction, including date, time, type of purchase, and amount of purchase. **Step 6.** The customer retrieves the can of soda.
ALTERNATE COURSES:	**Step 2.** The customer inserts a $1 bill into the machine's dollar bill receptacle. **Step 3.** If the dollar bill is not legitimate, reject bill out of receptacle. **Step 4A.** If customer depresses change return lever, release held coin(s) to change receptacle or reject dollar bill out of dollar bill receptacle. **Step 5A.** If customer has not deposited enough money for the soda selected, display message "Not enough money deposited" for five seconds, then display total amount deposited. Return to **Step 2.** **Step 5B.** If the soda the customer has selected is out of stock, display message "Out of Stock, Please Make Another Selection" for five seconds, then display total amount deposited. Return to **Step 4.** **Step 5C.** If change is due, release appropriate coins to the change receptacle.
PRECONDITION:	None
POSTCONDITION:	Customer has purchased soda.
ASSUMPTIONS:	The soda machine is operating normally and is able to give change.

7. Analyze the use case in problem 6 to identify behaviors.
8. Using the results of problems 6 and 7, develop a CRC card for one of the use case objects.
9. Using the results of problems 6, 7, and 8, construct an ideal object diagram.
10. Using the results of problems 6, 7, and 8, construct a sequence diagram.
11. Construct an activity diagram based on the scenario of withdrawing money from an ATM.

PROJECTS AND RESEARCH

1. Surf the Internet and make a list of links to sites containing information about object-oriented design.
2. Research the new Unified Modeling Language (UML) sponsored by the Rational Software Corporation. How are the diagrams that are used in the UML similar to the diagrams that we used in this module? How are they different?
3. Perform role playing using the use case in problem 6 above as the script.

MINICASES

1. Schedule an interview with the business analyst responsible for your school's registration system. Based on the information provided perform the following:
   a. Draw a context model diagram.
   b. Identify the actors and the use cases they initiate.
   c. Complete a use case description for the event of a student enrolling for a course. For a student dropping a course.
   d. Use the use cases you have previously prepared and construct an object class diagram.
   e. Transform the use cases you previously prepared into design use cases.
   f. Identify the design objects in each of the design use cases.
   g. Construct ideal object model diagrams for each of the design use cases.
   h. Using the previously prepared design use cases, identify the required responsibilities and behaviors, and document them on CRC cards.
   i. Construct sequence diagrams for each of the design use cases.
   j. Refine the object class diagram of step *d* to include attributes and methods.

SUGGESTED READINGS

Booch, G. *Object-Oriented Design with Applications*. Redwood City, CA: Benjamin Cummings, 1994. Many Booch concepts were integrated into the UML.

Booch, G. *Object Solutions—Managing the Object-Oriented Project*. Reading, MA: Addison-Wesley, 1996.

Coad, P., and E. Yourdon. *Object-Oriented Analysis*. 2nd ed. Englewood Cliffs, NJ: Prentice Hall, 1991. This book provides a very good overview of object-oriented concepts. However, the object model techniques are somewhat limited by comparison to UML and other object-oriented modeling approaches.

Eriksson, Hans-Erik, and Magnus Penker. *UML Toolkit*. New York: John Wiley & Sons, 1998. This book provides detailed coverage of the UML.

Fowler, Martin, and Kendall Scott. *UML Distilled—Applying the Standard Object Modeling Language*. Reading, MA: Addison-Wesley, 1997. A good short guide introducing the concepts and notation of the UML.

Goldberg, Adele, and Kenneth S. Rubin. *Succeeding with Objects*. Reading, MA: Addison-Wesley, 1995.

Harman, Paul, and Mark Watson. *Understanding UML—The Developer's Guide*. San Francisco: Morgan Kaufmann Publishers, 1997. This book is an excellent reference book. The examples were prepared using Popkin's *System Architect*.

Jacobson, Ivar; Magnus Christerson; Patrik Jonsson; and Gunnar Overgaard. *Object-Oriented Software Engineering—A Use Case Driven Approach*. Wokingham, England: Addison-Wesley, 1992. This book presents detailed coverage of use case modeling and how it's used to identify objects.

Larman, Craig. *Applying UML and Patterns—An Introduction to Object-Oriented Analysis and Design*. Englewood Cliffs, NJ: Prentice Hall, 1997. This is an excellent reference book explaining the concepts of OO development utilizing the UML.

Rumbaugh, James; Michael Blaha; William Premerlani; Frederick Eddy; and William Lorensen. *Object-Oriented Modeling and Design*. Englewood Cliffs, NJ: Prentice Hall, 1991. This book presents detailed coverage of the object modeling technique (OMT) and its application throughout the entire systems development life cycle. Many OMT constructs are now in the UML.

Taylor, David A. *Object-Oriented Information Systems—Planning and Implementation*. New York: John Wiley & Sons, 1992. This book is a very good entry-level resource for learning the concepts of object-oriented technology and techniques.

Wilkinson, Nancy M. *Using CRC Cards*. New York: SIGS Books, 1995.